HARNESSING MICROSTATION® V8 XM

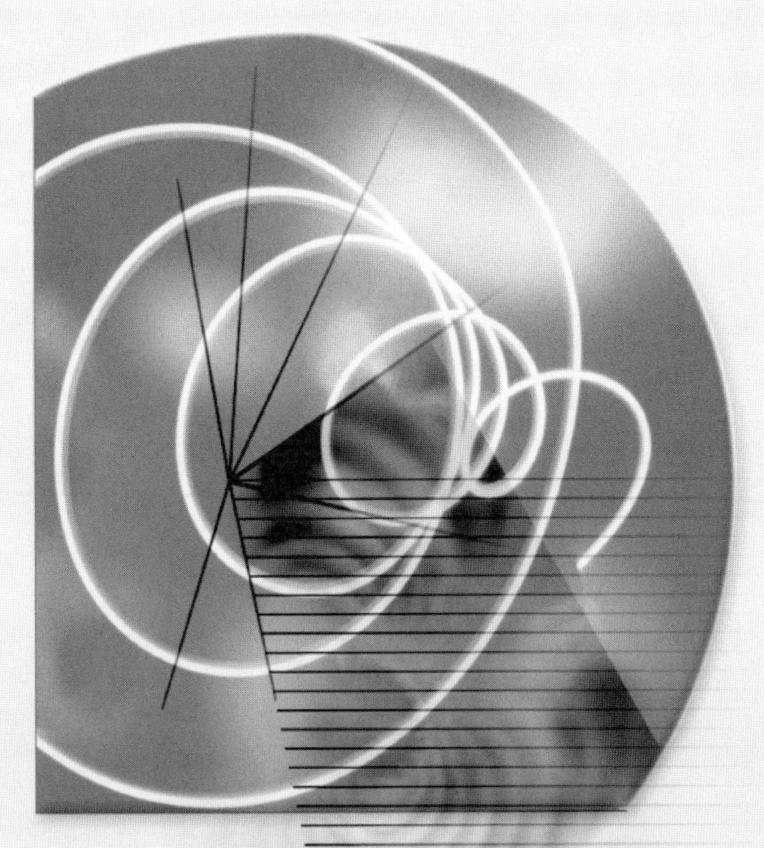

HARNESSING MICROSTATION® V8 XM

G.V. KRISHNAN
JAMES E. TAYLOR

Australia • Brazil • Japan • Korea • Mexico • Singapore • Spain • United Kingdom • United States

Harnessing Microstation® V8 XM
G.V. Krishnan, James E. Taylor

Vice President, Technology and Trades SBU: David Garza

Directorof Learning Solutions: Sandy Clark

Senior Acquisitions Editor: James Gish

Managing Editor: Tricia Coia

Marketing Director: Deborah S. Yarnell

Academic Marketing Manager: Guy Bakaran

Senior Content Project Manager: Stacy Masucci

Editorial Assistant: Niamh Matthews

Library of Congress Control Number: 2006031724

ISBN-13: 978-1-4180-5314-7

ISBN-10: 1-4180-5314-7

Delmar
Executive Woods
5 Maxwell Drive
Clifton Park, NY 12065
USA

Cengage Learning is a leading provider of customized learning solutions with office locations around the globe, including Singapore, the United Kingdom, Australia, Mexico, Brazil, and Japan. Locate your local office at **international.cengage.com/region**

Cengage Learning products are represented in Canada by Nelson Education, Ltd.

For your lifelong learning solutions, visit **www.cengage.com/delmar**

Visit our corporate website at **www.cengage.com**

Printed in Canada
2 3 4 5 6 7 11 10 09

CONTENTS

INTRODUCTION

Harnessing MicroStation V8 XM gives you the necessary skills to start using a powerful CADD (Computer Aided Drafting and Design) program, MicroStation V8 XM. We have created a comprehensive book providing information, references, instructions, and exercises for people of varied skill levels, disciplines, and requirements, for applying this powerful design/drafting software.

Now in its fifth edition, *Harnessing MicroStation V8 XM* was written and updated as a comprehensive tool for the novice and the experienced MicroStation user, both in the classroom and on the job. The book includes new features introduced in MicroStation V8 XM.

Readers immediately gain a broad range of knowledge of the elementary CADD concepts necessary to complete a simple design. We do not believe the user should be asked to wade through all components of every tool or concept the first time that tool or concept is introduced. Therefore, we have set up the early chapters to cover and practice fundamentals as preparation for the advanced topics covered later in the book.

Harnessing MicroStation V8 XM is intended to be both a classroom text and a desk reference. If you are already a user of earlier versions of MicroStation, you will see an in-depth explanation provided in the corresponding chapters for all new features. The new features in MicroStation V8 XM give personal computer-based CADD even greater depth and breadth.

IN THIS BOOK

Chapter 1—Getting Started The beginning of this chapter describes the hardware you need to get started with MicroStation. The balance of the chapter explains how to start MicroStation V8 XM; the salient features of dialogs, settings boxes, and MicroStation applications windows; input methods, design planes, and working units; saving changes and exiting the design file; and a summary of the enhancements in MicroStation V8 XM.

Chapters 2, 3, and 4—Fundamentals These chapters introduce the basic element placement and manipulation tools needed to draw a moderately intricate design. All tools discussions are accompanied by examples. Ample exercises are designed to give students the chance to test their level of skill and understanding.

Chapter 5—AccuDraw and SmartLine An in-depth explanation is provided for AccuDraw and SmartLine, two powerful drawing tools.

Chapter 6—Manipulating Groups of Elements Introduces the Power Selector and Fence tools that allow you to manipulate groups of elements.

Chapter 7—Placing Text, Data Fields, and Tags Provides an in-depth explanation of various methods for placing text, data fields that serve as place holders for text to be placed later, and tags that allow you to add non-graphical information to the design.

Chapter 8—Element Modification This chapter introduces various tools for modifying the elements of your design. All tools are accompanied by examples.

Chapter 9—Measurement and Dimensioning Introduces tools that allow you to display the length, angle, or area of elements; place dimensions on elements; and customize the way dimensions are placed.

Chapter 10—Plotting This chapter introduces all the features related to plotting (creating paper copies of your design).

Chapter 11—Cells and Cell Libraries Introduces the powerful set of tools available in MicroStation for creating and placing symbols—called cells—and for storing them in Cell libraries. These tools permit you to group elements under a user-determined name and perform manipulations on the group as though they were a single element.

Chapter 12—Patterning Introduces the set of tools available in MicroStation to place repeating patterns to fill regions in a design, such as the hatch lines in a cross-section.

Chapter 13—References A powerful and timesaving feature of MicroStation is its ability to view other models from an active design file (the one you currently have open for editing). MicroStation lets you display the contents of unlimited number of models from the active design file. When you view a model in this way, it is referred to as a Reference. This chapter describes all the tools used to manipulate references .

Chapter 14—Special Features MicroStation provides some special features that, though less often used than the tools described in chapters 1 through 13, add power and versatility to the MicroStation tool set. This chapter introduces several such features.

Chapter 15—Internet Utilities MicroStation provides utilities that allow you to take part in collaborative engineering projects over the World Wide Web. With these utilities you can share design information over the Web and insert engineering data directly into your design from the Web.

Chapter 16—Customizing Introduces several tools for customizing MicroStation, such as creating multi-line definitions, custom line styles, and workspaces; creating and modifying tool boxes, tool frames, and function key menus; installing fonts; and using the archive utility.

Chapter 17—3D Design and Rendering Provides an overview of the tools and specific tools available for 3D design.

Appendices provide additional valuable information to the user.

Harnessing MicroStation V8 XM provides a sequence suitable for learning, ample exercises, examples, review questions, and thorough coverage of the MicroStation program, that should make it a must for multiple courses in MicroStation, as well as self-learners, everyday operators on the job, and operators aspiring to customize MicroStation.

The project exercises and drawing exercises are contained in PDF files found on the CD in the back of this book. The PDF files correspond to the material presented in Chapters 1 through 13 and 17.

STYLE CONVENTIONS

In order to make this text easier for you to use, we have adopted certain conventions that are used throughout the book:

Convention	Example
Menu names appear with the first letter capitalized.	Element menu
Tool box names appear with the first letter capitalized.	Linear Elements tool box
All interface object names are indicated in boldface.	Click the **OK** button to save the changes and close the Design File settings box. Next, select **Tools** from the main menu.
User input is indicated by boldface.	Click in the **Angle** field, key-in **180**, and press ENTER.
Instructions are indicated by italics and are enclosed in parentheses.	Enter first point: *(Place data point or key-in coordinates)*

HOW TO INVOKE TOOLS

Throughout the book, instructions for invoking tools are summarized in tabular form, similar to the example shown below:

Menu Element	> Text Styles
Place Text Settings window	Click the magnifying glass icon
Key-in Window	**textstyle dialog open** (or **texts di o**) ENTER

The left column tells you where the action will take place (such as, by selecting a menu), and the right column tells you specifically what the action is (such as, by selecting the Text Styles option from the Element menu).

EXERCISE ICONS

A special icon is used to identify step-by-step Project Exercises. Exercises that give you practice with types of drawings that are often found in a particular engineering discipline are identified by icons that indicate the discipline. Exercises that are cross-discipline—that is, the skills used in the exercise are applicable for most or all disciplines—do not have a special icon designation. The following table presents all the exercise icons:

Type of Exercise	Icon	Type of Exercise	Icon
Project Exercises		Civil	
Electrical		Mechanical	
Piping		Architectural	

ACKNOWLEDGMENTS

This book was a team effort. We are very grateful to many people who worked very hard to help create this book. We are especially grateful to the following individuals at Thomson Delmar Learning, whose efforts made it possible to complete the project on time: Ms. Sandy Clark, Director of Learning Solutions; Mr. James Gish, Senior Acquisitions Editor; Ms. Tricia Coia, Managing Editor; Ms. Stacy Masucci, Senior Content Project Manager; and Ms. Niamh Matthews, Editorial Assistant. The authors and Thomson Delmar Learning would also like to acknowledge Mr. Michael Boyd and staff at ATLIS Publishing Services.

In addition, the authors would like to acknowledge the following individuals who reviewed the previous edition of this book: Dennis C. Jackson, Malcolm A. Roberts, Jr., and Michael J. White. The authors and Thomson Delmar Learning also gratefully acknowledge the thorough and thoughtful technical editing provided by Dennis Catalloni.

And, last but not least, special appreciation to Bentley Systems, Inc. for providing the MicroStation V8 XM software.

Getting Started

The beginning of this chapter describes how to get started with MicroStation V8 XM Edition. Users who are not familiar with the computer operating system (files, drives, directories, operating system commands, etc.) may wish to review rudimentary Windows operating system references, refer to the Installation Guide that came with this program, or consult the dealer from whom MicroStation V8 XM was purchased.

The balance of this chapter explains an overview of the screen layout and the salient features of dialog and settings boxes. Detailed explanations and examples are provided for the concepts and commands throughout the chapters that follow.

At the end of the chapter, a list of enhancements for MicroStation V8 XM is provided. Users already familiar with MicroStation V8, can check the list to see the new and improved features in MicroStation V8 XM. Throughout the book these new features are explained in greater detail. Features introduced in MicroStation V8 XM give personal computer–based CAD even greater depth and breadth.

STARTING MICROSTATION V8 XM

Designing and drafting is what MicroStation (and this book) is all about. To get started with MicroStation V8 XM, choose the Start button (Windows 2003 and XP operating systems), navigate to the Bentley program group, and then select the MicroStation V8 XM program. On some systems a MicroStation V8 XM startup icon will appear on the Windows desktop after installation. Upon launch, MicroStation displays the MicroStation Manager dialog box, similar to Figure 1–1. MicroStation dialog boxes have features that are similar to Windows file management dialog boxes.

Note: *All the screen captures shown in this textbook are taken from MicroStation V8 XM running in the Windows XP operating system.*

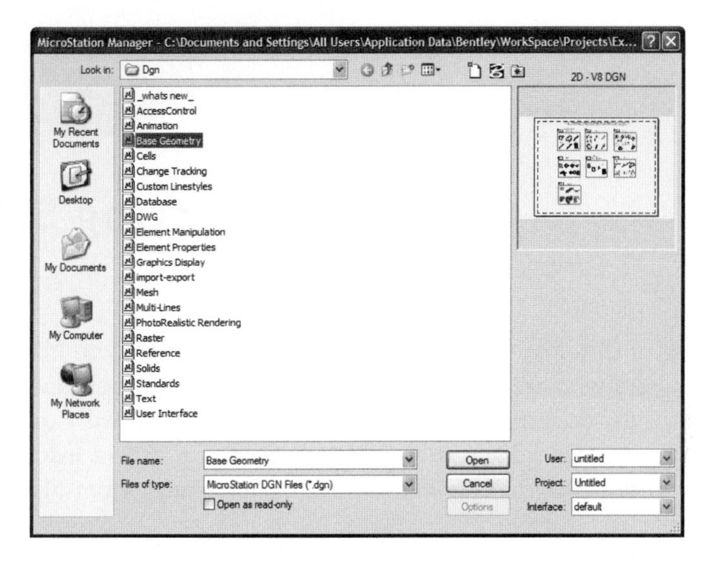

Figure 1–1 *MicroStation Manager dialog box*

MicroStation Manager is used to open MicroStation design (DGN) files and AutoCAD (DWG) files, create new MicroStation design files, manage folders and control viewing options. The MicroStation Manager is displayed when MicroStation is started without a DGN file specification or when Close is chosen from MicroStation Application Window File menu.

BEGINNING A NEW DESIGN

To begin a new design, select the **New** tool as shown in Figure 1–2. The New dialog box opens, as shown in Figure 1–3.

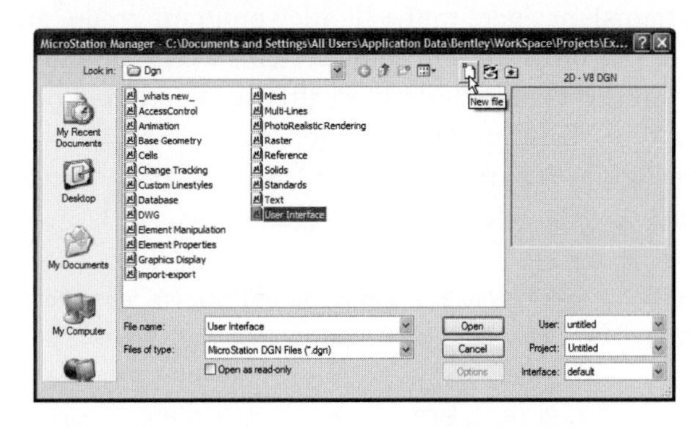

Figure 1–2 *Invoking the New tool from the MicroStation Manager*

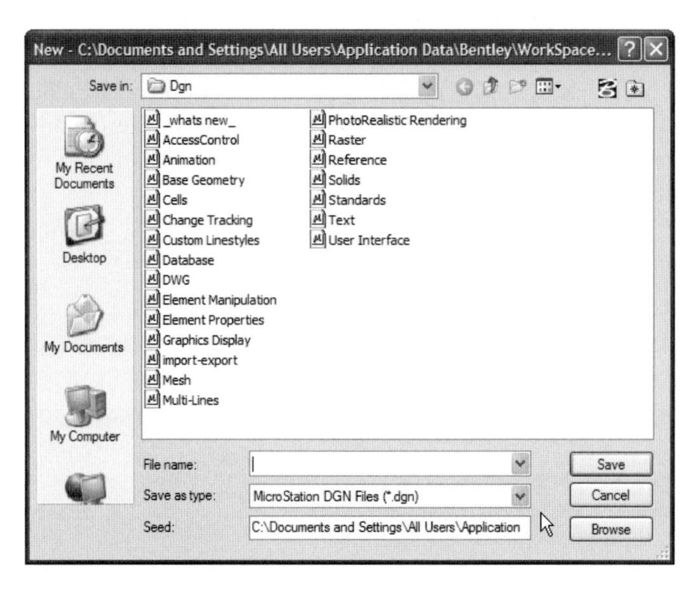

Figure 1–3 *New dialog box*

Navigate to the appropriate seed file (more about this in the section on Seed Files) by choosing the Browse button. Specify a file name for the new design in the **File name** field, select the appropriate folder where the newly created design file will be saved and click the **Save** button. The New dialog box is closed, and control is passed to the MicroStation Manager. MicroStation by default highlights the name of the file just created in the **Files** list box.

Note: *If the design file name is the same as the name of an existing file name, MicroStation displays an Alert Box asking if the existing file should be replaced. Click the* ***OK*** *button, or* ***Cancel*** *to reissue a new file name.*

Before clicking the Open button to open the newly created design file, make sure the appropriate **User**, **Project**, and **Interface** are selected from the option menu located at the bottom of the MicroStation Manager dialog box.

A **User** is a customized workspace drafting environment that permits the user to set up MicroStation for specific purposes. Multiple workspaces can be set up as needed. A workspace consists of "components" and "configuration files" for both the user and the project. By default, MicroStation selects the untitled workspace. In addition, workspaces can be created by modifying an existing workspace. Refer to Chapter 15 for a detailed description of creating or modifying workspaces.

The selection of the project sets the location and names of data files associated with a specific design project. Refer to Chapter 15 for setting up the project. By default, MicroStation selects the Untitled in the **Project** option menu.

The selection of the Interface sets a specific look and feel of MicroStation's tools and general on-screen operation. The interface selection can be changed from the Interface option menu. By default, MicroStation selects the default in the **Interface** option menu. Refer to Chapter 15 for a detailed explanation on creating and modifying the MicroStation interface.

To open the new design file, click the Open button. The screen will look similar to the one in Figure 1–4.

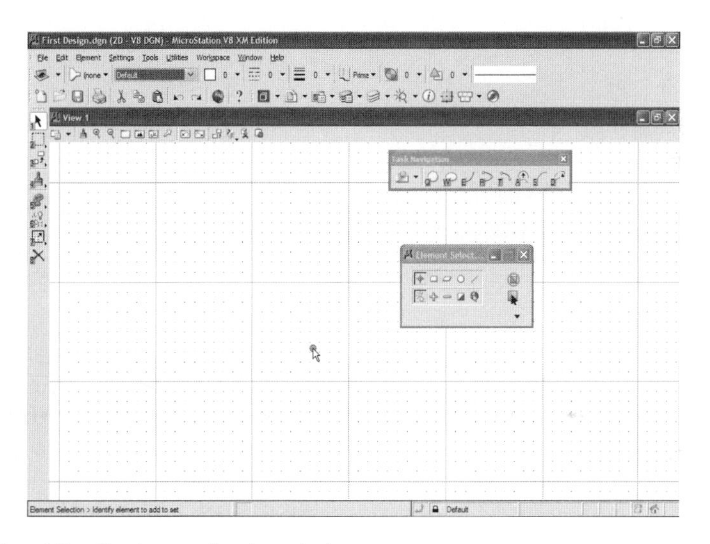

Figure 1–4 *MicroStation application window*

FILE NAMES

The name entered in the **File name:** field of the New dialog box will be the name of the file in which information about the design is stored. It must satisfy the requirements for file names as specified by the computer's particular operating system.

File names and file extensions can contain up to 255 characters. Names may be made up of combinations of uppercase and lowercase letters, numbers, the under_ score (_), the hyphen (-), embedded spaces, and punctuation. Valid examples include:

this is my first design.dgn
first house.dgn
machine part one.dgn
PART_NO5.wrk

When MicroStation prompts for a design name, just type in the file name and MicroStation will append the extension *.dgn* by default. For instance, keying-in the design name FLOOR1, will result in a file with the name *FLOOR1.DGN.* If a

different extension is required, it can be keyed-in with the file name. Before a file name is specified, make sure to navigate to the appropriate folder the where the new design file is created.

***Note:** The installed MicroStation directory structure for the MicroStation V8 XM Edition conforms to Microsoft Windows standards. In MicroStation V8 XM, there are separate default locations for program files and document files. Document files are installed by default to "C:\Documents and Settings\All Users\Application Data\Bentley," while program files are installed by default to "C:\Program Files\Bentley."*

SEED FILES

Whenever a new MicroStation design (DGN) or AutoCAD (DWG) file is created, a seed file is required. The contents of the seed file are copied into the new DGN or DWG file. Generally, seed files only contain settings and attributes. Some companies also incorporate graphic data such as logos and drawing sheets into a seed file.

The new DGN file is actually a copy of the seed file. Seed files do not (necessarily) contain elements, but, like other DGN files, they do contain at least one (default) model, settings, and view configurations. Having a seed file with customized settings frees the user from having to adjust settings each time a new DGN file is created. It may enhance workflow to have a different seed file for each type of drawing frequently created.

In a situation where DGN files are frequently converted to DWG format, it is helpful to set up a special DGN seed file that contains the standard level (layer), text style, dimension style, line style (linetype), and units settings required for the DWG deliverable. The easiest way to do this is to start with a DWG file containing the settings, then create a DGN seed file from it.

MicroStation programs come with several seed files. Depending on the discipline, use an appropriate seed file. For instance, if working on an architecture floor plan, use the architecture seed file (*2dEnglishArch.DGN*). Using the seed DWG file to create new DWG files, will take advantage of the standard level (DWG layer) configuration, text and dimension style, working units (DWG units), and line style (DWG linetype) settings. The default *seed.dwg* file delivered with MicroStation contains one design model and two sheet models named Layout1 and Layout2. The default seed file also contains view groups for each model.

The New dialog box displays the name of the default seed file as shown in Figure 1–5. If necessary, choose a different seed file by clicking the Browse button. MicroStation displays a list of available seed files as shown in Figure 1–6. Select the desired file from the list and click the Open button.

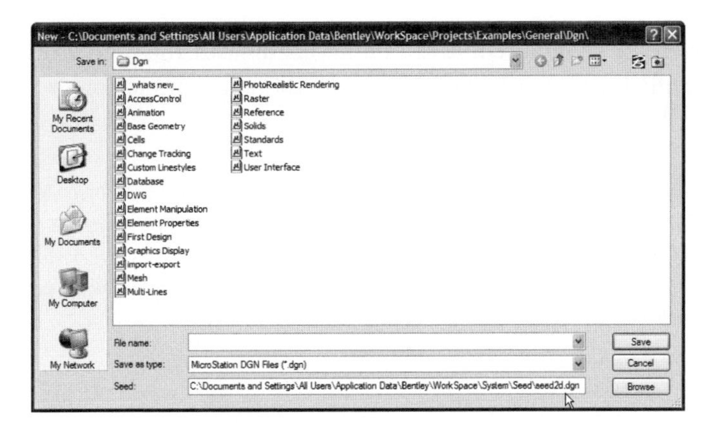

Figure 1–5 *New dialog box displaying the name of the default seed file*

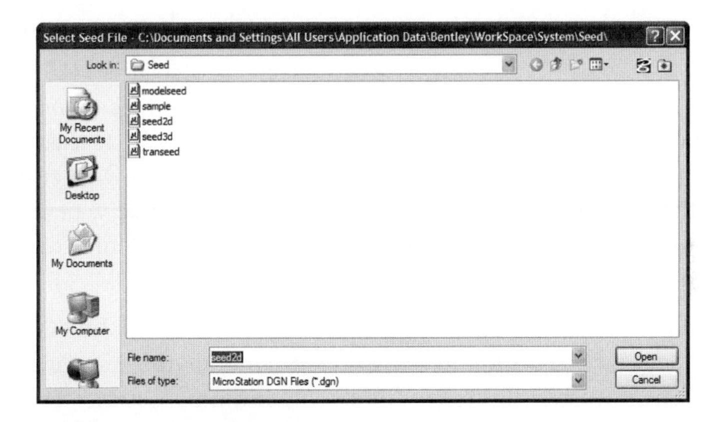

Figure 1–6 *Select Seed File dialog box*

OPENING AN EXISTING DESIGN FILE

To open an existing design file in MicroStation, simply click the name of the file in the Files list box item of the MicroStation Manager dialog box, and then click the Open button, or double-click the name of the file. If the design file is not in the current folder, change to the appropriate drive and folder from the Directories list box, then select the appropriate design file. In addition, MicroStation displays the names (including the path) of the last ten design files opened from the **Bentley** icon as shown in Figure 1–7. To open one of these ten design files, click on the file name.

Note: *MicroStation also allows to create a new design file or open an existing design file by selecting the New or Open tool, respectively, from the File menu located in the MicroStation Application Window.*

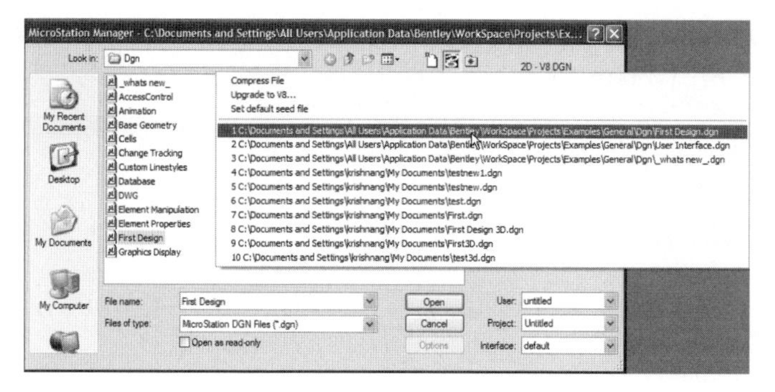

Figure 1–7 *Bentley menu from the MicroStation Manager dialog box, displaying the names of the last ten design files opened*

OPENING A DESIGN IN READ-ONLY MODE

Open a design in read-only mode by turning ON the check box for **Read-Only**, located in the MicroStation Manager dialog box. When the **Read-Only** option is chosen, MicroStation opens the active design file in a read-only state and displays a disk icon with a large red X in the lower right corner of the application window. Any changes made to the design will not be saved as part of the design file.

MICROSTATION APPLICATION WINDOW

The MicroStation application window consists of menus (also called *menu bars*), the status bar, tool boxes, AccuDraw window, and view windows with the View Control bar (see Figure 1–8).

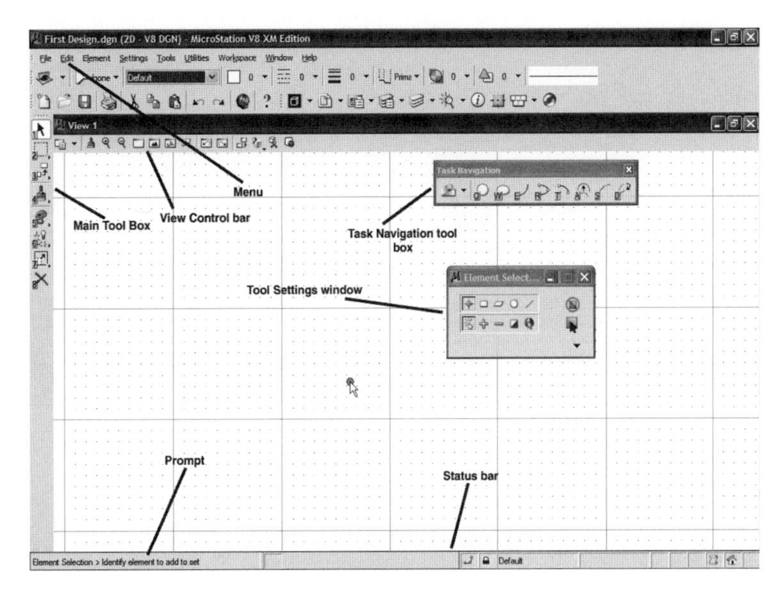

Figure 1–8 *MicroStation application window*

MENUS

The top menu bar of the MicroStation Application Window contains several drop-down menus. Tool boxes, dialog boxes, and settings boxes are available from these drop-down menus. Select one of the top menu items, and MicroStation displays the list of options available under that item (see Figure 1–9). Selecting from the list is a simple matter of moving the cursor down until the desired item is highlighted and then pressing the Data button on the pointing device. If a menu item has an arrow to the right, it has a cascading submenu. To display the submenu, just select the name of the submenu. Menu items that contain an ellipsis (...) display dialog boxes. When a dialog box is displayed, no other action is allowed until that dialog box is dismissed or closed.

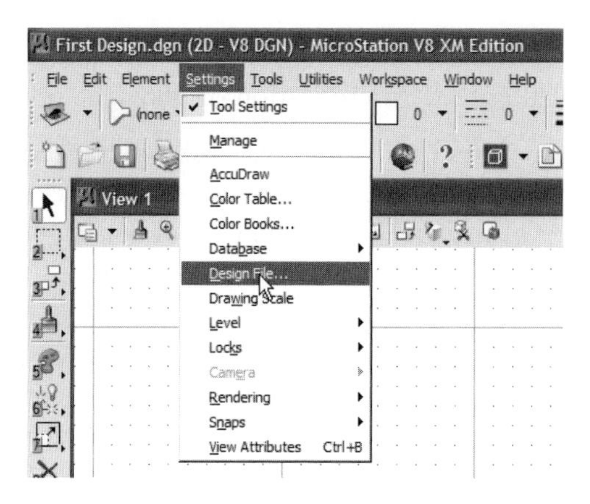

Figure 1–9 *Menu selection in the MicroStation Application Window*

The **Tools** drop-down menu displays the list of available tool boxes in MicroStation. To select one of the available tool boxes, just select the name of the tool box and it will be displayed on the screen. Check marks are placed in the **Tools** menu to indicate open tool boxes. Choosing an item in the **Tools** drop-down menu toggles the state of the corresponding tool box. If the tool box was closed, it opens; if the tool box was open, it closes. The tool box can be placed anywhere on the screen by dragging it with the pointing device. There is no limit to the number of the tool boxes that can be displayed on the screen. To open multiple tool boxes at the same time, open the Tool Boxes dialog box from:

Menu	Tools > Tool Boxes

MicroStation displays the Tool Boxes dialog box, similar to Figure 1–10.

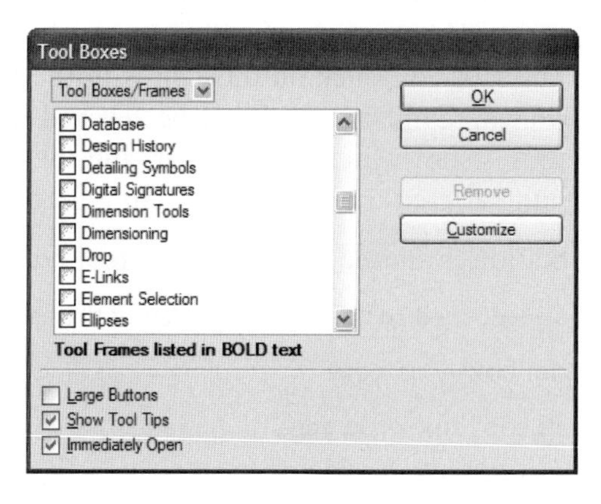

Figure 1–10 *Tool Boxes dialog box*

Select the desired tool boxes by turning ON their check boxes, then click the OK button to close the dialog box. MicroStation displays the selected tool boxes.

If working with two monitors, the tool boxes cab be dragged to the second monitor to unclutter the drawing screen. Close a tool box by clicking on the "X" located in the top right corner of the tool box. When a tool box is opened, it is displayed at the same location where it was previously on the screen.

TOOL BOXES

MicroStation has dozens of drawing tools (or tools, for short). Each open tool box is either floating in its own window (see Figure 1–11) or docked to an edge of the application window. The arrangement of tools in a floating tool box can be changed by resizing its window.

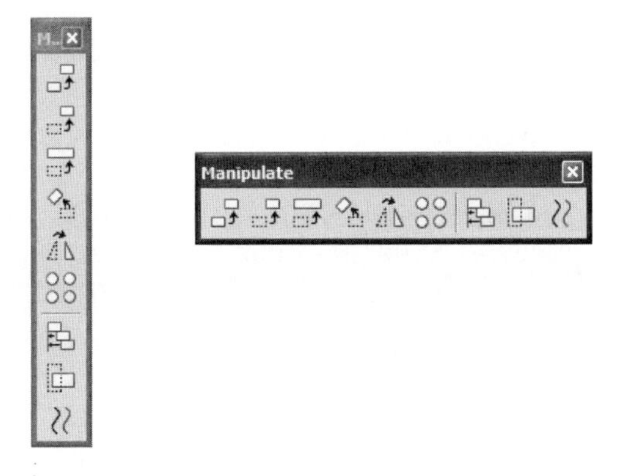

Figure 1–11 *Floating tool box*

Tool boxes consist of various tools. Tools are represented in tool boxes by icons. For simplicity, the term "tool" is used to refer both to a tool and its icon. To access one of the tools from the tool box, click on the icon with the Data button, and the appropriate tool is invoked. Only one tool can be selected at a given time. The name of the selected tool is shown in the status bar, and the tool is highlighted in the tool box. The default selected tool is the Element Selection tool.

The first time MicroStation is started, the following tool boxes are open:

- Main tool box—docked to the left-hand edge of the application window
- Attributes tool box—top edge
- Primary Tools tool box—top edge
- Standard Tool box—top edge
- Task Navigation tool box—floating tool box

Main Tool Box

The Main tool box (see Figure 1–12) is open when MicroStation is started for the first time. By default, the Main tool box is docked to the left-hand edge of the MicroStation window. It can be undocked or docked to the right-hand edge.

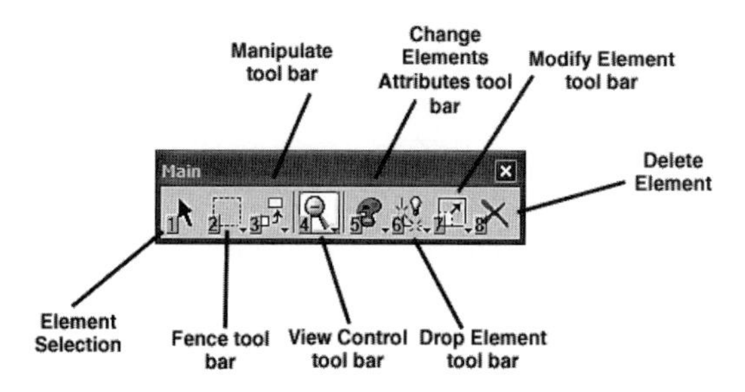

Figure 1–12 *Main tool box*

With the exception of Element Selection and Delete Element, the tools in the Main tool box also contain a "child" tool box. When one of the main tools is selected with the Data button held down, a drop-down menu opens (see Figure 1–13) from which a tool in the child tool box can be selected. The child tool box can be "torn off" and floated by choosing **Open As ToolBox** from the drop-down menu (see Figure 1–14).

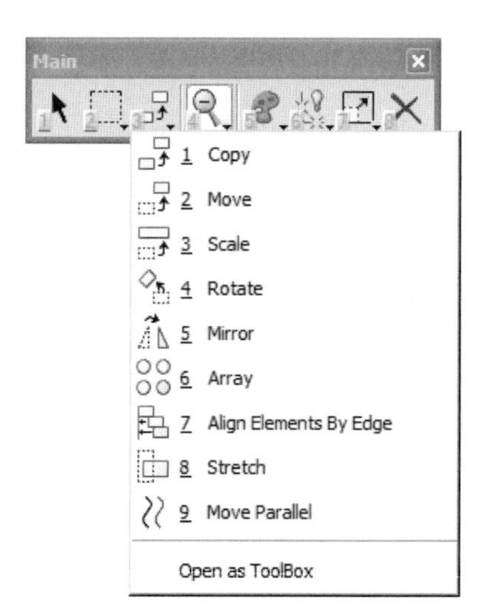

Figure 1–13 *Drop-down menu from the Main tool box*

Figure 1–14 *Select Open As ToolBox from the drop-down menu to float a child tool box.*

Task Navigation Tool Box

The Task Navigation tool box (see Figure 1–15) contains the Task List and the tools of the active task. In the as-delivered application window layout, the Task Navigation tool box is docked to the right-hand edge of the application window, and the active task is the Drawing task (see Figure 1–16).

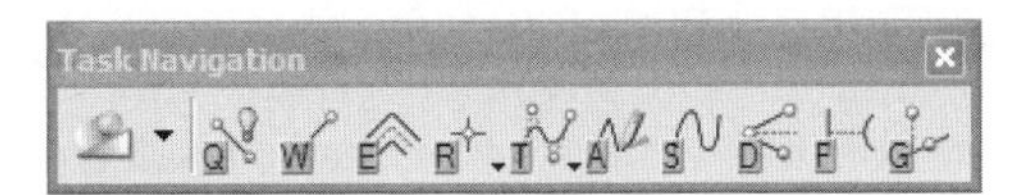

Figure 1–15 *Task Navigation tool box*

A task is a set of tools grouped to facilitate a particular workflow. The tools grouped into a given task can be standard MicroStation tools, custom tools, or a combination of both. The list of available tasks is generated by combining the tasks specified by MS_DGNLIBLIST and system DGN libraries.

To switch between the tasks, click the arrow on the Task List icon (see Figure 1–16) and the task list appears below it. If necessary, click the plus sign to expand the Task List and then click the task to use. The tools related to the selected task are displayed in the Task Navigation tool box. Selecting one of the available icons invokes the appropriate tool. The Task List can also be displayed at the cursor location by pressing the F2 function key (default function key menu) as shown in Figure 1–17. The available tools in the selected task list can be displayed at the cursor location by pressing the F4 function key (default function key menu) as shown in Figure 1–18.

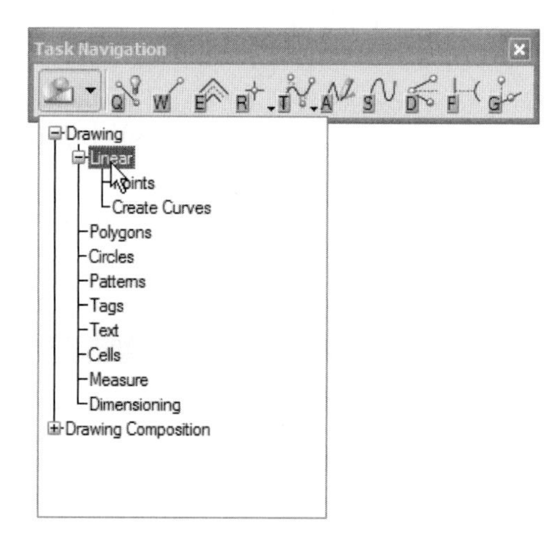

Figure 1–16 *Task list from the Task Navigation tool box*

1 Drawing
2 Drawing Composition

Figure 1–17 *Use F2 to invoke the task list at the cursor location*

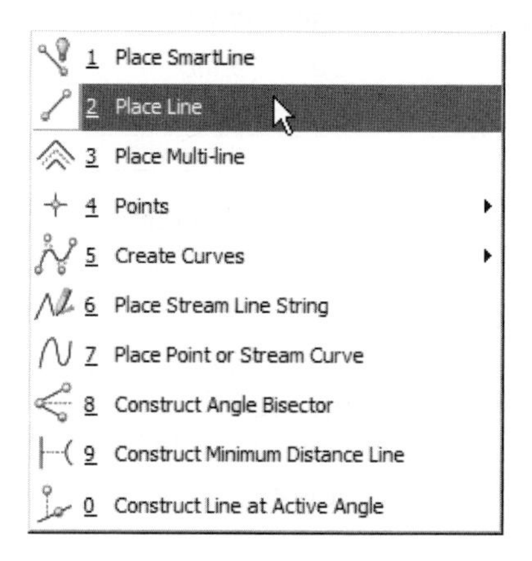

Figure 1–18 *Use F4 to invoke a list of available tools for the selected task at the cursor location*

Standard Tool Box

By default, the Standard tool box (see Figure 1–19) is docked just below the menu bar. The Standard tool box contains icons that enable quick access to many commonly used **File** and **Edit** drop-down menu items. Table 1–1 lists the tools available in the Standard tool box and the corresponding items in the menus.

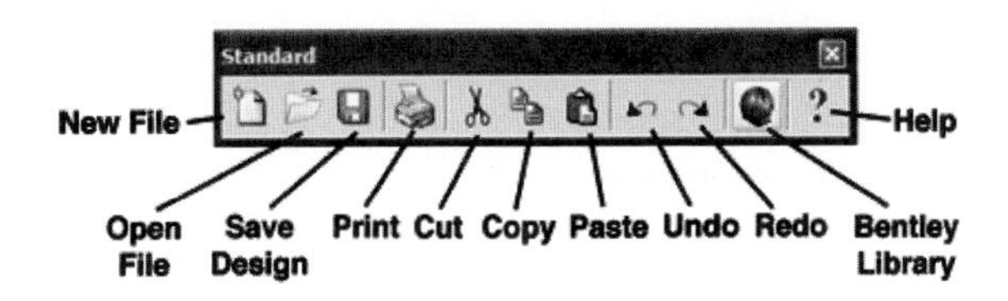

Figure 1–19 *Standard tool box*

Table 1–1 Tools Available in the Standard Tool Box

STANDARD TOOL BOX ITEM	MENU ITEM
New File	File > New
Open File	File > Open
Save Design	File > Save
Print File	File > Print/Plot
Cut	Edit > Cut
Copy	Edit > Copy
Paste	Edit > Paste
Undo	Edit > Undo
Redo	Edit > Redo
Bentley Library	————
Help	Help > Contents

Attributes Tool Box

By default, the Attributes tool box (see Figure 1–20) is docked along the side of the Standard tool box. The Attributes tool box contains icons that provide access to the more frequently accessed element symbology settings. Table 1–2 lists the tools available in the Attributes tool box.

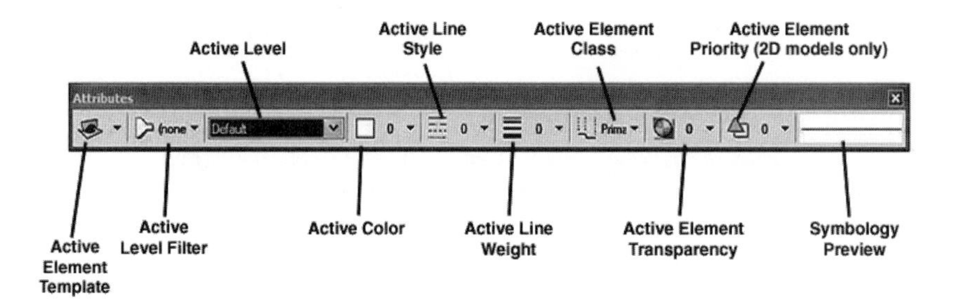

Figure 1–20 *Attributes tool box*

Table 1–2 Tools Available in the Attributes Tool Box

ATTRIBUTES TOOL BOX	ITEM FUNCTION
Active Element Template	Sets the Active Element Template.
Active Level Filter	Sets the Active Level Filter.
Active Level	Sets the Active Level.
Active Color	Sets the Active Color.
Active Line	Style Sets the Active Line Style.
Active Line Weight	Sets the Active Line Weight.
Active Element Class	Set the Active Element Class.
Active Element Transparency	Sets the Active Element Transparency.
Active Element Priority (2D models only)	Sets the Active Element Priority (2D models).
Symbology Preview	Demonstrates the active symbology.

Primary Tools Tool Box

By default, the Primary Tools tool box (see Figure 1–21) is docked along the side of the Standard and Attributes tool boxes. The Primary Tools tool box contains icons that provide access to the more frequently used tools. Table 1–3 lists the tools available in the Primary Tools tool box.

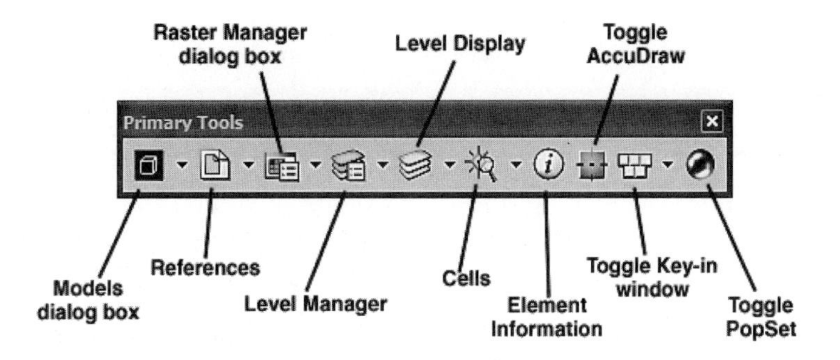

Figure 1–21 *Primary Tools tool box*

Table 1–3 Tools Available in the Primary Tool Box

PRIMARY TOOL BOX ITEM	FUNCTION
Models dialog box	Opens the Models dialog box to manage models.
References	Opens the References settings box.
Raster Manager dialog box	Opens the Raster Manager dialog box.
Level Manager	Opens the Level Manager dialog box.
Level Display	Opens the Level Display dialog box.
Cells	Opens Cell Library dialog box.
Element Information	Provides information about an element.
Toggle AccuDraw	Opens and closes AccuDraw window.
Toggle Key-in window	Opens and closes Key-in window.
Toggle PopSet	Controls the display of Tool Settings window.

Tool Tips

When the pointer is moved over any icon, MicroStation displays a tool tip, as shown in Figure 1–22. In addition, MicroStation displays a brief description of the function of the tool in the status bar. This is very helpful in learning which icon goes with which tool.

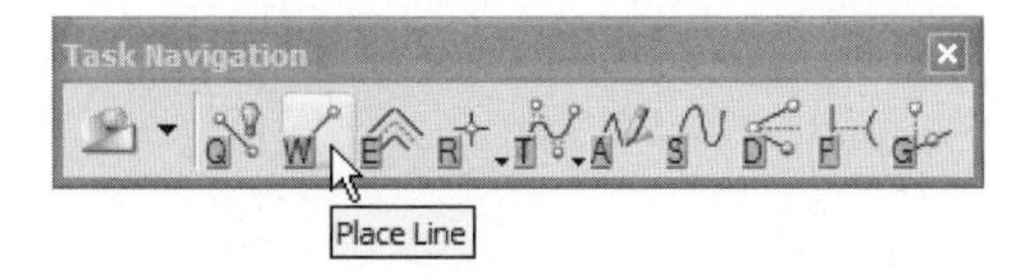

Figure 1–22 *Invoking the Place Line tool from the Task Navigation tool box (active task set to Linear)*

Tool tips can be disabled by turning off the option via the Help drop-down menu.

TOOL SETTINGS WINDOW

When a tool is invoked, MicroStation displays the controls required for adjusting the settings in the Tool Settings window. For example, if the Place Arc tool is selected, the **Method, Radius, Length, Start Angle, Sweep Angle** and **Direction** options are displayed in the Tool Settings window, as shown in Figure 1–23. If the Tool Settings window is closed, it opens automatically when a tool with settings is selected.

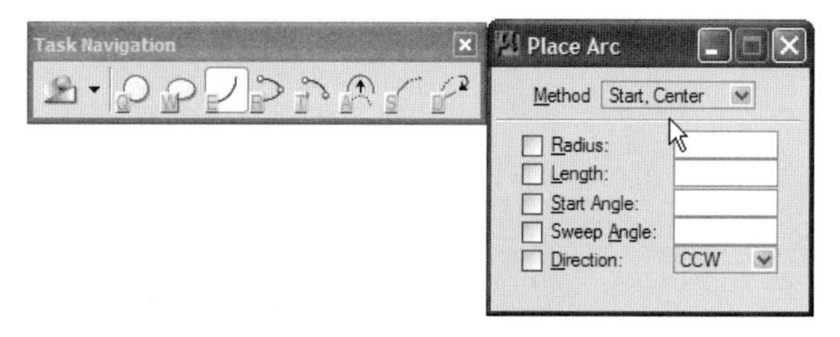

Figure 1–23 *Place Arc tool box and Tool Settings window*

The PopSet toggle is used to automatically prevent the display of Tool Settings window. When enabled, the Tool Settings window will close automatically when the controls are completed. This is a great way to reclaim valuable screen "real estate" and reduce pointer movement. Invoke the PopSet toggle from:

Primary Tools tool box	Select the Popset toggle (see Figure 1–24).
Key-in window	**popset on/off** (ENTER)

Figure 1–24 *Invoking the PopSet toggle from the Primary Tools tool box*

By default, the PopSet is disabled.

KEY-IN WINDOW

Key-ins are typed instructions entered into the Key-in window to control MicroStation. Any MicroStation tool can be invoked by typing the name of the tool in full or in abbreviated form in the Key-in window. Open the Key-in window from:

Primary Tools tool box	Select the Key-in toggle (see Figure 1–25).
Menu	Utilities > Key-in

Figure 1–25 *Invoking the Key-in toggle from Primary Tools tool box*

MicroStation displays the Key-in window, similar to Figure 1–26.

As characters are typed, they are matched to keywords in the list box below the Key-in window and are automatically selected in the list box. When the right key-in command is selected, press SPACE to complete the key-in, then click the Key-in button or press ENTER to complete the constructed key-in.

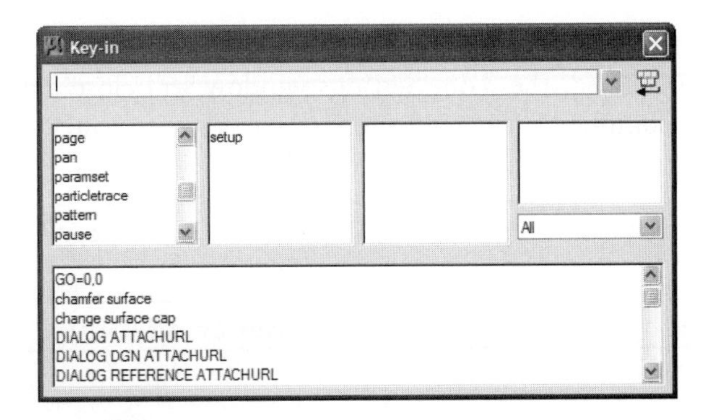

Figure 1–26 *Key-in window*

The list box in the Key-in window can be used to find and build key-ins. Scroll through the list of first words of key-ins in the left-most list box and select the keyword; it is then displayed in the key-in field. The subordinate, second-level keywords are shown in the key-in window's next list box. Select the desired keyword, and subsequently third-level keywords, if any, are shown. Select additional keywords, one per list box from left to right, until the desired key-in is constructed. To enter the constructed key-in, click the Key-in button or press ENTER.

MicroStation stores submitted key-ins in a buffer so they can easily be recalled or edited. Press the UP ARROW or DOWN ARROW on the keyboard repeatedly until the desired key-in text appears in the key-in field of the Key-in window. Make any necessary changes. In addition, MicroStation lists the submitted key-ins in the list box at the bottom of the Key-in window, from which they can be selected.

STATUS BAR

The status bar, located at the bottom of the MicroStation application window, as shown in Figure 1–27, displays a variety of useful information, including prompts, messages, and the name of the selected tool.

The Status bar is divided into two sections.

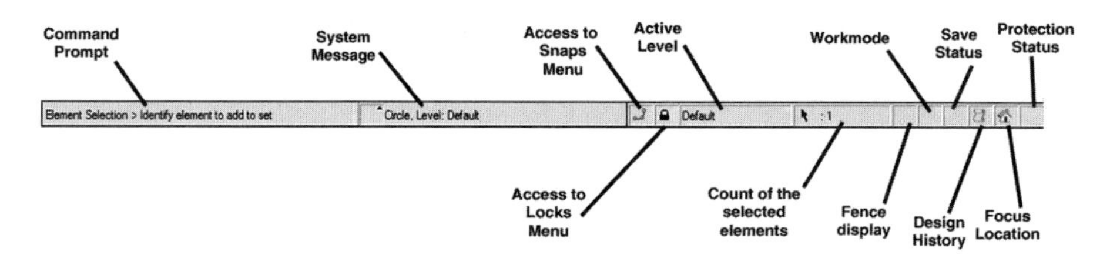

Figure 1–27 *Status bar*

Left-hand Section

The left-hand section of the status bar shows the name of the selected tool, followed by either a "greater than" symbol (>) or a colon (:) and message text. Message text that follows ">" is the selected tool's prompts. For example, with the Place Line tool selected, MicroStation prompts as follows:

Place Line > Identify start of line

The command prompts act as step-by-step guides through an operation with a tool. Message text that follows a colon indicates a possible problem.

In addition, as the pointer is moved over the tools in a tool box, the name of the selected tool and the associated message text are replaced with a description of the tool over which the pointer is located. This is intended as a form of online assistance.

Right-hand Section

The right-hand section of the status bar consists of a series of fields. Following are the available fields, from left to right:

- The first field shows system message information. Clicking in the Message Center portion of the status bar opens the Message Center dialog box. This dialog box contains a running log of system messages and any further description about the message if applicable.
- The second field indicates the Snap Mode setting.
- The Locks icon in the third field opens the Setting menu's Locks submenu.
- The fourth field indicates the current level.
- The fifth field indicates the count of the selected elements. If this field is blank, no elements are selected.

- The sixth field indicates whether there is a fence in the design. If this field is blank, no fence is placed.
- The seventh field indicates which work mode (DGN or DWG) is in effect. In DWG ("DWG") mode, certain functionality is disabled by default in order to restrict MicroStation to creating only information that can be stored in DWG format.
- The eighth field indicates whether changes to the active design file are unsaved. If the field is blank, there are no unsaved changes. If the field has a red icon with an "X" through it, the active design file is open for "read-only" access.
- The ninth field provides the status of the design history. If design history is not initialized, the icon is dimmed. If design history is initialized but there are uncommitted changes, a crayon is superimposed on the scroll.
- The tenth field indicates the focus location related to actions and attempts to move input focus for common operations.
- The last field indicates if the file is protected and digitally signed.

VIEW WINDOWS

MicroStation displays the elements drawn in the view windows. The portion of the design that is displaying in the view window is referred to as a *view*. In this part of the screen, all of the various commands entered will construct the design. Zoom In and Zoom Out can be used to control the visual size of the design's display. The scroll bars located both on the right side and at the bottom can be used to move the design in the view. A view typically shows a portion of the design, but may show it in its entirety, as in Figure 1–28.

Eight view windows can be open (ON) at the same time, and all view windows are active. An operation can begin in one view and be completed it in another. Move a view window by pressing and holding the cursor on the Title bar and dragging it to anywhere on the screen. Resize the display window by clicking and dragging on its surrounding border, and shrink and expand the window by clicking the push buttons located at the top right corner of the window. To close the view window, click the "X" located at the top right corner of the view window.

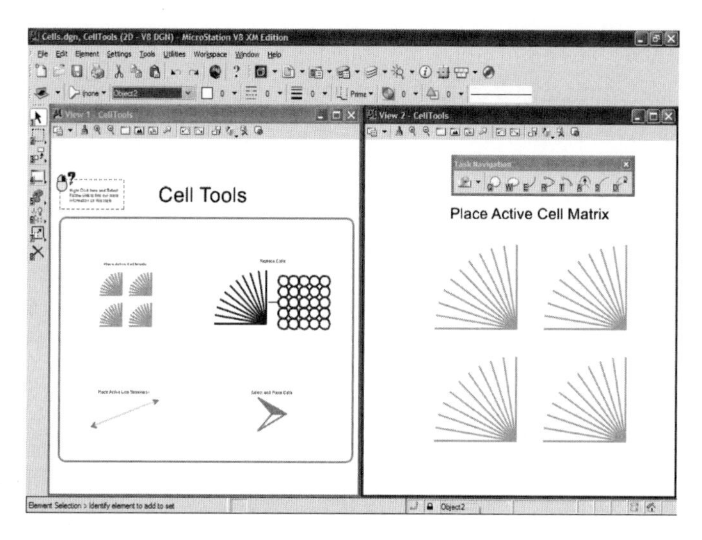

Figure 1–28 *Two view windows displaying different portions of a design*

MicroStation also provides, by default, a set of tools (View tool box) at the top left of the view window to control the view. View controls operate much like drawing tools; many even have "tool" settings. Detailed explanations of the View controls are provided in Chapter 3. The view tool box can be swapped with the Task Navigation tool box from the Windows menu.

INPUT METHODS

Input method refers to the manner in which MicroStation is instructed what tool to use and how to operate the tool. As mentioned earlier, in addition to the keyboard, the two most popular input devices are the mouse and the digitizing tablet.

KEYBOARD

To enter a command from the keyboard, simply key-in (type) the tool name in the Key-in window and click the Key-in button or press ENTER. *Key-in* is the name given to the function of providing information via the keyboard to MicroStation. MicroStation key-in language is much like plain English. For example, keying-in **PLACE LINE** selects the line command; **DELETE ELEMENT** selects the delete command, and so on. See Appendix B for a list of the key-in commands available in MicroStation.

Positional Keyboard Navigation

Positional keyboard navigation is a technique that utilizes a position-mapped keyboard. Position mapping is the mapping of keyboard zones to logical collections of controls in the user interface. MicroStation position maps is ON by default. Use the appropriate keys on the keyboard to invoke a tool when the focus is at home.

In Microsoft Windows and other graphical computer interfaces, the focus—sometimes called the input focus or keyboard focus—refers to the window or control to which keyboard input is directed.

MicroStation has a hierarchical focus model. The top level is called home. An icon in the status bar indicates the focus location. Figure 1–29 shows the display of the icons at various stages of the focus. To return to the **Home** from any of the focus stage, press the ESC key or the F12 function key.

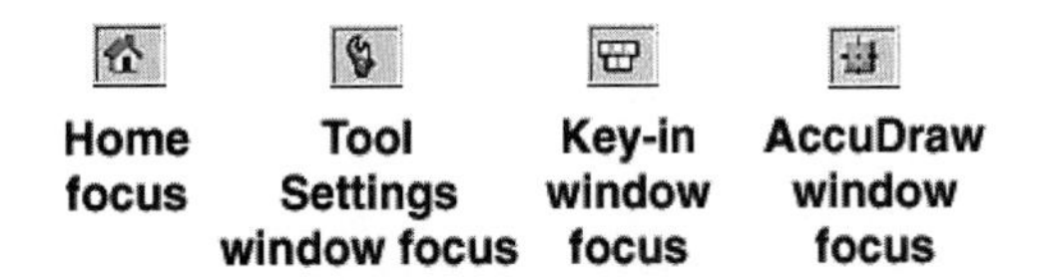

Figure 1–29 Icons at various stages of the focus

When the focus is home, positional keyboard navigation can be used to invoke tool box, task, and tool settings commands without using the mouse. Tools can be quickly selected by pressing keys on the keyboard.

The following keys are mapped to the icons in the Main tool box:

(1, 2, 3, 4, 5, 6, 7, 8, 9, and 0)

Following keys are mapped to the icons in the Task Navigation tool box:

(Q, W, E, R, T, A, S, D, F, G, Z, X, C, V, and B)

Following keys are mapped to the controls in the Tool Settings window:

(Y, U, I, O, P, H, J, K, L, ";", N, M, ",", ".", and/)

Figure 1–30 shows an example of the PC keyboard with symbolic representation of the controls.

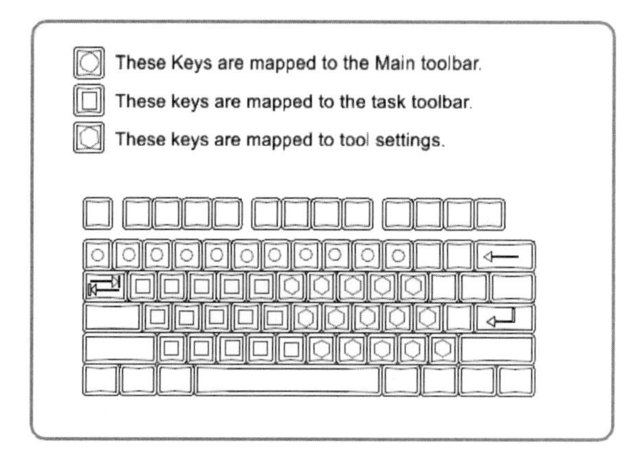

Figure 1–30 *PC Keyboard with symbolic representation of the positional keyboard navigation*

For example, to select the Mirror tool in the Manipulate tool box using positional keyboard navigation, follow these steps:

1. Set the focus to Home.
2. Press **3** to open the Manipulate tool box (the numbers are displayed in the Main tool box as shown in Figure 1–31).
3. A pop-up menu opens at the location of the pointer with the listing of the available tools (see Figure 1–32). Press **5** to select the Mirror tool.

Figure 1–31 *Main tool box*

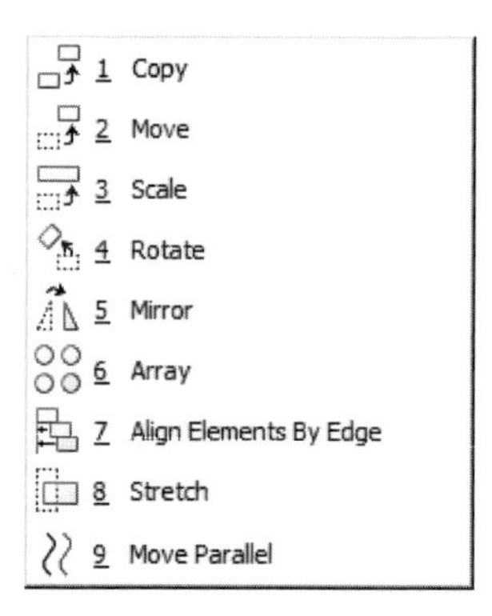

Figure 1–32 *Listing of available tools from the Manipulate tool box*

Following is another example in which the polygon tool is selected from the Task Navigation tool box with active task set to Polygons:

1. Set the focus to Home.
2. Press **R** to invoke the Polygon tool (The letter is displayed in the Task Navigation tool box as shown in Figure 1–33).

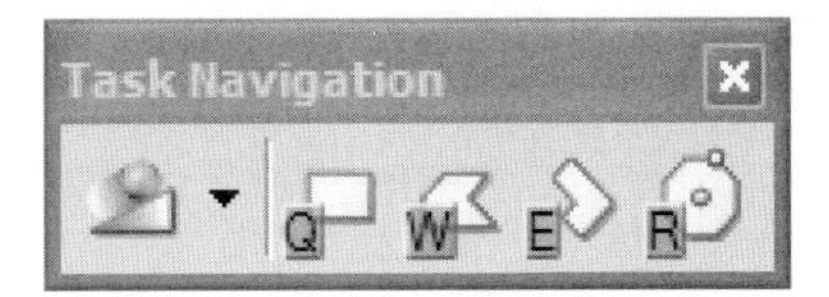

Figure 1–33 *Task Navigation tool box (active task set to Polygons)*

The Preferences dialog box (Position Mapping category) contains controls used to enable, disable, and customize positional keyboard navigation (see Figure 1–34). For a detailed explanation, refer to Chapter 15.

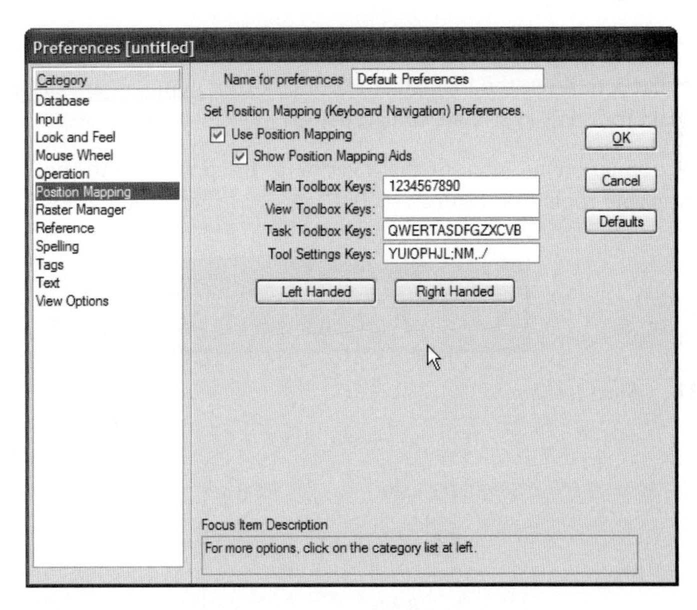

Figure 1–34 *Preferences dialog box (Position Mapping category selection)*

***Note:** Throughout the book, instructions for invoking tools are shown with Positional Keyboard Navigation in addition to accessing from the tool bar.*

POINTING DEVICES

MicroStation supports a two-button mouse, a three-button mouse (see Figure 1–35), and a digitizer. Depending on needs and available hardware, select one of the three pointing devices.

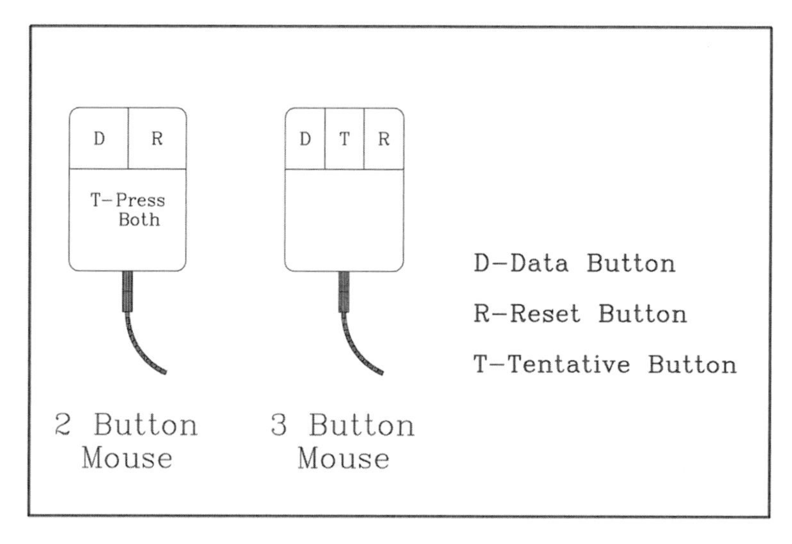

Figure 1–35 *Pointing devices—two-button mouse and three-button mouse*

See Table 1–4 for the specific functions that are programmed by default for the two- and three-button mouse.

Table 1–4 Button Functions on Two- and Three-Button Mice

BUTTON FUNCTION	BUTTON POSITION	
	Two-Button Mouse	**Three-Button Mouse**
Data button	Left (first button)	Left (first button)
Reset button	Right (second Button)	Right (third Button)
Tentative button	Left and right simultaneously	Center (Second Button)

Data Button

This is the most-used button on the mouse/puck. The Data button is used to do the following:

- Select tools from the menus and tool boxes
- Define location points in the design plane
- Identify elements that are to be manipulated

In addition, it is used to accept tentative points, and generally tell the computer "yes" (accept) whenever it is prompted to do so. The Data button is also referred to as the *Identify button* or the *Accept button.*

Reset Button

The Reset button stops the current operation and resets MicroStation to the beginning of the current command sequence. For instance, in the Place Line command, a series of lines can be drawn by using the Data button. To stop the sequence, press the Reset button. MicroStation will stop the current operation and reset the Line sequence to the beginning. In addition, the Reset button also can reject a prompt, and generally tell the computer "no" (reject) whenever it is prompted to do so. The Reset button is also referred to as the *Reject button.*

Tentative Button

The tentative point is one of MicroStation's most powerful features. The Tentative button places a tentative (temporary) point on the screen. If the location of the point is correct, accept it with the Data button. In other words, the tentative point enables experimentation in placement before actually selecting the final resting point for the data point. The Tentative button also can snap to elements at specific locations—for instance, the center and four quadrants of a circle, when the Snap lock is set to ON. For a detailed explanation, see Chapter 3.

Command Button

The Command button is only available on the digitizer tablet puck and is used to select commands from the tablet menu. To do so, look down the tablet menu, place the puck crosshairs in the box that represents the desired command, and then press the Command button. The corresponding command is invoked.

Button Mappings dialog box as shown in Figure 1–36 which is opened from Button Assignments dialog box (Workspace > Button Assignments) allows for reviewing/changing the current physical button mappings of the pointing device. Each physical button (or button combination) can be mapped to a logical button, which performs an action.

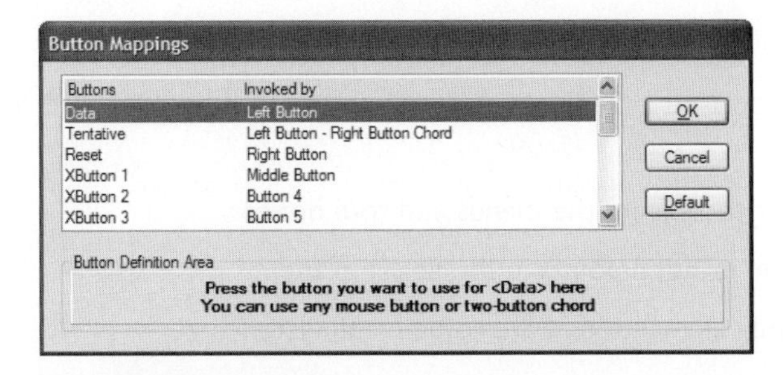

Figure 1–36 *Button Mappings dialog box*

Once a logical button is selected in the list box above, move the pointer to the Button Definition Area and press the physical button or button combination that will be used to invoke this logical button. A physical button (or button combination) cannot have more than one action assigned to it.

CURSOR MENU

MicroStation has a cursor menu that can be made to appear at the location of the cursor by pressing the designated button on the pointing device. Two cursor menus are available, one for View Control tools and another for Snap Mode options.

To invoke the View Control tools (Figure 1–37), press SHIFT plus the Reset button. The menu includes all the available tools for controlling the display of the design.

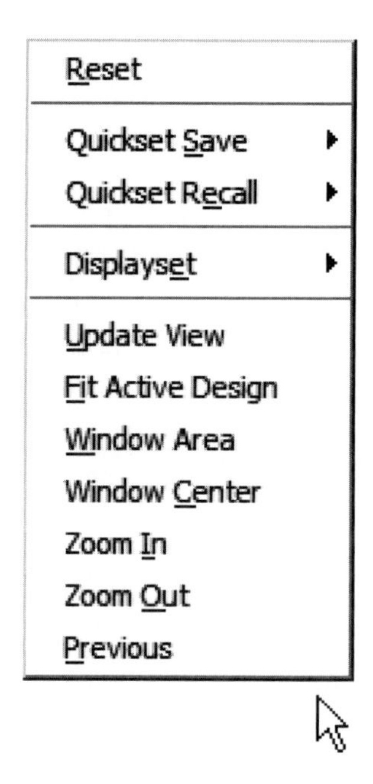

Figure 1–37 *Cursor menu (SHIFT + Reset button)—View Control tools*

To invoke the Snap Mode options (Figure 1–38), press SHIFT plus the Tentative button. The menu includes all the available Snap Mode options.

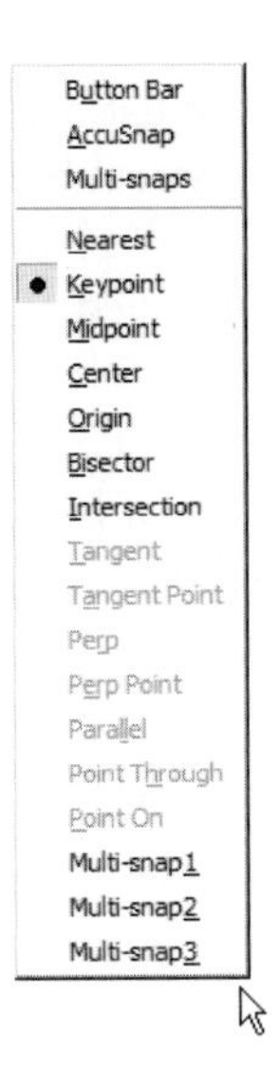

Figure 1–38 *Cursor menu (SHIFT + Tentative button)—Snap Mode options*

THE DESIGN PLANE

In conventional drafting, the drawing is normally done to a certain scale, such as 1′ = 1″. But in MicroStation, drawing is done full scale: all lines, circles, and other elements are drawn and measured as full size. For example, if a part is 150 feet long, it is drawn as 150 feet actual size. For plotting the part, MicroStation scales the design to fit a given sheet size. Alternatively, a scale factor can be specified for plotting on a given sheet size.

When a new *two-dimensional* design is started, it is given a design plane—the electronic equivalent of a sheet of paper on a drafting table. Unlike a sheet of drafting paper, however, the design plane (or cube in 3D) in a design file is extremely large, letting models be drawn at full scale. To draw various elements in the model, data points are entered. The center of the design plane is the global origin and is assigned coordinates (0,0). Any point to the right of the global origin has a positive *X* value; any point to the left has a negative *X* value. Any point above the global origin has a positive *Y* value; any point below has a negative *Y* value.

When a data point is entered, MicroStation saves its coordinates in 64-bit floating point format. The 3D design cube is similar to the 2D design plane, but with a third axis Z (depth). Points in 2D models are stored as coordinate values expressed in the form (X,Y), while those for 3D models are stored as (X,Y,Z).

If necessary, the location or coordinates of the global origin can be changed. For example, an architect might want all coordinates to be positive in value, so he or she would set the global origin at the bottom left corner of the design plane. To relocate the Global Origin, key-in **GO=0,0** (ENTER) and click the Reset button.

The Global Origin will be relocated to the lower left corner of the Design Plane and assigns the coordinate 0,0. To relocate the Global Origin to any location on the Design Plane, key-in **GO=0,0** (ENTER) and specify a data point to relocate the Global Origin.

WORKING UNITS

Working units are the real-world units used in creating models in a design file. Typically, the working units are defined in seed files, from which design files are created. Normally, they will not require any adjustment.

Working Units are comprised of Master or Major units (MU) and Sub Units (SU). The Master Unit is the largest unit being used in the design, such as feet and meters. The fractional parts of a Master Unit are called Sub Units, such as inches or centimeters. Sub Units cannot be larger than Master Units. Working units can be changed without affecting the size of elements in the design. For example, initial settings of feet and inches for Master units and Sub units can be changed to meters and centimeters to get the equivalent metric measurements. Existing elements and new elements to be drawn will reflect the change in units. Distances, such as measurements, that are input in design files are typically expressed in either of two forms:

- As a standard decimal number, such as 3.750
- As two numbers separated by a colon which represent Master Units:Sub Units (MU:SU). For example, 3:4 means 3 master units and 4 sub units.

The following are the options available to key-in 3 1/8 feet when the Working Units are set as feet for Master Units and inches for Sub Units:

3.125 *(In terms of feet)*

3:1.50 *(In terms of feet and inches.)*

0:37.50 *(In terms of inches.)*

3:1 1/2 *(In terms of feet and inches in fractions.)*

If the appropriate seed file is used to create the new design file, then the suitable Working Units will be set. To draw an architectural floor plan, use *2DEnglishArch.DGN* (architectural seed file) as a seed file; MicroStation sets the Working Units to feet and inches for Master Units and Sub Units respectively.

If necessary, the current Working Units can be changed. To do so, open the Design File Settings dialog box from:

Menu	Settings > Design File

MicroStation displays the Design File Settings dialog box, similar to Figure 1–39.

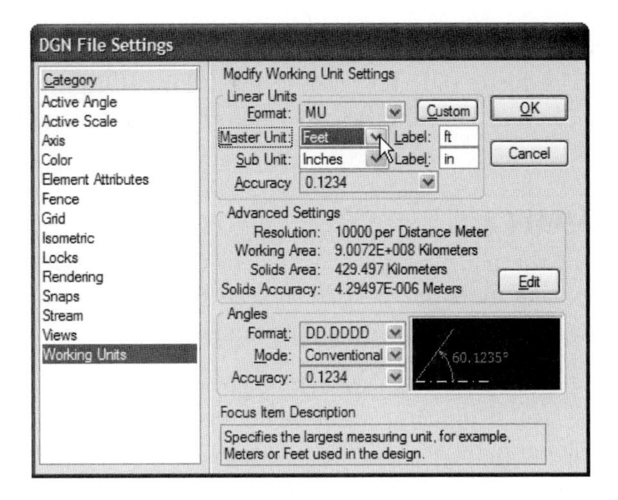

Figure 1–39 *Design File Settings dialog box*

Select **Working Units** in the **Category** list box. MicroStation displays the appropriate controls needed to modify the Working Units parameters (see Figure 1–39).

Select the **Master Units** from the **Master Unit** list box and appropriate label appears in the **Label** edit field. If necessary, change the label. Select the Sub Units from the **Sub Unit** list box and appropriate label appears in the **Label** edit field.

If the Master Unit is changed from Metric to English or vice versa, the Sub Unit is changed to a suitable unit. Similarly, if a Master Unit is specified that is smaller than the current Sub Unit, the Sub Unit is changed to a suitable unit.

The **Format** and **Accuracy** settings control how MicroStation displays coordinates, distances, and angles in the status bar and dialog boxes. Setting the Format and Accuracy does not affect the accuracy of calculations, only the accuracy with which the results are displayed.

- The **Format** option menu sets which units are displayed. Master Units selection displays only master units, Sub Units selection displays Master and Sub units in MU:SU format, and Working Units selection displays in Master, Sub and positional units in MU:SU:PU format.
- **Accuracy** sets decimal accuracy up to six decimal places for coordinates, 8 decimal places for angles or fractional accuracy to 1/2, 1/4, 1/8, 1/16, 1/32, or 1/64.

The **Angles** section contains controls that are used to set the format, direction, and accuracy of angle readout.

- **Format** sets the angle readout format.
- **Mode** sets the manner in which angles are measured.
- **Accuracy** sets decimal accuracy up to four decimal places.

CUSTOM UNITS

Custom Units is used to define customized settings for Master Unit and Sub Unit. Click the Custom button in the Design File Settings dialog box to open the Define Custom Units dialog box (see Figure 1–40).

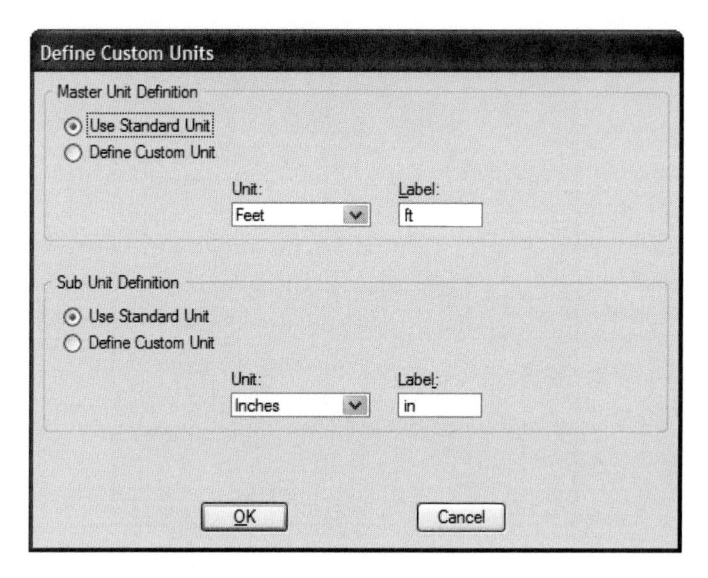

Figure 1–40 *Define Custom Units dialog box*

The **Master Unit Definition** section contains controls for selecting a Standard Unit or defining a Custom Unit to be the design Master Unit.

- **Use Standard Unit** provides options to select standard Master Unit settings recognized by MicroStation.
- **Define Custom Unit** creates a custom Master Unit, which is defined relative to one of the recognized Metric or English units.

The **Sub Unit Definition** section contains controls for selecting a Standard Unit or defining a Custom Units to be the design Sub Unit.

- **Use Standard Unit** provides options to select standard Sub Unit settings recognized by MicroStation.
- **Define Custom Unit** create a custom Sub Unit, which is defined relative to one of the recognized Metric or English units.

After making necessary changes, click the OK button to close the Define Custom Units dialog box.

ADVANCED UNITS SETTINGS

The Advanced Units Settings dialog box contains controls for setting the resolution of the design environment, which sets its size and accuracy. Click the Edit button in the Design File Settings dialog box to open the Advanced Unit Settings dialog box (see Figure 1–41).

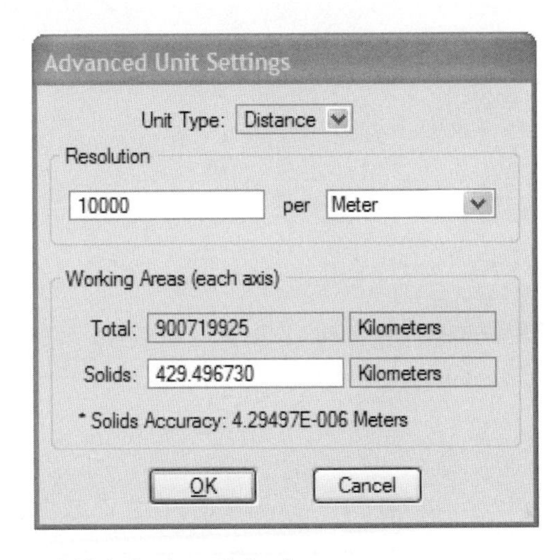

Figure 1–41 *Advanced Unit Settings dialog box*

The **Unit Type** option menu determines if the units in this design file are units of measurement or unitless representation.

- **Distance** selection set units are a unit of measurement.
- **Unitless** selection set units are other than a unit of linear measurement. For example, Longitude or Latitude.

The **Resolution** section determines the accuracy of the design plane/cube. The **Resolution** setting defines the worst-case accuracy for the design environment, which occurs at the very outer limits of the (very large) working area/volume. For example, working to a "worst-case" accuracy of 0.0001 meters, the size of the design plane/cube is 900 million kilometers along each axis. In almost all cases, therefore, there is no need to change the **Resolution** setting.

The **Working Areas (each axis)** section displays the length of each axis of the working environment (expressed in Miles or Kilometers when **Unit Type** is Distance) depending on the resolution. This area is recalculated, automatically, if the resolution is changed.

Note: *Do not change the* ***Resolution*** *if elements have already been placed in the design. Changing the* ***Resolution*** *alters the size of existing elements.*

Click the OK button to save any changes and close the dialog box.

Note: *The working units can also be viewed/changed from the Drawing Scale window (see Figure 1–42) available from the Settings menu.*

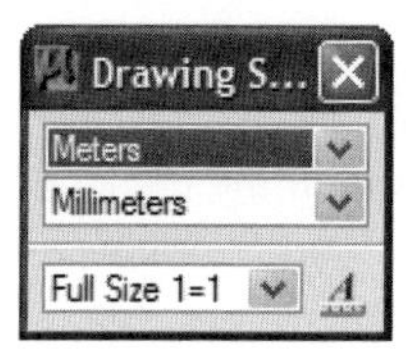

Figure 1–42 *Drawing Scale window*

CREATING MODELS

Each model in a design file is a set of design elements (such as lines, circles, etc.) that are unique to that model. A model has its own set of eight views and serves as a container for the geometry forming the model. Each model in a design is totally independent; in fact they may even have different working units. There is no limit to the number of models in a design file. For example, one Design model may be used for a house plan and another for its site details. One may have feet and inches for working units, and other may have decimal feet. Only one model can be active at any time.

MicroStation can create two types of models: Design and Sheet. As mentioned earlier, Design models contain the actual designs, Sheet models are used to compose drawings. Sheet models may contain reference designs (refer to Chapter 13 for detailed explanation), annotation and dimension as required, borders, title blocks etc. Printouts will normally be created from Sheet models.

The Models settings box is used to create, manage and switch between models in the open DGN file. Open the Models settings box from:

Primary Tools tool box	Select the Models (see Figure 1–43).
Menu	File > Models

MicroStation displays the Model settings box, similar to Figure 1–44.

Figure 1–43 *Invoking the Models from the Primary Tools tool box*

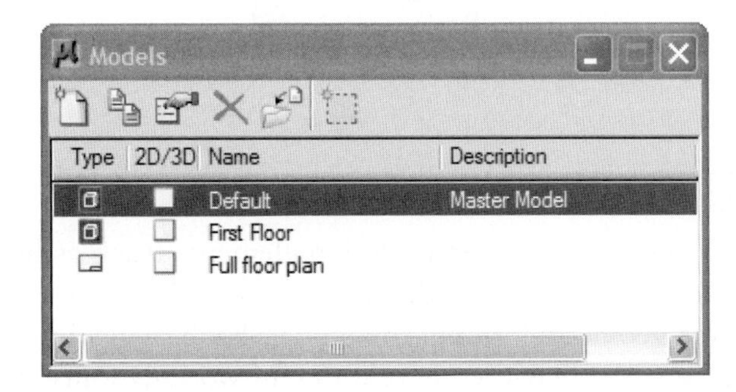

Figure 1–44 *Models settings box*

A model can be created as a 2D or 3D Design type, or as a Sheet. Icons at the top of the Models settings box are used to access its various functions. To switch to an active model or sheet, double-click on the name in the list box. The selected name becomes the active model and name of the active Model or Sheet is displayed as part of the View Window name.

CREATE A NEW MODEL/SHEET

To create a new model or sheet, click the Create a New Model icon. MicroStation opens a Create Model dialog box similar to Figure 1–45.

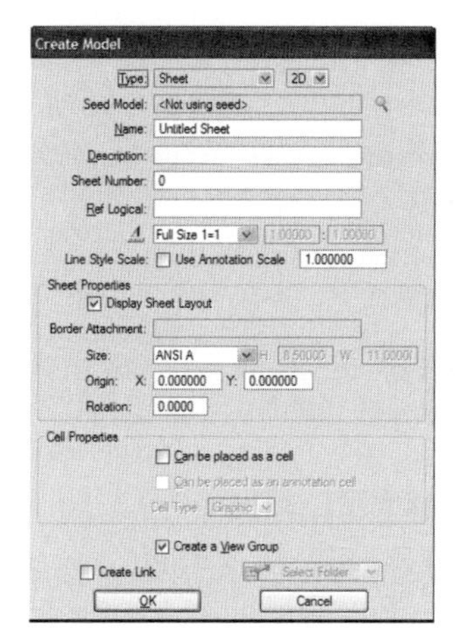

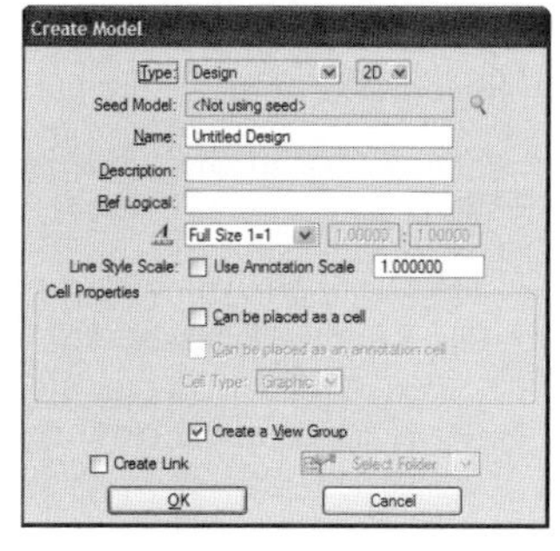

Figure 1–45 *Create Model dialog box (Model and Sheet creation dialog boxes)*

The **Type** option menu sets the type of model created.

Design selection is used to create a design model in either 2D or 3D.

Sheet selection is used to create a sheet model and attach references to create a drawing. This sheet model can be 2D or 3D.

Design From Seed selection creates a design model using the set of eight views from a model in the selected seed DGN file. The selected model also determines whether the new design model is 2D or 3D. The **Seed Model** menu allows selecting the DGN file.

Sheet From Seed selection creates a sheet model from a model in the selected seed DGN file. It includes any attached references present in the seed model. The selected model also determines whether the new sheet model is 2D or 3D. The Seed Model menu allows selecting the DGN file.

Name field is used to enter a name for the model.

Description field is used to add a description of the model.

Sheet Number (available Sheet selection only) assigns a sheet number to sheet models. This makes it easy to control the order of sheet models for presentation, printing, cataloging in a project, or generating PDFs.

Ref Logical fields is used to set the Reference Logical name for the model. The logical name identifies the model when it is attached to another model as a reference.

Annotation Scale icon displays the status of the Annotation Scale lock: ON or OFF. When placing text, the lock must be on to ensure that text is placed at the defined scale (Refer to Chapter 7 for detailed explanation). The option menu sets the scale factor for text and dimensioning in the model. Select from a list of common scales, or select CUSTOM and input scale information in the fields immediately to the right.

Sheet Properties section (available Sheet selection only) allows customizing the sheet properties, such as, size, origin, rotation, and controlling the display of the sheet elements.

Cell Properties section provides options for placing the design/sheet as a cell in the design file. (Refer to Chapter 11 for detailed explanation on Cells and Cell Libraries.)

Create a View Group check box sets whether a View Group is created with the model.

Create Link check box when set to ON creates a link to this model in the currently active Project Explorer link set.

Click the OK button to create a new model or sheet, as the case may be, and close the dialog box.

COPYING AN EXISTING MODEL/SHEET

To copy an existing model in the open DGN file, click the **Copy a Model** icon. MicroStation opens the Copy Model dialog box similar to Figure 1–46.

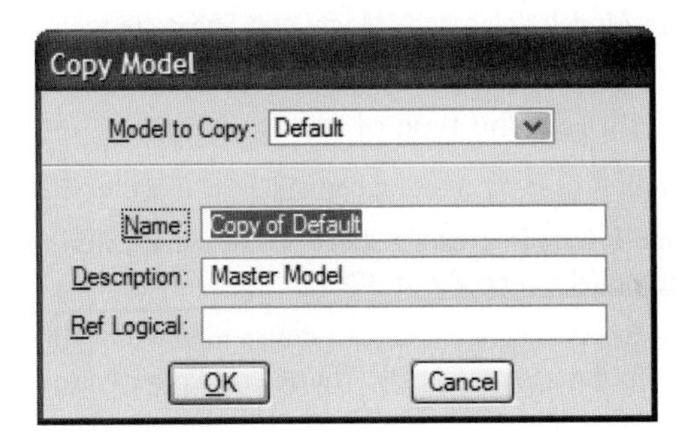

Figure 1–46 *Copy Model dialog box*

Model to Copy option menu is used to choose a model contained in the open DGN file (or the selected DGN file if importing a model).

Name field is used to enter a name for the copy of the model.

Description field is used to enter a description of the copied model.

Click the OK button to create a copy of the selected model and close the dialog box.

EDIT MODEL/SHEET PROPERTIES

To edit properties of a selected model/sheet contained in the open DGN file, click the **Edit Model Properties** icon. MicroStation opens the Model Properties dialog box where changes to the model/sheet properties can be made. After making necessary changes, click the OK button to accept the changes and close the dialog box.

DELETE A MODEL/SHEET

To delete a selected model/sheet, click the Delete Model icon. MicroStation deletes the selected model.

IMPORT A MODEL/SHEET

MicroStation can import a model/sheet into the open DGN file from another DGN or DWG file. To import a model/sheet, click the Import a Model icon. MicroStation opens the Import Model From File dialog box, which is similar to the Open dialog box. Choose a DGN or DWG file from which to select the model to import. Click the Open button the Select Models dialog box opens, from which a model/sheet is selected to copy into the open DGN file.

DRAWING PROPERTIES

The Properties dialog box is used to review the active DGN file's general properties and usage statistics and change the file's design properties. Open the Properties dialog box from:

Drop-down menu	File > Properties

MicroStation displays the Properties dialog box similar to Figure 1–47.

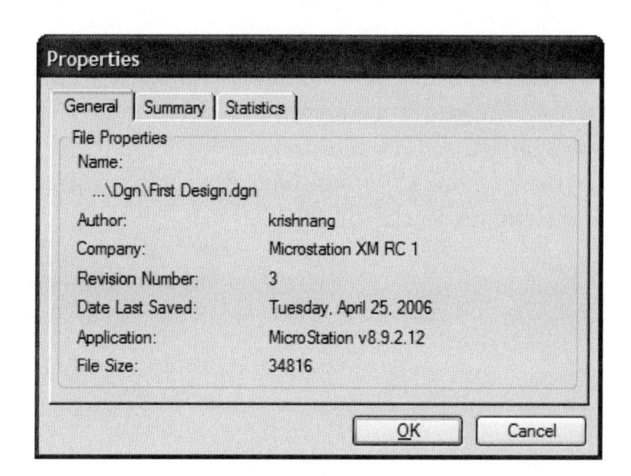

Figure 1–47 *Properties dialog box (General tab)*

The **General** tab of the dialog box displays general information (see Figure 1–47) about the DGN file; such as name of the design file with full path, name of the person who last edited the DGN file, company of the person who last edited the DGN file, number of times the DGN file has been edited and saved, date the last time the DGN file was edited and saved, software version on which the design was created, and size of the DGN file.

The **Summary** tab (see Figure 1–48) of the dialog box contains controls for editing a file's properties. Helpful information can be stored here, such as the project to which the file is related, explanation of the drawing or the client to which this

drawing is related, query keyword to find the file if it is in a database, comments associated with the file and the name of the manager associated with the file.

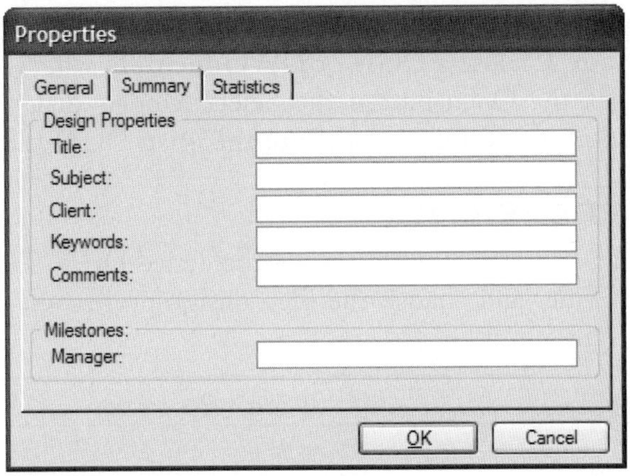

Figure 1–48 *Properties dialog box (Summary tab)*

The **Statistics** tab (see Figure 1–49) contains statistical information about the DGN file. Information includes total amount of time the DGN file has been edited, number of levels in the DGN file, number of levels used in the DGN file, number of models in the DGN file, number of references attached to the DGN file, and number of elements in the DGN file.

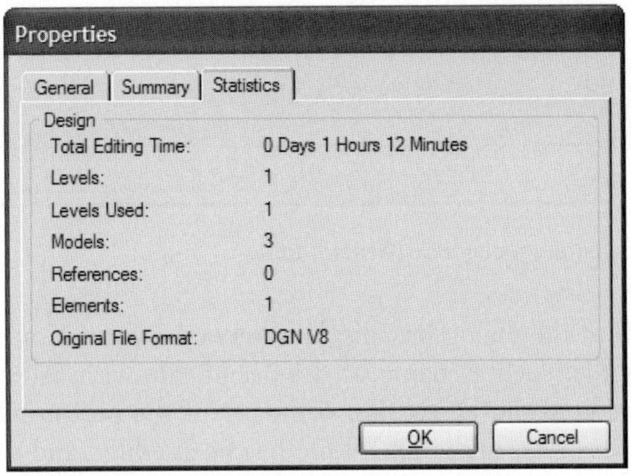

Figure 1–49 *Properties dialog box (Statistics tab)*

Click the OK button to accept the changes and close the Properties dialog box.

SAVING CHANGES AND EXITING THE DESIGN FILE

Before we discuss how to place elements in the design file, let's discuss how to save the current design file. By default, MicroStation saves all elements in a design file as they are drawn. There is no separate Save command. Even if there is a power failure during a design session, most of the design file will have been saved without significant loss.

If desired, set the **Immediately Save Design Changes** check box to OFF. Then MicroStation provides a Save tool in the **File** menu to save the design file. No other automatic save feature is provided. In order to save the design file, **Save** must be invoked from the **File** menu.

To change the status of the check box for **Automatically Save Design Changes**, open the Preferences dialog box from:

Menu	Workspace > Preferences

MicroStation displays the Preferences dialog box, similar to Figure 1–50.

Select the **Operation** option from the **Category** list box and set the check box to ON/OFF for **Automatically Save Design Changes**.

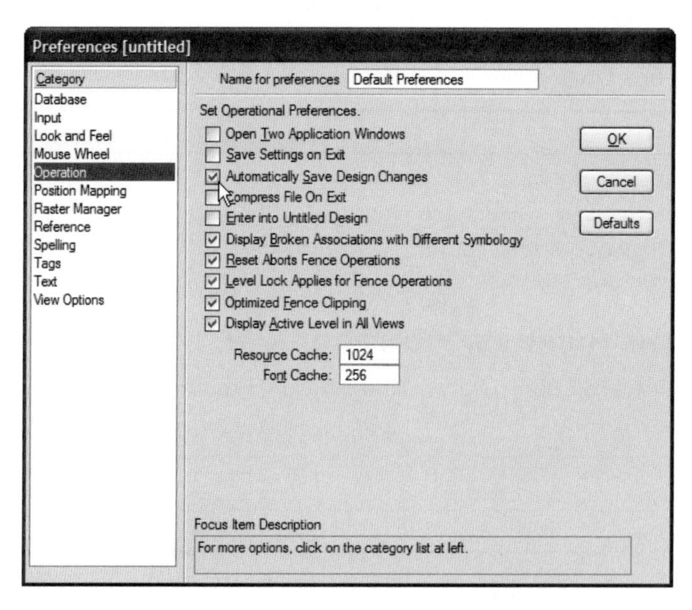

Figure 1–50 *Preferences dialog box with Operation category selection*

To save the current design file to a different file name, invoke the Save As command from:

Menu	File > Save As

MicroStation displays the Save As dialog box. Select the folder in which to save the design file, and key-in the name of the design file in the Files edit box. Click the **OK** button to save the file.

SAVING THE DESIGN FILE SETTINGS

Design file settings such as working units, grid spacing, or view settings, must be saved explicitly. To do so, invoke the Save Settings tool from:

Menu	File > Save Settings
Key-in window	**file design** (or **fil**) (ENTER)

MicroStation saves the current settings.

If the settings are not saved, time will be wasted adjusting the design and view settings to match the original intent.

To automatically save the settings from exiting the design file, set the check box for **Save Settings on Exit** to ON in the Preferences dialog box. By default it is set to OFF.

To change the status of the check box for **Save Settings on Exit**, open the Preferences dialog box from:

Menu	Workspace > Preferences

MicroStation displays the Preferences dialog box. Select the **Operation** option from the **Category** list box, and set the check box to ON/OFF for **Save Settings on Exit.**

EXITING THE MICROSTATION PROGRAM

To exit the MicroStation program and return to the operating system, invoke the Exit tool from:

Menu	File > Exit
Key-in window	**exit** (or **exi**) (ENTER)

MicroStation exits the program and returns to the operating system.

To return to the MicroStation Manager dialog box, invoke the Close tool from:

Menu	File > Close
Key-in window	**close design** (or **clo d**) (ENTER)

MicroStation returns to the MicroStation Manager dialog box.

GETTING HELP

MicroStation provides an online help facility available from the Help drop-down menu. Online help is provided through the Help window by specific topics, by searching for a text string within help topic names or help articles, or by browsing key-ins.

CONTENTS

The Contents window lists the top-level topics, as shown in Figure 1–51. To see a list of more specific subtopics related to a topic in the list, select the topic. MicroStation displays a list of subtopics, and by selecting a subtopic, MicroStation displays the available help information on that topic. Use text strings to search for relevant help (under the **Search** tab). Then click the **List Topics** button and MicroStation will display all the related topics.

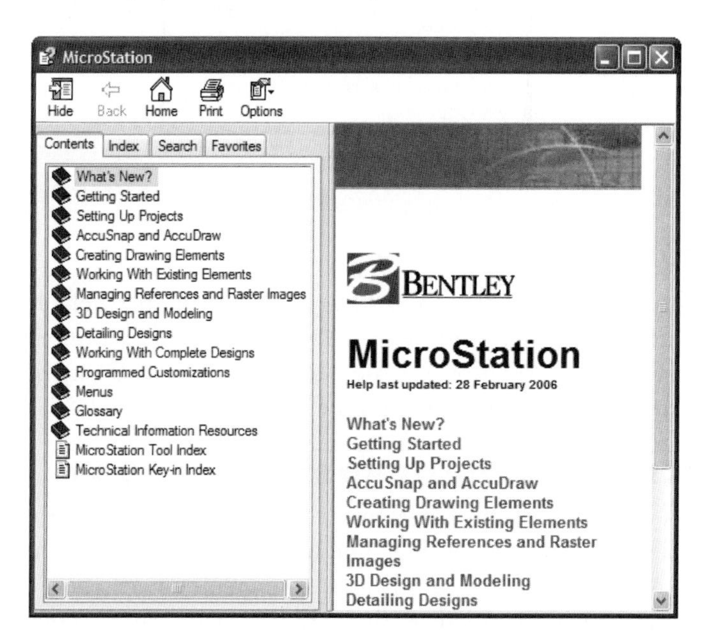

Figure 1–51 *Help Contents window*

Display the Help file's previous article by clicking the Back button.

TOOL INDEX

Tool Index from the Help menu opens the Tool Index window in the online help system as shown in Figure 1–52. Search for a tool with the **Find** form or select a tool from the alphabetical list. When the desired tool is selected, the associated tool documentation is displayed in the Help window.

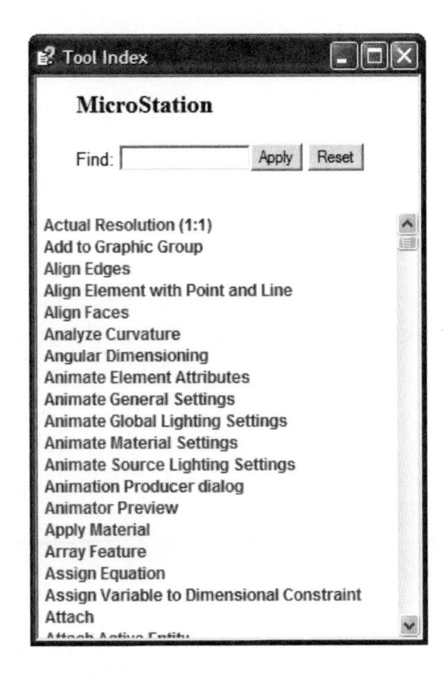

Figure 1–52 *Tool Index window*

ONLINE SUPPORT

Choosing **Online Support** from the Help menu opens the default browser and connects to the MicroStation resource website. MicroStation on the Web offers tools, resources and information for the MicroStation user from within a MicroStation session. It's a window to the online world of MicroStation.

ENHANCEMENTS IN MICROSTATION V8 XM

MicroStation V8 XM Edition represents a major step forward in the evolution of the entire MicroStation product line. The unparalleled combination of power and simplicity in MicroStation V8 XM Edition is driven by major new innovations in what is already widely regarded as the AEC industry's most powerful software platform.

Following are some of the features/benefits that were added to MicroStation V8 XM:

Updated GUI—An updated user interface is introduced in MicroStation V8 XM Edition with improved customization tools to simplify design tasks. Keyboard

mapping gives each user the flexibility to configure their entire keyboard, giving them immediate access to any MicroStation command with the touch of a key.

New Display Subsystem—MicroStation V8 XM Edition includes a new, more powerful, and faster graphics system which significantly increases view and navigation speed in 2D and 3D designs. This system leverages Microsoft DirectX technology and includes dialog and element transparency, integration with PANTONE colors, and display priority.

Structured Workflows—MicroStation V8 XM Edition dynamically applies a specific set of tools, standards, and interface elements to a particular task in a work process, enabling users to create consistent work. This innovation results in the standardization of core processes, enhanced efficiency, and fewer errors throughout design.

Structured Content—Managing and organizing project content is made simpler by providing users with an index of their project-specific information. This logical, structured view of project content enables users to easily navigate project content, define project organization, and establish and navigate links within and between designs, drawings, and supporting documentation.

3D in PDF—With the touch of a button, an entire AEC project—including 3D models, MicroStation and AutoCAD drawings, specifications, and PDF renditions of Microsoft Office files—can be packaged in a single PDF document.

CHAPTER 2

Fundamentals I

OBJECTIVES

Topics explored in this chapter:

- Drawing lines, blocks, shapes, circles, and arcs
- Dropping blocks and shapes and deleting elements
- Using Precision Input

PLACEMENT COMMANDS

MicroStation provides various tools for drawing objects. This section explains in detail the various tools for drawing lines, blocks, shapes, circles, and arcs.

PLACE LINE

The primary drawing element is the line, and the Place Line tool is used to draw series of lines. Invoke the Place Line tool from:

Task Navigation tool box (active task set to Linear)	Select the Place Line tool (see Figure 2–1).
Keyboard Navigation (Task Navigation tool box with active task set to Linear)	**w**

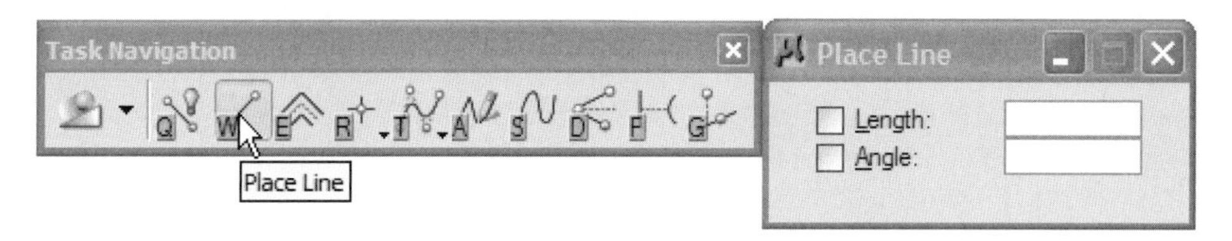

Figure 2–1 Invoking the Place Line tool from the Task Navigation tool box (active task set to Linear)

MicroStation prompts:

Place Line > Identify start of line

Specify the first point by providing a data point via a pointing device (mouse or puck) or by Precision Input coordinates (see the discussion, later in this chapter, on "Precision Input" for a more detailed explanation). After the first point is specified, MicroStation prompts:

Place Line > Identify end of line

Specify the end of the line by placing a data point via a pointing device (mouse or puck) or by Precision Input. MicroStation repeats the prompt:

Place Line > Identify end of line

Place a data point via the pointing device or by Precision Input to continue. To save time, the Place Line tool remains active and prompts for a new endpoint after each point specified. After placing a series of lines, press the Reset button or invoke another tool to terminate the Place Line tool.

When placing data points with a pointing device to draw a series of lines, a rubber-band line is displayed between the starting point and the crosshairs. This indicates where the resulting line will be placed. In Figure 2–2 the dotted lines represent previous cursor positions. To specify the endpoint of the line, click the Data button. Continue placing lines, then press the Reset button or select another tool to terminate the Place Line tool.

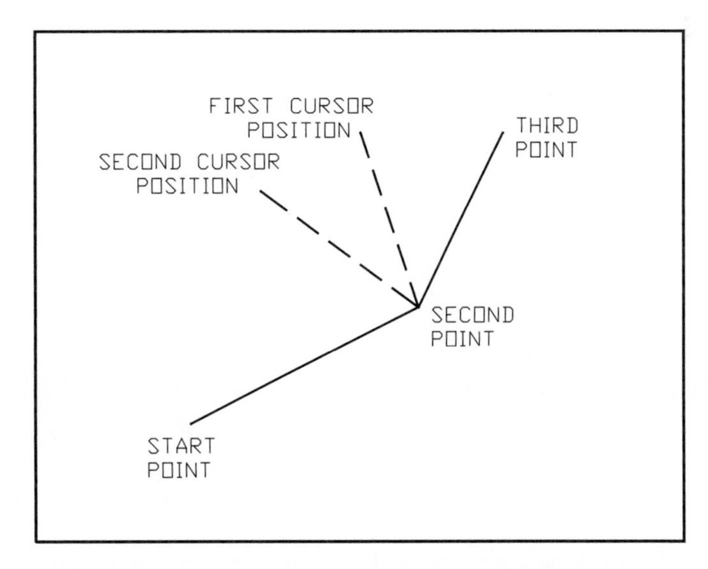

Figure 2–2 Placing data points with the cursor rather than with coordinates

Place Line of a Specified Length

To place a line of a specified length, select the Place Line tool in the Navigation tool box (active task set to Linear) and set the check box for **Length** to ON in the Tool Settings window. Type the distance, in MU:SU:PU format, in the **Length**

edit field. The prompts are similar to those for the Place Line tool, and any number of line segments of specified length can be placed.

Place Line at an Angle

To place a line at a specified angle, select the Place Line tool in the Navigation tool box (active task set to Linear) and set the check box for **Angle** to ON in the Tool Settings window. Key-in the angle in the **Angle** edit field. The prompts are similar to those for the Place Line tool, and any number of line segments at a specified angle can be placed.

If necessary, the check boxes for both **Length** and **Angle** edit fields may be turned ON to place lines with specific length and constrained angle.

PLACE BLOCK

Rectangular blocks may be drawn by two different methods: orthogonal and rotated.

Orthogonal Method

The Place Block tool (orthogonal) creates rectangular block using two points that define the diagonal corners of the shape. Place the two diagonal corners by specifying data points via a pointing device or by keying-in *two-dimensional (2D)* coordinates (see the discussion later in this chapter on "Precision Input"). Invoke the Place Block (orthogonal) tool from:

Task Navigation tool box (active task set to Polygons)	Select the Place Block tool and Orthogonal from the Method option menu located in the Tool Settings window (see Figure 2–3).
Keyboard Navigation (Task Navigation tool box with active task set to Polygons)	**q** (select Orthogonal from the Method option menu located in the Tool Settings window).

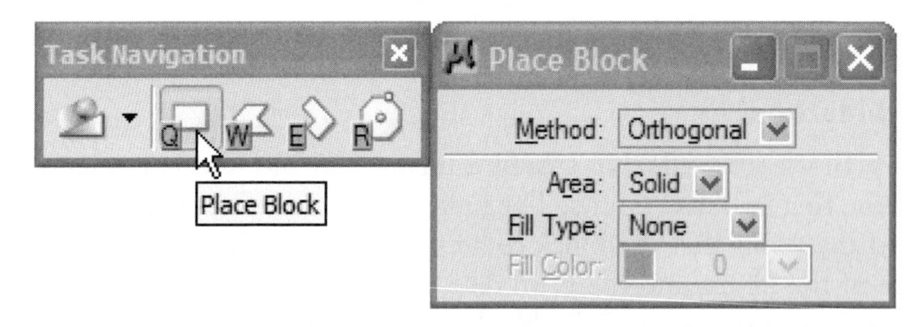

Figure 2–3 Invoking the Place Block (orthogonal) tool from the Task Navigation tool box (active task set to Polygons)

MicroStation prompts:

> Place Block > Enter first point *(Place a data point or key-in coordinates to define the start point of the block.)*
>
> Place Block > Enter opposite corner *(Place a data point or key-in coordinates to define the opposite corner of the block.)*

A block is a single element, and element manipulation tools such as Move, Copy, and Delete manipulate a block as one element. If necessary, blocks can be made into individual line elements with the Drop Line String tool.

For example, the following command sequence shows placement of a block by placing two data points diagonally opposite each other, as shown in Figure 2–4, using the Place Block (orthogonal) tool.

> Place Block > Enter first point *(Place a data point as shown in Figure 2–4.)*
>
> Place Block > Enter opposite corner *(Place a data point diagonally opposite to the first point.)*

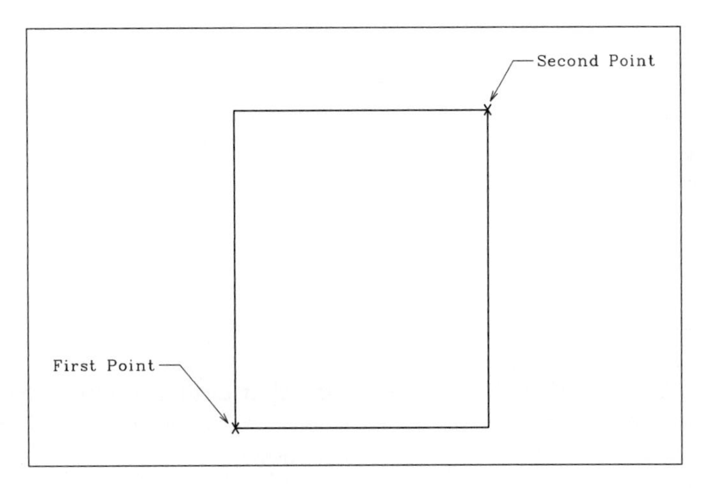

Figure 2–4 An example of placing an orthogonal block using the Place Block (orthogonal) tool

Rotated Method

The Place Block (rotated) tool creates a rectangular block at any angle that is defined by the first two data points. The first data point defines the first corner of the block and the point that the block rotates around. The second data point defines the angle of the block, and the third data point, entered diagonally from the first, defines the opposite corner of the block. Invoke the Place Block (rotated) tool:

Task Navigation tool box (active task set to Polygons)	Select the Place Block tool and Rotated from the Method option menu located in the Tool Settings window (see Figure 2–5).
Keyboard Navigation (Task Navigation tool box with active task set to Polygons)	**q** (select Rotated from the Method option menu located in the Tool Settings window)

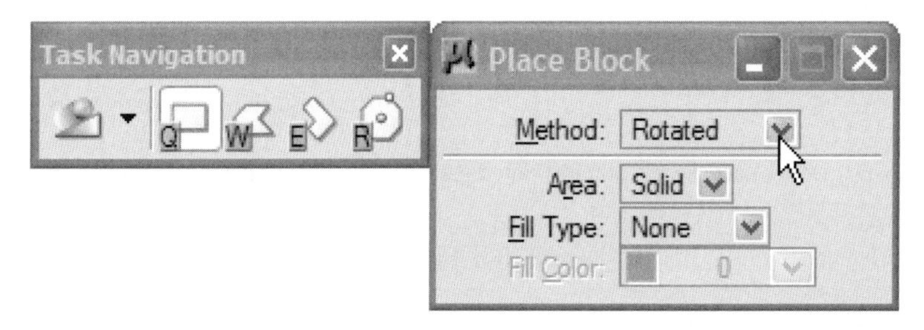

Figure 2–5 Invoking the Place Block (rotated) tool from the Task Navigation tool box (active task set to Polygons)

MicroStation prompts:

> Place Rotated Block > Enter first base point *(Place a data point or key-in coordinates to define the start point of the block.)*
>
> Place Rotated Block > Enter second base point *(Place a data point or key-in coordinates to define the angle of the block.)*
>
> Place Rotated Block > Enter diagonal point *(Place a data point or key-in coordinates to define the opposite corner of the block.)*

See Figure 2–6 for an example of placing a rotated block with the Place Block (rotated) tool by providing three data points.

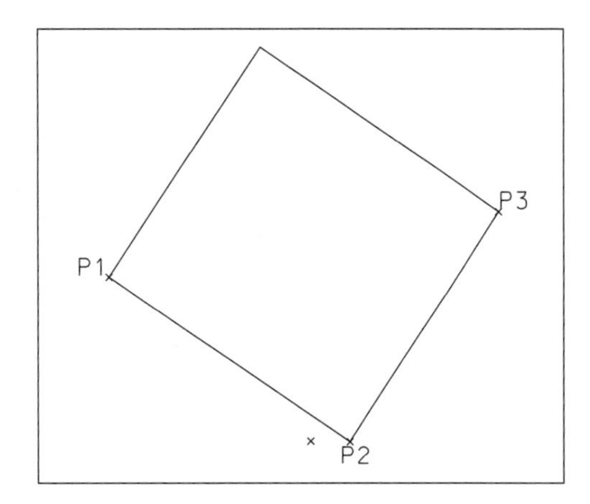

Figure 2–6 An example of placing a rotated block with the Place Block (rotated) tool

As with the orthogonal block, a rotated block is also a single element.

PLACE SHAPE

The Place Shape tool is used to place a multi-sided shape defined by a series of data points (3 to 100) that indicates the vertices of the polygon. To complete the polygon shape, the last data point should be placed on top of the starting point. The starting point and subsequent points are specified via Precision Input or by using a pointing device. Invoke the Place Shape tool from:

Task Navigation tool box (active task set to Polygons)	Select the Place Shape tool (see Figure 2–7).
Keyboard Navigation (Task Navigation tool box with active task set to Polygons)	**w**

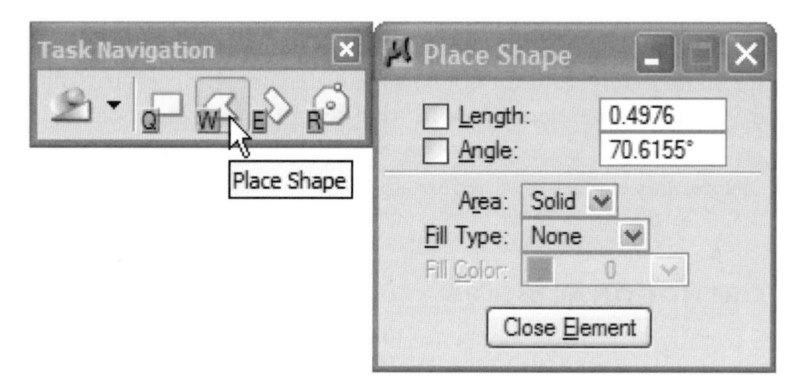

Figure 2–7 Invoking the Place Shape tool from the Task Navigation tool box (active task set to Polygons)

MicroStation prompts:

> Place Shape > Enter first point *(Place a data point or key-in coordinates to define the starting point of the shape.)*
>
> Place Shape > Enter vertex or Reset to cancel *(Place a data point or key-in coordinates to define the vertex, or press Reset button to cancel.)*

Continue placing data points. To complete the polygon shape, place the last data point on top of the starting point or click the **Close Element** button located in the Tool Settings window.

Length and/or Angle constraints may also be utilized with the Place Shape tool by turning on the check boxes for **Length** and **Angle**, appropriately located in the Tool Settings window.

See Figure 2–8 for an example of placing a closed shape with the Place Shape tool by providing six data points.

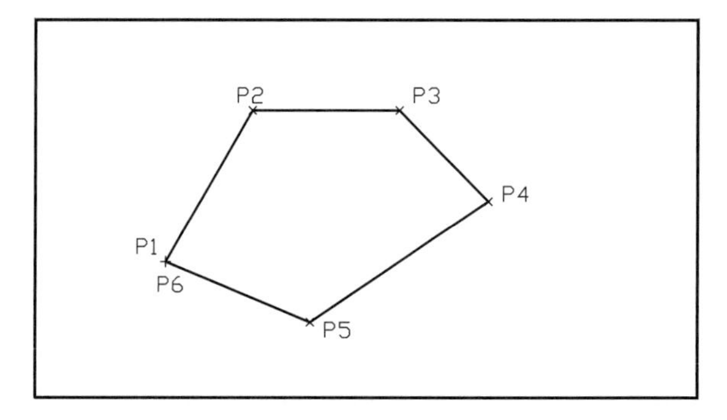

Figure 2–8 An example of placing a closed shape with the Place Shape tool

A shape is considered a single element. Element manipulation tools such as Move, Copy, and Delete manipulate the shape as one element. If necessary, shapes can be made into individual line elements with the Drop Line String tool.

Note: *Area and Fill type options are explained in the Patterning section of Chapter 12*

PLACE ORTHOGONAL SHAPE

The Place Orthogonal Shape tool is used to create a multi-sided shape that has adjacent sides at right angles. As with the Place Block (rotated) tool, the first two points define the vertices of the orthogonal shape. The additional points define the corners of the shape. To complete the polygon shape, the last data point should be placed on top of the starting point. The starting point and subsequent points may be specified with absolute or relative coordinates (see "Precision Input," later) or by using a pointing device. Invoke the Place Orthogonal Shape tool from:

Task Navigation tool box (active task set to Polygons)	Select the Place Orthogonal Shape tool (see Figure 2–9).
Keyboard Navigation (Task Navigation tool box with active task set to Polygons)	**e**

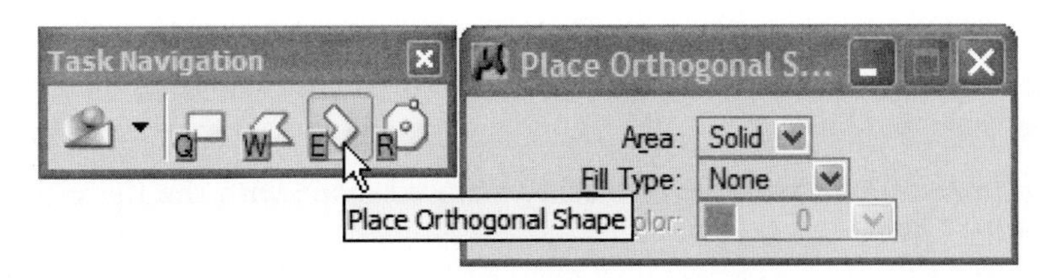

Figure 2–9 Invoking the Place Orthogonal Shape tool from the Task Navigation tool box (active task set to Polygons)

MicroStation prompts:

> Place Orthogonal Shape > Enter shape vertex *(Place a data point or key-in coordinates to define the start point of the shape.)*
>
> Place Orthogonal Shape > Enter shape vertex *(Place a data point or key-in coordinates to define the vertex.)*

MicroStation prompts for additional shape vertices. Continue placing data points. To complete the polygon shape, the last data point should be placed on top of the starting point.

An orthogonal shape is also a single element. Element manipulation tools such as Move, Copy, and Delete manipulate the orthogonal shape as one element. If necessary, the shape can be made into individual line elements with the Drop Line String tool.

See Figure 2–10 for an example of placing an orthogonal shape with the Place Orthogonal Shape tool by providing nine data points.

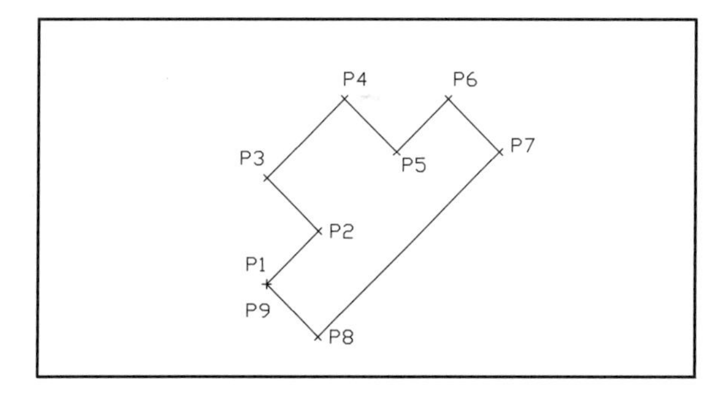

Figure 2–10 An example of placing an orthogonal shape with the Place Orthogonal Shape tool

***Note:** Area and Fill type options are explained in the Patterning section of Chapter 12.*

PLACE CIRCLE

MicroStation offers several methods for drawing circles. These include Place Circle By Center, Place Circle By Edge, and Place Circle By Diameter. The appropriate method is selected from the **Method** option menu located in the Tool Settings window.

Place Circle By Center

With the Place Circle By Center tool, a circle is drawn by defining the center point and a point on the circle. Invoke the Place Circle By Center tool from:

Task Navigation tool box (active task set to Circles)	Select the Place Circle tool and Center from the Method option menu located in the Tool Settings window (see Figure 2–11).
Keyboard Navigation (Task Navigation tool box with active task set to Circles)	**q** (select Center from the Method option menu located in the Tool Settings window)

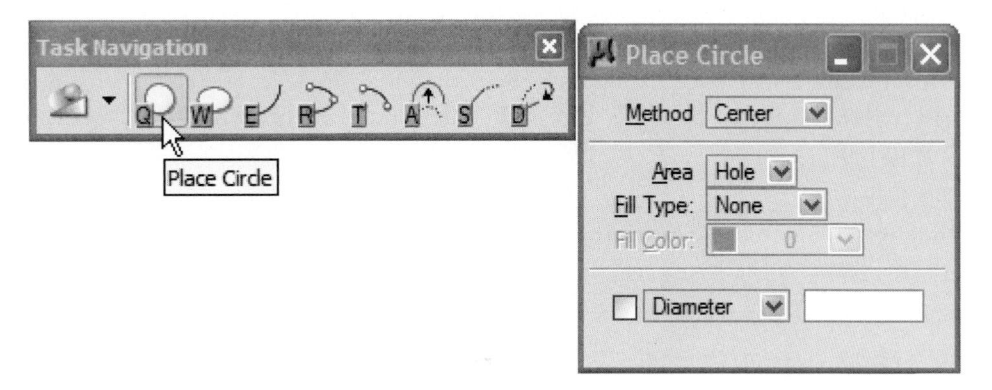

Figure 2–11 Invoking the Place Circle By Center tool from the Navigation tool box (active task set to Circles)

MicroStation prompts:

> Place Circle By Center > Identify center point *(Place a data point or key-in coordinates to define the center of the circle.)*
>
> Place Circle By Center > Define radius *(Place a data point or key-in coordinates to define the radius of the circle.)*

Note: *After placement of the first data point, a dynamic image of the circle drags with the screen pointer.*

To save time, the Place Circle By Center tool remains active and prompts for a new center point. To terminate the Place Circle By Center tool, invoke a different tool.

For example, the following command sequence shows placement of a circle with the Place Circle By Center tool (see Figure 2–12).

> Place Circle By Center > Identify center point *(place a data point to define the center point as shown in Figure 2–12)*
>
> Place Circle By Center > Define radius *(place a data point to draw a circle)*

In the last example, MicroStation used the distance between the center point and the point given on the circle for the radius of the circle.

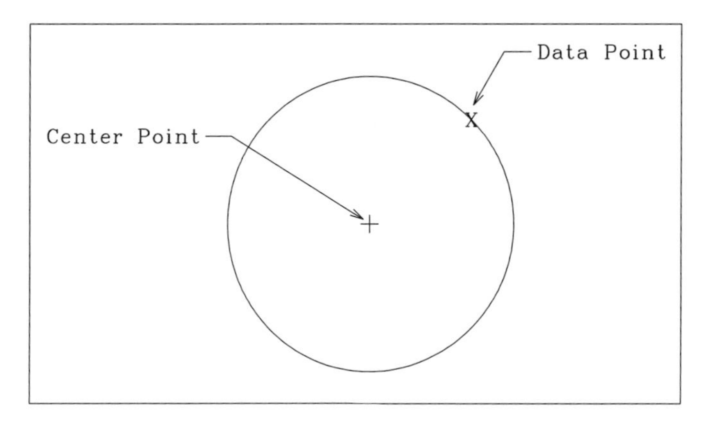

Figure 2–12 An example of placing a circle with the Place Circle By Center tool

A circle can also be placed by its center by keying-in a diameter or radius. Select **Diameter** or **Radius** from the options menu located in the Tool Settings window, turn on the check box, key-in the value in working units, then press (ENTER) or TAB. A circle of that diameter/radius appears at the screen cursor. When prompted to identify the center point of the circle, position the cursor appropriately, and place a data point. Continue placing circles with this same diameter/radius, or press the Reset button to change the diameter/radius of the circle. Invoke another tool to terminate the Place Circle By Center tool.

Place Circle By Edge

The Place Circle By Edge tool is used to draw a circle by defining three data points on the circle. Invoke the Place Circle By Edge tool from:

Task Navigation tool box (active task set to Circles)	Select the Place Circle tool and Edge from the Method option menu located in the Tool Settings window (see Figure 2–13).
Keyboard Navigation (Task Navigation tool box with active task set to Circles)	**q** (select Edge from the Method option menu located in the Tool Settings window)

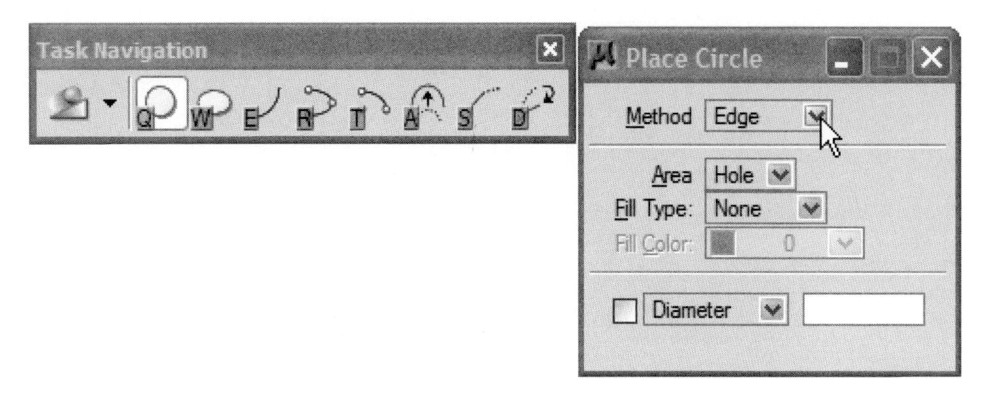

Figure 2–13 Invoking the Place Circle By Edge tool from the Navigation tool box (active task set to Circles)

MicroStation prompts:

> Place Circle By Edge > Identify first point on circle *(Place a data point or key-in coordinates to define the first edge point of the circle.)*
>
> Place Circle By Edge > Identify second point on circle *(Place a data point or key-in coordinates to define the second edge point of the circle.)*
>
> Place Circle By Edge > Identify third point on circle *(Place a data point or key-in coordinates to define the third edge point of the circle.)*

For example, the following command sequence shows placement of a circle via the Place Circle By Edge tool (see Figure 2–14).

> Place Circle By Edge > Identify first point on circle *(Place a data point to define the first point of the circle as shown in Figure 2–14.)*
>
> Place Circle By Edge > Identify second point on circle *(Place a data point to define the second point of the circle.)*
>
> Place Circle By Edge > Identify third point on circle *(Place a data point to define the third point of the circle.)*

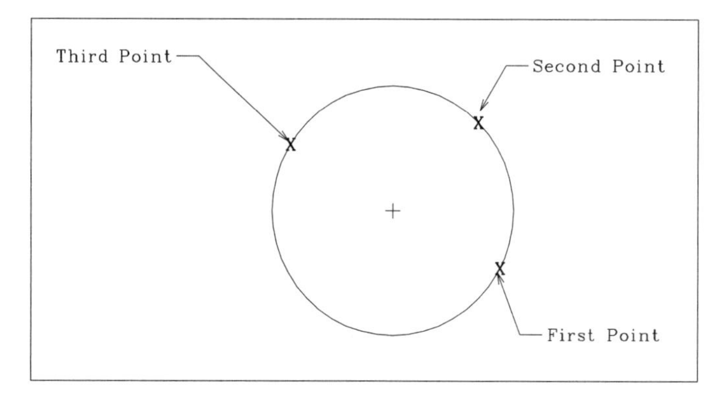

Figure 2–14 An example of placing a circle with the Place Circle By Edge tool

A circle can also be placed by its edge by keying-in a diameter or radius. Select **Diameter** or **Radius** from the options menu located in the Tool Settings window, turn on the check box, key-in the value, then press (ENTER) or TAB. When a diameter/radius is specified, MicroStation prompts for two data points, instead of three, to place a circle by edge.

Place Circle By Diameter

With the Place Circle By Diameter tool, a circle is drawn by defining two data points: the two endpoints of a diameter. Invoke the Place Circle By Diameter tool from:

Task Navigation tool box (active task set to Circles)	Select the Place Circle tool and Diameter from the Method option menu located in the Tool Settings window (see Figure 2–15).
Keyboard Navigation (Task Navigation tool box with active task set to Circles)	**q** (select Diameter from the Method option menu located in the Tool Settings window)

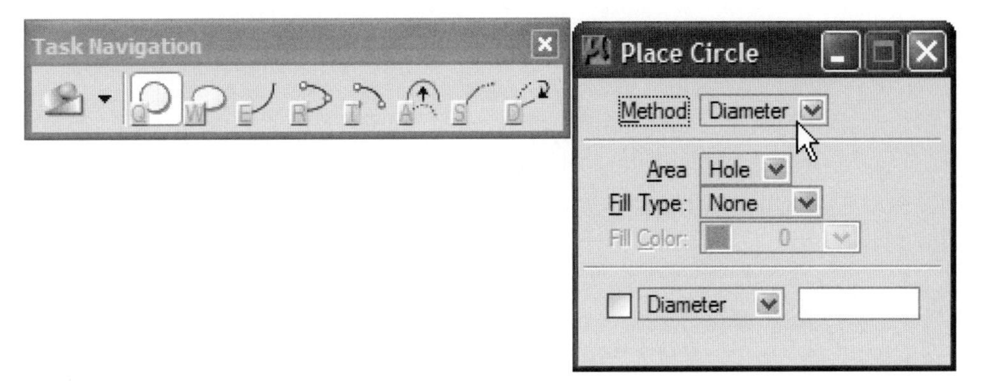

Figure 2–15 Invoking the Place Circle By Diameter tool from the Navigation tool box (active task set to Circles)

MicroStation prompts:

Place Circle By Diameter > Enter First Point on Diameter *(Place a data point or key-in coordinates to define the first endpoint of one of its diameters.)*

Place Circle By Diameter > Enter Second Point on Diameter *(Place a data point or key-in coordinates to define the second endpoint of one of its diameters.)*

For example, the following command sequence shows placement of a circle via the Place Circle By Diameter tool (see Figure 2–16).

Place Circle By Diameter > Enter First Point on Diameter *(Place a data point to define the first point to draw a circle as shown in Figure 2–16.)*

Place Circle By Diameter > Enter Second Point on Diameter *(Place a data point to define the second point to draw a circle.)*

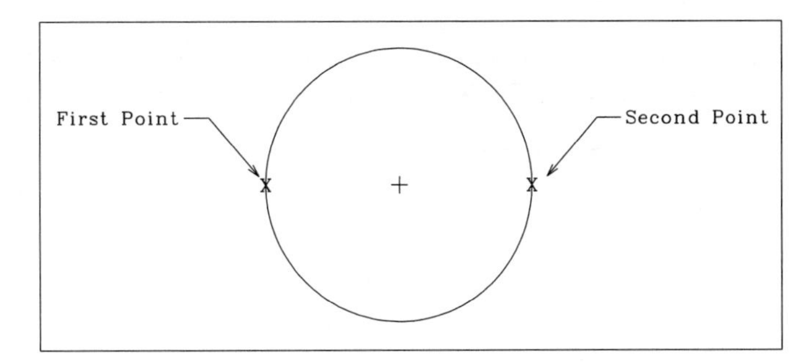

Figure 2–16 An example of placing a circle with the Place Circle By Diameter tool

PLACE ARC

Place Arc tool is used to place a circular arc. Arcs can be placed either clockwise or counterclockwise. MicroStation offers four different methods for placing arcs: Start Center, Center Start, Start Mid End, and Start End Mid. The appropriate

method is selected from the **Method** option menu located in the Tool Settings window.

Place Arc By Start, Center

The Place Arc By Start, Center tool is used to draw an arc by placing its start point, center point, sweep angle and direction. Invoke the Place Arc By Start, Center tool from:

Task Navigation tool box (active task set to Circles)	Select the Place Arc tool and Start, Center from the Method option menu located in the Tool Settings window (see Figure 2–17).
Keyboard Navigation (Task Navigation tool box with active task set to Circles)	**e** (select Start, Center from the Method option menu located in the Tool Settings window)

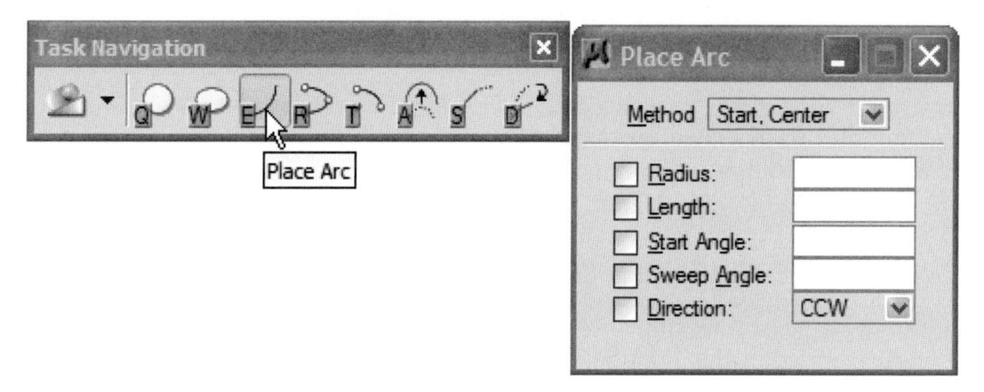

Figure 2–17 Invoking the Place Arc By Start, Center tool from the Navigation tool box (active task set to Circles)

MicroStation prompts:

> Place Arc By Center > Identify start of arc *(Place a data point or key-in coordinates to define the first arc start point.)*
>
> Place Arc By Center > Identify center point *(Place a data point or key-in coordinates to define the arc center.)*
>
> Place Arc By Center > Define arc sweep angle *(Place a data point or key-in coordinates to define the sweep angle.)*

Note: *After the first data point is placed, a dynamic image of the arc drags with the screen pointer. After the first and center points are placed, the second arc end point can be placed clockwise or counterclockwise.*

For example, the following command sequence shows placement of an arc with the Place Arc By Start, Center tool (see Figure 2–18).

Place Arc By Center > Identify start of arc *(Place a data point to define the first arc start point as shown in Figure 2–18.)*

Place Arc By Center > Identify center point *(Place a data point to define the arc center.)*

Place Arc By Center > Define arc sweep angle *(Place a data point to define the arc end-point.)*

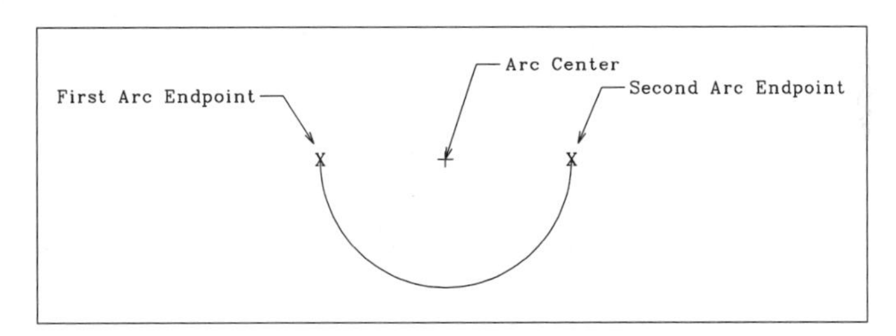

Figure 2–18 An example of placing an arc by the Place Arc By Start, Center tool

An arc can also be drawn by its center by keying-in a radius. To do so, turn on the check box for **Radius**, located in the Tool Settings window; key-in the appropriate value in the **Radius** edit field; and press (ENTER) or TAB. The prompts are similar to those shown above, except the First Arc Endpoint and Second Arc Endpoint define the starting and ending directions of the arc.

To draw an arc by its center using a specific length, turn on the check box for **Length**, located in the Tool Settings window; key-in the appropriate value in the **Length** edit field; and press (ENTER) or TAB. MicroStation prompts for the First Arc Endpoint and center point of arc and draws an arc to the specified arc length.

The **Start Angle** and **Sweep Angle can be constrained** by keying-in appropriate angles in the respective edit fields. The MicroStation prompts depend on the number of constraints turned ON. For example, if **Radius** and **Start Angle** are preset, MicroStation prompts for the center point of the arc and the sweep angle; if **Radius, Start Angle**, and **Sweep Angle** are preset, MicroStation prompts only for the center of the arc. In addition, the direction of the arc can be constrained by choosing CW for clockwise direction and CCW for counter clockwise direction in the **Direction** setting.

Place Arc By Center, Start

The Place Arc By Center, Start tool draws an arc based on its center point, start point, sweep angle and direction. Invoke the Place Arc By Center, Start tool from:

Task Navigation tool box (active task set to Circles)	Select the Place Arc tool and Center, Start from the Method option menu located in the Tool Settings window (see Figure 2–19).
Keyboard Navigation (Task Navigation tool box with active task set to Circles)	**e** (select Center, Start from the Method option menu located in the Tool Settings window)

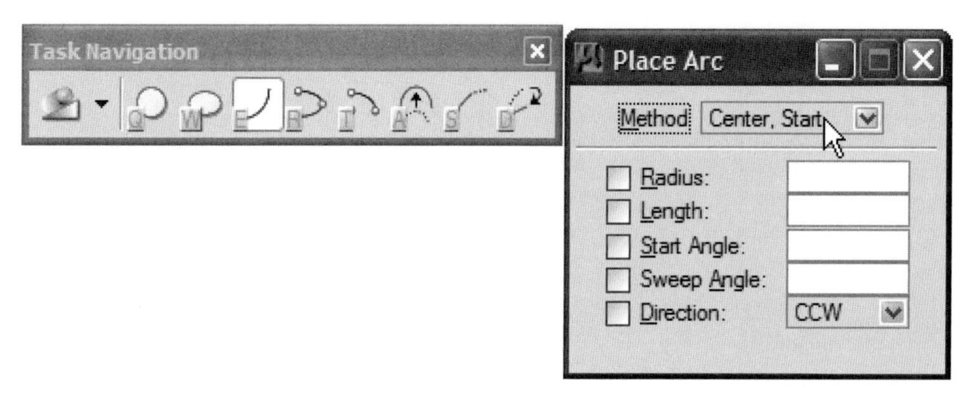

Figure 2–19 Invoking the Place Arc By Center, Start tool from the Navigation tool box (active task set to Circles)

MicroStation prompts:

> Place Arc By Center > Identify center point *(Place a data point or key-in coordinates to define the first arc center.)*
>
> Place Arc By Center > Identify start of arc *(Place a data point or key-in coordinates to define the arc start point.)*
>
> Place Arc By Center > Define arc sweep angle *(Place a data point or key-in coordinates to define the sweep angle.)*

Arc by Center, Start method can be constrained to **Radius, Length, Start Angle, Sweep Angle,** and **Direction.**

Place Arc By Start, Mid, End

The Place Arc By Start, Mid, End tool draws an arc based on three points defining its start point, a second point on the arc, and its end point. Invoke the Place Arc Start, Mid, End tool from:

Task Navigation tool box (active task set to Circles)	Select the Place Arc tool and Start, Mid, End from the Method option menu located in the Tool Settings window (see Figure 2–20).
Keyboard Navigation (Task Navigation tool box with active task set to Circles)	**e** (select Start, Mid, End from the Method option menu located in the Tool Settings window)

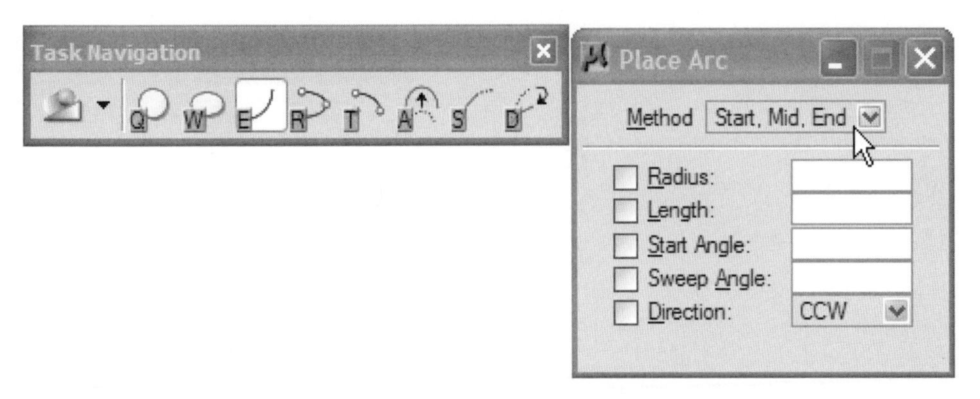

Figure 2–20 Invoking the Place Arc By Start, Mid, End tool from the Navigation tool box (active task set to Circles)

MicroStation prompts:

> Place Arc By Edge > Identify start of arc *(Place a data point or key-in coordinates to define the start point of arc.)*
>
> Place Arc By Edge > Identify point along an arc *(Place a data point or key-in coordinates to define a point along the arc.)*
>
> Place Arc By Edge > Identify end of arc *(Place a data point or key-in coordinates to define the end point of arc.)*

For example, the following command sequence shows the placement of an arc with the Place Arc Start, Mid, End tool (see Figure 2–21).

> Place Arc By Edge > Identify start of arc *(Place a data point to define the first arc start point as shown in Figure 2–21.)*
>
> Place Arc By Edge > Identify point along an arc *(Place a data point to define the point along the arc.)*
>
> Place Arc By Edge > Identify end of arc *(Place a data point to define the arc endpoint.)*

An arc can also be drawn by its edge by keying-in a radius. To do so, turn on the check box for **Radius**, located in the Tool Settings window; key-in the appropriate value in MU:SU:PU format in the **Radius** edit field; and press (ENTER) or TAB. The prompts are similar to those for the Place Arc Start, Mid, End tool, except

the First Arc Endpoint and Second Arc Endpoint define the starting and ending directions of the arc.

To draw an arc by its edge using a specific length, turn on the check box for **Length**, located in the Tool Settings window; key-in the appropriate value in the **Length** edit field; and press (ENTER) or TAB. MicroStation prompts for the First Arc Endpoint and Second Arc Endpoint and draws an arc to the specified arc length.

The **Start Angle** and **Sweep Angle** can be constrained by keying in appropriate angles in the respective edit fields. The MicroStation prompts depend on the number of constraints turned ON. For example, if **Radius** and **Start Angle** are preset, MicroStation prompts for the First Arc Endpoint and Second Arc Endpoint. If **Radius, Start Angle**, and **Sweep Angle** are preset, MicroStation prompts only for the First Arc Endpoint.

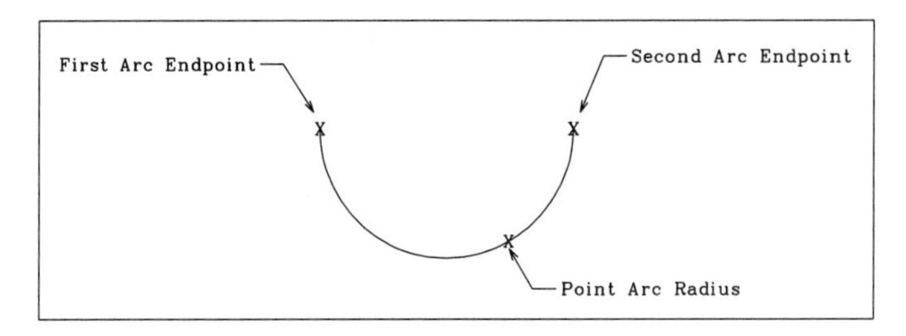

Figure 2–21 An example of placing an arc by the Place Arc By Edge tool

Place Arc By Start, End, Mid

The Place Arc By Start, End, Mid tool creates an arc defined by three points: start point, end point, and a third point on the arc. Invoke the Place Arc Start, End, Mid tool from:

Task Navigation tool box (active task set to Circles)	Select the Place Arc tool and Start, End, Mid from the Method option menu located in the Tool Settings window (see Figure 2–22).
Keyboard Navigation (Task Navigation tool box with active task set to Circles)	**e** (select Start, End, Mid from the Method option menu located in the Tool Settings window)

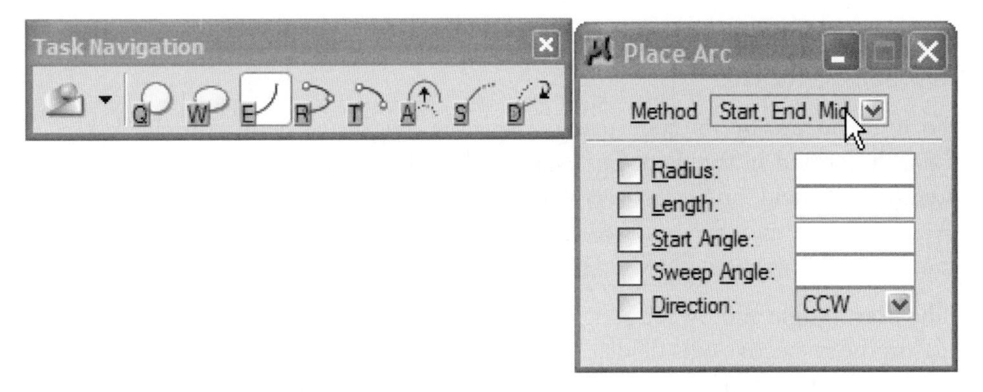

Figure 2–22 Invoking the Place Arc By Start, End, Mid tool from the Navigation tool box (active task set to Circles)

MicroStation prompts:

> Place Arc By Edge > Identify start of arc *(Place a data point or key-in coordinates to define the start point of arc.)*
>
> Place Arc By Edge > Identify end of arc *(Place a data point or key-in coordinates to define the end point of arc..)*
>
> Place Arc By Edge > Identify point along arc *(Place a data point or key-in coordinates to define a point along arc.)*

Arcs can be constrained to **Radius, Length, Start Angle,** and **Sweep Angle.**

MANIPULATING ELEMENTS

In addition to the drawing tools, MicroStation provides tools to facilitate manipulation of the created elements.

MicroStation's AccuSnap Settings provide valuable assistance in identifying elements and determining a tool's intent. Elements are located and identified automatically at the pointing device. If an element identified for manipulation is inappropriate for the selected tool, (e.g., it is the wrong type for the tool, it is locked, or in a reference, etc.), MicroStation reports the reason the element cannot be selected.

This feature is enabled by default. To disable it, open the AccuSnap Settings box from:

Menu	Settings > Snaps > AccuSnap

MicroStation opens the AccuSnap Settings box, as shown in Figure 2–23.

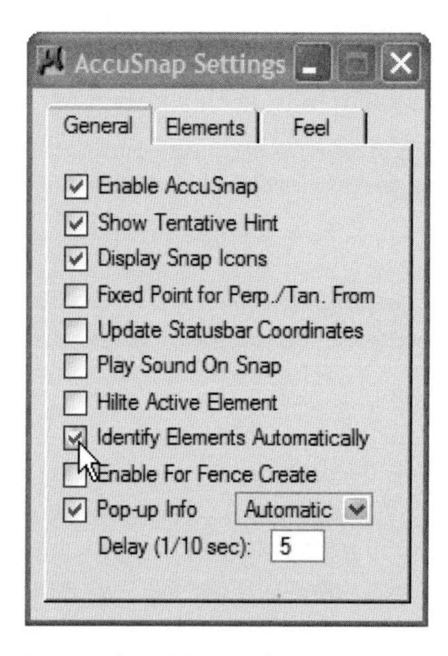

Figure 2–23 AccuSnap Settings box (General)

Set **Identify Elements Automatically** check box to OFF to disable the locating elements feature. By default, it is set to ON.

DELETE ELEMENT

Of the many manipulation tools available, the Delete Element tool is one of the most frequently used. Everyone makes mistakes, but with MicroStation, it is easy to delete them. Invoke the Delete Element tool from:

Main tool box	Select the Delete Element tool (see Figure 2–24).
Keyboard Navigation	**8**

Figure 2–24 Invoking the Delete Element tool from the Main tool box

MicroStation prompts:

Delete Element > Identify element *(Identify the element to delete.)*

The selected element is deleted. If the **Identify Elements Automatically** check box is set to OFF (in the AccuSnap Settings box), then MicroStatation prompts:

Delete Element > Accept/Reject (select next input) *(Click the Accept button to delete the selected element, select another element to delete, or click the Reject button to terminate the command sequence.)*

Note: *The Delete Element tool deletes only one element at a time. To delete a group of elements, use the Fence Delete tool, explained in Chapter 6.*

DROP ELEMENT

The Drop Element tool is used to break up an element(s) into simpler components, such as blocks and shapes into a series of connected individual lines that can be manipulated as individual elements. Tool settings are used to specify the element types on which the tool operates. Invoke the Drop Element tool from:

Groups tool box	Select the Drop Element tool and set Line Strings/ Shapes check box to ON in the Tool Settings window (see Figure 2–25).
Keyboard Navigation	**6** and select Drop Element from the drop-down menu and set Line Strings/Shapes check box to ON in the Tool Settings window

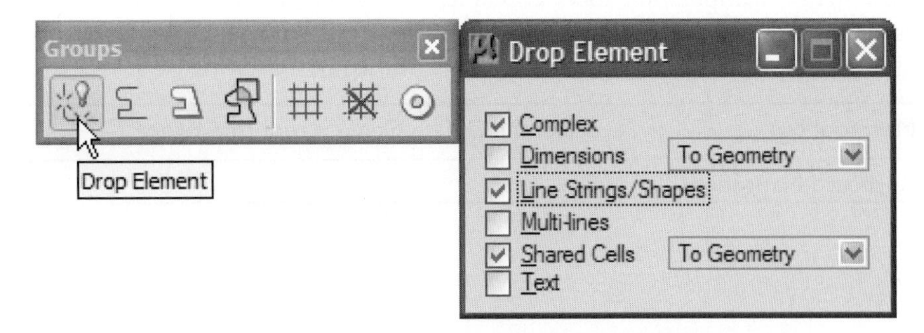

Figure 2–25 Invoking the Drop Element tool from the Groups tool box

MicroStation prompts:

Drop Element > Identify element *(Identify the block or shape to be dropped.)*

The selected element is dropped. If **Identify Elements Automatically** check box is set to OFF (in the AccuSnap Settings box), then MicroStatation prompts:

Drop Line String/Shape Status > Accept/Reject (select next input) *(Click the Accept button to accept, select another element to drop, or click the Reject button to reject.)*

PRECISION INPUT

In MicroStation, objects can be drawn at its true size and then the border, title block, and other non-object–associated features made to fit the object. The completed combination is scaled to fit the plotted sheet size specified when plotting.

Drawing a not-to-scale schematic does not take advantage of MicroStation's full graphics and computing potential. But even though the symbols and distances between them have no relationship to any real-life dimensions, the sheet size, text size, line widths, and other visible characteristics of the drawing must be considered to give the schematic maximum readability. Some planning, including sizing, should be undertaken for all drawings.

When MicroStation prompts for the location of a point, instead of providing the data point with a pointing device, using Precision Input commands enable more precise placement. The Precision Input methods based on Rectangle Coordinate system include Absolute Rectangular coordinates, Relative Rectangular coordinates, and Relative Polar coordinates.

The rectangular coordinates system is based on a point's distance from two intersecting perpendicular axes for *two-dimensional (2D)* points, or from three intersecting perpendicular planes for *three-dimensional (3D)* points. Each data point is measured along the *X* axis (horizontal) and *Y* axis (vertical) for *2D* design and along the *X* axis, *Y* axis, and *Z* axis (toward or away from the viewer) for *3D* design. The intersection of the axes, called the *origin* (XY=0,0) as shown in Figure 2–26.

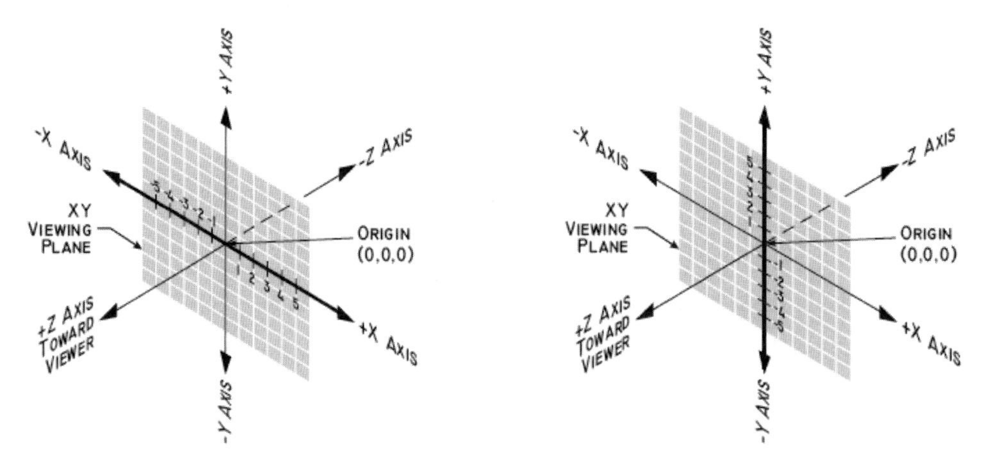

Figure 2–26 A 3D coordinate system

ABSOLUTE RECTANGULAR COORDINATES

Points are located by Absolute Rectangular Coordinates at an exact *X,Y* intersection on the design plane in relation to the Global Origin. By default, the Global Origin is located at the center of the design plane, as shown in Figure 2–27. The horizontal distance increases in the positive *X* direction from the origin, and the vertical distance increases in the positive *Y* direction from the origin. To enter an absolute coordinate, key-in:

XY=<X coordinate>**,**<Y coordinate> (ENTER)

or

POINT ABSOLUTE <X coordinate>**,**<Y coordinate> (ENTER)

The <X coordinate> and <Y coordinate> are the coordinates in working units in relation to the Global Origin. For example:

XY=2,4 (ENTER)

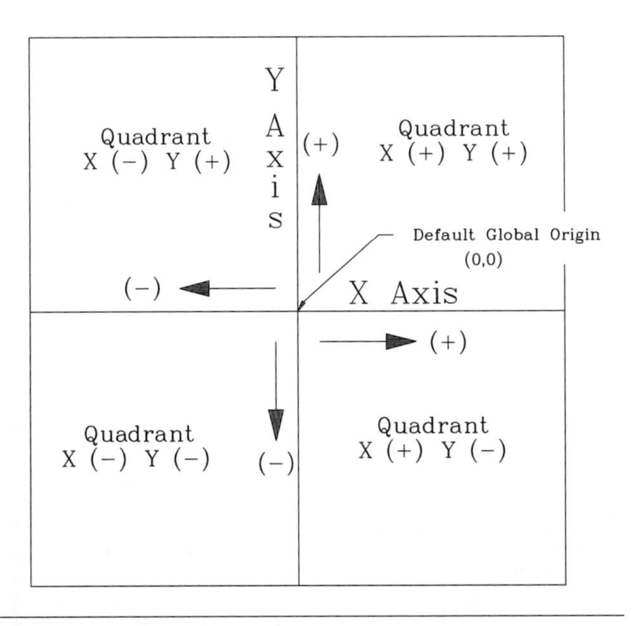

Figure 2–27 Showing Global Origin in a 2D design

The data point is located 2 master units from the origin along the positive *X* axis and 4 master units from the origin along the positive *Y* axis.

The Global Origin can be relocated anywhere on or off the design plane. To do so, key-in GO=MU:SU:PU,MU:SU:PU, and MicroStation will prompt:

Global Origin > Enter monument point *(Specify a point anywhere in the design plane to define the new global origin, or click the Reset button to automatically assign the coordinates to the lower-left corner of the design plane.)*

RELATIVE RECTANGULAR COORDINATES

Points are located by Relative Rectangular Coordinates in relation to the last specified position or point, rather than in relation to the origin. This is similar to specifying a point as an offset from the last point entered. To enter a Relative Rectangular Coordinate, key-in:

DL=<X coordinate>,<Y coordinate> (ENTER)

or

POINT DELTA <X coordinate>,<Y coordinate> (ENTER)

The <X coordinate> and <Y coordinate> are the coordinates in relation to the last specified position or point. For example, if the last point specified was *X,Y*=4,4, the command:

DL=5,4 (ENTER)

is equivalent to specifying the Absolute Rectangular Coordinates *X,Y*=9,8 (see Figure 2–28).

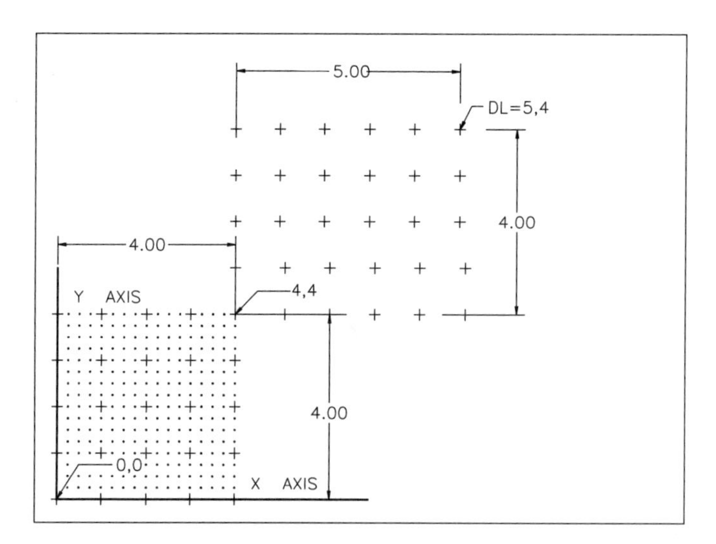

Figure 2–28 An example of placing a point by Relative Rectangular Coordinates

RELATIVE POLAR COORDINATES

Relative Polar Coordinates are based on a distance from a fixed point at a given angle. In MicroStation, a Relative Polar Coordinate is determined by the distance and angle measured from the previous data point. By default, the angle is mea-

sured in a counterclockwise direction relative to the positive *X* axis. It is important to remember that points located by Relative Polar Coordinates are always positioned relative to the previous point, not to the Global Origin (0,0). To enter a Relative Rectangular Coordinate, key-in:

DI=<distance>,<angle> (ENTER)

or

POINT DISTANCE <distance>,<angle> (ENTER)

The <distance> and <angle> are specified in relation to the last specified position or point. The distance is specified in current working units, and the direction is specified as an angle in current angular units relative to the *X* axis. For example, to specify a point at a distance of 6.4 Master Units from the previous point and at an angle of 39 degrees relative to the positive *X* axis (see Figure 2–29), key-in:

DI=6.4,39 (ENTER)

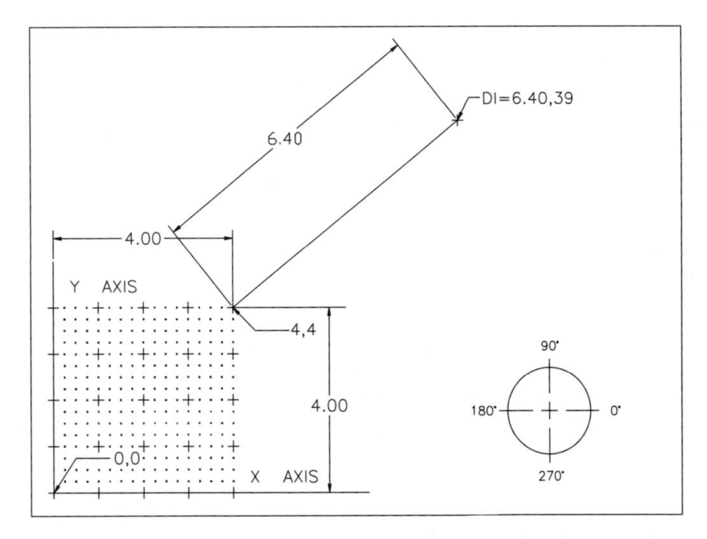

Figure 2–29 An example of placing a line by Relative Polar Coordinates

For example, the following key-ins show the placement of connected lines for the drawing shown in Figure 2–30 by the Place Line tool with absolute coordinates (see Figure 2–31):

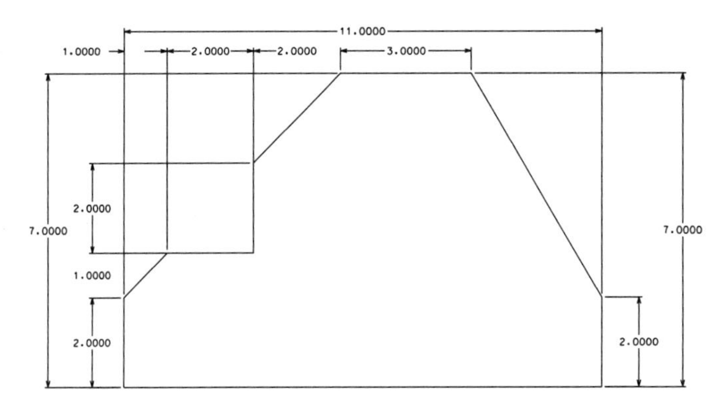

Figure 2–30 An example of placing connected lines

XY=2,2 (ENTER)

XY=2,4 (ENTER)

XY=3,5 (ENTER)

XY=5,5 (ENTER)

XY=5,7 (ENTER)

XY=7,9 (ENTER)

XY=10,9 (ENTER)

XY=13,4 (ENTER)

XY=13,2 (ENTER)

XY=2,2 (ENTER)

(click the Reset button to terminate the line sequence)

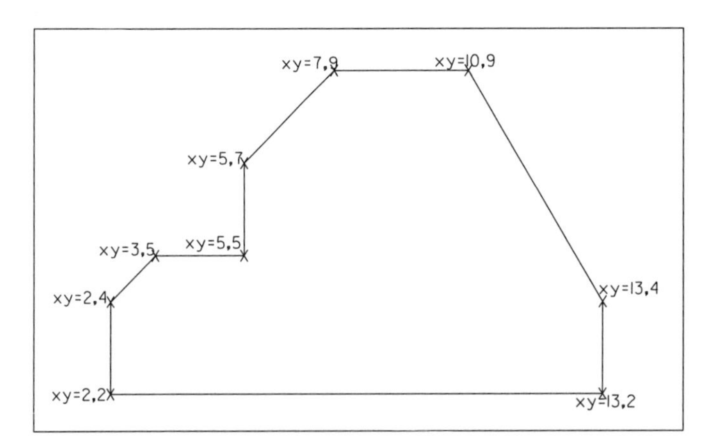

Figure 2–31 Placing connected lines using absolute coordinates

The following key-ins show the placement of connected lines for the drawing shown in Figure 2–30 by the Place Line tool with relative rectangular coordinates (see Figure 2–32).

XY=2,2 (ENTER)

DL=0,2 (ENTER)

DL=1,1 (ENTER)

DL=2,0 (ENTER)

DL=0,2 (ENTER)

DL=2,2 (ENTER)

DL=3,0 (ENTER)

DL=3,-5 (ENTER)

DL=0,-2 (ENTER)

XY=2,2 (ENTER)

(click the Reset button to terminate the line sequence)

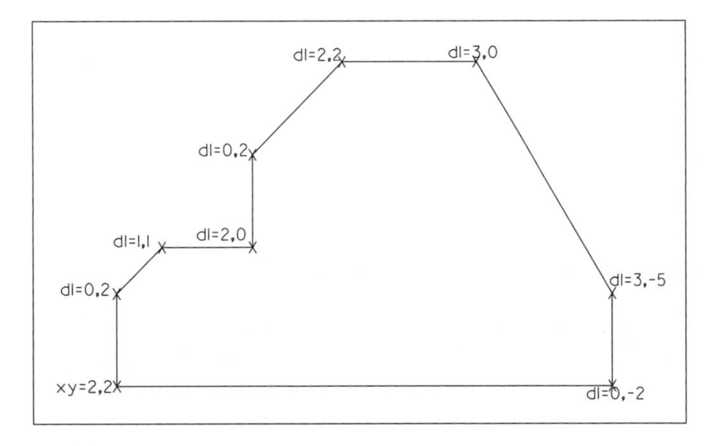

Figure 2–32 Placing connected lines using relative rectangular coordinates

The following key-ins show the placement of connected lines for the drawing shown in Figure 2–30 by the Place Line tool with relative polar coordinates and relative rectangular coordinates (see Figure 2–33):

XY=2,2 (ENTER)

DI=2,90 (ENTER)

DL=1,1 (ENTER)

DI=2,0 (ENTER)

DI=2,90 (ENTER)

DL=2,2 (ENTER)

DL=3,0 (ENTER)

DL=3,-5 (ENTER)

DI=2,270 (ENTER)

XY=2,2 (ENTER)

(click the Reset button to terminate the line sequence)

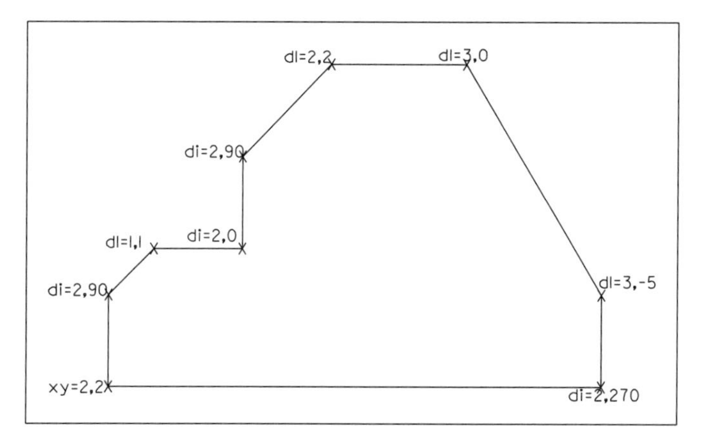

Figure 2–33 Placing connected lines using relative polar coordinates and relative rectangular coordinates

Open the Exercise Manual PDF file for Chapter 2 on the accompanying CD for project and discipline-specific exercises.

REVIEW QUESTIONS

Write your answers in the spaces provided.

1. Name the tool that will make individual elements from a shape.

   ____

2. Explain briefly the differences between the absolute rectangular coordinates and relative rectangular coordinates precision key-ins.

   ____

3. Name the three key-ins that are used in Precision Input for absolute rectangular coordinates, relative rectangular coordinates, and polar relative coordinates.

   ____

4. When MicroStation displays the information regarding the size of an element or coordinates, it does so in the following format: ________:________:________

5. Name the three methods by which circles can placed circles in MicroStation.

   ____

6. To draw a circle by specifying three known points on the circle, invoke the ______________________ tool.

7. The tool related to placing arcs is in the ____________________ tool box.
8. The Place Arc Edge tool places an arc by identifying ____________ points on the arc.
9. Number of vertices allowed when using the Place Shape tool range from ___ to ___.
10. Number of data points MicroStation prompts to draw a rotated block when using the Place Block Rotated tool is ____________.
11. Key-in to redefine the Global origin is ____________________.
12. What is the purpose of using the Save Settings command?
____

CHAPTER 3

Fundamentals II

OBJECTIVES

Topics explored in this chapter:

- Using drawing tools: grid, axis, units, and tentative snap
- Controlling and viewing levels
- Setting element attributes
- Matching element attributes
- Using View Controls: update, zoom in, zoom out, window area, fit, and pan
- Using View windows and view attributes
- Using Undo and Redo tools

DRAWING TOOLS

MicroStation provides several drawing tools to make drafting and design layout easier.

THE GRID SYSTEM

The grid system is a visual tool for measuring distances precisely and placing elements accurately. The grid appears in view windows as a matrix of evenly spaced dots and lines; it is similar to a sheet of linear graph paper. Tools allow the user to turn Grid display ON or OFF and change the spacing of the dots and lines. There are two parts to the grid system. The first is the Grid Reference, which appears on the screen as evenly-spaced horizontal and vertical lines. By default, the spacing between lines is one Master Unit. The second part of the grid system is the Grid Unit, which appears as dots on the screen. By default, the spacing between dots is one Sub Unit. For example, if the Master Units are feet and the Sub Units are inches, the default is one foot between reference lines and one inch between grid dots.

The grid is for visual reference only and does not plot. The grid serves two purposes: it provides a visual indication of distances and, with Grid lock ON, forces data points placed with the Data button to snap to grid points. This is useful for

keeping lines straight, ensuring that distances are exact, and making sure elements meet when they are expected to.

Note: *Using Precision Input, (entering the location of a point with keyed-in numeric information), overrides Grid lock.*

Grid Display

Grid display can be set to ON or OFF. When it is ON, the Grid appears on the screen. When it is OFF, the grid is not visible. Change the status of the Grid display from the View Attributes setting box. Invoke the View Attributes settings box from:

Menu	Settings > View Attributes

MicroStation displays the View Attributes settings box as shown in Figure 3–1.

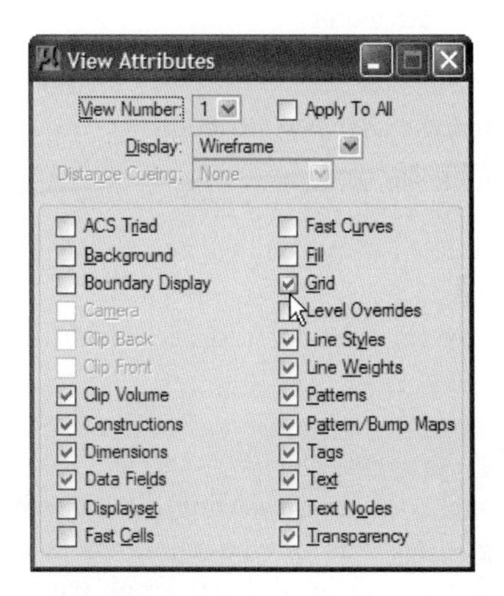

Figure 3–1 View Attributes settings box

To display the grid set the **Grid** check box to ON. Setting changes are applied immediately to the view window selected by the **View Number** or to all view windows if **Apply to All** is selected. (For a detailed explanation of Views and the View Attributes settings box, see the View Windows and View Attributes section later in this chapter.)

Note: *If Grid display is ON but the grid is not visible in the view window, the window may be zoomed out too far from the design plane (the Zoom tool is discussed later in this chapter). MicroStation, by default, displays a maximum of 90 grid dots and 46 reference line intersections. If the view window is zoomed out so that more than 90 dots are in the view area, Micro-*

Station does not display the grid dots. If the zoom factor is such that more than 46 reference line intersections are in the view area, the lines are not displayed.

Grid Spacing

The spacing between both the grid dots and the reference lines can be changed at any time as required for the design by making appropriate changes to the Grid settings in the Design File dialog box. Invoke the Design File dialog box from:

Menu	Settings > Design File

MicroStation displays a Design File dialog box, similar to the one shown in Figure 3–2.

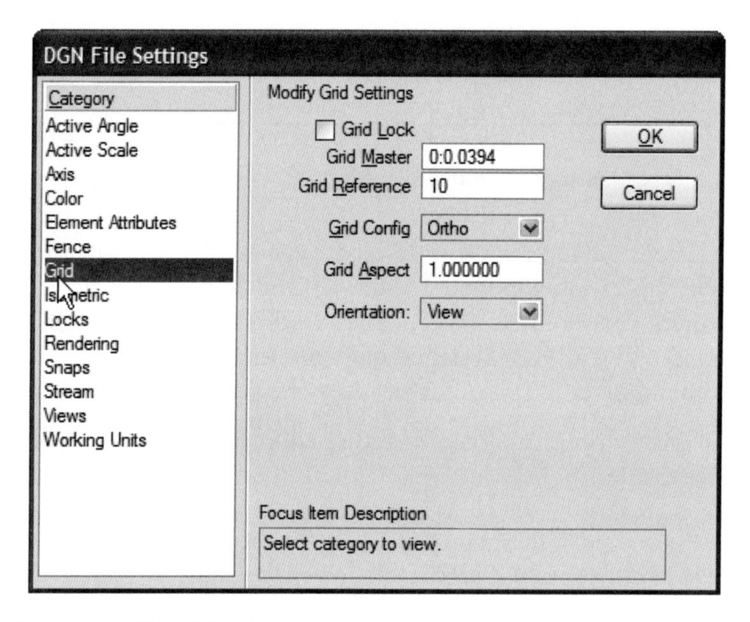

Figure 3–2 Design File dialog box

Select **Grid** from the **Category** list and MicroStation displays controls for adjusting the grid spacing. The **Grid Master** defines the distance between the grid dots and is specified in terms of the design's Master Units. Change the spacing by entering the number of Master Units or fraction of a Master Unit. The **Grid Reference** is an integer number that MicroStation multiplies by the **Grid Master** to determine the space between the reference lines.

For example, assume that in the design, the Master Units are feet and the Sub Units are inches. Set the **Grid Master** to 0.25 and the **Grid Reference** to four to put three inches between the dots (Master Unit x 0.25) and one foot between the

lines (0.25 x 4). Or, set the **Grid Master** to 5.0 and the **Grid Reference** to 20 to put five feet between the dots and 100 feet between the lines.

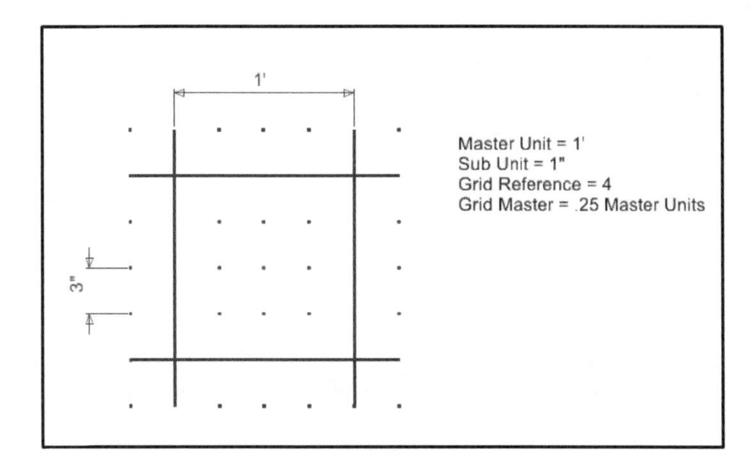

Figure 3–3 Screen display for the grid settings

The Grid Master unit and the Grid Reference can also be set by manually entering the spacing in the key-in window. To set the Grid Master unit, key-in **GU=**<distance> and press ENTER. The <distance> is specified in Master Units. To set the Grid Reference, key-in **GR=**<integer number> and press ENTER. The <integer number> is the number of grid units between the grid reference lines.

To keep the grid settings in effect for future editing sessions, select **Save Settings** from the **File** menu.

Grid Configuration

MicroStation provides three options for controlling the orientation of the Grid display: **Ortho, Isometric,** and **Offset.** Select an option from the Design File dialog box under the **Grid Config** menu. **Ortho** aligns the grid points orthogonally (the default option). **Isometric** aligns the grid points isometrically. **Offset** offsets the rows by half the distance between the horizontal grid points.

Grid Aspect Ratio (*Y/X*)

Grid Aspect is the ratio of vertical (*y)* grid points to horizontal (*x)* grid points. The default is 1.000, in which case the vertical and horizontal grid spacing is identical. If the ratio is 5.000, the space between rows of grid dots will be five times the space between columns of grid dots.

Grid Orientation

Grid Orientation is the orientation of the grid in relation to the design. The **Top, Right,** and **Front** orientation options are for 3D designs and align the grid with the 3D drawing axes. **View** and **ACS** (Auxiliary Coordinate System) are available

to both 2D and 3D drawings. **View** is usually selected for 2D drawings to align the grid parallel to the plane of the X and Y axes. For more information on 3D design and ACS, see Chapter 17.

Grid Lock

The **Grid Lock** check box controls where data points are placed in the design with the pointing device. When the lock is ON, MicroStation forces all the data points placed with the pointing device to be placed on the grid. Setting the lock ON allows quick placement of data points with the Data button, letting MicroStation ensure precise placement. The lock can be overridden by keying-in absolute or relative coordinate points.

Note: *Grid Lock is effective regardless of the status of Grid display. It still locks to grid points even if the grid is not visible on the view window.*

MicroStation displays the current Grid Lock setting on the Status bar. Toggle Grid Lock ON and OFF by clicking the lock icon on the Status bar. Grid Lock can also be toggled from two other locations:

1. The **Settings > Locks** submenu, as shown in Figure 3-4.
2. The **Toggles** settings box (open from **Settings > Locks** menu), as shown in Figure 3–5.

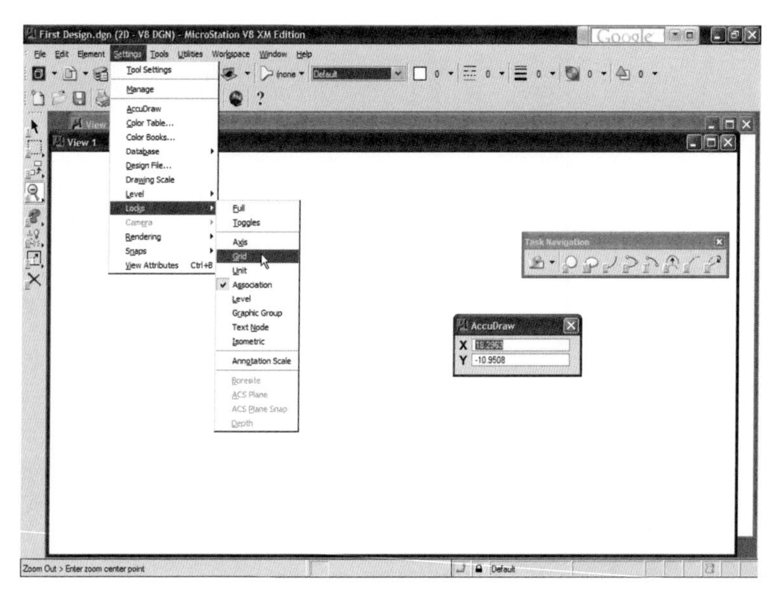

Figure 3–4 Locks submenu

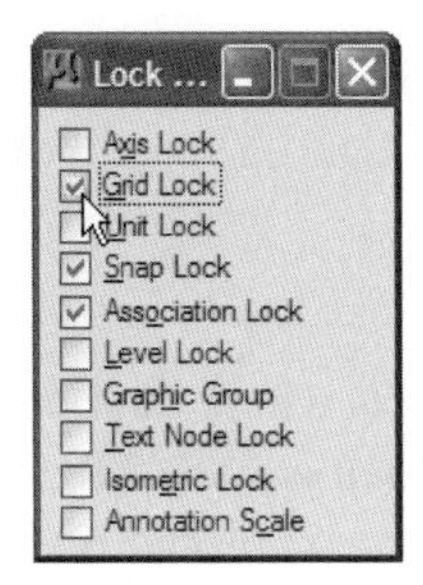

Figure 3–5 Lock Toggles settings box

To keep the grid settings (including Grid Lock) in effect for future editing sessions, choose **Save Settings** from the **File** menu.

AXIS LOCK

When the **Axis Lock** check box is ON, each data point placed with the pointing device is forced to lie at an incremental angle (or multiples of the angle) from the previous data point. Axis Lock settings are available in the Design File dialog box. Invoke the dialog box from:

Menu	Settings > Design File

MicroStation displays the Design File dialog box as shown in Figure 3–6.

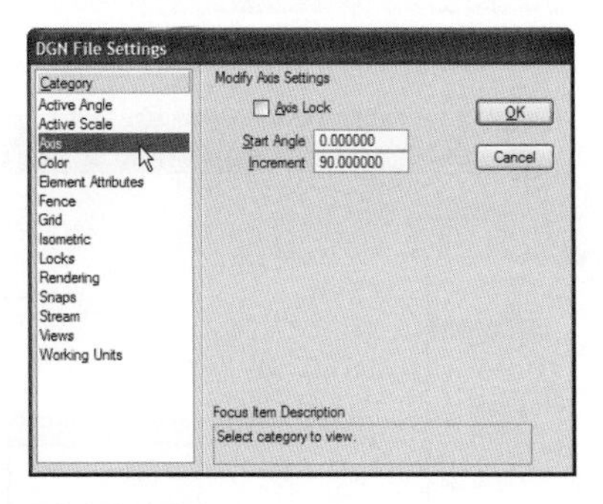

Figure 3–6 The Design File dialog box

Select **Axis** from the **Category** list and MicroStation displays controls for adjusting the Axis Lock settings. Key-in the appropriate Axis Start Angle in the **Start Angle** field and the Axis Increment angle in the **Increment** field. Turn ON the **Axis Lock** using the check box to constrain data points.

For example, to constrain data points to draw only horizontal and vertical lines, key-in a **Start Angle** of 0 degrees and an **Increment** angle of 90 degrees

SNAP LOCK

Snap Lock controls the placement of tentative snap points at specific locations on elements in the design. Tentative snapping provides a way to preview the location of a data point before entering it in the design and to place data points at precise locations on existing elements.

When a tentative snap point is placed (by clicking the Tentative button), a large cross appears to identify the point. If it is snapped to an element, the element is highlighted and MicroStation displays the absolute coordinates of the point on the Status bar. If the point is at the correct location, click the Accept button (same as the Data button) to confirm it. If, however, the point is not at the correct location, move the pointer closer to the correct location and click the Tentative button again. Selecting another tentative snap point rejects the last tentative snap point and selects a new one. When the tentative snap point is accepted with the Accept button, the large cross disappears and a data point is placed in the design. The tentative snap point can be canceled by clicking the Reset button.

The pointing device buttons used to place tentative snap points vary. Pucks that are used with a digitizing tablet usually have a specific Tentative button. On a three-button mouse, the middle button is usually the Tentative button. Tentative snapping with a two-button mouse requires pressing both buttons simultaneously (which may take some practice to learn to do successfully).

Snap Lock Mode

If the **Snap Lock** check box is ON, snaps can be made to a specific point on an element, depending on the snap mode selected. For example, if the snap mode is **Center** and **Snap Lock** ON, the tentative button snaps to the center of circles and blocks, and the midpoint of lines and segments of line strings. A tentative snap point can be placed while executing any MicroStation tool that requests a data point, such as Place Line, Place Circle, Place Arc, Move, and Copy. If **Snap Lock** is OFF, tentative points do not snap to elements.

Two settings boxes are provided for selecting Snap Lock settings:

Menu	Settings > Locks > Toggles
	Settings > Locks > Full

Select or clear the **Snap Lock** check box from either the Toggles settings box or the Full settings box, as shown in Figure 3–7.

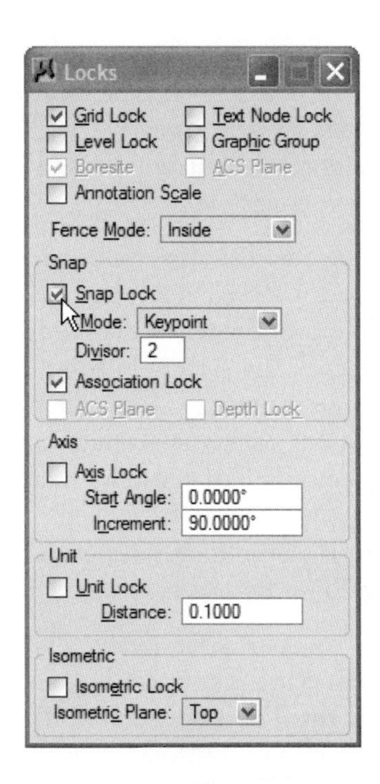

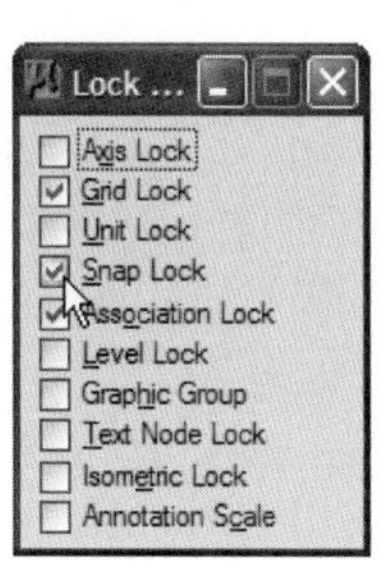

Figure 3–7 Locks Full setting box and Lock Toggles settings box

Selecting a Snap Mode

An active (default) Snap mode that always stays in effect can be selected. When a different one is occasionally needed, an override Snap mode can be selected that applies only to the next tentative point.

To set the Snap active mode:

Menu	Settings > Locks > Full (Select the desired Snap mode from the Snap Mode option menu. See Figure 3–8.)
Menu	Settings > Snaps > Button Bar (Double-click the desired Snap mode. See Figure 3–9.)

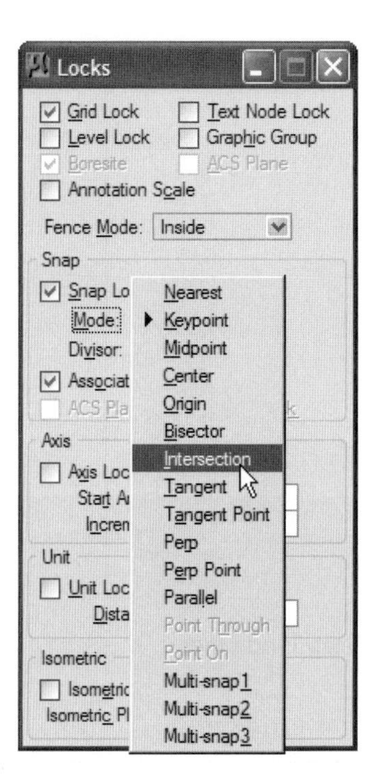

Figure 3–8 Locks settings box displaying the Snap Mode option menu.

Figure 3–9 Snap mode button bar

The default Snap mode can also be selected from the **Snap Mode** pop-up menu:

- Invoke the pop-up menu from the View window by holding down SHIFT and tapping the Tentative button. The pop-up menu is shown in Figure 3–10.
- Invoke the pop-up menu from the Status bar by positioning the pointer over the Active Snap Mode icon on the Status bar and clicking either the Data button or the Reset button.

Set the default Snap mode from the pop-up menu by positioning the pointer over the name of the desired Snap mode, holding down SHIFT, and tapping either the Data button or Reset button.

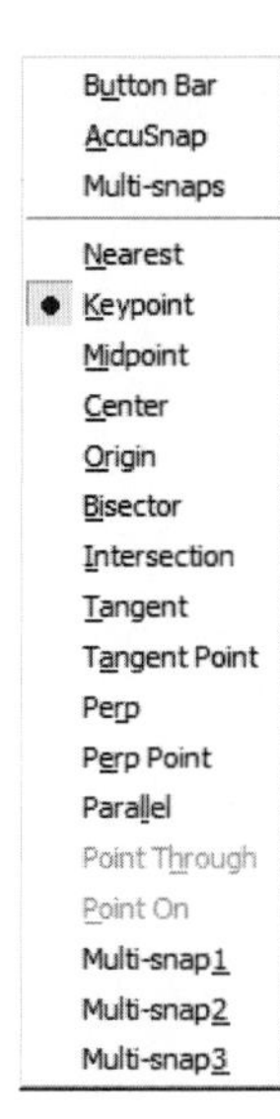

Figure 3–10 The pop-up Snap Mode menu

Set a temporary Snap mode override from:

Button bar	Click the desired Snap mode.
Menu	Settings > Snaps (Choose the desired Snap mode override.)

The override Snap mode can also be selected from the pop-up Snap Mode menu. Hold down SHIFT and press the Tentative button to display the pop-up menu and select the desired override snap mode from the menu.

By default, MicroStation displays the active Snap mode icon on the Status bar. MicroStation also displays a diamond-shaped icon to the left of the default active Snap mode in the Snaps menu. If an override mode is selected, a square appears to the left of the default Snap active mode and a diamond shaped icon appears to the left of the current override Snap mode.

Note: *The number of Snap modes included in the Snaps menu varies. The menu shows only the Snap modes that are available for the active placement or manipulation command. Some commands do not support all Snap modes.*

Using the Tentative Button

To use the Tentative button in placing and manipulating elements:

1. Select the placement or manipulation tool.
2. Select the desired Snap mode (the modes are described later).

3. Point to an element and click the Tentative button.
4. Click the Data button to accept the tentative point.
5. Continue using Snap modes and tentative points, as necessary, to complete the placement or manipulation.

For example, to start a line in the exact center of a block, select the Place Line tool, set the Snap Mode to **Center**, and click the Tentative button on the block. The block is highlighted and a large tentative cross appears at the exact center of the block. Place a data point to start the line at the center of the block (see Figure 3–11).

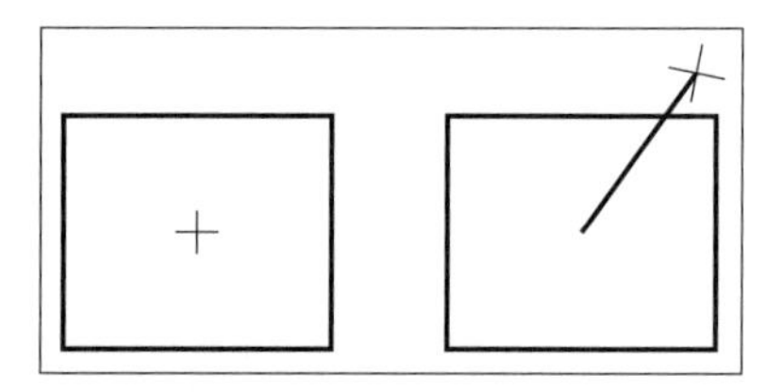

Figure 3–11 Example of using tentative snap

Keep in mind the following points when using the Tentative button.

- Snapping to elements is only in effect if Snap Lock is ON.
- If Grid Lock is ON and the element is placed between grid points, the Tentative button may snap to grid points rather than to the element. If that happens, turn Grid Lock OFF.
- When the Tentative button snaps to an element, the element is highlighted and the tentative cross appears at the tentative snap point. If the cross appears but the element is not highlighted, the snap may have attached to a grid point close to the element.
- When the Tentative button is pressed, MicroStation starts searching for elements in the area immediately around the screen pointer. It normally selects elements in the order they were placed in the design. If the tentative snap point appears on the wrong element, click the Tentative button again and the next element is found; there is no need to move the screen pointer location or to press the Reset button. If the Tentative button cycles through all the elements in the area without finding the required tentative snap point, move the screen pointer closer to the element and press the Tentative button again. For example: a block is placed, and then a line is placed starting very near one corner of the block. A snap to the end of the line is required for the next command, but the tentative point snaps to the corner of the block. Click the Tentative button again, and it should snap to the end of the line. If

the second snap also does not find the end of the line, move the screen pointer a little closer to it and snap again.

- The screen pointer does not have to be placed exactly on the point of the element to snap to it, just near it. In fact, to lessen the chance of snapping to the wrong element, it is best to move back along the element, away from other elements.

Types of Snap Modes

The following sections explain the available Snap modes. Availability of modes depends on the currently active command.

Nearest Mode

When active, Nearest mode places tentative snap points on the point of an element that is closest to the pointer. This rule remains the same among all element types, except text where the only snap point is the justification point. With Nearest mode any point on any type of element can be snapped.

To pick a specific point on an element, position the pointer close to the point to be selected, ensure that the Snap Lock is ON and the Snap mode is Nearest. When the Tentative button is pressed, the element is highlighted and the tentative cross appears at the closest point on the element. If the tentative cross appears, but the element does not highlight, there is no snap to the element. Continue pressing the Tentative button until a point is located, and then press the Accept button.

Keypoint Mode

Keypoint mode allows tentative points to snap to keypoints on elements. For a line, the keypoints are at the endpoints and center of the line; for a circle, they are at the center and four quadrants; for a block, they are at the four corners, and so on. See Figure 3–12 for the keypoint snap points for various element types. To snap to a keypoint on an element, position the pointer close to the keypoint (make sure Snap Lock is ON and the Keypoint mode is selected) and click the Tentative button. The tentative cross appears on the element's keypoint and the element is highlighted. If the tentative cross appears but the element does not highlight, element's snap point has not been found. Continue pressing the Tentative button until the correct snap point is located, and then press the Accept button.

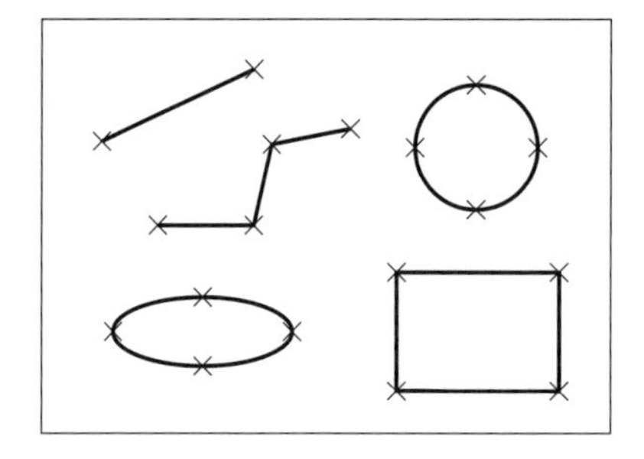

Figure 3–12 Keypoint snap locations on various element types

The Keypoint snap Divisor works with Keypoint snap mode to define additional snap points on an element. For example, setting the Keypoint snap Divisor to 5 divides an element into five equal divisions. Figure 3–13 shows the keypoint snaps for divisor values of two and five. The divisor can be set in the Full Locks settings box by keying-in the value in the Divisor edit field. The value can also be set by keying-in at the key-in window: KY=<number of divisors> and pressing ENTER.

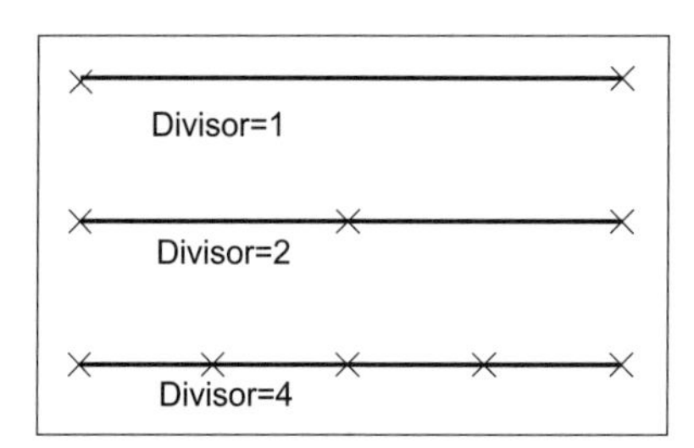

Figure 3–13 Example of various key point snaps divisors

Midpoint Mode

Midpoint mode, when active, places tentative snap points at the midpoint of an element or segment of a complex element (see Figure 3–14). The location of the midpoint varies with different types of elements:

- It bisects a line, arc, or partial ellipse.
- It bisects the selected segment of a line string, block, multi-sided shape, or regular polygon.
- It snaps to the 180-degree (9 o'clock) position of a circle or an ellipse.

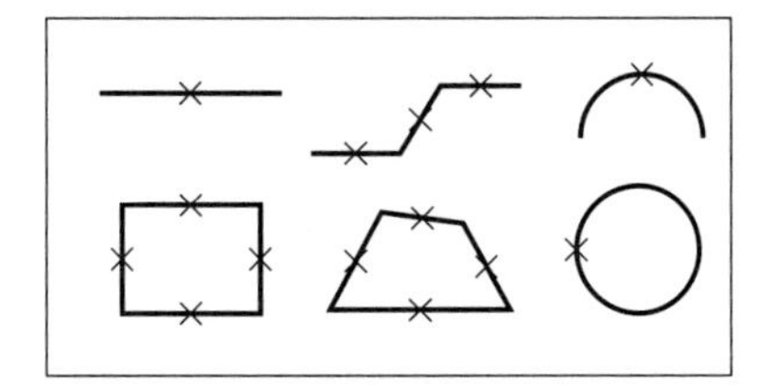

Figure 3–14 Midpoint snaps for various element types

Center Mode

Center mode causes tentative snap points to snap to the center of the space occupied by an element (such as a circle, block, or arc) as shown in Figure 3–15.

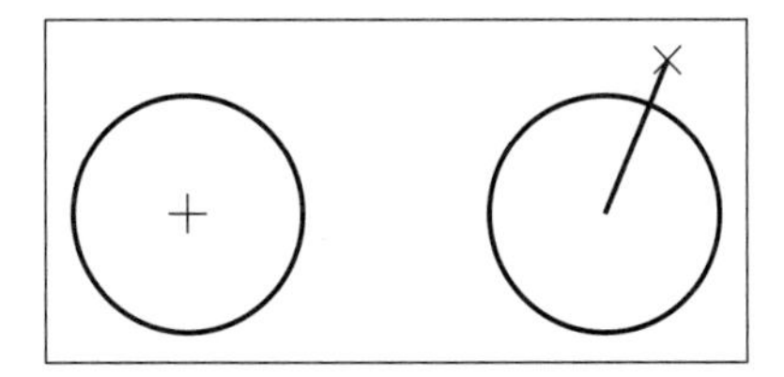

Figure 3–15 Example of snapping to the center of the circle to start a line

Origin Mode

Origin mode snaps to the center of an arc, circle, origin of the text, or cell.

Bisector Mode

Bisector mode sets the snap mode to bisect an element; the snap point varies with different types of elements.

Intersection Mode

Intersection mode causes tentative snap points to snap to the intersection of two elements. To find the intersection, snap to one of the intersecting elements close to the point of intersection. One or both of the elements may be highlighted and displayed with dashes to indicate that the intersection has been found. When the intersection is found the large tentative cross appears at the intersection (see Figure 3–16).

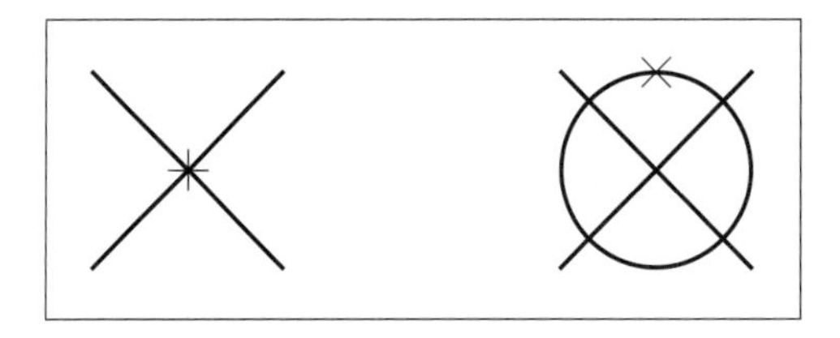

Figure 3–16 Example of snapping to the intersection of two elements for placing a circle

If the elements do not actually intersect, the tentative cross appears at the intersection of an imaginary extension of the two elements. If the two elements cannot be extended to an intersection, an error message appears in the Status bar and a tentative cross is not placed.

Tangent Mode

Tangent mode forces the element being created to be tangent to a nonlinear element (such as a circle, ellipse, or arc). The actual point of tangency varies, depending on how the element is placed (see Figure 3–17).

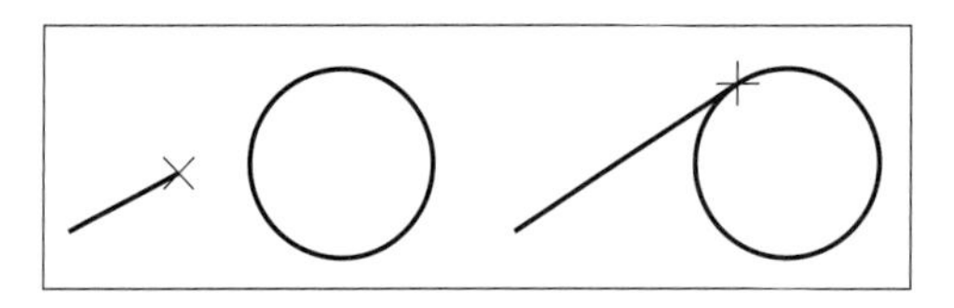

Figure 3–17 Example of snapping to a tangent point

Tangent Point Mode

Tangent Point mode forces the element being placed to be tangent to an existing nonlinear element (such as a circle, ellipse, or arc) at the point where the tentative snap point was placed (see Figure 3–18).

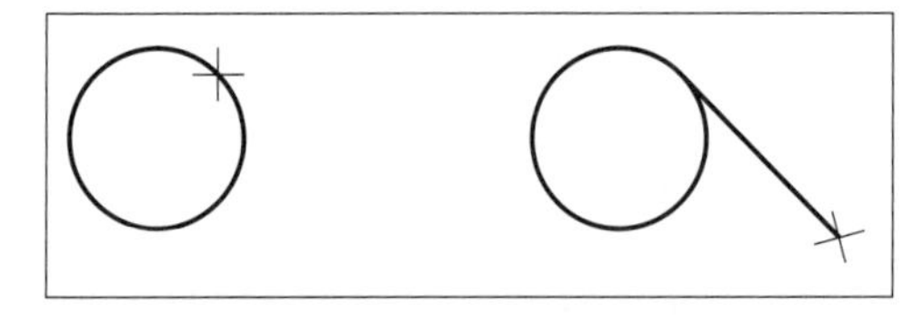

Figure 3–18 Example of snapping to a tangent on an existing nonlinear element

Perp Mode

Perp mode forces the element to be perpendicular to an existing element. The actual perpendicular point depends on the way the element is placed.

Perp Point Mode

Perp Point mode forces the element to be perpendicular to an existing element at the point where the tentative snap point was placed.

Parallel Mode

Parallel mode forces the line or segment of the line string to be parallel to a linear element.

Point Through Mode

Point Through mode causes tentative snap points to define a point on an existing element through which the element being placed must pass (see Figure 3–19).

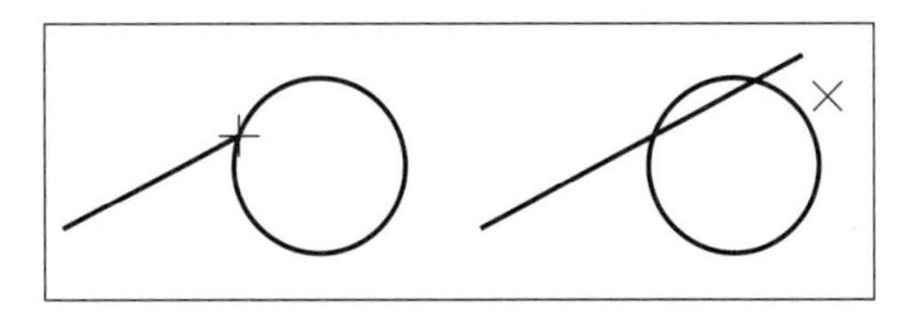

Figure 3–19 Example of snapping to Point Through

Point On Mode

Point On mode snaps to the nearest element (after the first element placement point) and constrains the next data point to lie on a closed element or anywhere on a linear element's line.

Multi-Snap Modes

Three Multi-Snap modes (Multi-Snap 1, Multi-Snap 2, and Multi-Snap 3) provide customizable groups of snap modes. The groups allow several snap modes to be active at the same time. For example, if one of the groups includes Midpoint and Intersection snap modes, positioning the screen pointer near the midpoint of an element causes a snap point to be indicated at the midpoint and positioning the screen pointer near the intersection of two elements causes a snap point to be indicated at the intersection of the two elements.

The three members of the Multi-Snap groups can be viewed and changed from the Multi-Snaps settings box. Invoke the Multi-Snap settings box from:

Menu	Settings > Snaps > Multi-Snaps

MicroStation displays the Multi-Snaps settings box, as shown in Figure 3–20.

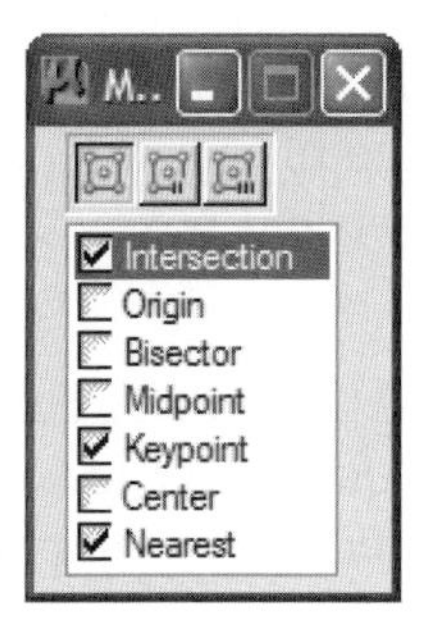

Figure 3–20 The Multi-Snap settings box

At the top of the Multi-Snap settings box are icons for each of the three Multi-Snap groups. Click an icon to see and change that group's snap modes. Below the icons are check boxes for all snap modes that can be included in the selected group. If a snap mode's check box is ON, that snap mode is part of the selected group. Modify the snap modes in the selected group by turning the snap mode check boxes ON and OFF. The group membership changes as soon as the setting of a snap mode check box is changed.

ACCUSNAP

AccuSnap improves the ability to quickly snap to elements by automatically placing tentative snap points and providing visual and audible clues to the location of the snap points. The Tentative button is not needed when AccuSnap is enabled (a timesaver for two-button mouse users).

Use AccuSnap to Snap to an Element

AccuSnap comes into play when selecting a tool and moving the pointer near a snap point on an element in the design. AccuSnap displays dashed crosshairs on the point. When the pointer is moved closer to the point, AccuSnap highlights the element containing the point and changes the crosshairs to an "X" with a different color to indicate that the screen pointer is at the point. To place a data point at the snap point, click the Data button, after the "X" appears. To reject it, move the pointer away from the point.

Control the Way AccuSnap Works

MicroStation provides an AccuSnap settings box that controls the behavior of AccuSnap. Invoke the AccuSnap settings box from:

Menu	Settings > Snaps > AccuSnap
Pop-up menu	While holding down SHIFT, click the Tentative button and select AccuSnap from the pop-up menu.

The AccuSnap settings box is divided into three groups of settings (**General, Element**, and **Feel**) that are available in separate tabs. To access one of the groups, click its tab near the top of the settings box.

General Settings

The **General** tab (see Figure 3–21) in the AccuSnap settings box toggles AccuSnap ON and OFF (enabled and disabled) and controls what AccuSnap does when the screen pointer approaches a tentative snap point on an element in the design.

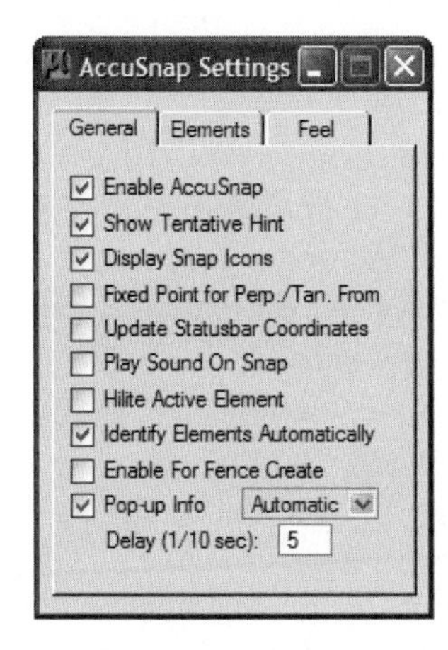

Figure 3–21 The General Tab on the AccuSnap settings box

If the Enable AccuSnap check box is ON, AccuSnap is ON and will automatically identify tentative snap points. If the check box is OFF, there is no AccuSnap action and the Tentative button must be used to place tentative snap points.

Note:** AccuSnap can also be turned ON and OFF from the Snap Mode tool box (see Figure 3–22). To open the tool box, select **Settings > Snaps > Button Bar.

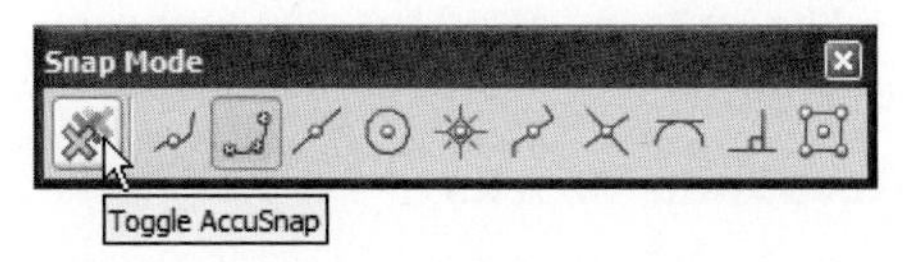

Figure 3–22 The AccuSnap button in the Snap Mode toolbar

Turning ON the **Show Tentative Hint** check box displays the cross when the pointer is near a tentative snap point. If the check box is OFF, AccuSnap only displays the "X" when the pointer is at a tentative snap point and there is no hint that the screen pointer is getting close to a tentative point.

Turning ON the **Display Snap Icons** check box displays a picture of the currently active snap mode's icon (example: Keypoint Snap) with the AccuSnap cross and "X." This setting provides a convenient reminder of what will happen when the tentative snap point is accepted by clicking the Data button.

The **Fixed Point for Perp./Tan. From** check box is used only with the Place SmartLine tool when Perpendicular or Tangent snap modes are in effect. The SmartLine tool is discussed in Chapter 5.

Turn ON the **Update Status Bar Coordinates** check box and AccuSnap will display the coordinate readout for each tentative snap point, in the Status bar.

Turn ON the **Play Sound On Snap** check box and AccuSnap make a clicking sound when it displays a tentative snap point (whenever the "X" appears).

Turn ON **Hilite Active Element** and AccuSnap highlights an element as soon as it finds a tentative snap point on the element (whenever the cross appears). Turn the check box OFF so that elements are highlighted only when AccuSnap selects a tentative snap on them (when the "X" appears).

Turn ON **Identify Element Automatically** and AccuSnap automatically identifies elements, as the pointer passes over them.

If **Enable For Fence Create is ON** (and AccuSnap is enabled) tentative snap points are identified when a fence is defined.

The **Pop-up Info** check box controls the display of Tips that appear next to the tentative snap points that AccuSnap finds or selects. When the check box is ON, an options list is available, to the right of the check box. The options are:

- **Automatic** – Causes the Tips to appear automatically.
- **Tentative** – Causes the Tips to appear only when the Tentative button is clicked to place a tentative snap point and then the screen pointer is allowed to hover over the point.

When the **Pop-up Info** check box is OFF, no tips appear.

Element Settings

The **Element** tab (see Figure 3–23) in the AccuSnap settings box provides check boxes for (B-spline) **Curves, Dimensions, Meshes,** and **Text.**

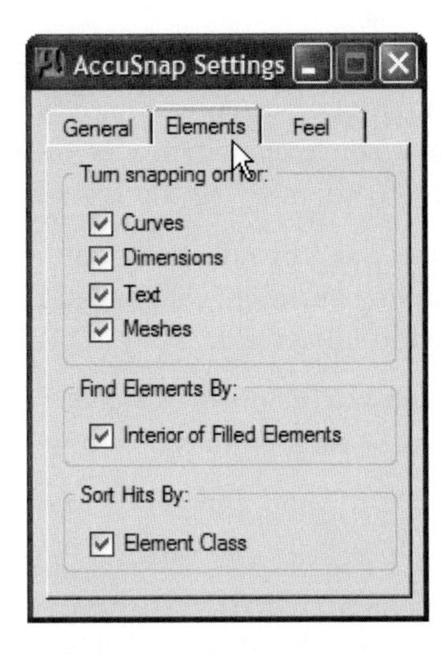

Figure 3–23 The Elements Tab on the AccuSnap settings box

If the check box for B-spline **Curves, Dimension, Text**, or **Meshes** is ON, AccuSnap snaps to that type of element. If the check box is OFF, AccuSnap does not snap to that type of element.

Note: *Override the Element settings by using the Tentative button to place tentative snap points.*

If the **Find Elements By Interior of Filled Elements** check box is ON, the interior of filled elements is included in the tentative snaps when the **Fill** view attribute is ON. For example, the entire filled area becomes a tentative snap point when Nearest snap mode is ON and the center of the filled area becomes a tentative snap point when Keypoint snap mode is ON.

If the **Sort Hits By Element Class** check box is on, the order in which AccuSnap snaps to overlapping elements is determined by their class: Primary elements first, followed, in order, by Construction, Pattern, and Dimension elements

Feel Settings

The **Feel** tab (see Figure 3–24) provides three slider bars used to control AccuSnap's sensitivity when locating and selecting tentative snap points.

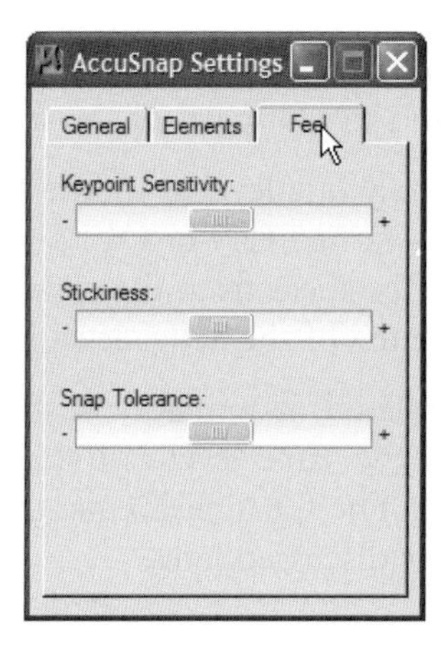

Figure 3–24 The Feel Tab on the AccuSnap settings box

The **Keypoint Sensitivity** slider bar adjusts how close the pointer must be to the snap point before AccuSnap will snap to it and display the "X". Moving the slider to the right (+) increases the allowable distance, while moving the slider to the left (-) reduces the allowable distance.

The **Stickiness** slider bar adjusts the sensitivity of AccuSnap to the current element when it has tentatively snapped to that element. Moving the slider to the right (+) increases the distance the pointer can be moved away from the active element before AccuSnap drops it and snaps to another element. Moving the slider to the left (-) decreases the distance.

The **Snap Tolerance** slider bar adjusts how close the pointer must be to an element before a tentative point can be snapped to it (the "snap tolerance"). Moving the slider to the right (+) increases the snap tolerance, moving to the left (-) decreases it.

ELEMENT ATTRIBUTES

There are several important attributes associated with the placement of elements: level, color, line style, line weight, transparency, and priority.

LEVELS

MicroStation offers a way to group elements on levels, in a manner similar to a designer drawing different parts of a design on separate transparent sheets. By stacking the transparent sheets one on top of another, the designer can see the complete drawing but can only draw on the top sheet. If the designer wants to show a customer only part of the design, he or she can remove from the stack the sheets that contain the parts of the design the customer does not need to see.

MicroStation has the ability to create an almost unlimited number of levels in each design file. For example, an architectural design might have the walls on one level, the dimensions on another level, electrical information on still another level, and

so on. Separating parts of the design by level, allows designers to turn on only the part they need to work on and to plot parts of the design separately.

Elements can be placed on only one level at a time (the active level), but the display of all other levels can be turned ON or OFF, in selected views. Elements on levels that are not displayed disappear from the view and do not plot, but they are still in the design file. The name of the active level is displayed on the Status bar.

When an element is manipulated, it remains on the same level. For instance, a copy of an element stays on the same level as the original element, regardless of what level is currently active. The Change Element Attributes tool, which moves elements to different levels, is discussed later.

Level Settings Boxes

Two settings boxes and a Menu are provided for managing levels. The Level Display settings box controls which levels are displayed and sets the active level (see Figure 3–25). The Level Manager settings box allows levels to be created, sets level symbology, and sets the active level (See Figure 3–26). Invoke the Level Display setting box from:

Menu	Settings > Levels > Display
Primary Tool box	Click the Level Display icon

Invoke the Level Manager settings box from:

Menu	Settings > Levels > Manager
Status Bar	Click the name of the active level.
Level Display settings box	Position the pointer over a level name, click the Reset button, and, in the resulting pop-up menu, click the Level Manager option.

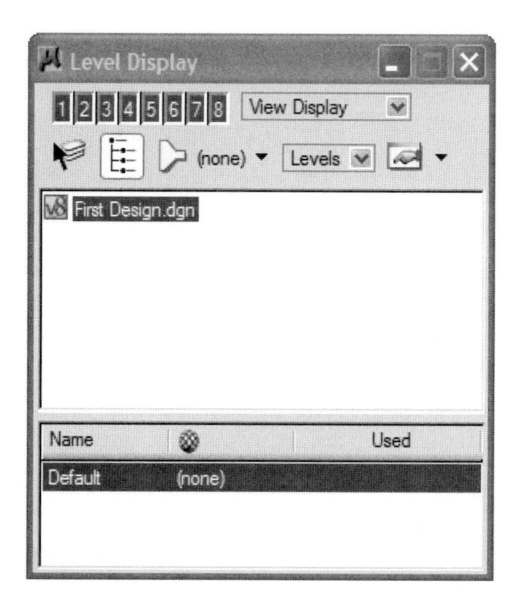

Figure 3–25 The Level Display settings box

Note: *The Level Display settings box can be docked against the left or right sides of the View windows.*

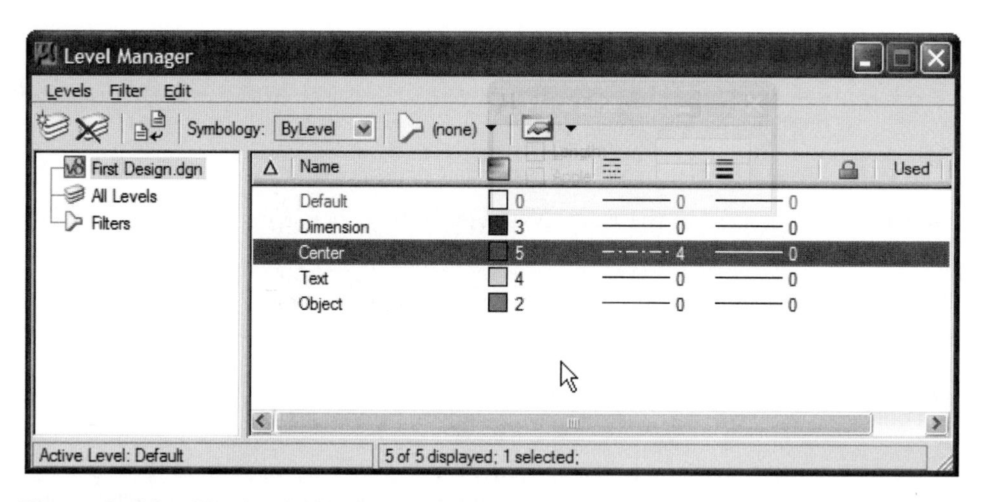

Figure 3–26 The Level Manager settings box

Setting the Active Level

There can be only one active level at any time, and most MicroStation placement tools place new elements on the active level. The name of the active level is dis-

played in the Status bar and is highlighted in the Level Manger and Level Display settings boxes.

Select the active level from the Level Manager or Level Display settings box by double-clicking the level's name. The selected name is highlighted in the settings boxes and displayed in the Status bar.

The Active Level can also be set from a Menu and with a key-in:

Attributes Tool box	Click the drop-down arrow on the Active Level menu to display the list of defined levels. Select the name of the level that is to be made the active level (see Figure 3–27).
Key-in window	**LV=**<level number> ENTER

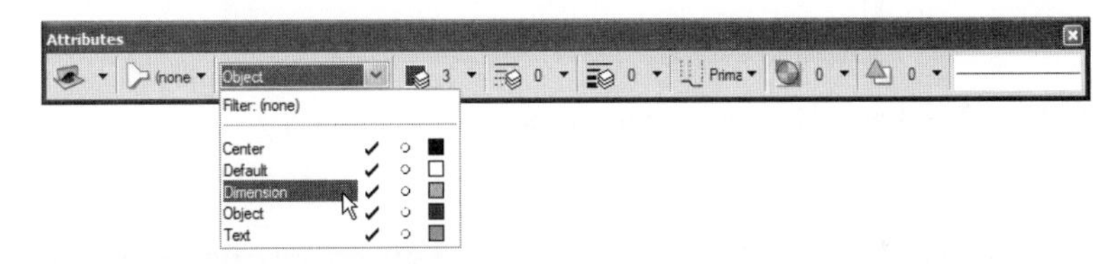

Figure 3–27 The Active Level menu

Note: *The active level remains in effect for the rest of the design session or until a new active level is selected. To save the current active level for the next design session, select **Save Settings** from the **File** menu.*

Creating New Levels

New levels are created in the Level Manager settings box. First create the new level and then set the new level's properties (such as the name).

To create a new level from the Level Manager settings box:

Menu	Levels>New
Tool box	Click the New icon
From an existing level name	Position the pointer over one of the existing level names, click the Reset button, and, in the resulting pop-up menu, select the New option.

When the new level is created, a description of its default settings is inserted at the bottom of the list of existing level names.

To change the name of the new level, click the default level name in the levels list, and type the new name. If the default name's background color does not change when clicked, click the name again before typing.

Setting Level Attributes

To the right of each named level in the Level Manager settings box are fields named **Color**, **Style**, and **Weight**. Make the necessary changes to the attributes for the newly created levels. The default attribute settings are all zero (white, solid, weight zero lines).

Note: *Make sure the Element Attributes are set to ByLevel (can also be set through Attributes tool box), so that the attribute settings that are set for active level will be applied when elements are placed.*

Changing the Properties of Existing Levels

To change the properties of an existing level in the Level Properties, select the level to be edited in the Level Manager settings box and invoke the Level Properties dialog box from:

Menu	Levels > Properties
From the new level name	Position the pointer over the name of the new level, click the Reset button, in the resulting pop-up menu, and select the Properties option.

MicroStation will open the Level Properties dialog box, as shown in Figure 3–28. The box contains options for changing the levels name and attributes. It also provides the Description field to record more information about the intended use of the level.

Note: *The level name, color, style, and weight can also be changed by clicking each setting in the level's row in the Level Manager settings box.*

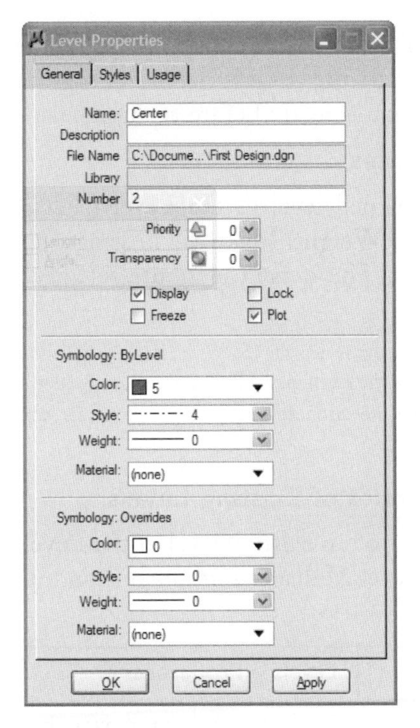

Figure 3–28 The Level Properties dialog box

Controlling the Display of Levels

The Level Display settings box controls the display ON or OFF status for one or all design views windows.

To control the display of levels, invoke the View Levels settings box from:

Menu	Settings > Level > Display
Key-in window	**ON=**<level numbers> ENTER **OFF=**<level numbers> ENTER

Displayed levels are shown in the Level Display settings box with a dark background and hidden levels are displayed with a light background.

Before changing the display of any levels, select the design view in which the display change is to occur. Click the drop-down arrow in the **View** menu and select one of the design views. (Only currently-open views can be selected). Any level display changes are applied only to the selected view. To apply the changes to all the design views, turn ON the **Use Global** check box.

To change the display state of a level, click it – if display of the level was ON, clicking it turns display OFF and vise-versa. To change the display state of a contiguous set of levels, click the first or last one in the list and, while continuing to

hold down the Data button, drag the pointer across all the levels. The resulting display state of all the selected levels is determined by the state of the first level selected. If display of the first level was OFF, all selected levels are turned ON (levels that were already ON remain ON). If the display of the first level was ON, all selected levels are turned OFF.

Notes: *Display of the active level cannot be turned OFF.*
The active level and the level display settings for each view remain in effect until they are changed or MicroStation is closed. To keep them in effect for the next editing session, select ***Save Settings****from the* ***File****menu.*

ELEMENT COLOR

Color helps to differentiate the various types of design elements. For example, the object being designed may be drawn using black while the dimensions are drawn using green. Before drawing elements, set the active color to ByLevel to use the color that is defined in the Level manager settings box for the Active Level. To override the color assigned to the active Level, select an alternate color from the color palette.

All new elements drawn are set to the active color. The active color remains in effect until it is changed to a different color or the design file is closed. To keep the active color in effect for the next editing session, select **Save Settings** from the **File** menu, after selecting the active color.

Note: *Element color is permanent, so the color of existing elements is not changed by changing the active color. Commands for changing an element's color are discussed later.*

- Three methods are provided for selecting the Active Color:
- Select the color from a color table that contains pre-defined color selections.
- Define a "True color" by selecting color values from a color model.
- Select a color name from a color book.

Set the Active Color from:

Attributes Tool box	Click the drop-down arrow on the Color menu. Select a color using one of the selection options on the resulting pop-up color options dialog box as described below.
Key-in window	**CO=**<name of the color or color number from 0 to 254> ENTER

The color options dialog box has three tabs that provide different methods for selecting or creating the active color. The first tab is normally selected when the dialog box is opened.

Color Table Tab

The first tab displays a Color Table that allows the Active Color to be selected from a set of numbered, predefined colors (see Figure 3-29).

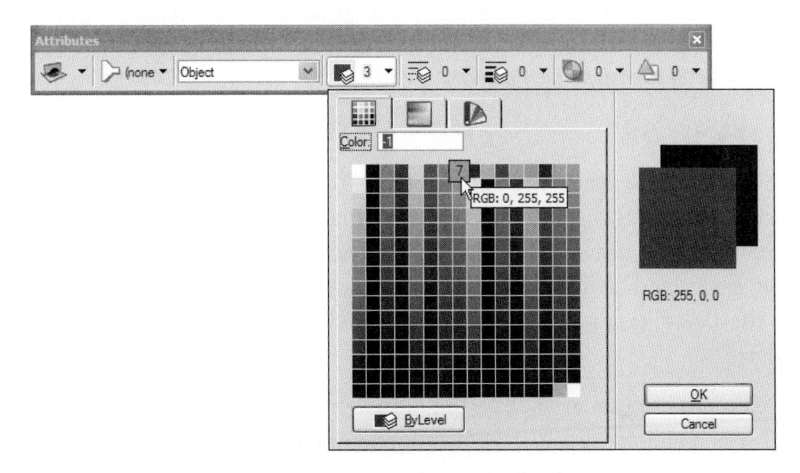

Figure 3–29 The Color options dialog box displaying the Color Table Tab

Select the Active Color from the color table by typing the color's ID number in the **Color** field or by clicking a color on the color palette. The color assigned to the Active Level can be set as the Active Color by clicking **ByLevel**. Selecting a color closes the dialog box and makes the selected color the active color. Close the dialog box without selecting an Active Color by clicking **Cancel** or by clicking away from the dialog box. The **OK** button is not used by the Color Table.

Note: *The colors assigned to each number can be customized. For more information, see Chapter 14.*

True Color Tab

The second tab displays True Color options that allow definition of a unique color to use as the Active Color (see Figure 3-30).

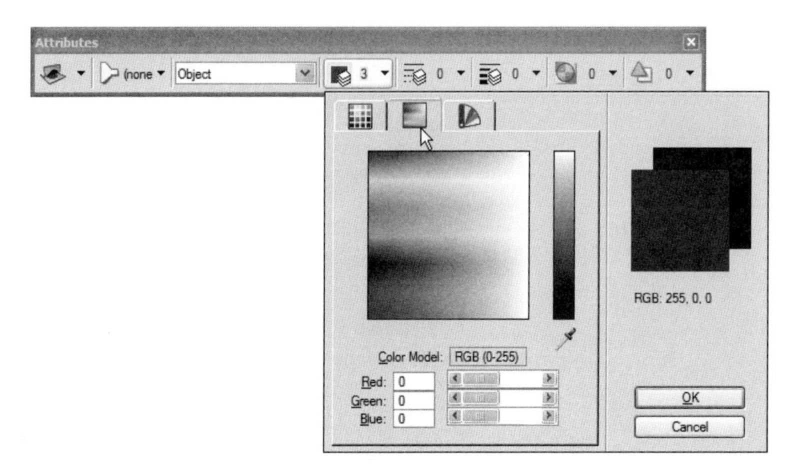

Figure 3–30 The Color options dialog box displaying the True Color options

The True Color options allow the Active Color to be interactively created. Define the Active Color by clicking a color in the color palette, by dragging the horizontal bar up and down the vertical color bar, or by selecting one of four color models and entering module values in the three text fields. For example, the **RGB** color model allows defining a color by entering red, green, and blue saturation values in the range of 0 to 255 (if all three values are zero, the color is black and if all three equal 255, the color is white).

The defined color is previewed on the right side of the dialog box (the second preview is the background color). Make the defined color the Active Color by clicking **OK**. Close the dialog box without defining an Active Color by clicking **Cancel** or by clicking away from the dialog box.

Color Book Tab

The third tab displays a color book that presents a set of named colors (see Figure 3-31).

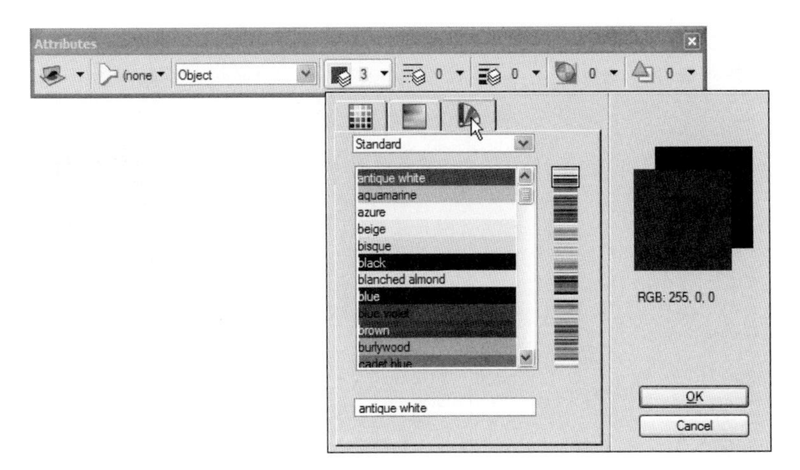

Figure 3–31 The Color options dialog box displaying the Color Book options

Select a color from the color book by clicking the color's name in the list of color names. Scroll through all the color names by dragging the rectangle located on the vertical color bar.

The defined color is previewed on the right side of the dialog box (the second preview is the background color). Make the defined color the Active Color by clicking **OK**. Close the dialog box without defining an Active Color by clicking **Cancel** or by clicking away from the dialog box.

Note: *The colors assigned to each number can be customized. For more information, see chapter 14.*

Note: *The actual colors shown depend on the monitor, graphics card, and what colors are defined in the MicroStation color table.*

ELEMENT LINE STYLE

In addition to color choices, MicroStation can assign elements a specific line style such as: solid or dashed. Select the active line style from a set of eight internal styles and several custom-made line styles, or elect to have the line style set to the Active Level's style attribute.

All new elements drawn are set to the active line style. The active line style remains in effect until it is changed it or the design file is closed. To keep the active line style in effect for the next editing session, select **Save Settings** from the **File** menu, after selecting the active style.

Note: *Element line style is not actively linked to the active line style, so the line style of existing elements is not changed by changing the active line style. Commands for changing an element's line style are discussed later.*

Set the Active Line Style from:

Attributes Tool box	Click the drop-down arrow on the Line Style menu. Select a style from the resulting pop-up menu or select ByLevel to use the Active Level Style attribute that is set in the Level Manager settings box (see Figure 3–32).
Key-in window	**LC=**<name of the line style or line style number> ENTER

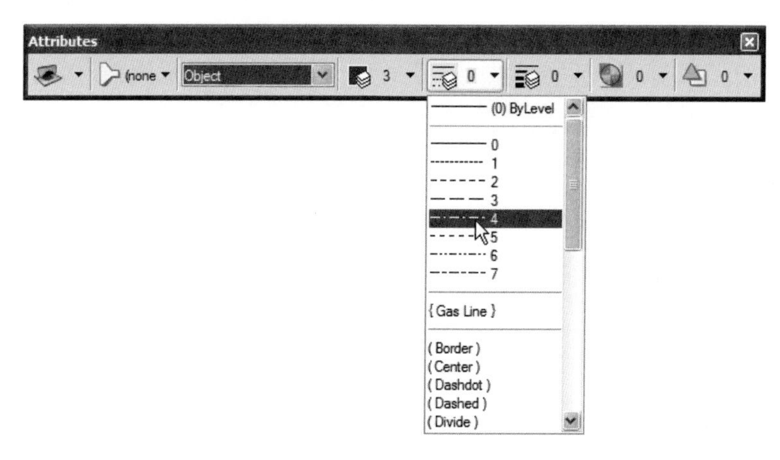

Figure 3–32 The Line Style options menu

ELEMENT LINE WEIGHT

In drafting, line width contributes to the "readability" or understanding of the design. For example, in a piping arrangement drawing, the line width for the pipe is the widest of the lines used on the drawing so that the pipe stands out from the equipment, foundations, and supports.

In MicroStation, line weight refers to the width of the element lines and elements are placed using the active weight. There are 32 line weights (numbered from 0 to 31), which is comparable to 32 different technical pens. The active weight can also be set to the Active Level Weight attribute.

All new element lines are set to the active weight.The active weight remains in effect until it is changed or the design file is closed.To keep the active line weight in effect for the next editing session, select **Save Settings** from the **File** menu, after selecting the active weight.

Note: *Element line weight is not actively linked to the active line style, so the weight of existing elements is not changed by changing the active weight. Commands for changing an element's line weight are discussed later.*

Set the Active Element Line Weight from:

Attributes Tool box	Click the drop-down arrow on the Line Weight menu. Select a weight from the resulting pop-up menu, or select ByLevel to use the Active Level Weight attribute that was set in the Level Manager settings box (see Figure 3–33).
Key-in window	**WT=**<line weight number anywhere from 0 to 32> ENTER

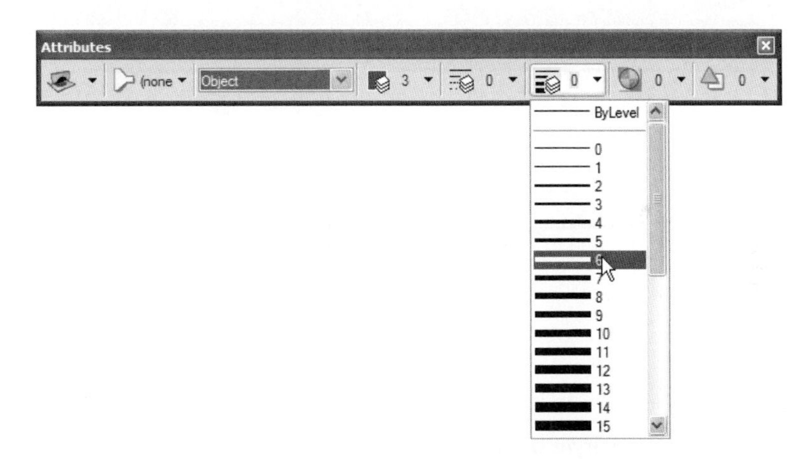

Figure 3–33 The Line Weight options Menu

ELEMENT TRANSPARENCY

Element Transparency can vary from 0% to 100%. A value of 0% indicates no transparency at all, while a value of 100% indicates almost complete transparency. If an element with 0% transparency is drawn over an existing element in the design, the new element covers the existing element. As transparency increases, the existing element will began to appear through the new element that was placed over it and the new element will appear to fade from view. Figure 3-34 shows an example of transparency. As discussed later in this chapter, there is a view attribute that controls the use of transparency in view windows. When the transparency attribute is off, all elements are displayed as if they had zero transparency.

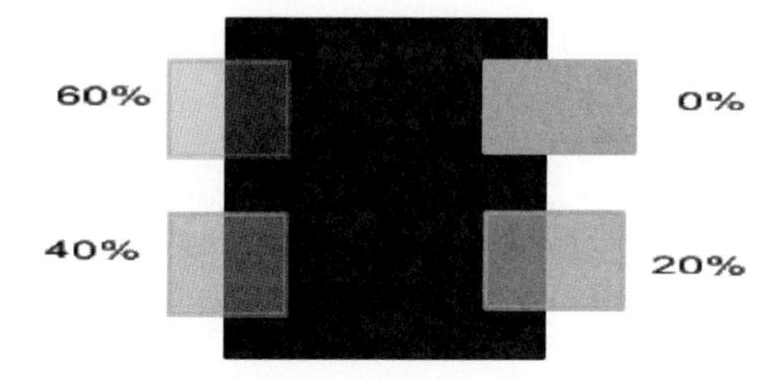

Figure 3–34 Element Transparency Examples

Set the Active Element Transparency from:

Attributes Tool box	Click the Transparency tool and select a transparency value from the resulting pop-up menu (see Figure 3–35). An alternate way to select a value is to drag the slider bar across the bottom of the menu until the required value is displayed.
Key-in window	ACTIVE TRANSPARENCY <transparency from 0 to 100> ENTER

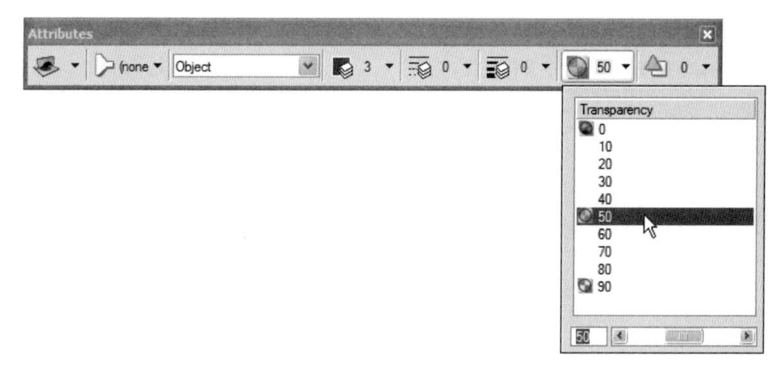

Figure 3–35 Element Transparency Options

ELEMENT PRIORITY

Elements are placed at Active Element Priority. Element Priority, which can be in the range of zero to plus or minus 500, determines how an element is displayed relative to other elements. Elements with a higher priority are displayed on top of elements with a lower priority.

Set the Active Element Priority from:

Attributes Tool box	Click the Priority tool and select a priority from the resulting pop-up menu (see Figure 3–36). An alternate way to select a value is to drag the slider bar across the bottom of the menu until the required value is displayed.
Key-in window	ACTIVE PRIORITY <priority from -500 to +500> ENTER

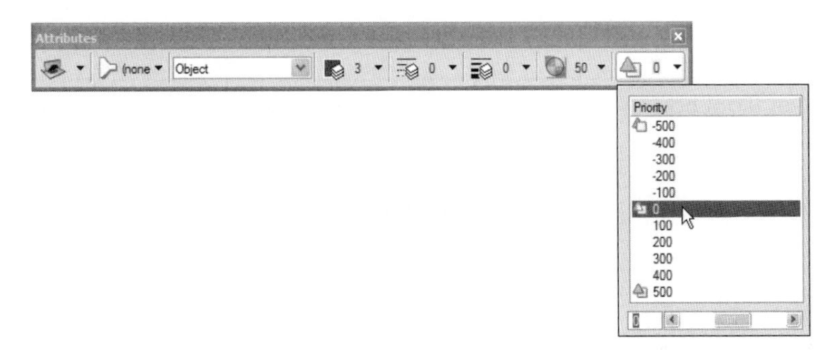

Figure 3–36 Element Priority Options

CHANGE ELEMENT ATTRIBUTES

The Change Element Attributes tool allows changing the level, color, style, and weight of elements already in the design file. Select the required attributes and then select elements that will be changed to the new attributes. Or, first select an element with the desired attributes and then select the elements that will share the attributes of the that element. Invoke the Change Attributes tool from:

Change Attributes tool box	Select the **Change Element Attributes** tool (see Figure 3–37).
Keyboard Navigation (Main tool box active)	**5 1**

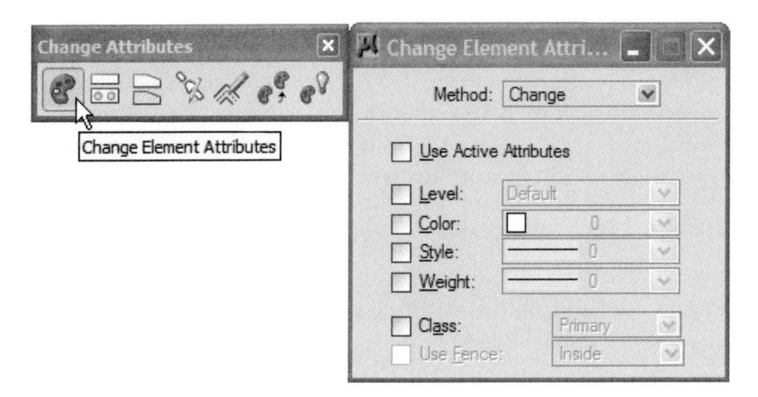

Figure 3–37 Invoke the Change Element Attributes tool from the Change Attributes tool box

Change Method

If no existing element uses the required attribute settings, select **Change** from the **Method** menu in the Tool Settings window.

Change the state of an attribute by clicking the attribute's check box. For example, to only change the level of an element, turn ON the **Level** check box and turn OFF the check boxes for **Color**, **Style**, **Weight**, and **Class** (Class is not discussed here).

To change elements to the current active element attributes (Active Level, Active Color, Active Style, and Active Weight), turn ON the **Use Active Attributes** check box to load the Tool Settings window with the current active settings.

Note: *When the **Use Active Attributes** check box is ON, any attribute setting changes made in the Tool Settings window also change the associated Active Attribute. For example, if the Active Color is red and the Color on the Tool Settings window is changed to green, the Active Color becomes green. If the check box is OFF, changes on the Tool Settings window do not change the Active Attributes.*

Instead of using the current Active Attribute settings, turn OFF the **Use Active Attributes** check box and select the attribute settings from the menus to the right of each attribute check box.

Once the attributes are set, start selecting the elements whose attributes are to be changed. MicroStation prompts:

> Change Element Attributes > Identify Element *(Identify the element whose attributes are to be changed.)*
>
> Change Element Attributes > Accept/Reject *(Click the Accept button to accept the changes for the selected element or click the Reject button to reject the element.)*

Note: *If the pointer is on another element when the Accept button is clicked, the first element is changed, the second element is selected, and MicroStation prompts to "Accept/Reject" the second element. If the pointer is not on another element when the Accept button is clicked, MicroStation prompts to "Identify Element" and another element can be selected to change. The Change Element Attributes tool remains active until another tool is selected.*

Match/ Change Method

If an existing element has the desired attribute settings, select the **Match/Change** option from the **Method** menu in the Tool Settings window. With this method, the attributes of the first selected element are set in the Tool Settings window and the subsequent elements selected are changed to the attribute settings of the first element. Before starting to select elements, turn ON the check boxes for the attributes to match and change (**Level**, **Color**, **Style**, and **Weight**).

To also change the Active Element Attributes to those of the first selected element, turn on the **Use Active Attributes** check box before selecting the first element.

MicroStation prompts:

Match Element Attributes > Identify Element to Match *(Identify the element that has the attributes to match.)*

Change Element Attributes > Accept/Reject Element to Match (Click the Accept button to accept the selected element and set its attribute settings in the Tool Settings window, or click the Reject button to reject the element.)

Change Element Attributes > Identify Element to Change *(Identify the element whose attributes are to be changed.)*

Change Element Attributes > Identify Element to Change *(Click the Accept button to accept the changes for the selected element or click the Reject button to reject the element.)*

MATCH ELEMENT ATTRIBUTES

The Match Element Attributes tool allows changing the active attributes (Active Level, Active Color, Active Style, and Active Weight) to the settings of an existing element. This tool provides a quick way to return to placing elements with the same attributes as elements previously placed in the design. Invoke the Match Element Attributes tool from:

Change Attributes tool box	Select the Match Element Attributes (see Figure 3-38)
Keyboard Navigation (Main tool box active)	**5 6**

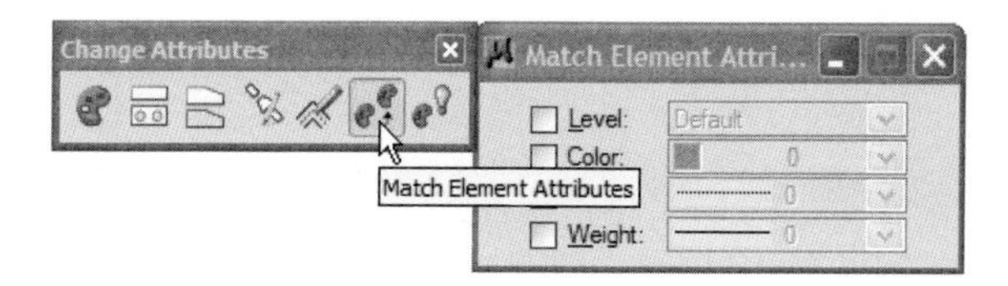

Figure 3–38 Invoke the Match element Attributes tool from the Change Attributes tool box

Before selecting the element, select the attributes to match (**Level, Color, Style,** and **Weight**) by turning on the attribute check boxes in the Tool Settings window. MicroStation prompts:

> Match Element Attributes > Identify Element *(Identify the element that has the attributes to be matched.)*
>
> Match Element Attributes > Accept/Reject (select next input) *(Click the Accept button to accept the changes for the selected element or click the Reject button to reject the changes.)*

VIEW CONTROL

The View Control tools allow selecting the portion of the drawing to be displayed in the View Window (for example zooming out to see more of the design in the View Window). By allowing the design to be viewed in different ways, MicroStation provides the means to draw more quickly, easily, and accurately. The View Control tools explained in this section are utility tools; they make the job easier and help with drawing more accurately, but they do not make any changes to the content of the design.

MicroStation provides eight separate view windows and the view control tools described here can manipulate any of the eight.

Access the view control tools from:

View Control Bar	Located on the upper-left corner of the View window, as shown in Figure 3–39.
View Control tool box	Tools > View Control (see Figure 3–40)
View Control pop-up menu	While holding down SHIFT, click the Reset button to display the pop-up menu shown in Figure 3–41.

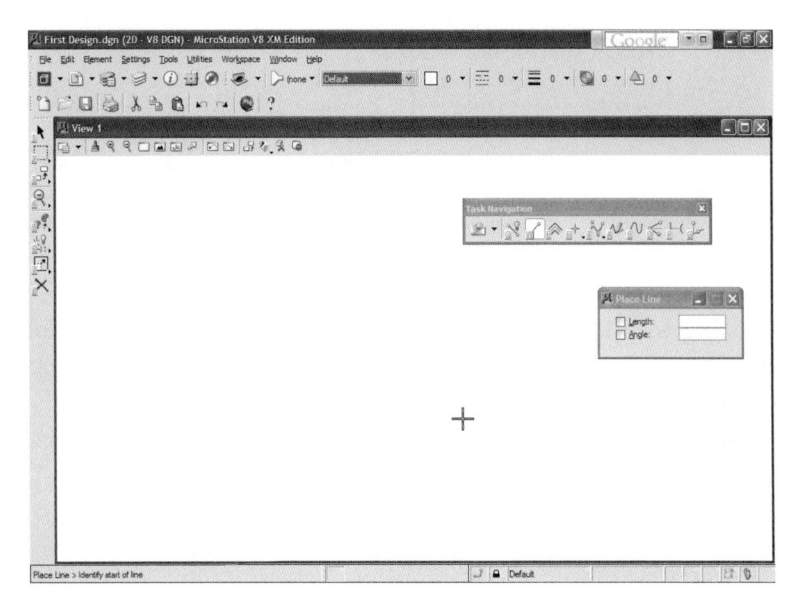

Figure 3–39 View Control Bar on the upper left corner of each view window

Figure 3–40 View Control tool box

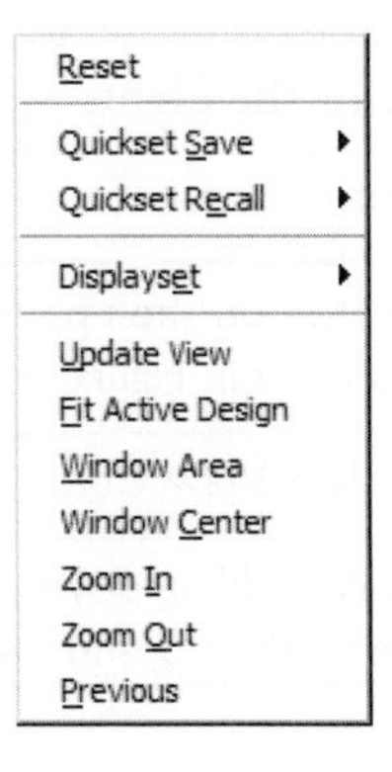

Figure 3–41 View Control pop-up menu

UPDATE VIEW

The Update View tool instructs the computer to repaint the contents of the selected view window making sure it correctly displays the design. Use this tool when

there is an incomplete image of the design in the view. For example, deleting elements can cause the false impression that there are gaps in elements that the deleted element crossed and gaps in the grid. Update the display to repaint the view and correctly display the design. Invoke the Update View tool from:

View Control tool box	Select the Update View tool (see Figure 3–42).
Key-in window	**update view** (or **up**). (ENTER) Then click with the mouse in the view.

Figure 3–42 Invoking the Update View tool from the View Control tool box

MicroStation immediately updates the active view window. There are no prompts. To update all open view windows, click **Update All Views** in the Update View settings window.

CONTROLLING THE DESIGN AREA DISPLAYED IN THE VIEW

The area of the design displayed in a view can be controlled in a way similar to using a zoom lens on a camera. Increase or decrease the viewing area by zooming in or out. Zooming in shows a smaller area of the design in greater detail. Zooming out shows a larger area of the design, in less detail. The design size is not changed, only the view of the design changes.

MicroStation provides three tools to control the area of the design that can be seen in a view window: **Zoom In**, **Zoom Out**, and **Window Area**.

Zoom In

The Zoom In tool moves the view window in closer to the design, showing a smaller area of the drawing in greater detail. Invoke the Zoom In tool from:

View Control tool box	Select the Zoom In tool (see Figure 3–43).
Key-in window	**zoom in extended** (or **z i e**). (ENTER)

Figure 3–43 Invoking the Zoom In tool from the View Control tool box

The Zoom In Tool Settings window contains one field, **Zoom Ratio,** that shows the active zoom in factor. The default ratio is two, but it can be set to range of one to 50 by typing the appropriate positive number in the field.

MicroStation prompts:

> Zoom In > Enter zoom center point *(Moving the pointer in the view displays a rectangular box that indicates the new view boundary. Click the Data button to define the center of the area of the design to be displayed in the view window (see Figures 3–44a and 3–44b) or click the Reset button to exit the Zoom In tool and return to the previously active tool.)*

Continue clicking the Accept button to zoom in closer. Zoom In remains active until the Reset button is clicked or another tool is invoked.

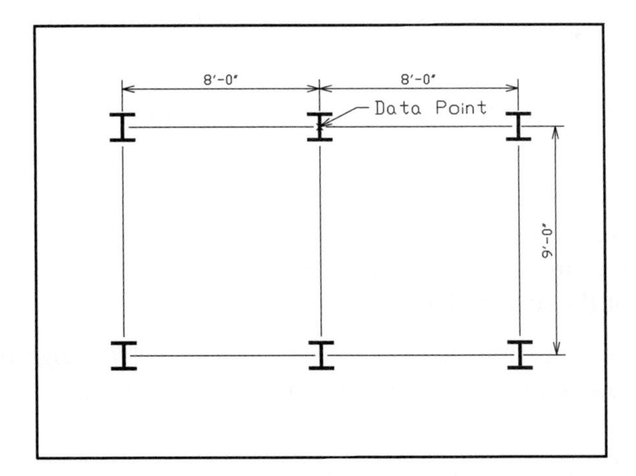

Figure 3–44a The design view before the Zoom In tool is invoked

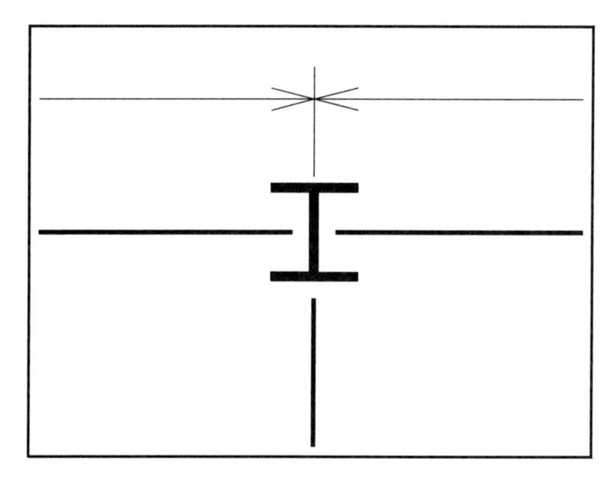

Figure 3–44b The design view after the Zoom In tool is invoked

Zoom Out

The Zoom Out tool moves the view window away from the design, showing a larger area of the drawing, but in less detail. Invoke the Zoom Out tool from:

View Control tool box	Select the Zoom Out tool (see Figure 3–45).
Key-in window	**zoom out extended** (or **z o e**).(ENTER)

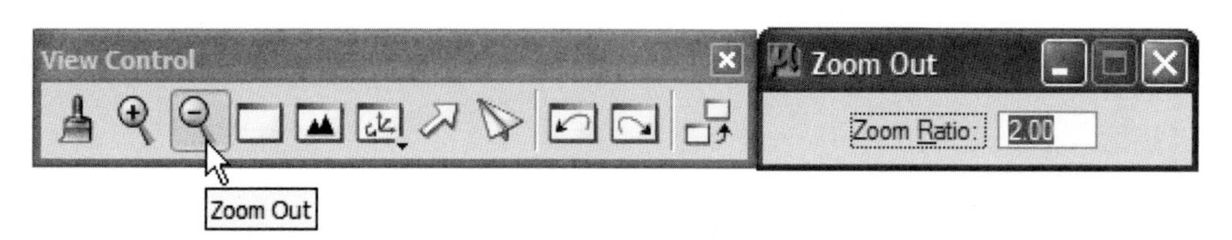

Figure 3–45 Invoking the Zoom Out tool from the View Control tool box

MicroStation immediately zooms the active view window. There are no prompts. Continue clicking the Data button in the view windows to zoom them out further. To exit the Zoom Out tool, click the Reset button or invoke another tool.

The Zoom Out Tool Settings window contains one field, **Zoom Ratio,** that shows the zoom out factor. The default ratio is two, but can be set to a range of one to 50, by typing the appropriate positive number in the field. Changing the Zoom Ratio and positioning the screen pointer over a view window causes MicroStation to prompt:

> Zoom Out > Enter zoom center point *(Position the screen pointer at the desired center point in the view and click the Data button to initiate zooming out by the factor entered in the settings box. (see Figures 3–46a and 3–46b) or click the Reset button to exit the Zoom In tool and return to the previously active tool.)*

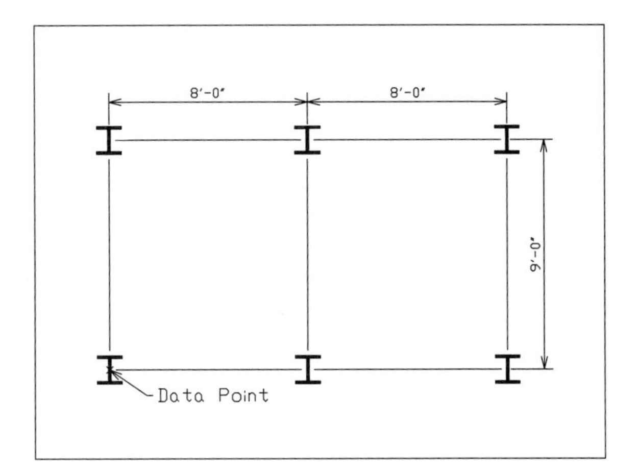

Figure 3–46a The design shown before the Zoom Out tool is invoked

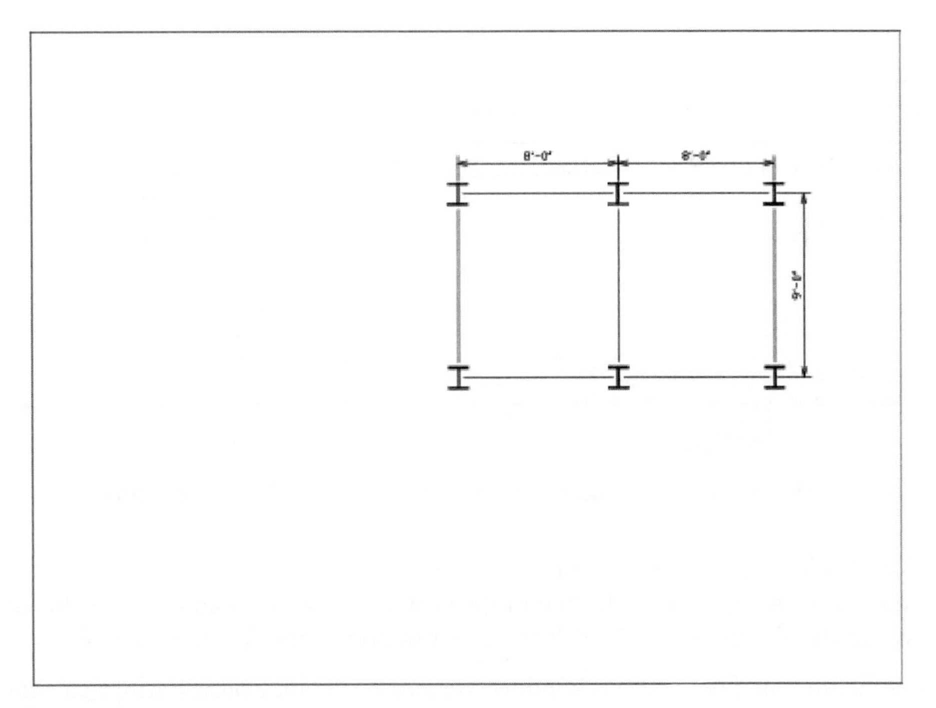

Figure 3–46b The design shown after the Zoom Out tool is invoked

Window Area

The Window Area tool allows placing the diagonally opposite points of a rectangle that defines the area of the design to be displayed in the view window. The center of the area selected is placed in the center of the view window and the design area

inside the rectangle is enlarged to fill the view window as completely as possible. Invoke the Window Area tool from:

View Control tool box	Select the Window Area tool (see Figure 3–47).
Key-in window	**window area extended** (or **w a e**). (ENTER)

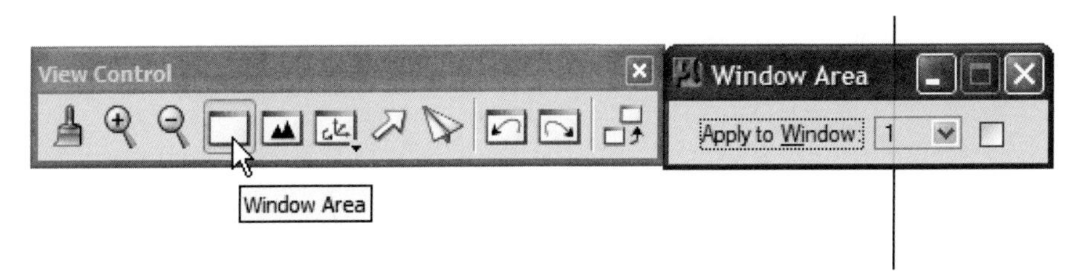

Figure 3–47 Invoking the Window Area tool from the View Control tool box

The Window Areas Tool Settings window contains the **Apply to Window** menu and check box. If the check box is set to ON, the window area definition is applied to the view window whose number is shown in the **Apply to Window** menu. If the check box is set to OFF, the new area is applied to the view window in which the area is defined. To set the destination view window, turn ON the check box and click the drop-down arrow in the **Apply to Window** menu to display the numbers of all currently open view windows. Select the destination view window number from the list. The new area can only be applied to open view windows.

MicroStation prompts:

> Window Area > Define first corner point *(A full screen crosshair appears. Position the pointer at the point in the view window that is to be one corner of a rectangle that defines the new displayed area and click the Data button, or click the Reset button to exit the Window Area tool and return to the previously active tool.)*
>
> Window Area > Define opposite corner point *(Position the pointer at the point in the view window that is to be the diagonally opposite point of the rectangle and click the Data button, or click the Reset button to exit the Window Area tool and return to the previously active tool.)*

MicroStation applies the new area to destination view window (see Figures 3–48a and 3–48b). Continue defining display areas or invoke another tool.

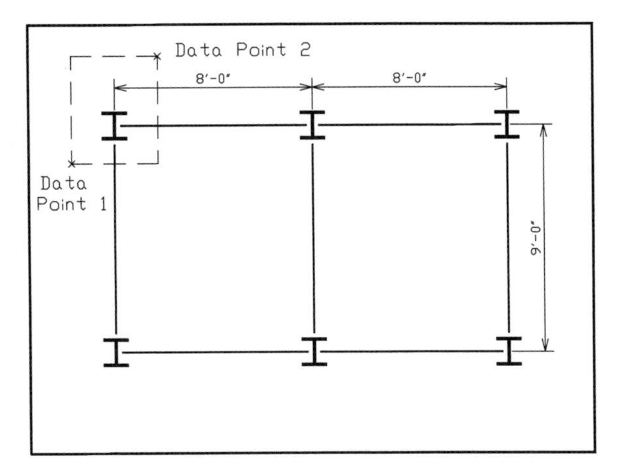

Figure 3–48a The design shown before the Window Area tool is invoked

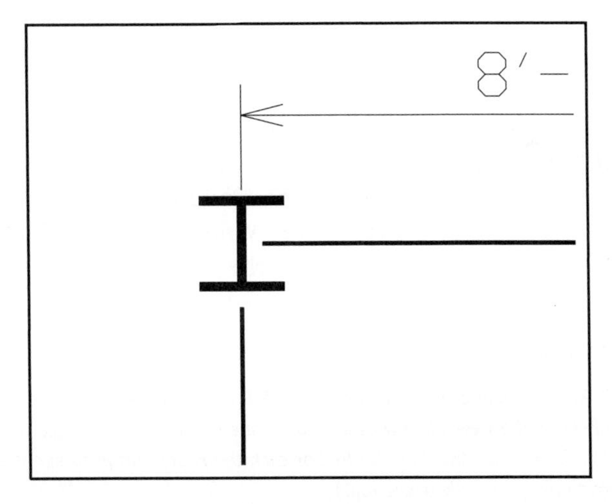

Figure 3–48b The design shown after the Window Area tool is invoked

FIT VIEW

The Fit View tool displays the entire design in the selected view window. Invoke the Fit View tool from:

View Control tool box	Select the Fit View tool (see Figure 3–49).
Key-in window	**fit view extended** (or **fit v e**). (ENTER)

Figure 3–49 Invoking the Fit View tool from the View Control tool box

MicroStation immediately fits the active view window. There are no prompts.

The Fit View Tool Settings window contains an option, **Files**, that allows selecting the types of files that MicroStation fits within the selected view window. The **All** option causes MicroStation to include all types of files, including the design, when it fits the view window. The **Active** option fits only the active design file. The **Reference** and **Raster** fit only the two types of references files. (Reference files are discussed in Chapter 13.) Click the drop-down arrow to display the file types to include in the fit and select an option by clicking on it. The default option is **All** and it is usually the selected option for fitting views.

Three check boxes are also on the Fit View Tool Settings window that apply to 3D designs.

MicroStation adjusts the view area and zoom factor as required to show all of the contents (elements) of the selected file types. (See Figures 3–50a and 3–50b for an example of fitting the active design file.

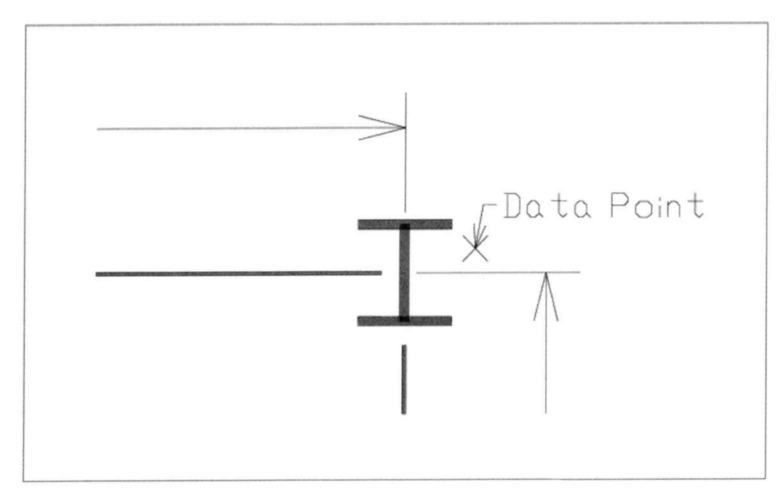

Figure 3–50a The view before the Fit View tool is invoked

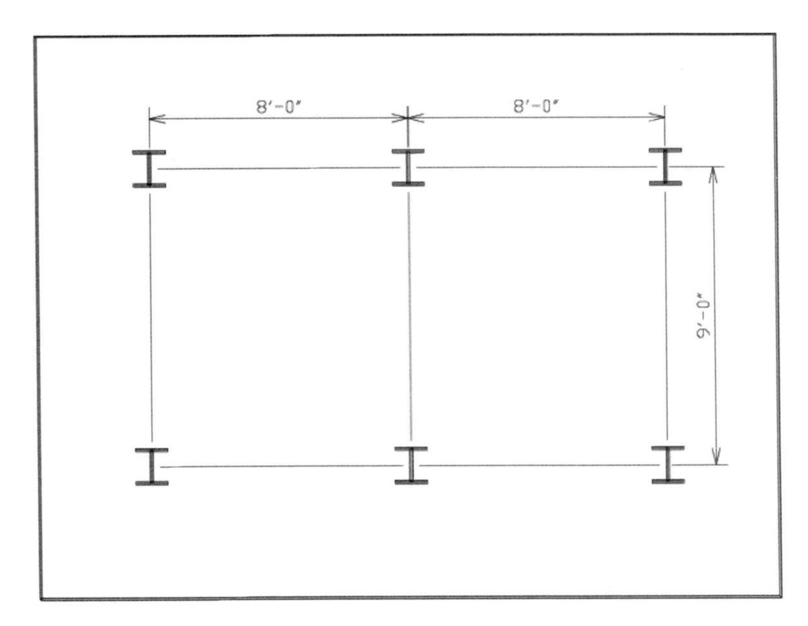

Figure 3–50b The view after the Fit View tool is invoked

ROTATE VIEW

The Rotate View tool allows rotating the view of the design in the view window about an imaginary Z axis that extends out from the view at a right angle to the X and Y axes. This tool rotates the X and Y axes in the view window, not the relation of the elements to the X and Y axes. Invoke the Rotate View tool:

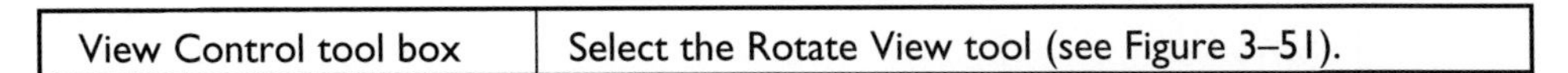

View Control tool box	Select the Rotate View tool (see Figure 3–51).

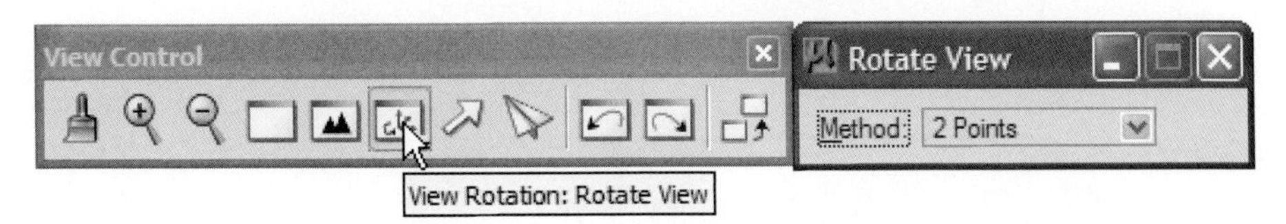

Figure 3–51 Invoking the Rotate View tool from the View Control tool box

Before placing rotation data points, select the desired rotation method from the Rotate View Tool Settings window. Three rotation methods are provided in the **Method** menu.

- The **Dynamic** option prompts for a rotation center point and rotates the image about the center point as the screen pointer is moved. Click the Data Button again to stop the rotation at the desired location. The tool remains in rotation mode so that the rotation angle can continue to be changed. Click the Reset button or select another command to complete the rotation.

- The **2 Points** option rotates the view by the number of degrees defined by center and axis datapoints. No rotation takes place until the second data point is placed.
- The **Unrotated** option returns the view to normal rotation (X axis horizontal and Y axis vertical).

2 Points Option

To rotate the design within the view window, select the **2 Points** option from the Method menu in the Tool Settings window.

MicroStation prompts:

> Rotate View > Define First Point *(Position the pointer in the view window at the desired center point for the rotation and click the Data button or click the Reset button to exit the Rotate View tool and return to the previously active tool.)*
>
> Rotate View > Define X Axis of View *(Move the pointer until the view is at the required rotation angle and click the Data button.)*
>
> Display Complete *(MicroStation rotates the design within the view window (see Figure 52a.)*

The tool is still active and the next data point defines a new center of view rotation. To exit the tool, either click the Reset button or invoke another tool.

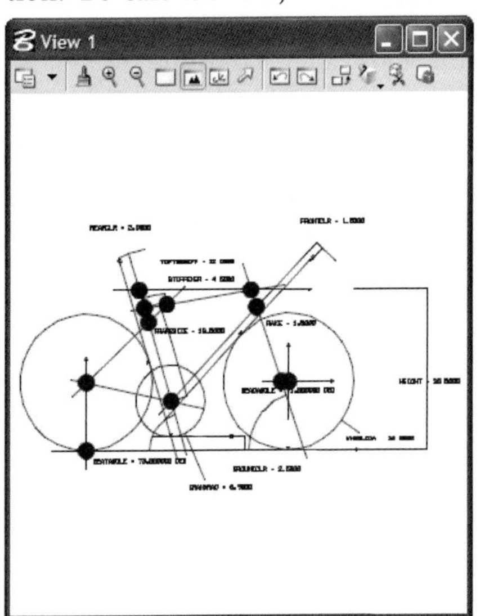

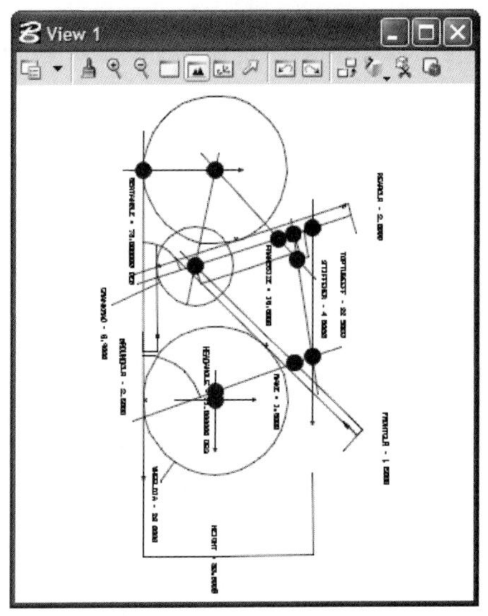

Figure 3–52a A view window - before and after view rotation

Unrotated Option

To set the rotation of a view to zero degrees (display the X axis horizontally), select the **Unrotated** option from the Method menu in the Tool Settings window menu. If only one view window is open, MicroStation immediately sets the view rotation

of the window to zero degrees. If more than one view window is open, MicroStation prompts:

> Top View > Select view *(Position the pointer anywhere in the view window and click the Data button, or click the Reset button to exit the Rotate View tool and return to the previously active tool.)*
>
> Display Complete *(MicroStation removed the rotation from the selected view.)*

PAN VIEW

The Pan View tool allows moving a view window over a part of the design that is not currently displayed in the view window. Invoke the Pan View tool from:

View Control tool box	Select the Pan View tool (see Figure 3–53).
Key-in window	**pan view** (or **pan v**). (ENTER)

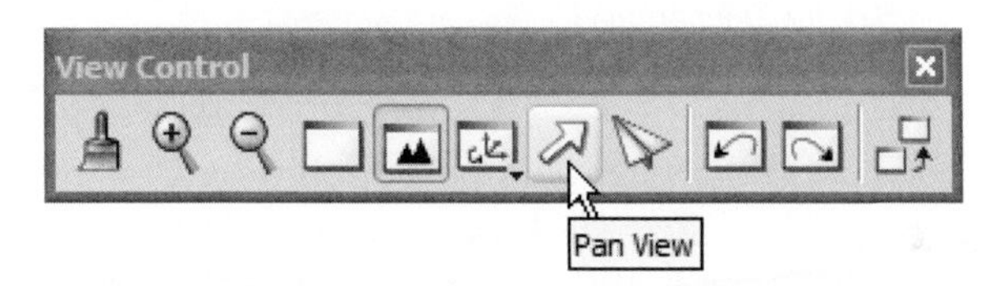

Figure 3–53 Invoking the Pan View tool from the View Control tool box

To pan the view, click the Data button to cause the view to move as the screen pointer is moved. The view continues to move with the screen pointer until a Data point is placed. To exit the tool, either click the Reset button or select another tool.

VIEW PREVIOUS

The View Previous tool displays the last displayed view. The tool can be invoked multiple times. For example, if the view was zoomed in two times, invoking the View Previous tool twice moves the view window back to the zoom setting before it was zoomed in two times. Invoke the View Previous tool:

View Control tool box	Select the View Previous tool (see Figure 3–54).
Key-in window	**view previous** (or **vi p**). (ENTER)

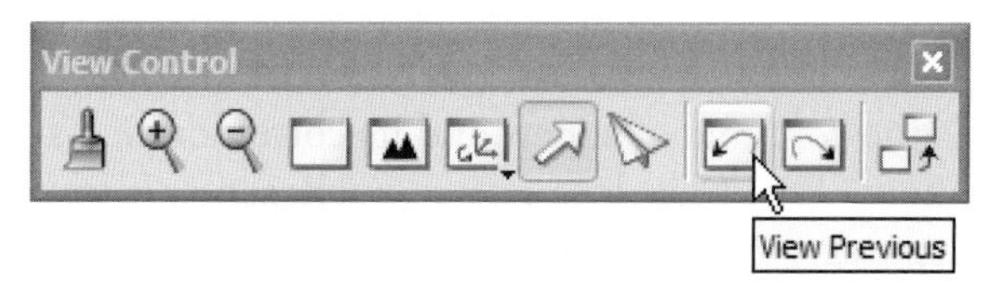

Figure 3–54 Invoking the View Previous tool from the View Control tool box

MicroStation prompts:

> View Previous > Select view *(Position the pointer anywhere in the view window and click the Data button to move back to the previous view. Continue clicking the Data button until the desired previous view is displayed or there are no more previous views.)*

The View Previous tool remains active and each Data button click restores a previous view. Click the Reset button to exit the tool and return to the previously active tool, or activate another tool.

VIEW NEXT

The View Next tool negates the view that was displayed by the View Previous tool. In other words, it moves forward through the previous views until the newest view is displayed again. Invoke the View Next tool:

View Control tool box	Select the View Next tool (see Figure 3–55).
Key-in window	**view next** (or **vi n**). (ENTER)

Figure 3–55 Invoking the View Next tool from the View Control tool box

MicroStation prompts:

> View Next > Select view *(Position the pointer anywhere in the view window and click the Data button to move forward to the next view. Continue clicking the Data button until the desired view is displayed or the newest view is reached.)*

The View Next tool remains active and clicking the Data button continues to restore previous views. Click the Reset button to exit the tool and return to the previously active tool, or activate another tool.

COPY VIEW

Copy View copies the contents of a view and its corresponding attributes to another open view window. This tool does not copy elements; it just sets up the target views to match the setup of the source view. Invoke the Copy View tool:

View Control tool box	Select the Copy View tool (see Figure 3–56).
Key-in window	copy view (ENTER)

Figure 3–56 Invoking the Copy View tool from the View Control tool box

MicroStation prompts:

> Copy View > Select source view *(Position the pointer anywhere in the view window to be copied and click the Data button.)*
>
> Copy View > Select destination view(s) *(Position the pointer anywhere in the view window that is to be set equal to the source view window and click the Data button.)*

The Copy View tool remains active and clicking the Data button in other view windows sets those views equal to the set up of the source view. Click the Reset button to exit the tool and return to the previously active tool, or activate another tool.

VIEW WINDOWS AND VIEW ATTRIBUTES

Thus far, all work has been done in one view window. That may have required spending extra time using the view commands to set up the view for the areas of the design. MicroStation provides eight separate view windows that allow working in different parts of the design at the same time.

Each view window is identified by its view number (1–8). The view windows are similar to having eight zoom lens cameras pointed at different parts of the design. For instance, one view window might display the entire drawing and two other view windows might be zoomed in close to different areas to show greater detail (see Figure 3–57). All eight view windows can be open at the same time on the monitor or on either monitor of a two-monitor workstation.

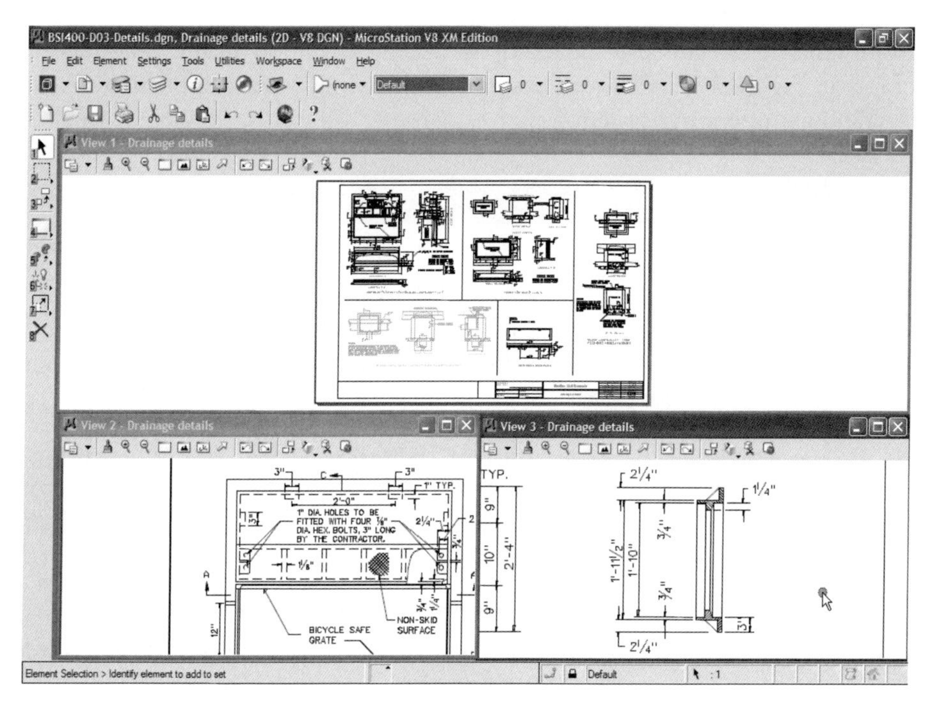

Figure 3–57 Three views showing different portions of a design

OPENING AND CLOSING VIEW WINDOWS

View windows can be opened and closed from the **Windows** > **Views** submenu, from the View Groups settings box, and by key-in commands.

To open or close a view window, select:

Menu	Window > Views > (Select a view number) (see Figure 3–57)
Key-in window	**view off** (or **vi off**) or **view on** (or **vi on**). (ENTER)

To open or close view windows from the View Groups settings box, select:

Menu	Window > Views > Dialog (see Figure 3–58)

Select the view number to open or close the one already open.

Figure 3–58 The View Groups settings box showing two view windows open

Most operating systems provide a tool for closing a window in the window's Control menu. Figure 3–59 shows an example of the window control menu available in the Windows operating system.

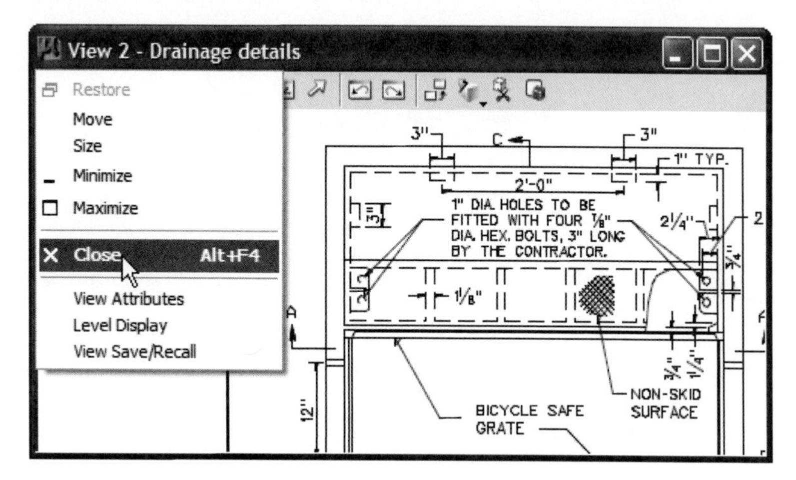

Figure 3–59 View Window close option in the Microsoft Windows XP view window

Note: *The view windows that are opened and closed apply only to the current editing session. To make the settings permanent, select the* ***Save Settings*** *option from the* ***File*** *menu.*

ARRANGING OPEN VIEW WINDOWS

The working area can become cluttered when several view windows are open and MicroStation provides three housekeeping tools for cleaning up the clutter: **Cascade, Tile**, and **Arrange**. The tools are provided in the Window menu.

Cascade

The Cascade tool stacks all open view windows in numerical order, with the lowest-numbered view window on top and the other view window title bars visible behind it.

To cascade the open view windows, select the Cascade tool:

Menu	Window > Cascade
Key-in window	**window cascade** (or **w c**). (ENTER)

The open view windows are cascaded (there are no MicroStation prompts). To work on a specific view, click its Title bar and it pops to the top of the stack.

Tile

The Tile tool arranges all open view windows side by side in a tiled fashion, with the lowest-numbered view window in the upper left.

To tile the open view windows, select the Tile tool:

Menu	Window > Tile
Key-in window	**window tile** (or **w t**). (ENTER)

The open view windows are tiled (there are no MicroStation prompts).

Arrange

The Arrange tool can size and move all open view windows as necessary to fill the MicroStation application window. The tool attempts to keep each view window as close to its original shape, size, and position as possible.

To arrange the open view windows, select the Arrange tool:

Menu	Window > Arrange
Key-in window	**window arrange** (or **w arr**). (ENTER)

The open view windows are arranged to fill the MicroStation application window (there are no MicroStation prompts).

In additional to opening and closing view windows, they can also be resized, minimized, and maximized just like any other open windows.

CHANGING THE TOOL BOX DISPLAYED ON THE VIEW WINDOWS

By default, the View Control box is docked to the top left on the individual view windows. The tool box docked to the windows can be swapped or removed using options available on the **Window** menu (**Task Navigation in Views** and **View Tool box**).

When **View Tool boxes** is docked (as indicated by a check mark on the option in the Window menu), the **Task Navigation in Views** tool from the Window menu can be used to switch the docked tool box

To switch between docking the View Control tool box and a drawing tool box (while **View Tool boxes** is selected), select:

Menu	Window > Task Navigation in Views

When **Task Navigation in Views** is selected, a check mark appears next to the option on the **Window** menu.

When Task Navigation in Views tool box is docked to the view windows, the menu for selecting drawing tool boxes is also docked, which allows different drawing tool boxes to be docked to each displayed view window. For example, one view window can be used to place polygon elements and another to place dimensions. The tool box selection menu is open on a view window in Figure 3-60.

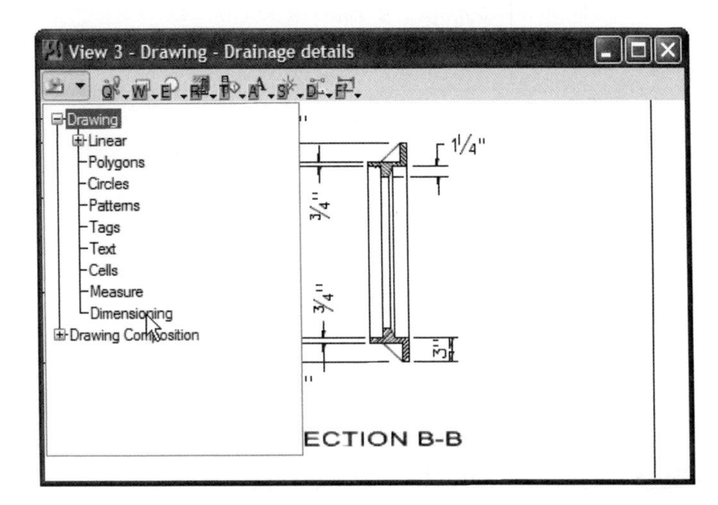

Figure 3-60 The Tool box Selection Menu on a View Window

CREATING AND USING VIEW WINDOW GROUPS

MicroStation provides tools for creating and recalling named groups of view windows in a model. When a group is created, it saves the currently open view windows and the alignment of each window. When a group is selected for display, the group's view windows open on the MicroStation desktop and all other view windows close. The open windows display the same areas of the design that were showing when the group was created.

View groups are useful for setting up optimal view windows for a particular type of design. For example, a floor plan might have three groups:

- A "Rooms" group that opens all eight view windows with each view window aligned to a different room in the plan.
- A "Notes" group that opens six view windows aligned to construction notes areas on the design.
- A "Border" group that opens three view windows aligned to areas of the design border.

The same view windows can be used in more than one group. For example, when the Rooms group is selected, view one displays the living room, when the Notes

group is selected, view one displays notes about electrical connections, and when the Border group is selected, view one displays the design's title block.

A view group is restored for viewing by clicking its name in the **View Groups** menu on the Views Dialog box (see Figure 3-61).

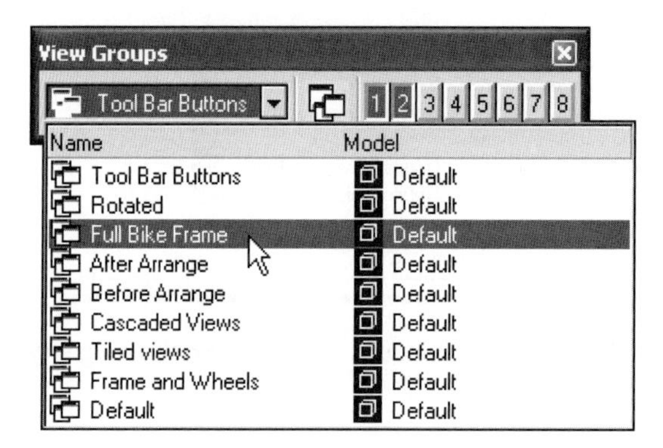

Figure 3–61 The View Groups settings box showing a typical View Groups menu

***Note:** If **File > Save Settings** is selected after opening or closing view windows, or changing the zoom factor and alignment of view windows, the current arrangement is applied to the displayed view group. Thereafter the view group displays the saved view window settings, not its original settings. Before saving settings, display the Default view group. There is always at least a "Default" view group (but the default view group can be deleted).*

MicroStation provides tools for creating, editing, and deleting view groups in the Manage View Groups settings box. Invoke the Manage View Groups settings box from:

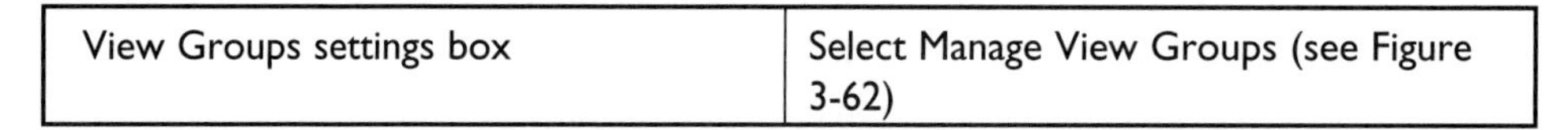

View Groups settings box	Select Manage View Groups (see Figure 3-62)

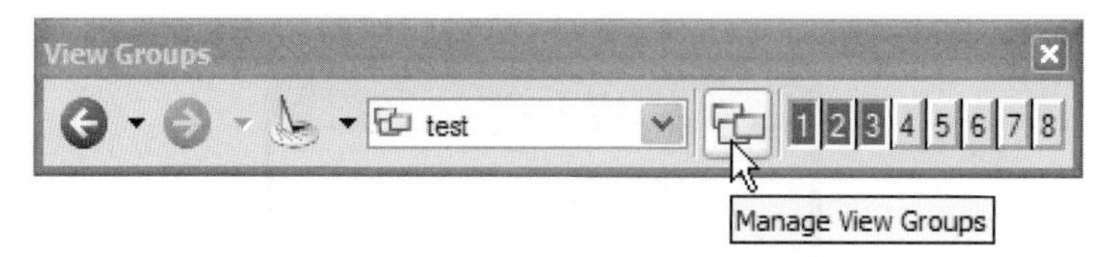

Figure 3–62 Invoke Manage View Groups from the View Groups settings box

MicroStation opens the Manage View Groups settings box with all existing groups displayed (see Figure 3–63). The menu bar in the box has options to **Create View Group, Edit View Group Properties**, and, **Delete View Group**, and **Apply.** There

is also an **Apply** button that defines an existing view group for the currently open view windows.

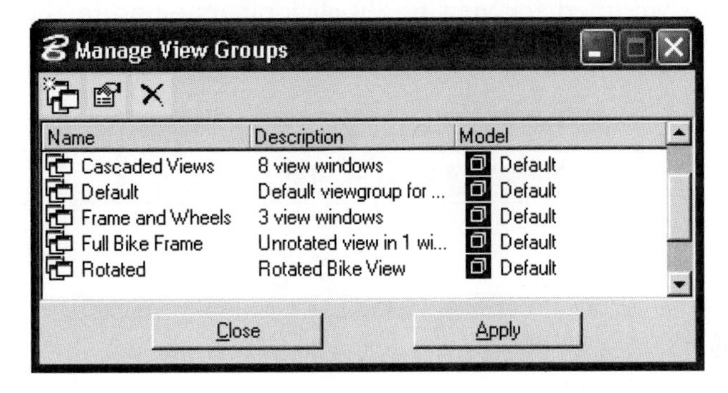

Figure 3–63 The Manage View Groups settings box showing a typical set of groups

Note: *New groups and changes to existing groups are saved automatically to the design file. It is not necessary to save settings to retain them.*

The **Create View Group** tool saves the size, alignment, and zoom level of each open view window. Before creating a new group, make sure the view windows are correctly set up in the group.

To create a new view group:

Manager View Groups settings box	Select Create View Group

MicroStation displays the Create View Group settings box (see Figure 3–64). To create the new View Group, enter a group name in the **Name** field, a description in the **Description** field, and click **OK** to save the new view group. The new group appears in the groups list of the Manage View Groups and View Groups settings boxes.

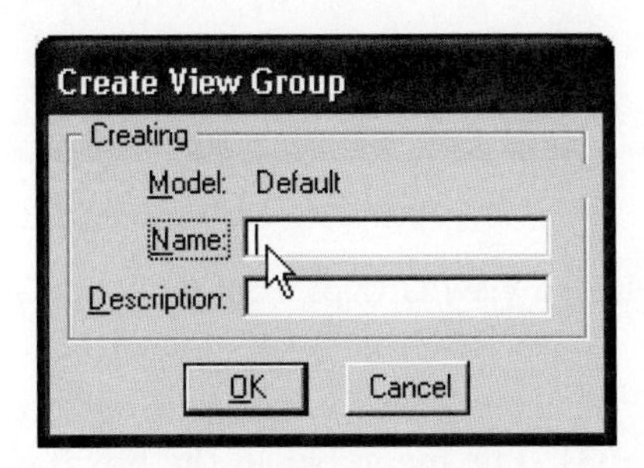

Figure 3–64 The Create View Group settings box

A view group's properties (**Name** and **Description**) can be changed from the View Group Properties dialog box. Invoke the dialog box from:

Manager View Groups settings box	Select the view group to be edited and select the Edit View Group Properties.

MicroStation displays the View Group Properties dialog box.

Edit the group name in the **Name** field, edit the group description in the **Description** field, and click **OK** to save the new properties, or click **Cancel** to close the dialog box without changing the properties.

To delete an existing view group, choose **Delete View Group** from the Manage View Groups settings box. Use this option to remove a selected view group from the design file. Invoke the **Delete View Group** from:

Manage View Groups settings box	Select the view group to be deleted and select the Delete View Group.

MicroStation displays an Alert message box that prompts for confirmation before deleting the view group. Click **OK** to delete the selected view group, or click **Cancel** to close the Alert box without deleting the view group.

To change the view windows definition (number, position, content, and shape) of an existing view group, make the required changes to the view windows, open the Manage View Groups settings box, select the name of the view group to be redefined, and click the **Apply** button. The open view windows are saved under the selected view group name.

SETTING VIEW ATTRIBUTES

MicroStation provides a View Attributes settings box that contains options to control the way elements and drawing aids appear in a view window. Turning OFF certain attributes can speed up the time to update a view and reduce clutter in a view. For example, if a view window contains a large amount of patterning, turn off the display of patterns to reduce the update time on a slow workstation. Invoke the View Attributes settings box from:

Menu	Settings > View Attributes

MicroStation displays the View Attributes settings box as shown in Figure 3–65, each view attribute is explained in Table 3–1.

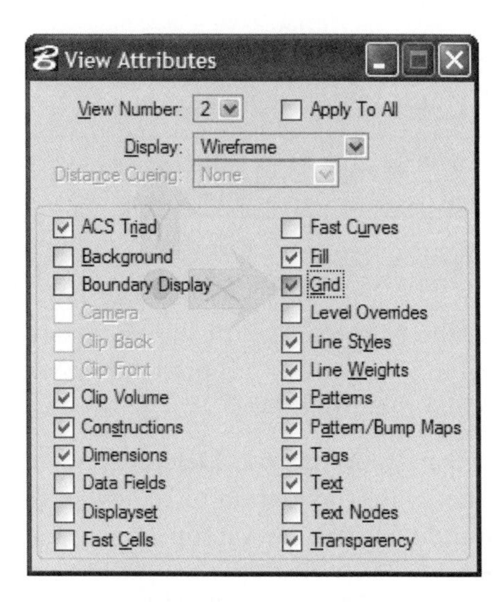

Figure 3–65 View Attributes settings box

The view attributes are changed immediately when an attribute check box is set or cleared. Before changing attributes, select the view to which the attributes are to be applied in the **View Number** menu or set the **Apply to All** check box to apply the attributes to all open view windows.

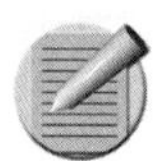

Note: *The **Display** menu and **Distance Queuing** field are both 3D tools.*

Follow these steps to use the settings box to change view attributes:

1. To set the attributes for all open view windows, turn **Apply to All** ON.
2. To set the attributes for one view window, select a view window number from the **View Number** menu.
3. Turn the appropriate attribute check box ON or OFF as required.

Table 3–1 View Attributes

ATTRIBUTE	TURNS ON AND OFF THE DISPLAY OF...
ACS Triad	The Auxiliary Coordinate System (ACS).
Background	The background image loaded with the Active Background command.
Boundary Display	Used in 3D designs to control whether the boundaries of the clip volume are displayed for a given view, as well as reference clip boundaries.
Camera	The 3D view camera.

ATTRIBUTE	TURNS ON AND OFF THE DISPLAY OF...
Clip Back	Used in 3D designs to toggle the display of elements, and parts of elements, located outside a 3D view's clipping planes. If ON, a back clipping plane is active in a view.
Clip Front	Used in 3D designs to toggle the display of elements, and parts of elements, located outside a 3D view's clipping planes. If ON, a front clipping plane is active in a view.
Clip Volume	Used in 3D designs to toggle the display of elements, and parts of elements, located outside a defined Clip Volume, for a given view. If ON, and a clip volume has been applied to the view, the view volume is restricted to the defined volume. If no clip volume has been applied to the view, it has no effect.
Constructions	Elements placed with Construction Class mode active.
Dimensions	Dimension elements.
Data Fields	Data Field placeholder characters.
Displayset	If ON, only elements in the current displayset are displayed in the view windows.
Fast Cells	The actual cells or a box indicating the area of the design occupied by cells.
Fast Curves	The actual curve string or straight line segments indicating the vertices.
Fill	The fill color in filled elements.
Grid	The grid (if the view is zoomed out far enough, the grid will be turned off even if this attribute is on).
Level Symbology Overrides	Elements according to the symbology table rather than the actual element symbology.
Line Styles	Elements with their actual line weights (when it is set to OFF, all elements are displayed with style 0, solid).
Line Weights	Elements with their actual line weights (when it is set to OFF, all elements are displayed at line weight 0).
Patterns	Pattern elements.
Tags	The tag information for tagged elements.
Text	The display of text elements (when it is set to OFF, no text elements are displayed).
Text Nodes	Text nodes as small crosses with numeric identifiers.
Transparency	If Transparency is ON, elements in the views are displayed using the transparency setting assigned to the elements. When Transparency is OFF, the elements are displayed with zero transparency regardless of their transparency settings.

***Note:** Changes to View Attributes setting remain in effect until they are changed or the design file is closed. To keep them in effect for future editing sessions, select **Save Settings** from the **File** menu.*

SAVING VIEWS

If work is regularly done in several specific areas of a design, MicroStation provides a way to return to those areas quickly by saving a view window's zoom factor, alignment, displayed levels, and view attributes under a user-defined name. To return to a saved view, provide the saved view name and the number of the view window in which to place the saved view.

The tools for displaying and saving views are in the Saved Views settings box. Invoke the settings box from:

Menu	Utilities > Saved Views
View window control menu	Select the View Save/Recall option

MicroStation displays the Saved Views settings box as shown in Figure 3–66.

Figure 3–66 Saved Views settings box

The name and description of all saved views are displayed in the settings box under the **Name** and **Description** columns. Above the list of saved views is a menu bar containing options for managing saved views. Below the list of saved views are options that affect the way saved views are displayed.

Display a Saved View

To display a saved view using the Saved Views settings box:

1. Position the pointer over the saved view's name and click the Data button to select it.
2. On the tool box select the number of the view window in the **View** option in which the saved view is to be displayed
3. In the **Apply Options** section, choose the options to apply to the selected view (See the following option descriptions).
4. Click the **Apply** button.

The saved view, whose name was selected, is displayed in the selected view window.

A view also can be restored by typing in at the key-in window field **vi=<name>**, and pressing (ENTER). Replace <name> with the name of the view to be restored. MicroStation prompts:

Select view *(Click the Data button in the view the view is to be restored.)*

Apply Options

The **Apply Options** check boxes on the Saved Views settings box affect the way the saved view appears in the target view window. Following are descriptions of each option:

- Turning ON the **Window** check box enables a menu of options that control the shape and size of the selected window:

1. Select the **Aspect Ratio** option to set the height and width of the selected view window, proportionate to the shape of the window from which the view was originally saved. For example, if the original window was twice as wide as it was tall, the window in which the saved view is placed will also be twice as wide as it is tall. The longer of the selected view window's two dimensions (horizontal or vertical) is adjusted to make the window proportional.
2. Select the **Size** check box to set the selected view window the same size and shape as the window from which the view was originally saved.
3. Select the **Size and Position** option to make the selected view window match the size, shape, and location on the MicroStation desktop, of the window from which the view was originally saved.

- Turning ON the **Camera Position** check box applies the camera settings, of the saved window, to the selected view window.
- Turning ON the **View Attributes** check box applies the view attributes settings, of the saved view, to the selected view.

- Turning ON the **Clip Volume** check box applies the clip volume, of the saved view, to the selected view.
- Turning ON the **Levels** check box applies the displayed levels, of the saved view, to the selected view.

Note: *The Active View is not impacted by the* ***Levels*** *check box. It remains at its current setting.*

- Turning ON the **Reference Settings** applies the reference file view settings, of the saved view, to the selected view. (Reference files are discussed in Chapter 13.)

Save View

Before saving a view, set up a view window to display the area of the design to be saved, set the window's view attributes (**Settings > View Attributes**) to the desired values. Next, turn on the display of the desired levels (**Settings > Level > Display**), and invoke the Save View settings box from:

Saved Views settings box	Select the Save View (the left-most button on the tool box that shows a pair of hands).

MicroStation displays the Save View settings box, as shown in Figure 3–67.

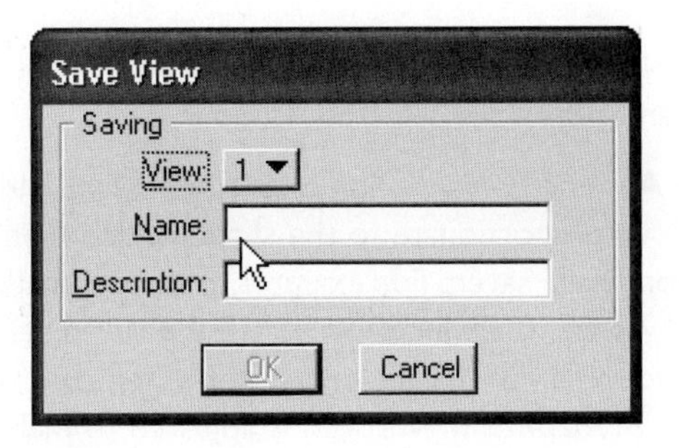

Figure 3–67 Save View settings box

In the Save View settings box:

1. Select the number of the view window in the **View** menu that contains the view to be saved (Only numbers for currently open views can be selected).
2. Type a name for the view in the **Name** field. The name can contain letters, numbers and the "$", ".", and "_" characters. It can be up to 511 characters long, but keep in mind that the Name field in the Saved Views settings box can only display the first 20, or so, characters of the name.
3. Optionally, type a description of the saved view in the **Description** field.

4. Click the **OK** button to save the view, or click the **Cancel** button to close the settings box without saving the view.

A view can also be saved by typing in the key-in window: **sv=<name>,description>** and pressing(ENTER). Replace <name> with the name of the saved view and <description> with a description of the saved view. MicroStation prompts:

> Save Named View > Select view *(Position the pointer in the view window to be saved and click the Data button.)*

Edit Saved View Properties

The properties of a Saved View include the view's name and description. Invoke the Edit Properties dialog box from:

Saved Views dialog box	Select the Edit Saved View Properties (second button from the left on the menu bar).

MicroStation displays the Edit Properties dialog box.

In the Edit Properties dialog box, edit the **Name** and **Description** as required. When the changes are complete, click the **OK** button to change the properties, or click **Cancel** to leave the properties unchanged.

Delete Saved View

To delete a saved view using the Saved Views settings box, click the name of the view to be deleted and invoke the **Delete Saved View** tool from:

Saved Views settings box	Select the Delete Saved View button (third button from the left on the menu bar).

MicroStation displays an Alert box that asks for confirmation of the deletion of the selected saved view. Click the **OK** button to delete the selected saved view, or click **Cancel** to cancel deleting the saved view.

A view can also be deleted by typing into the key-in window: **dv=<name>** and pressing (ENTER). Replace <name> with the name of the saved view to be deleted. The saved view is immediately deleted.

Update Saved View Settings

Updates a saved view to include changes to the settings. For example, modify the levels or view attributes for the saved view and use this icon to save the changes to the saved view. Invoke the Update Saved View Settings tool from:

Saved Views settings box	Select the Update Saved View settings (fourth button from the left on the menu bar).

MicroStation immediately updates the selected Saved View with the attributes from the active view window.

Import Saved View

MicroStation can Import saved views from other design files into the current design file. Imported views can be recalled and managed the same as any other saved views contained in the current design file. Invoke the Saved Views settings box from:

Saved Views settings box	Select the Import Saved View (fourth button from the left on the menu bar).

MicroStation displays the Open File dialog box. Select the design file from which to import the saved views.

MicroStation displays a list of the saved views that can be imported, in the Import Saved Views settings box.

In the Import Saved Views dialog box, the saved views are listed in the **Select Views to Import** section (If the design file does not contain any views, the list will be empty). Use the following methods to select the saved views to be imported:

1. To select one saved view for import, click that saved view's name.
2. To select a series of saved views, click on the name of the first view in the series, hold down SHIFT, and click on the name of the last view in the series. The selected views, and all saved views between the two selections, are selected for import.
3. To select several individual saved views, click on the name of the first view, hold down CTRL while clicking the name of each subsequent saved view. Only the clicked views are selected for import.
4. To select all saved views in the list, hold down CTRL and tap A.
5. To unselect a selected saved view, hold down CTRL while clicking the view name.

Click the **OK** button to import the selected saved views, or click the **Cancel** button to cancel the import operation. After clicking **OK**, the selected saved views are imported and appear in the Saved Views settings box of the active design file.

UNDO AND REDO

The Undo tool cancels the effects of the previous commands. The Redo tool reverses the effects of the previous Undo. For example, if a circle is placed in the design and the Undo tool is selected, the circle is removed from the design. If the Redo tool is clicked next, the circle is placed back in the design.

Commands can be undone because all the steps executed for each command are stored in an Undo buffer. The Undo tool goes to that buffer to get the information necessary to put things back the way they were before the tool was invoked. The last command executed is the first one undone, the next-to-last command is the next one undone, and so on.

Note: *The Undo tool can undo an unlimited number of commands that were executed during the current editing session (limited to amount of computer memory). Closing the design file clears the Undo buffer. Therefore, the next time the design file is opened for editing undo cannot undo the effects of commands issued in the previous editing session – there are other manipulation tools are used to alter changes made during previous editing sessions.*

UNDO TOOL

The Undo tool permits removing the effects of the last command.

To undo the last command, invoke the **Undo** tool from:

Standard tool box	Select the Undo tool (see Figure 3–68).
Key-in window	**undo** (or **und**). (ENTER)

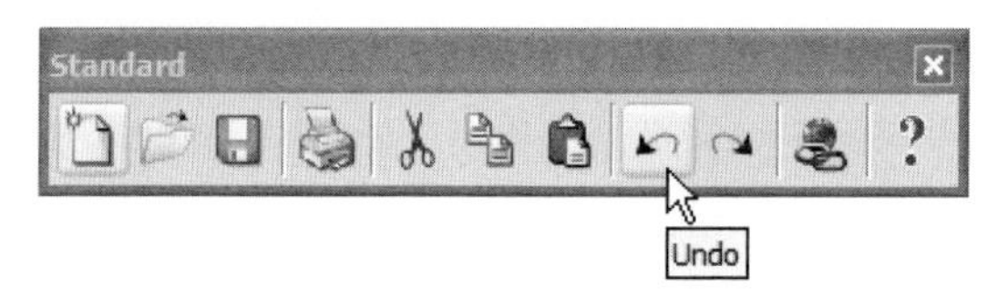

Figure 3–68 Invoking the Undo tool from the Standard tool box

To undo the last drawing operation, the **Undo (action)** option can also be selected from the **Edit** menu. MicroStation displays the name of the last command operation that was performed in place of **(action)**.

Set Mark and Undo Mark

To experiment with the design and be able to undo the experiment, place a Mark in the design before starting and, if necessary, undo back to the Mark.

To place a Mark, invoke the **Set Mark** tool from:

Menu	Edit > Set Mark
Key-in window	**mark** (or **mar**). (ENTER)

To undo all the steps back to when the Mark was placed, invoke the **Undo Mark** tool from:

Menu	Edit > Undo Other > To Mark
Key-in window	**undo mark** (or **und m**). (ENTER)

All the commands, after the Mark was placed, are undone and the Mark is also removed.

Undo All

The **Undo All** tool negates all drawing commands recorded in the Undo buffer.

To undo all drawing commands recorded in the Undo buffer, invoke the **Undo All** tool from:

Menu	Edit > Undo Other > All
Key-in window	**undo all** (or **und a**). (ENTER)

MicroStation displays an Alert box that asks for confirmation of undoing all changes. Click the **OK** button to undo all commands issued, or click the **Cancel** button to cancel the **Undo All** operation.

REDO TOOL

Invoking the **Redo** tool negates the action of the last Undo. Redo a series of Undo operations by repeatedly invoking **Redo**. Invoke the **Redo** tool from:

Standard tool box	Select the Redo tool (see Figure 3–69).
Key-in window	**redo.** (ENTER)

Figure 3–69 Invoking the Redo tool from the Standard tool box

THINGS TO CONSIDER BEFORE UNDOING

Following are two points to consider before invoking the Undo or Redo tools.

1. The Undo commands back up through the Undo buffer. Sometimes, Undo is not the best way to clean up a problem. For example, if a circle was placed five commands ago and it needs to be removed, the intervening four commands must be undone before the circle placement can be undone. In this case, a better way to remove the circle is with the **Delete Element** tool.

2. The Undo commands negates the effects of a previous command not an element. If the command manipulated multiple elements, Undo negates the manipulation of all of those elements. For example, the Fence commands can manipulate hundreds of elements at one time. Undoing a **Fence Contents Delete**, puts back all the elements that the fence deleted.

Open the Exercise Manual PDF file for Chapter 3 on the accompanying CD for project and discipline specific exercises.

REVIEW QUESTIONS

Write your answers in the spaces provided.

1. Briefly explain the difference between the Zoom In and Zoom Out tools.

   ____

   ____

2. Which tool will get in closer to a portion of the design by a factor of 2?

   ____

3. Panning affects the view by

   ____

4. To pan in a view, press the ____________ button while moving the pointer to drag the view window over the design.
5. What is the purpose of Grid lock?

   ____

6. The __________ Category in the ______________________ dialog box controls the display of the grid.
7. The Grid Master unit defines the distance between the _________________________ and is specified in terms of the design's __________ ___________.
8. The Grid Reference defines the distance between the

   ____

9. To keep the grid settings in effect for future editing sessions for the current design file, select __________________ from the ____________________ menu.
10. The Grid Aspect Ratio field in the Design File dialog box allows setting the

    ____

11. Name the two settings boxes in which the Snap Lock mode can be set: ____________ and ____________.

12. Name four of the Snap modes available in MicroStation.

    __________________________, ______________________,
    _____________________, and

    ____

13. Keypoint mode allows tentative points to snap to __________________________ _____________.on elements.
14. The Midpoint mode snaps to the _________________________________ of a circle and an ellipse.
15. Name three ways AccuSnap can indicate that it has snapped to a point?.______________________, _______________________, and

    _______________________.

16. What steps are used to accept an AccuSnap snap point?.____
17. The Axis Lock forces each data point

    ____

18. List four attributes associated with the placement of elements:

    _________________________,

    ______________________, _______________________, and

    ______________________.

19. How many levels can be defined in a design file?

    ____

20. How many levels can be active at any time?

    ____

21. How many levels can be turned ON or OFF at any time in a specific view window?

    ____

22. How many colors or shades of colors are available on the color table?

    ____

23. What two-letter key-in makes a color active? ___________
24. How many internal Line Styles are available in MicroStation?

    ____

25. In MicroStation, Line Weight refers to the ____________________ of lines.
26. How many line weights are available in MicroStation? ___________________

Fundamentals III

OBJECTIVES

Topics explored in this chapter:

- Drawing ellipses, polygons, point curves, curve streams, and multi-lines
- Modifying elements using fillets, chamfers, trim, and partial delete
- Manipulating elements using move, copy, move and copy parallel, scale original and copy, rotate original and copy, mirror original and copy, and array
- Placing text: Setting text parameters and placing text by origin

PLACEMENT TOOLS

In this chapter, four more placement tools are explained: Ellipse, Curve (Place Point Curve and Place Curve Stream), and Multi-line tools. This adds to the placement tools already described in Chapter 2.

PLACE ELLIPSE

MicroStation offers two methods for placing an ellipse: **Center** and **Edge**. The method is selected from the **Method** option menu in the Tool Settings window when the Place Ellipse tool is active.

Place Ellipse By Center

The **Center** method places an ellipse by defining three points: the center, one end of the primary (major) axis, and one end of the secondary (minor) axis. Invoke the Place Ellipse By Center tool from:

Task Navigation tool box (active task set to Circles)	Select the Place Ellipse tool and select Center from the Method menu on the Tool Settings window (see Figure 4–1).
Keyboard Navigation (Task Navigation tool box with active task set to Circles)	**w** (Select Center from the Method menu on the Tool Settings window.)

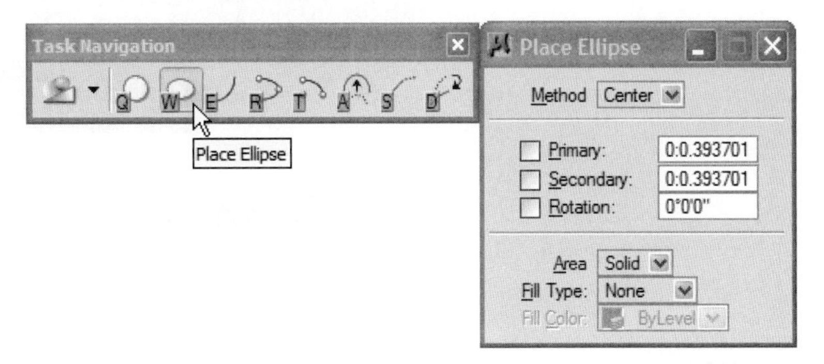

Figure 4–1 Invoke the Place Ellipse By Center and Edge tool from the Navigation tool box (active task set to Circles)

MicroStation prompts:

Place Ellipse By Center and Edge > Identify Ellipse Center *(Place a data point or key-in coordinates to define the center of the ellipse.)*

Place Ellipse By Center and Edge > Identify Ellipse Primary Radius *(Place a data point or key-in coordinates to define the one end of the primary axis.)*

Place Ellipse By Center and Edge > Identify Ellipse Secondary Radius *(Place a data point or key-in coordinates to define the one end of the secondary axis.)*

For example, the following tool sequence shows how to place an ellipse by Center (see Figure 4–2).

Place Ellipse By Center and Edge > Identify Ellipse Center (place a data point, P1)

Place Ellipse By Center and Edge > Identify Ellipse Primary Radius (place a data point P2)

Place Ellipse By Center and Edge > Identify Ellipse Secondary Radius (place a data point P3)

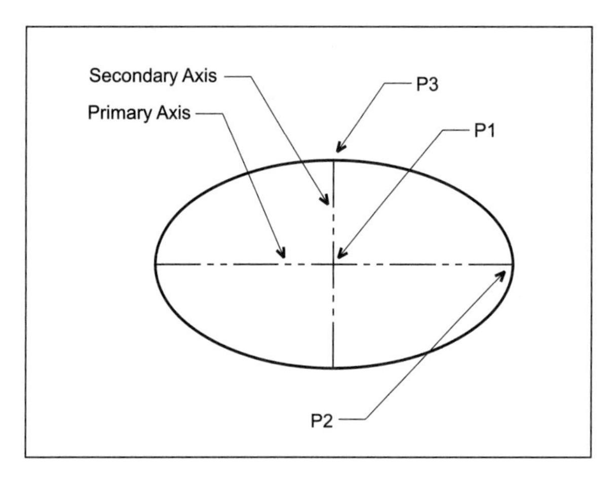

Figure 4–2 Example of placing an ellipse by means of the Place Ellipse by Center and Edge tool

Options in the Ellipse Tool Settings window constrain the **Primary** axis, **Secondary** axis, and **Rotation** angle key-in. To set a constraint, turn ON the appropriate check boxes and key-in the constraining value in the Tool Settings window. For example, to constrain the **Primary** axis, turn ON the **Primary** check box, key-in a value for the radius of the axis in the **Primary** text field and press either (ENTER) or TAB.

Constraining one or more of the three values, MicroStation adjusts its prompts to handle the constraint. For example, if the **Primary** and **Secondary** radii are constrained, MicroStation prompts for the ellipse center and rotation data points. If the **Primary** radius, **Secondary** radius, and the **Rotation** are all constrained, MicroStation only prompts for the ellipse center point.

Place Ellipse By Edge Points

The **Edge** method of the Place Ellipse tool (Place Ellipse By Edge Points) draws an ellipse by defining three points on the ellipse. Invoke the Place Ellipse By Edge Points tool from:

Task Navigation tool box (active task set to Circles)	Select the Place Ellipse tool, and select Edge from the Method menu on the Tool Settings window (see Figure 4–3).
Keyboard Navigation (Task Navigation tool box with active task set to Circles)	**w** (Select Edge from the Method menu on the Tool Settings window.)

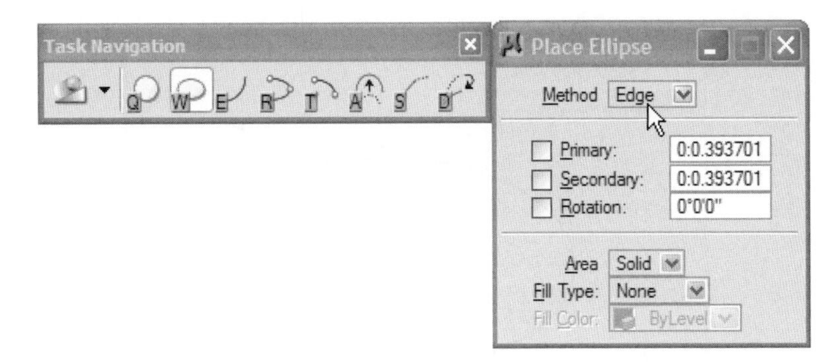

Figure 4–3 Invoke the Place Ellipse By Edge Points tool from the Navigation tool box (active task set to Circles)

MicroStation prompts:

Place Ellipse By Edge Points > Identify Point on Ellipse *(Place a data point or key-in coordinates to define the first point on the ellipse.)*

Place Ellipse By Edge Points > Identify Point on Ellipse *(Place a data point or key-in coordinates to define the second point on the ellipse.)*

Place Ellipse By Edge Points > Identify Point on Ellipse *(Place a data point or key-in coordinates to define the third point on the ellipse.)*

For example, the following tool sequence shows how to place an ellipse with the Place Ellipse by Edge Points tool (see Figure 4–4).

Place Ellipse By Edge Points > Identify Point on Ellipse (place a data point, P1)

Place Ellipse By Edge Points > Identify Point on Ellipse (place a data point, P2)

Place Ellipse By Edge Points > Identify Point on Ellipse (place a data point, P3)

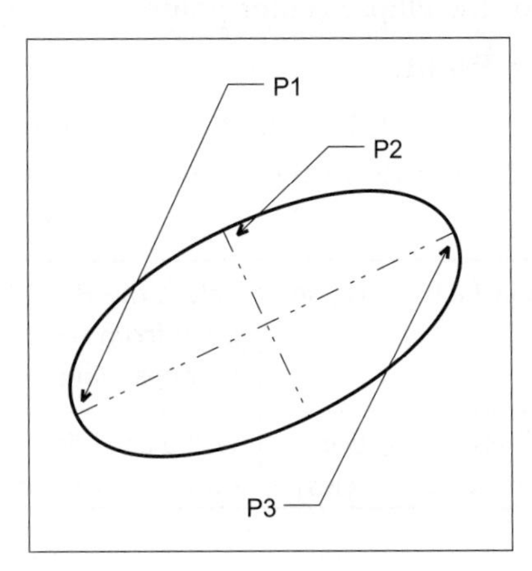

Figure 4–4 Example of placing an ellipse by edge

The Edge method provides the same placement constraints as the Center method.

Note: *Half and quarter ellipse can also be placed. The tools are located in the Navigation tool box (active task set to Circles).*

PLACE REGULAR POLYGON

With the Place Regular Polygon tool, regular polygons (all edges are equal length, all vertex angles are equal) with 3 to 4,999 edges can be drawn. MicroStation provides three methods for placing regular polygons: **Inscribed**, **Circumscribed**, and **By Edge**. The appropriate method is selected from the **Method** option menu on the Tool Settings window.

Place Polygon By Inscribed

The **Inscribed** method places a polygon that is inscribed inside an imaginary circle, whose diameter is equal to the distance across opposite polygon vertices (see Figure 4–5).

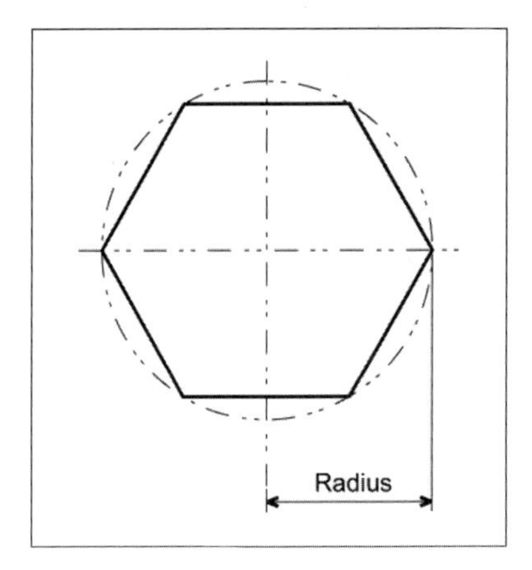

Figure 4–5 Example of displaying a polygon inscribed inside an imaginary circle

Invoke the Place Inscribed Polygon tool from:

Task Navigation tool box (active task set to Polygons)	Select the Place Regular Polygon tool and select Inscribed from the Method menu located on the Tool Settings window (see Figure 4–6).
Keyboard Navigation (Task Navigation tool box with active task set to Polygons)	**r** (Select Inscribed from the Method menu located on the Tool Settings window.)

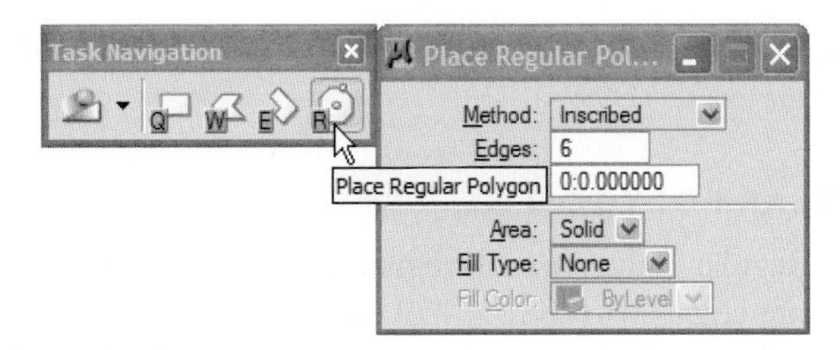

Figure 4–6 Invoke the Place Polygon Inscribed tool from the Navigation tool box (active task set to Polygons)

Key-in the number of polygon edges (from 3 to 4,999) in the **Edges** field on the Tool Settings window. Optionally, key-in the radius of the imaginary circle in the **Radius** edit field. If the **Radius** is set to 0, it can be defined graphically with a data point or by key-in from the Key-in window.

MicroStation prompts:

> Place Inscribed Polygon > Enter point on axis *(Place a data point or key-in coordinates to define the center of the polygon.)*
>
> Place Inscribed Polygon > Enter first edge point *(Place a data point or key-in coordinates to define the radius of the imaginary circle, the polygon's rotation, and one vertex.)*

Place Polygon By Circumscribed

The **Circumscribed** method places a polygon circumscribed around the outside of an imaginary circle having the same diameter as the distance across the opposite polygon edges (see Figure 4–7).

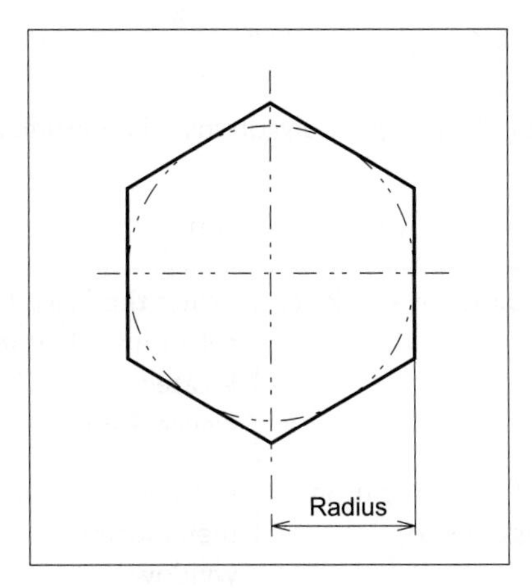

Figure 4–7 Example of displaying a polygon circumscribed around the outside of an imaginary circle

Invoke the Place Circumscribed Polygon tool:

Task Navigation tool box (active task set to Polygons)	Select the Place Regular Polygon tool and select Circumscribed from the Method option menu on the Tool Settings window (see Figure 4–8).
Keyboard Navigation (Task Navigation tool box with active task set to Polygons)	**r** (Select Circumscribed from the Method option menu on the Tool Settings window.)

Key-in the number of polygon edges (from 3 to 4,999) in the **Edges** field on the Tool Settings window. Optionally, key-in the radius of the imaginary circle in the **Radius** edit field. If the **Radius** is set to 0, it can be defined graphically with a data point or by key-in from the Key-in window.

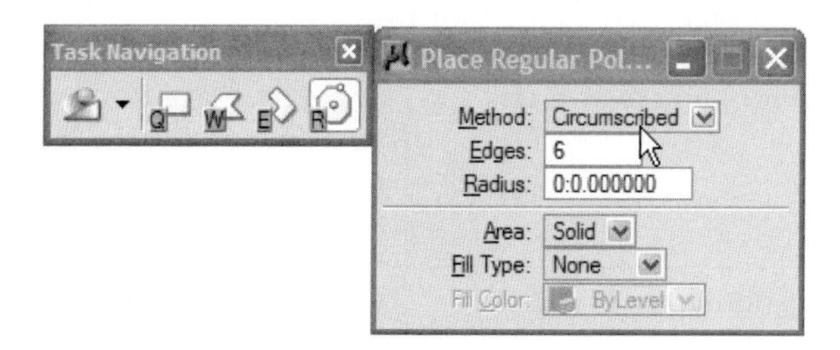

Figure 4–8 Invoke the Place Polygon Circumscribed tool from the Navigation tool box (active task set to Polygons)

MicroStation prompts:

> Place Circumscribed Polygon > Enter point on axis *(Place a data point or key-in coordinates to define the center of the polygon.)*
>
> Place Circumscribed Polygon > Enter radius or pnt on circle *(Place a data point or key-in coordinates to define the radius of the imaginary circle, the polygon's rotation, and one vertex.)*

Place Polygon By Edge

The **By Edge** method (Place Polygon By Edge) places a polygon by defining two endpoints of a polygon edge (see Figure 4–9).

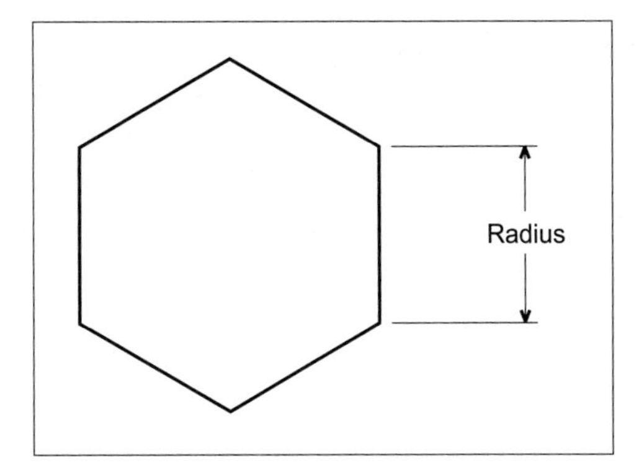

Figure 4–9 Example of a polygon placed by defining two endpoints of its side

Invoke the Place Polygon By Edge tool from:

Task Navigation tool box (active task set to Polygons)	Select the Place Regular Polygon tool and select Edge from the Method menu located in the Tool Settings window (see Figure 4–10).
Keyboard Navigation (Task Navigation tool box with active task set to Polygons)	**r** (Select Edge from the Method option menu on the Tool Settings window.)

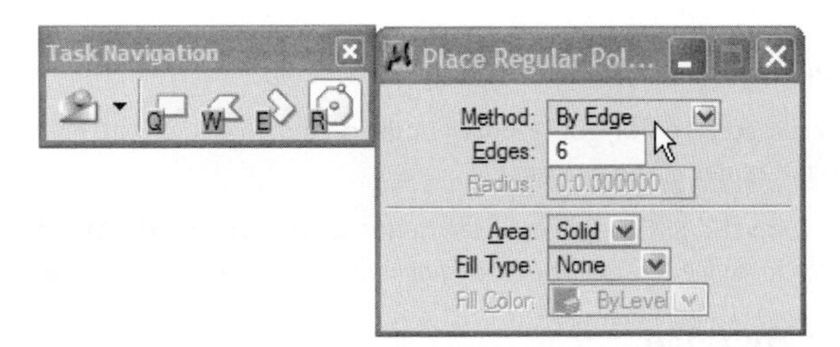

Figure 4–10 Invoke the Place Polygon by Edge tool from the Navigation tool box (active task set to Polygons)

Key-in the number of polygon edges (from 3 to 4,999) in the Edges field on the Tool Settings window.

MicroStation prompts:

> Place Polygon by Edge > Enter first edge point *(Place a data point or key-in coordinates to define the vertex of the edge.)*
>
> Place Polygon by Edge > Enter next (CCW) edge point *(Place a data point or key-in coordinates to define the second edge point.)*

PLACE POINT OR STREAM CURVE

The Place Point or Stream Curve tool places a curve element by defining a series of data points through which the curve passes. The curve element can contain from 3 to 4,994 vertices. If more than 4,994 vertices are placed, MicroStation starts a new Point Curve element and hooks it to the previous one as a complex chain (the elements in a chain act as if they are one element).

The **Points** method draws the curve through data points that have been defined by either clicking the Data button or keying-in precision coordinates. Invoke the Place Point Curve tool from:

Task Navigation tool box (active task set to Linear)	Select the Place Point or Stream Curve tool and select Points from the Method menu on the Tool Settings window (see Figure 4–11).
Keyboard Navigation (Task Navigation tool box with active task set to Linear)	**s** (**S**elect Points from the Method menu on the Tool Settings window.)

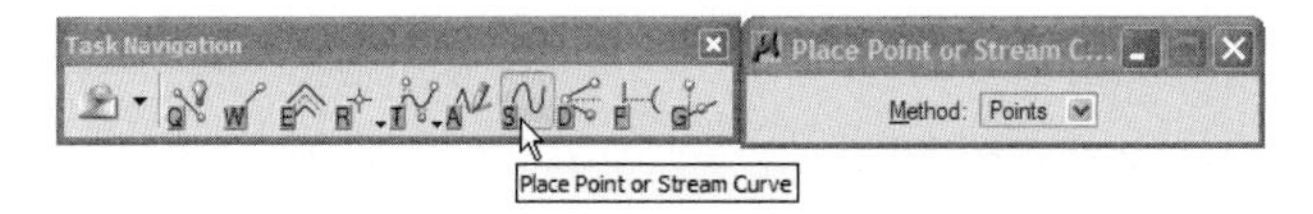

Figure 4–11 Invoke the Place Point Curve tool from the Navigation tool box (active task set to Linear)

MicroStation prompts:

> Place Point Curve > Enter first point in curve string *(Place a data point or key-in coordinates to define the starting point of the curve.)*
>
> Place Point Curve > Enter point or RESET to complete *(Place data points or key-in coordinates to define the curve vertices, or click the Reset button to complete the curve.)*

At least three points are required to describe a curved element. After defining curve data points or key-in coordinates, pressing the Reset button will terminate the tool sequence. See Figure 4–12 for an example of placing a curve using the Place Point Curve tool by providing five data points.

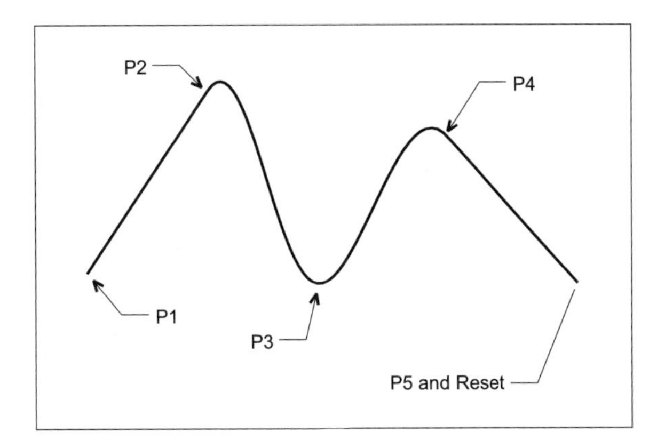

Figure 4–12 Example of placing an arc by the Place Point Curve tool

The **Stream** method draws the curve stream by following the movement of the pointer. As the pointer is moved, MicroStation defines data points based on the following streaming control settings:

- **Delta** sets the minimum distance, in working units, between points.
- **Tolerance** sets the maximum distance, in working units, between points.
- **Angle** sets the angle, in degrees, that the direction must change before a new vertex point is defined.
- **Area** sets the area that, when exceeded, causes a new vertex point to be defined.

Invoke the Place Curve Stream tool from:

Task Navigation tool box (active task set to Linear)	Select the Place Point or Stream Curve tool, then select Stream from the Method option menu located in the Tool Settings window (see Figure 4–13).
Keyboard Navigation (Task Navigation tool box with active task set to Linear)	**s** (Select Stream from the Method menu on the Tool Settings window.)

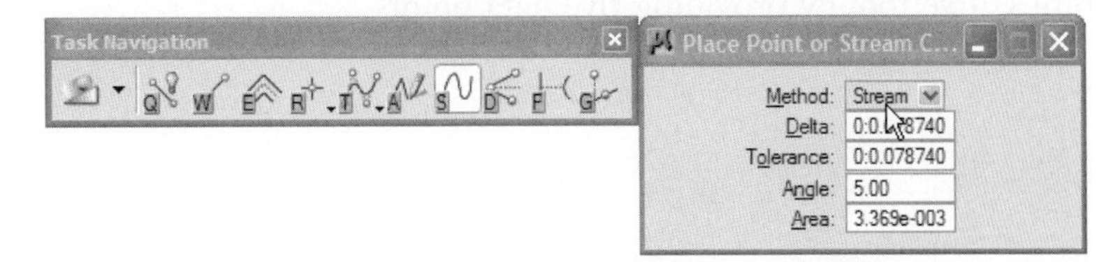

Figure 4–13 Invoke the Place Stream Curve tool from the Navigation tool box (active task set to Linear)

MicroStation prompts:

> Place Stream Curve > Enter first point in curve string *(Place a data point or key-in coordinates to define the starting point of the curve stream.)*
>
> Place Stream Curve > Enter point or RESET to complete *(Move the cursor to define the curve stream. Press the Reset button to complete the curve stream.)*

See Figure 4–14 for an example of placing a curve stream by means of the Place Curve Stream tool.

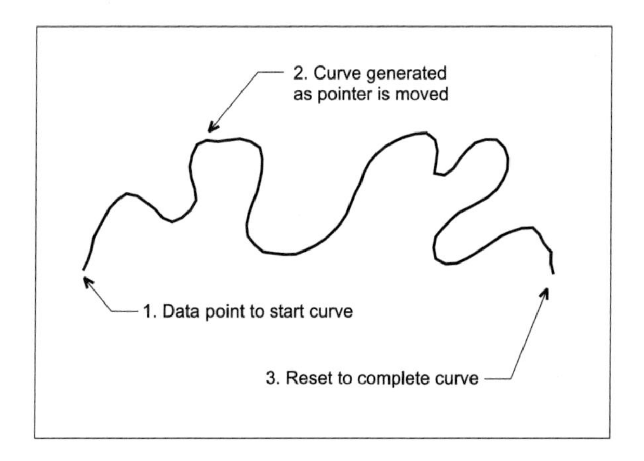

Figure 4–14 Example of placing a curve stream by means of the Place Curve Stream tool

PLACE MULTI-LINE

With the Place Multi-line tool, multiple parallel line segments can be drawn, which are then treated as one element. A multi-line can consist of as many as 16 separate, parallel lines of various line styles, weights, and colors. Multi-line definitions can be created or modified with the help of the Multi-line settings box invoked from the **Element** drop-down menu. (Refer to Chapter 16 for a detailed description of creating or modifying an existing multi-line definition.) One of the available multi-line definitions can be set as the active definition from the Tool Settings window.

To place a multi-line element, place a data point to identify the starting location and place additional data points to define each segment of the element. The following options in the Tool Settings window will constrain each multi-line segment to a specific length and angle:

1. To place segments of a specific length, turn ON the **Length** check box and key-in the length, in working units, in the **Length** text field.

2. To place segments at a specific angle, turn ON the **Angle** check box and key-in the angle in the **Angle** text box. Rotation is counter-clockwise and an angle of zero draws the segment horizontally to the right.

Note: *If both **Length** and **Angle** are ON, each data point defines a separate, one-segment multi-line element.*

Invoke the Place Multi-line tool from:

Task Navigation tool box (active task set to Linear)	Select the Place Multi-line tool (see Figure 4–15).
Keyboard Navigation (Task Navigation tool box with active task set to Linear)	**e**

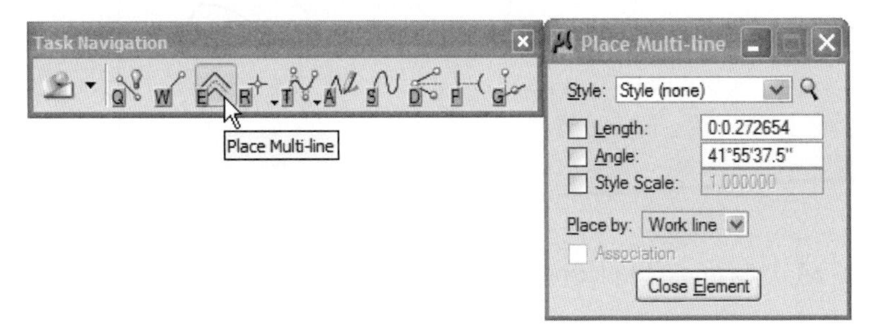

Figure 4–15 Invoke the Place Multi-line tool from the Navigation tool box (active task set to Linear)

MicroStation prompts:

Place Multi-line > Enter first point *(Place a data point or key-in coordinates to define the starting point of the multi-line.)*

Place Multi-line > Enter vertex or Reset to complete *(Place a data point or key-in coordinates to define a vertex, or press the Reset button to complete.)*

Note: *MicroStation has a set of tools to edit Multi-lines called Multi-line Joints. Refer to Chapter 8 for details about using Multi-line Joints.*

ELEMENT MODIFICATION

MicroStation not only facilitates placement of elements, but also provides tools to modify them as needed. This section discusses four important tools that will streamline modification of elements: Fillet, Chamfer, Trim, and Partial Delete.

CONSTRUCT CIRCULAR FILLET

The Construct Circular Fillet tool joins two elements (lines, line strings, circular arcs, circles, or shapes), two segments of a line string, or two sides of a shape with an arc of a specified radius. The arc is placed tangent to the two elements it connects.

The way MicroStation constructs a circular fillet depends on the option selected from the **Truncate** menu on the Tools Settings window.

- The **None** option places the fillet arc, but does not truncate the selected sides.
- The **Both** option places the fillet arc and truncates the elements at their point of tangency to create a smooth transition.
- The **First** option places the fillet and only truncates the first side identified.

For examples of placing a fillet using the three options, see Figure 4–16.

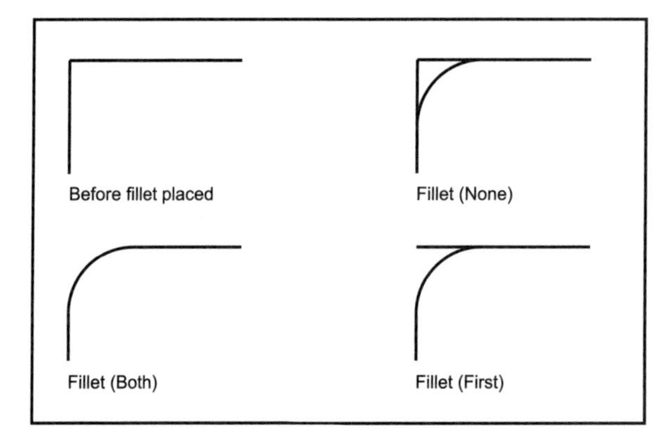

Figure 4–16 Examples of placing a fillet using the three truncation methods

Invoke the Construct Circular Fillet (no truncation) tool from:

Modify tool box	Select the Construct Circular Fillet tool. In the Tool Settings window, key-in the Radius in working units and select None from the Truncate menu (see Figure 4–17).
Keyboard Navigation (Main tool box active)	**7 0** (Key-in the Radius in working units and select None from the Truncate menu in the Tool Settings window.)

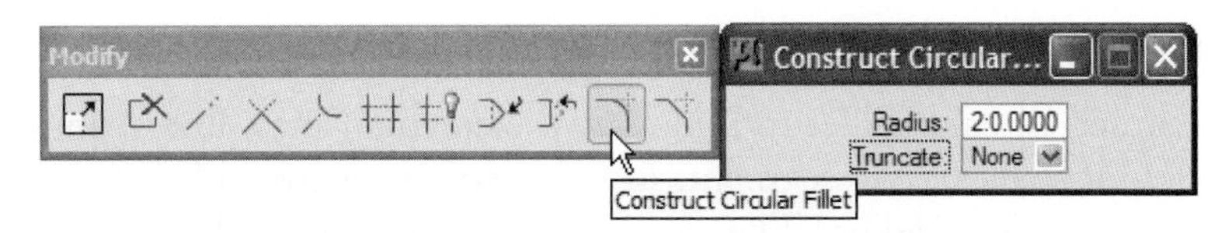

Figure 4–17 Invoke the Circular Fillet (no truncation) tool from the Modify tool box

MicroStation prompts:

> Circular Fillet (no truncation) > Select first segment *(Identify the first element or segment.)*
>
> Circular Fillet (no truncation) > Select second segment *(Identify the second element or segment.)*

MicroStation constructs a fillet of a specified radius. Invoke the Construct Circular Fillet (Truncate Both) tool from:

Modify tool box	Select the Construct Circular Fillet tool. In the Tool Settings window, key-in the Radius in working units and select Both from the Truncate menu.
Keyboard Navigation (Main tool box active)	**7 0** (Key-in the Radius in working units and select Both from the Truncate menu in the Tool Settings window.)

MicroStation prompts:

> Circular Fillet and Truncate Both > Select first segment *(Identify the first element or segment.)*
>
> Circular Fillet and Truncate Both > Select second segment *(Identify the second element or segment.)*

MicroStation constructs a fillet arc and truncates the elements at their point of tangency. Invoke the Construct Circular Fillet (Truncate Single) tool from:

Modify tool box	Select the Construct Circular Fillet tool. In the Tool Settings window, key-in the Radius in working units and select First from the Truncate menu.
Keyboard Navigation (Main tool box active)	**7 0** (Key-in in the Radius in working units and select First from the Truncate menu in the Tool Settings window.)

MicroStation prompts:

> Circular Fillet and Truncate Single > Select first segment *(Identify the first element or segment – the one that is to be truncated.)*
>
> Circular Fillet and Truncate Single > Select second segment *(Identify the second element or segment.)*

MicroStation constructs a fillet and truncates the first element or segment identified.

CONSTRUCT CHAMFER

The Construct Chamfer tool places an angled corner between two lines or between adjacent segments of a line string or shape.

The Tool Settings window provides the **Distance 1** and **Distance 2** text boxes that control the distance, in working units, that each end of the chamfer is placed from the corner. **Distance 1** is applied to the first element selected and **Distance 2** is applied to the second element. To place a 45-degree chamfer, enter equal distances in the two fields. Invoke the Construct Chamfer tool from:

Modify tool box	Select the Construct Chamfer tool. In the Tool Settings window, key-in the appropriate Distance 1 and Distance 2, in working units (see Figure 4–18).
Keyboard Navigation (Main tool box active)	**7** and select Construct Chamfer from the drop-down menu. In the Tool Settings window, key-in the appropriate Distance 1 and Distance 2, in working units

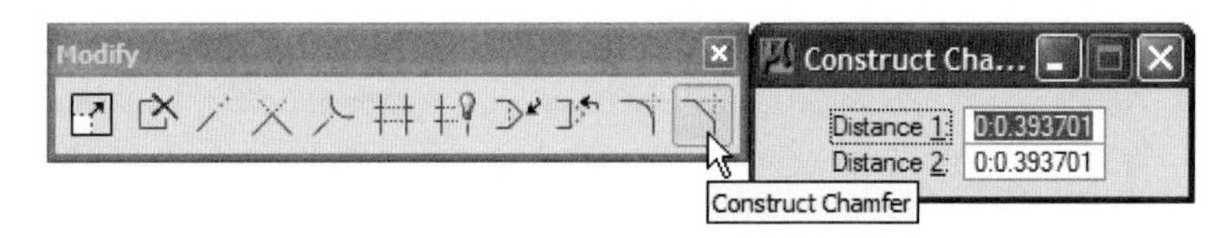

Figure 4–18 Invoke the Construct Chamfer tool from the Modify tool box

MicroStation prompts:

> Construct Chamfer > Select first chamfer segment *(Identify the first element or segment, as shown in Figure 4–19.)*
>
> Construct Chamfer > Select second chamfer segment *(Identify the second element or segment, as shown in Figure 4–19.)*
>
> Construct Chamfer > Accept-Initiate construction *(Click the Accept button to accept the placement of the chamfer, or click the Reject button to reject the placement of the chamfer.)*

Note: *The chamfer appears at the location selected immediately after selection of the second segment, but it is only a tentative placement. It does not become a permanent placement until the Accept button is clicked in response to the third prompt.*

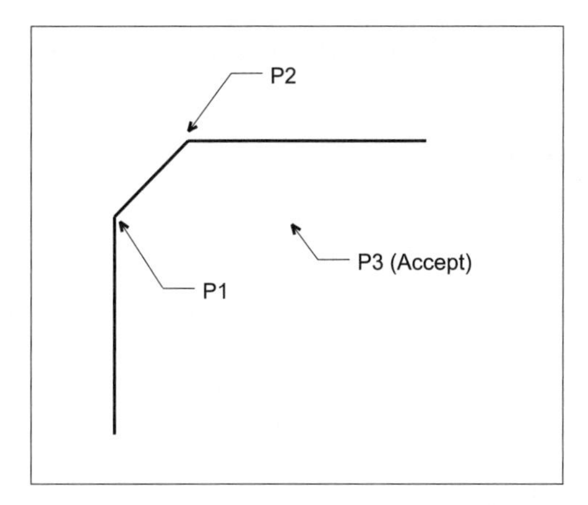

Figure 4–19 Example of placing the chamfer with the Construct Chamfer tool

TRIM ELEMENTS

IntelliTrim combines into one tool the ability to: **Trim** the part of selected elements that overlap a cutting element; **Extend** elements to a cutting element; and **Cut** elements into individual segments. In addition, the tool provides a **Quick Mode** that allows selecting one cutting element and an **Advanced Mode** that allows selecting one or more cutting elements.

- As elements are selected to be trimmed, extended, or cut, square guide posts appear at the point where the operation will take place (such as the point to which the elements will be extended).
- Elements can be extended to the elements, such as arcs, cell headers, complex shapes, complex strings, curves, ellipses, lines, line strings, shapes, and text nodes.
- Elements can be cut or trimmed, such as arcs, b-spline curves, complex shapes, complex strings, curves, ellipses, lines, line strings, and shapes.
- Elements can be extended, such as b-spline curves, complex chains that end with a line or line string, lines, and line strings. Elements that cannot be extended will usually be deleted.

Quick Trim

When using the **Quick Mode** and **Trim Operation,** IntelliTrim prompts for the cutting element. Next, it prompts the user to draw temporary lines across all elements that are to be trimmed to the first element selected. The parts of the elements that the temporary line touches are trimmed. Only elements that the cutting element crosses are trimmed. The cutting element is not modified. To execute a quick trim, invoke the IntelliTrim tool from:

Modify tool box	Select the IntelliTrim Elements tool and select the Quick Mode and the Trim Operation from the Tool Settings window (see Figure 4–20).
Keyboard Navigation (Main tool box active)	**7 7** (Set the Mode to Quick and the Operation to Trim in the Tool Settings window.)

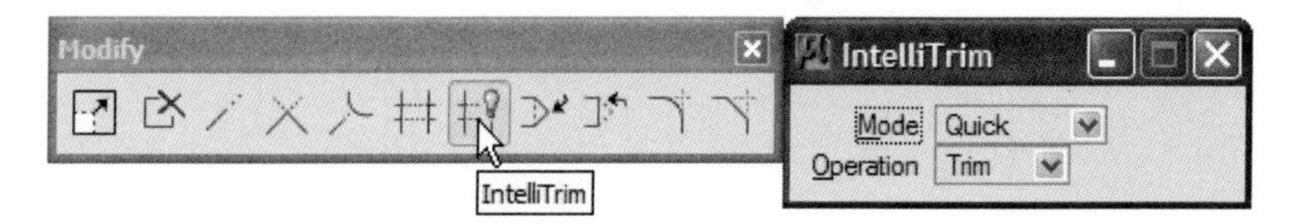

Figure 4–20 Invoke the IntelliTrim tool from the Modify tool box, then select the Quick Mode and the Trim Operation

MicroStation prompts:

IntelliTrim > Identify element to trim to *(Identify the cutting element.)*

IntelliTrim > Create line(s) crossing the element(s) to be trimmed *(Place a data point to identify one end of the line that will define the part of the elements to be trimmed.)*

IntelliTrim > Enter endpoint of the line *(Place a data point to identify the other end of the temporary line that will define the part of the elements to be trimmed. Make sure the temporary line crosses the part of the elements to be removed.)*

IntelliTrim > Create line(s) crossing the element(s) to be trimmed *(Continue identifying the start and endpoints of temporary lines until all elements to be trimmed are identified, and then click the Reset button to complete the trimming operation.)*

For example, the following tool sequence shows how to use the IntelliTrim tool's **Trim Operation** with **Quick Mode** to trim the parts of a set of lines that are outside of an ellipse. The sequence of commands is illustrated in Figure 4–21.

IntelliTrim > Identify Element *(Identify the ellipse that will be the cutting element.)*

IntelliTrim > Create line(s) crossing the element(s) to be trimmed *(Place a data point on the left side of the ellipse, above the lines, to start the temporary line that will identify the lines to be trimmed on the left side of the ellipse.)*

IntelliTrim > Enter endpoint of the line *(Place a data point below the lines to identify the end of the temporary line so that it passes across each of the three lines on the left side of the ellipse.)*

IntelliTrim > Create line(s) crossing the element(s) to be trimmed *(Place a data point on the right side of the ellipse, above the lines, to start the temporary line that will identify the lines to be trimmed on the right side of the ellipse.)*

IntelliTrim > Enter endpoint of the line *(Place a data point to identify the end of the temporary line so that it passes across each of the three lines on the right side of the ellipse.)*

IntelliTrim > Create line(s) crossing the element(s) to be *trimmed (Click the Reset button to complete trimming the part of the lines outside of the ellipse.)*

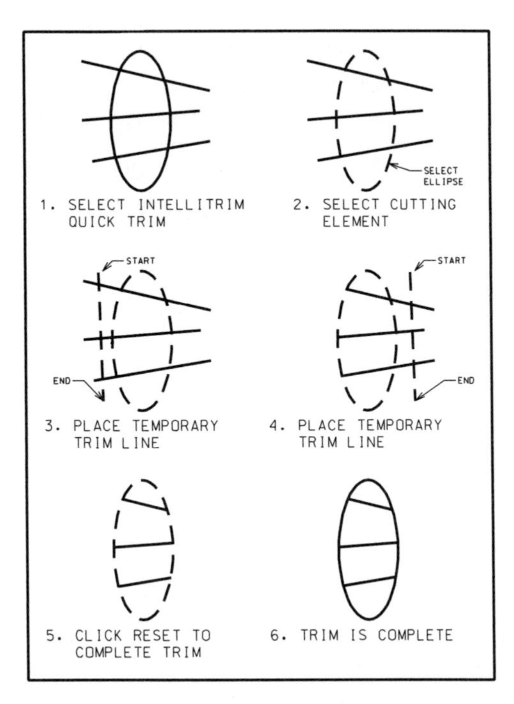

Figure 4–21 Example of trimming the parts of three lines that are outside of an ellipse

Quick Extend

When IntelliTrim is set to the **Quick Mode** and **Extend Operation**, an element to which other elements are to be extended is selected and then temporary lines are drawn across all the elements that are to be extended. If a selected element cannot be extended to actually touch the first selected element, it is not modified. If an element extends beyond the first selected element, it is not modified. If the first selected element is a closed shape, such as a circle, the extension is made to the near side of the element, unless an element already extends beyond the near side. If an element extends beyond the near side, it is extended to the far side. To execute a quick extend, invoke the IntelliTrim tool from:

Modify tool box	Select the IntelliTrim Elements tool and select the Quick Mode and Extend Operation from the Tool Settings window (see Figure 4–22).
Keyboard Navigation (Main tool box active)	**7 7** (Then set the Mode to Quick and the Operation to Extend in the Tool Settings window.)

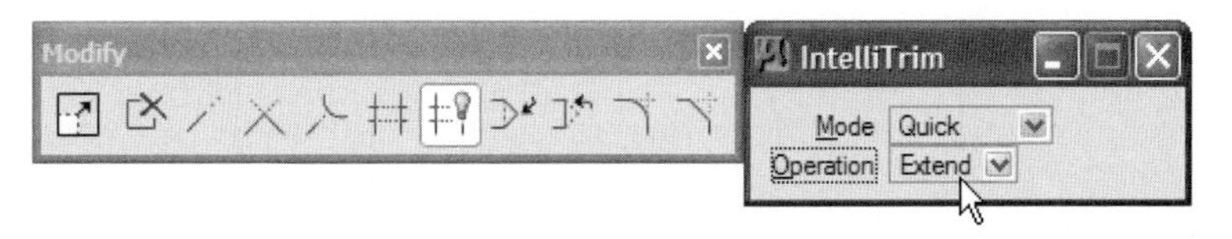

Figure 4–22 Invoke the IntelliTrim tool from the Modify tool box, then select the Quick Mode and Extend Operation

MicroStation prompts:

> IntelliTrim > Identify element to extend to *(Identify the element to which the other elements are to be extended.)*
>
> IntelliTrim > Create line(s) crossing the element(s) to be extended *(Place a data point to identify the location of the start of the line that will identify the elements to be extended.)*
>
> IntelliTrim > Enter endpoint of the line *(Place a data point to identify the location of the end of the temporary line such that it extends across the elements to be extended.)*
>
> IntelliTrim > Create line(s) crossing the element(s) to be extended *(Continue identifying temporary lines until all elements to be extended are identified, and then click the Reset button to complete the extension operation.)*

For example, the following tool sequence shows how to use the IntelliTrim tool **Quick Mode** and **Extend Operation** to extend three lines to the left side of an ellipse. The sequence of commands is illustrated in Figure 4–23.

> IntelliTrim > Identify element to extend to *(Identify the ellipse.)*
>
> IntelliTrim > Create line(s) crossing the element(s) to be extended *(Place a data point above the elements to be extended.)*
>
> IntelliTrim > Enter endpoint of the line *(Place a data point to below the lines to be extended making sure the temporary line crosses all the lines to be extended.)*
>
> IntelliTrim > Create line(s) crossing the element(s) to be extended *(Click the Reset button to complete extending the lines to the ellipse.)*

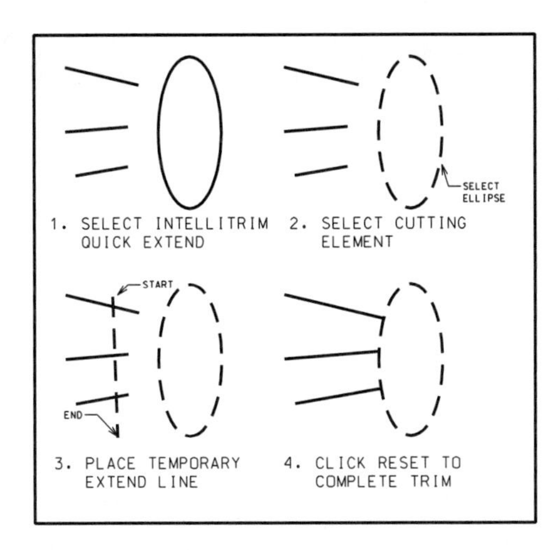

Figure 4–23 Example of extending three lines that are outside of an ellipse

Quick Cut

When IntelliTrim is set to the **Quick Mode** and **Cut Operation**, temporary lines are drawn across all elements that are to be cut. The temporary lines define the position of the cuts on each element.

The cut elements are broken into separate elements whose endpoints are where temporary lines cross the original elements. Changes will not be evident until one of the new elements is selected (such as by selecting one of them for deletion). To execute a quick cut, invoke the IntelliTrim tool from:

Modify tool box	Select the IntelliTrim Elements tool, then select the Quick Mode and the Cut Operation from the Tool Settings window (see Figure 4–24).
Keyboard Navigation (Main tool box active)	**7 7** (Set the **Mode** to **Quick** and the **Operation** to **Cut** in the Tool Settings window.)

Figure 4–24 Invoke the IntelliTrim tool from the Modify tool box, then select the Quick Mode and the Cut Operation

MicroStation prompts:

Cut elements > Create line which defines cut *(Place a data point to identify the location of the start of the line that will define the position of one of the cuts to be made on the elements it crosses.)*

Cut elements > Create line which defines cut *(Place a data point to identify the location of the end of the temporary line that will define the cut points. The temporary line should cross the elements to be cut at the points where they are to be cut.)*

Cut elements > Create line which defines cut *(Continue identifying temporary lines until all cut points are identified, then click the Reset button to complete the cutting operation.)*

For example, the following tool sequence shows how to cut two segments out of an ellipse. The sequence of commands is illustrated in Figure 4–25.

Cut elements > Create line which defines cut *(Place a data point on the left side of the ellipse to start the temporary line that will identify the top of the segment to be cut out of the ellipse.)*

Cut elements > Create line which defines cut *(Place a data point on the right side of the ellipse to identify the end of the temporary line that will identify the top of the segment to be cut out of the ellipse.)*

Cut elements > Create line which defines cut *(Place a data point on the left side of the ellipse to start the temporary line that will identify the bottom of the segment to be cut out of the ellipse.)*

Cut elements > Create line which defines cut *(Place a data point on the right side of the ellipse to identify the end of the temporary line that will identify the bottom of the segment to be cut out of the ellipse.)*

Cut elements > Create line which defines cut *(Click the Reset button to complete making the cuts in the ellipse.)*

Note: *The example procedure made two cuts each on the left and right sides of the ellipse. After the cuts were completed, the Delete Element tool was used to remove the pieces of the ellipse between the cuts as shown in Figure 4–25.*

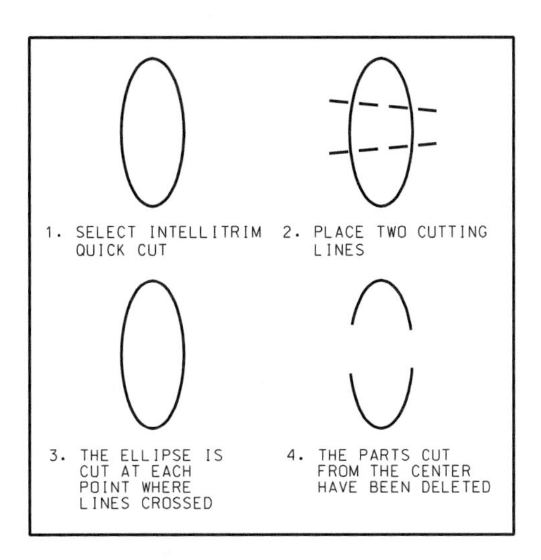

Figure 4–25 Example of making four cuts on an ellipse

Advanced Trim

When IntelliTrim is set to the **Advanced Mode** and **Trim Operation**, one or more cutting elements can be selected, and one or more elements can be trimmed to the cutting elements. Options on the Tool Settings window are **Select Elements to Trim** or **Select Cutting Elements**. To execute an advanced trim, invoke the IntelliTrim tool:

Modify tool box	Select the IntelliTrim Elements tool and then select the Advanced Mode and the Trim Operation from the Tool Settings window. Decide which set of elements to start with by choosing Select Elements to Trim or Select Cutting Elements (see Figure 4–26).
Keyboard Navigation (Main tool box active)	**7 7** (Set the **Mode** to **Advanced** and the **Operation** to **Trim** in the Tool Settings window and choose either Select Elements to Trim or Select Cutting Elements.)

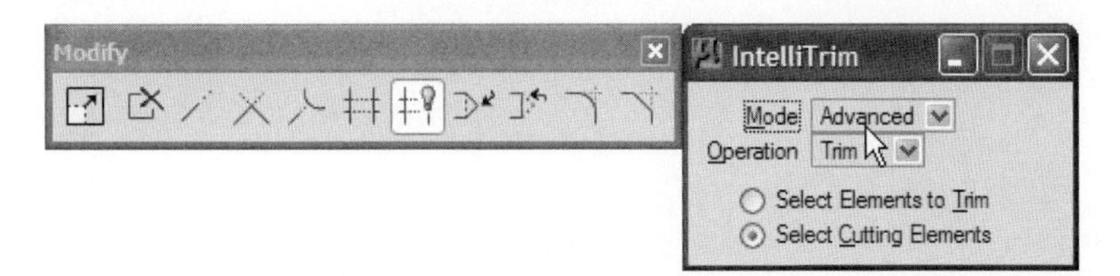

Figure 4–26 Invoke the IntelliTrim tool from the Modify tool box, then select the Advanced Mode and the Trim Operation

MicroStation prompts when the **Select Cutting Elements** option is selected first:

IntelliTrim > Identify cutting elements, reset to complete step *(Select each cutting element by clicking a data point on it. When all cutting elements are selected, click the Reset button.)*

IntelliTrim > Identify elements to trim/extend, reset to complete step *(Select each element to be trimmed by clicking a data point on it. When all elements to be trimmed are selected, click the Reset button.)*

IntelliTrim > Enter points near portions to keep, reset to complete command *(If the wrong part of any element has been trimmed, click a data point near it to switch the part that is trimmed. When the correct part of each element is shown trimmed, click the Reset button to make the trimming operation permanent.)*

For example, the following command sequence shows (see Figure 4-27) how to use the IntelliTrim tool's **Advanced Mode** and **Trim Operation** to use two orthogonal blocks as cutting elements for trimming two shapes with the Select Cutting Elements option selected first.

IntelliTrim > Identify cutting elements, reset to complete step *(Select one of the orthogonal blocks by clicking a data point on it.)*

IntelliTrim > Identify cutting elements, reset to complete step *(Select the other orthogonal block by clicking a data point on it, and then click the Reset button.)*

IntelliTrim > Identify elements to trim, reset to complete step *(Select one of the shapes by clicking a data point on it.)*

IntelliTrim > Identify elements to trim, reset to complete step *(Select the other shape by clicking a data point on it, and then click the Reset button.)*

IntelliTrim > Enter points near portions to keep, reset to complete command *(If the part of the two shapes between the inner and outer blocks is tentatively removed, click the Reset button to complete the operation. If the wrong parts of the shapes are trimmed, click a data point near the trimmed parts to switch the trimming to a different part of the elements. When the correct part of each shape is selected, click the Reset button to complete the trimming operation.)*

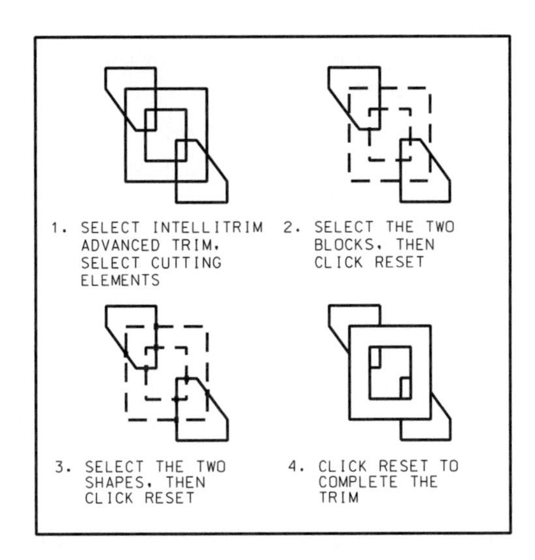

Figure 4–27 Example of using two blocks as cutting elements for trimming two shapes

Advanced Extend

When IntelliTrim is set to the **Advanced Mode** and **Extend Operation**, one or more boundary elements can be selected, and one or more elements can be extended to the boundary elements. Options on the Tool Settings are **Select Elements to Trim** (start by selecting the elements to be extended) or **Select Cutting Elements** (start by selecting the elements to which the other elements will be extended). To execute an advanced extend, invoke the IntelliTrim tool from:

Modify tool box	Select the IntelliTrim Elements tool, and then select the Advanced Mode and the Extend Operation from the Tool Settings window. Decide which set of elements to start with by choosing Select Elements to Trim or Select Cutting Elements (see Figure 4–28).
Keyboard Navigation (Main tool box active)	**7 7** (Set the **Mode** to **Advanced** and the **Operation** to **Extend** in the Tool Settings window and choose either Select Elements to Trim or Select Cutting Elements.)

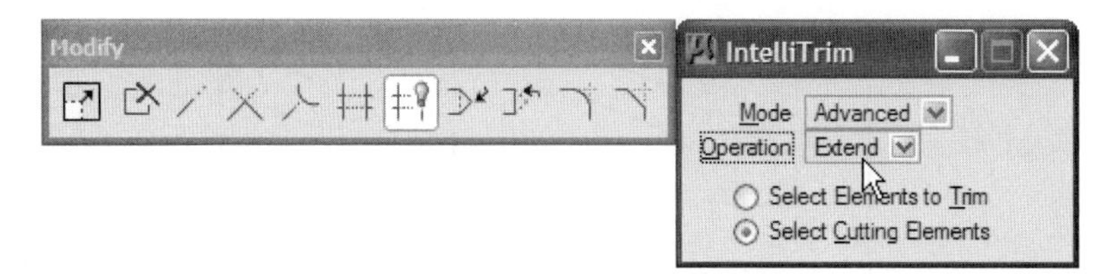

Figure 4–28 Invoke the IntelliTrim tool from the Modify tool box, then select the Advanced Mode and the Extend Operation

MicroStation prompts when the **Select Cutting Elements** option is selected first:

IntelliTrim > Identify cutting elements, reset to complete step *(Select each boundary element by clicking a data point on it, and then click the Reset button.)*

IntelliTrim > Identify elements to trim/extend, reset to complete step *(Select each element to be extended by clicking a data point on it, and then click the Reset button.)*

IntelliTrim > Enter points near portions to keep, reset to complete command *(Click the Reset button to make the extend operation permanent.)*

For example, the following command sequence shows (see Figure 4-29) how to use the IntelliTrim tool's **Advanced Mode** and **Extend Operation** to extend one line to one cutting block and another line to a second cutting block with Select Cutting Elements option selected first.

IntelliTrim > Identify cutting elements, reset to complete step *(Select one of the orthogonal blocks by clicking a data point on it.)*

IntelliTrim > Identify cutting elements, reset to complete step *(Select the other orthogonal block by clicking a data point on it, and then click the Reset button.)*

IntelliTrim > Identify elements to trim/extend, reset to complete step *(Select one of the lines by clicking a data point on it.)*

IntelliTrim > Identify elements to trim/extend, reset to complete step *(Select the other line by clicking a data point on it, and then click the Reset button.)*

IntelliTrim > Enter points near portions to keep, reset to complete command *(Click the Reset button to complete the trimming operation.)*

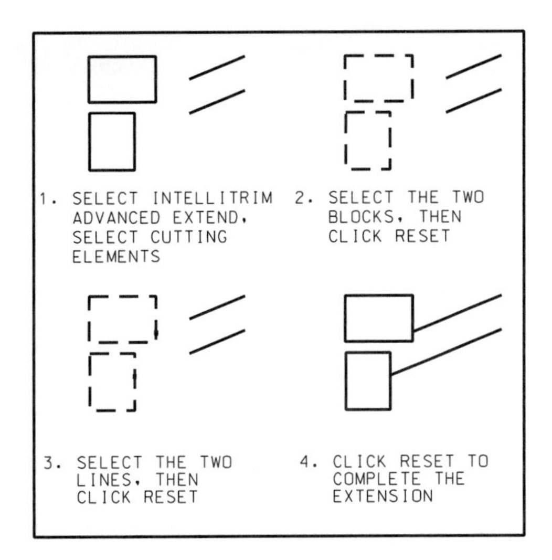

1. SELECT INTELLITRIM ADVANCED EXTEND, SELECT CUTTING ELEMENTS
2. SELECT THE TWO BLOCKS, THEN CLICK RESET
3. SELECT THE TWO LINES, THEN CLICK RESET
4. CLICK RESET TO COMPLETE THE EXTENSION

Figure 4–29 Example of using two blocks as cutting elements for extending two lines

PARTIAL DELETE

The Partial Delete tool deletes part of an element. In the case of a line, line string, multi-line, curve, or arc, the Partial Delete tool removes part of the element, and the element is divided into two elements of the same type. A partially deleted ellipse or circle becomes an arc, and a shape becomes a line string. There are no Tool Settings options for this tool. Invoke the Partial Delete tool from:

Modify tool box	Select the Partial Delete tool (see Figure 4–30.)
Keyboard Navigation (Main tool box active)	**7 2**

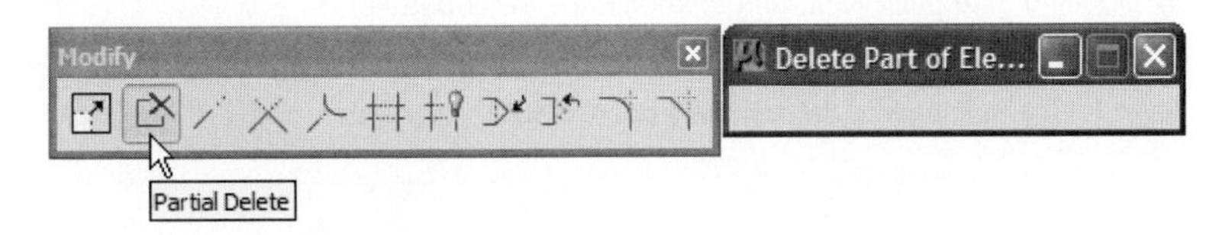

Figure 4–30 The Partial Delete tool

MicroStation prompts:

> Delete Part of Element > Select start pnt for partial delete *(Identify the element at the point where the partial delete should start.)*
>
> Delete Part of Element > Select direction of partial delete *(This prompt appears only when a closed element, such as a circle, ellipse, or polygon is selected. Move the pointer a short distance in the direction of the cut, then click the Data button.)*

Delete Part of Element > Select end pnt for partial delete *(Place a data point or key-in coordinates to identify the location of the end point of the partial delete.)*

See Figure 4–31 for examples of the Partial Delete tool.

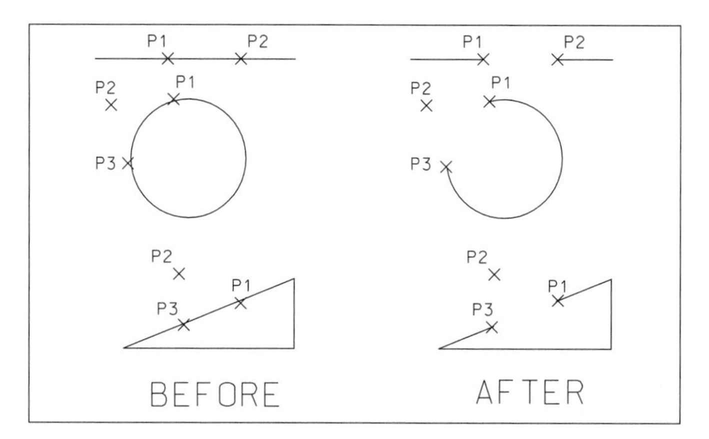

Figure 4–31 Examples of deleting part of an element with the Partial Delete tool

ELEMENT MANIPULATION

MicroStation offers two main categories of manipulation tools: single-element manipulation and multi-element manipulation. Single-element manipulation tools allow manipulation of one element at a time; multi-element manipulation tools manipulate groups of elements. This section discusses single-element manipulation tools. Multi-element manipulations are performed with Element Selection and Fence manipulation tools (see Chapter 6).

Mastering the element manipulation tools and learning when and how to apply them are key to appreciating the power and capability of MicroStation. For example, one element can be drawn, and then element manipulation tools used to quickly make copies, saving time and effort. Planning ahead to utilize these powerful tools will become second nature after a little practice.

All the manipulation tools described here require identification of the element to be manipulated To identify an element, position the pointer on the element and click the Data button. MicroStation indicates that an element is selected by displaying it in the highlight color. If the highlighted element is the correct one, continue following the tool prompts shown in the prompt field. If the element that is highlighted is not the correct one, click the Reject button to reject the element.

Note: *Be sure to check the status of lock settings before modification. A good rule to follow is to turn OFF all of the locks that are not being used with the exception of the Snap Lock. It is very frustrating to try to select an element that does not lie on the grid when the Grid Lock is*

turned ON. The cursor bounces around from grid dot to grid dot, and it may be impossible to identify an element if it is not on the grid. It is simple to toggle the grid lock OFF quickly, identify the object, and then toggle the grid lock back ON again, if needed.

COPY ELEMENT

The Copy tool places a copy of the selected element at a specified location and leaves the original element intact. The copy is identical to the original and as many copies of the original as needed may be made. Each copy is independent of the original and can be manipulated and modified like any other element. The data point entered identify the element to copy is the (base) point on the element to which the pointer is attached. Select this point with care and use the tentative snap, if needed, to snap to the element at a precise location (with the appropriate snap mode selected).

Click the Reset button to reset the command sequence. If desired, another element can be selected to copy, or another tool invoked to continue working on the design file. Invoke the Copy Element tool from:

Manipulate tool box	Select the Copy tool, then specify number of Copies (default is set to 1) in the Tool Settings window (see Figure 4–32).
Keyboard Navigation (Main tool box active)	**3 1** (Specify number of copies in the Tool Settings window.)

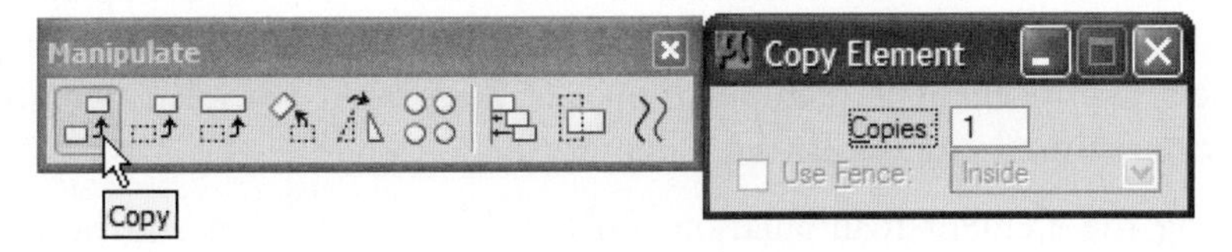

Figure 4–32 The Copy tool

MicroStation prompts:

Copy Element > Identify element *(Identify an element to copy.)*

Copy Element > Enter point to define distance and direction *(Provide the location of the copy by clicking the Data button or by keying-in coordinates.)*

Copy Element > Enter point to define distance and direction *(If necessary, copy to another location by providing a data point or by keying-in coordinates. Click the Reset button to terminate the tool sequence.)*

For example, the following tool sequence shows how to copy a line to the center of a circle using the Copy Element tool (see Figure 4–33).

Copy Element > Identify element *(Identify the line by tentative snapping to the end of the line and then clicking the Data button.)*

Copy Element > Enter point to define distance and direction *(Tentative snap to the center of the circle and then click the Data button.)*

Copy Element > Enter point to define distance and direction *(Click the Reset button to place the element at the new location.)*

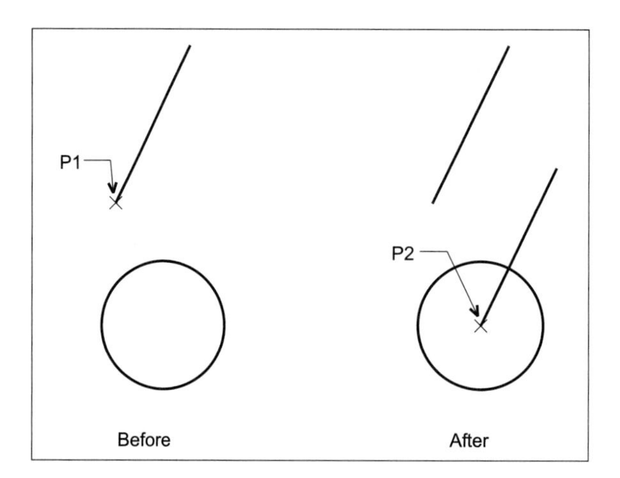

Figure 4–33 Example of copying an element by means of the Copy tool

MOVE ELEMENT

The Move tool moves an element to a new location without changing its orientation or size. The data point used to identify the element is the (base) point by which the element will be moved. Careful selection of this point combined with the tentative snap (with the appropriate snap mode selected) will facilitate precise element location. Invoke the Move Element tool from:

Manipulate tool box	Select the Move tool (see Figure 4–34).
Keyboard Navigation (Main tool box active)	**3 2**

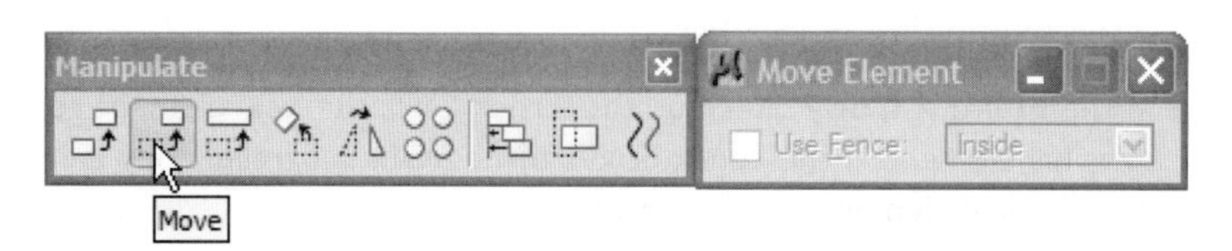

Figure 4–34 Invoke the Move tool from the Manipulate tool box

MicroStation prompts:

Move Element > Identify element *(Identify an element to move.)*

Move Element > Enter point to define distance and direction *(Reposition the element to its new location by providing a data point or keying-in coordinates.)*

For example, the following tool sequence shows how to move a line to the center of a circle using the Move Element tool (see Figure 4–35).

Move Element > Identify element *(Identify the line by snapping to the end point of the line and clicking the Data button.)*

Move Element > Enter point to define distance and direction *(Snap to the center of the circle and click the Data button.)*

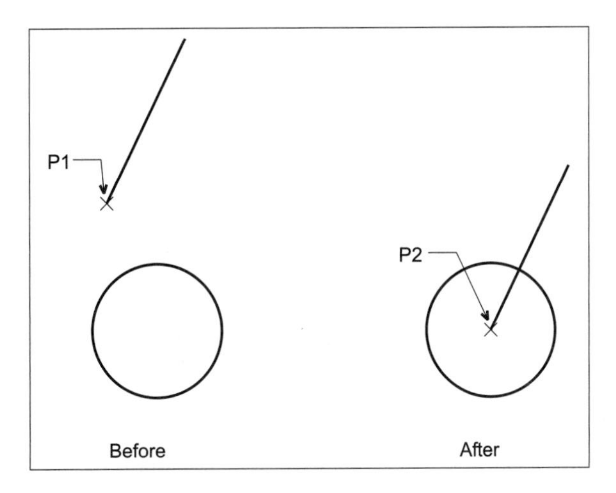

Figure 4–35 Example of moving an element

MOVE PARALLEL ELEMENT

The Move Parallel tool moves or copies an element (such as a line, line string, multi-line, circle, curve, arc, ellipse, shape, complex chain, or complex shape) parallel to the original location of the element. All points on the original element are moved an equal distance.

The Tool Settings window for the parallel tool provides the following options for controlling the manipulation results:

- If the **Make a Copy** check box is turned OFF, the new element replaces the original element. If the check box is turned ON, a parallel copy is made and the original element remains in the design.
- If the **Distance** check box is turned OFF, the position of the moved or copied element is graphically determined with a data point or precision key-in. If the check box is turned ON, the distance of the new element from the original is determined by the value (in working units) typed in the associated text box.

- The **Gap Mode** menu provides three options for controlling the shape of the corners of linear objects such as orthogonal blocks. The **Miter** option maintains a linear element's sharp corners. The **Round** option converts a linear element's sharp corners to fillets. The **Original** option creates the resulting element as the same type as the original element. Figure 4–36 shows the result of the **Miter** and **Round** Gap Modes on a shape that is copied parallel.
- If the **Use Active Attributes** check box is turned OFF, the moved or copied element retains the attributes of the existing element.
- If the **Use Active Attributes** check box is turned ON, the moved or copied element takes on the active attributes.

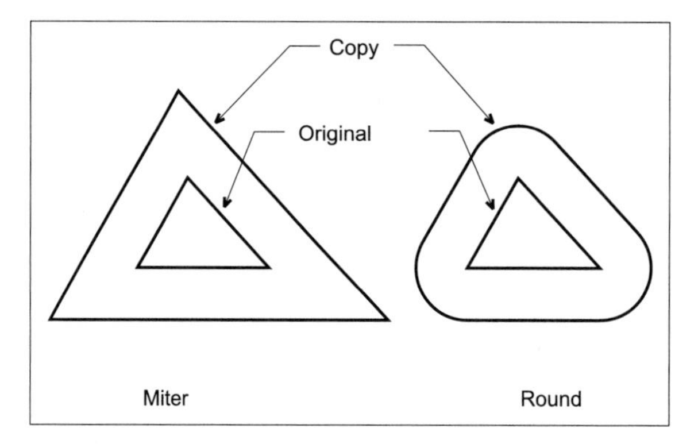

Figure 4–36 Examples of parallel copies

Invoke the Move Parallel tool from:

Manipulate tool box	Select the Move Parallel tool and select the desired options in Tool Settings window (see Figure 4–37).
Keyboard Navigation (Main tool box active)	**3 9** (Set appropriate options in Tool Settings window.)

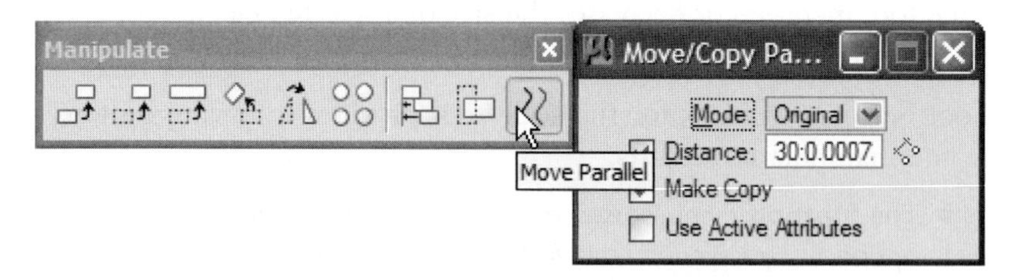

Figure 4–37 Invoke the Move Parallel tool from the Manipulate tool box

MicroStation prompts:

Move Parallel > Identify element *(Identify an element to move parallel.)*

Move Parallel > Accept/Reject *(Identify the new parallel location of the element or a copy of the element by clicking a data point or by keying-in coordinates.)*

Move Parallel > Accept/Reject *(If necessary, continue moving or copying the element in parallel by providing data points or by keying-in coordinates. When copying and moving is completed, click the Reset button.)*

For example, the following sequence shows how to move a line, in parallel to its original location, using the Move Parallel tool. The **Distance** and **Make Copy** check boxes are set to OFF. The **Gap Mode** options are not used for lines. (see Figure 4–38).

Move Parallel by Distance > Identify element *(Identify the line.)*

Move Parallel by Distance > Accept/Reject (Select next input) *(Place a data point to define the location of the parallel copy of the line.)*

Move Parallel by Distance > Accept/Reject (Select next input) *(Click the Reset button to place the element at the new location.)*

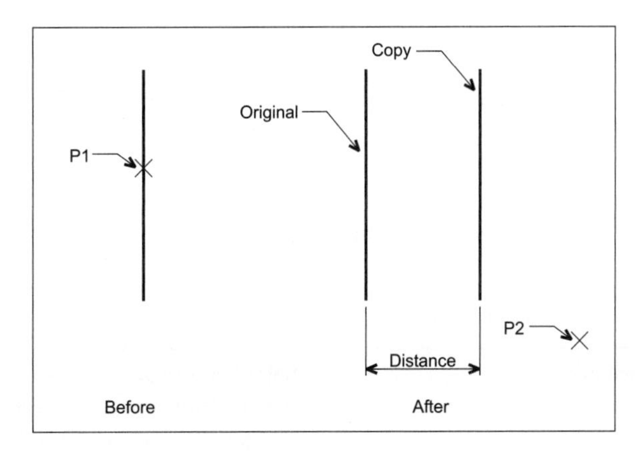

Figure 4–38 Example of moving an element by a fixed distance

SCALE ELEMENT

The Scale tool increases or decreases the size of a selected element or a copy of the element. Options in the Tool Settings window control the way the tool operates.

The Tool Settings window for the scale tool provides following options for controlling the manipulation results:

- The **Method** menu is used to scale the element graphically by the **3 Points** option or by keyed-in scale factors, using the **Active Scale** option. With the **Active Scale** option, **X Scale** and **Y Scale** text fields appear on the Tool Settings window and the element is scaled using these values. In the **3 Point** method, the scale text fields disappear from the Tool Settings window and a

Proportional check box appears. This check box forces the graphically scaled element to be proportional to the original element.

- If the **Copies** check box is set to OFF, then the original element is scaled, If it set to ON, then the selected element is copied and the copy(s) are scaled; the original is not manipulated.
- If the **About Element Center** is set to OFF, then the selected element is scaled with reference to the point where the element was identified. If it is set ON, then the selected element is scaled about its center point instead of a selected point. Cells and text elements are scaled about their origins.

Active Scale Method

To increase the size of the element using Active Scale method, select **Active Scale** from the method menu and key-in **X Scale** and **Y Scale** factors that are greater than one. To decrease the size of the element, key-in **X Scale** and **Y Scale** factors that are between 0 and 1. For example, a scale factor of 3 makes the element three times larger and a scale factor of 0.75 shrinks the selected element to three-quarters of its original size.

The scale factors can be positive or negative (-) numbers:

- Positive X and Y scale factors produce a scaled element that has the same orientation as the original element.
- A negative X and positive Y scale factor produces a scaled element that is a mirror (backward) image of the original element around the Y axis.
- A positive X and negative Y scale factor produces a scaled element that is a mirror image around the X axis.
- Negative X and Y scale factors produce a scaled element that is a mirror image around both axes. Figure 4–39 shows the effect of positive and negative scale factors.

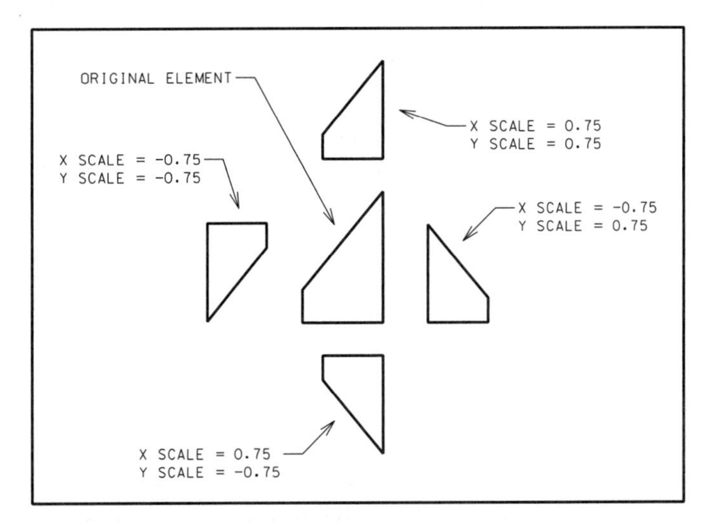

Figure 4–39 Example of the effect of positive and negative scale factors

To the right of the **X Scale** and **Y Scale** text fields is a lock symbol that can be opened and closed by clicking it with the Data button. When the lock is open, different factors can be keyed-in in the scale text fields. When the lock is closed, entering a new value in one field automatically changes the value of the other field to be equal.

To scale an element by keying-in the X and Y scale factors, invoke the Scale tool from:

Manipulate tool box	Select the Scale tool and in the Tool Settings window, select the Active Scale from the Method option menu. Key-in the appropriate scale factors in the X Scale and Y Scale text fields. To scale a copy of the original element, turn on the Copies check box (see Figure 4–40).
Keyboard Navigation (Main tool box active)	**3 3** (Select the Active Scale from the Method option menu and set appropriate scale factors in the X Scale and Y Scale text fields. To scale a copy of the original element, turn on the Copies check box.)

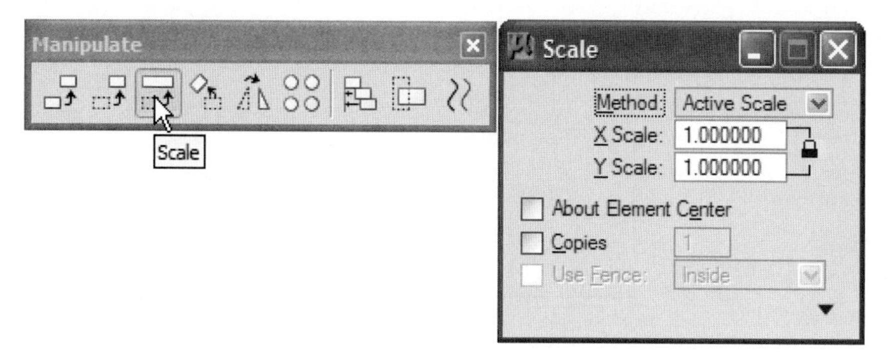

Figure 4–40 The Scale tool

MicroStation prompts:

Scale Element > Identify element *(Identify an element to scale.)*

Scale Element > Enter origin point (point to scale about) *(Identify the location of the scaled element by providing a data point or by keying-in coordinates.)*

***Note:** If the **Copies** check box is ON, the word "Copy" is added to the MicroStation prompts. For example, "Scale Element (Copy) > Identify element."*

The following tool sequence example shows how to scale a copy of an element to half its original size (see Figure 4–41). Select Scale from the Manipulations tool box. In the Tool Settings window, set the **X Scale** and **Y Scale** to 0.5, and set the **Copies** check box to ON.

MicroStation prompts:

Scale Element (Copy) > Identify element *(Identify the element by selecting a point on the bottom of the element.)*

Scale Element (Copy) > Enter origin point (point to scale about) *(Move the pointer to the left in order to move the dynamic, scaled image of the element to the right, and click the Data button to place it.)*

Scale Element (Copy) > Enter origin point (point to scale about) *(Click the Reset button and place the scaled element.)*

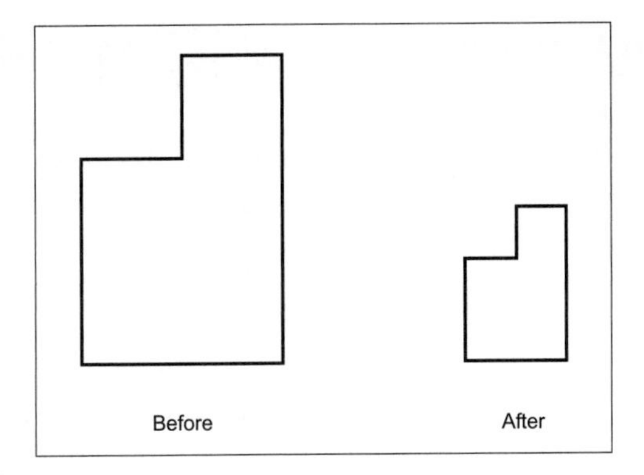

Figure 4–41 Example of scaling a copy of an element with the Active Scale tool

3-Points Method

Scaling graphically involves providing three data points or keying-in their coordinates. The scale factors are computed by dividing the distance between the first and third points by the distance between the first and second points. To scale an element graphically, invoke the Scale tool from:

Manipulate tool box	Select the Scale tool and then select 3 points from the Method option menu (see Figure 4–42).
Keyboard Navigation (Main tool box active)	**3 3** (Select 3 points from the Method option menu.)

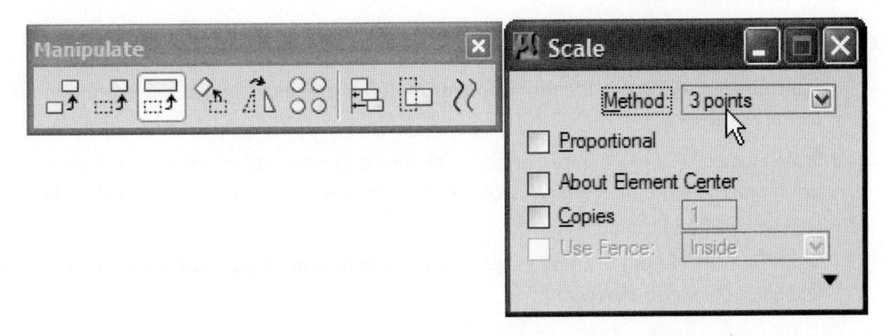

Figure 4–42 Select the Scale by 3 Points Method from the Manipulate tool box

MicroStation prompts:

> Scale Element by 3 Points > Identify element *(Identify an element to scale.)*
>
> Scale Element by 3 Points > Enter origin point (point to scale about) *(Place a data point or key-in coordinates to define the origin point.)*
>
> Scale Element by 3 Points > Enter reference point *(Place a data point or key-in coordinates to define the reference point.)*

Scale Element by 3 Points > Enter point to define amount of scaling *(Place a data point or key-in coordinates to define the amount of scaling.)*

Depending on the relationship of the three data points, the different scale factors may be applied to the vertical and horizontal size of the element. To maintain the proportionality of the selected element, turn the **Proportional** check box ON in the Tool Settings window.

Note: *If the* **Copies** *check box is ON, the word "Copy" is added to the MicroStation prompts. For example, "Scale Element by 3 Points (Copy) > Identify element."*

ROTATE ELEMENT

The Rotate tool rotates a selected element, or copy of the element, about a specified pivot point. Options on the Tool Settings window change the way the tool operates.

The Tool Settings window for the scale tool provides the following options for controlling the manipulation results:

- The **Method** menu rotates the element by an **Active Angle**; graphically, by defining **2 Points**; and graphically, by defining **3 Points**. Descriptions of each method follow.
- If the **Copies** check box is set to OFF, then the original element is rotated, If it set to ON, then the selected element is copied and the copy(s) are scaled; the original is not manipulated.
- If the **About Element Center** is set to OFF, then the selected element is scaled with reference to the point where the element was identified. If it is set ON, then the selected element is scaled about its center point instead of a selected point. Cells and text elements are scaled about their origins.

Active Angle Method

To rotate selected element using the Active Angle method, select **Active Angle** from the method menu and key-in the rotation angle in the text box or click the scroll arrows on the right side of the text box, to select standard angles such as 180 or 270 degrees. The selected element is rotated for the chosen angle in a counter-clockwise direction, starting at an imaginary line from the point on the element where it was originally identified to the rotation point.

Note: *The active angle that is set will become the new default value for subsequent commands that use the active angle. If necessary, reset the active angle to zero degrees.*

To rotate an element by the Active Angle, invoke the Rotate Element tool from:

Manipulate tool box	Select the Rotate tool. In the Tool Settings window, select Active Angle from the Method option menu, and key-in the rotation angle in the text field. To rotate a copy of the element, turn ON the Copies check box. (See Figure 4-43.)
Keyboard Navigation (Main tool box active)	**3 4** (Select the Active Angle from the Method option menu and set appropriate rotate angle. To rotate a copy of the original element, turn on the Copies check box.)

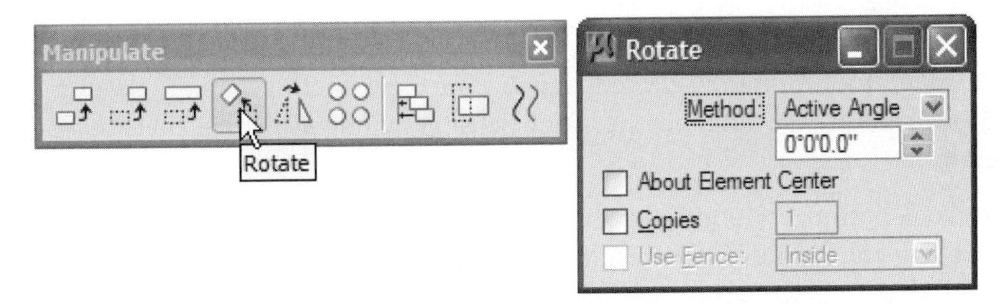

Figure 4–43 Invoke the Rotate tool from the Manipulate tool box

MicroStation prompts:

Rotate Element > Identify element *(Identify an element to rotate.)*

Rotate Element > Enter pivot point (point to rotate about) *(Define the point about which the element will rotate by providing a data point or by keying-in coordinates.)*

Note: *If the* ***Make Copy*** *check box is ON, the word "Copy" is added to the MicroStation prompts. For example, "Rotate Element (Copy) > Identify element."*

For example, the following tool sequence shows how to rotate a copy of an element to a keyed-in Active Angle of 45 degrees, using the Rotate tool (see Figure 4–44).

Rotate Element (Copy) > Identify element *(Identify the element to rotate.)*

Rotate Element (Copy) > Enter pivot point (point to rotate about) *(Reposition the rotated element to its new location by a data point.)*

Rotate Element (Copy) > Enter pivot point (point to rotate about) *(Click the Reset button to place the rotated element.)*

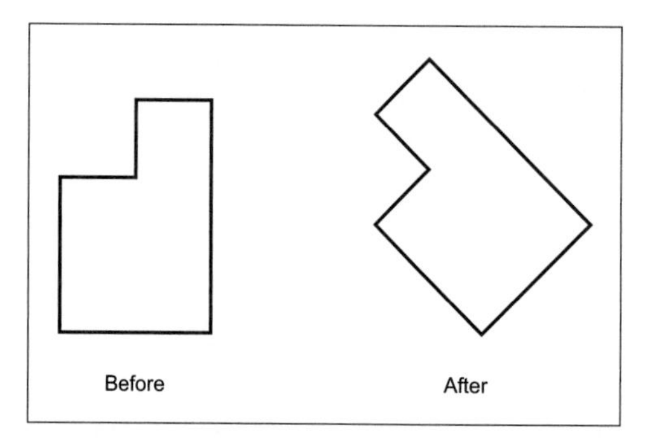

Figure 4–44 Example of rotating a copy of an element using the Active Angle Method

2 Points Method

With the **2 Points** rotation method, the element is identified with the first data point and a second data point defines the point about which the element is to be rotated. After the rotation point is defined, a dynamic image of the element spins in a circle, around the point, as the pointer moves. The radius of the circle is equal to the distance from the selection point on the element to the rotation point. The third data point determines the angle of rotation and completes the rotation operation, and the element, or a copy of the element, is placed at the rotated position if Copies option is selected. For an example of rotating an element by 2 points, see Figure 4–45.

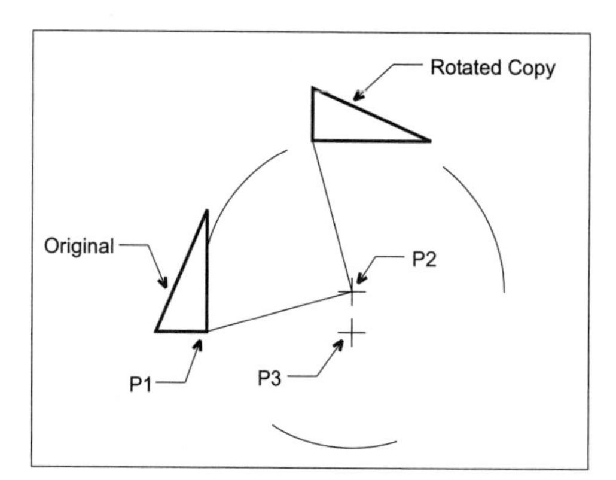

Figure 4–45 Example of rotating a copy of an element using the 2 Points Method

To rotate an element by 2 points, invoke the Rotate Element tool from:

Manipulate tool box	Select the Rotate tool. From the Tool Settings window, select the 2 Points Method. To rotate a copy of the original element, turn the Copies check box ON (see Figure 4–46).
Keyboard Navigation (Main tool box active)	**3 4** (Select the 2 Points from the Method option menu.To rotate a copy of the original element, turn on the Copies check box.)

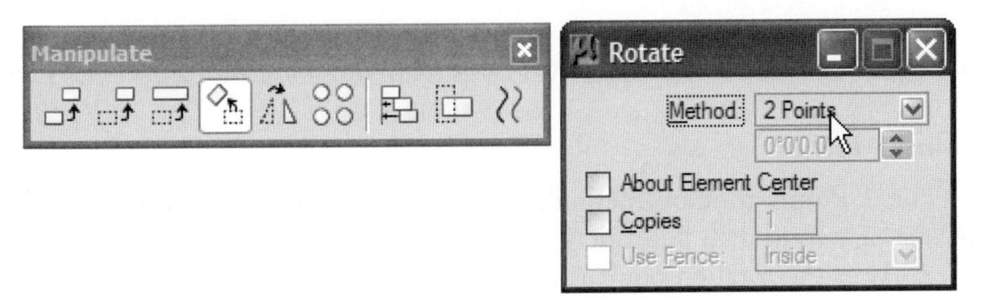

Figure 4–46 Invoke the Rotate by 2 Points Method from the Manipulate tool box

MicroStation prompts:

Rotate Element > Identify element *(Identify an element to rotate.)*

Spin Element > Enter pivot point (point to rotate about) *(Place a data point or key-in coordinates to define the pivot point.)*

Spin Element > Enter point to define amount of rotation *(Place a data point or key-in coordinates to define the amount of rotation and place the rotated element.)*

Note: *If the* ***Copies*** *check box is ON, the word "Copy" is added to the MicroStation prompts. For example, "Rotate Element (Copy) > Identify element."*

3 Points Method

With the **3 Points** rotation method, an element is identified with the first data point, a second data point defines the point about which the element is to be rotated, and a third data point starts the rotation. After the third data point, a dynamic image of the element spins around the second (rotation) data point, as the pointer moves. The fourth data point determines the angle of rotation and completes the rotation operation. The element, or a copy of the element, is placed at the rotated position if the Copies option is selected.

The rotation angle is determined by the relationship of the four data points. The angle formed by data point one and data point three, with data point two as the angle vertex, is equal to the angle formed by data point three and data point four, with data point two as the angle vertex (see Figure 4–47).

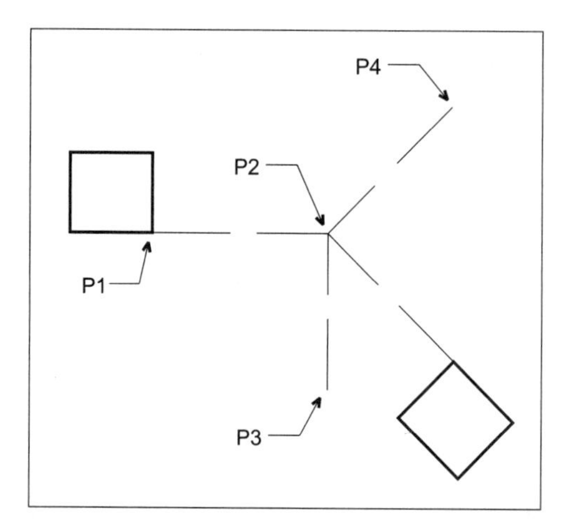

Figure 4–47 Example of rotating a copy of an element using the 3 Points Method

To rotate an element by 3 points, invoke the Rotate Element tool from:

Manipulate tool box	Select the Rotate tool. From the Tool Settings window, select the 3 points Method. To rotate a copy of the original element, turn the Copies check box ON (see Figure 4–48).
Keyboard Navigation (Main tool box active)	**3 4** (Select the 3 Points from the Method option menu. To rotate a copy of the original element, turn on the Copies check box.)

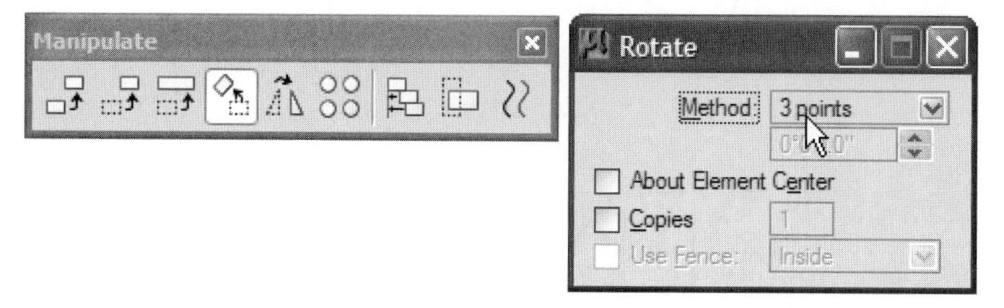

Figure 4–48 Invoke the Rotate Element by 3 Points tool from the Manipulate tool box

MicroStation prompts:

Rotate Element by 3 Points > Identify element *(Identify an element to rotate.)*

Rotate Element by 3 Points > Enter pivot point (point to rotate about) *(Place a data point or key-in coordinates to define the pivot point.)*

Rotate Element by 3 Points > Enter point to define start of rotation *(Place a data point or key-in coordinates to define the starting point of rotation.)*

Rotate Element by 3 Points > Enter point to define amount of rotation *(Place a data point or key-in coordinates to define the amount of rotation.)*

Note: *If the* **Copies** *check box is ON, the word "Copy" is added to the MicroStation prompts. For example, "Rotate Element (Copy) > Identify element."*

MIRROR ELEMENT

The Mirror Element tool creates a mirror (backward) image of an element. Options in the Tool Settings window control the way the tool operates.

To mirror the original element, set the **Copies** check box to OFF. To mirror a copy of the original element (and leave the original element unchanged), set the **Copies** check box to ON. If the element contains text to be mirrored, set the **Mirror Text** check box to ON. To mirroring a multi-line element with the offsets mirrored, set the **Mirror Multi-line Offsets** check box to ON.

Note: *The* **Mirror Text** *check box is useful for mirroring groups of elements that include text elements. Group manipulation tools are discussed in Chapter 6.*

The **Mirror About** menu mirrors the element about a **Horizontal** axis, about **Vertical** axis, or about an imaginary user-defined **Line**. Examples of the three mirror orientations are shown in Figure 4–49.

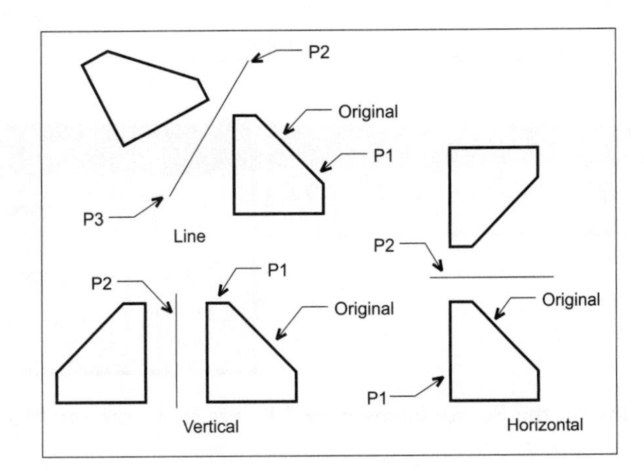

Figure 4–49 Examples of the mirroring an element using each of the Mirror About methods

Horizontal Method

To mirror an element about a horizontal axis, invoke the Mirror tool from:

Manipulate tool box	Select the Mirror tool. In the Tool Settings window, select Horizontal from the Mirror About menu. To mirror a copy of the element, turn the Copies check box ON (see Figure 4–50).
Keyboard Navigation (Main tool box active)	**3 5** (Select the Horizontal from the Mirror About menu. To mirror a copy of the original element, turn on the Copies check box.)

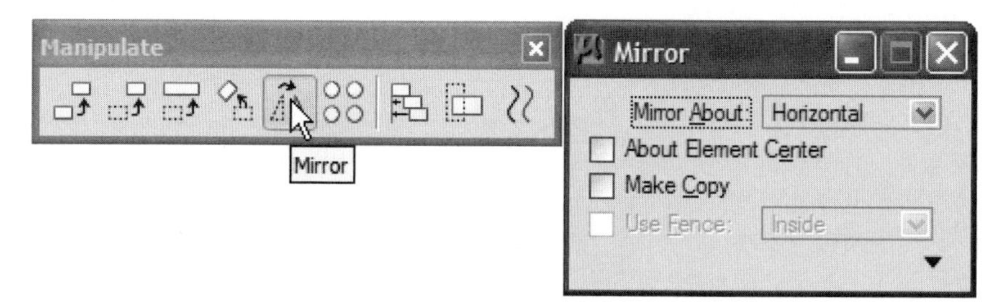

Figure 4–50 The Tool Settings window set to mirror an element about a horizontal axis

MicroStation prompts:

Mirror Element > Identify element *(Identify an element to mirror.)*

Mirror Element > Accept/Reject Selection *(Identify the distance of the horizontal axis from the element by placing a data point or keying in coordinates.)*

Note: *If the* ***Copies*** *check box is ON, the word "Original" is replaced by "Copy" in the MicroStation prompts. For example, "Mirror Element About Horizontal (Copy) > Identify element."*

Vertical Method

To mirror an element about a vertical axis, invoke the Mirror tool from:

Manipulate tool box	Select the Mirror tool. In the Tool Settings window, select Vertical from the Mirror About menu. To mirror a copy of the element, turn the Copies check box ON.
Keyboard Navigation (Main tool box active)	**3 5** (Select the Vertical from the Mirror About menu. To mirror a copy of the original element, turn on the Copies check box.)

MicroStation prompts:

Mirror Element > Identify element *(Identify an element to mirror.)*

Mirror Element About > Accept/Reject Selection *(Identify the distance of the vertical axis from the element by placing a data point or keying-in coordinates.)*

Note: *If the* **Copies** *check box is ON, the word "Original" is replaced by "Copy" in the MicroStation prompts. For example, "Mirror Element About Vertical (Copy) > Identify element."*

Line Method

The **Mirror About Line** method mirrors an element about an imaginary ser-defined line. It requires one more data point than the other mirroring methods. After selecting the element to be mirrored, two more data points are placed to define the location and angle of the line about which the element is to be mirrored, as shown in Figure 5-51

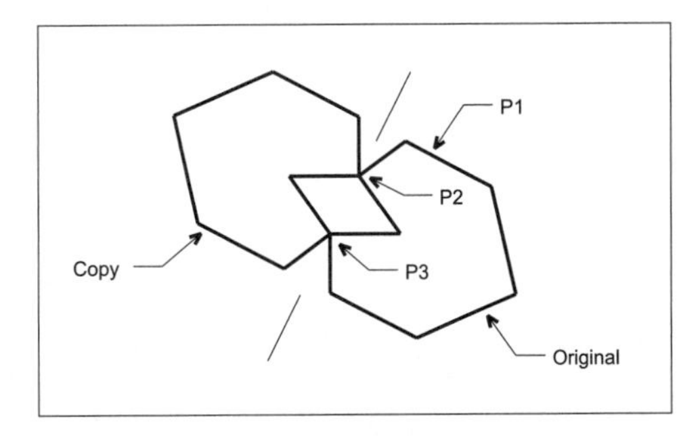

Figure 4–51 Example of mirroring a copied element about a line

To mirror an element about a line, invoke the Mirror tool from:

Manipulate tool box	Select the Mirror tool. In the Tool Settings window, select Line from the Mirror About menu. To mirror a copy of the element, turn the Copies check box ON.
Keyboard Navigation (Main tool box active)	**3 5** (Select the Line from the Mirror About menu. To mirror a copy of the original element, turn on the Copies check box.)

Note: *If the* **Copies** *check box is ON, the word "Original" is replaced by "Copy" in the "MicroStation prompts." For example, "Mirror Element About Line (Copy) > Identify element."*

Mirror Element > Identify element *(Identify an element to mirror.)*

Mirror Element > Enter first point on mirror line *(Place a data point or key-in coordinates to define one end of the line about which the element will be mirrored.)*

Mirror Element > Enter second point on mirror line *(Place a data point or key-in coordinates to define the angle of the line about which the element will be mirrored.)*

CONSTRUCT ARRAY

The Construct Array tool makes multiple copies of a selected element in a rectangular or polar array, as shown in Figure 4–52. Use the **Method** menu on the Tool Settings window to select the type of array and control the way the array is placed.

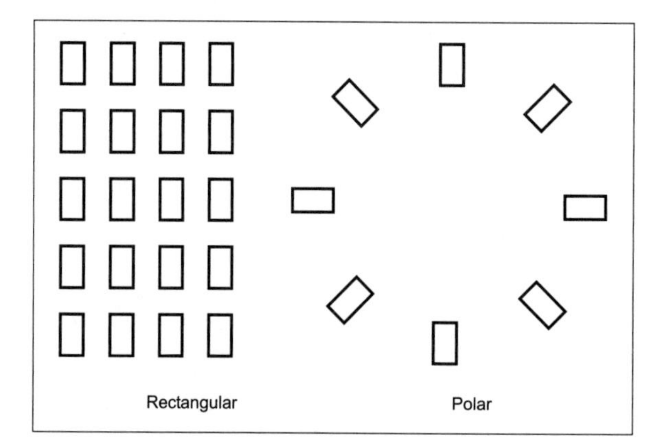

Figure 4–52 Example of a rectangular and a polar array

Rectangular Array

When **Rectangular** is selected from the **Method** menu, a set of options for controlling the rectangular array structure appears in the Tool Settings tool box:

- In the **Active Angle** text field, key-in the counter clockwise rotation angle of the array rows. Alternately, the up and down scroll arrows can be used to select a standard rotation angle, such as 180 or 270 degrees.
- In the **Rows** and **Columns** text fields, key-in the number of rows and columns to include in the array. The numbers include the row and column that contain the original element, so a one by one array does not place any copies of the selected element.
- In the **Row Spacing** and **Column Spacing** text fields, key-in the row and column spacing (in working units). The spacing values can be positive or negative. Positive row spacing builds the rows upward and a negative number builds them downward. Positive column spacing builds the columns to the right and negative row spacing builds the columns to the left.

Note: *The spacing is not between the copies of the element, but rather from a point on one element to the same point on adjacent elements in the array.*

Invoke the Construct Array tool from:

Manipulate tool box	Select the Construct Array tool. In the Tool Setting box, select the Rectangular Array Type and specify the angle, number of columns and rows, and the row and column spacing. (See Figure 4–53).
Keyboard Navigation (Main tool box active)	**3 6** (Select the Rectangular Array type and specify the angle, number of columns and rows, and the row and column spacing.)

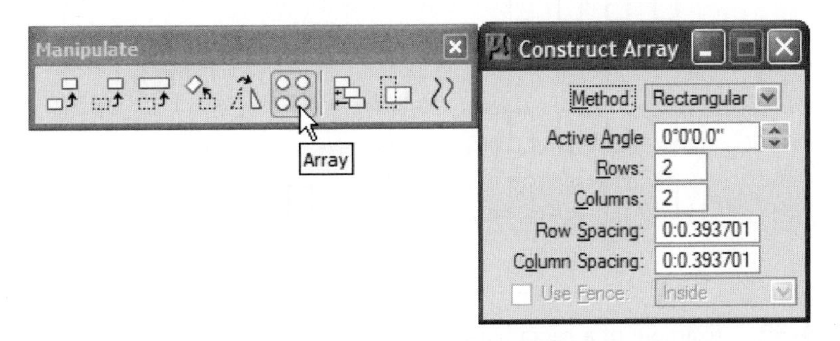

Figure 4–53 Invoke the Construct Array (Rectangular) tool from the Manipulate tool box

MicroStation prompts:

Array Element (Rectangular) > Identify element *(Identify an element to array.)*

Array Element (Rectangular) > Accept/Reject Selection *(Click the Accept button to place copies, or click the Reject button to disregard the selection.)*

Figure 4–54 shows an example of placing a rectangular array.

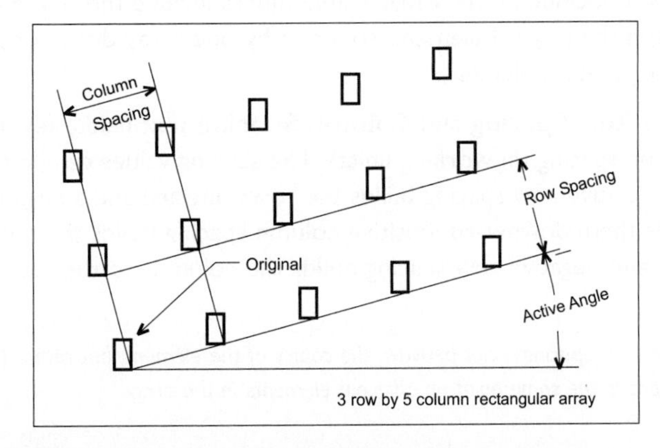

Figure 4–54 Example of placing a rectangular array

Polar Array

When **Polar** is selected from the **Method** menu, a set of options for controlling the polar array structure appear in the Tool Settings window:

- In the **Items** field, key in the number of items (elements) that are to be in the array. The number includes the original element (for example, if seven is keyed in, the polar array will consist of the original element and six copies of the element).
- In the **Delta Angle** field, key in the degrees of rotation between each element in the polar array. The array is constructed in a counter clockwise direction from the point on the element where it was identified to that same point on the each copy of the element.
- To copy each of the elements in the array to be rotated by the **Delta Angle**, turn the **Rotate Items** check box ON. Turn the check box OFF if all elements in the array are to be placed at the same orientation as the original element. Figure 4–55 shows difference between rotating and not rotating the array elements.

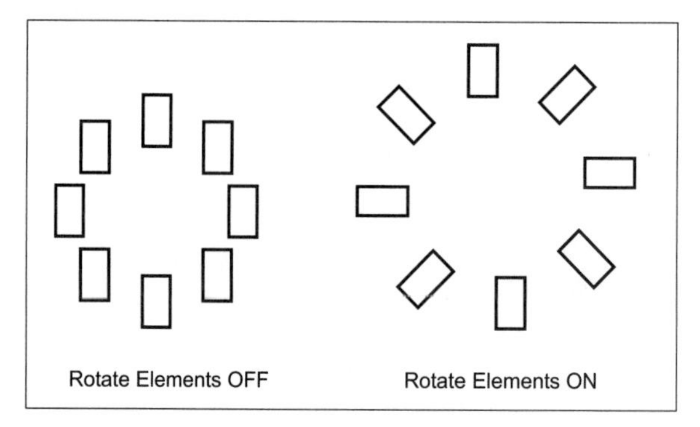

Figure 4–55 Example of rotating and not rotating the polar array elements

Invoke the Construct Array tool from:

Manipulate tool box	Select the Construct Array tool. In the Tool Setting box, select the Polar Array Type and specify the number of Items and Delta Angle (see Figure 4–56).
Keyboard Navigation (Main tool box active)	**3 6** (Select the Polar Array type and specify the number of Items and Delta Angle.)

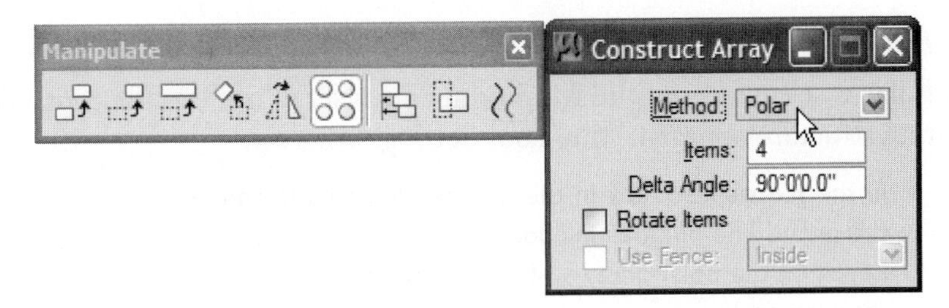

Figure 4–56 Invoke the Construct Array (Polar) tool from the Manipulate tool box

MicroStation prompts:

Array Element (Polar) > Identify element *(Identify the element to array.)*

Array Element (Polar) > Accept/Reject Selection *(Specify the center point for the array by pressing the Data button or by keying-in coordinates, or click the Reject button to cancel the selection.)*

The following command sequence shows an example of using the Construct Array tool to place eight, evenly spaced bolt holes in a pipe flange (The flange and center lines were already drawn). Figure 4–57 shows the flange before and after the polar array was constructed.

After selecting the tool and the **Polar Array** Type, the **Items** are set to eight and the **Delta Angle** to 45 degrees. MicroStation prompts:

Array Element (Polar) > Identify element *(Identify the shape.)*

Array Element (Polar) > Accept, select Center/Reject *(Place a tentative point at the center of the circle and accept it.)*

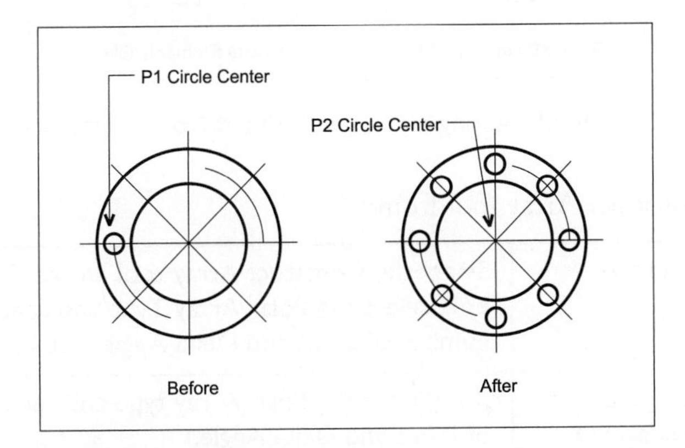

Figure 4–57 Example of placing a polar array

TEXT PLACEMENT

When drawings are created by hand on paper, adding text such as design component specifications and shop and fabrication notes is a time-consuming, tedious process. MicroStation provides several text placement tools that greatly reduce the time and tedium of text placement.

MicroStation provides tools to place text, create "fill-in-the-blank" forms, import text from other applications, edit text, and change the appearance of text. In this chapter, we describe several text placement tools and settings. Additional text tools and settings are described later in Chapter 7.

Text is placed by typing it into the Text Editor window and then defining the location(s) where the text should be placed in the design. When placed, all the keyed-in text becomes one text element called either a "text string" or a "text node." The element is a text string when all the text is in one line. The element is a text node when the text is placed as more than one line by pressing (ENTER) while keying in the text.

PLACE TEXT

The Place Text tool provides several text placement methods and the associated Tool Settings window provides settings that control the appearance of the text and where it is placed, in relation to the placement data point. Invoke the Place Text tool from:

Task Navigation tool box (active task set to Text)	Select the Place Text tool. MicroStation displays the Text Editor word Processor window, as shown in Figure 4–58.
Keyboard Navigation (Text tool box active)	q

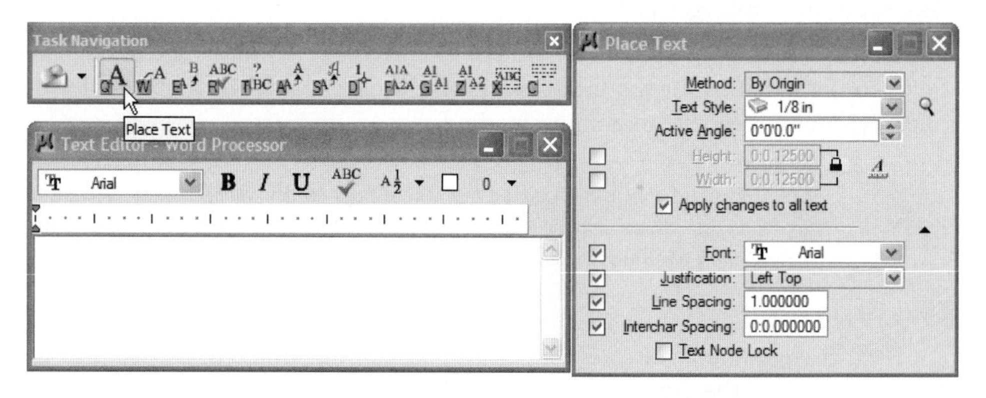

Figure 4–58 Invoking the Place Text tool from the Text tool box

MicroStation prompts:

> Place Text > Enter more characters or position text *(Make the required text settings on the Tool Settings and Text Editor – Word Processor window. Type the text in the Text Editor window and place a data point in the design to define the text placement point. Continue defining points where additional copies of the text should be placed, or click the Reset button to drop the current text string. Details on these steps are provided in the following topics.)*

The Tool Settings window provides settings for controlling the way text is placed and its appearance. The Text Editor window provides the text field in which to type the text and additional settings for controlling the appearance of the text.

TOOL SETTINGS WINDOW

Settings on the Tool Settings window control style, placement method, angle, size, font, line spacing, interchar spacing, and justification. Additional tools can be accessed by clicking the expansion arrow in the lower right corner of the window, as shown in Figure 4–59.

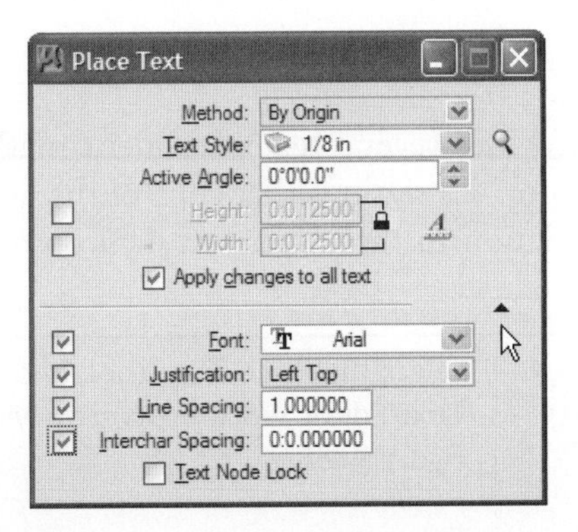

Figure 4–59 Expanding the Place Text Tool Settings window

Methods for Placing Text

The **Method** menu provides nine methods for placing text. In this section detailed discussion is provided for the first method, **By Origin**. Refer to Chapter 7 for detailed explanation for the remaining methods.

Note: *The set of options available on the Tool Settings window changes when different text placement methods are selected. The settings discussed here are the options available when the* ***By Origin Method*** *is selected.*

Text Style

The Text **Style** menu offers pre-defined text styles from which to choose. For this discussion, select the **Style (none)** option. (Text style creation is described in Chapter 7.)

Active Angle

The **Active Angle** field rotates the text when it is placed. The default rotation angle is zero degrees. The angle may be set by keying-in the degrees of rotation, or by using the up and down scrolling arrows to select standard rotation angles, such as 90, 180, and 270 degrees. Entering a positive number will rotate the text in a counter-clockwise direction, while a negative number will rotate the text in a clockwise direction. Figure 4–60 shows examples of various rotation angles.

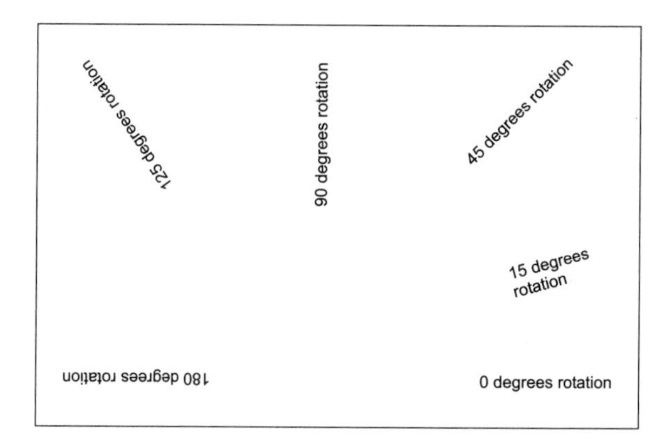

Figure 4–60 Examples of placing text at various rotation angles

Note: *The active angle that is set will become the new default value for subsequent commands that use the active angle. If necessary, reset the active angle to zero degrees.*

Line Spacing and Inter-character Spacing

The **Line Spacing** and **Interchar Spacing** text fields are not available with the **Place Text By Origin Method**. Tools that use these settings are described in Chapter 7.

Text Size

The **Height** and **Width** text fields control the size of the text placed it in the design. Enter values in these fields in working units. For example, if the Master Unit is inches, enter 1.5 or 1:5 to set the text height to 1.5 inches.

To the right of the **Height** and **Width** fields is a picture of a lock. If the lock is closed, the two fields will be forced to the same value: Key-in a value in one field and the other fields is automatically set equal to the value keyed in. If the lock is open, the **Height** and **Width** can be different values.

When drawing an unscaled schematic, or if the design will be printed full size, selecting a text size is simple—key-in the actual size the text will be when printed.

If the design must be scaled when printed, selecting a text size is a little more complicated. As mentioned earlier, in MicroStation objects are drawn full size, and MicroStation is told what scale to use when it plots the design to paper. MicroStation scales everything in the design to fit the size of paper chosen, including the text. Therefore the text must be scaled by the *inverse* of the print scale so it will be the correct size on paper.

For example, if the design will be printed at 1″ = 10′ and the text size must be 0.1 inch on paper, the text height in the design must be 1 foot. Below is a formula that can be used to calculate this value:

Text height in design = design scale ÷ plot scale × printed text size

(where all values are in the same measurement units, such as inches or centimeters).

For example, to output 1/4″ tall text on paper when plotting at 1/8″ = 1′ using a design scale of one foot or 12 inches and a plot scale of 1/8 inch, the formula yields a text height of 24-inch text, as shown below.

24″ = 12″ ÷ 1/8″ × 1/4″

Text Font

The **Font** menu provides character sets of different typeface designs. To display the menu, click the drop-down arrow in the right side of the field. Use the scroll bar on the right side of the menu to scroll through the font names and click the name to select it, as shown in Figure 4–61.

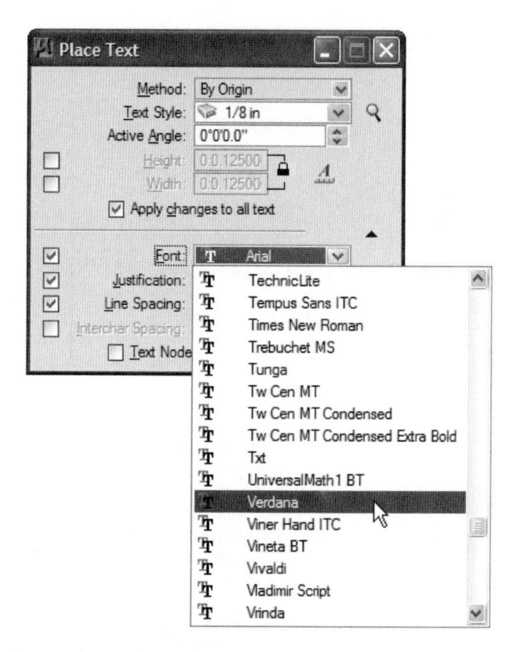

Figure 4–61 The Fonts drop-down menu

This menu can contain the names of up to 255 internal MicroStation fonts (such as Engineering or Architectural) and fonts that are in the operating system font library on your workstation (such as Arial and Times New Roman). Only a few of the internal fonts are normally installed in MicroStation. The number of system fonts varies.

***Note:** This method of selecting a font does not show a preview of the typeface that the font provides, so results will not be visible until the text is placed in the design. Chapter 7 describes a method for defining sets of text styles that includes a way to preview the typeface of each font.*

Justification

The **Justification** menu provides options that control where the data appears in relation to the location of the data points defined to place the text. Display the menu by clicking the drop-down arrow in the **Justification** field and then click the option, as shown in Figure 4–62.

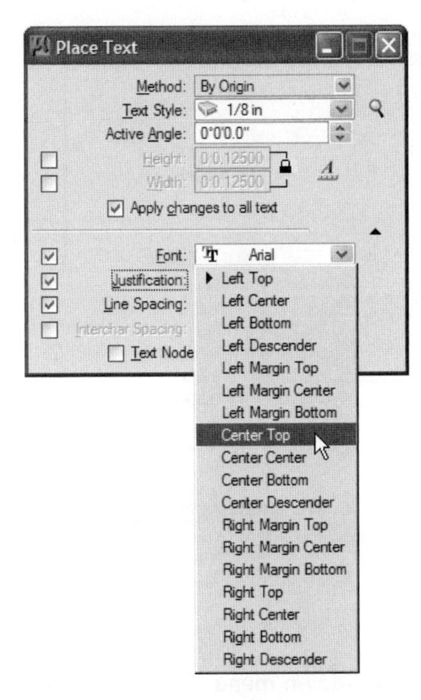

Figure 4–62 The Justification drop-down menu

The name of the justification option describes the location of the data point in relation to the text, as shown in Figure 4–63.

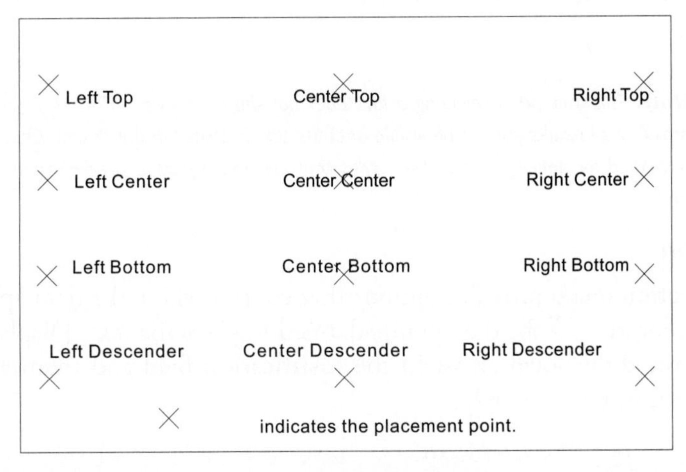

Figure 4–63 Examples of data point location for various text justifications

Text Editor Window

The Text Editor window includes a settings bar that provides additional text appearance settings and a field for keying-in the text string to place in the design.

Font

On the left end of the bar is the same Font menu that is available in the Tool Settings window when the Place Text tool is invoked. To change the selected font, click the drop-down arrow and select a font name from the menu.

Bold, Italics, and Underline

When these settings are available text can be placed using the Bold, Italic, and Underline versions of the selected font.

If there is no version of a font for one of these settings, the button is unavailable. For example, there is no bold version of the MicroStation fonts, so when a MicroStation font is selected, the Bold button cannot be selected.

Spelling

Click the **Spelling** icon on the settings bar to invoke the MicroStation spelling check utility. Each time the utility finds a spelling error, in the text currently keyed into the text field of the Text Editor window, it displays the Spell Checker window. The Spell Checker is described in chapter 7.

Stacked Fractions

Stacked fractions are discussed in Chapter 7.

Color

When first invoked the Place Text tool, the text color is set to the Active Color that is currently selected in the Attributes tool box. To change the text color without changing the Active Color, click the drop-down arrow next to the **Color** button to display the Color palette and then select a color from the palette.

Note: *When using one of the MicroStation internal fonts, the Active Line Weight is applied to the text. If a heavy line weight is selected, the lines and curves that make up the text string are wider. True Type fonts do not use the line weight. To change the Active Line Weight, select it in the Attributes tool box.*

PLACE TEXT BY ORIGIN

Using the settings discussed above to control the appearance of the text start placing text in the design. In this chapter text is placed using the **By Origin Method**.

1. Invoke the Place Text tool and select **By Origin** from the **Method** menu in the Tool Settings window.
2. In the Tool Settings and Text Editor windows, select the other settings needed for the text about to placed.
3. Click in the text field of the Text Editor window and type the text string. To place the text string on multiple lines, press (ENTER) at any point in the text string to start a new line.

Notes: The Text Editor window can be resized, if needed, to provide a larger area in which to type the text. To resize it, point to the box border, press and hold the Data button while drag the box to the new size, and then release the Data button.

If ENTER is not pressed, as when creating multiple lines, the text will wrap to a new line at the right edge of the Text Editor window, but all the text will be on one line when placed in your design.

4. After typing the text, move the pointer onto the design surface and the pointer drags a dynamic image of the text.
5. To place a copy of the text string in the design, move the pointer to the desired location and click the Data button.
6. To place additional copies of the same text string, continue placing data points at each location.
7. To edit or replace the text string at any time during the placement process, click in the Text Editor window and edit the text. When the pointer is moved back onto the design surface, it will drag a dynamic image of the new text string.
8. To drop the current text string and clear the Text Editor window, click the Reset button.

Notes: To change any of the text settings, first drop the current text string from the pointer by clicking the Reset button. Changes to settings usually do not go into effect while there is a dynamic image of a text string on the pointer.

Each copy of the text placed becomes a single element that can be manipulated like any other element. The only key point in a text string or multi-line element is the placement point.

KEYBOARD SHORTCUTS FOR THE TEXT EDITOR WINDOW

The keyboard shortcuts listed in Table 4–1 move the text pointer to specific positions in the current line and between lines when the box contains more than one line of text.

Table 4–1 Positioning the Text Cursor

PRESS:	TO MOVE THE TEXT CURSOR:
LEFT	Left one character.
RIGHT	Right one character.
CTRL + LEFT	Left one word.
CTRL + RIGHT	Right one word.
HOME	To the beginning of the current text line.
END	To the end of the current text line.

PRESS:	TO MOVE THE TEXT CURSOR:
UP	Up to the previous line of text.
DOWN	Down to the next line of text.
PAGE UP	Straight up into the first text line.
PAGE DOWN	Straight down into the last text line.

The key-ins described in Table 4–2 delete characters from the text in the Text Editor box.

Table 4–2 Keys that Delete Text

PRESS:	TO DELETE:
BACKSPACE	The character to the left of the text cursor.
DELETE	The character to the right of the text cursor.
ALT + DELETE	All characters from the text cursor to the end of the word.
CTRL + DELETE	All characters from the text cursor to the end of the current word.
CTRL + BACKSPACE	All characters from the text cursor to the beginning of the current word.

The key-ins described in Table 4–3 select or deselect text in the Text Editor box. Selected text is shown with a dark background, and it can be moved, copied, or deleted.

Table 4–3 Selecting Text with Key-Ins

PRESS:	TO SELECT (OR DESELECT IF ALREADY SELECTED):
SHIFT + LEFT	The character to the left of the text cursor.
SHIFT + RIGHT	The character to the right of the text cursor.
CTRL + SHIFT + LEFT	The characters from the text cursor to the start of a word.
CTRL + SHIFT + RIGHT	The characters from the text cursor to the end of a word.
SHIFT + END	The characters from the text cursor to the end of a line.
SHIFT + HOME	The characters from the text cursor to the start of a line.
CTRL + a	To select all text in the Text Editor box.
LEFT or RIGHT	To deselect all previously selected text.

The pointing device actions described in Table 4–4 select or deselect text in the Text Editor box.

Table 4–4 Selecting Text with the Pointing Device

POINTING DEVICE ACTION	RESULT
Press the Data button and drag the screen cursor across the text	Selects all the text dragged across.
Double-click the Data button	Selects the word the cursor is in.
Triple-click the Data button	Selects all text in the Text Editor window.
Hold down SHIFT + Data button	Adds more text to the text already selected and drag across the text.
Click the Data button in an area	Deselects all previously selected text where there is no text.

The actions that replace, delete, and copy previously selected text are shown in Table 4–5.

Table 4–5 Replacing, Deleting, and Copying Selected Text

ACTION	RESULT
Start typing characters	Replaces the selected text with the text typed.
Press BackSpace	Deletes all the selected text.
Press DELETE	Deletes all the selected text.
Press CTRL + INSERT	Copies the selected text to a buffer.
Press SHIFT + INSERT	Pastes the previously copied or deleted text at the text cursor position.

Open the Exercise Manual PDF file for Chapter 4 on the accompanying CD for project and discipline-specific exercises.

REVIEW QUESTIONS

Write your answers in the spaces provided.

1. The Place Polygon tool places polygons that can have a maximum of ____________ sides.
2. The Points Method for the Place Point or Stream Curve tool is used to place a curve string by

   ____.

3. The Stream Method of the Place Point or Stream Curve tool is used to place a curve string by
   ______________________________.
4. The Multi-line tool allows placing up to ________ separate lines of various ____________, ____________, and ____________ with a single tool.
5. The Place Fillet tool joins two lines, adjacent segments of a line string, arcs, or circles with an ____________ of a specified radius.
6. Name the three methods for constructing a fillet: ____________, ____________, and ____________.
7. The Chamfer tool allows drawing a ________________ instead of an arc.
8. The purpose of the Trim tool is to
   ______________________________.
9. What is the name of the tool that deletes part of an element?
   ____
10. Name the two categories of manipulation tools available in MicroStation.
    ______________________ and
    ______________________.
11. The Copy tool is similar to the Move tool, but it ________________.
12. Name at least three element manipulation tools available in MicroStation:
    ____________, ____________, and ____________
13. To rotate an element by a keyed in angle, select the ________________ method.
14. The Array tool makes multiple copies of a selected element in either a ________________ or ________________ array.
15. List four of the settings that control the shape and size of a rectangular array:
    ____________, ____________, ____________, and ____________.
16. How are additional settings displayed on the Place Text Tool Settings window?
    ______________________.
17. Name three text settings that can be changed in the Place Text Tool Settings window:
    ____________, ____________, and ____________.
18. Fonts can be selected from the ________________ and ________________windows.
19. A new Height value typed in the Place Text Tool Settings window makes the Width value change to be equal to the Height. What caused that to happen?
    ____

20. To place a text element that must be divided into two lines of text, what must be done while typing the text?

____

AccuDraw and SmartLine

OBJECTIVES

Topics explored in this chapter:

- Setting up AccuDraw
- Using AccuDraw to place elements with fewer data points and less typing
- Using SmartLine to draw complex models quickly with one tool

GETTING TO KNOW ACCUDRAW

AccuDraw is a powerful MicroStation feature that increases drawing productivity by tracking previous drawing steps and anticipating the next steps.

When AccuDraw is active, the AccuDraw settings box appears on the MicroStation desktop and the AccuDraw compass appears on data points. The AccuDraw settings box provides a faster and easier way to enter precise coordinates than the key-in method, and the compass helps place data points accurately when using the pointing device. The AccuDraw Settings box controls the way AccuDraw interacts with the design. Figure 5–1 shows the boxes and compass.

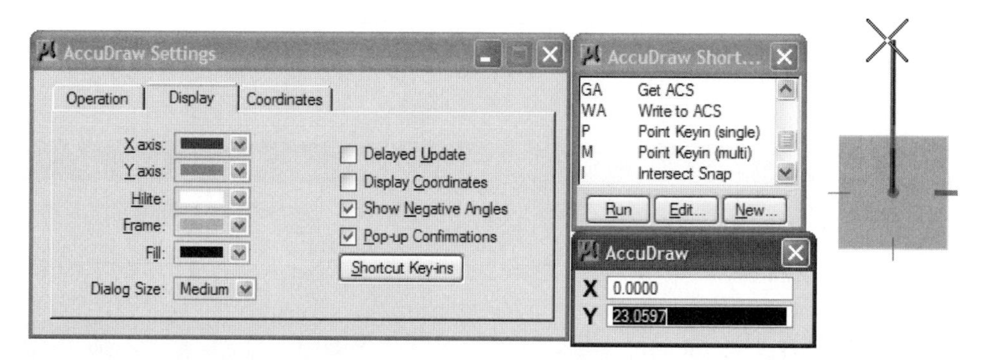

Figure 5–1 AccuDraw tools

START AND STOP ACCUDRAW

Opening a design file in MicroStation starts AccuDraw but there are options to stop AccuDraw while working in a design file.

Stop and Start AccuDraw from:

Primary tool box	Select the Start AccuDraw tool (see Figure 5–2).
Key-in window	**accudraw activate** (or **a a**) (ENTER)

Figure 5–2 Invoking the AccuDraw tool from the Primary tool box

Note: *A check box on the AccuDraw Settings box* ***Operation*** *tab enables/disables the automatic start of AccuDraw when design files open (discussed later in this chapter).*

Stopping AccuDraw removes the AccuDraw settings box compass from the MicroStation desktop. Starting AccuDraw opens the AccuDraw settings box and, and placing the next data point displays the compass.

The AccuDraw settings box initially appears floating, but it can be docked to the top or bottom edges of the MicroStation desktop. Figure 5–3a shows the box floating and Figure 5–3b shows the box docked on the top edge of the desktop.

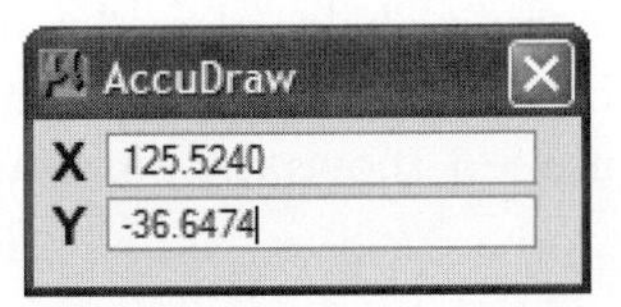

Figure 5–3a AccuDraw settings box floating in the View window

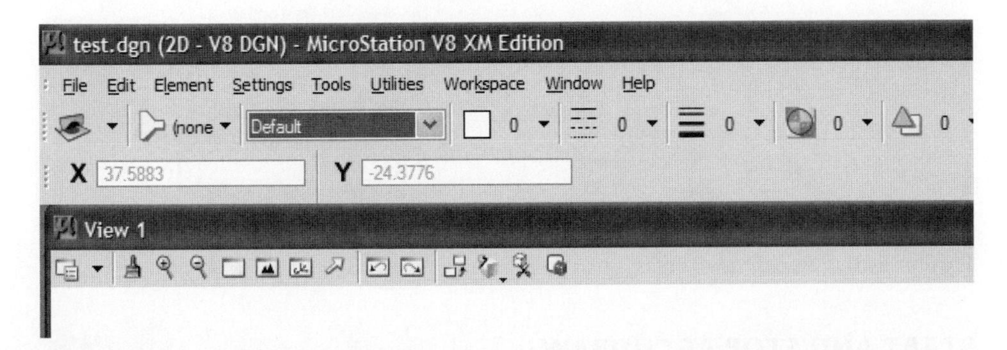

Figure 5–3b AccuDraw settings box docked at the top of the MicroStation application window

KEY-IN SHORTCUTS

The AccuDraw tools include one and two-character shortcut key-ins that control AccuDraw actions and select AccuDraw features. When focus is on the AccuDraw settings box, MicroStation checks all keyed in characters to see if they are AccuDraw shortcuts.

THE ACCUDRAW COMPASS

When AccuDraw is active, the AccuDraw compass appears on data points placed in the design and on the selection point of elements for manipulation. The compass is the center of the AccuDraw plane and is the focus for input.

Coordinate System

The AccuDraw drawing plane includes rectangular and polar coordinate systems for locating points. The systems are the same as those provided by the precision key-in tools, except that all values displayed in the AccuDraw coordinate systems are from the AccuDraw compass origin, not the design plane's origin point.

The rectangular coordinate system has fields, labeled **X** and **Y**, that display the rectangular offset of the pointer from the compass center (See Figure 5–4a).

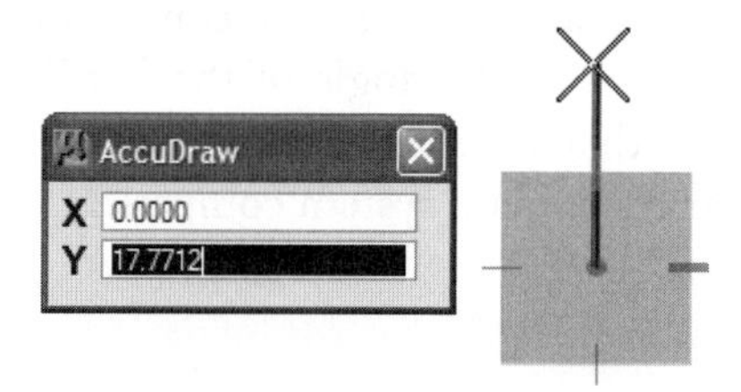

Figure 5–4a Compass with the rectangular coordinate system active

The polar coordinate system has fields that display the distance and rotation of the pointer from the compass center. A linear dimension icon indicates the distance field and an angular dimension icon indicates the angle field (see Figure 5–4b).

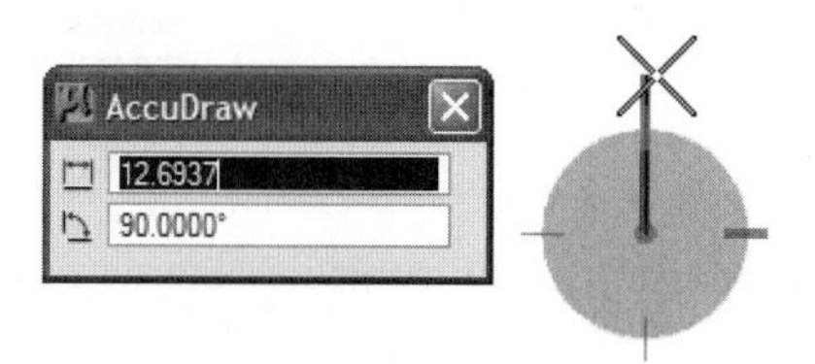

Figure 5–4b Compass with the polar coordinate system active

To switch from one coordinate system to the other, use one of these methods:

Keyboard shortcut	SPACEBAR (when focus is on the AccuDraw settings box)
Key-in window	**accudraw mode** (or **a m**) (ENTER)
AccuDraw Settings box	On the **Coordinates** tab, select **Polar** or **Rectangular** from the **Type** menu.

The coordinate system is switched (there are no MicroStation prompts).

Orthogonal Axes

A dot in the center of the compass indicates the center of the AccuDraw design plan. Short tick marks (lines) on the compass rectangle or circle indicate the orientation of the AccuDraw drawing plane *X-axis* and *Y-axis*.

To aid in distinguishing the two axes, the positive *X-axis* tick mark is a red line and the positive *Y-axis* tick mark is a green line. The AccuDraw Settings box (discussed later in this chapter) has options for changing the colors.

Compass Movement

If the **Floating Origin** check box (discussed later in this chapter) is ON, the compass moves to and centers on each new data point. If the **Context Sensitivity** check box (discussed later in this chapter) is on, the compass also rotates so that its *X-axis* (or zero rotation) parallel to the angle of the last line segment drawn and is pointing to the data point that completed the segment. Figure 5–5 shows an example of the Rectangular coordinate system compass at the end of a line drawn at 45 degrees of rotation.

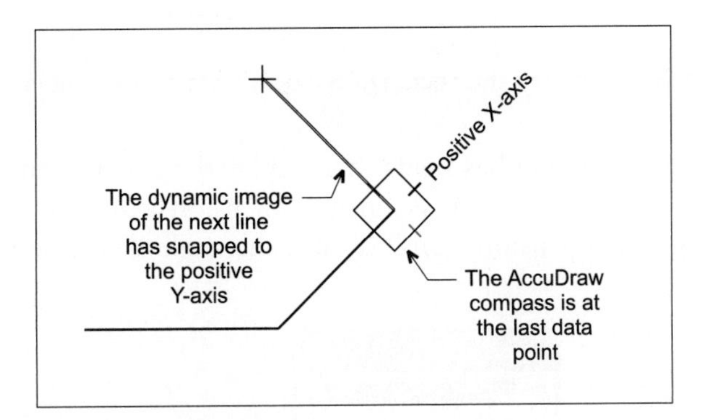

Figure 5–5 Example of the AccuDraw compass rotated to match the rotation of the last line placed in the design

Key-in any of the following one or two letter shortcuts to change the position and rotation of the compass:

- **O** (Set Origin) moves the compass from its current location to the position of the pointer.

- **T** (Top Rotation) sets the compass rotation to zero degrees (the red positive *X-axis* tick mark points to the right).
- **RZ** (Rotate about Z) rotates the compass 90 degrees in a counterclockwise direction.
- **RQ** (Rotate Quick) allows dynamically setting the compass rotation by moving the pointer. The compass follows the pointer by rotating around its center point until the Data button is clicked.

The AccuDraw Settings box has options to lock the compass at its current location so it no longer jumps to the last data point. Also, the compass can be locked at its current rotation, so it does not assume the rotation angle of the last element segment placed.

Axis Indexing

If the **Indexing** Axis check box (discussed later in this chapter) is on, AccuDraw tracks the pointer location in relation to the compass *X*- and *Y*-axes for rectangular coordinates and in relation to the degrees of rotation for polar coordinates. AccuDraw indexes (snaps) to the axis when the pointer is almost lined up with the *X* or *Y-axis* (or at 90 degree increments). Figure 5–6 shows an example of the pointer indexed to the *X-axis* while placing a circle by center (indexing is indicated by the wide line extending from the center of the compass to the pointer).

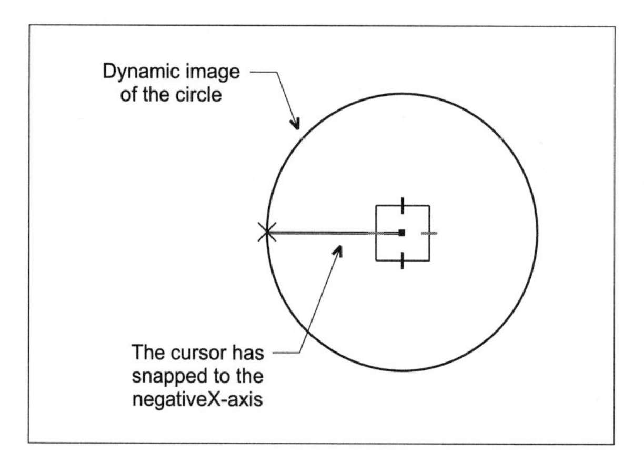

Figure 5–6 The AccuDraw compass indexed to the X-axis

Indexing makes it easy to use the pointer to place a new element in the same direction or at a right angle to the previous element. Indexing does not lock the pointer to the axis and the pointer can be moved away from the axis after it snaps to the axis.

THE ACCUDRAW SETTINGS BOX

The AccuDraw settings box provides a place to key-in *X*- and *Y-axis* numeric values in the Rectangular coordinate system and Distance and Angle offsets in the Polar coordinate system (see Figures 5–4a and 5–4b). This box almost eliminates the need to key in precision input codes such as **DL** and **DI**.

As the screen pointer moves, the coordinate box input fields update automatically with the pointer's position, relative to the current AccuDraw origin.

Note: *For 3D designs, the Rectangular coordinate system also includes the Z-axis.*

Selecting Fields

If the **Smart Key-ins** check box is ON, AccuDraw focuses on the AccuDraw settings box field that is most likely to be used next, so there is often no need to select the field first. A change in the appearance of the field indicates focus. In rectangular mode, the axis closest to the direction of the dynamic image of the line from the center of the AccuDraw axis and the pointer is in focus. Thus, if the dynamic image is close to the *X*-axis, the **X** field has focus. In polar mode, the distance field is always in focus.

If the wrong edit field has focus, press TAB to move focus to the other edit field. Keying characters into a field replaces the characters currently in the field. The up and down arrow keys also switch focus between the two AccuDraw settings box fields. Focus cannot be switch between the fields if either field is locked (discussed later in this chapter).

Keyed-in characters appear in the field that has focus and lock the field until the next data point is placed. For rectangular coordinates, the pointer locks on an imaginary line that is perpendicular to the locked axis at the keyed in distance from the axis. For polar coordinates, a lock on the distance field forces the next data point to be on an imaginary circle whose center is the AccuDraw compass center and whose radius equal to the keyed in value. In both cases, the next data point is forced to be somewhere on the imaginary line or circle even if the screen point is moved away from the imaginary line.

Figure 5–7a shows the rectangular compass with a locked *X*-axis during placement of a line element. The imaginary locking line appears on the screen as a dashed line perpendicular to the locked axis. The dynamic image of the line element extends from the compass center to a point on the imaginary locking line. The screen pointer only moves the dynamic image of the line element along the locking line. Note that the *Y*-axis field has focus because it is still unlocked.

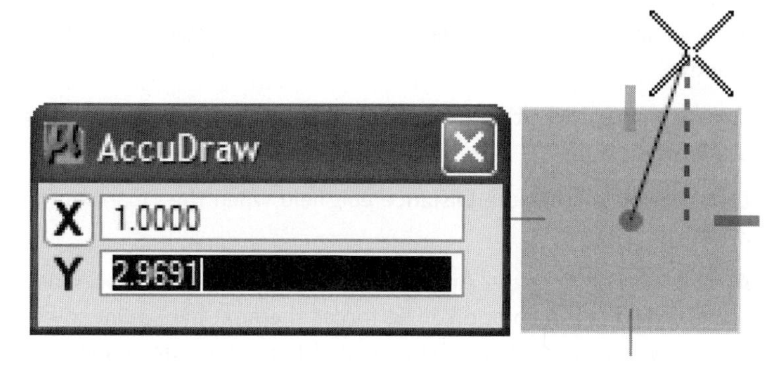

Figure 5–7a The AccuDraw rectangular compass with the X-axis locked at I master unit

Figure 5–7b shows the polar compass with a locked distance during placement of a line element. The imaginary locking circle does not appear on the screen. The dynamic image of the line element extends from the compass center to a point on the imaginary locking circle. The screen pointer only moves the dynamic image of the line element around the locking circle. Note that the angle field has focus because it is still unlocked.

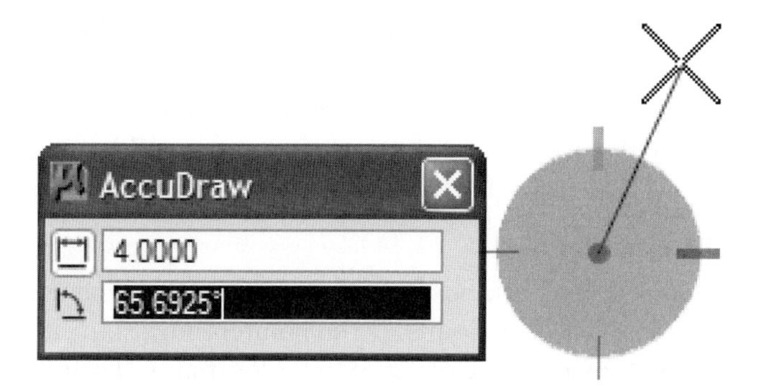

Figure 5–7b The AccuDraw polar compass with the distance locked at 4.0000 master units

Accepting Field Contents

When the AccuDraw settings box fields contain the correct offset from the compass origin, click the Data button to place the data point at the offset values. Placing the data point unlocks all locked AccuDraw fields.

Manually Locking Fields

As discussed earlier, keying characters into an AccuDraw field locks the field and placing the next data point unlocks the locked fields. There are key-in shortcuts that also lock and unlock fields.

- **X** locks and unlocks the X edit field when the Rectangular coordinate system is active.
- **Y** locks and unlocks the Y edit field when the Rectangular coordinate system is active.
- **D** locks and unlocks the Distance edit field when the Polar coordinate system is active.
- **A** locks and unlocks the Angle edit field when the Polar coordinate system is active. Locking the rotation angle forces the next data point to be on an imaginary line drawn from the center of the polar compass at the current rotation angle.
- **L** locks and unlocks indexing for both coordinate systems. When indexing is unlocked, AccuDraw does not indicate snapping the axes or 90-degree rotation increments. When the lock is on, AccuDraw indexes as described earlier in this chapter.
- Pressing the right or left arrow key locks the axis closest to the dynamic image of the line from the compass center to the pointer.

Negative Distances

If the **Smart Key-ins** check box (discussed later in this chapter) is ON, the direction of the screen pointer from the compass origin indicates the direction of input for the rectangular compass, so there is usually no need to key-in the negative sign for distances that are to the left of or down from the AccuDraw origin. Move the screen pointer near the area where the next data point will be placed to establish the correct sign.

Indexing Distance

If the **Indexing Distance** check box is ON, AccuDraw remembers the linear distance between the last two data points. If the pointer indexes to either Rectangular coordinate axis, a tick mark at the screen pointer location indicates when the linear distance from the last data point to the current pointer location is equal to the linear distance between the last two data points. For Polar coordinates, the tick mark appears at all degrees of rotation. Figure 5–8 shows the tick mark at the pointer position for rectangular coordinates.

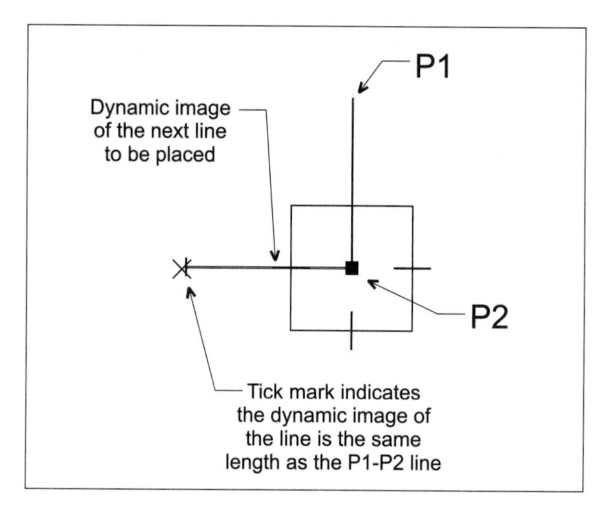

Figure 5–8 Current distance is equal to previous distance, as indicated by tick mark at pointer position

Recalling Previous Values

AccuDraw remembers the offsets used to place each data point. All distance and angle values are stored in a buffer. Press PAGE UP to load the last value used in the AccuDraw field that is in focus. Press PAGE UP again and the next-to-last value is loaded. Continue pressing PAGE UP to load more saved values into the field.

THE POPUP CALCULATOR

MicroStation provides a calculator that works with AccuDraw to do calculations using the values in the AccuDraw settings box fields. To use the calculator on the value in the field with focus:

1. Key-in a symbol for the required mathematical operation: + (add), – (subtract), * (multiply),/ (divide), or = (equals).
2. Key-in a value to complete the calculation.
3. Place a data point to accept the calculation, or click ESC to reject the calculation.

Keying in a mathematical symbol opens the calculator box below the AccuDraw field that has focus. The calculation result appears at the bottom of the calculator box. Figure 5–9 shows an example of using the calculator to multiply the contents of the *X* field by two.

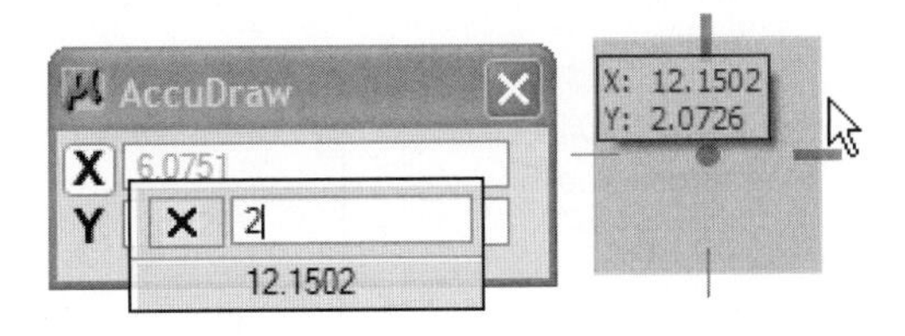

Figure 5–9 The calculator box

Replace Current Value

Starting the calculation with an equal sign replaces the existing AccuDraw value rather than operating on it. For example, in Figure 5–10, the equal sign was typed while focus was on the Y field, 2 times (*) 3.5 was typed in the calculator, and the calculation result (7.0) was placed in the Y field. The original contents of the Y field were not used.

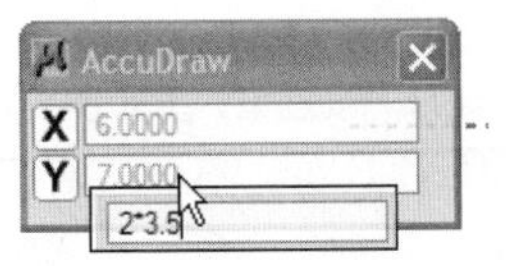

Figure 5–10 The calculator box when typing begins with an equal sign (=)

Complex Calculations

Use parentheses to create complex calculations. AccuDraw calculates functions enclosed in parenthesis first.

Figure 5–11a shows that keying in **=10*(3+4)** from the **Y** field places 70 in the field (3 and 4 are added and the result is multiplied by ten).

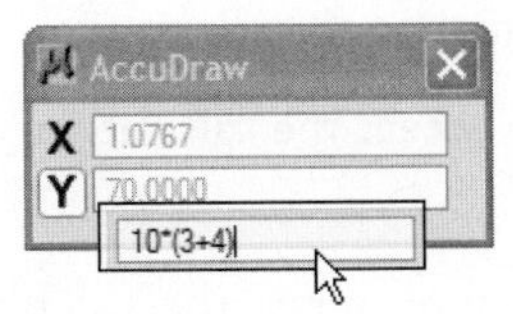

Figure 5–11a The use of parentheses in a calculation

Figure 5–11b shows that keying in **=sin(45)** from the Angle field places 0.7071 in the field.

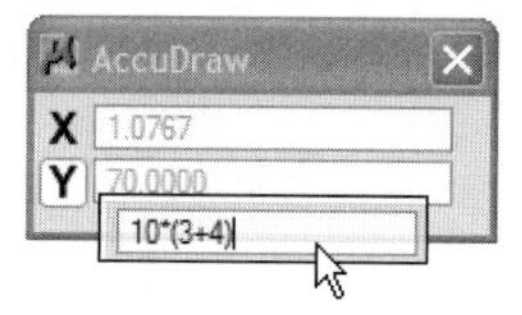

Figure 5–11b The use of parentheses in a calculation

SMART LOCK

Smart Lock is an AccuDraw key-in shortcut that allows for constraining the next data point to the nearest axis. To use Smart Lock to constrain the placement of the next data point, move the pointer near to the desired constraining axis and direction and press ENTER. If the rectangular compass is active, it locks the X or **Y-axis.** If the polar compass is active, it locks the **Angle** at 0, 90, 180, or -90 (270) degrees.

To remove the constraint from the next data point, press ENTER again.

CHANGE ACCUDRAW SETTINGS

Use the AccuDraw settings box to change the appearance of the compass and the way AccuDraw works. The settings box contains three tabs: **Operation, Display,** and **Coordinates.**

Invoke the AccuDraw settings box from:

Pull-down menu	Settings > AccuDraw
Key-in shortcut	**GS** (with focus on the AccuDraw settings box)
Key-in window	**accudraw dialog settings** (or **a d se**) (ENTER)

MicroStation opens the AccuDraw settings box. Following are descriptions of the options available on each tab.

OPERATION

The **Operation** tab provides check boxes and a menu that change the way AccuDraw performs, as shown in Figure 5–12.

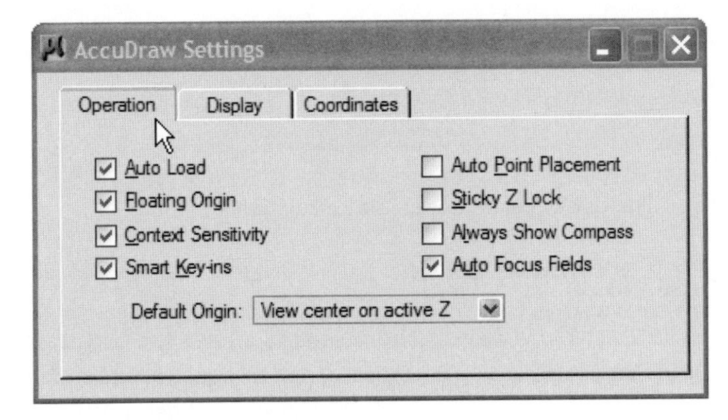

Figure 5–12 The AccuDraw settings box Operation tab

Auto Load

If the Auto Load check box is ON, AccuDraw starts automatically when a design file opens in MicroStation. Turning the **Auto Load** check box OFF prevents AccuDraw from starting when any design file opens unless the design file had AccuDraw on when **Save Settings** was selected in a previous editing session.

Floating Origin

When the **Floating Origin** check box is ON, the AccuDraw compass always moves to the location of the last data point. If the check box is OFF, the compass stays at its current location and does not jump to the location of new data points.

Context Sensitivity

When the **Context Sensitivity** check box is ON, tools can override AccuDraw default behavior to insure smoother operation of AccuDraw with the tool. For example, when placing lines or line segments, the AccuDraw compass rotates to the angle of the last placed line or line segment. If the last line or line segment was rotated 45 degrees, the rectangular coordinate compass is rotated 45 degrees.

If the check box is OFF, tools do not override AccuDraw default behavior. For example, if the check is OFF, the compass does not rotate to match the rotation of the last line or line segment.

Smart Key-ins

When the **Smart Key-ins** check box is ON, AccuDraw interprets the numbers typed into the coordinate fields as positive or negative based on the pointer position in relation to the compass. If the compass is in rectangular coordinate mode, focus also switches to the field of the axis closest to the pointer.

If the check box is OFF, key-in a dash to indicate a negative distance and press TAB to switch focus between the rectangular coordinate axis fields.

Auto Point Placement

When the **Auto Point Placement** check box is ON, AccuDraw places data points automatically when both the X and Y values are locked or when one or the other was locked while the pointer is indexed to zero.

Sticky Z Lock

The **Sticky Z Lock** check box is used in 3D designs to keep the *Z-axis* locked when data points are placed (Locks normally turn off when a data point is placed). Keeping the *Z-axis* locked makes it easy for designers to draw in one plane.

Always Show Compass

If the **Always Show Compass** check box is ON, turning on AccuDraw makes the compass appear at the location of the last data point in the current operation. For example, invoking the Copy tool, selecting an element, and turning on AccuDraw, places the compass at the selection point on the element. If the check box is OFF, placing the next data point causes the compass to appear.

Default Origin

Default Origin is a menu with three options for determining where AccuDraw starts operation in a 3D design:

- **View Center on Active Z** starts with origin at the center of the view and at the Active Z depth of the view.
- **Global Origin** starts with the origin at the Global Origin of the design file.
- **Global Origin on Active Z** starts with the origin at the Global Origin and at the Active Z depth of the view.

DISPLAY

The **Display** tab contains menus and check boxes that control the compass colors and the display of AccuDraw information, as shown in Figure 5–13.

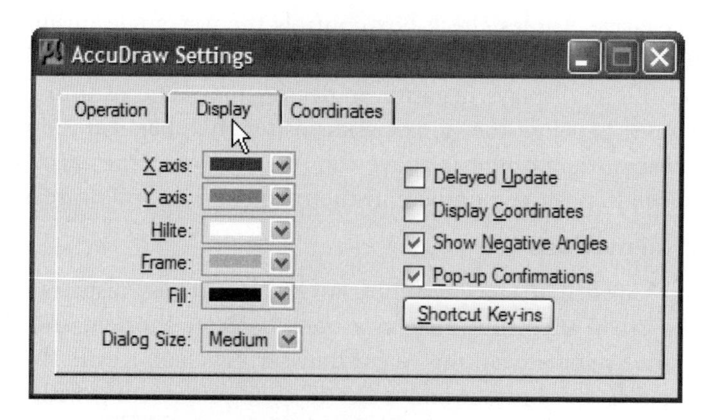

Figure 5–13 The AccuDraw settings box Display tab

The Color Menus

The color menus allow changing the colors used for the following AccuDraw compass parts:

- **X-axis** sets the color of the positive *X-axis* tick mark on the compass.
- **Y-axis** sets the color of the positive *Y-axis* tick mark on the compass.
- **Hilite** sets the color of dynamic line from the compass center to the pointer when the pointer indexes to a compass axis.
- **Frame** sets the color of the compass center and the negative axis tick marks.
- **Fill** sets the color that fills the interior of the compass.

Delayed Update

If the **Delayed Update** check box is set to OFF (default), the fields in the AccuDraw settings box update continuously as the screen pointer moves. If it is set to ON, the fields are not updated when the screen pointer moves.

Display Coordinates

If the **Display coordinates** check box is ON, MicroStation dynamically displays AccuDraw drawing plane coordinates with the pointer (see Figure 5–14)

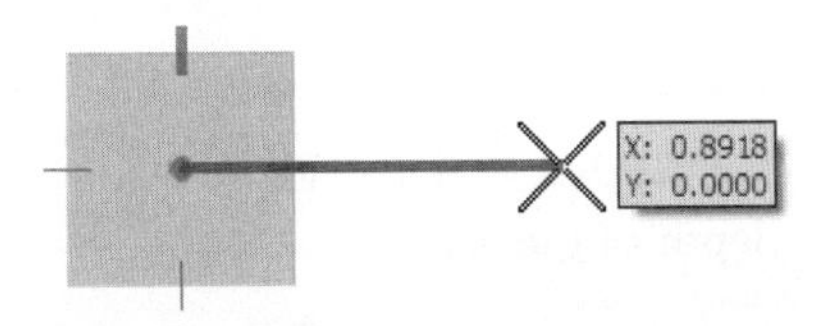

Figure 5–14 Drawing plane coordinates displayed with the pointer

Show Negative Angles

The **Show Negative Angles** check box controls the way angles display in the AccuDraw settings box when the Polar coordinate system is active. With the check box ON, AccuDraw displays negative angles (in the range of 0 to -180 degrees) when the pointer is below the *X-axis* and positive numbers (in the range of 0 to 180 degrees) when the pointer is above the *X-axis*. When the check box is OFF, AccuDraw displays positive numbers in the range of 0 to 360 degrees.

Pop-up Confirmations

If the **Pop-up Confirmations** check box is ON, AccuDraw displays pop-up boxes to confirm keyed-in shortcuts. Key in a one-character shortcut and the name of the shortcut briefly appears on the AccuDraw settings box. Key in the first character of a two-character shortcut and a list of shortcuts starting with the keyed in character appears (see Figure 5–15). If the check box is OFF, no messages appear.

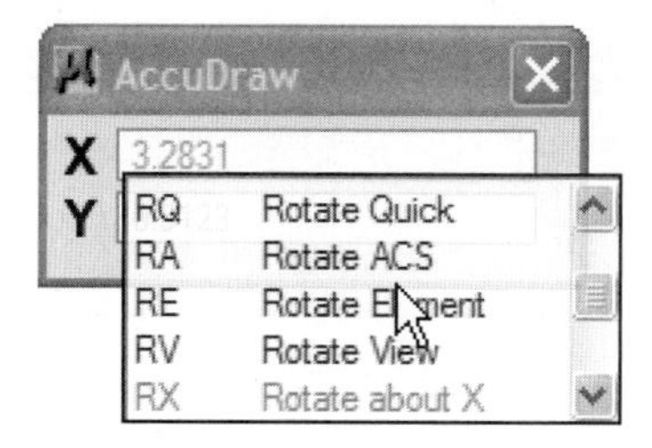

Figure 5–15 The pop-up box that appears when the first character of a two-letter shortcut is pressed

Shortcut Key-ins

Click **Shortcut Key-ins** to open the AccuDraw Shortcuts box, as shown in Figure 5–16.

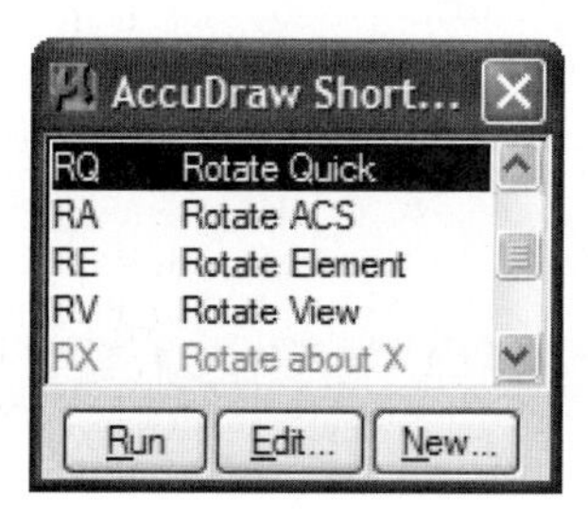

Figure 5–16 The AccuDraw Shortcuts box

The AccuDraw Shortcuts box displays all defined AccuDraw keyboard shortcuts. To invoke a shortcut from the box, select the shortcut and click **Run**. The box is rather small when initially opened. Change the box size by dragging the box's top or bottom border up or down (the width is fixed).

Clicking **Edit** opens the Edit Shortcut box. Options on the Edit Shortcut box allow changing the shortcut's name, description, and the MicroStation command it executes. To edit a shortcut, select it in the AccuDraw Shortcuts box and click **Edit**. The Edit Shortcut box opens displaying the selected shortcut's current settings, as shown in Figure 5–17. After editing the shortcut, click **OK** to close the box and apply the changes or click **Cancel** to close the box without saving the changes.

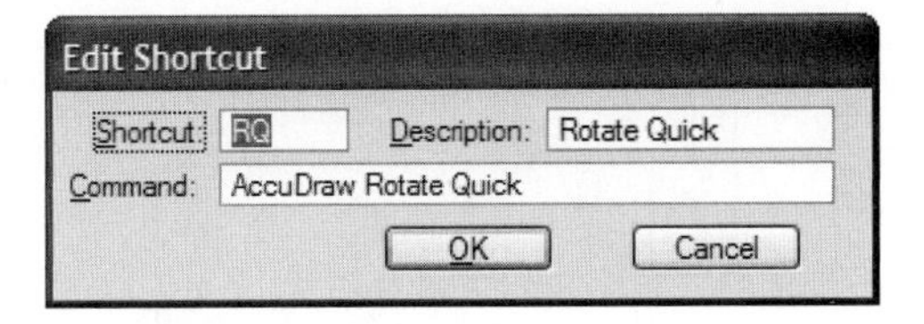

Figure 5–17 The Edit Shortcut box

Clicking **New** opens the New Shortcut box, which allows the creation of custom shortcuts by entering a one or two-character shortcut name, a shortcut description, and the command the shortcut is to execute. Custom shortcuts provide a quick way to invoke commonly used tools, such as Place Line and Copy Element. For example, Figure 5–18 shows the New Shortcut box with a Place Line shortcut. Click **OK** to close the box and create the new shortcut or click **Cancel** to close the box without creating the shortcut.

Note: *AccuDraw can support up to 400 keyboard shortcuts.*

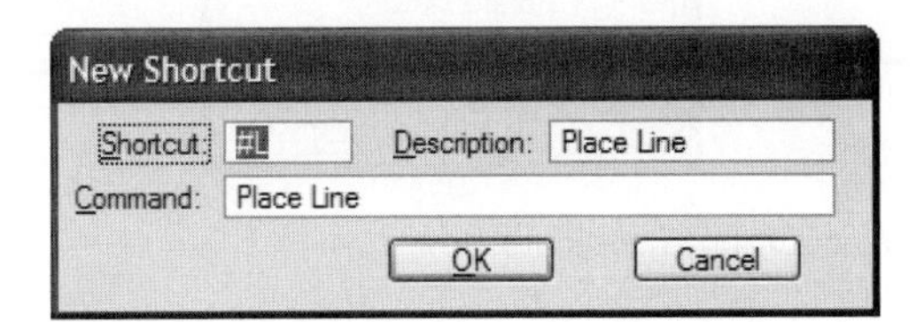

Figure 5–18 The New Shortcut box

COORDINATES

The **Coordinates** tab contains menus and check boxes that change the way the AccuDraw coordinate system works, as shown in Figure 5–19.

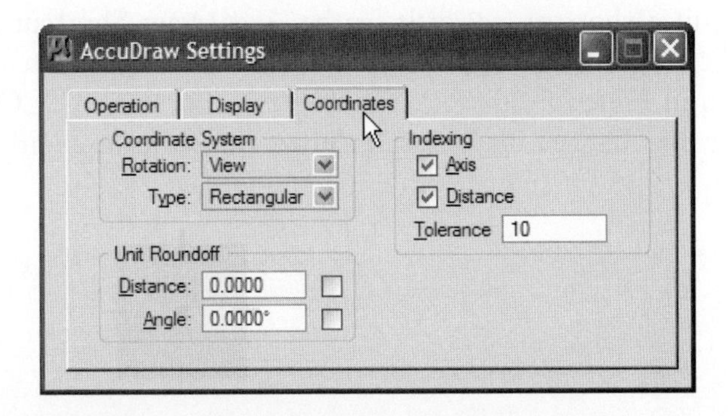

Figure 5–19 The AccuDraw settings box Coordinates tab

Rotation

The Coordinate System **Rotation** menu allows for aligning the compass to the following settings:

- **Top**—Aligns the compass to the top view in *three dimensions*, or the current view axes in *two dimensions* (same as View).
- **Front**—Aligns the compass to the front view in 3D only.
- **Side**—Aligns the compass to the side view in 3D only.
- **View**—Aligns the compass to the current view axes.
- **Auxiliary**—Aligns the compass to the last defined auxiliary coordinate system.
- **Context**—A temporary orientation affected by several factors, including the RQ keyboard shortcut.

Type

The Coordinate System **Type** menu allows switching the compass between the **Polar** and **Rectangular** coordinate systems (the same as pressing the Spacebar when focus is on the AccuDraw settings box).

Distance

The Unit Roundoff **Distance** text field and checkbox allow forcing distances, defined by moving the pointer, to be limited to the roundoff value or multiples of the value. To lock a roundoff distance value, key-in the roundoff value in the **Distance** text field and turn the **Distance** checkbox ON. For example, a **Distance** roundoff value of 0.25 Master Units forces the line length to change in increments of 0.25 Master Units (0.25, 0.5, 0.75, 1.0, and so forth), as the pointer moves in the design.

The roundoff value applies to the distance, when AccuDraw is using the polar coordinate system, and the *X* and *Y-axis value, when AccuDraw is using the rectangular coordinate system.*

Angle

The Unit Roundoff **Angle** text field limits angles, defined by moving the pointer, to the roundoff value or multiples of the value. To lock a roundoff angle value, key-in the roundoff value in the **Angle** text field and turn the **Angle** check box ON.

The roundoff value applies to the rotation angle when AccuDraw is using the polar coordinate system. For example, an **Angle** roundoff value of 30 degrees forces all angles to increments of 30 degrees (0, 30, 60, 90, and so forth) as the pointer moves in the design.

Note: *Keying values into to the AccuDraw settings box edit fields overrides the roundoff values.*

Axis

The Indexing **Axis** check box turns compass indexing ON and OFF. If the check box is ON, moving the pointer close to one of the axis marks on the AccuDraw compass causes the dynamic image to snap to the mark. If the checkbox is OFF, no snapping occurs.

Distance

The Indexing **Distance** check box turns ON and OFF distance indexing. When the check box is ON, AccuDraw places a tick mark on the end of the dynamic image of the line being placed when the line is snapped to a compass axis mark and is the same length as the previously placed line. If the check box is OFF, there is no indication of equal length lines.

Tolerance

The Indexing **Tolerance** text field allows for controlling how close the pointer must be to the Axis and Distance indexing points before AccuDraw triggers indexing. The **Tolerance** is in screen pixels and can be in the range of 1 to 99. A pixel is the smallest addressable unit on a display screen.

ACCUDRAW SHORTCUTS

As was discussed earlier in this chapter, AccuDraw includes one and two-character keyboard shortcuts that allow for changing the action AccuDraw is about to perform. Invoke a shortcut by typing the shortcut while focus is in the AccuDraw settings box. Table 5–1 lists the shortcuts available in 2D designs (additional shortcuts are available for 3D designs).

Note: *AccuDraw can support up to 400 keyboard shortcuts.*

Table 5–1 AccuDraw Shortcuts for 2D design.

KEY-IN	ACCUDRAW DIRECTIVE
(ENTER)	Turn Smart Lock ON and OFF.
SPACE	Toggle between the Rectangular and Polar coordinate compass.
O	Move the origin point to the current screen pointer location.
V	Rotate the drawing plane to align with the view (its normal rotation).
T	Rotate the drawing plane to align with the top view.

KEY-IN	ACCUDRAW DIRECTIVE
B	Rotate the compass between its current rotation and the drawing's top view. A toggle switch alternates between rotating the compass between the two positions.
X	Turn ON and OFF the Rectangular coordinate *X* value lock.
Y	Turn ON and OFF the Rectangular coordinate *Y* value lock.
D	Turn the Polar coordinate Distance value lock.
A	Turn ON and OFF the Polar coordinate Angle value lock.
L	Turn ON and OFF the Index lock. If the Index lock is OFF, the only way to index the pointer position to an axis is to use Smart Lock.
RQ	Temporarily rotate the drawing plane about the compass origin point. The next data point turns off this lock.
RA	Permanently rotate the drawing plane. It stays active after the current tool terminates.
RE	Rotate the drawing plane to match the orientation of a selected element.
RV	Rotate the active view to match the current drawing plane.
RZ	Rotate the drawing plane 90 degrees about its *Z-axis*. In a 2D drawing, the *Z-axis* is perpendicular to the drawing plane.)
?	Open the AccuDraw Shortcuts box.
~	Bump the selected item in the top menu in the Tool Settings window.
GT	Open or set focus to the AccuDraw Tool Settings window.
G K	Open, or move focus, to the Key-in window (same as choosing Key-in from the Utility menu).
G S	Open, or move focus, to the AccuDraw settings box.
GA	Get a saved ACS (Auxiliary Coordinate System)
WA	Save the drawing plane alignment as an ACS.
P	Open or set focus to the Data Point Key-in window that can be used for precision input (*XY*=, DL=, DI=, etc.).
M	Open or set focus to the Data Point Key-in box that can be used for precision input (*XY*=, DL=, DI=, etc.).
I	Activate Intersect snap mode.
N	Activate Nearest snap mode.
C	Activate Center snap mode.

KEY-IN	ACCUDRAW DIRECTIVE
K	Open the Keypoint Snap divisor setting box so the snap divisor can be set.
HA	Suspend AccuDraw for the current tool operation. Selecting a new tool, or entering a Reset re-enables AccuDraw.
HS	Turn ON and OFF AccuSnap.
HU	Suspend AccuSnap for the current tool operation. Selecting a new tool, or entering a Reset, re-enables AccuSnap.
Q	Close the AccuDraw tool.

WORKING WITH ACCUDRAW

The AccuDraw parts and their customization were described. Now it is time to see how to use AccuDraw to enhance the function of other MicroStation tools.

Note: *The following examples assume Axis lock is OFF, Snap lock is ON, the distance and angle can be set to any value, and angles are entered using positive degrees of rotation (0-360).*

EXAMPLE OF SIMPLE PLACEMENT

Figure 5–20 shows AccuDraw enhancing the drawing of an arrowhead. A description of the steps used to draw the arrowhead follow the figure.

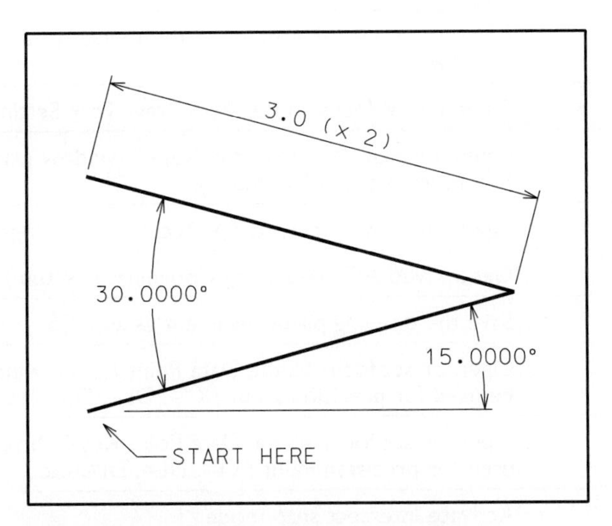

Figure 5–20 Simple Arrowhead drawn with AccuDraw

1. Open the AccuDraw Settings box, click the **Operation** tab, and turn ON the **Floating Origin** and **Context Sensitivity** check boxes.

2. Invoke AccuDraw and the Place Line tool.
3. Place a data point to define the lower left end of the arrowhead.
4. If the AccuDraw Rectangular coordinates compass appears, press the SPACEBAR to switch to the Polar coordinates compass.
5. Drag the pointer a short distance to the right and upwards.
6. If the AccuDraw settings box focus is on the **Angle** text field, press TAB to focus on the **Distance** text field.
7. Key-in **3** to lock the line length in the **Distance** text field.
8. Press TAB to move focus to the **Angle** edit field, and key-in **15** to lock the rotation angle in the **Angle** text field.
9. Click the Data button to draw the bottom half of the arrowhead.
10. Drag the pointer to the left and up until the distance indexing tick mark appears at the screen pointer (the **Distance** should be 3).
11. If the AccuDraw window focus is not on the **Angle** text field, press TAB to focus on it.
12. Key-in **150** to place and lock the angle of the second line in the **Angle** text field.
13. Click the Data button to draw the top half of the arrowhead.

MOVING THE COMPASS ORIGIN

In almost all uses of AccuDraw, the compass origin is on the last placed data point, but it can be moved without placing a data point. Defining a tentative point and then keying in the letter O causes the compass to jump to the tentative point. Key in O, without first defining a tentative point, and the compass jumps to the current pointer location.

USING TENTATIVE POINTS WITH ACCUDRAW

Tentative points in combination with the compass allow placing elements in precise relationships to other elements. For example, to start a line 2 Master Units to the right of the corner of an existing element, do the following:

1. Invoke the Place Line tool.
2. Tentative snap to the corner of the existing element from which the offset is to be measured.
3. Key-in the letter **O** to move the compass to the tentative point.
4. If the Polar coordinate system is not active, press the SPACEBAR to switch to it. (This example uses Polar coordinates, but it can be done in Rectangular coordinates as well.)

5. Using TAB as necessary, key-in **2** in the **Distance** field and **0** in the **Angle** field.
6. Click the Data button to place the first point of the line 2 units to the right of the tentative point.
7. Finish drawing the line.

Note: *AccuDraw provides keyboard shortcuts for selecting some of the tentative snap modes (see Table 5–1).*

ROTATING THE ACCUDRAW PLANE

In 2D designs, the AccuDraw plane can be rotated about the *Z-axis* (which is perpendicular to the 2D plane) any time the compass is visible. To rotate the plane, key-in one of the two-letter rotation shortcuts (see Table 5–1).

Note: *The compass rotates automatically to the same angle as the previously placed line segment when the context sensitivity check box is set to ON. The rotation shortcuts allow overriding that rotation.*

PLACING ELEMENTS WITH ACCUDRAW ACTIVE

Here are two examples of using AccuDraw to enhance placement of elements in a design.

Drawing an Ellipse

The following example uses the Place Ellipse by center method and AccuDraw to place an ellipse with a major axis 8 Master Units long, rotated 30 degrees, and a minor axis 4 Master Units long. Place a data point to define one end of the major axis and use the AccuDraw settings box to complete the ellipse.

1. Invoke the Place Ellipse tool and select **Center** from the **Method** menu.
2. Place the first data point to define one end of the major axis.
3. If the rectangular compass is active, press the SPACEBAR to switch to the polar compass.
4. In the AccuDraw settings box, use TAB as necessary to set the **Distance** to **4** and the **Angle** to **30**.
5. Click the Data button to define the major ellipse axis.
6. In the AccuDraw window, press TAB if necessary to focus on the Distance field and key-in **2**.
7. Click the Data button to complete the ellipse by defining its minor axis.

Drawing a Block

The following example uses the Place Block tool with the **Rotated Method** and AccuDraw to place a 3 x 5 block, rotated 15 degrees. Place a data point to define one corner of the block and use the AccuDraw settings box to define the rotation angle and place the block.

1. Invoke the Place Block tool and select **Rotated** from the **Method** menu.
2. Place a data point to define the lower left corner of the rotated block and move the pointer slightly above and to the right of the data point.
3. If the rectangular compass is active, press the SPACEBAR to switch to the polar compass.
4. In the AccuDraw settings box, press TAB as required to key-in **3** in the **Distance** field and **15** in the **Angle** field.
5. Click the Data button to define the bottom edge of the rotated block (the compass switches to rectangular coordinates) and move the pointer up a short distance.
6. In the AccuDraw settings box, key-in **5** in the **Y** field.
7. Click the Data button to complete the rotated block.

MANIPULATING ELEMENTS WITH ACCUDRAW ACTIVE

AccuDraw also enhances the manipulation of elements. The following example draws the lower left block in Figure 5–21, and uses AccuDraw with AccuSnap to place two copies of the left block.

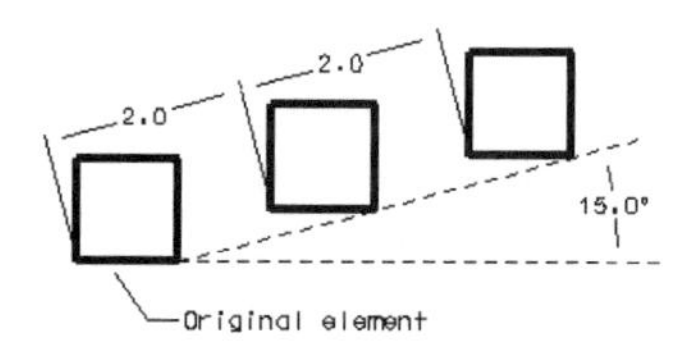

Figure 5–21 Example of using AccuDraw with the Copy Element tool

If AccuDraw is not active, invoke it.

1. Place a 1 by 1 block.
2. Open the AccuSnap Settings box, turn the **Enable AccuSnap** check box ON and use the other check boxes to set AccuSnap operation as desired.
3. Invoke the Copy Element tool and turn the **Make Copy** check box ON.

4. Open the AccuDraw Settings box, set the Coordinate System **Type** to **Polar**, turn the Unit Roundoff **Distance** check box ON, and key-in **2** in the **Distance** text field.
5. Turn the Unit Roundoff **Angle** check box ON and key in **15** in the **Angle** field.
6. Select the lower right corner of the block by moving the pointer toward the corner until AccuSnap snaps to the corner and then click the Data button to accept it.
7. Drag the pointer up and to the right until the AccuDraw settings box shows an **Angle** of **15** and a **Distance** of **2**.
8. Click the Data button to place the copy.
9. Repeat steps 7 and 8 for the second copy.
10. Click the Reset button to drop the element.

PLACE SMARTLINE TOOL

The SmartLine tool places a chain of connected lines and arcs as individual elements, or as a line string, shape, complex chain, or complex shape. The segment vertices can be sharp points, tangent arcs (rounded), or chamfers.

SmartLine, AccuDraw, and AccuSnap, in combination, provide a powerful toolset for creating designs in MicroStation.

Invoke SmartLine from:

Task Navigation tool box (active task set to Linear)	Select the Place SmartLine tool (see Figure 5–22) and make the initial tool settings in the Tool Settings window.
Keyboard Navigation (Task Navigation tool box with active task set to Linear)	**Q**

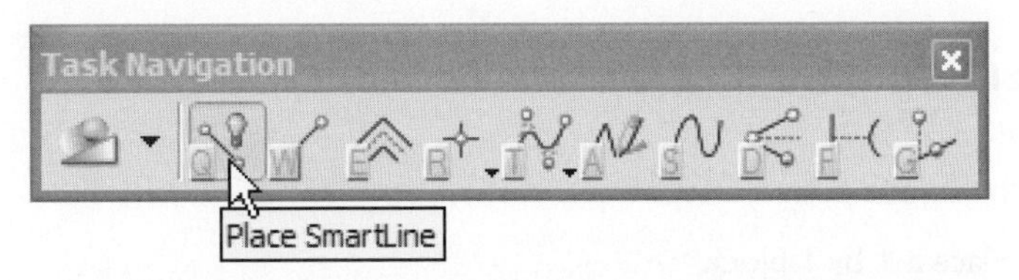

Figure 5–22 Place SmartLine tool icon in the Linear Elements tool box

MicroStation prompts:

Place SmartLine > Enter first vertex *(Place the first data point to start the SmartLine.)*

Place SmartLine > Enter next vertex or reset to complete *(Place data points to define SmartLine segments and change tool settings as required to complete the design. After completing drawing all segments, click the Reset button to end the SmartLine.)*

THE SMARTLINE TOOL SETTINGS BOX

When the SmartLine tool is active, the options available in the Tool Settings window vary depending on the selected options and the current use of the tool. Following is a discussion of the options and their impact on SmartLine operation.

Choosing the Type of Segment to Place

The **Segment Type** menu allows for choosing between placing **Lines** and **Arcs**. Placing **Line** segments is similar to the Place Line tool. Placing **Arcs** is similar to the Place Arcs by the Center tool. Segment types can be switched while the SmartLine tool is active, as shown in Figure 5–23.

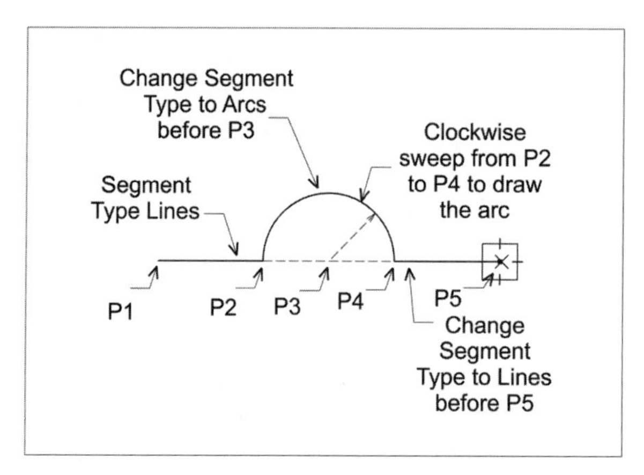

Figure 5–23 Using the SmartLine tool to place lines and arcs

Switch to placing arcs and the next data point defines the center of the arc. As the pointer moves around the center point, a dynamic image shows arc placement after the next data point is defined. The direction in which the pointer moves after defining the arc center point controls the direction in which the arc sweeps.

Controlling the Shape of Line Segment Edges

Choose a **Vertex Type** to change the way the intersections of the segments are drawn:

- Choose **Sharp** to join the two line segments at the data point and form a sharp edge.
- Choose **Rounded** to round off the intersection of the two line segments with an arc and form a smoothly curved edge. Set the radius of the rounding arcs by keying-in a radius in the **Rounding Radius** field. A rounded vertex joins line segments or line and arc segments.

- Choose **Chamfered** to cut off the edge formed by the intersection of the two line segments and join them with a line segment. To set the cut amount, enter a distance in the **Chamfer Offset** field. SmartLine cuts the offset value from each line segment and joins them with a line segment. Chamfered edges only join line segments. If one of the segments is an arc, a sharp corner is created.

Note: *SmartLine only creates the Rounded or Chamfered vertex if the adjacent segments are long enough to contain the vertex. If the segments are too short, a sharp vertex is placed.*

Figure 5–24 shows examples of vertex types.

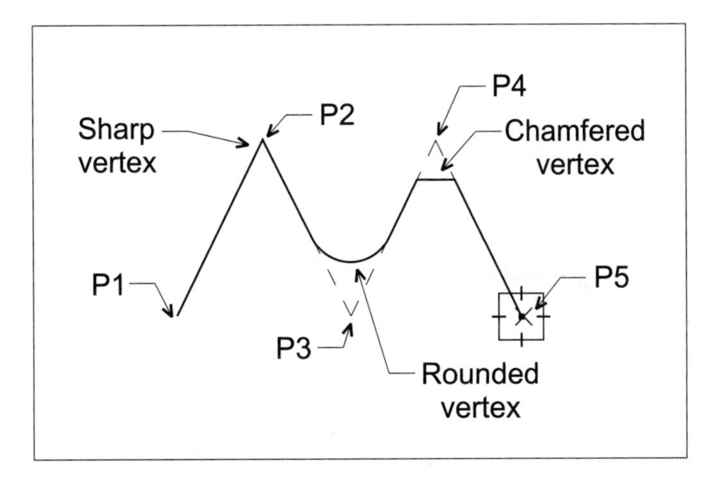

Figure 5–24 The SmartLine tool vertex types

Joining Elements

The **Join Elements** check box controls placement of the SmartLine. If the check box is ON, all the SmartLine segments form a single element. If the check box is OFF, SmartLine segments are separate elements, including the rounded and chamfered corners.

Closing SmartLines

If the **Join Elements** check box is ON, the SmartLines are closed by moving the pointer over the first data point of the SmartLine, snapping to the point and clicking the Data button. Snapping to the starting point expands the Tool settings window to show the **Join Element** check box and other close options, as shown in Figure 5–25.

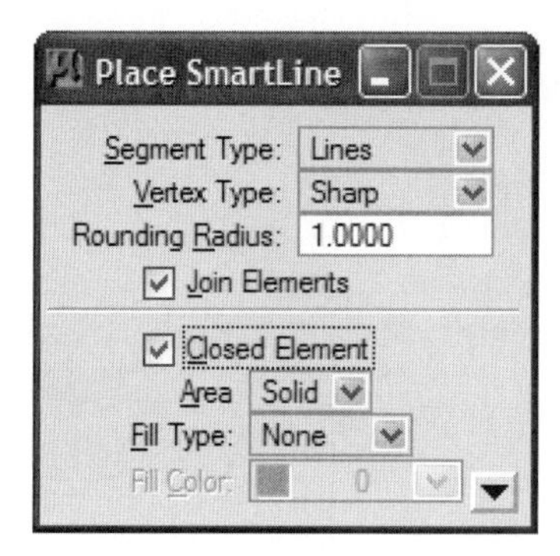

Figure 5–25 The Expanded SmartLine Tool settings window

The additional options in the Tool settings window control the completion of the SmartLine and the type of element created by closing it. To change these options, click the Tentative button to snap to the starting point. Otherwise, the extra options disappear when the pointer moves away from the starting point. The AccuSnap snap, alone, does not make the options "stick."

If the **Closed Element** check box is ON, the SmartLine clicking the Data button to accept the tentative snap completes the SmartLine as a closed element. If the check box OFF, the SmartLine is not completed and clicking the Reset button completes drawing the SmartLine.

If the **Closed Element** check box is ON, the **Area, Fill Type**, and **Fill Color** options allow filling the interior of the closed element with color or a pattern. Chapter 12 discusses fill and patterning tools.

Note: *If the* ***Closed Element*** *check box is OFF when SmartLine element closing is signaled by tentative snapping to the starting point of the element, the only additional option that appears in the Tool Settings window is the* ***Closed Element*** *checkbox. Turn the check box ON for the additional options.*

SmartLine Placement Settings

On the bottom, right corner of the SmartLine Tool Settings window is a down-pointing arrowhead that opens the **SmartLine Placement Settings** box, shown in Figure 5–26. The box contains two check boxes that change the way SmartLine operates.

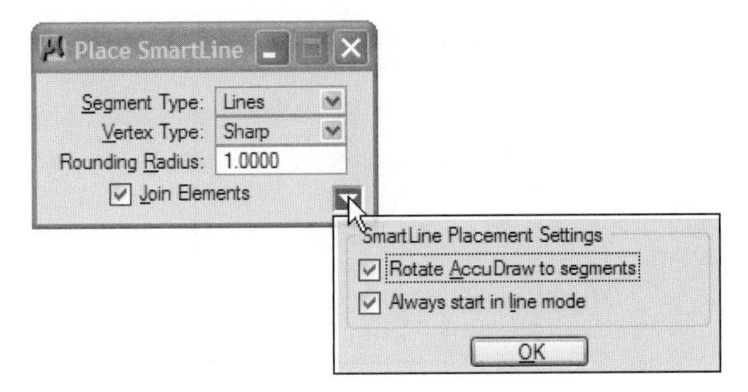

Figure 5–26 The SmartLine Placement Settings box

If the **Rotate AccuDraw to Segments** check box is ON, the AccuDraw compass rotates so that its *X-axis* is in line with the angle of the segment to which it is attached. If the check box is OFF, the compass rotation is always zero, with its *X-axis* parallel to the design plane's *X-axis*. This check box overrides rotations in the AccuDraw Settings box.

If the **Always Start in Line Mode** checkbox is ON, SmartLine always starts with the **Segment Type** set to **Lines** when it is selected from the Linear Elements tool box This occurs even if it was left set for Arcs the last time it was used.

USING SMARTLINE WITH ACCUDRAW

The following example illustrates the use of SmartLine and AccuDraw by creating the simple design shown in Figure 5–27. The letters in the figure point to the locations of all data points. It assumes the design file has its Master and Sub Units both set to inches and the active view window is set to display an area of about six by eight inches.

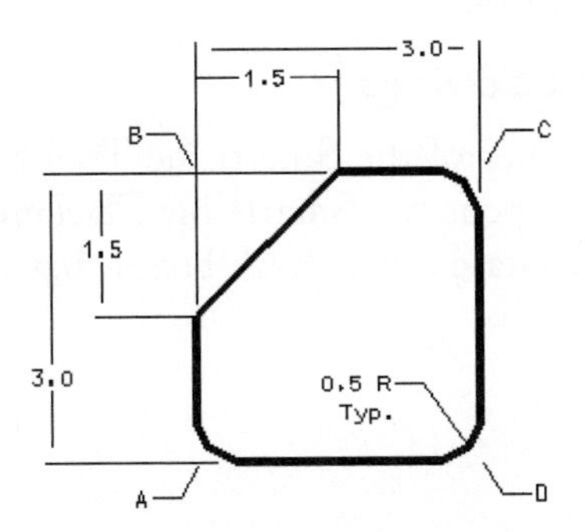

Figure 5–27 Example of a drawing made with the SmartLine tool

1. If AccuDraw is not active, invoke the AccuDraw tool.

2. Open the AccuDraw Settings box.
3. On the **Coordinates** tab, lock the **Distance** roundoff value at **0.5** inch, lock the **Angle** roundoff value at **90** degrees, select the **Polar** Coordinate System **Type**, and close the AccuDraw Settings box.
4. Invoke SmartLine from the Lines tool box.
5. In the Tool Settings window, select the **Lines Segment Type**, select the **Chamfered Vertex Type**, and key-in a **Chamfer Offset** of 1.5 inches.
6. Define a point near the bottom left of the view window to start the object (Point A).

Note: *Because the AccuDraw distance roundoff value also forces the first data point to a multiple of the roundoff value on the design plane grid, the point may not be placed exactly where the Data button was clicked.*

7. Move the pointer up until the AccuDraw settings box **Distance** field displays 3 inches and click the Data button (Point B).
8. Move the pointer to the right until the AccuDraw settings box **Distance** field displays 3 inches and click the Data button (Point C).

Note: *The 1.5 inch offset chamfer does not appear until there is room for it on the line segment (when the pointer has moved at least 1.5 inches to the right).*

9. In the Tool Settings window, select the **Rounded Vertex Type** and set the **Rounding Radius** to 0.5 inch.
10. Move the pointer down until the AccuDraw settings box **Distance** field displays 3 inches and click the Data button (Point D).
11. Slide the pointer to the left until the AccuDraw settings box **Distance** field displays 3 inches and touches the starting point of the SmartLine (Point A) and click the Tentative button.
12. If the **Closed Element** check box is OFF, turn it ON in the Tool Settings window
13. Click the Data button to close the element and complete the SmartLine operation.

The completed element should be identical to the element shown in Figure 5–27 (without the dimensions and letters).

Open the Exercise Manual PDF file for Chapter 5 on the accompanying CD for project and discipline specific exercises.

REVIEW QUESTIONS

Write your answers in the spaces provided.

1. How is AccuDraw activated?

   ____

2. Name the two coordinate systems that can be used with AccuDraw.

   ____

3. Name three settings that can be adjusted in the AccuDraw Settings box **Operation** tab.

   ____

4. What is the purpose of rounding off AccuDraw distances and angles?

   ____

5. How are previous values recalled in the AccuDraw settings box?

   ____

6. What is the shortcut key-in that will move the compass origin from the previous data point?

   ____

7. Explain briefly the benefits of manipulating elements with AccuDraw active.

   ____

   ____

   ____

8. Explain the difference between the Place Line tool and the Place SmartLine tool.

   ____

   ____

9. Name the two segment types that can be placed with the SmartLine tool.

   ____

   ____

   ____

   ____

10. What can cause SmartLine to place a sharp vertex when SmartLine is set to place 1.5-inch radius rounded vertices between line segments?

    ____

Manipulating a Group of Elements

OBJECTIVES

Topics explored in this chapter:

- Using Element Selection tools to select and manipulate elements
- Placing fences and manipulating fence contents

ELEMENT SELECTION TOOLS

The element manipulation tools presented in the previous chapters invoked an element manipulation tool and then identified the element to manipulate. This chapter introduces the Element Selection tools that are used to select or deselect one element, several elements, or all elements in the design. Element manipulation tools manipulate selected elements and some manipulations (such as deleting selected elements) can be performed without actually invoking a tool. Invoke the Element Selection tool from:

Main tool box	Select the Element Selection tool (see Figure 6-1).
Keyboard Navigation	1

The Element settings box displays the Element Selection tools.

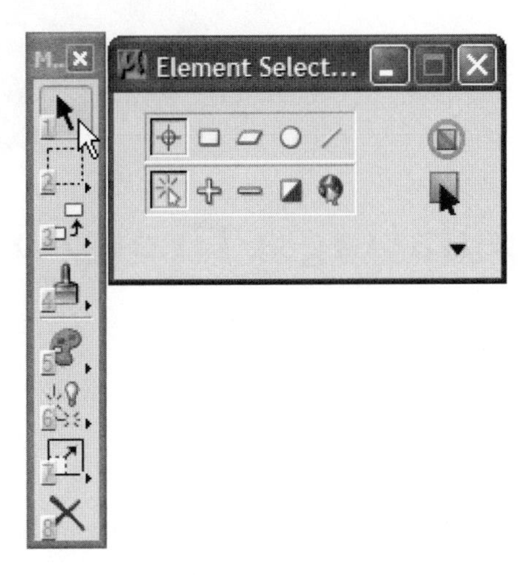

Figure 6–1 Invoking Element Selection from the Main tool box

The Element Selection settings box contains two rows of tools in recessed areas of the box and two additional tools on the right side of the box. The top recessed area provides the Method tools for selecting and unselecting elements. The bottom recessed area provides Mode tools that determine how the Method tools select and unselect elements.

MODE TOOLS

As was stated earlier, the Mode tools influence the action of the Method tools when using them to select elements, so they are presented before the methods.

***Note:** A change in element color or the appearance of "handles" indicates the selection of an element. Handles are discussed later in this chapter.*

New Mode Tool

The left-most Mode tool on the Element Selection settings box is **New**, as shown in Figure 6–2.

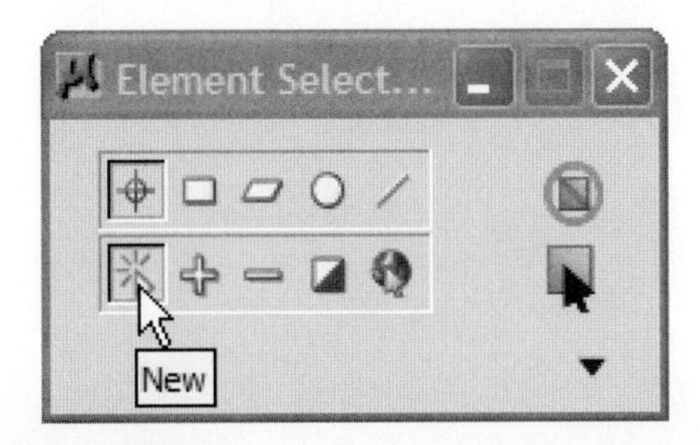

Figure 6–2 The New mode tool

New mode is used to select one element, or group of elements. It is possible to override the single selection action of the **New** mode by holding down CTRL while selecting the additional elements, or group of elements. The CTRL key can also remove the selection of one or more elements in a group of selected elements. To unselect elements in the group, hold down CTRL while selecting the previously selected elements.

Adding elements to, or removing them from the current selection may be easier with the **Add**, **Subtract**, and **Invert** modes because there is no need to hold down CTRL while selecting and unselecting elements—forgetting to hold down CTRL one time unselects all previously selected elements.

Add Mode Tool

The second Mode tool on the Element Selection settings box is **Add**, as shown in Figure 6–3.

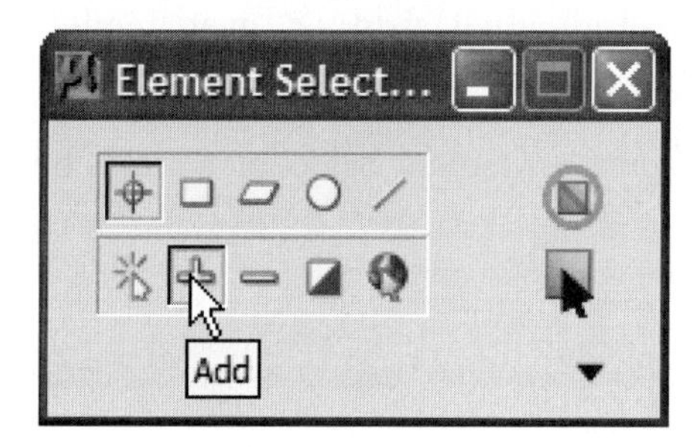

Figure 6–3 The Add Mode tool

Use **Add** mode when more than one selection is required to select a group of elements. Each additional selection adds the elements to the current set of selected elements. **Add** mode eliminates the need to hold down CTRL while adding element to the selection. **Add** mode can also start a new selection.

Add mode does not allow for removing elements from the current selection. To remove elements, switch to **Individual**, **Subtract**, or **Invert** mode.

Subtract Mode Tool

The third Mode tool on the Element Selection settings box is **Subtract**, as shown in Figure 6–4.

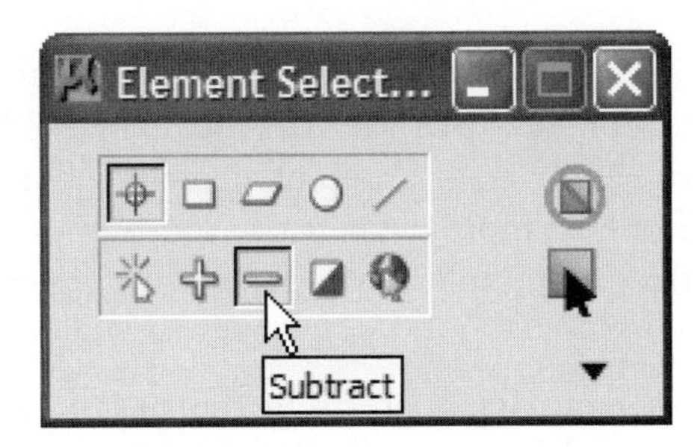

Figure 6–4 The Subtract Mode tool

Use **Subtract** mode to remove elements from the current set of selected elements. This mode eliminates the need to hold down CTRL while removing elements from the selection. **Subtract** mode is only useful when elements are already selected.

Subtract mode does not allow for adding elements from the current selection. To add elements, switch to **Individual**, **Add**, or **Invert** mode.

Invert Mode Tool

The fourth Mode tool on the Element Selection settings box is **Invert**, as shown in Figure 6–5.

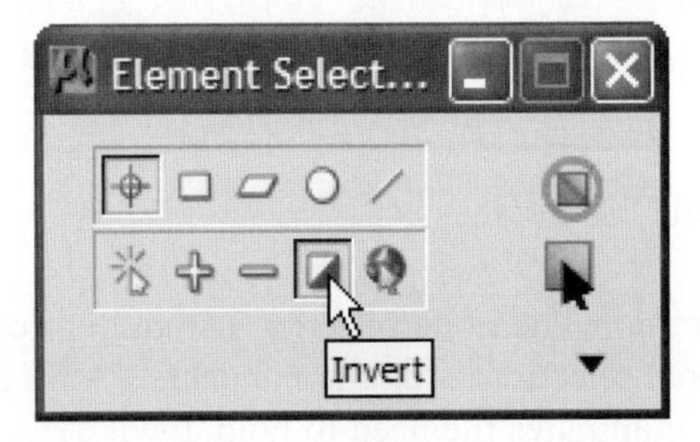

Figure 6–5 The Invert Mode tool

Invert mode combines **Add** and **Subtract** modes into one tool. Selecting a previously selected element, or group of elements, removes the element, or group of elements, from the selection. Selecting an unselected element, or group of elements, adds the element, or group of elements to the selection. There is no need to hold down CTRL while adding or removing elements. **Invert** mode can also start a new selection.

Select All/Clear Mode Tool

The fifth Mode tool on the Element Selection settings box is actually two modes and clicking it causes it to switch to the other mode (see Figure 6–6).

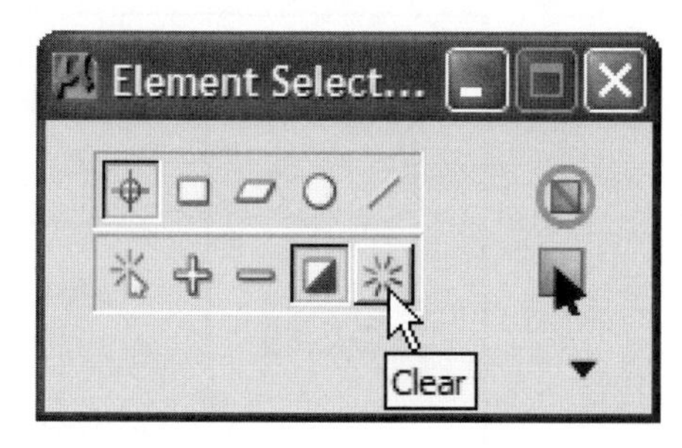

Figure 6–6 The Select All and Clear All mode selection tool

Select All mode selects all elements in the design and switches the icon to **Clear** mode. Conversely, **Clear** mode unselects all elements in the design and switches the icon to **Select All** mode.

METHOD TOOLS

Method tools provide various options for selecting and unselecting elements.

Individual Method

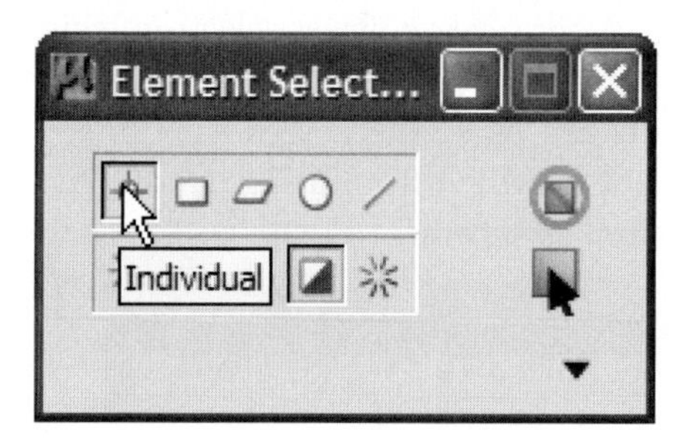

Figure 6–7 The Individual method tool

The **Individual** method (see Figure 6-7) is used to select a single element by clicking the element or several elements by dragging a box around the elements. While dragging over elements, a dynamic image of a box appears as the cursor moves. Release the data button to select all elements completely inside the image of the box. The active Mode tool controls the exact action of the **Individual** method.

Selecting the first element by clicking it places "selection handles" on the element to indicate selection. If selection handles appear on the first selected element, they appear on all subsequently selected elements as well. Figure 6–8 shows selection handles that appear on typical elements.

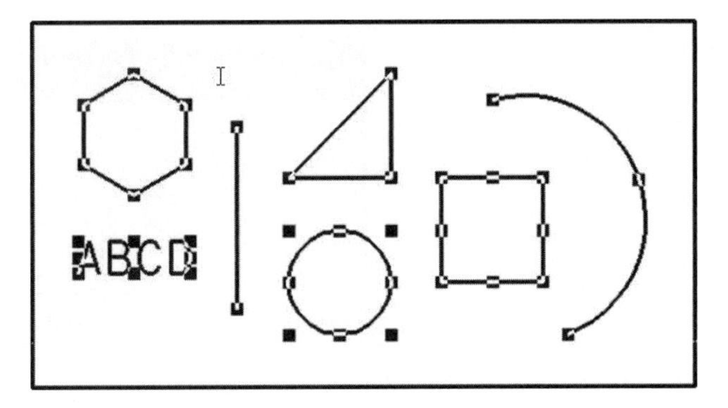

Figure 6–8 An example of selection handles on typical elements

Selecting the first element by dragging a box around it causes the elements to change to a specific color that indicates a selection. If a change in element color indicates the first selection, all following selections use color change to indicate selection rather than selection handles.

Control the Display of Handles

As was discussed in the previous section, selecting the first element by clicking it when the **Individual** method is active places selection handles on the element. The handles can be used to make modifications to selected elements but their appearance on several elements can make the design hard to see. MicroStation provides the **Disable Handles** tool to switch from using handles to always using a highlight color to indicate selection, as shown in Figure 6–9.

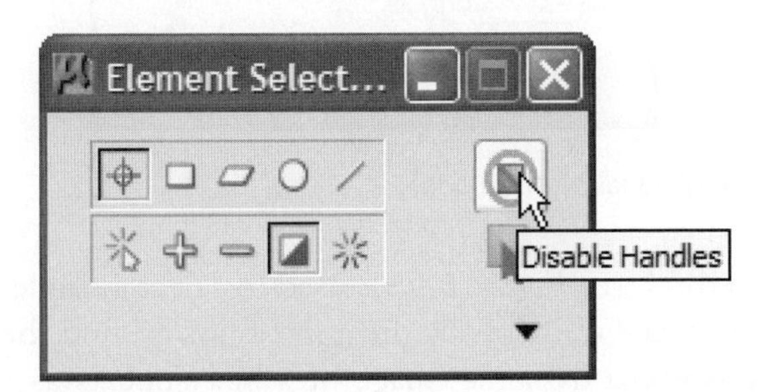

Figure 6–9 The Disable Handles tool

Disable Handles is a toggle switch that turns selection handles ON and OFF. Selecting the tool when handles are in use replaces the handles with the highlight color. Selecting the tool when handles are OFF, turns them ON.

Block Method

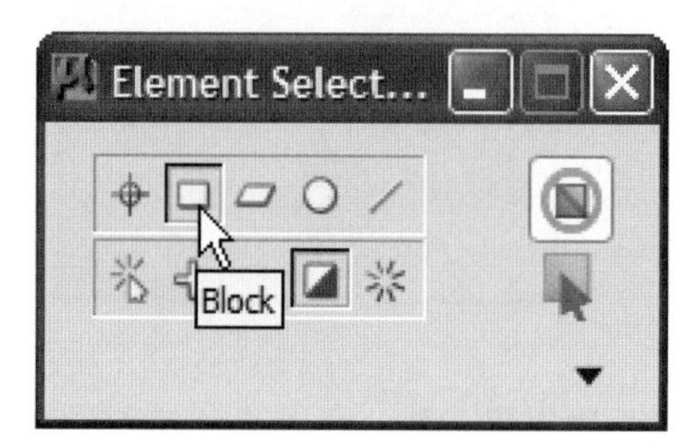

Figure 6–10 The Block method tool

The **Block** method (see Figure 6-10) selects elements by dragging the dynamic image of a block around the elements. Releasing the data button selects all elements *completely* inside the block. The tool does not select elements that merely touch or overlap the edges of the block. The active Mode tool controls the exact action of the **Block** method.

Shape Method

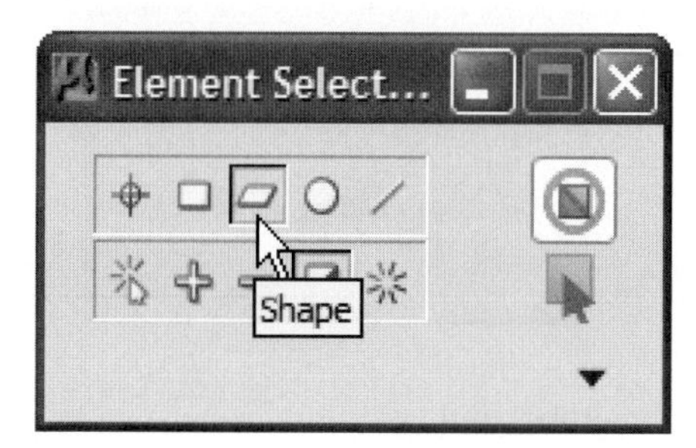

Figure 6–11 The Shape method tool

The **Shape** method (see Figure 6-11) selects elements by defining a multi-sided, closed shape that completely encloses the elements. Releasing the data button selects all elements *completely* inside the shape. The tool does not select elements that merely touch or overlap the edges of the shape. The active Mode tool controls the exact action of the **Shape** method.

Circle Method

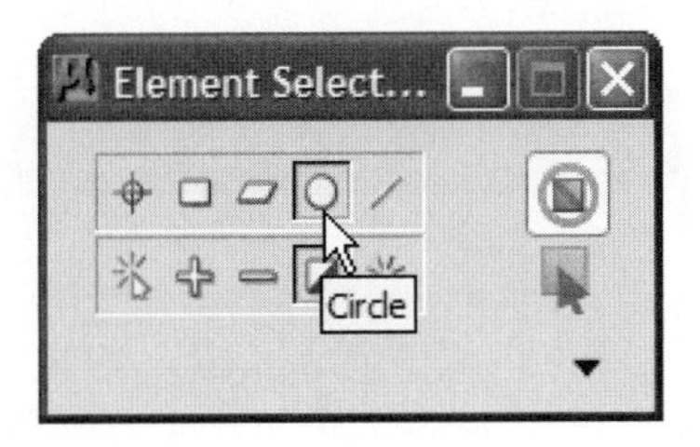

Figure 6–12 The Circle method tool

The **Circle** method (see Figure 6-12) selects elements by defining the center and a point on the circumference of a circle that completely encloses the elements. Releasing the data button selects all elements *completely* inside the circle. The tool does not select elements that merely touch or overlap the edges of the circle. The active Mode tool controls the exact action of the **Circle** method.

Line Method

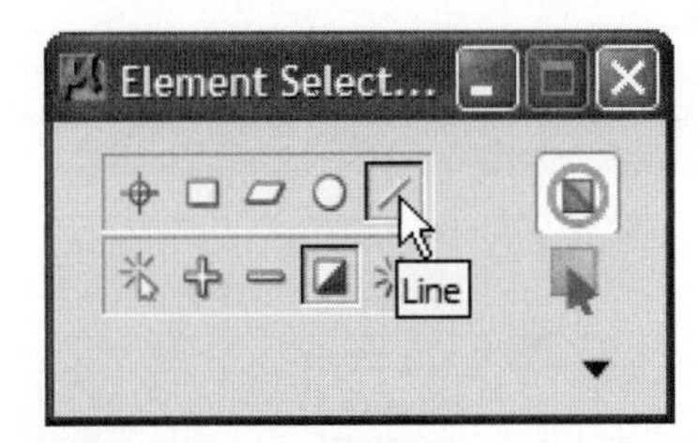

Figure 6–13 The Line method tool

The **Line** method (see Figure 6-13) selects elements by drawing a line through the elements. Placing the second data point selects all elements that touch the line. The active Mode tool controls the exact action of the **Line** method.

Following is an example of selecting elements using the **Block** Method. Invoke the Element Selection settings box from:

Main tool box	Select the **Element Selection** tool and select **Block** method and **New** mode from the Element Selection settings box (see Figure 6–14).
Keyboard Navigation	1 (Select **Block** method and **New** mode from the Element Selection settings box.)

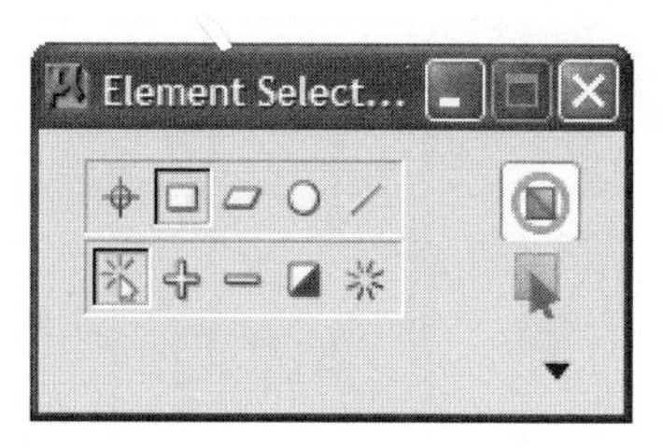

Figure 6–14 Invoking the Block method and New mode

Move the cursor off the settings box and MicroStation prompts:

> Element Selector > Place Shape for elements to add to set *(Place a data point to define one corner of the selection block and drag the cursor to enclose the elements, as shown in Figure 6–15.)*

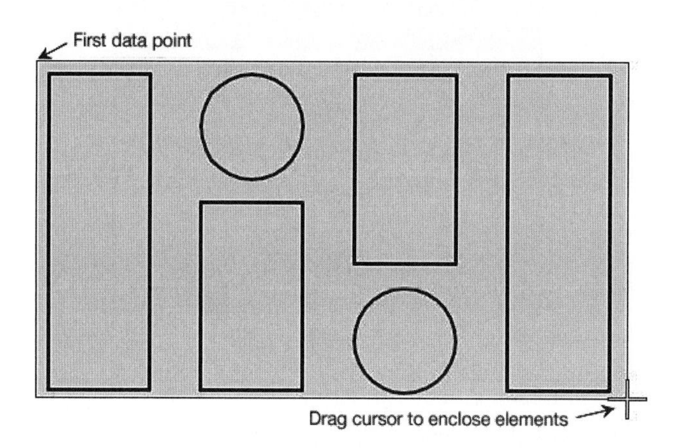

Figure 6–15 Defining the selection block

Placing the second data point removes the dynamic image of the selection block and selects all elements completely inside the area defined by the block. The elements change color to indicate selection.

SELECT ELEMENTS BY ATTRIBUTES

The **Show Extended Settings** option in the lower, right corner of the Element Selection settings box extends the height of the settings box to reveal options for selecting elements based on one or more element attributes (see Figure 6–16).

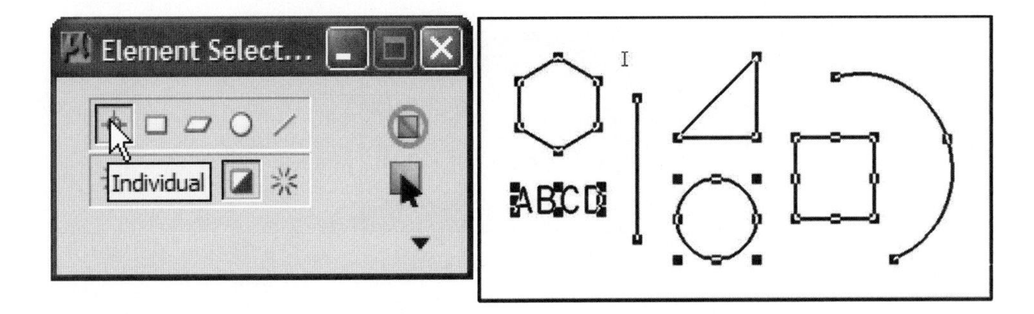

Figure 6–16 The Show Extended Settings option on the Element Selection settings box and the extended Element Selection settings box

The seven tabs at the top of the extended area of the settings box allow for selection of elements based on different types of attributes, as shown in Figure 6–17.

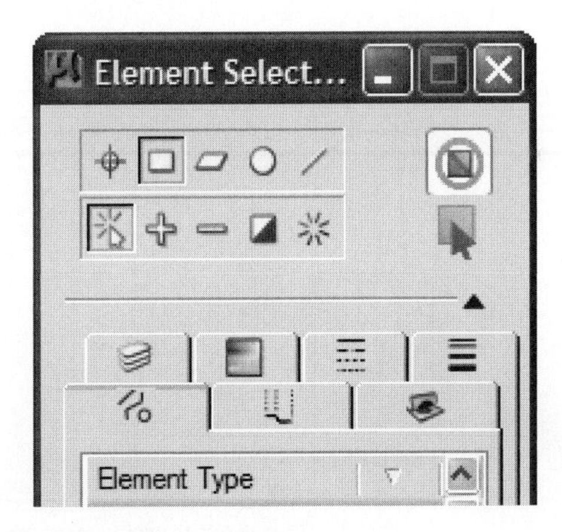

Figure 6–17 The extended Element Selection tabs

Clicking on an attribute on the settings box selects all elements in the design that have that attribute and shows the attribute as selected on the settings box. Attributes selected on the settings box are cumulative. For example, selecting line style 1 on the settings box selects all elements in the design that have the line style 1 attribute. Selecting line style 3 on the settings box next adds to the selection all elements in the design that have the line style 3 attribute. The result is that all line style 1 and 3 elements are selected in the design and the two line styles show as selected on the settings box. Attributes only show on the settings box as being selected when the attribute is used by one or more elements in the design.

Selecting an element selects all of the element's attributes on the settings box. Selections on the settings box made by selecting elements are not cumulative. For example, selecting a red element selects the red color on the Colors tab of the

settings box. Selecting a blue element next unselects the color red on the setting box and selects the color blue.

CONSOLIDATE ELEMENTS INTO A GROUP

The MicroStation Group option consolidates all selected elements into a permanent group that acts like a single element when manipulated. To create a group, select the elements and invoke the Group tool from:

Menu	Edit > Group
Key-in window	**group selection** (or **gr s**) (ENTER)

The selected elements immediately consolidate into a group (there are no Micro_Station prompts). If selection of the elements is indicated by selection handles, grouping them replaces the individual element handles with handles on an imaginary box that encloses the grouped elements, as shown in Figure 6–18. If selection of the individual elements is shown by a selection color, grouping does not change the appearance of the selection (the selection color remains on the elements in the group).

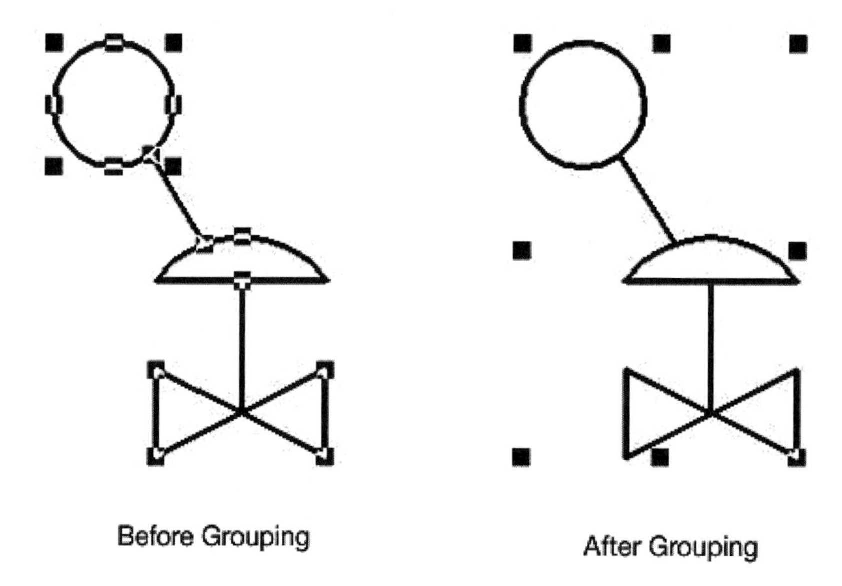

Figure 6–18 Example of grouping elements whose selection is indicated by handles

Note: *If no elements are selected, the Group tool is dim on the Edit drop-down menu and is unavailable.*

UNGROUP CONSOLIDATED ELEMENTS

The MicroStation Ungroup option permanently breaks all selected element groups into separate elements. To ungroup a group of elements, select the group (or groups) and invoke the Ungroup tool from:

Menu	Edit > Ungroup
Key-in window	**ungroup** (or **ung**) (ENTER)

The selected group (or groups) immediately breaks into separate elements (there are no MicroStation prompts).

If the group selection is indicated by selection handles, ungrouping will restore display handles for each element. If the group selection is indicated by color, ungrouping will not cause a change in appearance (the selection color remains on each element that was in the group).

Note: *If no elements are selected, the Ungroup tool is dim on the Edit drop-down menu and is unavailable.*

LOCKING SELECTED ELEMENTS

Locking elements prevents them from being manipulated. Locking is an easy way to protect completed parts of a design from accidental changes. To lock elements, select the elements and invoke the Lock tool from:

Drop-down menu	Edit > Lock
Key-in window	**change lock** (or **chan lo**) (ENTER)

The selected elements immediately lock (there are no MicroStation prompts). If selection handles indicate element selection, locking the elements changes the handles to the selection color. If a selection color indicates element selection, locking the elements makes no change in appearance.

Attempting to select a locked element for manipulation causes a locked symbol to appear by the pointer (as shown in Figure 6–19), and the element cannot be manipulated.

Note: *If no elements are selected, the Lock tool is shown dim on the Edit drop-down menu and is unavailable.*

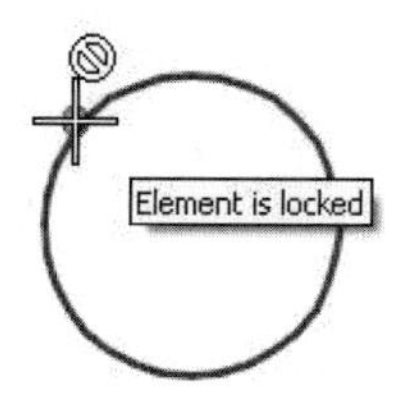

Figure 6–19 Attempting to select a locked element changes the appearance of the pointer

UNLOCK SELECTED ELEMENTS

Unlocking elements allows them to be manipulated again and removes the selection color from the elements. To unlock elements, select the elements to unlock and invoke the Unlock tool from:

Menu	Edit > Unlock
Key-in window	**change unlock** (or **chan u**) (ENTER)

The selected elements immediately unlock (there are no MicroStation prompts). Selecting locked elements always highlights them using the selection color. Unlocking elements leaves the element highlighted until the next action in MicroStation.

Note: *If no elements are selected, the Unlock tool is shown dim on the Edit drop-down menu and is unavailable.*

DRAG SELECTED ELEMENTS TO A NEW POSITION

The Individual Method with New Mode allows for dragging of selected elements to a new position in the design without using the Move element tool.

1. Select the elements using any of the Element Selection methods and modes.
2. Switch to the Individual Method and New Mode.
3. Position the pointer over one of the selected elements so that the selection circle that appears on the pointer touches the outline of the element (see Figure 6–20).
4. Press and hold the Data button while moving the pointer to the new location (the selected elements move with the pointer).
5. Release the Data button to place the selected elements at the new location.

The elements remain selected at the new location.

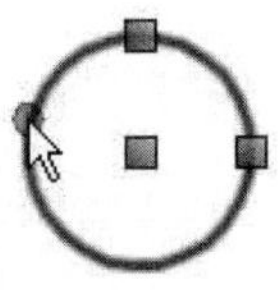

Figure 6–20 The Individual Method selection circle on an element

Note: *The other element selection method and mode combinations do not drag elements because pressing the Data button causes them to start a new selection.*

DRAG AN ELEMENT HANDLE TO CHANGE ITS SHAPE

When selection handles are on an element, the Individual Method with New Mode allows changing the shape of the element by dragging one of the selection handles.

1. Select the element using the Individual Method and New Mode.
2. Position the pointer over one of the element handles with the selection circle on the handle.
3. Press and hold the Data button while moving the handle to the new location (the selected element changes shape as the pointer moves).
4. Release the Data button to complete changing the element's shape.

The reshaped element remains selected. Figure 6–21 shows an example of dragging the middle, left handle to the right on a multi-sided shape.

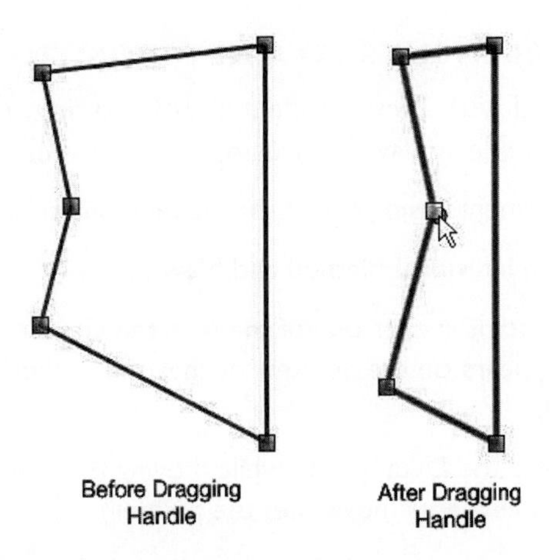

Figure 6–21 Example of modifying an element's shape

***Note:** This drag method only works on one element at a time. To change the shape of several elements at the same time, first group them.*

MANIPULATION TOOLS WITH PREVIOUSLY SELECTED ELEMENTS

Most manipulation tools work with selected elements. The tools that recognize selected elements include Align Elements by Edge, Array, Change Element Attributes, Delete, Mirror, Move and Copy, Move and Copy Parallel, Rotate, and Scale.

These manipulation tools affect all selected elements as if they were one element. To manipulate the elements, select them and invoke the appropriate manipulation tool. The tools immediately begin manipulating the elements so fewer data points are normally required. Some of the prompts are also slightly different than they are when working with individual elements.

The Move and Copy tools still require two data points for moving previously selected elements, but the points define the relative distance to move or copy the selected elements. In other words, the new location of the elements has the same relationship to the second data point as the original elements had to the first data point (see Figure 6–22).

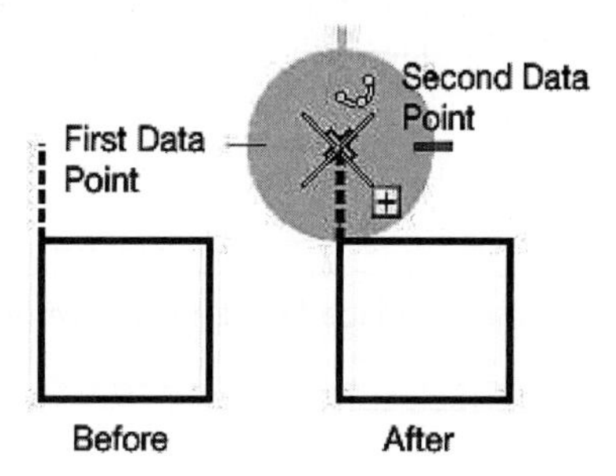

Figure 6–22 Example of using the Move tool with a selected element

***Note:** An alternate method for copying or moving selected elements is to drag the elements while holding down the Data button and release the Data button when the elements are at the desired location.*

Quickly delete selected elements by pressing DELETE or BACKSPACE; or by selecting the Delete tool on the Main tool box. To delete selected elements:

1. Use any element selection method to select the elements to delete.
2. Press DELETE or select the Delete tool from the Main tool frame.

All selected elements are deleted (there are no MicroStation prompts).

CHANGE THE ATTRIBUTES OF SELECTED ELEMENTS

Change element attributes by selecting the elements, selecting the desired attribute in the Change Element Attributes tool box, and clicking anywhere in the design. For example:

1. Select the elements whose attributes are to be changed.
2. Activate the Change Element Attributes tool box and select the desired attributes (e.g., change the line weight from 2 to 4 and the color from green to red).
3. Click the Data button anywhere in the design.

The selected elements immediately change to the selected attributes.

Selected elements can also be manipulated by right-clicking with the Element Selection tool. A pop-up menu appears that provides convenient, contextual access to tools for manipulating that element: Copy Element, Move Element, Scale, Rotate, Mirror, and Delete Element.

FENCE MANIPULATION

The Fence manipulation tools provide another way to manipulate groups of elements. Place a fence around the elements and use the fence content manipulation tools to manipulate the fenced elements. Only one fence at a time is active in the design plane and it remains active until it is removed. Select the Place Fence tool from:

Fence tool box	Select Place Fence (as shown in Figure 6–23).
Keyboard Navigation	**2** (Select Place Fence from the drop down menu.)

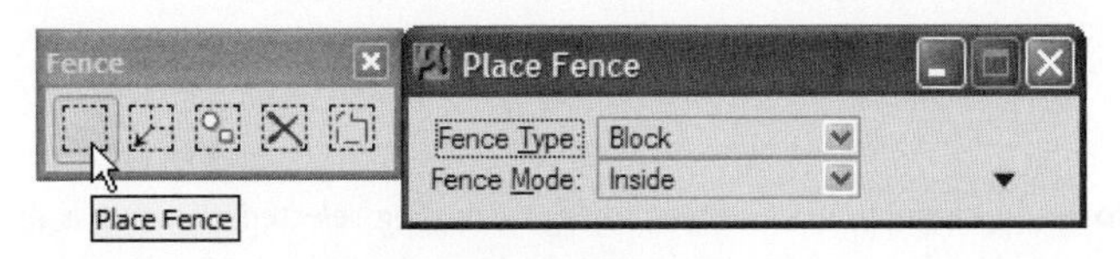

Figure 6–23 Place Fence tool on the Fence toolbar

FENCE TYPES

The **Fence Type** menu on the Place Fence settings box provides options for placing seven types of fences on the design, as shown in Figure 6–24.

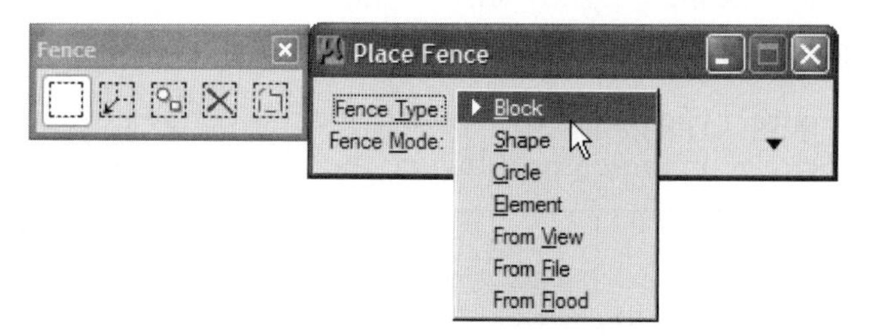

Figure 6–24 The Fence Type menu on the Place Fence Settings Box

Following are descriptions of each Fence Type.

Place a Block Fence

The **Block** Fence Type places the fence as a rectangular block by defining data points that are the diagonally opposite corners of the block. Invoke the Place Fence Block tool from:

Fence tool bar	Select the **Place Fence** tool and select **Block** from the **Fence Type** menu on the Place Fence settings box.
Key-in window	**place fence block** (or **pl f b**) (ENTER)

MicroStation prompts:

Place Fence Block > Enter first point *(Define one corner of the block in the design plane.)*

Place Fence Block > Enter opposite corner (Define the diagonally opposite corner of the block in the design plane.)

The result is an outlined, lightly shaded block fence.

Place a Shape Fence

The **Shape** Fence Type places the fence as a closed, multi-sided shape by defining the location of each shape vertex. The shape can have up to 5000 vertices. Invoke the Place Fence Shape tool from:

Fence tool bar	Select the Place Fence tool and select **Shape** from the **Fence Type** menu on the Place Fence settings box.
Key-in window	**place fence shape** (or **pl f s**) (ENTER)

MicroStation prompts:

Place Fence Shape > Enter Fence Points

Define the location of each fence vertex in the design plane. To complete the fence shape, either place the last vertex data point on top of the first fence data point or click **Close** on the Place Fence Settings box.

The result is an outlined, lightly shaded multi-sided shape fence.

Place a Circle Fence

The **Circle** Fence Type places the fence as a circle defined by a center point and a point on the circle circumference. Invoke the Place Fence Circle tool from:

Fence tool bar	Select the Place Fence tool and select **Circle** from the **Fence Type** menu on the Place Fence settings box.
Key-in window	**place fence circle** (or **pl f c**) (ENTER)

MicroStation prompts:

> Place Fence Circle > Enter circle center *(Define the location of the fence center in the design plane.)*
>
> Place Fence Circle > Enter edge point *(Define a data point on the circumference of the fence circle.)*

The result is an outlined, lightly shaded circle fence.

Place Fence on an Element

The **Element** Fence Type uses a selected element as the fence. Fence content manipulation tools manipulate the selected element with the rest of the elements in the fence depending on the active Fence Mode (discussed later). Invoke the Place Fence Element tool from:

Fence tool bar	Select the Place Fence tool and select **Element** from the **Fence Type** menu in the Tool settings box.
Key-in window	**place fence from shape** (or **pl f e**) (ENTER)

MicroStation prompts:

> Create Fence From Element > Identify element *(Select a closed element.)*

The resulting fence's perimeter exactly matches the element's perimeter and the area enclosed by the fence is lightly shaded.

Place a Fence From View

The **From View** Fence Type places a block fence that is the size of the selected View Window. Invoke the Place Fence From View tool from:

Fence tool bar	Select the Place Fence tool and select **From View** from the **Fence Type** option menu on the Tool settings box
Key-in window	**place fence view** (or **pl f v**) ENTER

MicroStation prompts:

> Create Fence From View > Select View *(Place a data point anywhere on the view to place a fence in that view.)*

MicroStation places an outlined, lightly shaded fence block that encloses the area of the design that is displayed in view window.

Place a Fence From File

The **From File** Fence Type places a fence block that encloses all elements in the design file. Invoke the Place Fence From File tool from:

Fence tool bar	Select the Place Fence tool and select **From File** from the **Fence Type** menu on the Tool settings box.
Key-in window	**place fence active** (or **pl f a**) (ENTER)

MicroStation prompts:

> Create Fence From Active Design File > Select View *(Click anywhere in the view window to place the fence.)*
>
> Create Fence From Active Design File > Fence placed - <Reset> to place again. *(Proceed with using the fence contents manipulation commands.)*

The result is an outlined, lightly shaded block fence that is just large enough to enclose all elements in the active design file.

Place a Fence From Flood

The **From Flood** Fence Type places a fence that outlines an area enclosed by existing elements in the design. Figure 6–25 shows an example of a Fence From Flood on an area enclosed by several block, circle, and line elements.

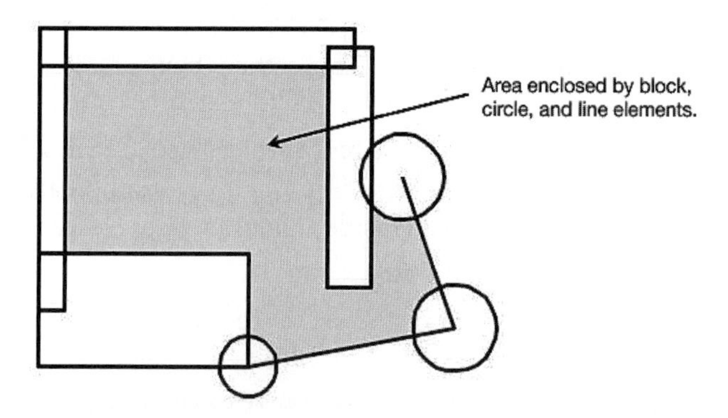

Figure 6–25 Example of a Fence From Flood in an enclosed area formed by several elements

Invoke the Place Fence From Flood tool from:

Fence tool bar	Select the **Place Fence** tool and select **From Flood** from the **Fence Type** menu on the Place Fence settings box.
Key-in window	**place fence active** (or **pl f a**) (ENTER)

MicroStation prompts:

Create Fence From Area Enclosing Point > Enter data point inside area *(Click anywhere within the area enclosed by elements. The flood area is outlined.)*

Create Fence From Area Enclosing Point > Accept/Reject (select next input) *(Click anywhere to accept the flood area and place the fence.)*

The result is an outlined, lightly shaded fence that outlines the closed area formed by the elements.

Note: *The Fence from Flood tool fails to create a fence when the flood area is not completely closed. It displays a message in the status bar that says, "Error – No enclosing region found."*

FENCE MODE

The **Fence Mode** menu (see Figure 6-26) on the Place Fence settings box provides options that control which elements the fence tools manipulate.

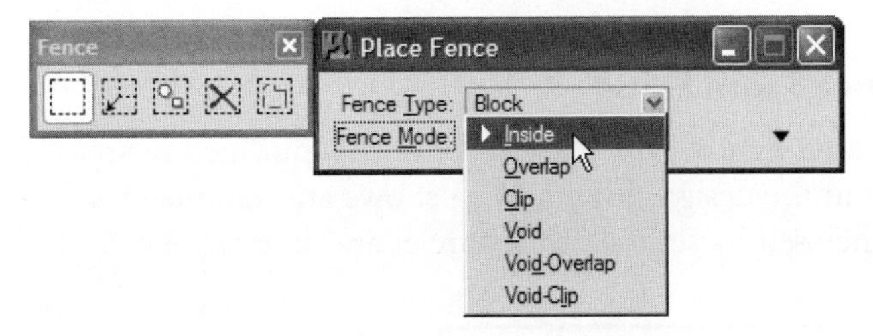

Figure 6–26 Fence Mode options in the Tool settings box

Before manipulating the contents of a fence, select the **Fence Mode** to control which elements the fence manipulates. For example, manipulate only elements that are completely inside the fence (Inside mode) or only elements that are completely outside the fence (Void mode).

Following are descriptions of each of the six fence modes.

Note: *The Fence Mode options are also available on the* ***Settings>Locks>Full*** *dialog box and on the settings box of tools that can manipulate fence contents.*

Inside Mode

Inside Fence Mode limits manipulation to the elements completely inside the fence. Figure 6–27 shows an example of deleting fence contents with Inside Mode active. Delete Fence Contents only deletes the blocks completely inside the fence.

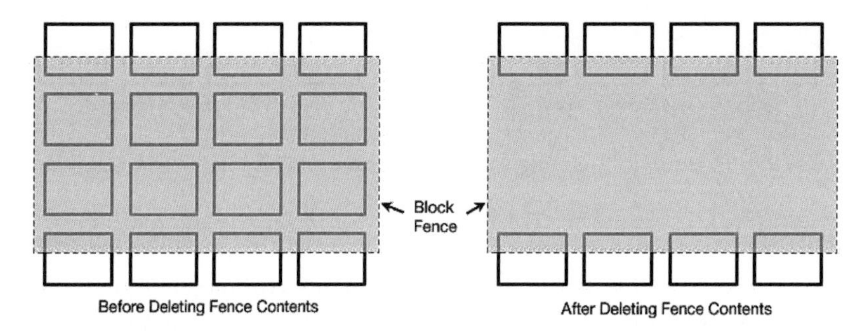

Figure 6–27 Example of deleting the fence contents with Inside mode active

Overlap Mode

Overlap Fence Mode allows manipulation of elements inside and overlapping the fence. Figure 6–28 shows an example of deleting a set of blocks with Overlap mode active. Delete Fence Contents deletes all the blocks, both inside and overlapping the fence.

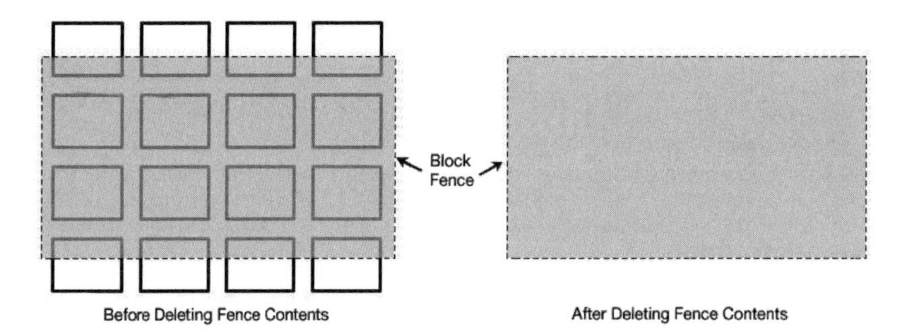

Figure 6–28 Example of deleting the fence contents with Overlap mode selected

Clip Mode

Clip Fence Mode limits manipulation to elements inside the fence and the inside part of elements overlapping the fence. Fence manipulation tools clip elements overlapping the fence at the fence boundary. Figure 6–29 shows an example of deleting fence contents with Clip mode. Delete Fence Contents clips the blocks overlapping the fence leaving the part of the blocks outside the fence area.

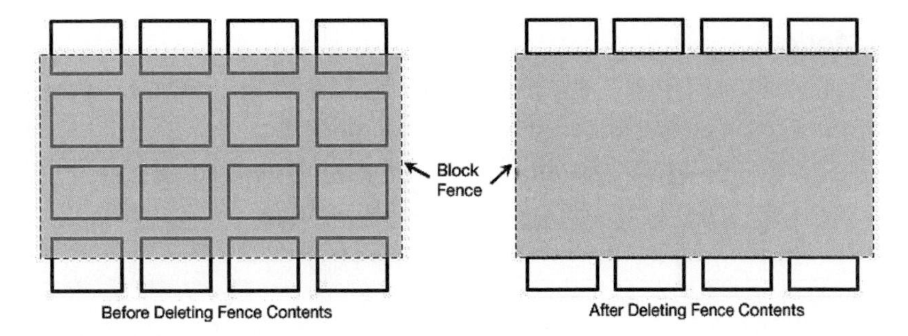

Figure 6–29 Example of deleting the fence contents with Clip mode selected

Void Mode

Void Fence Mode limits manipulation to elements completely outside the fence. Figure 6–30 shows an example of deleting fence contents with Void mode active. The Delete Fence Contents tool deletes the blocks completely outside the fence, leaving the blocks completely inside the fence.

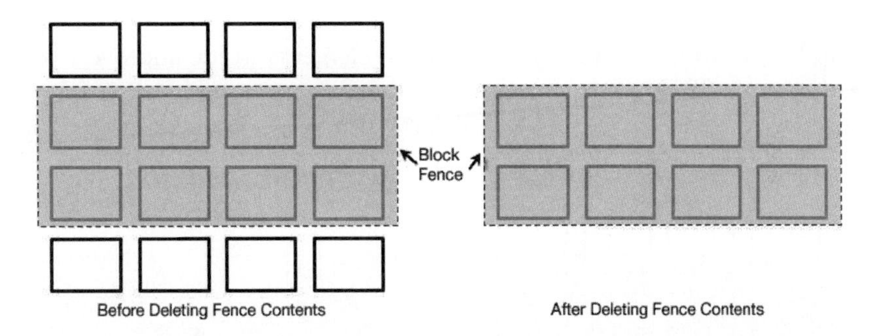

Figure 6–30 Example of deleting the fence contents with Void mode selected

Void-Overlap Mode

Void-Overlap Fence Mode limits manipulation to elements outside and overlapping the fence. Figure 6–31 shows an example of deleting fence contents with Void-Overlap mode active. The Fence Contents Delete tool deletes the blocks outside and overlapping the fence, leaving only the blocks completly inside the fence.

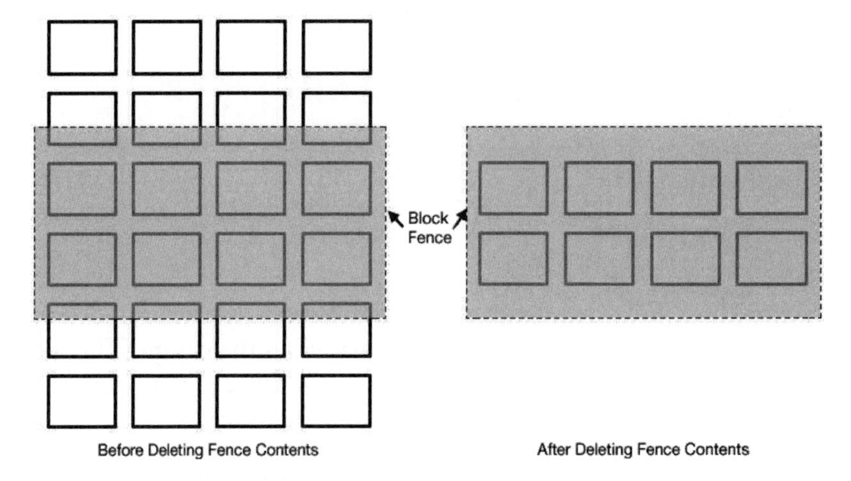

Figure 6–31 Example of deleting the fence contents with Void-Overlap mode selected

Void-Clip Mode

Void-Clip Fence Mode limits manipulation to elements outside the fence and the parts of the overlapping elements that are outside the fence. Figure 6–32 shows an example of deleting fence contents with Void-Clip mode active. The Delete Fence Contents tool clips the clips the blocks overlapping the fence, leaving only the part of the blocks inside the fence and the blocks that are completely inside the fence.

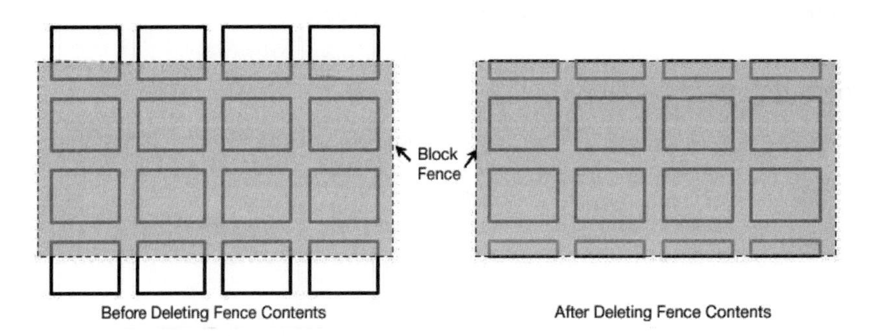

Figure 6–32 Example of deleting the fence contents with Void-Clip mode selected

MODIFY A FENCE'S SHAPE OR LOCATION

The **Modify Fence** tool modifies the shape of a fence already placed in the design or moves the existing fence to a new position. For example, if a fence shape does not include required elements, use the Modify Fence tool to drag the fence shape across the elements rather than completely redrawing the fence shape. Figure 6–33 shows the Modify Fence tool on the Fence tool box.

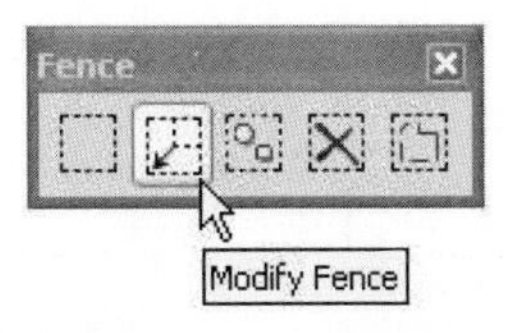

Figure 6–33 The Modify Fence tool on the Fence tool box

Modify a Fence Vertex

The Modify Fence **Vertex** mode moves a fence vertex to a new position in order to modify the size and shape of a fence already placed on the design. The tool modifies the fence itself, not the elements selected by the fence. To modify a fence vertex, invoke the **Modify Fence** tool and **Vertex** mode from:

Fence tool box	Select the **Modify Fence** tool and select **Vertex** from the **Modify Mode** menu on the Modify Fence settings box (see Figure 6–34).
Key-in window	**modify fence** (or **modi f**) (ENTER)

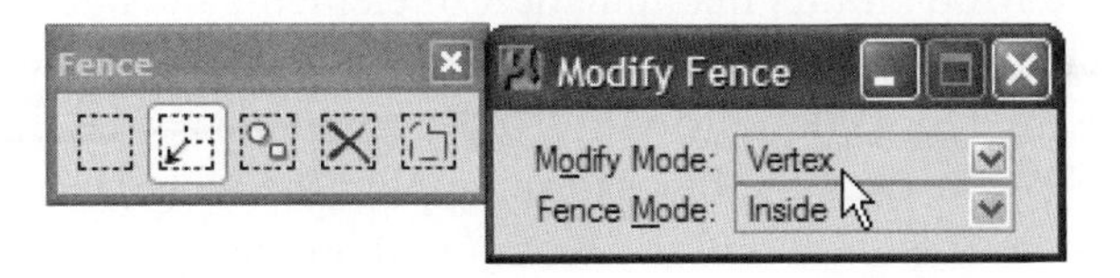

Figure 6–34 Invoking the Modify Fence Vertex tool

MicroStation prompts:

> Modify Fence Vertex > Identify vertex *(Click the Data button on the fence outline near the vertex to be modified, drag the vertex to the new position, and click the Data button again. Click the Reset button to complete the modification.)*

Figure 6–35 shows an example of modifying a vertex on a fence block.

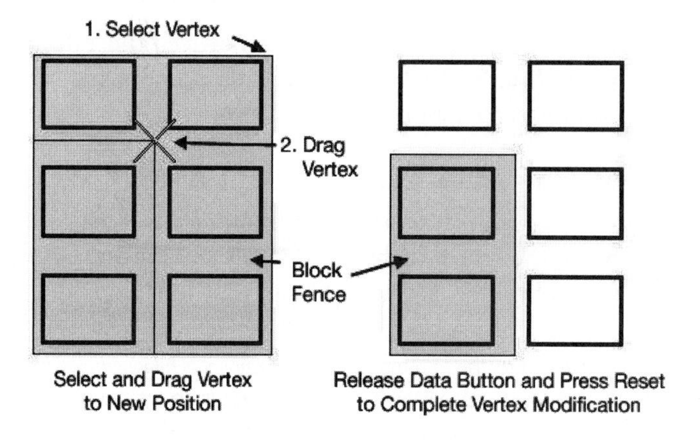

Figure 6–35 Example of modifying a block fence vertex

***Note:** Placing the first data displays a dynamic image of the fence that resizes as the cursor moves.*

Move a Fence

The Modify Fence **Position** mode moves an existing fence to a new location. The fence itself moves, not the elements in the fence. To move a fence to a new location in the design plane, invoke the **Move Fence** tool and Position mode from:

Fence tool box	Select the **Modify Fence** tool and select **Position** from the **Modify Mode** menu on the Tool settings box (see Figure 6–36).
Key-in window	**move fence** (or **mov f**) (ENTER)

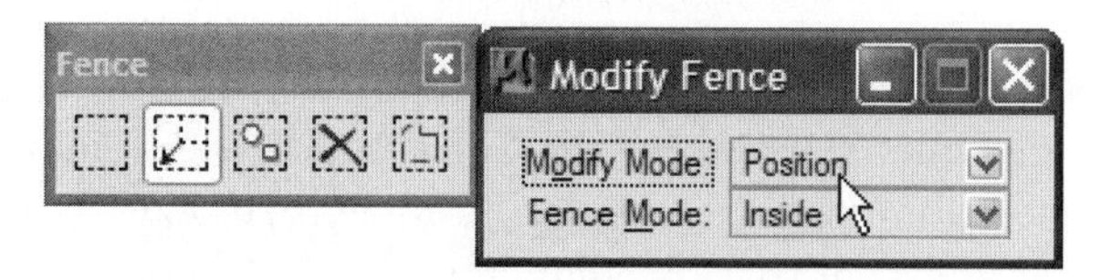

Figure 6–36 Invoking the Modify Fence tool (Move Position) from the Fence tool box

MicroStation prompts:

> Move Fence Block/Shape > Define origin *(Click the Data button in the design plane to identify the relative starting position of the move.)*
>
> Move Fence Block/Shape > Define distance *(Click the Data button in the design plane to identify the relative position to which the fence is to be moved. Click the Reset button to complete the move.)*

The fence moves a distance equal to the distance between the two data points, and the relationship of the final fence position to data point two is the same as the original fence position was to data point one.

REMOVE THE ACTIVE FENCE

Selecting the Place Fence tool removes the currently active fence. There is no separate tool for removing the fence. If the goal is only to remove the fence, select another tool after selecting the Place Fence tool.

***Note:** It is a good practice to remove the fence when it is no longer needed to protect against accidental manipulation. For example, when a fence is on the design, selecting Delete Fence Contents rather than the element delete tool could delete all the elements selected by the fence.*

MANIPULATE FENCE CONTENTS

Manipulation tools are available from the Fence and Manipulate tool boxes.

Manipulate Fence Contents from the Fence Tool Box

Click the **Manipulate Fence Contents** tool on the Fence tool box to display the **Fence Mode** menu and seven manipulation tools on the Manipulate Fence Contents settings box, as shown in Figure 6–37.

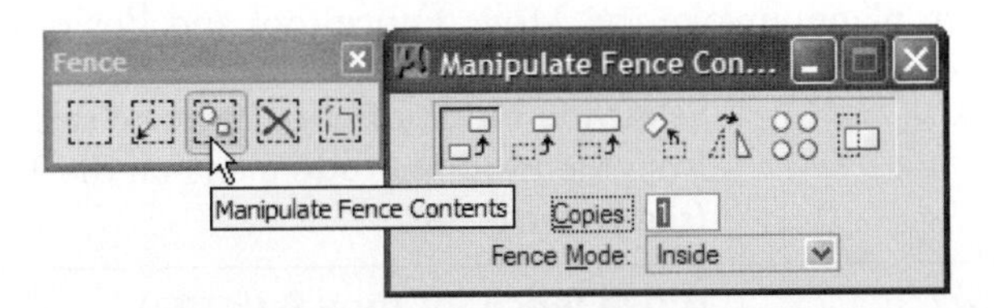

Figure 6–37 Invoking the Manipulate Fence contents tool from the Fence tool box

The manipulation tools (**Copy, Move, Scale, Rotate, Mirror, Array,** and **Stretch**) manipulate the elements selected by the active fence under control of the active Fence Mode. Operation of the manipulation tools is similar to the operations described in Chapter 4, except that the elements are selected by the fence rather than individually. For example, to move the contents of the fence to a new location in the design plane, select the Manipulate Fence Contents tool from:

Fence tool box	Select the **Manipulate Fence Contents** tool and select **Move** from the Manipulate Fence Contents settings box (see Figure 6–38).
Key-in window	**fence move** (or **f mo**) (ENTER)

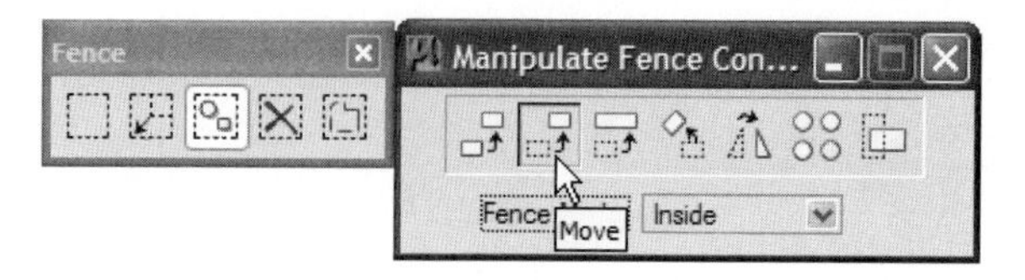

Figure 6–38 Invoking the Manipulate Fence Contents Move tool from the Fence tool box

MicroStation prompts:

> Move Fence Contents > Enter first point *(Locate the starting position of the move in the design plane.)*
>
> Move Fence Contents > Enter point to define distance and direction *(Locate the final destination in the design plane.)*

Manipulate Fence Contents from the Manipulate Tool Box

The manipulation tools on the Manipulate tool box can manipulate individually selected elements or all the elements selected by the active fence.

The settings box for each manipulation tool contains a **Use Fence** check box and a **Fence Mode** menu. Turning the check box ON switches the tools from manipulating individually selected elements to manipulating elements selected by the fence. The check box and menu are only available for use when a fence is defined on the design. In Figure 6–39, the **Use Fence** check box is on in the Copy Element settings box.

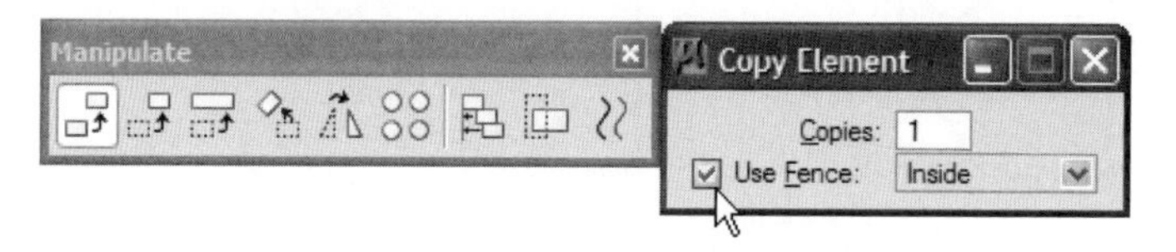

Figure 6–39 The Use Fence check box set to ON for the Copy tool

***Note:** Turning the check box ON for one manipulation tool turns it ON for all tools that contain the check box.*

Stretch Fence Contents

The **Stretch** tool is available from the Fence and Manipulate tool boxes but the tool only manipulates fence contents. If the **Use Fence** check box is OFF, the tool places a fence as the first step. If the **Use Fence** check box is ON, a fence must already be active on the design. Figure 6–40 shows the **Stretch** tool selected on the Manipulate tool box.

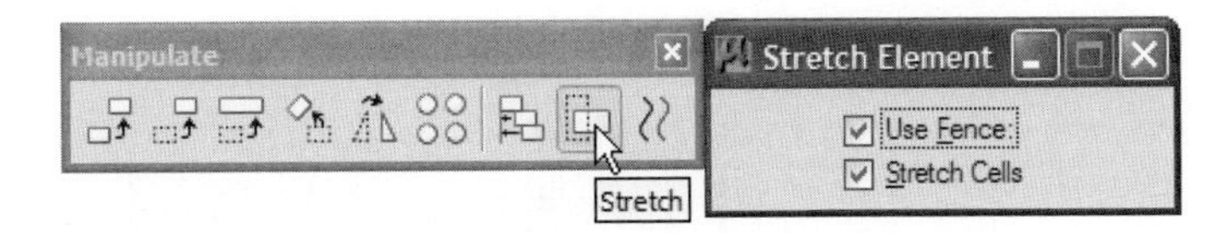

Figure 6–40 Stretch tool selected on the Manipulate tool box

The **Stretch** tool stretches the contents of a Block fence. The tool's impact on elements is as follows:

- It stretches Line, Line String, Multi-line, Curve String, Shape, Polygon, and Arc, elements that overlap the fence.
- It stretches cells that overlap the fence when the **Stretch Cells** check box is ON (for more information on cells, see Chapter 10).
- It ignores Circle and Ellipse elements that overlap the fence.
- It moves elements completely inside the fence.

There is no **Fence Mode** menu on the settings box when **Stretch** is active because the tool forces the mode to **Overlap**. There is also no dynamic image of the stretching elements as the pointer moves toward the final location. Figure 6–41 shows an example of stretching the contents of a fence horizontally to the right.

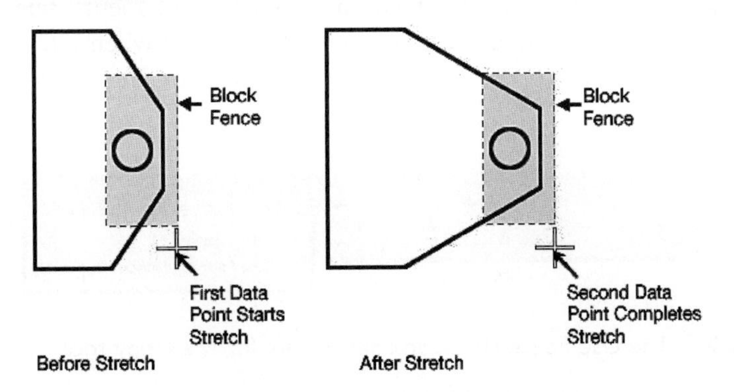

Figure 6–41 Example of stretching the contents of a fence

To stretch a group of elements, place a fence that overlaps the elements and select the Fence Stretch tool from:

Fence tool box	Select the **Manipulate Fence Contents** tool on the Fence tool box and select **Stretch** from the Fence Stretch settings box.
Key-in window	**fence stretch** (or **f st**) (ENTER)

MicroStation prompts:

Fence Stretch > Enter first point *(Locate a point in the design plane to start the stretch.)*

Fence Stretch > Enter point to define distance and direction *(Locate a point in the design plane to end the stretch.)*

The new fence location has the same relationship to the second data point as the original fence location had to the first data point.

Change the Attributes of Elements Selected by a Fence

The **Use Fence** check box and **Fence Mode** menu are also on the Change Element Attributes settings box. Turning the check box ON switches the Attributes tool from changing the attributes of individually selected elements to changing the attributes of elements selected by a fence. The check box and menu are only available for use when a fence is active on the design. Figure 6–42 shows the **Use Fence** check box selected for the Change Attributes tool.

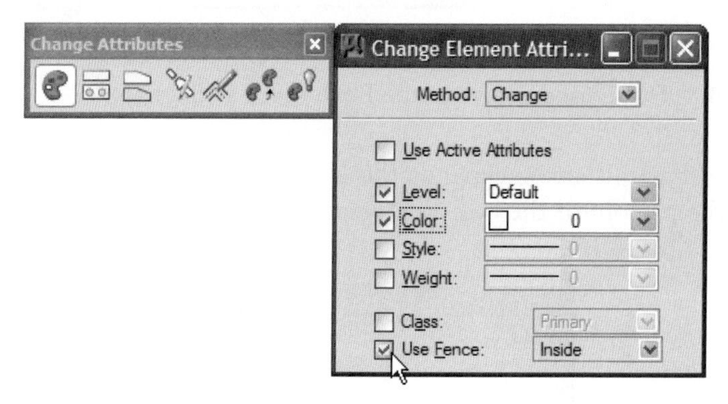

Figure 6–42 Use Fence check box selected for the Change Attributes tool

Drop Fence Contents

The Drop Fence Contents tool is invoked from the Fence tool box (as shown in Figure 6–43) and drops all complex elements within the contents of the fence.

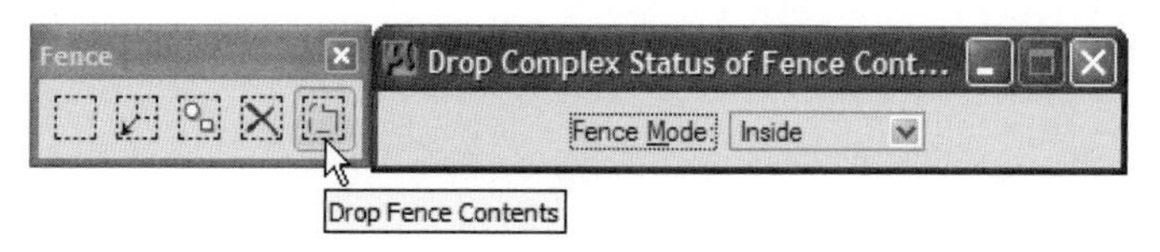

Figure 6–43 The Drop Fence Contents tool selected on the Fence tool box

MicroStation prompts:

Drop Complex Status of Fence Contents > Accept/Reject Fence contents *(Click the Data button to initiate dropping of the elements.)*

Grouped elements selected by the fence become individual elements.

NAMED FENCES

The Place Fence tool on the Fence tool bar includes options for creating and recalling named fences. The options allow placing fences and saving their location for later recall. The options allow saving multiple named fences in the design. Display the named fences options from:

Fence tool box	Select the **Place Fence** tool and click the **Expand** arrow on the Place Fence settings box (see Figure 6–44.)

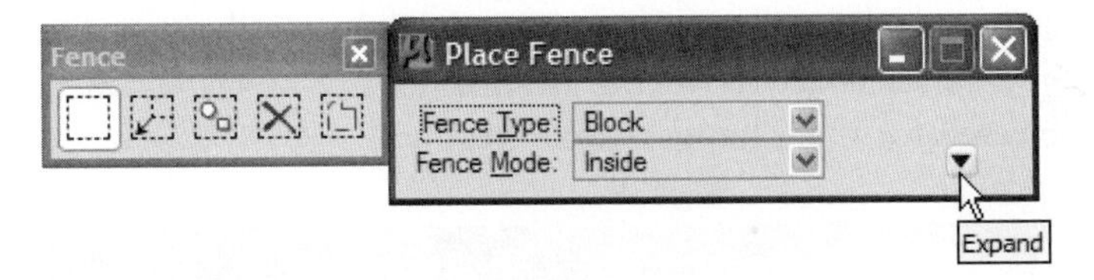

Figure 6–44 The Expand arrow on the Place Fence settings box

Clicking the Expand arrow expands the settings box to display the named fence options at the bottom of the box, as shown in Figure 6–45.

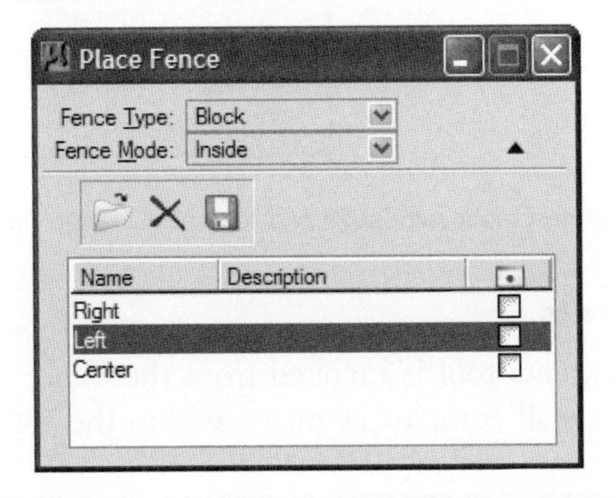

Figure 6–45 The Place Fence settings box expanded to show the Named Fences options

Create a Named Fence

Use the Place Fence tool and settings box to create the fence. After creating the fence, expand the Place Fence settings box and click the **Create Named Fence From Active Fence** tool (see Figure 6–46) to create the new named fence.

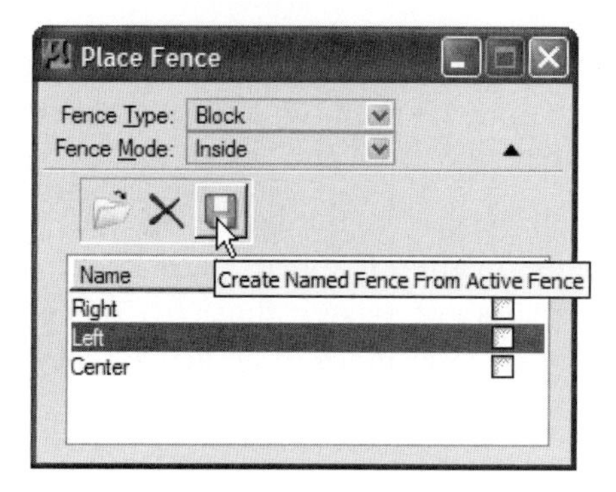

Figure 6–46 The Create Named Fence From Active Fence option

Clicking **Create Named Fence From Active Fence** saves the fence, names it "Untitled," places the named fence in the fences list on the settings box, and selects the name for editing (see Figure 6–47). MicroStation saves the fence location, type, mode, name, and description.

To complete saving the fence, replace "Untitled" in the **Name** field with a meaningful fence name and optionally provide a brief description of the fence in the **Description** field. Press ENTER after typing each entry to complete the entry. Select the **Name** or **Description** field for editing by clicking the field with the Data button.

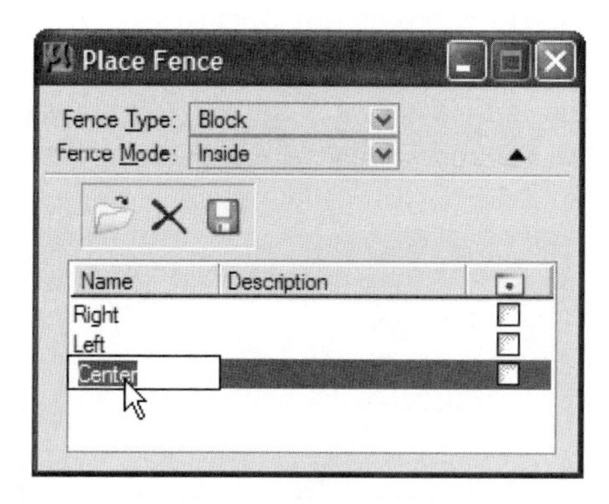

Figure 6–47 A saved fence name selected on the Place Fence settings box

Display the Location of Named Fences

The named fences options allow viewing the location of named fences on the design plane.

If there is only one named view in the list, an outline of it is always visible on the design plane when the Place Fence settings box is expanded to show the named fences options, even if the named fence is not the active fence. If the named fence is not the active fence, the display is only an indication of where the fence is locat-

ed and it cannot be used for fence manipulations (although, as discussed later, it can be made the active fence). Closing the Create Fence settings box removes the view of the selected named view from the design plane.

If more than one named fence is in the list, select one to display on the design plane by clicking its name in the expanded Place Fence settings box. Displaying one of the named fences does not make it the active fence. See Figure 6–48 for an example of the difference in appearance of an active fence and a displayed named fence that is not active.

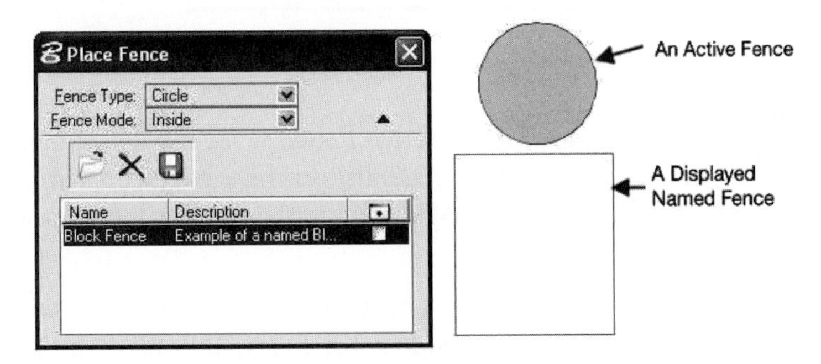

Figure 6–48 Example of an active fence and a displayed, but not active, named fence

To keep the location of a named view visible on the design plane even when the expanded Place Fence settings box is not open, turn ON the named view's check box. If there are multiple named views in the design, the check box can be turned ON for one to all of the named views. Figure 6–49 shows all named fence check boxed turned ON in the expanded Place Fence settings box.

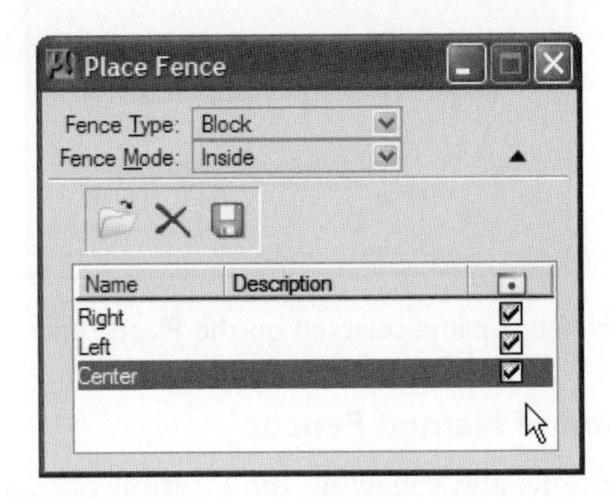

Figure 6–49 Named fence check boxes turned ON in the named view list on the expanded Create Fence settings box

Make a Named Fence the Active Fence

To make a named fence the active fence in the design, select the named fence in the list of named fences on the expanded Create Fence settings box and click the **Activate Named Fence** tool (see Figure 6–50).

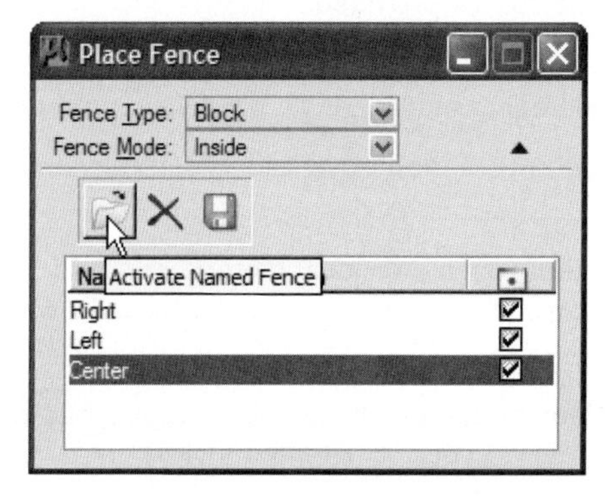

Figure 6–50 The Activate Named Fence tool on the expanded Create Fence settings box

The selected named fence becomes the active fence and is available for use in fence content manipulations. Its appearance is differentiated from an unnamed active fence by a wide, solid outline.

Delete a Named Fence

To delete a named fence, select the named fence in the list of named fences on the expanded Create Fence settings box and click the **Delete Named Fence** tool (see Figure 6–51).

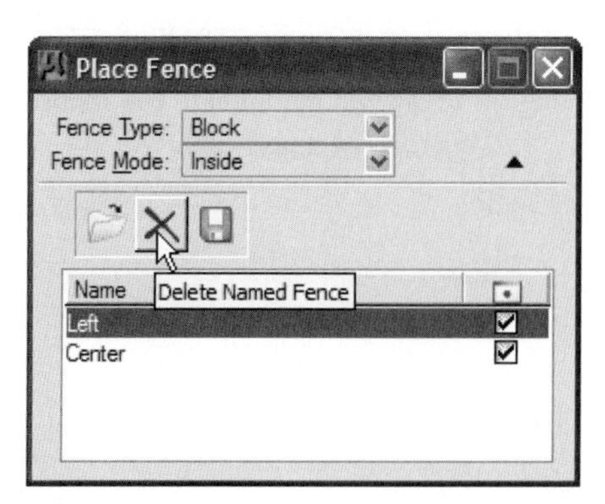

Figure 6–51 The Delete Named Fence tool on the expanded Create Fence settings box

The tool immediately deletes the selected named fence from the list.

Open the Exercise Manual PDF file for Chapter 6 on the accompanying CD for project and discipline specific exercises.

REVIEW QUESTIONS

Write your answers in the spaces provided.

1. The Element Selection tool, ____________ Method, and ____________ Mode are used to select all elements by dragging a block around them.
2. The Element Selection tool Mode that allows selecting and unselecting elements at one time is the ____________ mode.
3. The Element Selection tool's Individual Method and Add Mode can be turned on by pressing two keyboard shortcuts. What two keys will turn them on? ________________________.
4. The shape of an element can be changed by dragging one of its handles only when the ____________________________ tool is active.
5. Briefly explain the purpose of locking individual elements.

   ____

   ____
6. List the six fence selection modes available in MicroStation.

   ____

   ____
7. Explain briefly the difference between the Overlap and Void-Overlap mode.

   ____

   ____
8. The Fence Stretch tool will stretch an arc. TRUE or FALSE
9. The element manipulation tools will manipulate the fence contents when the ______________________ button is turned ON.

10 How is a Fence removed?

   ____

   ____

Placing Text, Data Fields and Tags

OBJECTIVES

Topics explored in this chapter:

- Placing single-character fractions
- Using several tools to place text elements
- Importing text from other computer applications
- Editing the content of text elements
- Manipulating the attributes of text elements
- Placing notes in the design
- Creating and using "fill-in-the-blanks" Text Node and Data Field elements
- Placing and managing Tags

PLACE TEXT

Chapter 4 described the Place Text tool's **By Origin** method for placing text strings in a design and using settings in the Tool Settings box to control the size and shape of the text strings. This section describes the other placement methods that the Place Text tool offers.

SETTING TEXT ATTRIBUTES

Chapter 4 discussed the text attributes available on the Place Text settings box and the Text Editor box. One additional tool on the settings box is used to create text strings with mixed attributes. The **Apply Changes to All Text** check box controls how attribute changes are applied.

- When the check box is ON, attribute changes apply to all text in the Text Editor box.
- When the check box is OFF, attribute changes apply only to text selected in the Text Editor box (For example, making part of a text string Bold and Italic and the rest of the text only Italic.)

THE PLACE TEXT METHODS MENU

The Place Text methods described in this section are all available from the **Methods** menu on the Place Text settings box, as shown in Figure 7–1.

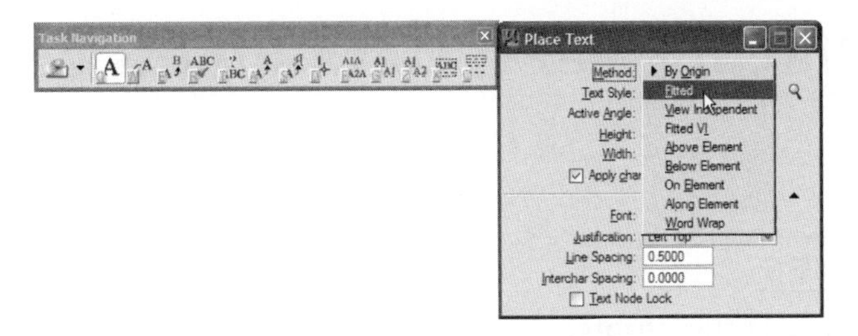

Figure 7–1 The Method menu options on the Place Text settings box

Note: *If the Place Text settings box does not display all setting fields, click the* ***Expand*** *arrow on the lower right corner of the box to display the additional fields. The settings box is expanded in Figure 7–1.*

Fitted Method

The Place Text tool's **Fitted** method places text keyed into the Text Editor box along an imaginary line between two placement data points. The text rotates and changes size as necessary to extend from the first data point to the second data point. Figure 7–2 shows two placements of fitted text and where the data points were placed for each one.

Before typing the text, select the text **Font** and **Justification** from the Place Text settings box. The top, center, and bottom justification options control text placement along the imaginary line between the two placement data points. For example, selecting **Left Bottom, Center Bottom,** or **Right Bottom,** places the text above the imaginary line. The text in Figure 7–2 is bottom justified.

Once placed, the text string is a normal text element and all the manipulation tools apply to it. The line string has one keypoint at the location of the first placement data point.

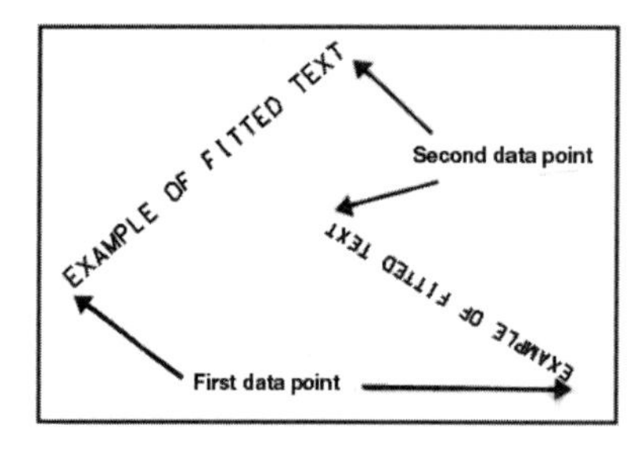

Figure 7–2 Example of placing fitted text

Invoke the Place Text tool from:

Task Navigation tool box (active task set to Text)	Select the Place Text tool. In the Tool Settings box, select Fitted from the Method menu and make necessary changes to text attributes.
Keyboard Navigation (Task Navigation tool box with active task set to Text)	**q** (In the Tool Settings box select Fitted from the Method menu and make necessary changes to text attributes.)

MicroStation prompts:

Place Fitted Text > Enter text before placing start point *(Type the text in the Text Editor box.)*

Place Fitted Text > Enter more text or position start point *(Place a data point to define the starting point of the text.)*

Place Fitted Text > Enter more text or position end point *(Place a data point to define the ending point of the text.)*

Place Fitted Text > Enter more text or position start point *(Continue placing fitted copies of the text, key-in different text in the Text Editor box, or click the Reset button to drop the text.)*

View Independent and Fitted VI Methods

The Place Text tool's **View Independent** method places text in the same way as the **Place Text** origin method, and the Place Text tool **Fitted VI** method places text in the same way as **Fitted**. Both of these methods are 3D tools for placing text that is readable regardless of the view orientation. For example, view-independent text placed on the top view of the 3D design is readable when viewed in the left, right, and bottom views.

Above Element Method

The Place Text tool's **Above Element** method places a text string above a selected linear element. Key-in the text string in the Text Editor box and identify the linear

element at the text placement point. The tool places the text parallel to and above the selected element, as shown in Figure 7–3.

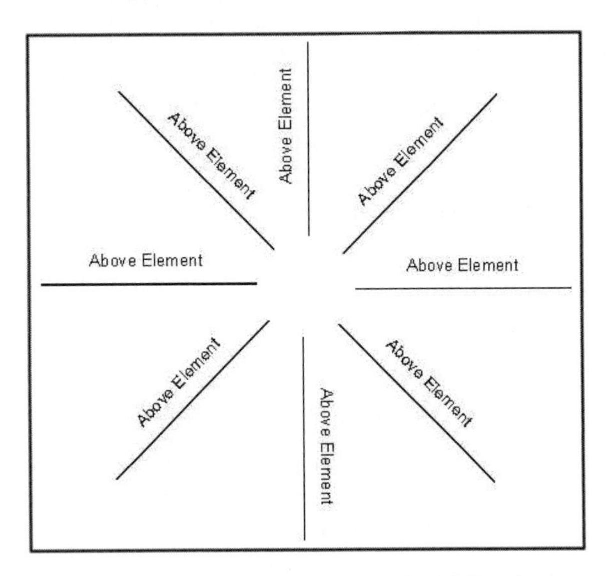

Figure 7–3 Examples of placing text above a linear element

Note: *The "above" side of the line or line segment depends on the line's rotation. The first data point that placed the line is the line's origin point. The lines in Figure 7–3 extend from the center out.*

The **Font, Height**, and **Width** fields in the Tool Settings window determine the text style and size. The **Line Spacing** field sets the space between the bottom of the text string and the selected element in working units. For example, a **Line Spacing** of zero places the text touching the selected element and a **Line Spacing** of 3.0 places the text three Master Units above the element.

The left, center and right **Justification** selections determine text string position in relation to the selection point on the element. For example, a left setting (**Left Top, Left Center**, or **Left Bottom**) places the text such that the selection point on the element is to the left of the text string.

Once placed, the text string is a normal element and all the manipulation commands apply to it. The line string has one keypoint at the left, center, or right of the element depending on the selected **Justification**. The text string has no connection to the selected element. For example, deleting the element does not delete the text string. Invoke the Place Text tool from:

Task Navigation tool box (active task set to Text)	Select the Place Text tool. In the Tool Settings box, select Above Element from the Method menu.
Keyboard Navigation (Task Navigation tool box with active task set to Text)	**q** (In the Tool Settings box select Above Element from the Method menu.)

MicroStation Prompts:

Place Text Above > Enter text *(Type the text in the Text Editor box.)*

Place Text Above > Identify element *(Identify the element at the point where the text is to be placed. A dynamic image of the text appears above the element.)*

Place Text Above > Accept/Reject (Select next input) *(Click the Data button to accept the element and place the text, or click the Reset button to reject it.)*

***Note:** If the Text Editor box contains text when the **Above Element** method is selected, MicroStation skips the "Enter text" prompt and goes straight to asking for the element to be identified.*

Below Element Method

The Place Text tool's **Below Element** method works the same way as the **Above Element** method, except that it places the text string below the element, as shown in Figure 7–4. Invoke the Place Text tool from:

Task Navigation tool box (active task set to Text)	Select the Place Text tool. In the Tool Settings box, select Below Element from the Method menu.
Keyboard Navigation (Task Navigation tool box with active task set to Text)	**q** (In the Tool Settings box select Below Element from the Method menu.)

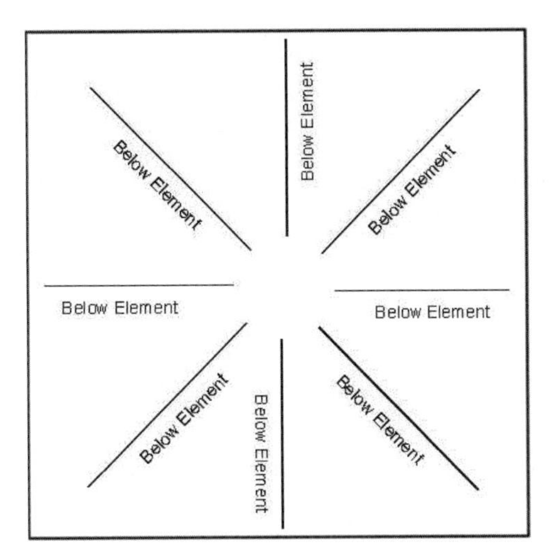

Figure 7–4 Examples of placing text below a linear element

On Element Method

The Place Text tool's **On Element** method works the same way as the **Above Element** and **Below Element** methods, except that it cuts a hole in the linear element that is slightly wider than the text string and places the text string in the hole, as shown in Figure 7–5.

The left, center, and right **Justification** options determine where to cut the hole on the line or line segment. For example, selecting **Left Top, Left Center,** or **Left Bottom** cuts the hole such that the data point is near the left end of the hole and to the left of the text string. Invoke the Place Text tool from:

Task Navigation tool box (active task set to Text)	Select the Place Text tool. In the Tool Settings box, select On Element from the Method menu.
Keyboard Navigation (Task Navigation tool box with active task set to Text)	**q** (In the Tool Settings box select On Element from the Method menu.)

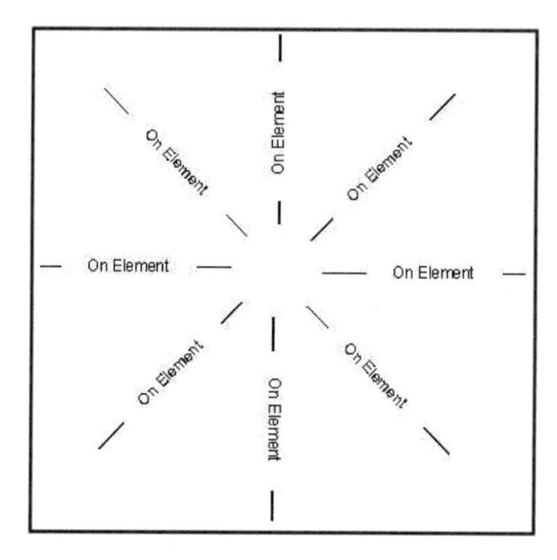

Figure 7–5 Examples of placing text on a linear element

Along Element

The Place Text tool's **Along Element** method places text that follows the contour of non-linear elements (such as curve strings). Figure 7–6 shows examples of placing text along elements.

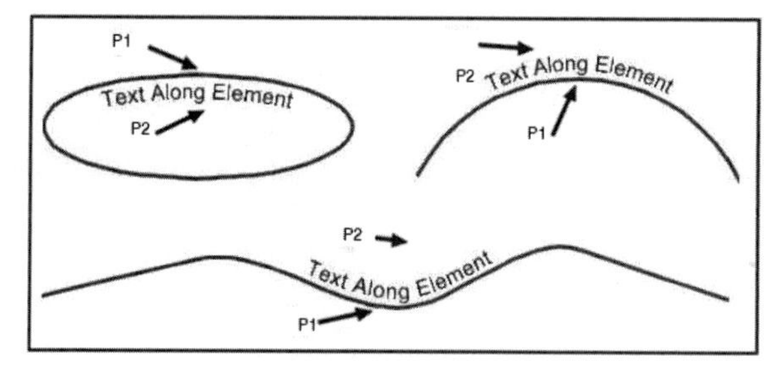

Figure 7–6 Examples of placing text along elements

The **Font, Height,** and **Width** fields in the Tool Settings window determine the text style and size. The left, center, and right text **Justification** options determine where the text lines up in relation to the element selection point. The examples in Figure 7–6 are center justified.

To compensate for tight curves, increase the space between characters in the string with the **Interchar Spacing** field. Move the text away from the element with the **Line Spacing** field.

The single character strings along the element form a "complex shape" and act like one text element when manipulated. The complex shape has one keypoint for tentative snapping at the left, center, or right of the element depending on the select-

ed **Justification.** The text string has no connection to the selected element For example, deleting or moving the element does not delete or move the text. Invoke the Place Text tool from:

Task Navigation tool box (active task set to Text)	Select the Place Text tool. In the Tool Settings box, select Along Element from the Method menu.
Keyboard Navigation (Task Navigation tool box with active task set to Text)	**q** (In the Tool Settings box select Along Element from the Method menu.)

MicroStation prompts:

> Place Text Along > Enter text *(Type the text in the Text Editor box.)*
>
> Place Text Along > Identify element, text location *(Define the placement point on the element. Dynamic images of the text appear both above and below the element, as shown in Figure 7–7.*
>
> Place Text Along > Accept, select text above/below *(Click the Data button on the side of the element where the text is to be placed, or click the Reset button to reject it.)*

The tool places the text on the selected side of the element and the dynamic image of the text disappears from the other side.

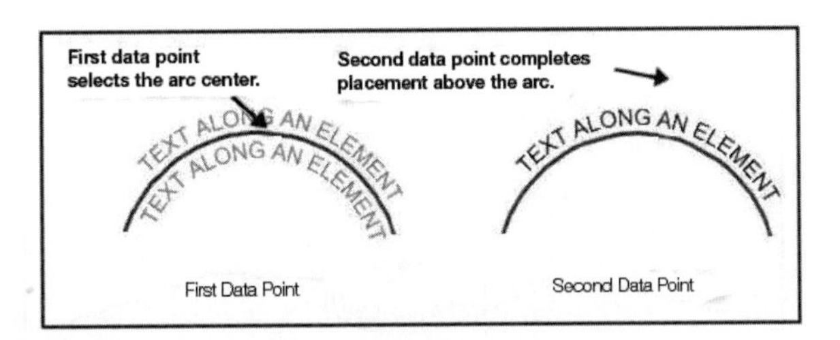

Figure 7–7 Example of placing text along an element

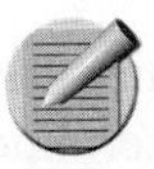

> **Note:** *If the text string contains line breaks, the tool removes them when it places the text along the selected element.*

Word Wrap Method

The Place Text tool's **Word Wrap** method places a text string within the area of an imaginary box using word wrapping as necessary to stay within the box width. Start the tool sequence by placing two data points to define the diagonally-opposite points of the imaginary box. The dynamic image of the imaginary box appears in the design using a dashed line style.

After defining the box, key-in the text in the Text Editor box. As the text is typed, a dynamic image of it appears in the box starting at the corner of the box defined

by the first data point. Wrapping occurs at word breaks. If the text string is too long to fit within the box, the lines wrap to the area outside the box.

After keying in the text, place a third data point anywhere in the design to complete the tool sequence and place the text in the design. The third data point also removes the imaginary box. To terminate the sequence without placing the text, click the Reset button.

Note: *Resetting the sequence should remove the dynamic image of the box and text. If it does not disappear, update the view.*

The location of the two data points that define the box is important. Placing the second data point to the left of the first data point places the text as a mirror image. Placing the second data point below the first data point places the text upside down. Figure 7–8 shows examples of placement with different data point relationships.

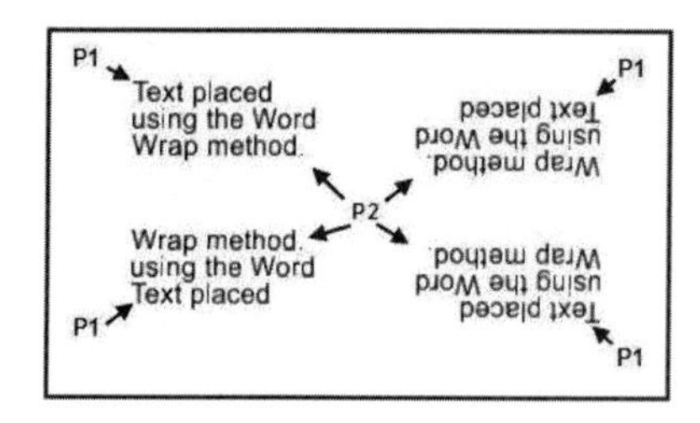

Figure 7–8 Examples of placing text using the Word Wrap method

The **Font**, **Height**, and **Width** fields determine the text style and size. The left, center, and right **Justification** options control placement of the lines of text within the area of the box. For example, **Left Top**, **Left Center**, and **Left Bottom** all create a smooth left text margin. The **Line Spacing** field determines the space between the wrapped lines of text. Invoke the Place Text tool from:

Task Navigation tool box (active task set to Text)	Select the Place Text tool. In the Tool Settings box, select Word Wrap from the Method menu.
Keyboard Navigation (Task Navigation tool box with active task set to Text)	**q** (In the Tool Settings box select Word Wrap from the Method menu.)

MicroStation prompts:

Place Word Wrapped Text > Place first corner point *(Place a data point to define one corner of the box.)*

Place Word Wrapped Text > Place second corner point *(Place a data point to define the diagonally opposite corner of the box.)*

Place Word Wrapped Text *(Key-in the text string in the Text Editor box.)*

Place Word Wrapped Text > Enter data point to accept text *(Click the Data button to place the text or click the Reset button to terminate the tool without placing the text.)*

Place Word Wrapped Text > Place first corner point *(Place a data point to define one corner of another placement box or select another tool.)*

PLACE NOTE

The **Place Note** tool places text as a dimension with a leader line and arrow. Figure 7–9 shows the tool selected and thc expanded Place Text settings box.

Note: *To expand the settings box, click the* **Expand** *arrow on the lower right corner of the box.*

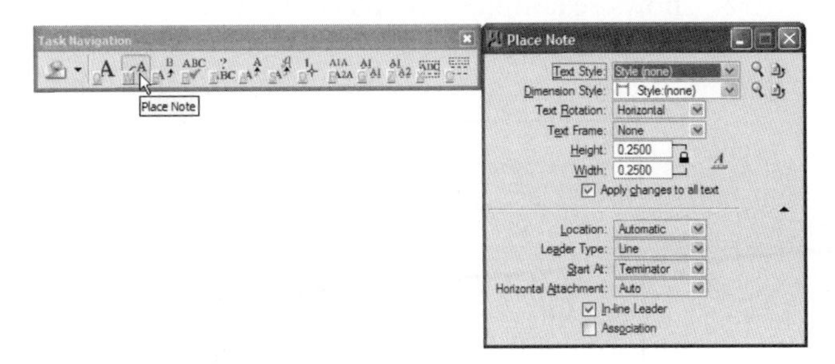

Figure 7–9 Invoke the Place Note tool

Place a note by typing the note text onto the Text Editor box and then defining two data points on the design: one data point defines the location of the arrow and the other defines the location of the text. The Start menu controls data point order (discussed later in this section).

PLACE NOTE SETTINGS

Note text uses the same settings (such as Height, Width, and Font) that the other Place Text tools use, but several additional settings customize note placement in the design.

Dimension Style

The Dimension Style menu allows for selecting a dimension style for the notes from a list of available styles that control the appearance of dimensioning elements such as the arrows on the end of the leader lines. For more information, see Chapter 9.

Text Rotation

The **Text** Rotation menu provides three options (**Horizontal**, **Vertical**, and **Inline**) that set the rotation of the text relative to the leader line. Settings in the Di-

mension Style determine the default Text Rotation value. Figure 7–10 shows examples of the rotation settings.

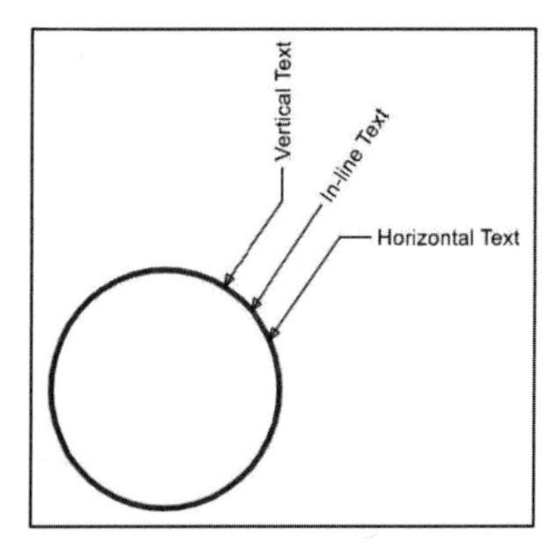

Figure 7–10 Notes placed using the three Text Rotation settings

Text Frame

The Text Frame menu provides several types of frames for enclosing the text in notes. Figure 7–11 shows example of each frame provided by the menu. The text on each note provides the name of the frame option.

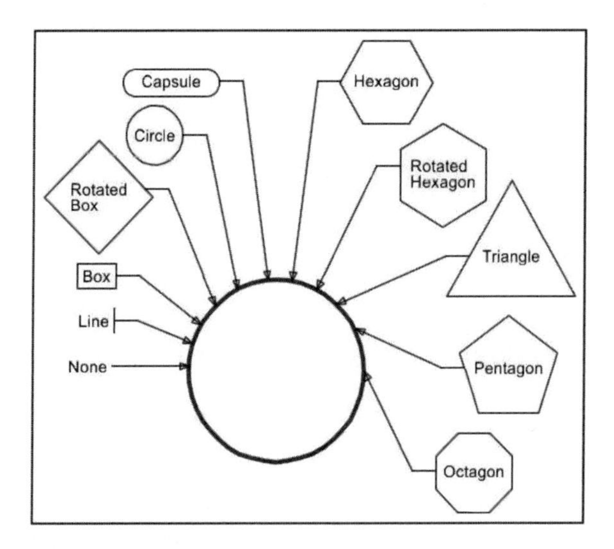

Figure 7–11 Notes placed using all of the available text frame options

Location

The **Location** menu controls the number of leader line segments allowed for placing a note. Notes placed with **Automatic** location active can have only one seg-

ment. Notes placed with **Manual** location active can have multiple segments. Figure 7–12 shows examples of notes placed with the two Location options.

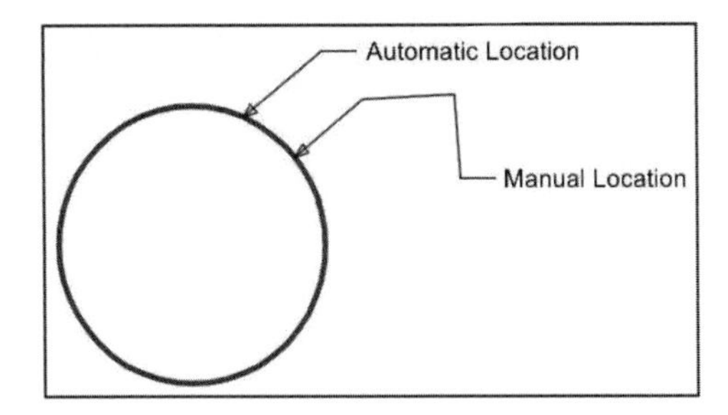

Figure 7–12 Notes placed using the two Location options

Leader Type

The **Leader Type** menu allows for placing Notes with a straight **Line** leader and a **Curve** leader. The top two Notes in Figure 7–13 are examples of the two leader types.

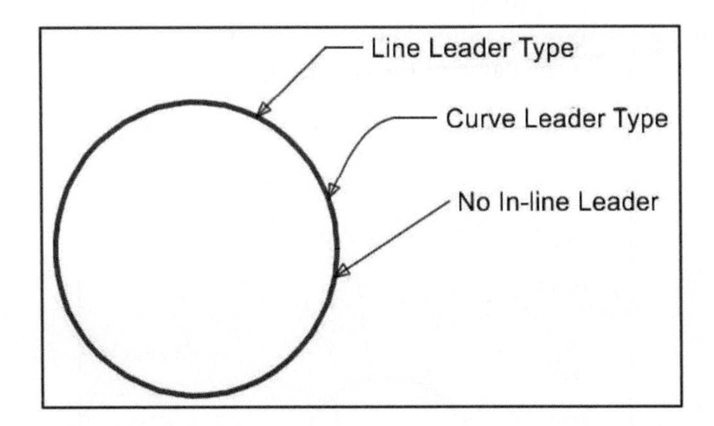

Figure 7–13 Notes placed with the Leader Type options and with the In-line Leader check box OFF

Start At

The **Start At** menu provides two options to control the action of the data points that place notes. Selecting **Terminator** causes the first data point to define the location on the design of the end of the terminator symbol and the second data point to define the location of the note text. Selecting **Text** causes the first data point to define the note text location and the second data point to define the terminator location.

Horizontal Attachment

The **Horizontal Attachment** menu controls the placement of the text in relation to the leader line for horizontally placed text. The **Automatic** option lets the tool decide on which side to place the text. The **Left** option forces the leader to be on the left side of the text and the **Right** option forces the leader to be on the right side of the text. Figure 7–14 shows examples of placement with each of the options.

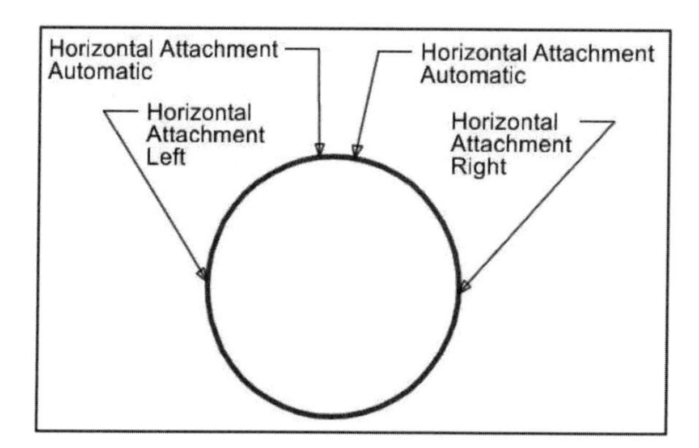

Figure 7–14 Notes placed with the Leader Type options and with the In-line Leader check box OFF

In-line Leader Check Box

If the **In-line Leader** check box is ON, the Place Note tool places notes that have a leader segment parallel to the note text. If the check box is OFF, there is no segment parallel to the note text. The first two notes in Figure 7–13 are examples of note placement with the check box ON and the third (bottom) note is an example of note placement with the check box OFF.

Association Check Box

Turn the **Association** check box ON to associate the note with another element by first snapping to the other element, and then selecting it, when defining the note terminator location in the design. If the **Association** check box is OFF, notes cannot be associated to other elements. When a note terminator is associated with an element, the terminator moves with the element, but the note text does not move. Invoke the Place Note tool from:

Task Navigation tool box (active task set to Text)	Select the Place Note tool.
Keyboard Navigation (Task Navigation tool box with active task set to Text)	**w**

MicroStation prompts:

Place Note > Define start point (*Type the note text in the Tool Editor box and define first note placement point).*

Place Note > Define next point or <Reset> to abort or (Ctrl) to create multiple leaders *(Define the second note placement point.)*

Place Note > Define next point, or <Reset> to abort or (Ctrl) to create multiple leaders *(Continue placing copies of the note or click the Reset button to drop the note and clear the Text Editor box.)*

PLACING A NOTE WITH MULTIPLE LEADERS

The Place Note tool provides for placing multiple leader lines attached to a single note text string. Figure 7–15 shows a note with multiple leader lines.

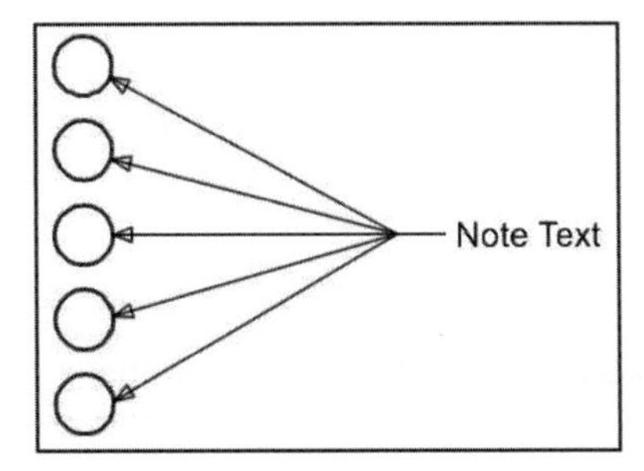

Figure 7–15 A note with multiple leader lines

To place a note with multiple leader lines, select **Text** from the **Start** menu before placing the note. Type the note text and place a data point to define the location of the note text. After defining the text position, hold down CTRL while defining the terminator end of each leader line. Releasing the CTRL key ends placement and completes the note.

SINGLE-CHARACTER FRACTIONS

Natural fractions are several characters long, which can take up a lot of space in the design. For example, 9/16 is four characters long. To reduce the space required for natural fractions, MicroStation provides "stacked fractions" in several fonts that place typical natural fractions as one, slightly taller than normal, character.

The Text Editor box provides a **Stacked Fractions** menu for controlling stacked fractions placement in the design. The menu options allow for aligning stacked fractions such that the bottom of the fraction is in line with the bottom of the text string, vertically centering the fraction within the text string, and aligning the top of the fraction with the top of the text string. Figure 7–16 shows the menu open in the Text Editor box.

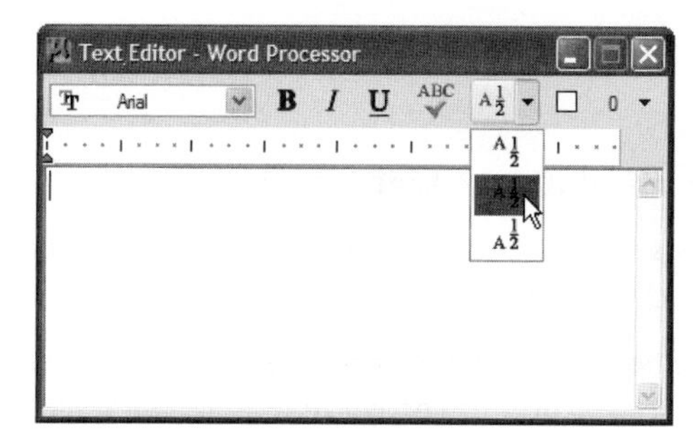

Figure 7–16 The Stacked Fractions option on the Text Editor box

Figure 7–17 shows an example of natural fractions placed both as separate characters and as single-character stacked fractions.

Single-Character Fractions OFF: 1/2, 1/4, 3/4, 7/64
Single-Character Fractions ON: ½, ¼, ¾, 7/64

Figure 7–17 Examples of natural fractions with Single-Character Fractions ON and OFF

The single-character fractions menu only appears in the word processor type of Text Editor box. If the option is not on the Text Editor box, change to the word processor type by doing the following:

1. Select **Preferences** from the Workspace menu on the MicroStation menu bar to open the Preferences dialog box.
2. Select **Text** from the **Category** list on the left side of the dialog box.
3. Select Word Processor from the Text Editor **Style** menu.
4. Click **OK** to close the dialog box and switch to using the Word Processor Text Editor box.

The **Fractions** check box must be ON to place Stacked fractions in the design. Turn the check box ON from the Text Style settings box:

1. Select **Text Style** from the **Elements** menu on the MicroStation menu bar to open the Text Style settings box.
2. Click the **General** tab on the settings box.
3. Turn the **Fractions** check box ON.

The check box setting goes into effect immediately and the settings box remains open. To close the settings box, select **Exit** from the **Style** menu on the menu bar.

Note: *For more information on the Text Styles settings box, see the next section.*

Things to Keep in Mind about Natural Fractions

- If a font does not include stacked fractions, MicroStation places the fractions as separate characters.
- There must be a space before and after a natural fraction in a string of text in order for MicroStation to recognize it as a natural fraction.
- Single-character fractions take up slightly more horizontal space than single characters, so extra space characters may be required to keep the fraction from running into the characters before and after it.

TEXT STYLES

With the text placement methods described thus far, changing the style of text requires making several setting changes in the Tool Settings window. For example, the switch from placing the names of rooms in a house floor plan to placing text for the details may require changing the text font, size, and justification. A complex drawing may require working with several different styles and require frequent changes in tool settings.

To eliminate the need for frequent and complex text tool setting changes, MicroStation provides options for building a library of text styles. Define the settings for each text setup in the text style library and change from one text setup to another by simply selecting the text style from the library.

SELECT A TEXT STYLE

A **Text Style** menu is available in the Tool Settings window for all text placement tools. The menu lists the names of all text styles in the design file's text styles library. Figure 7–18 shows a typical **Text Style** menu.

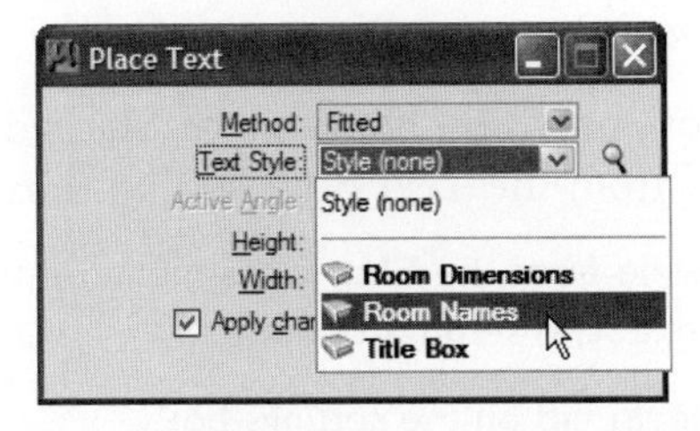

Figure 7–18 The Text Style menu in the Place Text Tool Settings window

Note: *Selecting a text style places check boxes next to some of the settings in the Place Text and Place Note tool settings boxes. Turning a check box ON overrides the selected text style for the setting associated with the check box and allows changing the setting in the tool settings box.*

CREATE AND MAINTAIN TEXT STYLES

Text Styles are created and maintained in the Text Styles box. Invoke the Text Style box from:

Menu	Element > Text Styles
Place Text Tool Settings window	Click the magnifying glass symbol.
Key-in window	**textstyle dialog open** (or **texts di o**) (ENTER)

Figure 7–19 shows a typical Text Styles box and the following discussion describes using the box to create and maintain text styles.

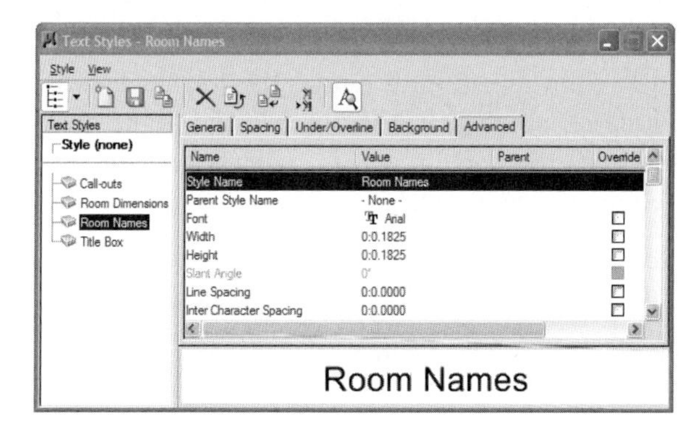

Figure 7–19 The Text Styles box

TEXT STYLES BOX PARTS

The Text Styles box contains the following parts.

Menu and Button Bars

The menu bar at the top of the settings provides options that initiate actions such as saving a new style. The button bar below the menu bar has icons for some of the more frequently used options in the menus.

Text Styles

The **Text Styles** area on the left side of the settings box below the bars is a hierarchy tree that displays all styles currently defined in the design file. To select a style for editing, click its name in this area.

Text Style Settings

The area to the right of the **Text Styles** area lists the settings for the selected style or the default settings if there are no defined styles. Five groups of settings are available on the settings box on separate tabs (**General, Spacing, Under/Overline, Background** and **Advanced**). Click a tab to access its options. The tabs are described in more detail later in this section.

Preview Area

The bottom area of the Text Styles box displays the name of the text style using the style's settings. It provides a preview of the way text appears when placed using the style. If this area is not visible in the Text Style box, display it by selecting the **Preview** option from the **View** menu on the dialog box's menu bar.

CREATE A NEW TEXT STYLE

To create a new style in the design, do the following:

Select **New** from the **File** menu on the Text Styles dialog box's menu bar. Micro_ Station creates a new text style that appears in the **Text Styles** field with "Untitled" as the style name. The new style's settings are set to default values.

Replace "Untitled" with a meaningful name. By default, creating a new text style selects the style name for editing. If it is not selected, right-click the current name and select **Rename** from the pop-up menu. After typing the new name, click ENTER to make the name permanent.

Make the required text settings on the five text style settings tabs. For more information, see the tab options descriptions later in this section.

Save the new style by selecting **Save** from the **File** menu on the Text Styles dialog box's menu bar.

Note: *If the Place Text or Place Note tools are active during the definition of a new style, the new style may not appear in the* **Text Style** *menu. To make it appear, select another tool and then select the Place Text or Place Note tool again.*

MODIFY AN EXISTING TEXT STYLE

To modify the settings of an existing text style in the design, do the following:

1. Select the style in the **Text Styles** area of the Text Styles box's menu bar.
2. Make the required text settings on the five text style settings tabs. For more information, see the tab options descriptions later in this section.

Save the new style by selecting **Save** from the **File** menu on the Text Styles dialog box's menu bar.

DELETE A STYLE

To delete a style from the design file, do the following:

1. Select the style in the **Text Styles** area of the Text Styles box's menu bar.
2. Select **Delete** from the **File** menu on the Text Style dialog box's menu bar.
3. In the **Alert** box that appears, click **Yes** to delete the text style **No** to cancel deleting the text style.

THE TEXT STYLE TABS

Following are descriptions of the options offered on each text style tab. All settings on these tabs become the active settings when the Text Style is active in the design.

General Settings

The General tab provides common text style settings, as shown in Figure 7–20.

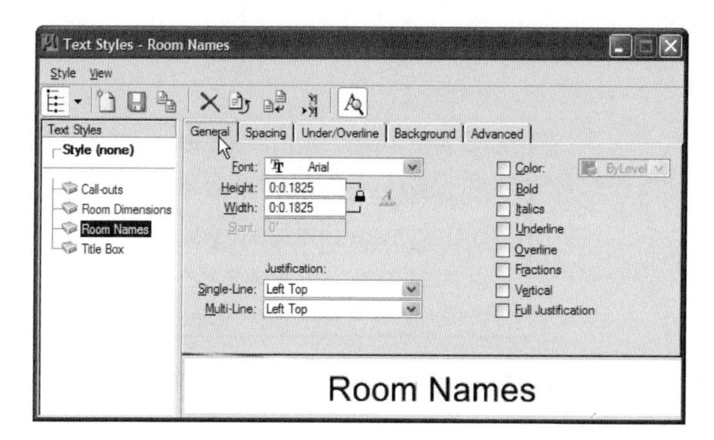

Figure 7–20 The General tab on the Text Style dialog box

Font

Select the active front for the Text Style from the **Font** menu.

Height and Width

Enter the active text height and width for the Text Style in the **Height** and **Width** fields. The text lock (located just to the right of the fields) controls the values entered into the fields. Typing a value in one of the fields with the lock closed sets the other field to the same value. Unlock the lock to allow entering different values in the fields.

Annotation Scale Lock

Turning **Annotation Scale Lock** on applies the Annotation Scale factor to all text placed in the design. The annotation scale factor is a multiplication factor that is

used for all the text that is placed when the Annotation Scale lock is set to ON. For example, if you are creating a design that is drawn full scale and set to plot at a scale of 1" = 200 " (or 1:200), any text that is placed would have to be 200 times larger than the size it was set in order to plot correctly. With Annotation Scale set to 200:1, the size of the text does not need to be calculated by the user. For example, with the Annotation Scale Lock set to a text size of .25", text will be placed in the drawing at 50" (200 x .25) high, when plotted at 1:200 scale the text will be plotted to .25" high. The Annotation Scale factor is set in the Drawing Scale window which is opened by selecting Drawing Scale from the Settings menu. The Drawing Scale window is a dockable window that contains controls for viewing and/or modifying working units, the annotation scale factor, and the Annotation Scale Lock.

Slant Angle

Enter the active text slant angle for the Text Style in the **Slant Angle** field. The **Slant Angle** is sets the slant of each character in text strings, as shown in Figure 7–21. A positive angle slants the text clockwise and a negative angle slants it counter clockwise.

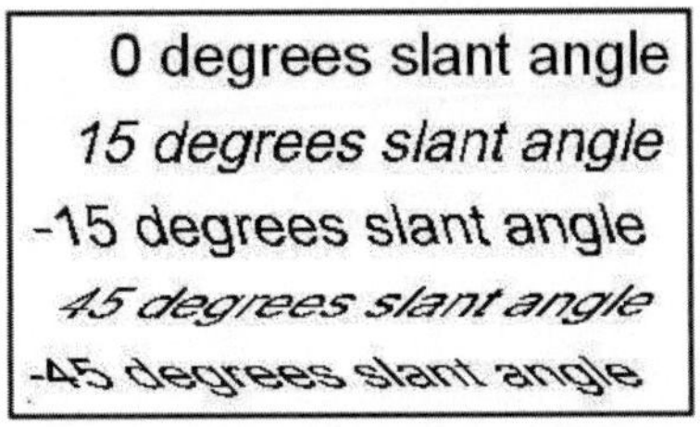

0 degrees slant angle
15 degrees slant angle
-15 degrees slant angle
45 degrees slant angle
-45 degrees slant angle

Figure 7–21 Examples of placing text with various slant angles

Justification – Single-Line

Select the active single-line text justification for the Text Style in the **Justification – Single-Line** menu. Justification controls the position of text strings in relation to the data point that places the text string. For example, with **Left Top Justification** the placement data point is at the top left corner of the text string. Figure 7–22 shows the options available in the menu.

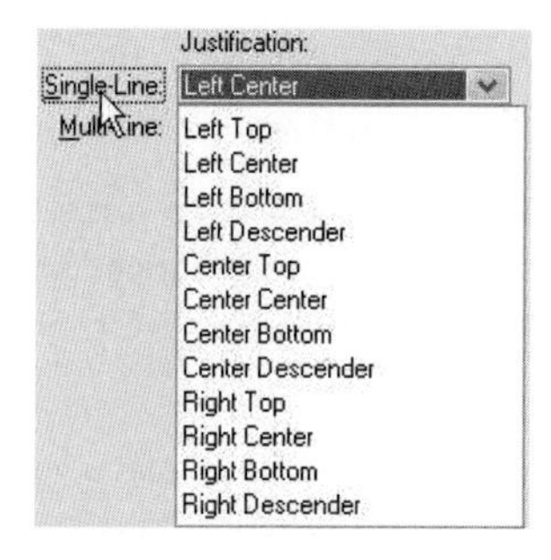

Figure 7–22 Justification – Single-Line menu options

Justification – Multi-Line

Select the active multi-line text (Text Node) justification for the Text Style in the **Justification – Multi-Line** menu. Figure 7–23 shows the options available on the menu.

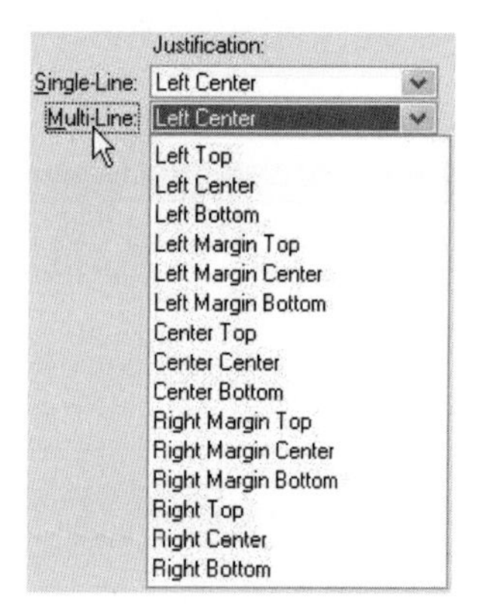

Figure 7–23 Justification – Multi-Line menu options

Check Boxes

The **General** tab contains a set of check boxes that select active settings for the Text Style.

- Turn the **Color** check box ON to activate the color menu located to the right of the check box. Select a color from the picker to set the active color for the Text Style.
- Turn the **Bold** check box ON to place bold text in the design while the Text Style is active. The **Bold** check box is only available for TrueType fonts.
- Turn the **Italics** check box ON to place italic text in the design while the Text Style is active.
- Turn the **Underline** check box ON to place underlined text in the design while the Text Style is active.

- Turn the **Overline** check box ON to place a line over text placed in the design while the Text Style is active.
- Turn the **Fractions** check box ON to place single-character natural fractions in the design while the Text Style is active. Single-character fractions are only placed if the active font supports them. Otherwise, multiple-character fractions are placed.
- Turn the **Vertical** check box ON to place vertically oriented text strings while the Text Style is active. Vertical text stacks the individual characters in the text string. If the **Vertical** check box is ON, MicroStation ignores the **Underline** and **Overline** check boxes, and places the text without underlining or overlining.
- If the **Full Justification** check box is ON, MicroStation adjusts word spacing in text strings so that no word is hyphenated to complete a line of text.

Spacing Settings

The **Spacing** tab contains settings for the space between characters and rows as shown in Figure 7–24.

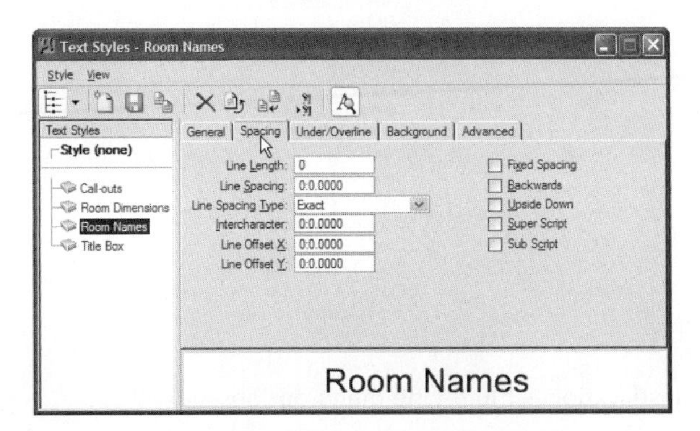

Figure 7–24 The Spacing Tab on the Text Style dialog box

Line Length

The **Line Length** field sets Active Line Length value for the Text Style—the maximum number of characters allowed in multi-line text strings (Text Nodes).

Line Spacing

The **Line Spacing** field sets the Active Line Spacing value for the Text Style. The Line Spacing value sets the space between lines in a multi-line string and the space between the text and the selected element for The Place Text tool's **Above Element, Below Element**, and **Along Element** methods.

Line Spacing Type

The **Line Spacing Type** menu contains options that calculate the vertical space between lines using the Line Spacing value and other character size values.

Exact—Calculates the distance to the next line as the Top of line + Line Spacing + the baseline, where Baseline is determined by the arrangement origin of the text (upper left) + the height of the tallest character. When the Vertical text style setting is on, the **Exact** method calculates the distance between lines of vertical text as the Current line origin + the maximum width of characters in the line + the Line Spacing.

Note: *The **Exact** method does not account for extreme variance in descender depth. To get the best line spacing, try adjusting the Line Spacing value. The descender is the part of text characters that extends below the text base such as g, q, p, and y.*

Automatic—Calculates the distance to the next line as the Top of line + Line Spacing + baseline where Baseline is determined by the arrangement origin of the text (upper left) + the height of the tallest character. When the Vertical text style setting is on, the **Automatic** method calculates the distance between lines of vertical text as the maximum of either Line Spacing or the maximum width of characters in the line.

Note: *The **Automatic** method does not account for extreme variance in descender depth. To get the best line spacing, try adjusting the Line Spacing value. The descender is the part of text characters that extends below the text base such as g, q, p, and y.*

From Line Top—For horizontal and vertical text, calculates the distance to the next line as the Top of line + Line Spacing.

Note: *This method is analogous to the AutoCAD Exactly line spacing setting. It provides rigid control over line spacing, and is generally used for table-based text. However, because the spacing does not vary, the lowest point on one line may overlap the top of the next line.*

At Least—Calculates the distance to the next line as 1/3 of the maximum height of a character + 1/3 of the node number height + the lowest point of the text Line Spacing. When the Vertical text style setting is on, the same rules apply, except that the distance is between the left side of one line of text and the left side of the following line of text.

Note: *This method is analogous to the AutoCAD At Least line spacing setting. It is variable, data dependent, and uses Line Spacing as a factor, not a distance.*

Intercharacter

The **Intercharacter** value sets the distance, in working units, between characters in text when placed. The user preference **Fixed-Width Character Spacing** controls

the manner in which the specified distance is measured (for more information on user preferences, see Chapter 16).

Line Offset X

The **Line Offset** X value controls the spacing, in working units, between placed text and the position from which the text is drawn in X.

Line Offset Y

The **Line Offset Y** value controls the spacing, in working units, between placed text and the position from which the text is drawn in Y.

Check Boxes

The **Spacing** tab contains a set of check boxes that affect the placement of text strings.

- Turn the **Fixed Spacing** check box ON to use the **Inter Character Spacing** value to set the distance from the start of one character to the start of the next character, rather then between characters.
- Turn the **Backwards** check box ON to reverse the orientation of text horizontally when placed.
- Turn the **Upside Down** check box ON to reverse the orientation of text vertically when placed.
- Turn the **Super Script** check box ON to set placed text with superscript characteristics.
- Turn the **Sub Script** check box ON to set placed text with subscript characteristics.

Under/Overline Tab Settings

The **Under/Overline** tab provides settings for placing lines over and under text styles, as shown in Figure 7–25.

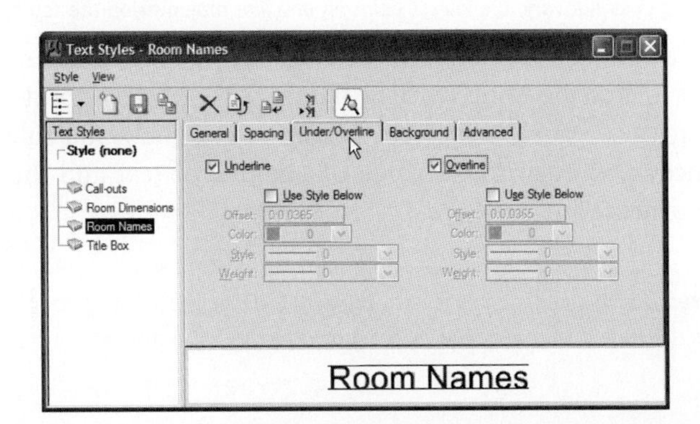

Figure 7–25 The Under/Overline Tab on the Text Style dialog box

Underline

Turn ON the **Underline** check box to place a line under all text placed in the design when the Text Style is active.

Overline

Turn ON the **Overline** check box to place a line over all text placed in the design when the Text Style is active.

Use Style Below

Turn ON the **Use Style Below** check box below Underline and/or Overline respectively to activate the **Offset, Color, Style,** and **Weight** settings for the line.

Offset

The **Offset** field sets the spacing, in working units, between the base of the text and the underline, for **Underline,** and between the height of the text and the overline, for **Overline**.

Color

The **Color** menu sets the line color.

Style

The **Style** menu sets the type of line used.

Weight

The **Weight** menu sets the thickness of the line.

Background Settings

The **Background** tab provides settings for controlling text backgrounds, as shown in Figure 7–26.

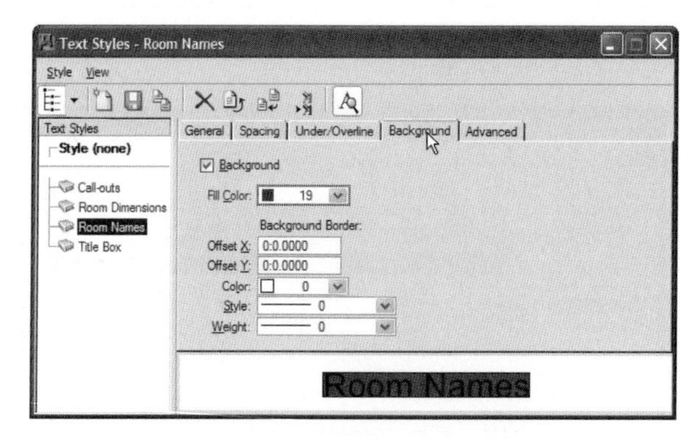

Figure 7–26 The Background Tab on the Text Style dialog box

Background

Turn ON the **Background** check box to define a background when the text style is active. The other settings on this tab are only enabled when the **Background** checkbox is ON.

Fill Color

The **Fill Color** sets the fill color of the background when text is placed.

Background Border Offset X

The **Background border Offset X** field sets the thickness of the border around the top and bottom of the text box when the text style is active.

Background Border Offset Y

The **Background border Offset Y** field sets the thickness of the border around the sides of the text box when the text style is active.

Background Border Color

The **Background Border Color** sets the color of the border around text when the text style is active.

Background Border Style

The **Background Border Style** sets the style of the border around text placed when the Text Style is active.

Background Border Weight

The **Background Border Weight** sets the weight of the border around text placed when the Text Style is active. Line weights range from 1 to 15 in working units.

Advanced Settings

The **Advanced** tab displays all the settings from the other tabs plus the **Parent Style Name** field, as shown in Figure 7–27.

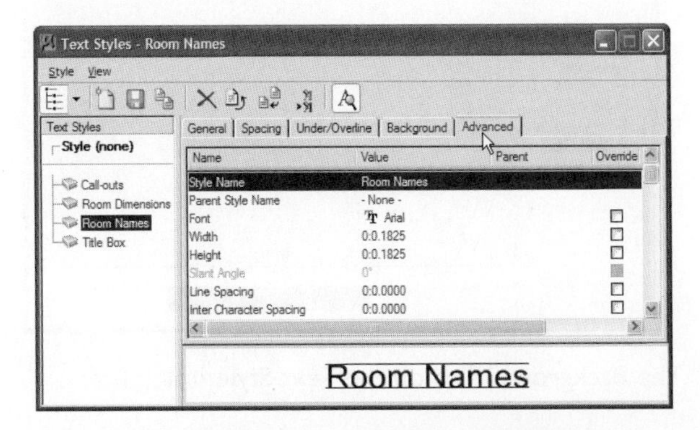

Figure 7–27 The Advanced Tab on the Text Style dialog box

Parent Style Name

The **Parent Style Name** row on the **Advanced** tab allows the currently selected Text Style to be turned into a subset (the "child") of another Text Style (the "parent"). When created, the child takes on the settings of the parent, but the child settings can be changed. The ability to create a hierarchical structure of parent and children is useful for ensuring uniform settings for all text in a design. For example, the only setting changes in the children might be text size for different uses of text (such as for dimensions construction notes, and so forth).

To declare the selected Text Style the child of another Text Style, click near the middle of the **Parent Style Name** and select the parent Text Style from the resulting menu. In Figure 7–26 "Room Names" is the parent Text Style and "Room Dimensions" is the child.

IMPORTING TEXT STYLES

In addition to creating text styles in the active design file, MicroStation provides a tool for importing text styles from other design files into the active design file. The tool adds the imported text styles to the text styles library of the active design file. Invoke the Textstyle Import tool from:

Text Styles box	**File > Import**

MicroStation opens the Text Style Import box.

1. In the **Look in** menu, navigate to the folder that contains the design file with text styles to be imported.
2. Select the file from the list of design files in the selected folder.
3. Click **OK** to import the text styles or click **Cancel** to close the box without importing anything.

IMPORT TEXT

The **Import Text** tool imports text from a text file created by another computer application. The file to be imported must contain only unformatted (ASCII) text. Invoke the Import Text tool from:

Menu	**File > Import > Text**
Key-in window	**include** (or **in**) (ENTER)

MicroStation displays the Include Text File dialog box. Find and select the file that contains the text to be imported and click on **OK**.

MicroStation prompts:

Import Text File > Enter text node origin *(Define the point in the drawing plane where the text is to be placed.)*

A dynamic image of the text follows the cursor. The text is placed using the currently active Multi-Line Justification to determine the text position in relation to the placement data point.

Text Attribute Settings

The text file can contain MicroStation element and text attribute setting key-ins to control the way the text appears when placed.

Keep the following in mind when adding attribute settings to a Text File that is to be imported:

- Use standard MicroStation attribute key-ins.
- Precede each key-in with a period.
- Put only key-ins on a line (text cannot be on the same line).
- The settings act on the text that follows them in the file.
- The key-ins are not placed in the imported text string.
- If an attribute setting is not included in the file, the drawing's current active setting applies.
- The settings in the imported file become the drawing's active settings after the text is imported.

Table 7–1 lists useful key-ins for importing text. Figure 7–28 shows an example of a text file that contains key-ins.

Table 7–1 Element Attribute Key-ins

KEY-IN	SETS
AA=	Rotation angle in degrees
CO=	Color name or number
FT=	Font name or number
LS=	Line Spacing in working units
LV=	Level name
TH=	Text Height in working units
TW=	Text Width in working units
TX=	Text Size (sets height and width equal) in working units
WT=	Weight number
Indent #	Indents each following line of text with "#" number of spaces

```
.AA=5
.LV=Dimensions
.FT=Architectural
.CO=Red
.WT=1
.TH=1.5
.TW=1.0
.LS=0.8
When this text is imported, it will be placed at an angle of
5 degrees on the Dimensions level using using the Architectural
font, the color red, line weight one, and a text size of 1.5 x 1.0
master units.
.Indent=5
NOTE: This note will be indented five spaces.
```

Figure 7–28 An example of a text file with MicroStation key-ins

TEXT MANIPULATION TOOLS

The manipulation tools (Move, Copy, and Rotate, among others) manipulate the text element but not the text itself. This section describes text manipulation tools that change the text in text elements.

SPELL CHECKER

The Spell Checker tool compares the words in a selected text element with the MicroStation dictionary and suggests possible corrections for words it does not find in the dictionary. Select one of the suggested words to replace the misspelled word in the text string. Invoke the Spell Checker tool from:

Task Navigation tool box (active task set to Text)	Select the Spell Checker tool (see Figure 7–29).
Keyboard Navigation (Task Navigation tool box with active task set to Text)	**r**

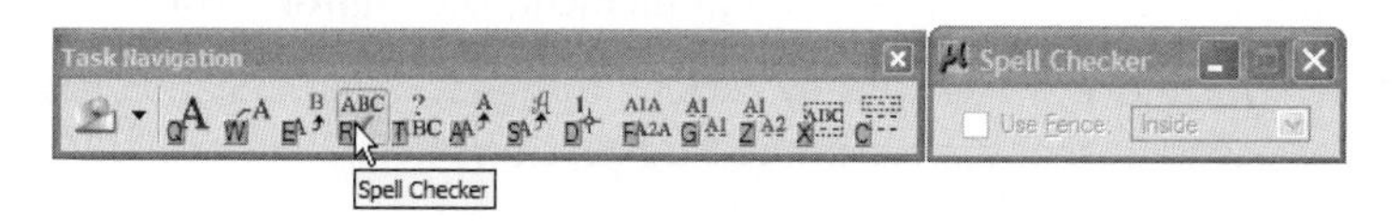

Figure 7–29 Invoking the Spell Checker tool

MicroStation prompts:

> Spell Checker > Identify element *(Click the Data button on the text element whose spelling is to be checked.)*

If the selected element contains words that are not in the MicroStation dictionary, the Spell Checker dialog box opens, as shown in Figure 7–30, and displays the first misspelled word.

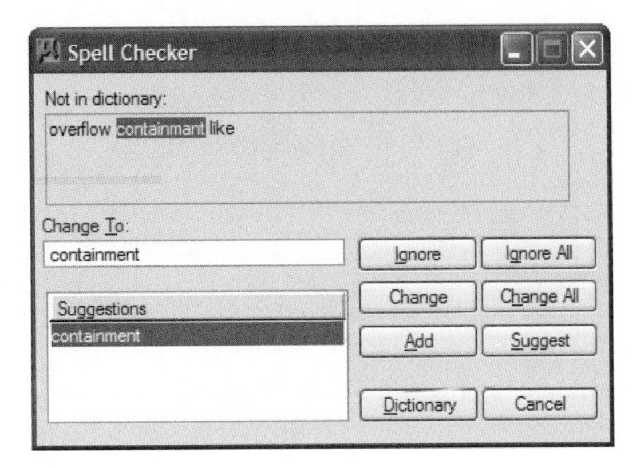

Figure 7–30 The Spell Checker box

The tool displays the selected text string with the misspelled word highlighted in the **Not in dictionary** field. The tool also displays the recommended spelling in the **Change To** field and alternate spellings in the **Suggestions** field.

- Replace the misspelled word with the current **Change To** content by clicking **Change**.
- Replace the misspelled word with another suggested spelling by clicking the spelling in the **Suggestions** list and clicking **Change To**.
- Change the content of the **Change To** field manually and then click **Change**.
- If the same misspelling occurs in more than one place in the text string, click **Change All** to replace all misspellings with the suggested change.
- To ignore the misspelled word and continue checking for misspellings in the selected text element, click **Ignore**, or to ignore all occurrences of the misspelled word in the selected text element, click **Ignore All**.
- Ask for more suggestions by clicking **Suggest**. If MicroStation has additional suggestions, they appear in the **Suggestions** field.
- If the spelling of the word is actually correct, add it to the dictionary by clicking **Add**.
- Close the Spell Checker box without checking any more words by clicking **Cancel**.

When no misspelled words remain in the selected text string, the Spell Checker box closes.

Note: *The* ***Dictionary*** *button displays the Edit User Dictionary settings box. This box provides options for adding and removing dictionary words and for defining special actions for words.*

***Note:** The **Spell Checker** tool is also on the Text Editor box to allow for checking spelling before placing a text string in the design.*

EDIT TEXT ELEMENTS

The Edit Text tool provides a way to change the text in an existing text element. Invoke the Edit Text tool from:

Task Navigation tool box (active task set to Text)	Select the Edit Text tool (see Figure 7–31).
Keyboard Navigation (Task Navigation tool box with active task set to Text)	e

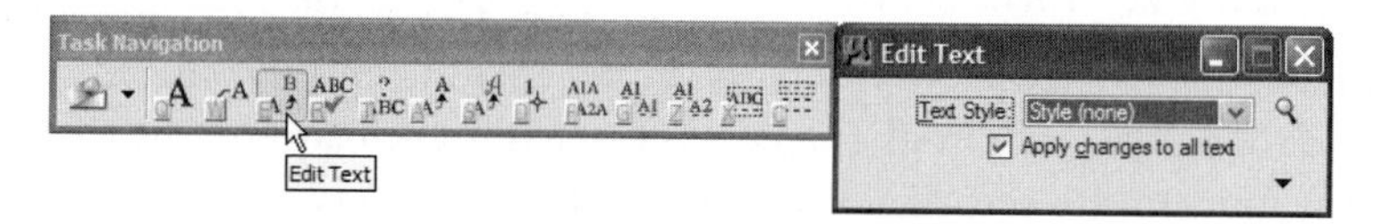

Figure 7–31 Invoking the Edit Text tool from the Text tool box

MicroStation prompts:

Edit Text > Identify element *(Select the text element to be edited.)*

Edit Text > Accept/Reject (Select next input) *(Click the Data button again to accept the text and place it in the Text Editor box for editing.)*

Edit Text > Accept/Reject (Select next input) *(Make the required changes to the text in the Text Editor box and click the Data button somewhere outside of the Text Editor box to replace the text in the design with the edited text, or click the Reset button to leave the text unchanged and clear the Text Editor box.)*

***Note:** Refer to the discussion of "Place Text Tools" in Chapter 4 for notes on keyboard shortcuts in the Text Editor box.*

The Edit Text Tool Settings window allows for changing the **Text Style, Height, Width,** and **Font** settings while editing the selected text. By default, only the **Text Style** menu is visible in the window. Click the **Expand** arrow on the right side of the window to display the additional settings. If the arrow is pointing up, the window is already expanded and the arrow's name is **Collapse.**

To change formatting for the text, select all of the text string in the Text Editor box and make each required setting change. The appearance of the selected text changes in the Text Editor box as the settings are changed. Accept the changes by clicking the Data button outside of the box.

The tool allows for adding text to a string using different settings than the original text. Select the desired options and key-in the new text. For example, if the text height and width settings are changed and text is keyed in at the end of the text string in the Text Editor window, the new text is placed at the new text size and the original text remains it the same size. The text string is still one element, even though it contains text of two different sizes.

The method of selecting text and changing its formatting only works when the selected text has uniform formatting. For example, if the selected text has two different font values, the text cannot be changed to new settings. Select the part of the text that has one font, make the changes, and then select and change the part that has the other font.

MATCH TEXT ATTRIBUTES

The Match Text Attributes tool sets the active text attributes to match those of an existing text element. The tool changes the active font, text size, line spacing, and text justification to the settings of the selected text element, and all text placed afterwards uses the new active settings. Invoke the Match Text Attributes tool from:

Task Navigation tool box (active task set to Text)	Select the Match Text Attributes tool (see Figure 7–32).
Keyboard Navigation (Task Navigation tool box with active task set to Text)	**a**

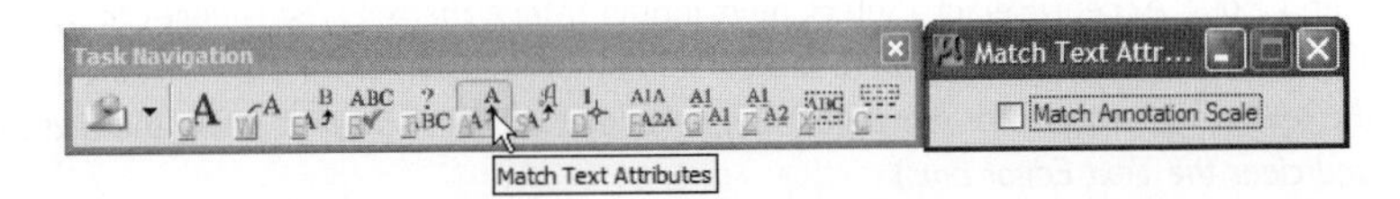

Figure 7–32 Invoking the Match Text Attributes tool from the Text tool box

MicroStation prompts:

Match Text Attributes > Identify text element *(Identify the text element to match.)*

Match Text Attributes > Accept/Reject (Select next input) *(Click the Data button to set the active settings to selected text element, or click the Reset button to reject it.)*

MicroStation sets the new active text attributes and displays them on the Status bar. To make these changes permanent, select **File** > **Save Settings**.

CHANGE TEXT ATTRIBUTES

The Change Text Attributes tool changes the attributes of an existing text element. Invoke the Change Text Attributes tool from:

Task Navigation tool box (active task set to Text)	Select the Change Text Attributes tool (see Figure 7–33). In the Change Text Attributes settings box, turn ON the check box for each attribute that needs to be changed and select or type the required attributes in the option menus and edit fields of the attributes that are ON.
Keyboard Navigation (Task Navigation tool box with active task set to Text)	**s**

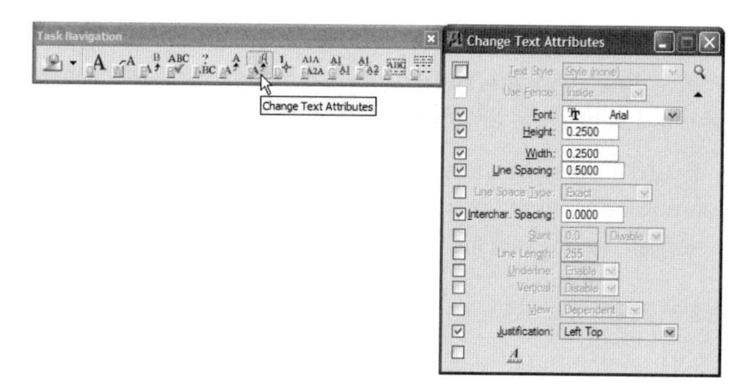

Figure 7–33 Invoking the Change Text Attributes tool from the Text tool box

MicroStation prompts:

Change Text Attributes > Identify text *(Select text element.)*

Change Text Attributes > Accept/Reject (Select next input) *(Click the Data button to change the attributes of the selected text element, or click the Reset button to reject it. This data point can also select another text element at the same time it accepts the current element.)*

DISPLAY TEXT ATTRIBUTES

Display Text Attributes is an information-only tool that displays the attributes of an existing text element. Invoke the Display Text Attributes tool from:

Task Navigation tool box (active task set to Text)	Select the Display Text Attributes tool (see Figure 7–34).
Keyboard Navigation (Task Navigation tool box with active task set to Text)	**t**

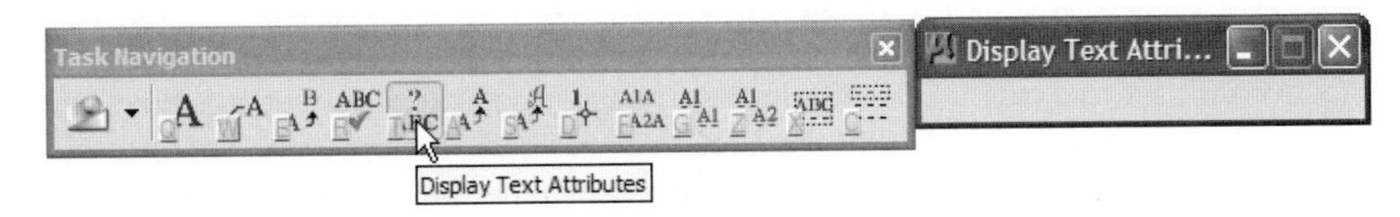

Figure 7–34 Invoking the Display Text Attributes tool from the Text tool box

MicroStation prompts:

Display Text Attributes > Identify text *(Identify the text element.)*

MicroStation displays the attributes in the left side of the Status bar. If desired, select another text element. The text attributes displayed in the Status bar are different for one-line text elements and multi-line text elements (Text Nodes):

- **For one-line text elements**, the displayed attributes include the text height and width, the level the element is on, and the font number.
- **For multi-line text elements**, the displayed attributes include the Text Node number, the maximum characters per line, the line spacing, the level the element is on, and the font number.

COPY/INCREMENT TEXT

Annotating a series of objects with an incremented identification (such as P100, P101, P102) would be a tedious job without the Copy/Increment Text tool. This tool copies and increments numbers in text strings. To make incremented copies, select the element to be copied and incremented, and place data points at each location where an incremented copy is to be placed.

The **Tag Increment** text field on the Copy and Increment settings box allows setting a positive or negative increment value. For example, an increment value of 10 causes each copy to be 10 greater than the previous one. A value of −10 causes each new copy to be 10 less than the previous one.

Only the numeric portion of a text string is incremented, and, if the string contains more than one numeric portion separated by nonnumeric characters, only the rightmost numeric portion is incremented. For example, only the 30 in the string P100-30 is incremented (P100-31, P100-32, P100-33, and so on).

To place a series of incremented text strings, place the starting text string and invoke Copy/Increment Text from:

Task Navigation tool box (active task set to Text)	Select the Copy/Increment Text tool and, optionally, set the Tag Increment value in the Copy/Increment Text settings box (see Figure 7–35).
Keyboard Navigation (Task Navigation tool box with active task set to Text)	**f**

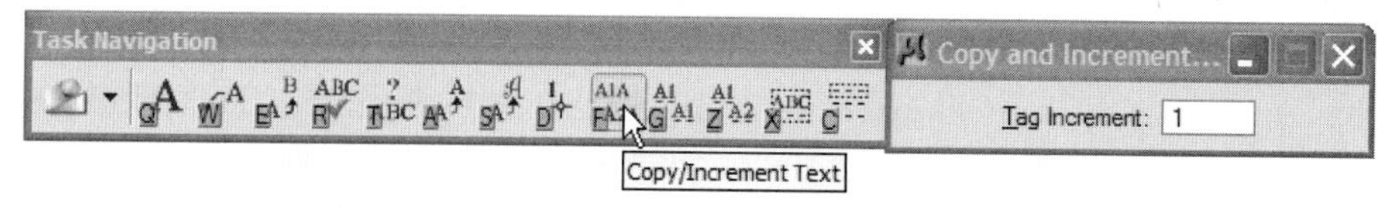

Figure 7–35 Invoking the Copy/Increment Text tool from the Text tool box

MicroStation prompts:

Copy and Increment Text > Identify element *(Identify the text string to be copied and incremented.)*

Copy and Increment Text > Accept/Reject (Select next input) *(Define the location of each incremented copy, or press the Reset button to reject the copy.)*

Note: *The Copy/Increment tool only accepts single-line text strings that contain numbers. Strings that do not contain numbers or multi-line strings are not accepted.*

TEXT NODES

The Text Node tool provides a way to reserve space in a design where text is to be placed later. When a Text Node is placed, it saves the active element and text attribute settings. When text is added to the node later, the text takes on those settings.

A common use of Nodes is for building "fill-in-the-blank" forms in a design that can be filled in later with information specific to the design. A common example is a title block form that has all the required fields held with Text Nodes. Use of the form provides a standard title block layout for all designs.

VIEW TEXT NODES

The visual indication of a text node is a unique identification number and a cross indicating the node origin point. Figure 7–36 shows examples of empty and filled-in Text Nodes on a design when the Text Node View Attribute is ON.

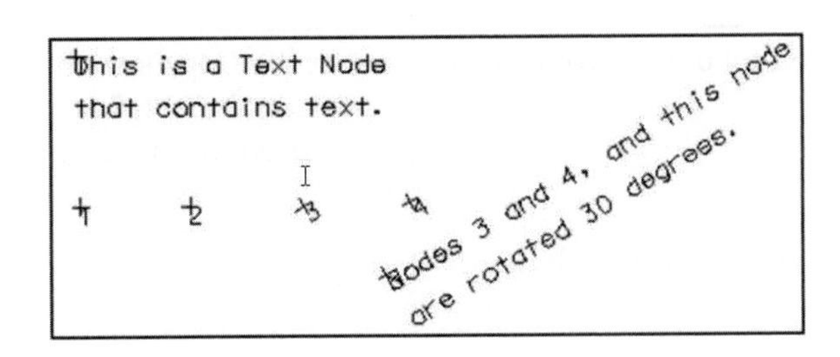

Figure 7–36 Examples of Text Node indicators

The Text Node view attribute controls the display and printing of Text Node indicators for selected views. Change the display of the Text Node view attribute from:

Settings menu	Select **View Attributes** (see Figure 7–37).

The Text Node check box is at the bottom of the right column in the dialog box.

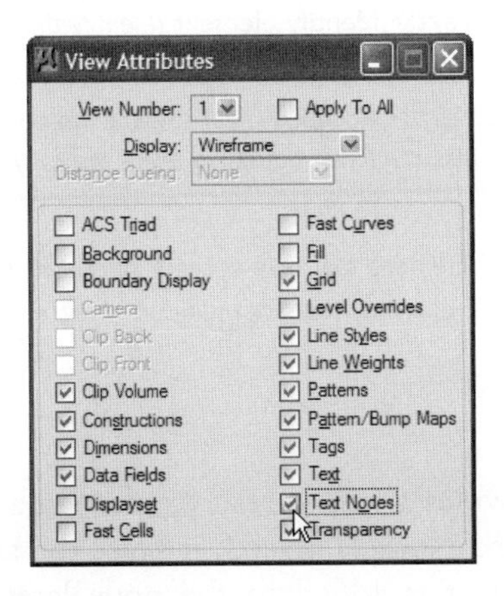

Figure 7–37 The View Attributes box

Turn the Text Node View Attribute check box ON to display the text node numbers and crosses. Turn it OFF to hide them.

Note: *Multi-line text strings are also text nodes. Placing a multi-line text string in the design also assigns a text node to the string. If the* ***Text Nodes*** *view attribute is ON, the node cross and number appear with the text element at the justification point.*

PLACE TEXT NODES

To place empty text nodes, invoke the Place Text Node tool from:

Task Navigation tool box (active task set to Text)	Select the Place Text Node Text tool and if necessary change the Active Angle (see Figure 7-38).
Keyboard Navigation (Task Navigation tool box with active task set to Text)	**d**

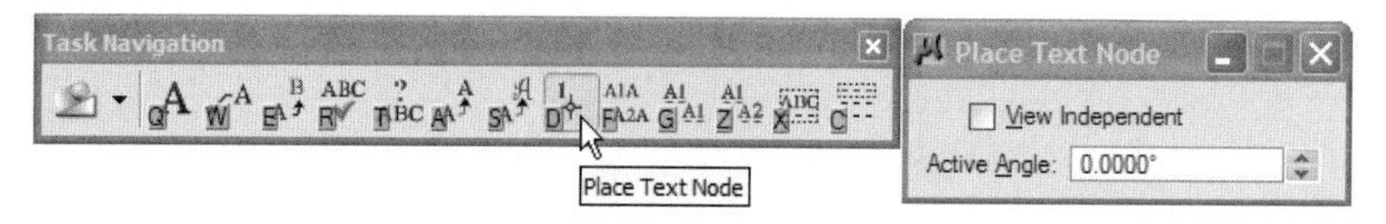

Figure 7–38 Invoking the Place Text Node tool from the Text tool box

MicroStation prompts:

Place Text Node > Enter text node origin *(Define the origin point for each node that is to be placed.)*

Note: *If the* ***Text Nodes*** *view attribute is OFF, nothing appears to happen when the Place Text Node tool is invoked to place empty nodes. Turn ON the* ***Text Nodes*** *view attribute to see the results of placing empty nodes.*

Note: *The Place Text Node settings include a* ***View Independent*** *check box that applies to 3D designs only. If the check box is ON, MicroStation prompts "Place View Independent Text Node," but there is no difference in the way the nodes are placed in a 2D design.*

FILL IN TEXT NODES

Text can only be placed on empty text nodes when the Text Node Lock check box is ON. Turn on the Text Node Lock from:

Settings menu	Select **Settings > Locks > Text Node.**
Settings menu	Select **Settings > Locks > Full** and turn the **Text Node Lock** check box ON in the Locks settings box.
Settings menu	Select **Settings > Locks > Toggles** and turn the **Text Node Lock** check box ON in the Toggles settings box as shown in Figure 7–39.
Key-in window	**lock textnode on** (or **lock t on**) (ENTER)

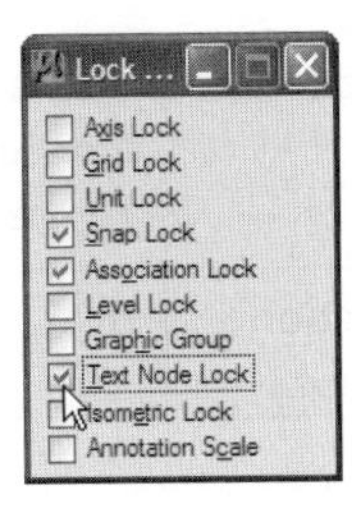

Figure 7–39 Setting the Text Node Lock check box to ON.

After turning the **Text Node Lock** ON, use the Place Text tool's **By Origin** method to place text on empty text nodes.

MicroStation prompts:

> Place Text > Enter text *(Enter the text in the Text Editor box and select the Text Node the text is to be placed on.)*
>
> Place Text > Enter more characters or position text *(Click the Data button again to accept the previous node and, optionally, select another node. More text can also be entered in the Text Editor box before clicking again. Click the Reset button to clear the Text Editor box.)*

The text is placed using the element and text attributes that were in effect when the node was created.

Note: *If the selection or acceptance points are placed in an empty space or on a node that already contains text, MicroStation displays the message "Text node not found" on the Status bar. When the* ***Text Node Lock*** *is set to ON, text can only be placed on empty Text Nodes.*

DATA FIELDS

Data Fields are similar to Text Nodes in that they create placeholders for text that will be filled in later. Data Fields, though, are more powerful than Text Nodes because there are tools available to automate filling in the fields.

A common use for Data Fields is to provide placeholders for descriptive text in cells. For example, a control valve cell might contain Data Fields for the valve type, size, and identification code. Cells are discussed in Chapter 11.

DATA FIELD CHARACTER

Create Data Fields by typing a contiguous string of underscores in a text string. For example, "_____" is a five-character Enter Data field. Use any of The Place Text tool's methods to create the data fields.

The number of underscores in a Data Field sets the limit on the number of characters that can be placed in the Data Field, so when creating the field, the number of text characters to be placed in the field must be anticipated. The Edit Text tool al-

lows adding and removing underscores in Data Fields that are already in the design.

A text string can contain more than one data field. For example, the string, "Pump ___ - _" contains two data fields. Inserting a line break in a string of underscores, creates separate data fields on each line.

***Note:** The underscore is the default Data Field character, but the reserved character can be changed from the MicroStation Preferences settings box (preferences are discussed in Chapter 16).*

DATA FIELD VIEW ATTRIBUTE

The View Attributes box includes a **Data Fields** check box for turning ON and OFF the display of the Data Field underscores for a selected view, as shown in Figure 7–40.

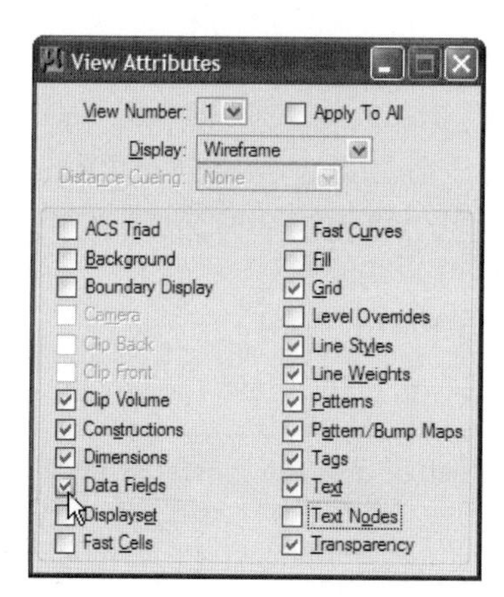

Figure 7–40 View Attributes box

When the **Data Fields** View Attributes check box is:

- **OFF**—The underscores disappear from the selected view. The text of filled-in data fields remains visible.
- **ON**—The underscores are visible in the selected view and they print. If the data fields contain characters, the underscores appear at the bottom of each fill-in character.

SET JUSTIFICATION FOR DATA FIELD CONTENTS

The Data Field contents can be justified Left, Right, or Center within the field when there are fewer characters in the field than underscores. Data Field content justification is different from text justification and is applied to the Data Field only after the text string is placed. Invoke the Data Field justification tool from:

Key-in window	**justify left** (or **ju l**) (ENTER)
	justify center (or **ju c**) (ENTER)
	justify right (or **ju r**) (ENTER)

The MicroStation prompt shown here appears when Center justification is selected (the MicroStation prompt includes the selected justification):

Center Justify Enter Data Field > Identify element *(Click on the Data Field to be justified.)*

No acceptance is required for this tool. Selecting the field applies the justification to it.

Note: *Applying data field justification to a Data Field that already contains text does not change the text position. Replacing the existing text after changing the justification applies the new justification.*

FILL IN DATA FIELDS

MicroStation provides tools for filling in text in Data Fields one at a time, automatically, by copying, and by copying and incrementing. There is also a tool for editing text already placed in Data Fields.

Fill In Single Enter Data Field Tool

Invoke the Fill In Single Enter Data Field tool from:

Task Navigation tool box (active task set to Text)	Select the Fill In Single Enter Data Field tool (see Figure 7–41).
Keyboard Navigation (Task Navigation tool box with active task set to Text)	**x**

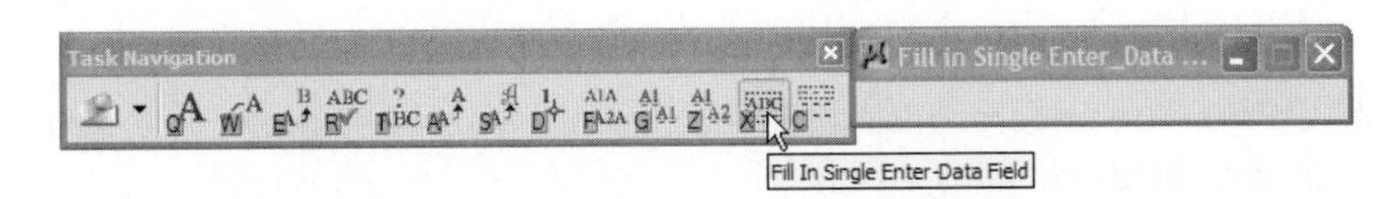

Figure 7–41 Invoking the Fill In Single Enter Data Field tool from the Text tool box

MicroStation prompts:

Fill in Single Enter_Data Field > Identify element *(Identify the Data Field.)*

Selecting the tool opens the Fill in Single Enter Data field settings box and the Text Editor box. Select a Data Field, type the text in the Text Editor box, and press (ENTER) to place the text in the selected Data Field. The text placed in the

data field is also displayed on the status bar. The tool remains active and additional Data Fields can be filled in.

Note: *The Data Field tools open the dialog box type of Text Editor box. The dialog box type does not have formatting tools on the menu bar as does the word processor version of the box. The formatting tools are not necessary because formatting is set when Data Fields are created.*

Auto Fill In Enter Data Fields Tool

Invoke the Auto Fill In Enter Data Fields tool from:

Task Navigation tool box (active task set to Text)	Select the Auto Fill In Enter Data Fields tool (see Figure 7–42).
Keyboard Navigation (Task Navigation tool box with active task set to Text)	**c**

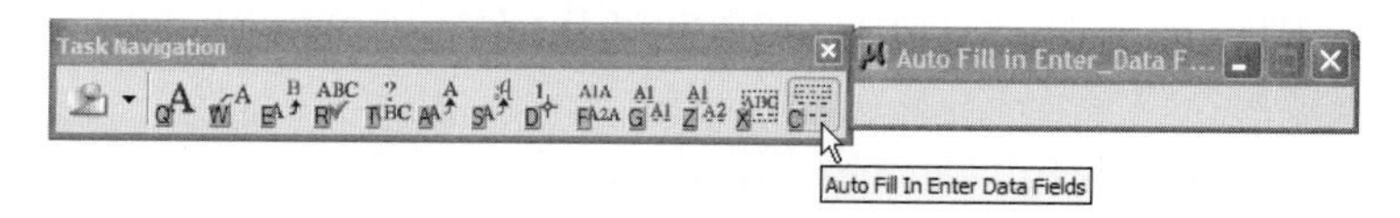

Figure 7–42 Invoking the Auto Fill In Enter Data Fields tool from the Text tool box

MicroStation prompts:

> Auto Fill in Enter Data Fields > Select view *(Click the Data button anywhere in the view containing the fields to be filled in.)*
>
> Auto Fill in Enter Data Fields ><CR> to fill in or DATA for next field *(Type the text in the Text Editor box and press* (ENTER)*, or click the Data button to skip the field.)*

The tool continues through the view selecting empty Data Fields in the order they were created. It skips Data Fields that already contain text. If there are no empty data fields in the view, nothing happens when the view is selected.

COPY DATA FIELDS

To copy the contents of one Data Field to another Data Field, invoke Copy Enter Data Field from:

Task Navigation tool box (active task set to Text)	Select the Copy Enter Data Field tool (see Figure 7–43).
Keyboard Navigation (Task Navigation tool box with active task set to Text)	**g**

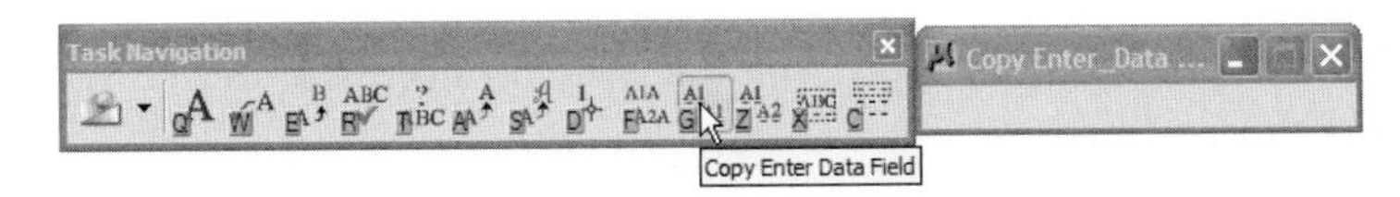

Figure 7–43 Invoking the Copy Enter Data Field tool from the Text tool box

MicroStation prompts:

> Copy Enter Data Field > Select enter data field to copy *(Select the Data Field containing the text to be copied and click each Data Field to which the text is to be copied.)*

Selecting the Data Field that contains the text to copy causes MicroStation to display the selected text in the Status bar. This is a useful aid for selecting the correct field when the view window is zoomed out so far that the text is too small to read.

COPY AND INCREMENT DATA FIELDS

The Copy and Increment Enter Data Field tool copies text from a filled-in Data Field and increments the numeric portion of the text before placing it in each selected empty Data Field.

The Copy and Increment Data Fields settings box contains one setting, the **Tag Increment** field. The default increment value is 1. MicroStation increments the copied text by the **Tag Increment** value each time a destination Data Field is selected. A positive increment value increases the number and a negative increment value decreases it. For example, an increment value of 10 causes each copy to be 10 greater than the previous one and a value of −10 causes each copy to be 10 less than the previous one.

To copy the contents of one Data Field to another Data Field and increment the numeric portion of the text, invoke Copy/Increment Enter Data Fields from:

Task Navigation tool box (active task set to Text)	Select the Copy/Increment Enter Data Fields tool (see Figure 7–44), and, optionally, set the **Tag Increment** value in the Tool Settings box.
Keyboard Navigation (Task Navigation tool box with active task set to Text)	**z**

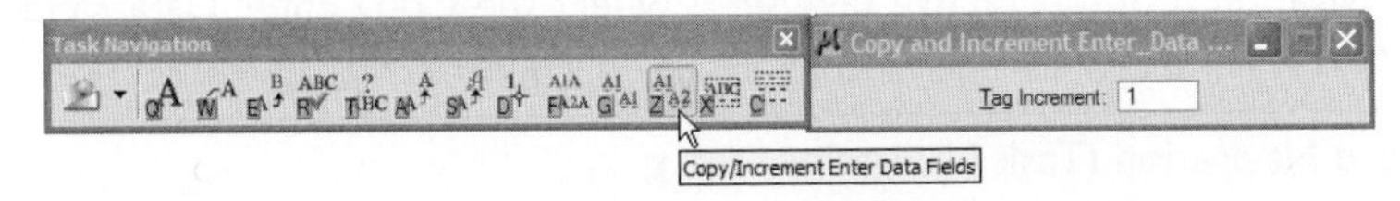

Figure 7–44 Invoking the Copy and Increment Enter Data Field tool from the Text tool box

MicroStation prompts:

Copy and Increment Enter Data Field > Select enter data field to copy *(Select the Data Field containing the text to be copied and click in each Data Field to which the text is to be copied and incremented.)*

EDIT TEXT IN A DATA FIELD

The Edit Text tool can edit the number of underscore characters in an existing Data Field and can change the text entered into a Data Field.

In the Text Editor box the underscores are represented by spaces enclosed in pairs of angle brackets. For example, the string "PUMP __-__" appears as "Pump <<   >>-<<   >>" in the Text Editor box.

- To shorten a Data Field, remove spaces from between the angle brackets.
- To lengthen a Data Field, insert spaces (or underscores) between the angle brackets.
- To delete a Data Field completely, delete the angle brackets and the spaces between them.
- To insert a new Data Field:

1. Position the cursor at the insertion point in the text string.
2. Type a pair of left angle brackets (<<).
3. Type spaces (or underscores) to define the length of the Data Field.
4. Type a pair of right angle brackets (>>).

TAGS

Engineering drawings have long served to convey more than how a design looks. Drawings, for example, must tell fabricators how to construct the design. This non-graphical information includes such things as the construction material, how many to make, colors, where to obtain materials, and what finishes to apply to the surface. Painstaking work was required to extract lists of this information from paper drawings. A major innovation of CAD is the ability to automate the creation of such lists.

One way that MicroStation provides this automation by attaching "Tags" to elements. Any element, or element group, can be tagged with descriptive information, and tag reports can be generated. For example, each electrical fixture in an architectural floor plan can be tagged with its rating, order number, price, and project name. An estimator can extract a fixture tag report from the design and insert the resulting data in a spreadsheet to obtain the total project cost for electrical fixtures. A purchasing agent can use the tag reports from several projects to order fixtures and take advantage of quantity discounts. Receiving clerks can employ the order numbers and project names to route the received fixtures to the correct projects.

MicroStation's Tag tools are helpful when the tagging requirements are simple and when the project must import or export drawings from other CAD packages that store non-graphical data inside the design files. For complex tagging requirements, MicroStation supports connections to databases (which is beyond the scope of this book).

All element types accept Tags. Figure 7–45 shows an example of tags. In the example, tags are assigned to a small point (actually a short line) in each tract of land in a plot plan. The points provide an element to which the tag can be hooked. Each "Tract" tag set includes the tract identification number, the purchase status, and the tract size.

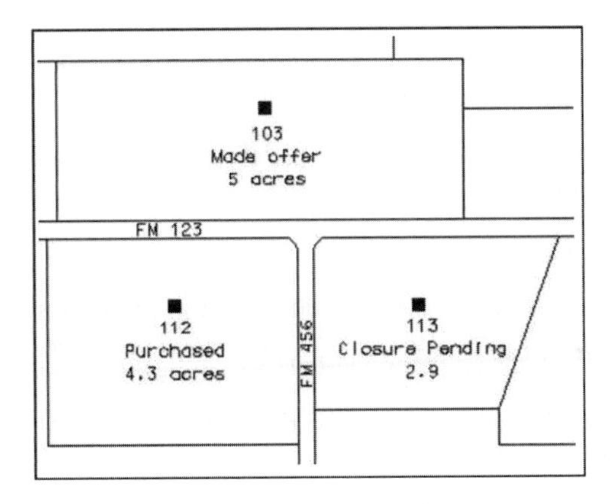

Figure 7–45 Tags displayed in a plot plan

TAG TERMS

Adding tags to a design requires the understanding of tag terminology. Table 7–2 defines important tagging terms.

Table 7–2 Tagging Terms

TERM	DEFINITION	EXAMPLES
Tag set	A set of associated tags.	Separate tag sets for doors, windows, and electrical fixtures.
Tag	Non-graphical attributes that may be attached to graphical elements	Part number, size, material of construction, vendor, price.
Tag report template	A file that specifies the tag set and the set's member tags to include on each line of reports that use the template.	For the fixtures set, the part number, rating, price, and project name of report.
		One tag set per each tagged element.

TERM	DEFINITION	EXAMPLES
Tag report	A list of all tags based on a tag report template.	F300-2, 220V, $300, New ABC, Inc. building.
Tag set library	Files containing tag sets exported from design files.	A library of architectural tag definitions for use in multiple sets.

Note: *Deleting an element with attached tags also deletes the tag attachments.*

CREATE A TAG SET AND TAGS

The first step in creating tags in a design is to create the tag set and define the tags in the set. Open the Tag Sets settings box from:

Menu	**Element > Tags > Define**
Key-in window	**mdl load tags define** (or **md l tags define**) (ENTER)

MicroStation displays the Tag Sets settings box as shown in Figure 7–46. All defined tag sets are on the left side of the window, and the tag names in the selected tag set are on the right side of the window.

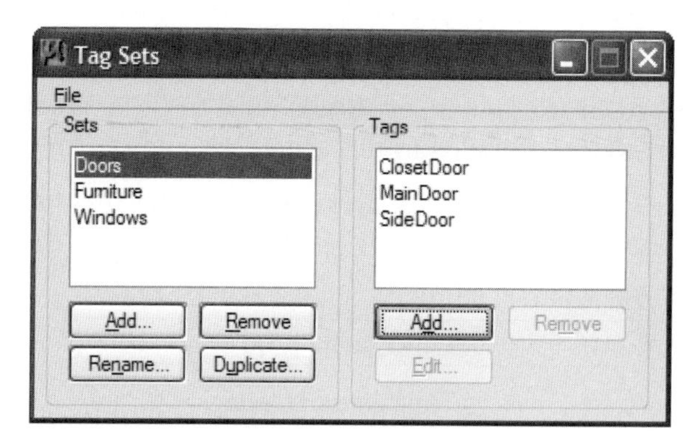

Figure 7–46 Tag Sets settings box

To create a new tag set name, click **Add** in the **Sets** area on the left side of the Tag Sets settings box. MicroStation displays the Tag Set Name dialog box, as shown in Figure 7–47.

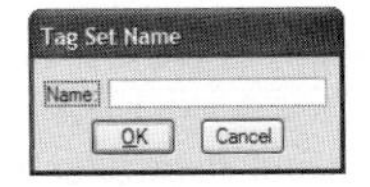

Figure 7–47 Tag Set Name dialog box

Key-in the Tag Set name in the **Name** field of the Tag Set Name dialog box and click **OK** to create the new Tag Set. The new Tag Set name appears in the **Sets** field in the Tag Sets settings box.

To create a new tag under a specific tag set, select the Tag Set name in the **Sets** list and click **Add** on the right side of the settings box under the **Tags** names. MicroStation displays the Define Tag dialog box, as shown in Figure 7–48.

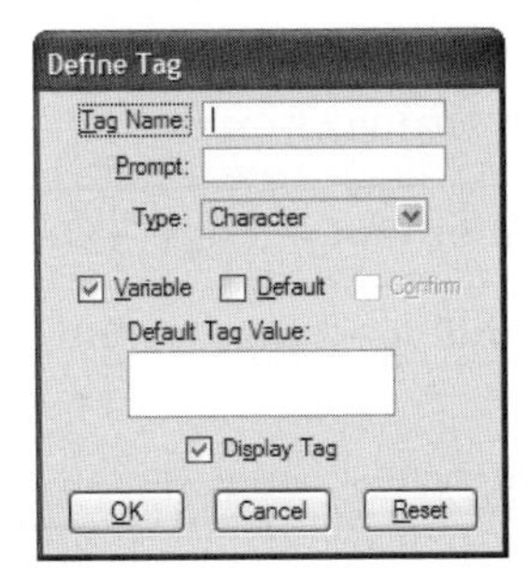

Figure 7–48 Define Tag dialog box

Key-in the appropriate information in the fields provided in the Define Tag dialog box and click **OK** to create the tag, or click **Cancel** to close the dialog box without creating the tag. To clear all fields in the dialog box and start over, click **Reset**.

Refer to Table 7–3 for a detailed explanation of the available fields in the Define Tag dialog box.

Table 7–3 Tag Attributes

ATTRIBUTE	DESCRIPTION
Tag Name	The name of the tag.
Prompt	A 32-character-maximum text string that will serve to tell the user what the tag is for when it is assigned to an element.
Type	A menu for selecting the type of information that will be placed in the tag: Character—a text string Integer—a whole number Real—a number with a fractional part

ATTRIBUTE	DESCRIPTION
Tag Name	The name of the tag.
Variable	A check box that, when OFF, prevents the tag value from being changed with the Edit Tags tool (discussed later). If ON, the value can be edited.
Default	A check box that, when OFF, uses the default tag value and prevents the tag value from being changed with the Edit Tags tool. If ON, the check box uses the default but allows editing of the value.
Default Tag Value	The default value for the tag when it is assigned to an element. The default value can be overridden.
Display Tag	Controls how the tags are displayed in the views and what can be done to them.

MAINTAIN TAG SET DEFINITIONS

The Tag Sets settings box provides options for maintaining existing tag and tag set definitions. The options include:

- **Remove**—Remove (delete) the selected tag set (with all its tags) or tag. A confirmation window opens and removal is initiated by clicking **OK**.
- **Rename**—Change a tag set's name. It opens the Tag Set Name window, in which the new tag set name is typed.
- **Duplicate**—Create a duplicate copy of a tag set. It opens the Tag Set Name window, in which the name to use for the duplicate tag set is typed.
- **Edit**—edit the attributes of the selected Tag. It opens the Define Tag window, in which the tag attributes are edited.

ATTACH TAGS TO ELEMENTS

To assign a tag to an element, invoke the Attach Tags tool from:

Task Navigation tool box (active task set to Tags)	Select the Attach Tags tool and select a **Tag Sets** name from the Attach Tags settings box (see Figure 7–49).
Keyboard Navigation (Task Navigation tool box with active task set to Tags)	**q**

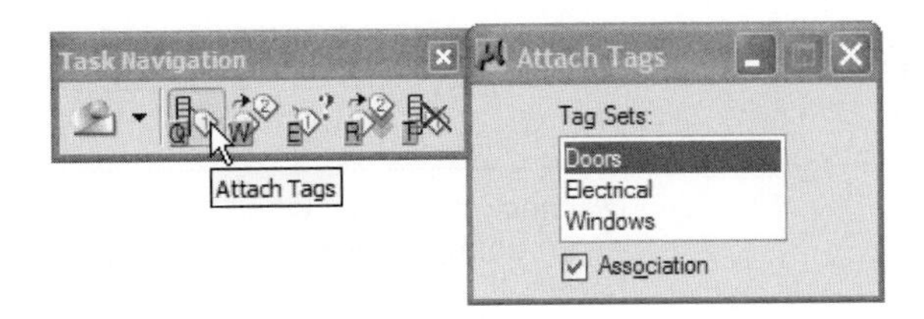

Figure 7–49 Invoking the Attach Tags tool from the Tags tool box

MicroStation prompts:

> Attach Tags > Identify element *(Select a Tag Set from the Tool Settings window and identify the element to which tags are to be attached .)*
>
> Attach Tags > Accept/Reject (Select next input) *(Click the Accept button to accept the selected element, and, if required, select the next element to which tags will be attached after the current one is completed.)*

If the selected Tag Set contains Tags, the Attach Tags window opens when the element is accepted. Figure 7–50 provides an example of the window displaying the tag names for a "windows" tag set; Table 7–4 describes each field.

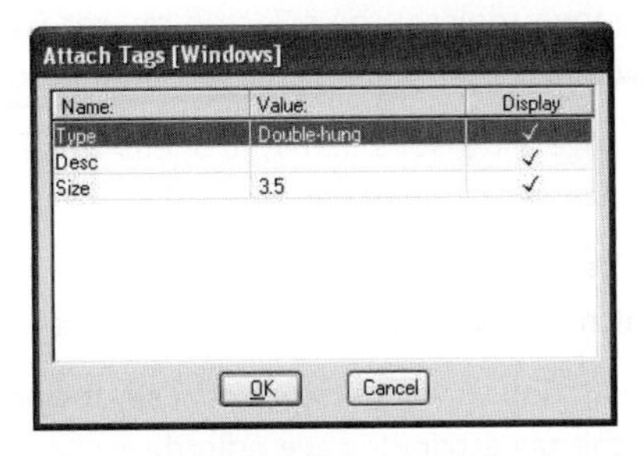

Figure 7–50 Typical Attach Tags dialog box

Table 7–4 The Fields in the Attach Tags Window

FIELD	DESCRIPTION
Name	The name of each tag in the set.
Value	The default value, if any, of each tag in the set.
Display	Turns the display of each tag ON or OFF .
Prompt	The input prompt for the selected tag.
Value Field	Next to the prompt is an input field for entering the values for selected tags. If a tag has a default value, the value appears here when the tag is selected.

For each tag to place on the selected element:

1. Placing a check under the **Display** column makes the tag visible in the design when placed. If a check is not present, the tag is not be visible in the design but is included in tag reports.
2. Type the selected tag's value in the entry field, unless the tag has an acceptable default value.
3. When all the tags for the selected element are ready, click **OK** to close the window and place the tags on the element, or click **Cancel** to discard the changes.

Clicking **OK** when there are selected tags with the **Display** check box ON, causes MicroStation to prompt:

> Attach Tags > Place Tag *(Click the Data button at the location in the design where the tags are to be placed.)*

The tag values are placed using the active element and text attribute settings. If no tags are to be displayed, there is no prompt for placing the tags.

EDIT TAGS

To make changes to the tag values attached to an element, invoke the Edit Tags tool from:

Task Navigation tool box (active task set to Tags)	Select the Edit Tags tool (see Figure 7–51).
Keyboard Navigation (Task Navigation tool box with active task set to Tags)	**w**

Figure 7–51 Invoking the Edit Tags tool from the Tags tool box

MicroStation prompts:

> Edit Tags > Identify element *(Select the element containing tags to be edited.)*
>
> Edit Tags > Accept/Reject (Select next input) *(Click Accept to accept the selected element.)*

***Note:** If more than one tag set is attached to the selected element, the Edit Tags dialog box appears and lists all tag sets attached to the selected element, (as shown in Figure 7–52) and the following prompt appears. Otherwise, the Edit Tags [tag set name] dialog box appears, as shown in Figure 7–53.*

Edit Tags > Select Tag to Edit (If, the Edit Tags dialog box appears, select one of the tag sets from the dialog box and click **OK**.)

Figure 7–52 Typical Edit Tags window

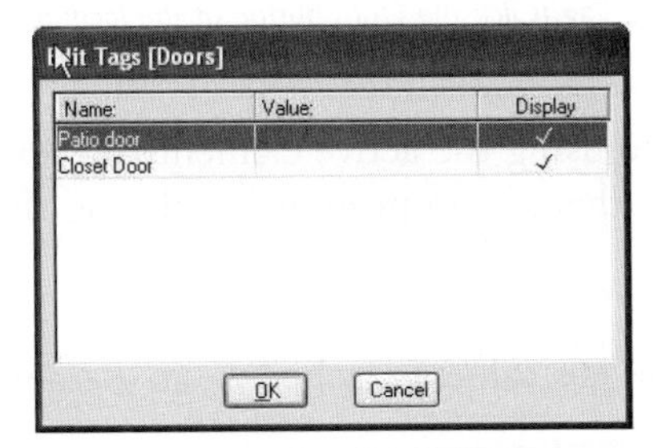

Figure 7–53 Typical Edit Tags [tag set name] window

The Edit Tags [tag set name] settings box shows all tags in the selected tag set. For each tag in the dialog box, the **Display** can be turned ON or OFF and the tag's value can be changed. After completing the required tag changes, click **OK** to apply the changes to the tags for the selected element, or click **Cancel** to discard the changes.

Note: *Tags created with the Variable option OFF cannot be edited.*

REVIEW TAGS

The Review Tags tool allows selecting an element and viewing the tags attached to the element. The tool works the same as the Edit Tags tool, except that it only allows views. Invoke the Review Tags tool from:

Task Navigation tool box (active task set to Tags)	Select the Review Tags tool (see Figure 7–54).
Keyboard Navigation (Task Navigation tool box with active task set to Tags)	**e**

Figure 7–54 Invoking the Review Tags tool from the Tags tool box

MicroStation prompts:

Review Tags > Identify element *(Identify the element.)*

Review Tags > Accept/Reject (Select next input) *(Click Accept to accept the selected element or click Reset to reject the element.)*

Note: *If more than one tag set is attached to the selected element, the Review Tags dialog box appears with a list of all tag sets attached to the selected element and the following prompt appears. Otherwise, the Edit Tags [tag set name] dialog box appears. Except for the box name, these two boxes are identical to the Edit Tag tool boxes.*

Review Tags > Select Tag to review *(If, the Review Tags dialog box appears, select one of the tag sets from the dialog box and click **OK**.*

The Review Tags [tag set name] settings box shows all tags in the selected tag set. Click **OK** to close the dialog box.

CHANGE TAGS

The Change Tags tool allows changing the values assigned to tags in the design. For example, change every "Door Material" tag with the value "Pine," to "White Oak." Invoke the Change Tags tool from:

Task Navigation tool box (active task set to Tags)	Select the Change Tags tool and select required change options in the Change Tags settings box (see Figure 7–55).
Keyboard Navigation (Task Navigation tool box with active task set to Tags)	**r**

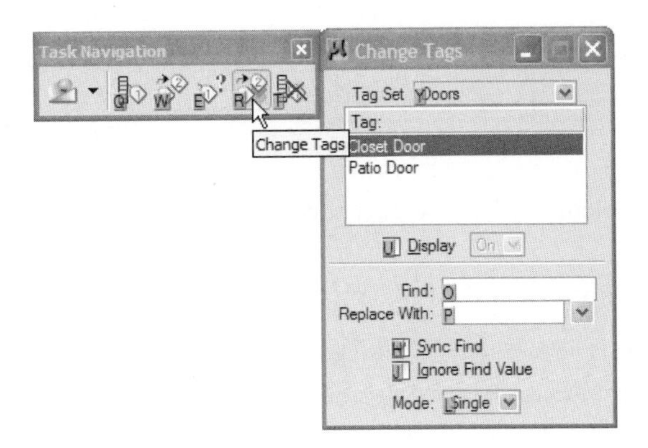

Figure 7–55 Invoking the Change Tags tool from the Tags tool box

When invoked, the Change Tags settings box contains several fields for controlling the way the tag changes are handled. Table 7–5 describes each option.

Table 7–5 The Fields in the Change Tags Window

FIELD	DESCRIPTION
Tag Set	An options menu that allows selecting the Tag Set that contains the Tag to be changed.
Tag	Lists all Tags for the selected Tag Set. Pick the Tag to be changed from this list.
Display	If the **Display** check box is ON, display of the selected Tag can be turned ON or OFF the.
Find	Provides an edit field typing the value to change.
Replace With	Provides an edit field typing the new value to assign to the selected Tag.
Sync Find	Provides a check box that enables and disables the function of the **Find** field
Mode	This option menu controls how the tool searches for occurrences of the selected Tag. **Single** mode requires selecting and changing individual Tags. **Fence** mode changes all occurrences of the selected Tag inside the fence (Place a fence before this Mode). **All** mode changes all occurrences of the selected Tag in the design.

For example, the following procedure changes all Material Tags in the Doors Tag Set with a value of "Pine" to "White Oak."

1. Invoke **Change Tags** from the Tags tool box.
2. In the Tool Settings window, enter the following settings:

- In the **Tag Set** menu, select the Doors Tag Set.

- In the **Tag** list, select the Material Tag.
- Turn the **Display** check box ON and select **On** from the **Display** menu to see all occurrences of the Material Tag in the design after the tool action is completed.
- In the **Find** edit field, type **Pine**.
- In the **Replace With** edit field, type **White Oak**.
- Turn on the **Sync Find** check box to enable searching for Material Tags with values equal to the contents of the **Find** edit field.
- Turn off the **Ignore Find Value** check box to make the tool use the contents of the **Find** edit field for the search.
- Set the **Mode** to **All** to cause the tool to find all occurrences of the Materials Tag in the design (MicroStation prompts, "Change Tag > Accept/Reject entire design file").

3. Click the Data button anywhere in any open View Window to initiate the change.

CREATE TAG REPORTS

As mentioned earlier, MicroStation creates a tag report that lists the non-graphical information required to fabricate the model created in the project's design files.

Creating a report is a two-step process. First, generate a tags template to control what is included in each type of report. Second, generate a report based on the tag template.

Generating a Tags Template

A tag template defines the data columns in a tag report. Each column is either a tag from one tag set or an element attribute. To create a tag template, open the Generate Template settings box from:

Menu	**Element > Tags > Generate Templates**
Key-in window	**mdl load tags template** (or **md l tags template**) (ENTER)

MicroStation displays the General Templates settings box, as shown in Figure 7–56; Table 7–6 describes the fields in the settings box.

Figure 7–56 Typical Generate Templates settings box

Table 7–6 Generate Templates Settings Box Fields

FIELD	DESCRIPTION
Tag Sets	Lists the tag sets attached to the design file. Select the one for which the template is to be created.
Tags	Lists all tags in the selected tag set and all element attributes that can be included in the report. The names in the list that start with a dollar sign ($) are element attributes.
Report Columns	Lists the tags and element attributes selected for inclusion in the report. Each name will be the name of a column in the report, and the order in the window determines the column order in the report.
Report File Name	
Report On Menu	Select the type of elements to include in the report: *Tagged elements*—Include only tagged elements in the report. *All elements*—Include all elements, both tagged and untagged. If all elements are included but no element attribute columns are included, the untagged elements will show up in the report as empty rows.
File Menu	*Open*—Displays the Open Template dialog box, from which an existing template file can be selected to open. *Save*—Saves the template information to the same file that was previously opened or saved as. *Save As*—Opens the Save Template As dialog box, from which the template information to any directory path can be saved with a file name. Use this tool to save new templates. Both the Open and Save Template As windows display the default template files directory path the first time they are opened.

Following are the steps to create a new template in the Generate Templates settings box.

1. Select a Tag Set from the **Tag Sets** menu.
2. Key-in the file name for the template in the **Report File Name** field.
3. For each Tag or element attribute to include in the report, select the name in the **Tags** column and click **Add**.
4. To remove a name added by mistake, select it in the **Report Columns** list and click **Remove**.
5. After creating all report columns, select **File > Save** to save the new template using the name currently in the **Report File Name** field.

Generating a Tags Report

Tag reports list tags and element data information based on Tags Templates. To generate a tag report, open the Generate Reports dialog box from:

Menu	**Element > Tags > Generate Reports**
Key-in window	**mdl load tags report** (or **md l tags report**) (ENTER)

MicroStation displays the Generate Reports dialog box. Following are the steps to generate a report from the Generate Templates dialog box:

1. If required, change the directory path and file type to display the required templates.
2. To generate each report, click the required template file name in the **Files** list and click **Add**. The file specification for each selected template appears in the **Templates for Reports** list.
3. To correct a mistake in selecting a template, select it in the **Templates for Reports** list and click **Remove**. The **Remove** button is dim unless a template is selected.
4. After selecting the required templates, click **Done** to generate the reports.

Tag reports are created using the template's file name and the *.rpt* extension. The reports are stored in MicroStation's default reports path:

Program Files\Bentley\Workspace\projects\untitled\out

The folder under which this path is found varies, depending on how MicroStation was installed and the computer's operating system. The MS_TAGREPORTS configuration variable defines complete path.

Accessing the Reports

Tag reports are in ASCII files (also called "flat files") that can be accessed in several ways. For example:

- View and print a report with one of the operating system's text viewers, such as NotePad in Microsoft Windows.
- Import a report into another application, such as Microsoft Office Excel, the Microsoft Windows spreadsheet application.

TAG LIBRARIES

MicroStation provides tools that allow collecting the tag sets from several design files into a tag library. Tag sets from the library can be imported into other design files. This allows quickly obtaining the required tag sets without having to go to several other design files. Tools for creating tag libraries, exporting sets to existing tag libraries, and importing tag sets from a tag library are available from the Tag Sets settings box's **File** menu. Open the Tag Sets settings box from:

Menu	**Element > Tags > Define**
Key-in window	**mdl load tags define** (or **md l tags define**) (ENTER)

Create a Tag Set Library

Following are the steps to create a tag library:

1. Invoke the Tag Sets settings box.
2. Select a tag set in the **Sets** list box.
3. From the settings box's menu bar, select **File > Export > Create Tag Library** to open the Export Tag Library dialog box.
4. In the **Directories** field, locate and select the folder that is to hold the new tag set library.
5. Key-in the library file name in the **Files** field.
6. Click **OK**.

MicroStation creates a tag library file and exports the selected tag set to the new library.

Append a Tag Set to an Existing Library

Following are the steps to export a tag set from the design to an existing tag library:

1. Invoke the Tag Sets settings box.
2. Select the tag set export from the **Sets** list box.

3. From the settings box's menu bar, select **File > Export > Append to Tag Library** to open the Export Tag Library dialog box.
4. In the **Directories** field, locate and select the folder that contains the tag library.
5. In the **Files** field, find and select the name of the library to which the selected tag set is to be imported
6. Click **OK**.

MicroStation appends the selected tag set to the tag library file.

Note: *If the library contains a tag set with the same name as the selected tag set, a message box appears asking for confirmation before replacing the existing set in the library. Click* **OK** *to replace it or click* **Cancel** *to keep the set already in the library.*

Import a Tag Set from a Library

Following are the steps to import a tag set from a selected library:

1. Invoke the Tag Sets settings box.
2. From the settings box's menu bar, select **File > Import > From Tag Library** to open the Open Tag Library dialog box.
3. In the **Directories** field, locate and select the folder that contains the tag library.
4. In the **Files** field, find and select the tag library that contains the tag set to import.
5. Click **OK** to open the Import Sets dialog box.
6. Find and select the tag set to import.
7. Click **OK**.

MicroStation imports the selected tag set to the design file.

Note: *If the design file contains a tag set with the same name as the selected tag set, a message box appears asking for confirmation before replacing the existing set in the design file. Click* **OK** *to replace it or click* **Cancel** *to save the set already in the design.*

Open the Exercise Manual PDF file for Chapter 7 on the accompanying CD for project- and discipline-specific exercises.

REVIEW QUESTIONS

Write your answers in the spaces provided.

1. An option to control the presentation of stacked fractions is available on the menu bar of the ____________________ box.
2. The Place Fitted Text tool fits the text between two ____________________.
3. When placing a text string above a line with the Place Text Above tool, the distance between the line and the text is controlled by the Text Style ____________________ setting.
4. Explain briefly why the Intercharacter Spacing attribute is useful in placing text along a curving element.

   ____

   ____

5. Under what circumstance might the Match Text Attributes tool be used?

   ____

   ____

6. What is the purpose of the Tag Increment?

   ____

7. To determine the text attributes of an existing text element in a design file, invoke the ____________________ tool.
8. Explain briefly the purpose of placing nodes in a design file.

   ____

   ____

9. Explain briefly the purpose of defining tags.

   ____

   ____

10. List the steps involved in generating a template and report file.

    ____

    ____

CHAPTER 8

Element Modification

OBJECTIVES

Topics explored in this chapter:

- Extending Elements
- Modifying vertices and arcs
- Creating complex shapes and chains
- Creating multi-line profiles
- Modifying multi-line joints

EXTENDING ELEMENTS

As described in earlier chapters, MicroStation tools allow easy placement of elements. MicroStation also provides tools that easily modify elements as needed. Three tools that modify the length of elements (Extend Line, Extend Elements to Intersection, and Extend Element to Intersection) are available from the Modify Element tool box.

EXTENDING LINES

The Extend Line tool extends or shortens a line, line string, or multi-line by a graphically defined length (with a data point) or by a keyed-in distance. The Extend Line settings box has two check boxes that modify the way the tool functions.

Distance sets the distance the tool shortens or extends the selected element. A negative number shortens the element and a positive number extends the element.

If **From End** is ON, the element shortening or extension starts at the end of the element closest to the element identification point. If **From end** is OFF, the shortening or extension of the element starts at an element keypoint regardless of the location of the element identification point. Figure 8–1a shows an example of selecting a line element with the From End check box OFF and Figure 8–1b shows an example of selecting a line element with the From End check box ON.

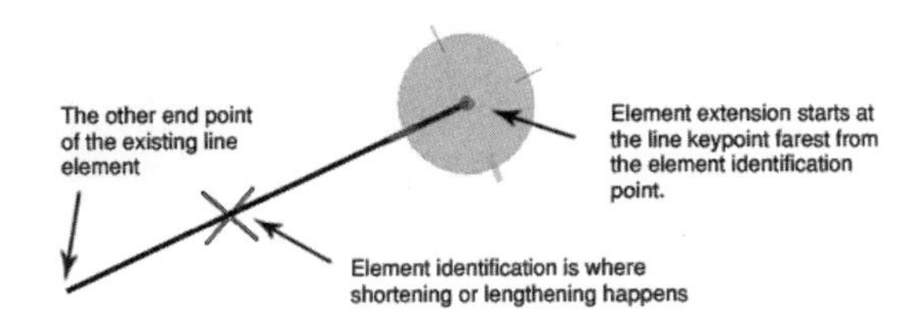

Figure 8–1a *Examples of selecting a line element with the Extend Line From End check box OFF*

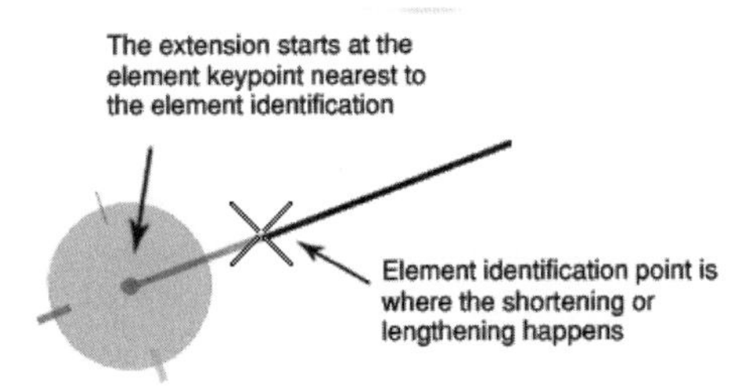

Figure 8–1b *Examples of selecting a line element with the Extend Line From End check box ON*

Note: *The tool does not use the* ***From End*** *check box when the* ***Distance*** *check box is ON.*

Extending Lines Graphically

To extend an element graphically, invoke Extend Line from:

Modify tool box	Select the Extend Line tool and if necessary set From End check box to ON in the Tool Settings window (see Figure 8-2).
Keyboard Navigation (Main tool box active)	**7 3** (If necessary set From End check box to ON in the Tool Settings window.)

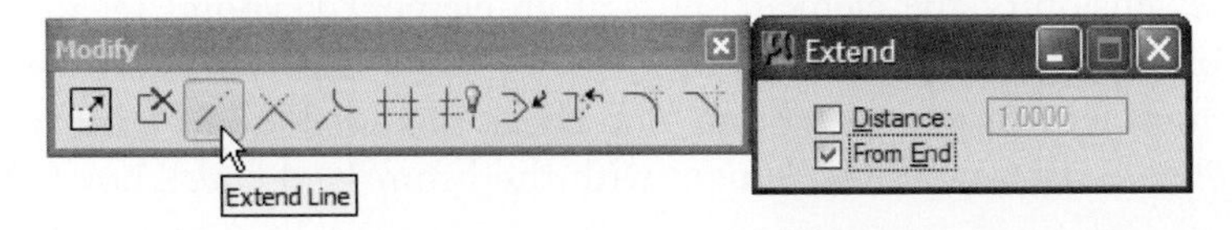

Figure 8–2 *Invoking the Extend Line tool from the Modify Element tool box*

MicroStation prompts:

Extend Line > Identify element *(Identify the element at the point were the shortening or extension is to start.)*

Extend Line > Accept or Reject (Select next input) *(Drag the element to the new length and click the Data button to accept, or click the Reject button to disregard the modification.)*

Extending Lines by Key-in

To extend an element by keying-in the distance, invoke Extend Element from:

Modify tool box	Select the Extend Line tool and set the Distance check box to ON and key-in the shortening (a negative value) or extension distance in the Distance field in the Tool Settings window (see Figure 8-3).
Keyboard Navigation (Main tool box active)	**7 3** (Set the Distance check box to ON and key-in appropriate distance in the Tool Settings window.)

Figure 8–3 *Invoking the Extend Element tool via key-in from the Modify Element tool box*

MicroStation prompts:

Extend Line by Key-in > Identify element *(Identify the element near the end to be extended or shortened.)*

Extend Line by Key-in > Accept/Reject (Select next input) *(Click the Data button again anywhere in the design plane to accept the extension.)*

Note: *To shorten the element, key-in a negative value in the Distance edit field.*

EXTENDING ELEMENTS TO INTERSECTION

The Extend Elements to Intersection tool extends or shortens two elements, as necessary so that the elements intersect and end at the point of intersection. The tool extends open elements, including lines, line strings, arcs, half ellipses, and quarter ellipses. The tool cannot extend closed elements, such as rectangles and circles to an intersection. Figures 8–4 shows examples extending elements to their intersection point.

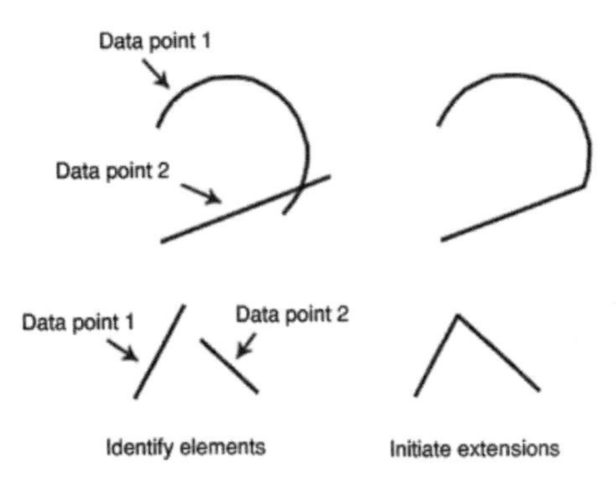

Figure 8–4 *Examples of extending elements to a common intersection*

To extend two elements to their common intersection, invoke Extend two Elements to Intersection from:

Modify tool box	Select the Extend Elements to Intersection tool (see Figure 8–5).
Keyboard Navigation (Main tool box active)	**7 4**

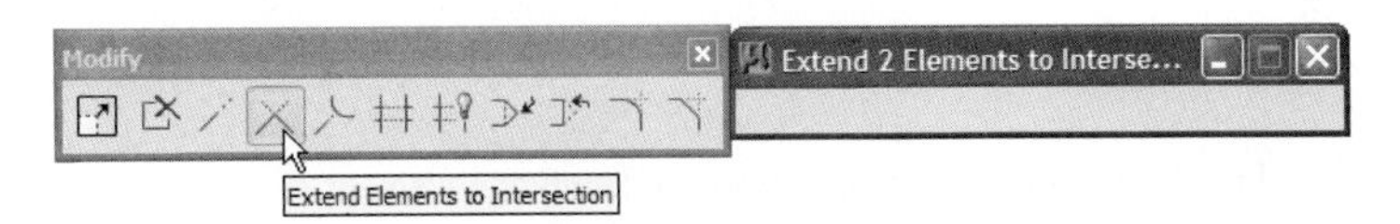

Figure 8–5 *Invoking the Extend Elements to Intersection tool from the Modify Element tool box*

MicroStation prompts:

Extend 2 Elements to Intersection > Select first element for extension *(Identify one of the two elements.)*

Extend 2 Elements to Intersection > Select element for intersection *(Identify the second element.)*

Note: *If an element overlaps the intersection, select it on the part to keep. The tool deletes the part of the element beyond the intersection. If dynamic update shows the wrong part of the element deleted, click the Reset button to back up and try again.*

EXTENDING ELEMENT TO INTERSECTION

The Extend Element to Intersection tool shortens or extends first selected element as necessary to intersect with, and end at, the second selected element. The tool extends open elements including lines, line strings, arcs, half ellipses, and quarter

ellipses. It does not extend closed elements such as rectangles or circles. Figure 8–6 shows two examples of extending an element to its contact point with another element.

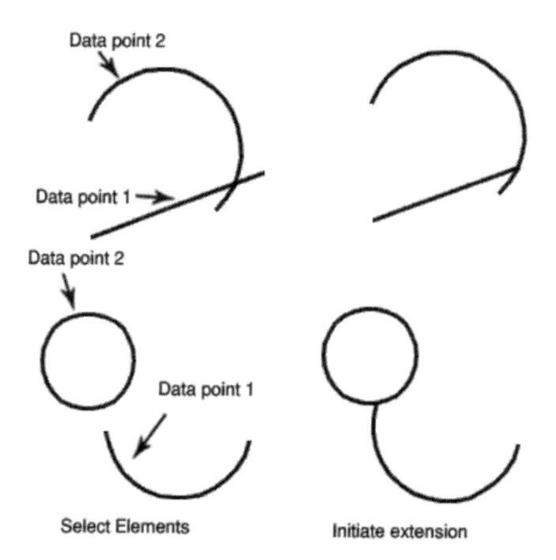

Figure 8–6 *Examples of extending an element to an intersection*

To extend an element to its intersection with another element, invoke Extend Element to Intersection from:

Modify tool box	Select the Extend Element to Intersection tool (see Figure 8–7).
Keyboard Navigation (Main tool box active)	**7 5**

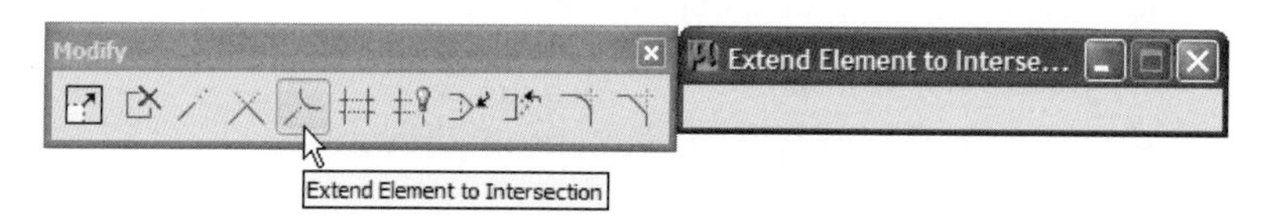

Figure 8–7 *Invoking the Extend Element to Intersection tool from the Modify Element tool box*

MicroStation prompts:

> Extend Element to Intersection > Select first element for extension *(Identify the element to extend.)*
>
> Extend Element to Intersection > Select element for intersection *(Identify the second element and initiate the extension of the first element to the second element.)*

***Note:** If the extended element overlaps the intersection, select it on the part to keep. The tool deletes the part of the element beyond the intersection.*

MODIFYING VERTICES

Several tools are available to modify the geometric shape of elements by moving, deleting, or inserting vertices. For example, change the size of a block by selecting and moving one of the vertices of the block, or turn the block into a triangle by deleting one of the vertices.

MODIFYING ELEMENTS

The Modify Element tool can modify the geometric shape of any type of element except text elements. Here are the possible types of modifications:

- Move a vertex or segment of a line, line string, multi-line, curve, B-spline control polygon, shape, complex chain, or complex shape.
- Scale a block about the opposite vertex.
- Modify rounded segments of complex chains and complex shapes created with the Place SmartLine tool while preserving their tangency.
- Change rounded segments of complex chains and complex shapes to sharp, and vice versa.
- Scale a circular arc while maintaining its sweep angle (use the Modify Arc Angle tool to change the sweep angle of an arc).
- Change a circle's radius or the length of one axis of an ellipse (if the ellipse axes are made equal, the ellipse becomes a circle and only the radius can be modified after that).
- Move dimension text or modify the extension line length of a dimension element.

Figure 8–8 shows examples of typical element modifications.

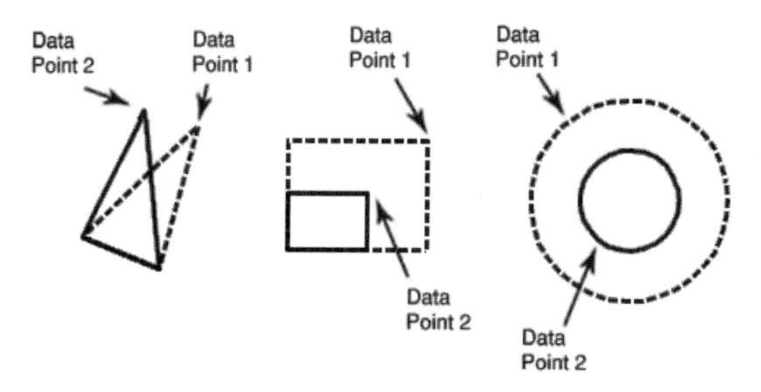

Figure 8–8 *Examples of element modifications*

Note: *Select a segment near its center to move the segment. Select it near a vertex to move the vertex.*

Invoke Modify Element from:

Modify tool box	Select the Modify Element tool (see Figure 8–9).
Keyboard Navigation (Main tool box active)	**7 1**

Figure 8–9 *Invoking the Modify Element tool from the Modify tool box*

MicroStation prompts:

Modify Element > Identify element *(Identify the element to modify.)*

Modify Element > Accept/Reject (Select next input) *(Move the selection point to the desired new location and place a data point to complete the modification, or click the Reset button to deselect the element.)*

Selecting a vertex for modification displays a set of options on the Modify Element tool settings box.

- The **Vertex Type** menu provides the **Sharp**, **Rounded**, and **Chamfered** options that specify the geometric shape of the selected vertex at its new location.

- If the **Vertex Type** is **Sharp** or **Rounded**, the next option on the settings box is the **Rounding Radius** field that is used by the **Rounded** option to set the radius of the vertex rounding in working units. The **Sharp** option does not use this field. Figure 8–10a shows the settings box when the **Vertex Type** is **Rounded**.

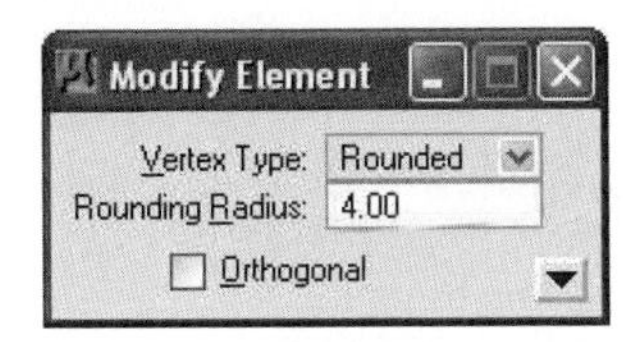

Figure 8–10a *Modify Element tool settings window with the Vertex Type set to Rounded*

- If the **Vertex Type** is **Chamfered**, the next option on the settings box is the **Chamfer Offset** field that is used to set the offset of each end of the chamfered vertex in working units. Figure 8–10b shows the settings box when the **Vertex Type** is **Chamfered**.

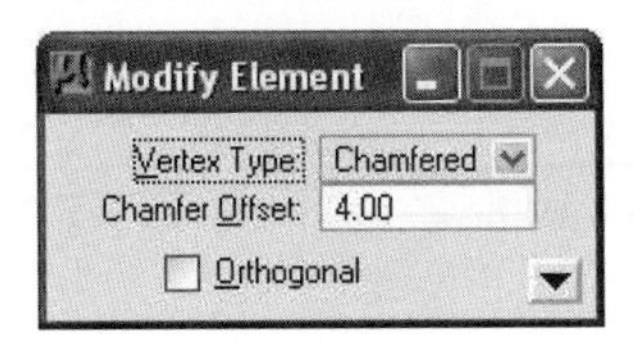

Figure 8–10b *Modify Element tool settings window with the Vertex Type set to chamfered*

- The **Orthogonal** check box is only visible on the settings box when the selected vertex is orthogonal. Turn the check box ON to maintain the orthogonal shape of the vertex.

Using Modify Element and AccuDraw Together

Turn on AccuDraw before invoking the Modify Element tool to benefit from the extra drawing aids AccuDraw provides. For example, locking the angle for a line makes it easy to adjust only the line length and locking the length makes it easy to change only the rotation angle.

DELETING VERTICES

The Delete Vertex tool removes a vertex from a shape, line string, or curve string. Figure 8–11 shows an example of deleting a vertex from a point curve element.

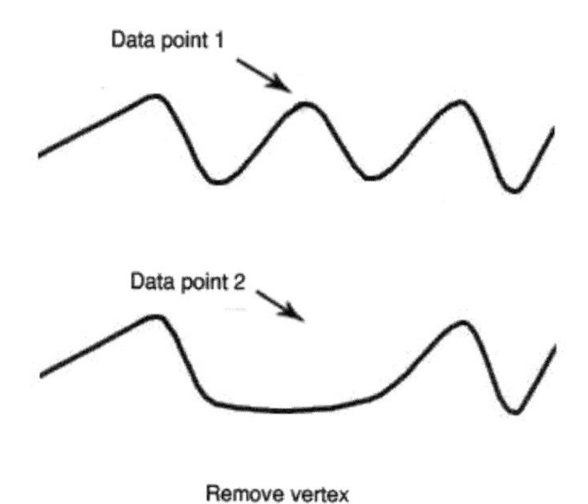

Figure 8–11 *Example of deleting a vertex from a line string*

To delete a vertex from an element, invoke the Delete Vertex tool from:

Modify tool box	Select the Delete Vertex tool (see Figure 8–12).
Keyboard Navigation (Main tool box active)	**7 9**

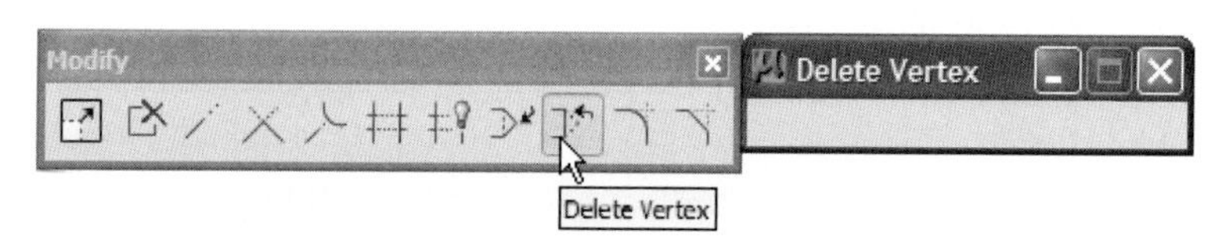

Figure 8–12 *Invoking the Delete Vertex tool from the Modify Element tool box*

MicroStation prompts:

> Delete Vertex > Identify element *(Identify the element near the vertex to delete.)*
>
> Delete Vertex > Accept/Reject (Select next input) *(Click the Data button to initiate deleting the vertex, or click the Reject button to cancel the tool action.)*

Selecting the vertex to delete displays a dynamic update that shows the element without the vertex, but is the second data point actually removes the vertex. The second data point can also select another vertex to delete.

Note: *The tool does not allow for deleting a vertex from an element that only has the minimum number of vertices required to define that type of element. In this situation the tool indicates it is deleting the vertex but does not delete it. For example, a minimum of three vertices is required to define a shape.*

INSERTING VERTICES

The Insert Vertex tool inserts a new vertex into a shape, line string, or curve string. Figure 8–13 shows an example of inserting a vertex into a point curve element.

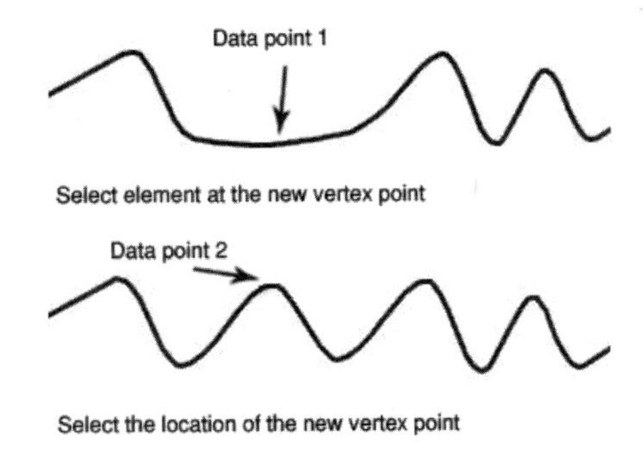

Figure 8–13 *Example of inserting a vertex*

To insert a vertex from an element, invoke the Insert Vertex tool from:

Modify tool box	Select the Insert Vertex tool (see Figure 8–14).
Keyboard Navigation (Main tool box active)	**7 8**

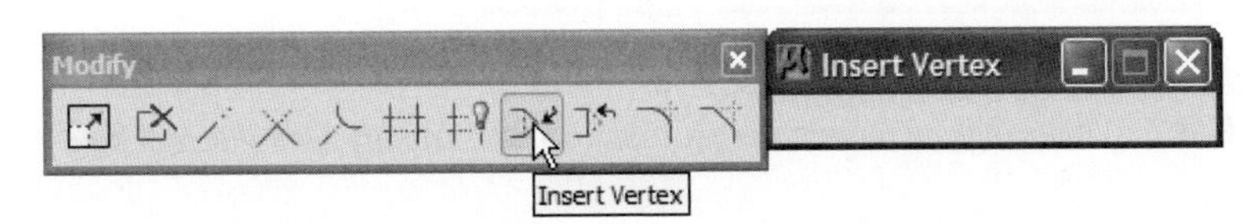

Figure 8–14 *Invoking the Insert Vertex tool from the Modify Element tool box*

MicroStation prompts:

> Insert Vertex > Identify element *(Identify the element at the vertex insertion point.)*
>
> Insert Vertex > Accept/Reject (Select next input) *(Drag the new vertex to in the desired location on the design plane and click the Data button to insert it, or click the Reject button to cancel the tool action.)*

MODIFYING ARCS

MicroStation provides tools for modifying an arc's radius, sweep angle, and axis. The tools are available in the Arcs tool box.

MODIFYING AN ARC RADIUS

The Modify Arc Radius tool changes the length of the radius of the selected arc. Figure 8–15 shows examples of modifying an arc radius.

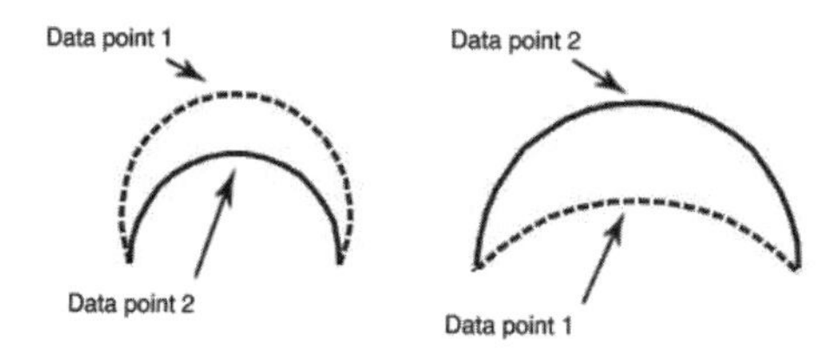

Figure 8–15 *Examples of modifying an arc radius*

To modify the radius of an arc, invoke the Modify Arc Radius tool from:

Task Navigation tool box (active task set to Circles)	Select the Modify Arc Radius tool (see Figure 8–16).
Keyboard Navigation (Task Navigation tool box with active task set to Circles)	**a**

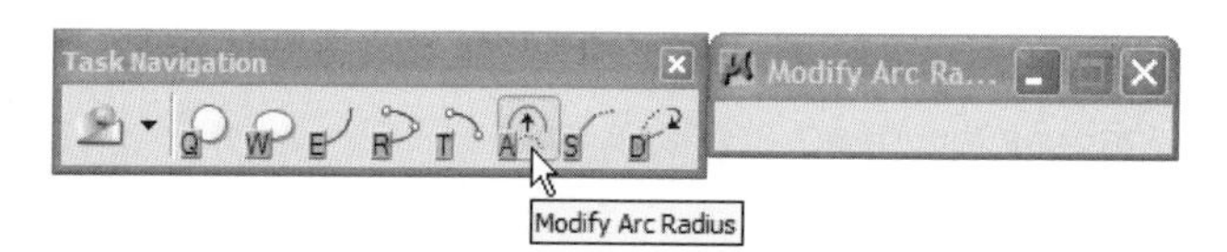

Figure 8–16 *Invoking the Modify Arc Radius tool*

MicroStation prompts:

> Modify Arc Radius > Identify element *(Identify the arc to modify.)*
>
> Modify Arc Radius > Accept/Reject (Select next input) *(Reposition the arc and click the Data button to place it, or click the Reject button to reject the modification.)*

Selecting the arch causes a dynamic image of it to follow the cursor but placing the second data point, actually modifies the arc.

MODIFYING ARC ANGLE

The Modify Arc Angle tool increases or decreases the sweep angle of a selected arc. Figure 8–17 shows an example of modifying an arc angle.

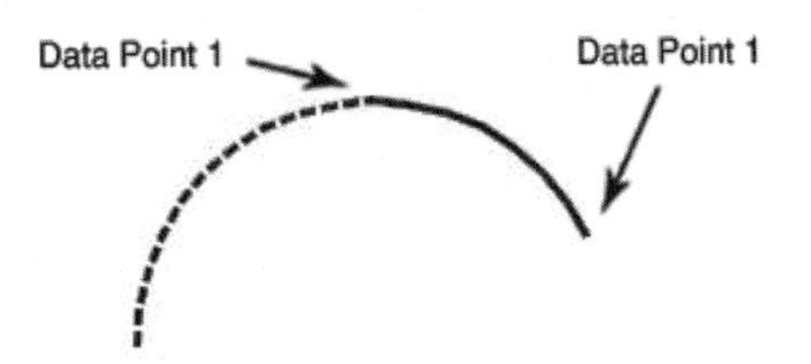

Figure 8–17 *Example of modifying an arc angle*

To modify the sweep angle of an arc, invoke the Modify Arc Angle tool from:

Task Navigation tool box (active task set to Circles)	Select the Modify Arc Angle tool (see Figure 8–18).
Keyboard Navigation (Task Navigation tool box with active task set to Circles)	**s**

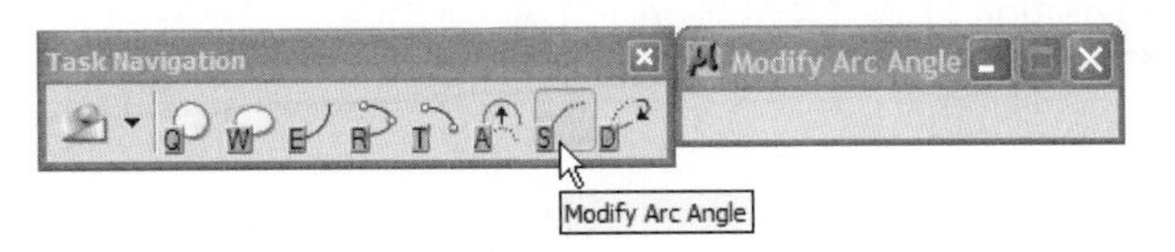

Figure 8–18 *Invoking the Modify Arc Angle tool*

MicroStation prompts:

> Modify Arc Angle > Identify element *(Identify the arc near the end whose sweep to be modified.)*
>
> Modify Arc Angle > Accept/Reject (Select next input) *(Reposition the end of the arc and click the Data button to initiate the change, or click the Reject button to reject the modification.)*

Dynamic update shows the arc following the cursor after selecting the arc but the second data point actually completes modifying the sweep angle.

Note: *Extending the sweep angle until the arc appears to be a circle does not change the arc to a closed circle: It is still an open arc. Arcs that look like circles can be confusing later when using tools like patterning. If the arc should have been a circle, place a circle and delete the arc.*

MODIFYING ARC AXIS

The Modify Arc Axis tool changes the major or minor axis radius of the selected arc. Figure 8–19 shows an example of modifying an arc axis.

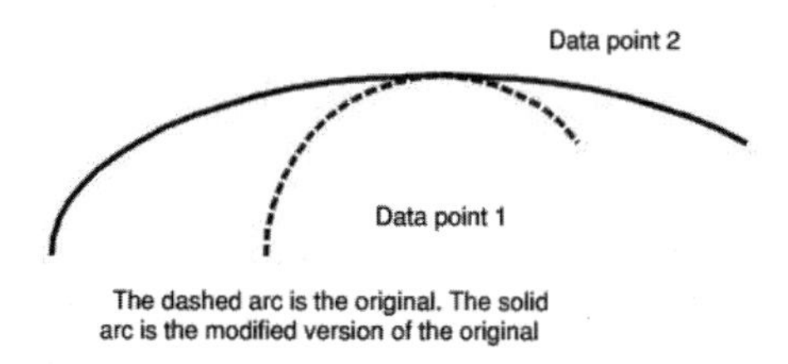
Data point 2
Data point 1
The dashed arc is the original. The solid arc is the modified version of the original

Figure 8–19 *Example of modifying an arc axis*

To modify an arc axis, invoke the Modify Arc Axis tool from:

Task Navigation tool box (active task set to Circles)	Select the Modify Arc Axis tool (see Figure 8–20).
Keyboard Navigation (Task Navigation tool box with active task set to Circles)	**d**

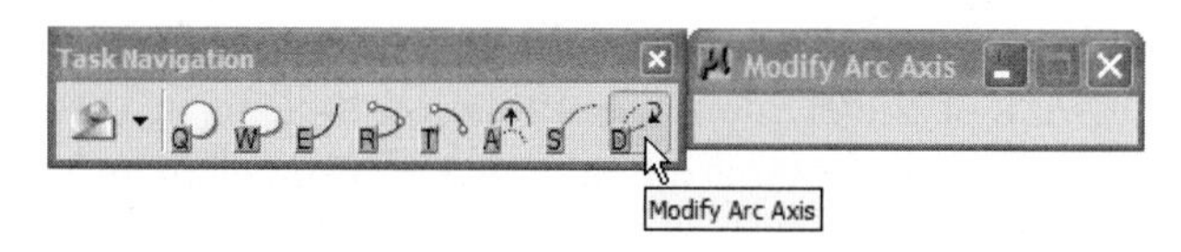
Task Navigation
Modify Arc Axis
Modify Arc Axis

Figure 8–20 *Invoking the Modify Arc Axis tool from the Arcs tool box*

MicroStation prompts:

> Modify Arc Axis > Identify element *(Identify the arc whose axis has to be modified.)*
>
> Modify Arc Axis > Accept/Reject (Select next input) *(Reposition the axis of the arc and click the Data button to place it, or click the Reject button to reject the modification.)*

Identifying the arc displays a dynamic image of the arc that follows the cursor until the second data point completes modifying the axis. The second data point actually modifies the arc.

Note: *The Modify Arc Radius tool cannot manipulate an arc with a modified axis.*

CREATING COMPLEX CHAINS

The Create Complex Chain tool turns groups of connected elements into one complex element. A complex chain is an open element (the "water" can flow out between the two ends of the chain). The element manipulation tools treat the elements in a complex group as one element.

When created, complex chains take on the current active element attributes and gaps between the selected elements are closed. Figure 8–21 shows four of individual elements and the complex shape that results from grouping the elements.

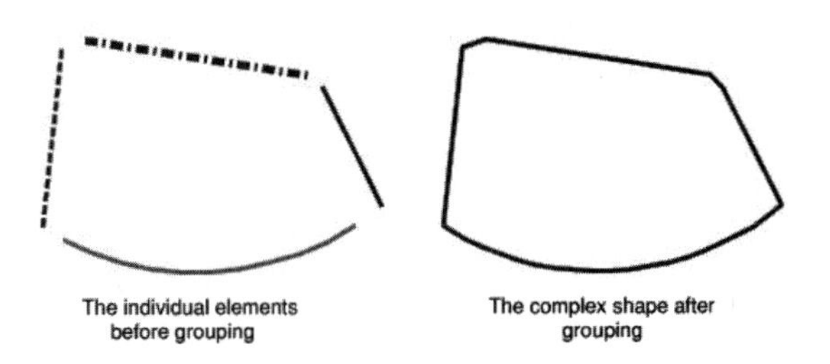

Figure 8–21 *Example of creating a complex shape from separate elements*

Complex chains and shapes are created manually by selecting each element to include or automatically by letting MicroStation find each element.

A quick way to check to see the tool actually created a complex group from the elements is to apply one of the element manipulation tools to it. If the elements are a group, the dynamic image contains all elements and the element type in the Status bar says it is either a Complex Chain or a Complex Shape.

Note: *It is easy to make a mistake in creating complex chains and shapes from a large number of elements. Correct a mistake by selecting the Undo tool and try again.*

MANUAL METHOD

Manually creating a complex chain requires selecting, in order, all the elements that are to make up the chain. The tool closes gaps between selected elements and, when completed, the chain takes on the currently active element attributes.

To create a complex chain manually, invoke the Create Complex Chain tool from:

Groups tool box	Select the Create Complex Chain tool and select Manual from the Method menu in the Tool settings box (see Figure 8–22).
Keyboard Navigation (Main tool box active)	**6 2** (Select Manual from the Method menu in the Tool settings box.)

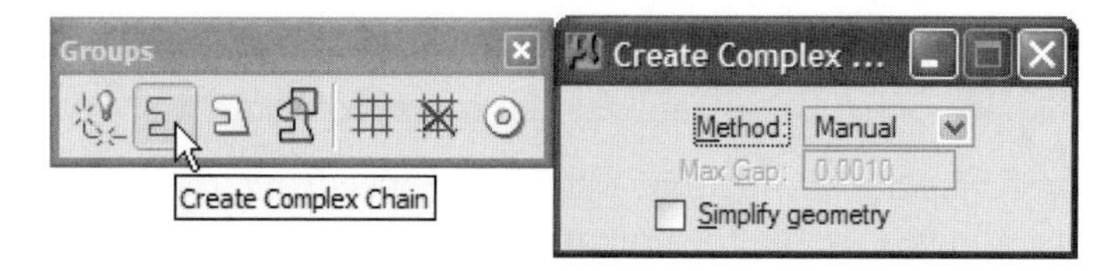

Figure 8–22 *Invoking the Create Complex Chain tool from the Groups tool box*

MicroStation prompts:

> Create Complex Chain > Identify element *(Identify the first element to include in the complex chain.)*
>
> Create Complex Chain > Accept/Reject (Select next input) *(Identify the next element to include in the complex chain and continue selecting elements in order until all elements are selected. After selecting and accepting the last element, click the Reset button to create the complex chain.)*

Turning the **Simplify Geometry** check box ON, adds connected lines as line strings. Identifying only connected lines produces a primitive line string element rather than a complex chain. Figure 8–23 provides an example of manually creating a complex chain.

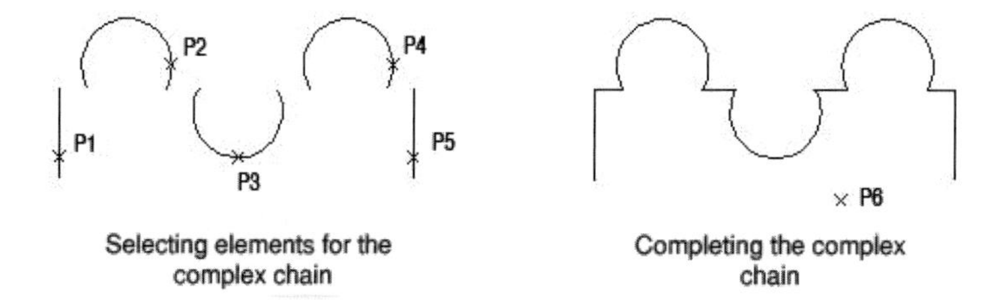

Figure 8–23 *Example of manually creating a complex chain*

***Note:** Sometimes updating the view that shows a complex chain causes the lines that close the gaps between elements in the complex chain to disappear. They are still there—update the view again and they should reappear.*

AUTOMATIC METHOD

To create a complex chain automatically, select and accept the first element. The tool finds the next element to include in the chain, selects it, and waits for selection acceptance. Click the Data button to accept the element or the Reset button to reject it and cause the tool to search for another element to include. Each time an element is accepted, the search usually continues in from the end of the previous element nearest to the location of the acceptance data point.

The process continues until the tool cannot find another element to include. When that happens, the tool completes the complex chain.

The automatic version of the tool allows for specifying the maximum gap, in working units, allowed between elements. If the maximum gap is zero, the ends of the elements must touch before the tool finds them.

If there are two or more possible elements at a junction, the tool displays a message saying there is a fork in the path and selects one of the possible elements. Accept the selected element or reject it and have tool highlight another possible element in the fork.

Note: *If the complex chain contains many elements, and there are not very many forks, the automatic method is probably faster than the manual method. If there are many fork points, creating the chain manually may be faster.*

To create a complex chain automatically, invoke Create Complex Chain from:

Groups tool box	Select the Create Complex Chain tool and select Automatic from the Method menu and, optionally, specify a Max Gap value in working units in the Tool Settings window (see Figure 8–24).
Keyboard Navigation (Main tool box active)	**6 2** (Select Automatic from the Method menu and, optionally, specify a Max Gap value in working units in the Tool Settings window.)

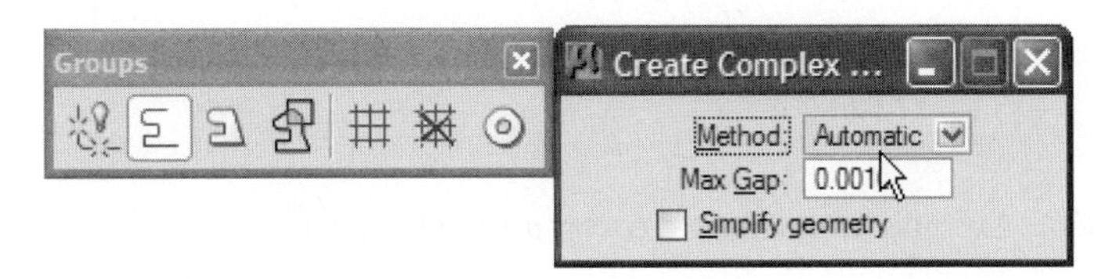

Figure 8–24 *Invoking the Create Complex Chain Automatic tool from the Groups tool box*

MicroStation prompts:

Automatic Create Complex Chain > Identify element *(Identify the first element to include in the complex chain.)*

Automatic Create Complex Chain > Accept/Reject (Select next input) *(Move the cursor near the end of the selected element from which the search should continue and click the Data button to accept the selection, or click the Reject button to reject it and start over.)*

If there are no forks in the path from the previous element, MicroStation prompts:

Automatic Create Complex Chain > Accept Chain Element *(Click the Data button to accept the element and continue the search, or click the Reset button to complete the chain with the previous element.)*

If there is a fork in the path from the previous element, MicroStation prompts:

> Automatic Create Complex Chain > FORK—Accept or reject to See Alternate *(To accept the element MicroStation selected, click the Data button; or click the Reject button to disregard the current selection and select another fork element.)*

The process continues until MicroStation cannot find another element to add or until a selection is rejected when there is no fork in the path. The search cannot end at a fork point. If the **Simplify Geometry** check box is set to ON, the tool adds connected lines as line strings. if only connected lines are identified, the tool produces a primitive line string element, rather than a complex chain. Figure 8–25 shows an example of creating a complex chain automatically.

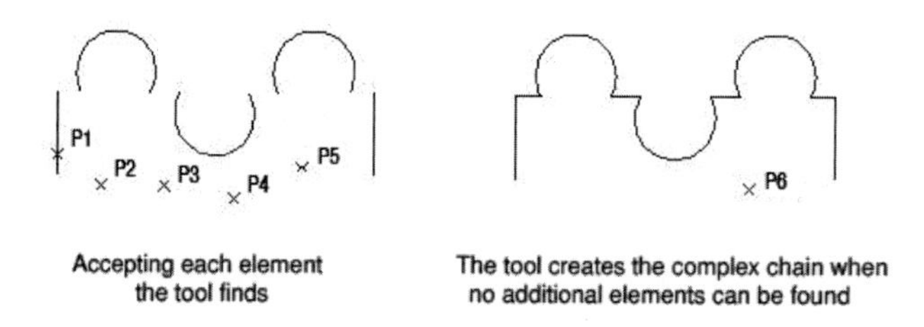
P1
P2
P3
P4
P5
P6
Accepting each element the tool finds
The tool creates the complex chain when no additional elements can be found

Figure 8–25 *Example of creating a complex chain automatically*

CREATING COMPLEX SHAPES

Create Complex Shape tools turn groups of connected elements into one complex element. A complex shape is a closed element (it "holds water"). The element manipulation tools treat the elements in a complex group as one element. When created, complex shapes take on the current active element attributes and gaps between the selected elements are closed.

MANUAL METHOD

Creating a complex shape manually requires selecting and accepting, in order, each element to include in the shape. The tool closes gaps between selected elements and the completed complex shape takes on the current active element attributes. To create a complex shape manually, invoke Create Complex Shape from:

Groups tool box	Select the Create Complex Shape tool and select Manual from the Method menu in the Tool settings box (see Figure 8–26).
Keyboard Navigation (Main tool box active)	**6 3** (Select Manual from the Method menu in the Tool settings box.)

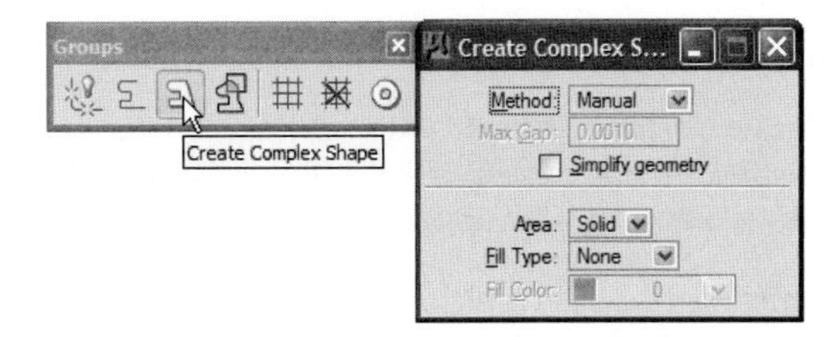

Figure 8–26 *Invoking the Create Complex Shape tool from the Groups tool box*

MicroStation prompts:

> Create Complex Shape > Identify element *(Identify the first element to include in the complex shape.)*
>
> Create Complex Shape > Accept/Reject (Select next input) *(Identify the next element to include in the complex shape and continue selecting elements in order until all elements are selected. When all elements are selected and the shape appears closed, click the Accept button to create the complex shape.)*

If the **Simplify Geometry** check box is set to ON, connected lines are added as line strings. If only connected lines are identified, the tool produces a primitive line string element, rather than a complex chain. Figure 8–27 provides an example of creating a complex shape manually.

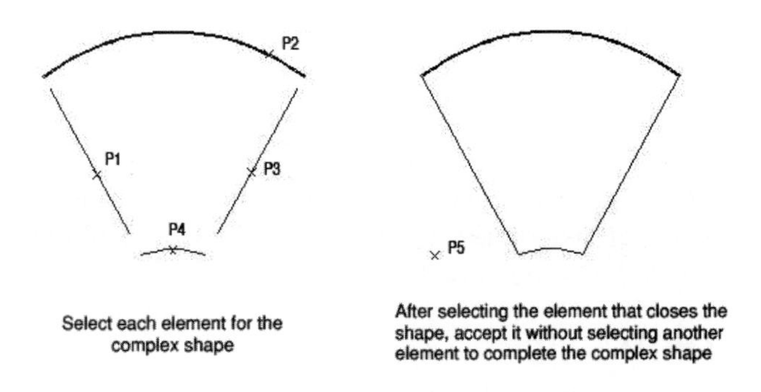

Figure 8–27 *Example of creating a complex shape manually*

AUTOMATIC METHOD

To create a complex shape automatically, start by selecting and accepting the first element in the shape. The tool then finds and selects another element and waits for acceptance or rejection. Accept the element by clicking the Data button, or reject it by selecting the Reset button. If the element is accepted, the tool searches for the next element to include. If the element is rejected, the tool searches for another element to replace the previous selection. This process continues until the tool considers the shape to be closed.

If there are two or more possible elements at a junction, the tool displays a message saying there is a fork in the path and picks one of the possible elements. Either accept the element or reject it to cause the tool to highlight another element at the fork.

The automatic version of the tool also allows for specifying a maximum gap that tells MicroStation how far away, in working units, from the end of the previous element it can search for another element. If the tolerance is set to zero, the ends of elements must touch one before MicroStation finds them.

Note: *If the complex shape contains many elements, and there are not very many forks, the automatic method is probably faster than the manual method. If there are many fork points, creating the shape manually may be faster.*

To create a complex shape automatically, invoke the Create Complex Shape tool from:

Groups tool box	Select the Create Complex Shape tool and select Automatic from the Method menu, and optionally, specify a Max Gap value in the Tool Settings window (see Figure 8–28).
Keyboard Navigation (Main tool box active)	**6 3** (Select Automatic from the Method menu and optionally, specify a Max Gap value in the Tool settings box.)

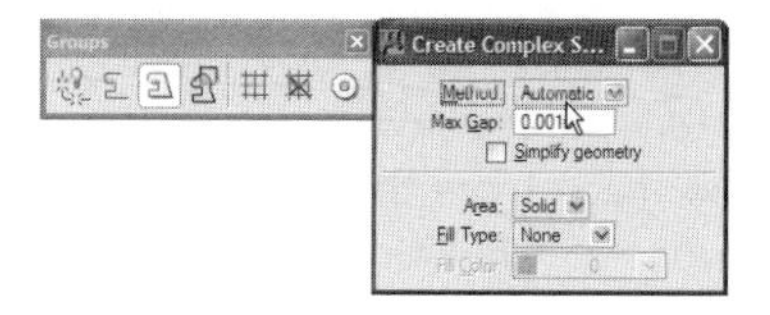

Figure 8–28 *Invoking the Create Complex Shape Automatic tool from the Groups tool box*

MicroStation prompts:

Automatic Create Complex Shape > Identify element *(Identify the first element to include in the complex shape.)*

Automatic Create Complex Shape > Accept/Reject (Select next input) *(Move the near the end of the selected element from which the search is to continue and click the Data button to accept the first element, or click the Reject button to reject it and start over.)*

If there are no forks in the path from the previous element, MicroStation prompts:

Automatic Create Complex Shape > Accept chain Element *(Click the Data button to accept the element and continue the search, or click the Reset button to complete the chain with the previous element.)*

If there is a fork in the path from the previous element, MicroStation prompts:

> Automatic Create Complex Chain > FORK—Accept or reject to See Alternate *(To accept the element MicroStation selected, click the Data button or click the Reject button to cause the too to select another fork element.)*

The process continues until the tool considers the shape closed or until a selection is rejected when there is no fork in the path. The search cannot end at a fork point. Turning the **Simplify Geometry** check box ON adds connected lines as line strings. Figure 8–29 shows an example of creating a complex shape automatically.

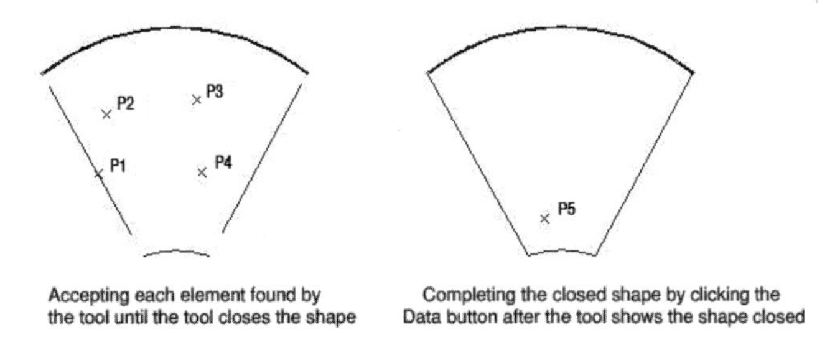
P2
P3
P1
P4
P5
Accepting each element found by the tool until the tool closes the shape
Completing the closed shape by clicking the Data button after the tool shows the shape closed

Figure 8–29 *Example of creating a complex shape automatically*

CREATING REGIONS

The Create Region tool is similar to the Create Complex Shape tool, in that it creates a complex shape, but it provides the following methods for creating the shapes:

- **Flood** creates the shape from the perimeter of an area enclosed by elements that either touch one another or that are no farther apart than a maximum gap value.
- **Union** creates the shape from the perimeter of the total area enclosed by two overlapping closed elements.
- **Intersection** creates the shape form the perimeter of the common area of two overlapping closed elements.
- **Difference** creates the shape from the area within two overlapping closed elements that is not common to both elements.

The tool provides the methods as icons on the Create Region settings box.

The **Keep Original** check box on the settings box controls the handling of the original elements for each of the methods. Turn the check box OFF to delete the original elements, leaving only the complex shape formed from the elements. Turn the check box ON to leave the original elements.

FLOOD METHOD

The **Flood** method creates a complex shape from the perimeter of an area enclosed by elements that either touch one another or are no farther apart than a maximum gap value, as shown in Figure 8–30.

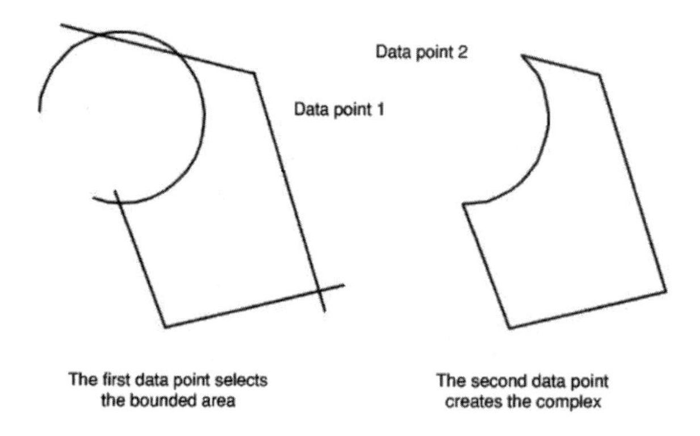
Data point 2
Data point 1
The first data point selects the bounded area
The second data point creates the complex

Figure 8–30 *Example of creating a complex shape using Flood and with Keep Original OFF*

The method also provides icons on the settings box for handling elements within the area enclosed by the elements:

- If **Ignore Interior Shapes** is ON, elements within the bounded area are ignored and do not become part of the complex shape.
- If **Locate Interior Shapes** is ON, elements within the bounded area become part of the complex shape.
- If **Identify Alternating Interior Shapes** is ON, alternating areas are flooded where shapes are nested inside one another.
- If **Locate Interior Text** is ON, the tool avoids any text or dimension text inside or overlapping the selected area.
- If **Dynamic Area Locate** is ON, the area to be included in the region displays dynamically as the cursor moves over the view.

Invoke the Create Region tool from:

Groups tool box	Select the Create Region tool and select Flood and set additional related settings as required in the Tool Settings window (see Figure 8–31).
Keyboard Navigation (Main tool box active)	**6 4** (Select Flood and set additional related settings required in the Tool settings box.)

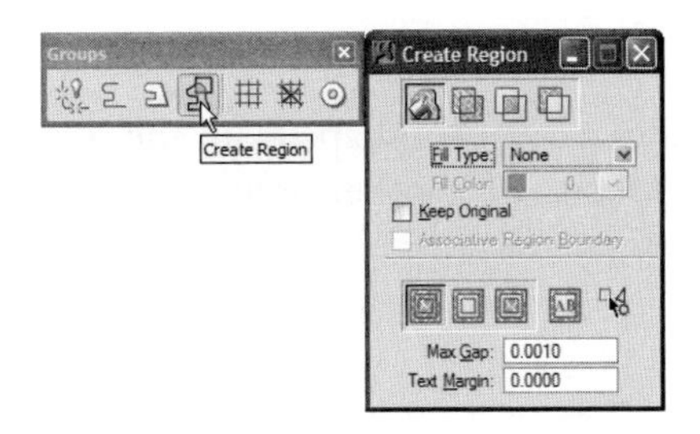

Figure 8–31 *Invoking the Create Region tool with Flood method*

MicroStation prompts:

Create Region From Area Enclosing Point > Enter data point inside area *(Place a data point inside the enclosed area.)*

Create Region From Area Enclosing Point > Accept-Create Region *(Click the Data button to accept the complex shape or click the Reset button to reject the area.)*

Note: *If the tool does not find an enclosing area, it displays the following error message in the status bar: "Error – No enclosing region found."*

UNION METHOD

The **Union** method creates a complex shape from a composite area formed in such a way that there is no duplication between two overlapping closed elements, as shown in Figure 8–32. The total resulting area can be equal to or less than the sum of the areas in the original closed elements.

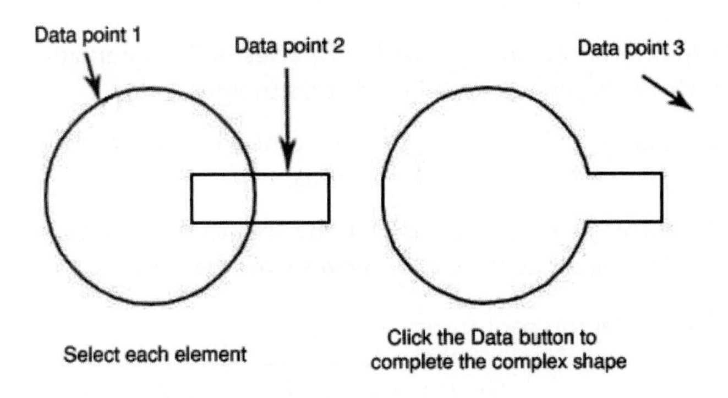

Figure 8–32 *Example of creating a complex shape using Union and with Keep Original OFF*

Invoke the Create Region tool from:

Groups tool box	Select the Create Region tool and select Union and set related settings as required in the Tool settings window (see Figure 8–33).
Keyboard Navigation (Main tool box active)	**6 4** (Select Union and set additional related settings required in the Tool settings box.)

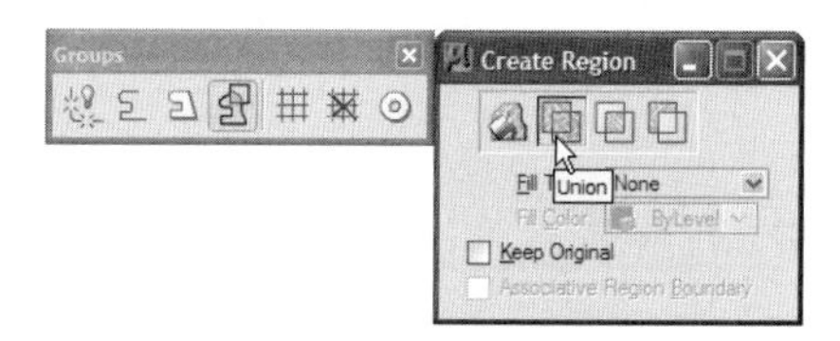

Figure 8–33 *Invoking the Create Region tool with Union method from Groups tool box*

MicroStation prompts:

Create Region From Element Union > Identify element *(Identify one of the elements.)*

Create Region From Element Union > Identify Additional/Reset to Complete *(Select the other element.)*

Create Region From Element Union > Accept/Reject (CTRL+DATA to select) *(Click the Data button again to initiate creating the complex shape from.)*

Note: *The tool normally creates a complex shape from the union of two elements, but the tool's third prompt, "Accept/Reject (CTRL+DATA to select)," allows for adding more elements to the set before initiating creation of the complex shape. To add additional elements, hold down the* CTRL *key while clicking the Data button on the additional element.*

INTERSECTION METHOD

The **Intersection** method creates a complex shape from the area common to two or more closed elements, as shown in Figure 8–34.

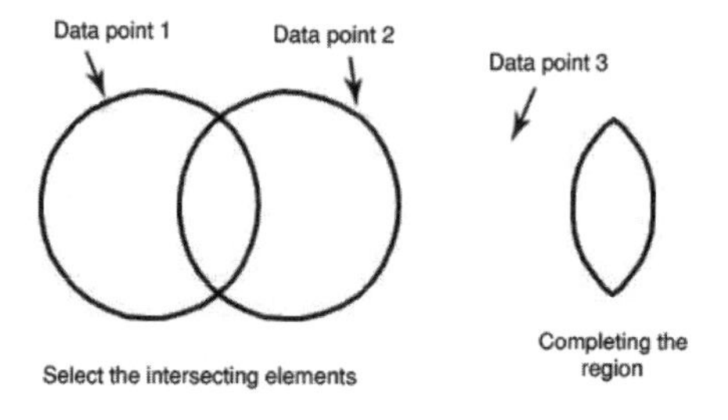

Figure 8–34 *Example of creating a complex shape using the Intersection method*

Invoke the Create Region tool from:

Groups tool box	Select the Create Region tool and select Intersection and set additional settings as required in the Tool Settings window (see Figure 8–35).
Keyboard Navigation (Main tool box active)	**6 4** (Select Intersection and set additional related settings required in the Tool settings box.)

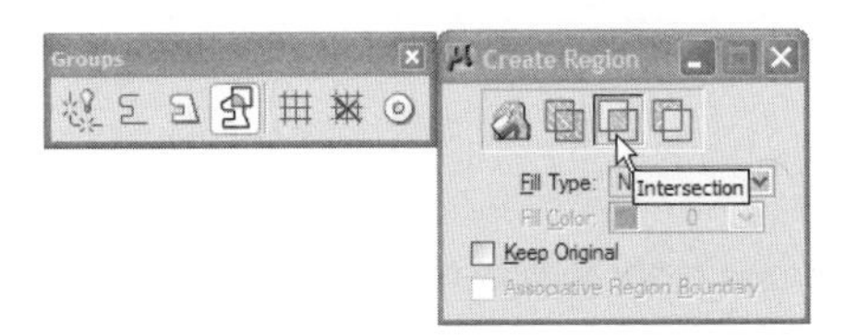

Figure 8–35 *Invoking the Create Region tool with Intersection method from the Groups tool box.*

MicroStation prompts:

Create Region From Element Intersection > Identify element *(Identify one of the two circles.)*

Create Region From Element Intersection > Identify additional/Reset to complete *(Identify the second circle.)*

Create Region From Element Union > Accept/Reject (CTRL+DATA to select) *(Click the Data button again to initiate creating the complex shape.)*

Note: *The tool normally creates a complex shape from the intersection of two elements, but the tool's third prompt, "Accept/Reject (CTRL+DATA to select)," allows for adding more elements to the set before initiating creation of the complex shape. To add additional elements, hold down the* CTRL *key while clicking the Data button.*

DIFFERENCE METHOD

The **Difference** method creates a complex shape from a closed element after removing from it any area it has in common with one or more additional elements, as shown in Figure 8–36.

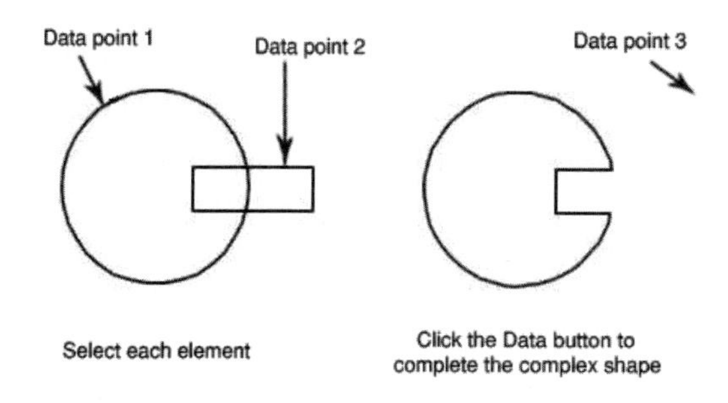

Figure 8–36 *Example of creating a complex shape using the Difference method*

Invoke the Create Region tool from:

Groups tool box	Select the Create Region tool and select Difference and set additional settings as required in the Tool Settings window (see Figure 8–37).
Keyboard Navigation (Main tool box active)	**6 4** (Select Difference and set additional related settings required in the Tool settings box.)

Figure 8–37 *Invoking the Create Region tool with Difference method from the Groups tool box*

MicroStation prompts:

Create Region From Element Difference > Identify element *(Identify the first element.)*

Create Region From Element Difference > Identify additional/Reset to complete *(Identify the second element.)*

Create Region From Element Union > Accept/Reject (CTRL+DATA to select) *(Click the Data button again to initiate creating the complex shape.)*

Note: *The tool normally creates a complex shape from two elements, but the tool's third prompt, "Accept/Reject (CTRL+DATA to select)," allows for adding more elements to the set before initiating creation of the complex shape. To add additional elements, hold down the* CTRL *key while clicking the Data button on the additional elements.*

DROPPING COMPLEX CHAINS AND SHAPES

Drop tools allow complex chains and shapes to be "dropped" or broken apart into the individual elements that made up the complex element. The Drop Element tool drops a selected complex shape. The Drop Fence Contents tool drops all complex shapes selected by a fence. Dropped complex elements return to being individual elements, but the elements keep the attributes (color, weight, and so forth) of the complex shape, they do not return to their original attributes.

Note: *For information on the Drop Fence Contents tool, see Chapter 6.*

To drop a complex chain or shape, invoke the Drop Element tool from:

Groups tool box	Select the Drop Element tool and set the Complex check box to ON, and set the remaining check boxes to OFF in the Tool Settings window (see Figure 8–38).
Keyboard Navigation (Main tool box active)	**6 1** (Set the Complex check box to ON and set the remaining check boxes to OFF in the Tool Settings window.)

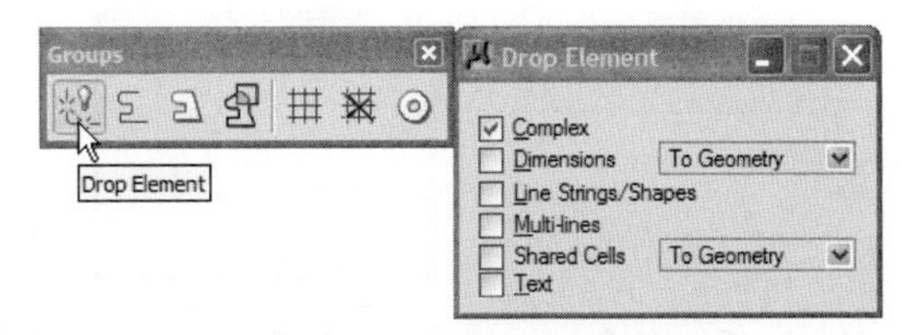

Figure 8–38 *Invoking the Drop Element tool from the Groups tool box*

MicroStation prompts:

Drop Complex Status > Identify element *(Identify the complex chain or shape to drop into individual elements.)*

Drop Complex Status > Accept/Reject (select next input) *(Click the Data button to accept the change in status from a complex element into individual elements, or click the Reset button to reject the change in the status.)*

Note: *If there were gaps between the elements in the complex shape or chain, dropping the complex shape or chain removes the lines that connected the gaps. If the gaps do not disappear, repaint the view window.*

MULTI-LINE STYLES

Chapter 4 described placing multi-lines using the active multi-line style (usually three parallel lines with a dashed centerline). This topic expands the value of multi-lines by describing how to create and maintain multi-line styles using the Multi-Line Styles settings box.

MULTI-LINE STYLE COMPONENTS

Multi-line styles consist of the following components:

- **Lines**—Multi-lines can contain from one to sixteen parallel lines. Each line can have unique attributes (such as color, weight, and style) and an offset (in working units) from the multi-line element's workline. A positive offset value positions the line above the multi-line element's workline when it is drawn horizontally from left to right and on the left when the element is drawn vertically from bottom to top. Drawing the element in the reverse directions reverses the offset.
- **Caps**—The ends of a multi-line can be capped by lines and arcs. **Start Caps** go at the start of the multi-line element and **End Caps** go at the end of the multi-line element. Lines and arcs can be placed between the two outer lines and arcs can be placed across interior pairs of lines. The cap attributes can be set separately from the line attributes.
- **Joints**—A line can be placed across each multi-line joint (vertex) and the joint line can have its own set of attributes.

Figure 8–39 show and example of a multi-line style that has caps and joints.

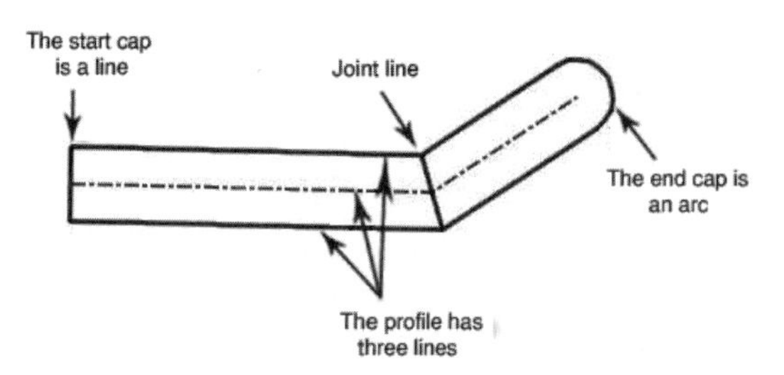

Figure 8–39 *The Multi-Line components*

THE MULTI-LINE STYLES SETTINGS BOX

MicroStation provides the Multi-lines Styles settings box for creating and maintaining multi-line styles and the profiles that define the styles. A style is the set of lines that the multi-line places when it is active. Each style's profile describes the components in the multi-line. Invoke the Multi-Lines settings box:

Menu	Element > Multi-lines Styles
Key-in window	**dialog mlinestyle open** (or **di mu o**) (ENTER)

MicroStation displays the Multi-lines settings box as shown in Figure 8–40.

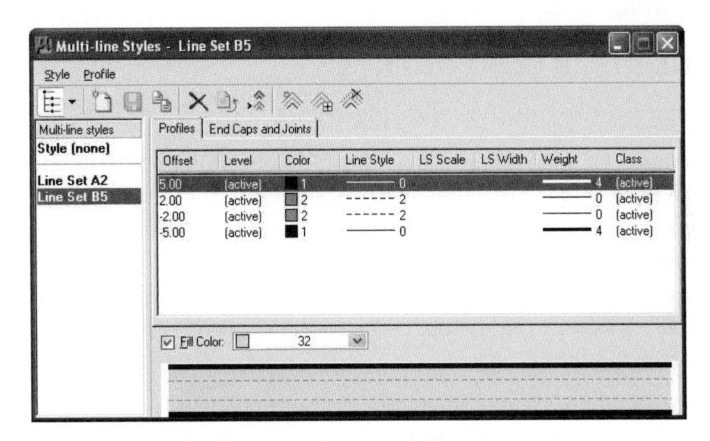

Figure 8–40 *The Multi-Line Styles settings box*

The Multi-Line List

The multi-line list on the left side of the settings box displays the names of all defined multi-line styles in the design. The **Style (none)** multi-line style contains the profile of the currently active multi-line style.

The Multi-Line Definition

The area of the dialog box under the two tabs provides options for creating and maintaining the definition of the multi-line style currently selected in the multi-line list.

- The **Profiles** tab provides controls for selecting and modifying the multi-lines in the selected multi-line style.
- The **End Caps and Joints** tab provides controls for specifying the appearance of the start cap, end cap and joints of the selected multi-line style.
- The **Fill Color** check box and menu allow for selecting a color to fill the entire area between the outermost component lines of the multi-line.
- The preview area across the bottom of the dialog box provides an example of the selected multi-line style.

The Menu Bar

The menu bar on the dialog box provides menus for inserting and removing multi-line styles and the lines in the selected multi-line style.

The **Style** menu manipulates styles and provides the following options:

- **New** creates a new multi-line style by copying the currently active multi-line style.
- **Save** saves the selected multi-line style.
- **Copy** places a copy of a selected style immediately after the selected style in the multi-line list pane.
- **Rename** opens the selected style's name for editing.
- **Delete** deletes the selected style. The option immediately deletes the selected style—there is no confirmation question.
- **Reset** resets the selected style to the settings it had at the last save and throws out all unsaved changes made on the two tabs for the selected style.
- **Save All** saves all multi-line styles listed in the Multi-line styles list box.
- **Import** opens the Multi-line Style Import dialog box that allows importing a multi-line style from another design file or design library.
- **Update from Library** updates the multi-line definition to match the definitions in the attached design library.
- **Exit** closes the Multi-Line Styles settings box.

The **Profile** menu adds and removes lines in the **Profile** tab and contains the following options:

- **Insert** adds a new line to the multi-line profile directly above the currently selected line.
- **Copy** copies the selected line in the profile and pastes it directly above the selected line.
- **Delete** deletes the selected line from the profile.

The Multi-line Tool Bar

The tool bar, located directly below the menu bar, provides icons for options in the two menus plus a menu that displays a list of all multi-line styles in design. Figure 8-41 shows the tool bar and names the icons on it.

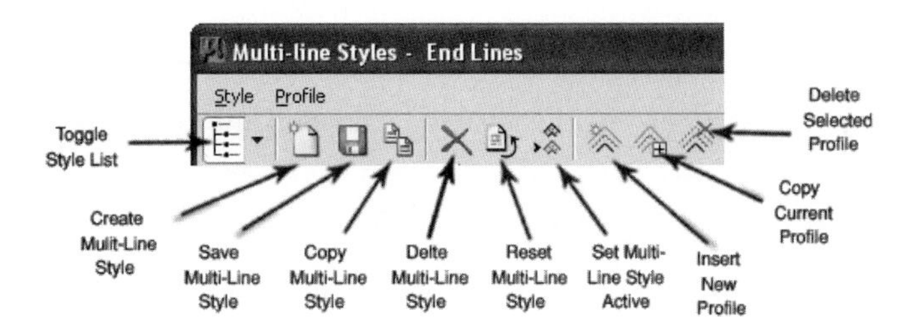

Figure 8–41 *Multi-lines settings box tool bar*

Profiles Tab

The Profile tab displays the lines contained in the selected multi-line style's profile. Each line is on a separate row and each row contains the current offset and attributes settings for the line. Figure 8–40 shows the profile of a typical multi-line style.

- The **Offset** field contains the line's distance, in working units, from the center of the multi-line style. A line with a positive offset value is above the center of the multi-line element when it is drawn horizontally from left to right and to the left of center when drawn vertically from bottom to top. A dash preceding the value indicates a negative offset (For example, -5.25).
- **Level**, **Color**, **Line Style**, **Weight**, and **Class** are the attribute settings for each line. They are also menus that, when clicked with the Data button, allow for changing the attribute setting. Figure 8–42 shows the **Weight** attribute menu open for one of the lines in a multi-line style.

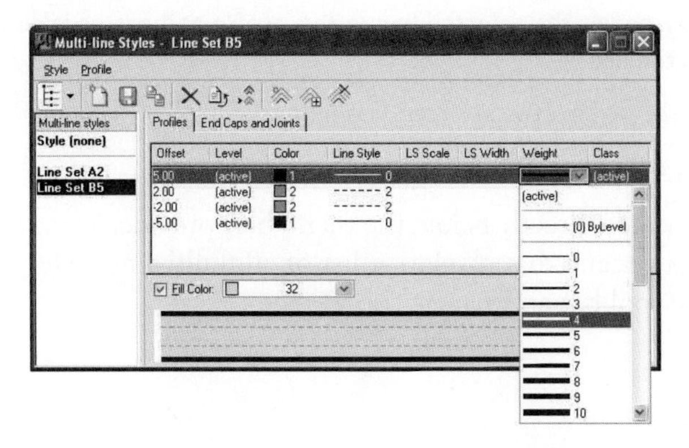

Figure 8–42 *The Weight menu open for a line in a Multi-Line Style*

- **LS Scale** sets the scale factor applied to all displayable characteristics (dash length and width, point symbol size) of the selected Line Style.

- **LS Width** sets the starting width, in master units, of each dash stroke (thereby modifying the widths specified for these strokes in the definition of the selected Line Style).

End Caps and Joints Tab

The **End Caps and Joints** tab contains options for managing the appearance of the ends and joints of multi-line styles. The tab displays three rows (one each for **Start Cap, End Cap,** and **Joints**), and twelve columns of settings (see Figure 8-43).

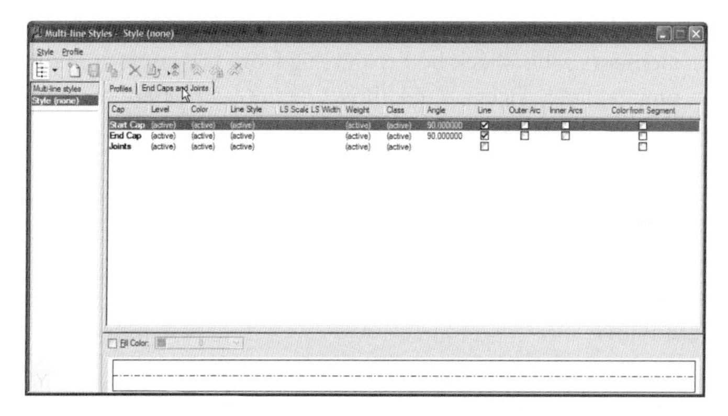

Figure 8–43 *The Multi-line Styles setting box with End Caps and Joints tab selection*

- **Cap** is the type of element. A **Start Cap** is placed on the starting end of multi-line elements (the location of the first placement data point). An **End Cap** is placed on the end of placed multi-line elements (the location of the last data point). **Joints** are placed at each multi-line vertex (at placement points between the first and last data point).
- **Level, Color, Line Style, Weight,** and **Class** are the attribute settings for each cap. They are also menus that, when clicked with the Data button, allow changing the attribute setting.
- **LS Scale** sets the scale factor applied to all displayable characteristics (dash length and width, point symbol size) of the selected Line Style.
- **LS Width** sets the starting width, in master units, of each dash stroke (thereby modifying the widths specified for these strokes in the definition of the selected Line Style).
- **Angle** controls the placement angle at which Start and End Caps on the ends of the multi-line element. A 90-degree angle places the end caps at right angles to the multi-line. A 45-degree angle places the end caps at 45 degrees relative to the lines in the multi-line element.
- **Line** is a check box that controls placement of lines as start caps, end caps, and joints. If the check box is ON, the caps are lines and joints are placed as

part of the multi-line element. Figure 4–44 shows an example of line caps and joints.

- **Outer Arc** are check boxes that control the placement of arcs as start and end caps. Outer arcs connect the outer pair of lines when placed with multi-line elements. The check box is missing from the Joints row because joints can only be lines. Figure 4–44 shows an example of outer arc start and end caps.
- **Inner Arcs** are check boxes that control the placement of arcs as start and end caps. Inner arcs connect interior pairs of lines when placed with multi-line elements. The check box is missing from the Joints row because joints can only be lines. Figure 4–44 shows examples of inner arc start and end caps.

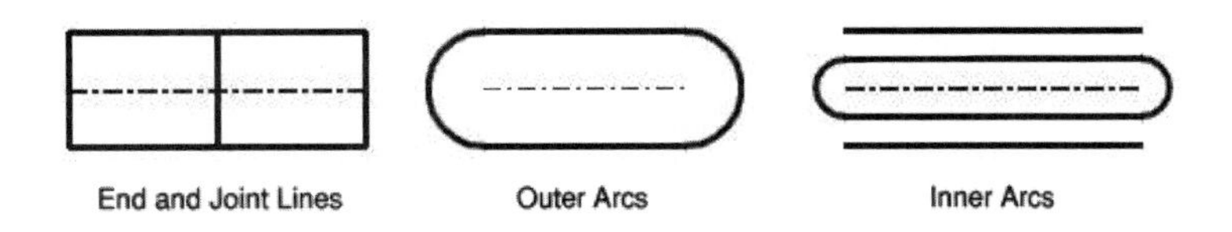

Figure 8–44 *Examples of Start Caps, End Caps, and Joints*

CHANGING A STYLE CHANGES PLACED ELEMENTS

Changing the profile of a style also changes all multi-line elements placed using that style. For example, two multi-line elements that use multi-line style "Roadways" are in the design. If a new line is added to the "Roadways" profile, the new line also appears in the two multi-line elements in the design.

CREATE A MULTI-LINE STYLE

To create a new multi-line style, invoke New from:

Multi-Line Styles settings box menus	**Style > New**
Multi-Line Styles settings box tool bar	**Create Multi-Line Style**

The tool creates a new style, inserts new style in the Styles List pane, names it "Untitled-#," and selects the name for editing. Type a short, descriptive name for the new style in place of the default name.

To save the new multi-line style, invoke Save from:

Multi-Line Styles settings box menus	Select **Style > Save**
Multi-Line Styles settings box tool bar	Click **Save Mu'ti-Line Style**

MicroStation immediately saves the multi-line style in the design file. The new style's profile matches the profile of the currently active multi-line style.

Add a Line to a Multi-Line Style

To add a new line to a multi-line style's profile, select the multi-line style's name from the Multi-Line Styles List pane and display the style's profile from:

Multi-Line Styles settings box	Click the **Profile** tab

MicroStation displays the contents of the **Profiles** tab on the settings box.

The tab displays the profile (lines) of the selected style with the first line in the profile selected. To add a new line to the profile, click the existing line above which the new line is to be inserted, and invoke Insert from:

Multi-Line Styles settings box menus	Select **Profile > Insert**
Multi-Line Styles settings box tool bar	Click **Insert New Profile**

The tool inserts the new line directly above the selected existing line and gives it a zero offset value. To change the offset value, right-click the Offset field (currently 0.00) and type the required offset from the multi-line element's workline. To enter a negative offset value, type a dash before the value.

To save the new line, invoke Save from:

Multi-Line Styles settings box menus	Select **Style > Save**
Multi-Line Styles settings box tool bar	Click **Save Multi-Line Style**

MicroStation immediately saves the multi-line style in the design file.

Change a Profile Line's Attributes

To change the attributes of lines in a multi-line style's profile, select the multi-line style's name from the Multi-Line Styles List pane and display the style's profile from:

Multi-Line Styles settings box	Click the **Profile** tab

MicroStation displays the contents of the **Profiles** tab on the settings box.

The tab displays the profile (lines) of the selected style with each line is on a separate row that contains the current attribute values for the line. To change an attribute value for a line, click attribute with the Data button to display a menu of all possible values for the attribute. Change the attribute value by selecting a new attribute from the menu.

After completing all required changes to the profile attributes, save the multi-line style by invoking Save from:

Multi-Line Styles settings box menus	Select **Style > Save**
Multi-Line Styles settings box tool bar	Click **Save Multi-Line Style**

MicroStation immediately saves the multi-line style in the design file.

Change a Profile Line's Offset

To change the offset value of a line in a multi-line style's profile, select the multi-line style's name from the Multi-Line Styles List pane and display the style's profile from:

Multi-Line Styles settings box	Click the **Profile** tab

MicroStation displays the contents of the **Profiles** tab on the settings box.

To change the offset field of a line, select the line, position the cursor over the **Offset** field and click the Reset button. Type the new value in the field and click ENTER.

After completing all required offset changes, save the multi-line style by invoking Save from:

Multi-Line Styles settings box menus	Select **Style > Save**
Multi-Line Styles settings box tool bar	Click **Save Multi-Line Style**

MicroStation immediately saves the multi-line style in the design file.

Copy a Profile Line

To create a copy of a line in a multi-line style's profile, select the style's name from the Multi-Line List pane and invoke the profile from:

Multi-Line Styles settings box	Click the **Profile** tab

MicroStation displays the contents of the **Profiles** tab on the settings box.

Select the line to copy on the **Profile** tab and invoke Copy from:

Multi-Line Styles settings box menus	Select **Style > Copy**
Multi-Line Styles settings box tool bar	Click **Copy Current Profile**

MicroStation places the copy directly above the copied line. The new line is identical to the copied line. Make the required changes to the new line's offset and attributes, and then invoke Save from:

Multi-Line Styles settings box menus	Select **Style > Save**
Multi-Line Styles settings box tool bar	Click **Save Multi-Line Style**

MicroStation immediately saves the multi-line style in the design file.

Delete a Profile Line

To delete a line from the profile of a multi-line style, select the style's name from the Multi-Line List pane and invoke the profile from:

Multi-Line Styles settings box	Click the **Profile** tab

MicroStation displays the contents of the **Profiles** tab on the settings box.

Select the line to delete and invoke Delete from:

Multi-Line Styles settings box menus	Select **Profile > Delete**
Multi-Line Styles settings box tool bar	Click **Delete Selected Profile**

MicroStation removes the line from the profile. To make the deletion permanent, invoke Save from:

Multi-Line Styles settings box menus	Select **Style > Save**
Multi-Line Styles settings box tool bar	Click **Save Multi-Line Style**

MicroStation immediately saves the multi-line style in the design file.

Modify a Multi-Line Style's Caps and Joints

To change the cap and joint settings for a multi-line style, select the multi-line style's name from the Multi-Line Styles List pane and display the style's cap and joints settings from:

Multi-Line Styles settings box	Click the **End Caps and Joints** tab

MicroStation displays contents of the **End Caps and Joints** tab on the settings box.

Turn caps and joints ON or OFF by clicking the appropriate check boxes on the **Start Cap**, **End Cap**, and **Joint** check boxes. Select the appropriate attribute settings from the attribute menus on each row.

After completing all required cap and joint changes, save the multi-line style by invoking Save from:

Multi-Line Styles settings box menus	Select **Style > Save**
Multi-Line Styles settings box tool bar	Click **Save Multi-Line Style**

MicroStation immediately saves the multi-line style in the design file.

Control Multi-line Fill

The **Fill** color check box controls the color of the background area within the multi-line's parallel lines.

Turn the **Fill** check box to ON and select a color from the menu to fill the multi-line with the selected color. Turn it OFF to have an unfilled multi-line.

MODIFYING MULTI-LINE JOINTS

The Modify Multi-Line Joints tools clean up intersections formed when multi-lines cross over other multi-lines. There are tools for cleaning up such intersections and for cutting holes in multi-lines.

CONSTRUCTING CLOSED CROSS JOINTS

The Construct Closed Cross Joint tool cuts all lines that make up the first selected multi-line at the point where it crosses the second selected multi-line, as shown in Figure 8–45.

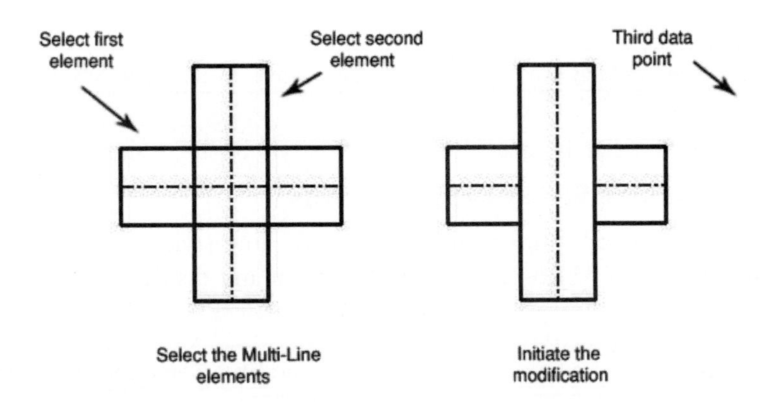

Figure 8–45 *Example of using the Construct Closed Cross Joint tool*

Invoke Construct Closed Cross Joint from:

Multi-line Joints tool box	Select the **Construct Closed Cross Joint** tool (see Figure 8–46).
Key-in window	**join cross closed** (or **jo cr c**) (ENTER)

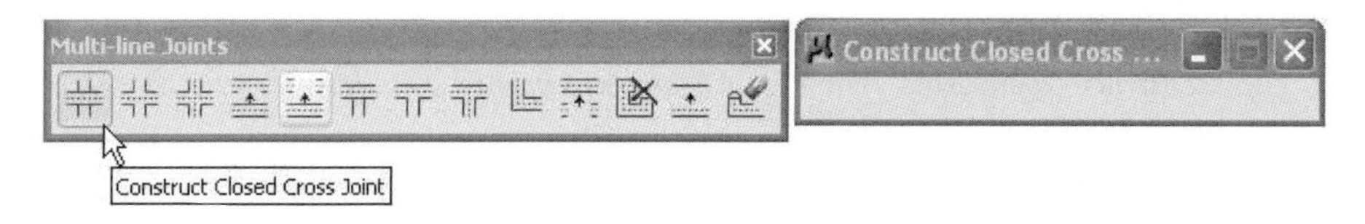

Figure 8–46 *Invoking the Construct Closed Cross Joint tool from the Multi-line Joints tool box*

MicroStation prompts:

Construct Closed Cross Joint > Identify element *(Identify the multi-line element to cut.)*

Construct Closed Cross Joint > Identify element *(Identify the other multi-line element.)*

Construct Closed Cross Joint > Identify element *(Click the Data button anywhere in the view to initiate cleaning up of the intersection or click the Reject button to reject the change.)*

CONSTRUCTING OPEN CROSS JOINTS

The Construct Open Cross Joint tool cuts all lines that make up the first selected multi-line element and cuts only the outside lines of the second multi-line element, as shown in Figure 8–47.

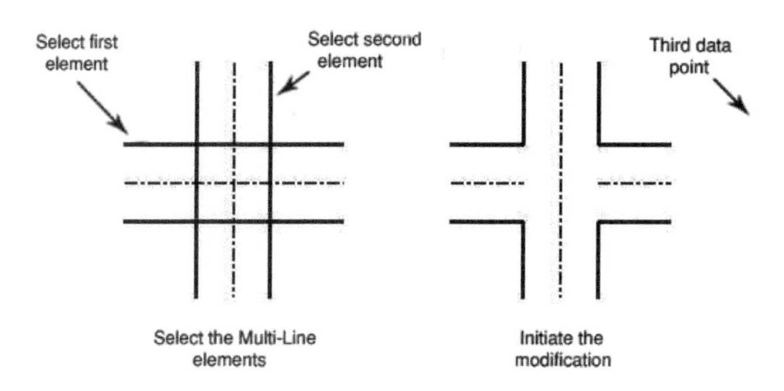

Figure 8–47 *Example of using the Construct Open Cross Joint tool*

Invoke the Construct Open Cross Joint tool from:

Multi-line Joints tool box	Select the **Construct Open Cross Joint** tool (see Figure 8–48).
Key-in window	**join cross open** (or **jo cr o**) (ENTER)

Figure 8–48 *Invoking the Construct Open Cross Joint tool from the Multi-line Joints tool box*

MicroStation prompts:

> Construct Open Cross Joint > Identify element *(Identify the multi-line element whose lines are to be cut)*
>
> Construct Open Cross Joint > Identify element *(Identify the other multi-line element.)*
>
> Construct Open Cross Joint > Identify element *(Click the Data button anywhere in the view to initiate modification or click the Reject button to reject the change.)*

The tool cuts all lines of the first selected multi-line element and cuts only the outer lines of the second selected multi-line element.

CONSTRUCTING MERGED CROSS JOINTS

The Construct Merged Cross Joint tool cuts only the outer lines from each of the selected intersecting multi-line elements, as shown in Figure 8–49.

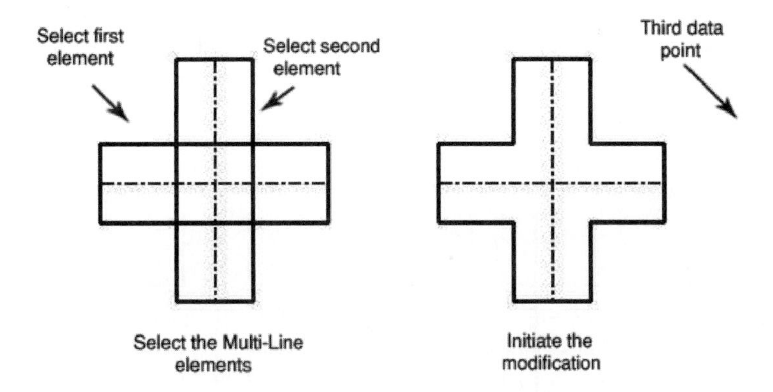

Figure 8–49 *Example of using the Construct Merged Cross Joint tool*

Invoke the Construct Merged Cross Joint tool from:

Multi-line Joints tool box	Select the **Construct Merged Cross Joint** tool (see Figure 8–50).
Key-in window	**join cross merge** (or **jo cr m**) (ENTER)

Figure 8–50 *Invoking the Construct Merged Cross Joint tool from the Multi-line Joints tool box*

MicroStation prompts:

> Construct Merged Cross Joint > Identify element *(Identify one of the intersecting multi-line elements.)*
>
> Construct Merged Cross Joint > Identify element *(Identify the other multi-line element.)*
>
> Construct Merged Cross Joint > Identify element *(Click the Data button anywhere in the view to initiate cleaning up of the intersection or click the Reject button to reject the change.)*

CONSTRUCTING CLOSED TEE JOINTS

The Construct Closed Tee Joint tool extends or shortens the first selected multi-line element to its intersection with the second multi-line element. The first multi-line element ends at the near side of the second multi-line element, as shown in Figure 8–51.

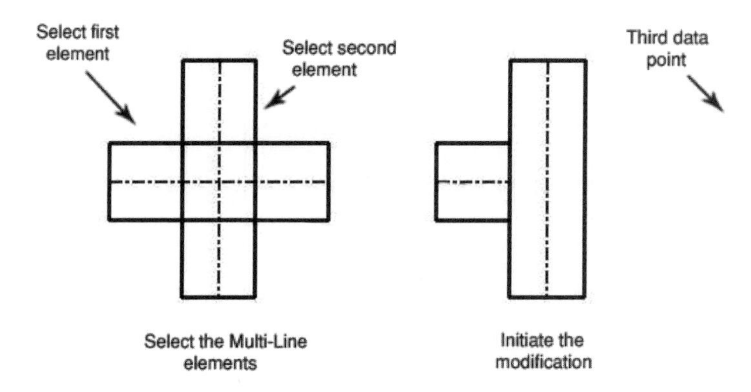

Figure 8–51 *Example of using the Construct Closed Tee Joint tool*

Invoke the Construct Closed Tee Joint tool from:

Multi-line Joints tool box	Select the **Construct Closed Tee Joint** tool (see Figure 8–52).
Key-in window	**join tee closed** (or **jo t c**) (ENTER)

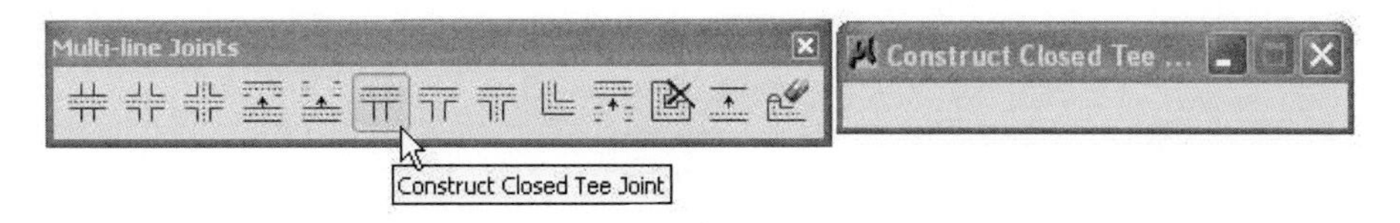

Figure 8–52 *Invoking the Construct Closed Tee Joint tool from the Multi-line Joints tool box*

MicroStation prompts:

> Construct Closed Tee Joint > Identify element *(Identify the multi-line element to trim.)*
>
> Construct Closed Tee Joint > Identify element *(Identify the other multi-line element.)*
>
> Construct Closed Tee Joint > Identify element *(Click the Data button anywhere in the view to initiate cleaning up of the intersection or click the Reject button to reject the change.)*

CONSTRUCTING OPEN TEE JOINTS

The Construct Open Tee Joint tool is similar to the Closed Tee Joint tool, except it cuts the touching outer line of the second element to form an open intersection, as shown in Figure 8–53.

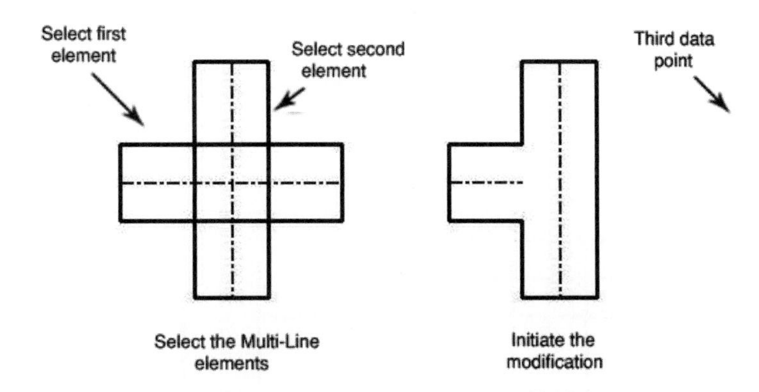

Figure 8–53 *Example of using the Construct Open Tee Cross Joint tool*

Invoke the Construct Open Tee Joint tool from:

Multi-line Joints tool box	Select the **Construct Open Tee Joint** tool (see Figure 8–54).
Key-in window	**oin tee open** (or **jo t o**) (ENTER)

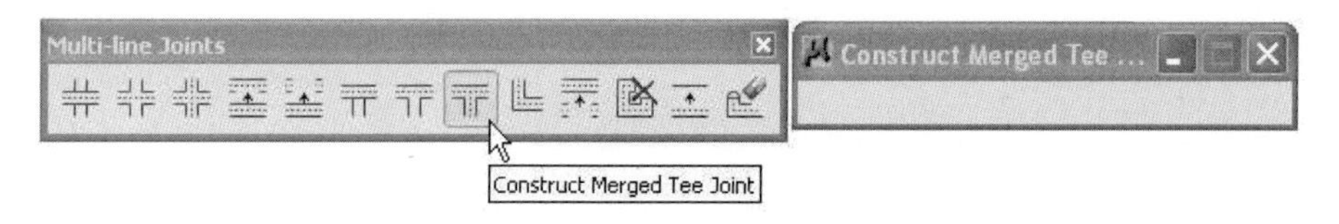

Figure 8–54 *Invoking the Construct Open Tee Joint tool from the Multi-line Joints tool box*

MicroStation prompts:

> Construct Open Tee Joint > Identify element *(Identify the multi-line element to trim.)*
>
> Construct Open Tee Joint > Identify element *(Identify the other multi-line element.)*
>
> Construct Open Tee Joint > Identify element *(Click the Data button anywhere in the view to initiate cleaning up of the intersection or click the Reject button to reject the change.)*

CONSTRUCTING MERGED TEE JOINTS

The Construct Merged Tee Joint tool is similar to the Open Tee Joint tool, except the inner lines of the first multi-line element extend to their intersection with the matching inner lines of the second multi-line element, as shown in Figure 8–55. If there are no matches in the second element for lines in the first element, the lines may extend to the far side of the second element.

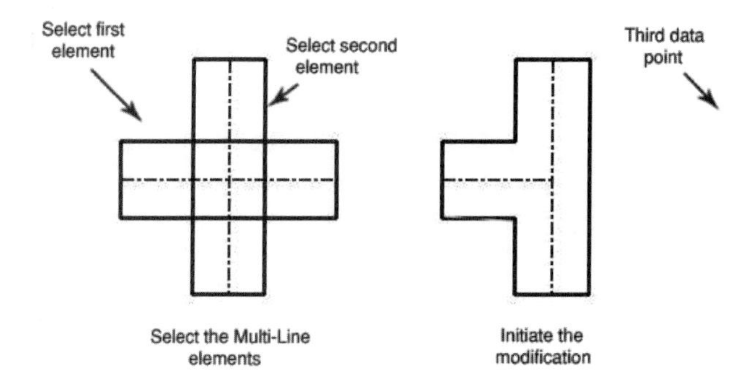

Figure 8–55 *Example of using the Construct Merged Tee Joint tool*

Invoke the Construct Merged Tee Joint tool from:

Multi-line Joints tool box	Select the **Construct Merged Tee Joint** tool (see Figure 8–56).
Key-in window	**join tee merge** (or **jo t m**) (ENTER)

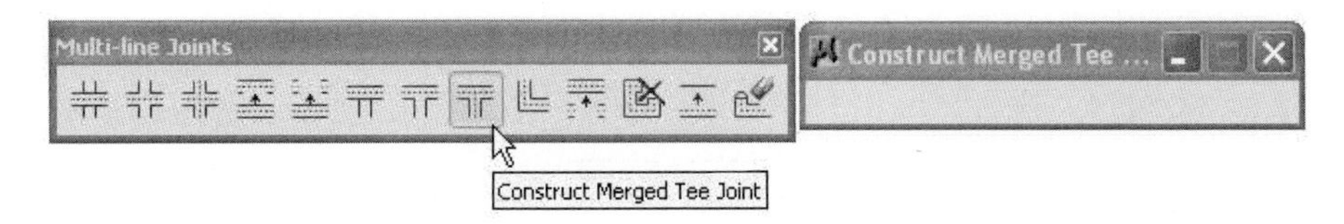

Figure 8–56 *Invoking the Construct Merged Tee Joint tool from the Multi-line Joints tool box*

MicroStation prompts:

> Construct Merged Tee Joint > Identify element *(Identify the multi-line element to trim.)*
>
> Construct Merged Tee Joint > Identify element *(Identify the other multi-line element.)*
>
> Construct Merged Tee Joint > Identify element *(Click the Data button anywhere in the view to initiate cleaning up of the intersection or click the Reject button to reject the change.)*

CONSTRUCTING CORNER JOINTS

The Construct Corner Joint tool lengthens or shortens each of the two selected multi-lines as necessary to create a clean intersection, as shown in Figure 8–57.

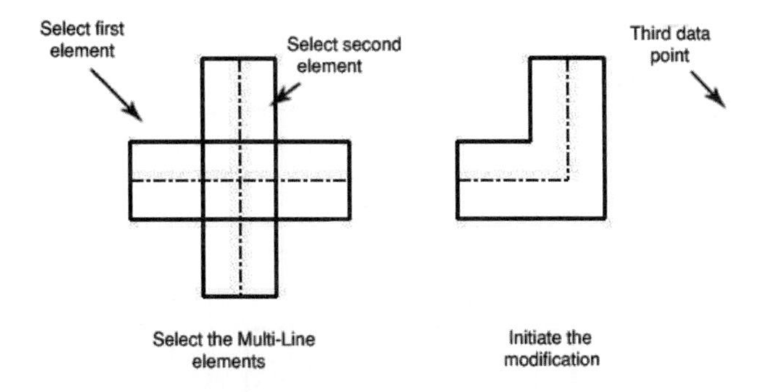

Figure 8–57 *Example of using the Construct Corner Joint tool*

Invoke the Construct Corner Joint tool from:

Multi-line Joints tool box	Select the **Construct Corner Joint** tool (see Figure 8–58).
Key-in window	**join corner** (or **jo c**) (ENTER)

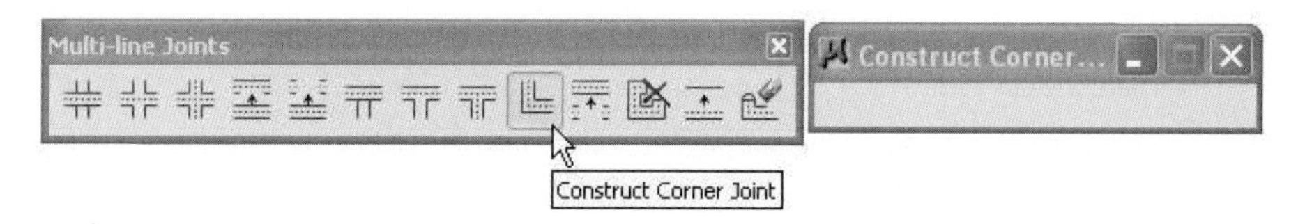

Figure 8–58 *Invoking the Construct Corner Joint tool from the Multi-line Joints tool box*

MicroStation prompts:

Construct Joint > Identify element (*Identify one of the multi-line elements.*)

Construct Joint > Identify element (*Identify the other multi-line element.*)

Construct Joint > Identify element (*Click the Data button anywhere in the view to initiate cle-aning up of the intersection or click the Reject button to reject the change.*)

CUTTING SINGLE COMPONENT LINES

The Cut Single Component Line tool cuts a hole in the selected line in a multi-line from the first data point to the second data point, as shown in Figure 8–59.

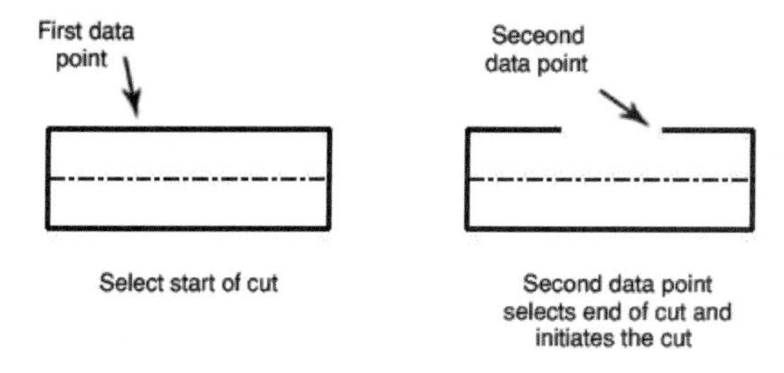

Figure 8–59 *Example of using the Cut Single Component Line tool*

Invoke the Cut Single Component Line tool from:

Multi-line Joints tool box	Select the **Cut Single Component Line** tool (see Figure 8–60).
Key-in window	**cut single** (or **cu s**) (ENTER)

Figure 8–60 *Invoking the Cut Single Component Line tool from the Multi-line Joints tool box*

MicroStation prompts:

Cut Single Component Line > Identify element *(Identify the first multi-line element the point where the cut starts.)*

Cut Single Component Line *(Identify the second multi-line element at the point where the cut ends or click the Reset button to reject the change.)*

The second data point initiates the cut from the first to the second data point.

CUTTING ALL COMPONENT LINES

The Cut All Component Lines tool cuts a hole in the selected multi-line element from the first data point to the second data point, as shown in Figure 8–61. The multi-line is still one element after this tool cuts a hole in it.

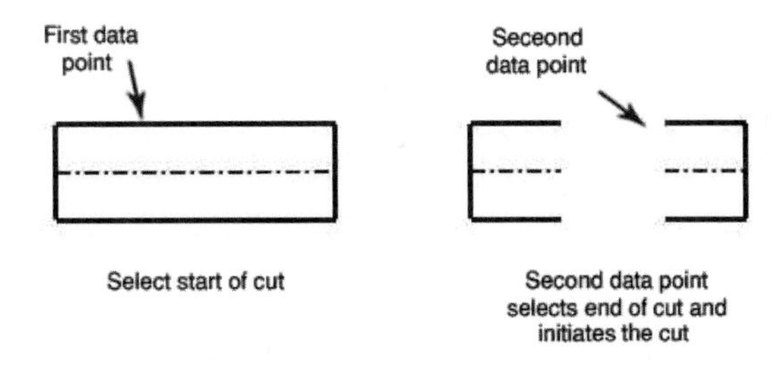

Figure 8–61 *Example of using the Cut All Component Lines tool*

Invoke the Cut All Component Lines tool from:

Multi-line Joints tool box	Select the **Cut All Component Lines** tool (see Figure 8–62.)
Key-in window	**cut all** (or **cu a**) (ENTER)

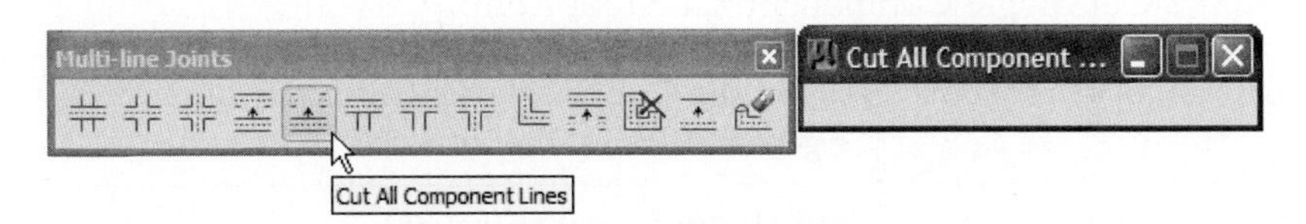

Figure 8–62 *Invoking the Cut All Component Lines tool from the Multi-line Joints tool box*

MicroStation prompts:

Cut All Component Lines > Identify element *(Identify the first multi-line element the point where the cut starts.)*

Cut All Component Lines *(Identify the second multi-line element at the point where the cut ends or click the Reset button to reject the change.)*

The second data point initiates the cut from the first to the second data point.

UNCUTTING COMPONENT LINES

The Uncut Component Lines tool allows for removing a cut in a multi-line element. It removes the cut from an individual line that was cut using the Cut Single Component Line tool and removes the cuts from all lines that were cut by the Cut All Component Lines tool. To remove a cut, keypoint snap to one end of the cut, click the data button to accept the snap, and click the Data button a second time to remove the cut, as shown in Figure 8–63.

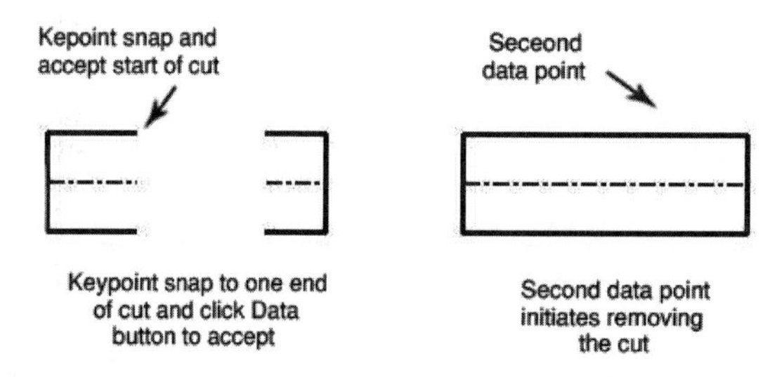

Figure 8–63 *Example of using the Uncut Component Lines tool*

Invoke the Uncut Component Lines tool from:

Multi-line Joints tool box	Select the **Uncut Component Lines** tool (see Figure 8–64).
Key-in window	**uncut** (or **un**) (ENTER)

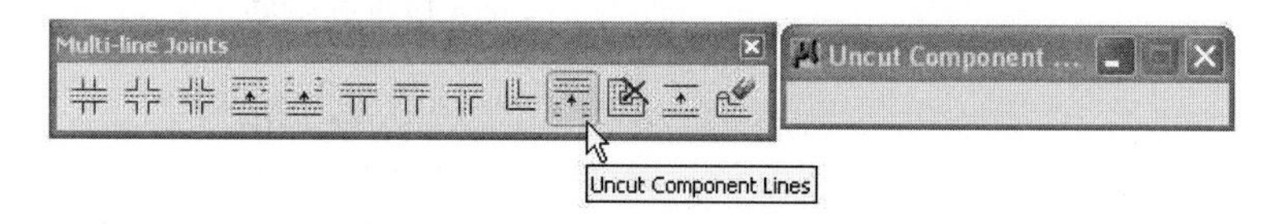

Figure 8–64 *Invoking the Uncut Component Lines tool from the Multi-line Joints tool box*

MicroStation prompts:

Uncut Component Lines > Identify element *(Keypoint snap to one end of the cut.)*

Uncut Component Lines *(Place a data point to accept the Keypoint snap or click the Reset button to reject the change.)*

Uncut Component Lines *(Place a second data point to remove the cut or click the Reset button to reject the change.)*

MULTI-LINE PARTIAL DELETE

The Multi-line Partial Delete tool reduces the length of a multi-line element by selecting one end of it or breaks it into two separate multi-line elements by cutting

a hole in it. The tool can also place caps on the new ends created by the partial delete, using one of the following options in the **Cap Mode** option menu on the Multi-line Partial Delete settings box:

- **None** does not place caps.
- **Current** places the start cap and end cap definitions on the multi-line element.
- **Active** places the active multi-line style's start and end cap definitions.
- **Joint** places a 90-degree joint line.

Invoke the Multi-line Partial Delete tool from:

Multi-line Joints tool box	Select the **Multi-line Partial Delete** tool and, on the Multi-Line Partial Delete settings box, select an option from the **Cap Mode** menu (see Figure 8–65).
Key-in window	**mline partial delete** (or **ml p d**) (ENTER)

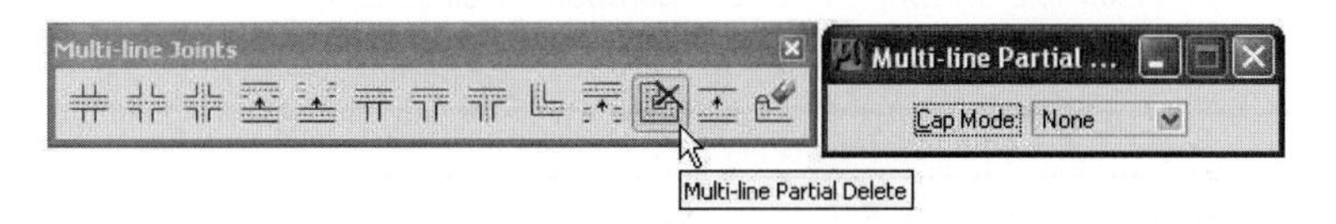

Figure 8–65 *Invoking the Multi-line Partial Delete tool from the Multi-line Joints tool box*

MicroStation prompts:

Multi-line Partial Delete > Identify multi-line at start of delete *(Identify the multi-line element at one end of the part to delete.)*

Multi-line Partial Delete > Define length of delete *(Place a data point to define the length of the delete or click the Reset button to reject the change.)*

MOVING MULTI-LINE PROFILES

The Move Multi-line Profile tool moves an individual line in multi-line element or reposition the working line of a multi-line without moving any component lines.

Move Component

The **Move Component** option allows for moving one selected line component in a multi-line element. The line moves in parallel to the other lines in the multi-line element, as shown in Figure 8–66.

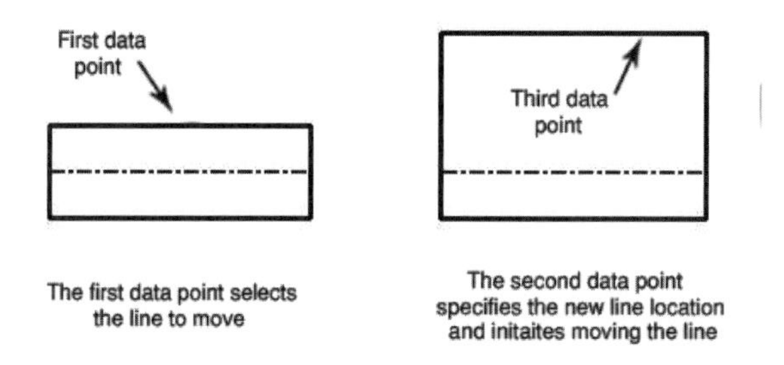
First data point
Third data point
The first data point selects the line to move
The second data point specifies the new line location and initaites moving the line

Figure 8–66 *Example of using the Move Multi-Line Profile tool to move a line*

To move a line component, invoke Move Multi-line Profile from:

Multi-line Joints tool box	Select the **Move Multi-line Profile** tool and, on the Move Multi-Line Profile settings box, select **Component** from the **Move** menu (see Figure 8–67).
Key-in window	**mline edit profile** (or **ml e p**) (ENTER)

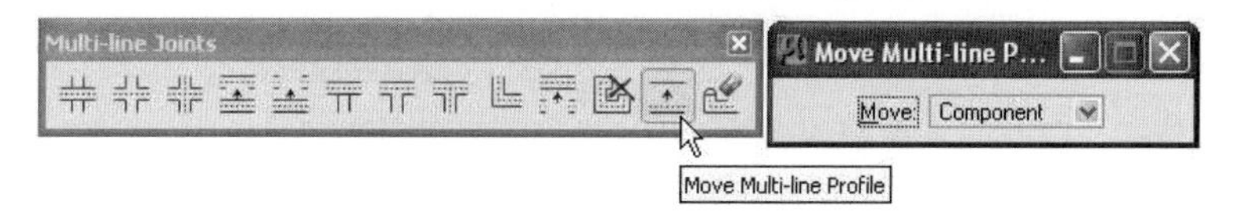
Multi-line Joints
Move Multi-line P...
Move:
Component
Move Multi-line Profile

Figure 8–67 *The Move Multi-line Profile tool set for a Component Move*

MicroStation prompts:

Move Multi-line Profile > Identify multi-line component to move *(Identify the line to move.)*

Move Multi-line Profile > Define component position (reset to reject element) *(Place a Data point to reposition the component or click the Reset button to reject the move.)*

Move Working Line

The **Workline Move** option allows for moving the location of the insertion point (working line) in a multi-line element. This option has the effect of changing the zero Offset position. Usually, no change in the multi-line element is visible after using this tool.

The result of a working line move becomes apparent when manipulating the multi-line element. For example, if the Modify Element tool to move one end of the multi-line element, the element may pivot about the new working line position.

To move the working line, invoke the Move Multi-line Profile from:

Multi-line Joints tool box	Select the **Move Multi-line Profile** tool and, from the Multi-Line Profile settings box, select **Workline** from the **Move** menu (see Figure 8–68).
Key-in window	**mline edit profile** (or **ml e p**) (ENTER)

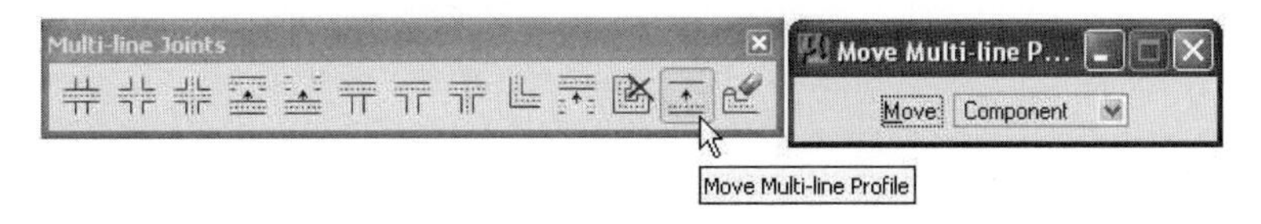

Figure 8–68 *The Move Multi-line Profile tool set for a Workline Move*

MicroStation prompts:

Move Multi-line Profile > Identify multi-line profile *(Identify the multi-line.)*

Move Multi-line Profile > Define new workline (reset to reject element) *(Place a Data point to reposition the working line, or click the Reset button to reject the move.)*

EDITING MULTI-LINE CAPS

The Edit Multi-line Cap tool changes the end cap of a multi-line. MicroStation provides four options under the **Cap Mode** option menu:

- **None** removes an end cap.
- **Current** does not change the end cap and is enabled only when **Adjust Angle** is turned ON.
- **Active** uses the active multi-line style definitions for the end cap.
- **Joint** uses the identified multi-line element's joint definition instead of the end cap definition and ensures that the end cap is always 90 degrees.

Invoke the Edit Multi-line Cap tool from:

Multi-line Joints tool box	Select the **Edit Multi-line Cap** tool and, from the Edit Multi-line Cap settings box, select an option from the **Cap Mode** menu (see Figure 8–69).
Key-in window	**mline edit cap** (or **ml e c**) (ENTER)

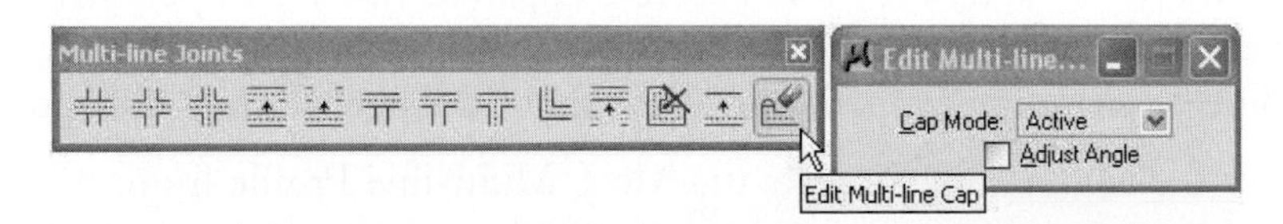

Figure 8–69 *Invoking the Edit Multi-line Cap tool from the Multi-line Joints tool box*

MicroStation prompts:

Edit Multi-line Cap > Identify multi-line near the end cap to modify *(Identify the multi-line element near the end cap.)*

Edit Multi-line Cap > Data to change end cap (reset to reject element) *(Place a data point to change the end cap or click the Reject button to reject the change.)*

Open the Exercise Manual PDF file for Chapter 8 on the accompanying CD for project and discipline specific exercises.

REVIEW QUESTIONS

Write your answers in the spaces provided.

1. Explain briefly the options available with the Extend Line tool.
   ____
2. List the element types that the Modify Element tool can modify.
   ____
3. List the tools available to modify an arc.
   ____
4. Explain the difference between creating a chain manually and creating one automatically.
   ____
5. To drop a complex chain, invoke the ______________________________ tool.
6. Explain the difference between the Construct Closed Cross Joint tool and the Construct Merged Cross Joint tool.
   ____
7. Explain the difference between the Construct Closed Tee Joint tool and the Construct Merged Tee Joint tool.
   ____
8. List the steps involved in moving the Multi-line profile.
   ____

CHAPTER 9

Measurement and Dimensioning

OBJECTIVES

Topics explored in this chapter:

- Using the measurement tools, such as Measure Distance, Measure Radius, Measure Angle, Measure Length, and Measure Area
- Using the dimensioning tools for linear, angular, and radial measurement
- Creating and modifying dimension styles

MEASUREMENT TOOLS

"Is that line really 12 feet long?" "What's the radius of that circle?" "What is the surface area of that foundation?" MicroStation can answer these questions with the measurement tools.

The measurement tools do nothing to the design, they simply display distances, areas, and angles in the Status bar/Tool Settings window.

All measurement tools are available from the Task Navigation tool box (active task set to Measure), shown in Figure 9–1.

Figure 9–1 *Task Navigation tool box (active task set to Measure)*

MEASURE DISTANCE

MicroStation provides four distance measurement options. Distance options include measuring the distance between user-defined points, the distance along an element between user-defined points, the perpendicular distance from an element, and the minimum distance between two elements.

Measure Distance Between Points

This tool measures the cumulative straight-line distance from the first data point, through successive data points, to the last data point defined. Invoke the Measure Distance Between Points tool from:

Task Navigation tool box (active task set to Measure)	Select the Measure Distance tool. In the Tool Settings window select Between Points from the Distance option menu (see Figure 9–2).
Keyboard Navigation (Task Navigation tool box with active task set to Measure)	**q** (Select Between Points from the Distance option menu located in the Tool Settings window.)

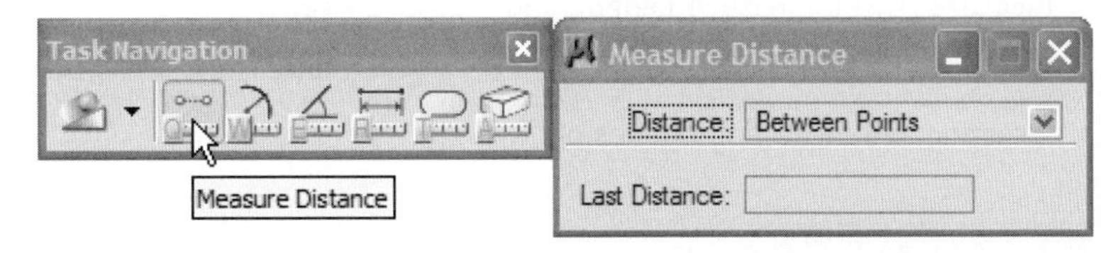

Figure 9–2 *Invoking the Measure Distance (Between Points) tool from the Task Navigation tool box (active task set to Measure)*

MicroStation prompts:

> Measure Distance Between Points > Enter start point *(Place a data point to start the measurement.)*
>
> Measure Distance Between Points > Define distance to measure *(Place a data point to define the distance to measure and continue placing data points to define additional measurement segments. Click the Reset button to terminate the tool sequence.)*

As each data point is placed, the cumulative linear distance between points is displayed in the Status bar/Tool Settings window.

Measure Distance Along Element

This tool measures the cumulative distance along an element, from the data point that selects it through successive data points on the element to the last data point defined. Invoke the Measure Distance Along Element tool from:

Task Navigation tool box (active task set to Measure)	Select the Measure Distance tool. In the Tool Settings window select Along Element from the Distance option menu (see Figure 9–3).
Keyboard Navigation (Task Navigation tool box with active task set to Measure)	**q** (Select Along Element from the Distance option menu located in the Tool Settings window.)

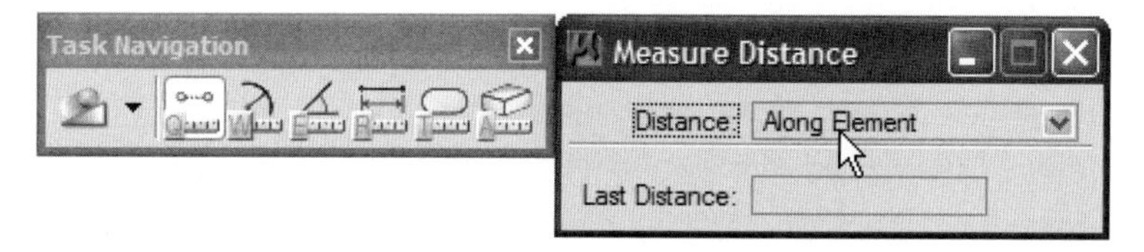

Figure 9–3 *Invoking the Measure Distance (Along Element) tool from the Task Navigation tool box (active task set to Measure)*

MicroStation prompts:

> Measure Distance Along Element > Identify Element @ first point *(Place a data point on the element to start the measurement.)*
>
> Measure Distance Along Element > Enter endpoint *(Place a data point on the element to end the measurement.)*
>
> Measure Distance Along Element > Measure more points/Reset to select *(Continue placing data points on the element to obtain the cumulative measurement from the first data point through the succeeding points, or click the Reset button to terminate the measurement.)*

As each data point after the first one is placed, the cumulative distance along the element is displayed in the Status bar/Tool Settings window.

Measure Distance Perpendicular From Element

This tool measures the perpendicular distance from a point to an element. Invoke the Measure Distance Perpendicular From Element tool from:

Task Navigation tool box (active task set to Measure)	Select the Measure Distance tool. In the Tool Settings window select Perpendicular (true) from the Distance option menu (see Figure 9–4).
Keyboard Navigation (Task Navigation tool box with active task set to Measure)	**q** (Select Perpendicular (true) from the Distance option menu located in the Tool Settings window.)

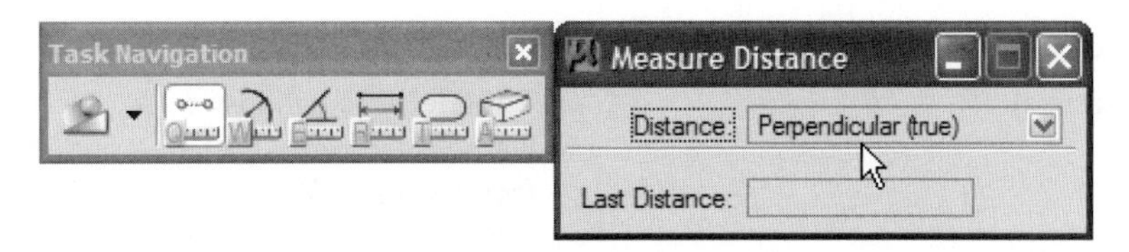

Figure 9–4 *Invoking the Measure Distance (Perpendicular (true)) tool from the Task Navigation tool box (active task set to Measure)*

MicroStation prompts:

> Measure Distance Perpendicular From Element > Enter start point *(Identify the element from which to measure the perpendicular distance.)*
>
> Measure Distance Perpendicular From Element > Enter end point *(Place a data point to measure the perpendicular distance to the element.)*
>
> Measure Distance Perpendicular From Element > Measure more points/Reset to reselect *(Continue placing data points to obtain additional perpendicular measurements from the element, or click the Reset button to terminate he measurement.)*

A line from the element to the cursor shows the location of the calculation. The line is a temporary image that disappears when the Reset button is clicked or another tool is selected.

Note: *If the measurement end point is placed beyond the end of a linear element (such as a line or a box), the perpendicular is calculated from an imaginary extension of the measured element.*

Measure Minimum Distance Between Elements

This tool measures the minimum distance between two elements. Invoke the Measure Minimum Distance Between Elements tool from:

Task Navigation tool box (active task set to Measure)	Select the Measure Distance tool. In the Tool Settings window select Minimum Between from the Distance option menu (see Figure 9–5).
Keyboard Navigation (Task Navigation tool box with active task set to Measure)	**q** (Select Minimum Between from the Distance option menu located in the Tool Settings window.)

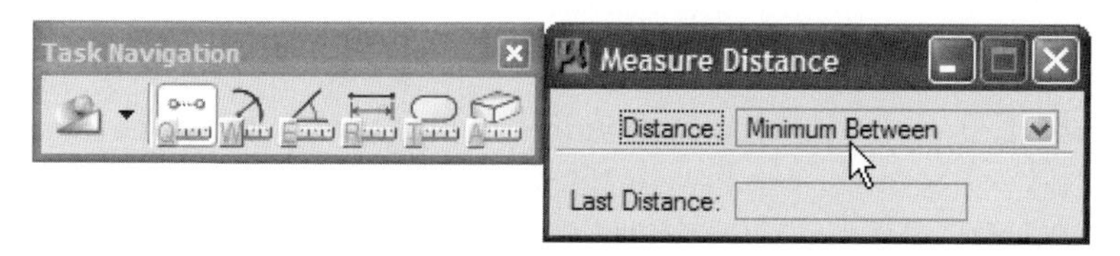

Figure 9–5 *Invoking the Measure Distance tool (Minimum Between Elements) tool from the Task Navigation tool box (active task set to Measure)*

MicroStation prompts:

> Measure Minimum Distance Between Elements > Identify first element *(Identify the first element.)*
>
> Measure Minimum Distance Between Elements > Accept, Identify 2nd element/Reject *(Identify the second element, or click the Reject button to start all over again.)*
>
> Measure Minimum Distance Between Elements > Identify first element *(Identify another element to continue.)*

After the second data point is placed, a line appears in the design to indicate where the minimum distance is, and the minimum distance is displayed in the Status bar/ Tool Settings window. The line is a temporary image that disappears when the Reset button is clicked or another tool is selected.

MEASURE RADIUS

The Measure Radius tool displays the radius of arcs, circles, partial ellipses, and ellipses in the Status bar/Tool Settings window. Invoke the Measure Radius tool from:

Task Navigation tool box (active task set to Measure)	Select the Measure Radius tool (see Figure 9–6).
Keyboard Navigation (Task Navigation tool box with active task set to Measure)	**w**

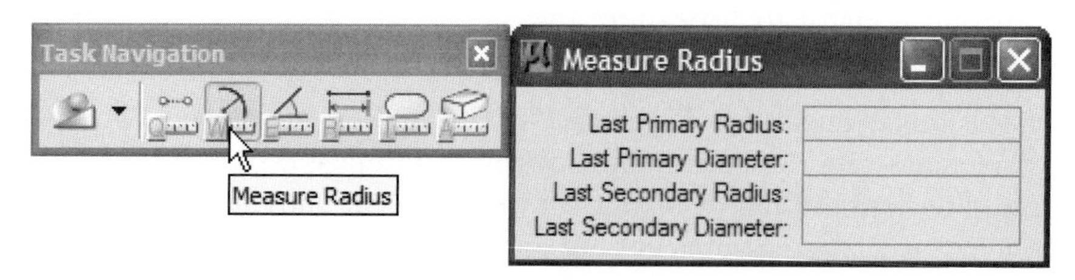

Figure 9–6 *Invoking the Measure Radius tool from the Task Navigation tool box (active task set to Measure)*

MicroStation prompts:

Measure Radius > Identify first element *(Identify the element to measure the radius.)*

MicroStation displays the element's radius in the current Working Units, in the Status bar/Tool Settings window. If the selected element is an ellipse or a partial ellipse, the major axis and minor axis radii are displayed.

MEASURE ANGLE

The Measure Angle Between Lines tool measures the minimum angle formed by two elements. Invoke the Measure Angle Between Lines tool from:

Task Navigation tool box (active task set to Measure)	Select the Measure Angle tool (see Figure 9–7).
Keyboard Navigation (Task Navigation tool box with active task set to Measure)	e

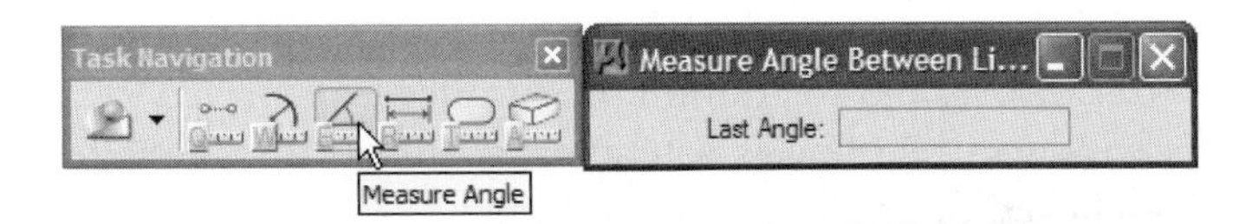

Figure 9–7 *Invoking the Measure Angle tool from the Task Navigation tool box (active task set to Measure)*

MicroStation prompts:

Measure Angle Between Lines > Identify first element *(Identify the first element.)*

Measure Angle Between Lines > Accept, Identify 2nd element/Reject *(Identify the second element to measure the angle.)*

The angle between the two elements is displayed in the Status bar/Tool Settings window.

MEASURE LENGTH

The Measure Length tool measures the total length of an open element or the length of the perimeter of a closed element.

When this tool is invoked, a Tolerance (%) field appears in the Tool Settings window. Tolerance sets the maximum allowable percentage of the distance between the true curve and the approximation for measurement purposes. A low value produces a very accurate measurement but may take a long time to calculate. The default value is sufficient in most cases. Invoke the Measure Length tool from:

Task Navigation tool box (active task set to Measure)	Select the Measure Length tool (see Figure 9–8).
Keyboard Navigation (Task Navigation tool box with active task set to Measure)	**r**

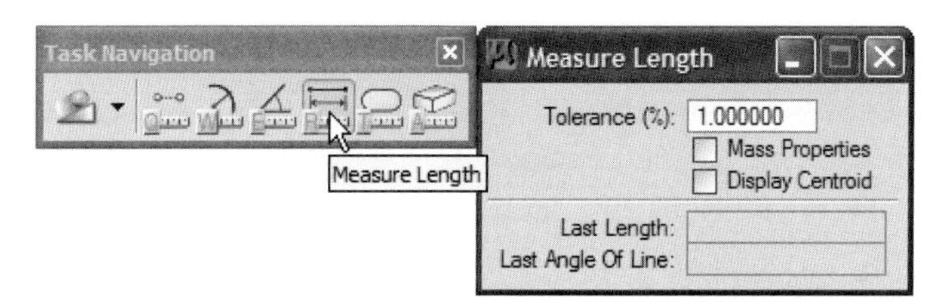

Figure 9–8 *Invoking the Measure Length tool from the Task Navigation tool box (active task set to Measure) MicroStation prompts:*

Measure Length > Identify element *(Identify the element to measure.)*

The total length of the element, or element perimeter, is displayed in the Status bar/Tool Settings window.

The tool can be used to measure the cumulative length of several elements by first invoking the Element Selection tool to select all the elements to include in the measurement. After selecting the elements, select Measure Length from the Measure tool box, and the cumulative length of all selected elements appears in the Status bar/Tool Settings window.

MEASURE AREA

MicroStation provides seven different ways to measure area. Area options include measuring the area of a closed element; a fence; the intersection, union, or difference of two overlapping closed elements; a group of intersecting elements; or a group of points.

When this tool is invoked, a Tolerance (%) field appears in the Tool Settings window. Tolerance sets the maximum allowable percentage of the distance between the true curve and the approximation for area calculation purposes. A low value produces a very accurate area but may take a long time to calculate. The default value is sufficient in most cases.

Measure Area of an Element

The Measure Area tool measures the area of a closed element, such as a circle, ellipse, shape, or block. Invoke the Measure Area tool from:

Task Navigation tool box (active task set to Measure)	Select the Measure Area tool. In the Tool Settings window, select Element from the Method option menu (see Figure 9–9).
Keyboard Navigation (Task Navigation tool box with active task set to Measure)	**t** (select Element from the Method option menu located in the Tool Settings window)

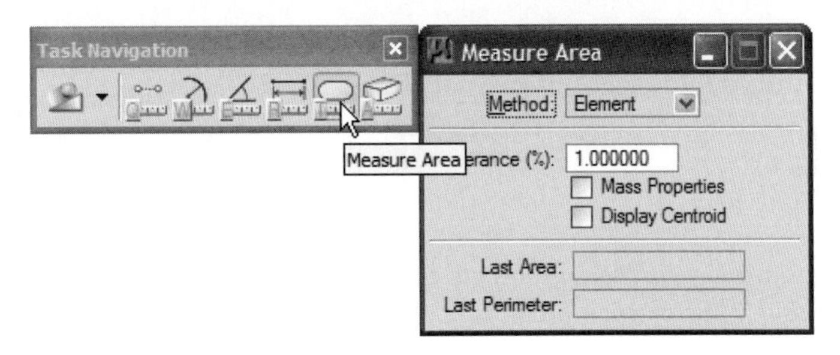

Figure 9–9 *Invoking the Measure Area (Element) tool from the Task Navigation tool box (active task set to Measure)*

MicroStation prompts:

Measure Area > Identify element *(Identify the closed element to measure the area.)*

The element's area and perimeter length are displayed in the Status bar/Tool Settings window.

This tool can be used to measure the cumulative area of several closed elements by first employing the Element Selection tool to select all the elements to include in the area measurement. After selecting the elements, select Measure Area from the Measure tool box, and the cumulative area of all selected elements then appears in the Status bar.

This tool also measures the area enclosed by a fence. Place a fence and then invoke the Measure Fence Area tool from:

Task Navigation tool box (active task set to Measure)	Select the Measure Area tool. In the Tool Settings window, select Fence from the Method option menu (see Figure 9–10).
Keyboard Navigation (Task Navigation tool box with active task set to Measure)	**t** (Select Fence from the Method option menu located in the Tool Settings window.)

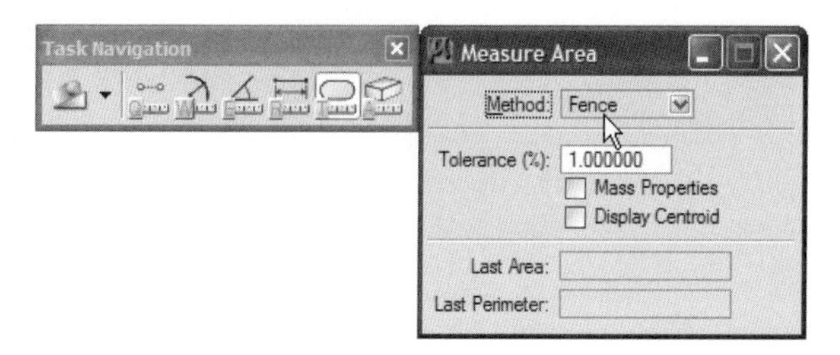

Figure 9–10 *Invoking the Measure Area tool (Fence) tool from the Task Navigation tool box (active task set to Measure)*

MicroStation prompts:

> Measure Fence Area > Accept/Reject Fence Contents *(Click the Data button to accept the fence contents to measure the area, or click the Reject button to disregard the measurement.)*

MicroStation displays the area of the fence in the Status bar/Tool Settings window.

Measure Area Intersection, Union, or Difference

These options measure areas formed by intersecting closed elements. The intersection option determines the area that is common to two closed elements. The union option determines the area in such a way that there is no duplication between two closed elements. The difference option determines the area that is formed from a closed element after removing from it any area that it has in common with the other selected element. Invoke the Measure Element Union (or Difference, or Intersection) Area tool from:

Task Navigation tool box (active task set to Measure)	Select the Measure Area tool. In the Tool Settings window select Union, Difference, or Intersection from the Method option menu (see Figure 9–11).
Keyboard Navigation (Task Navigation tool box with active task set to Measure)	**t** (Select Union, Difference, or Intersection from the Method option menu located in the Tool Settings window.)

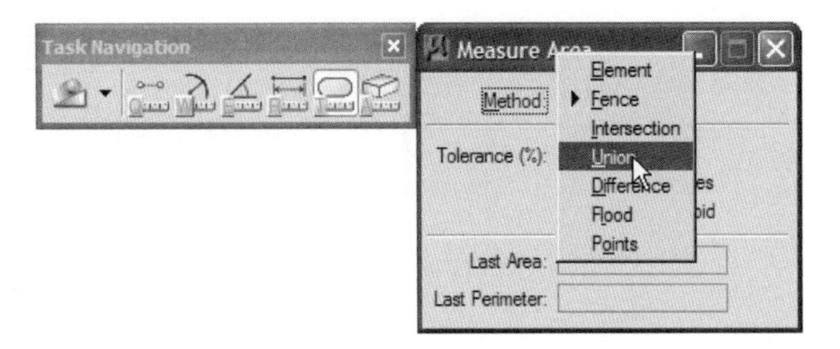

Figure 9–11 *Invoking the Measure Area (Union) tool from the Task Navigation tool box (active task set to Measure)*

MicroStation prompts:

> Measure Element Union Area > Identify the element *(Identify the first element.)*
>
> Measure Element Union Area > Identify additional/reset to complete *(Identify the second element.)*
>
> Measure Element Union Area > Accept/Reject (Select next input) *(Click the Data button to accept the element, or identify another element.)*
>
> Measure Element Union Area > Identify Additional/Reset to terminate *(Identify additional elements, or click the Reset button to terminate and initiate the measurement.)*

After all the elements are selected, and the last data point placed in space, MicroStation displays an image of only the part of the elements that are included in the type of area selected. Click the Reset Button to cause the elements to reappear, and MicroStation displays the area and perimeter length in the Status bar/Tool Settings window.

Measure Area Flood

This measures the area enclosed by a group of elements. Invoke the Measure Area Enclosing Point tool from:

Task Navigation tool box (active task set to Measure)	Select the Measure Area tool. In the Tool Settings window select Flood from the Method option menu (see Figure 9–12).
Keyboard Navigation (Task Navigation tool box with active task set to Measure)	**t** (Select Flood from the Method option menu located in the Tool Settings window.)

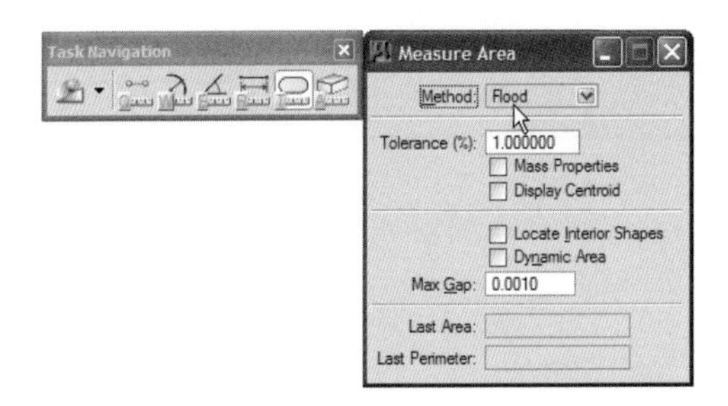

Figure 9–12 *Invoking the Measure Area (Flood) tool from the Task Navigation tool box (active task set to Measure)*

MicroStation prompts:

> Measure Area Enclosing Point > Enter data point inside area *(Click the Data button inside the area enclosed by the elements.)*
>
> Measure Area Enclosing Point > Accept, Initiate Measurement *(Click the Data button to accept and initiate measurement.)*

After the Data button is clicked, a small spinner will appear in the Status bar. The spinner spins to indicate that MicroStation is determining the area enclosed by the elements. As MicroStation traces the area, it highlights the elements. When the area has been determined, the spinner stops spinning and the area and perimeter length appear in the Status bar/Tool Settings window.

Note: *The* ***Max Gap*** *setting controls the space that is allowed between the elements that define the flood area. If the gap is zero, the elements must be touching.*

Measure Area Points

This tool measures the area formed by a set of data points specified. It assumes the perimeter of the area is formed by straight lines between the data points. An image of the area is displayed as the data points are entered. Invoke the Measure Area Defined By Points tool from:

Task Navigation tool box (active task set to Measure)	Select the Measure Area tool. In the Tool Settings window select Points from the Method option menu (see Figure 9–13).
Keyboard Navigation (Task Navigation tool box with active task set to Measure)	**t** (Select Points from the Method option menu located in the Tool Settings window.)

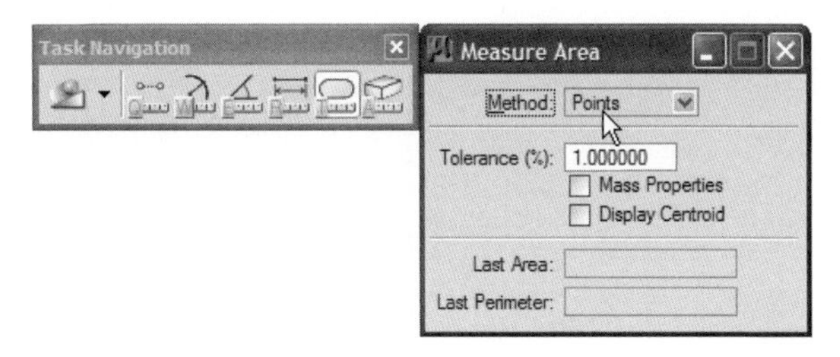

Figure 9–13 *Invoking the Measure Area (Points) tool from the Task Navigation tool box (active task set to Measure)*

MicroStation prompts:

> Measure Area Defined By Points > Enter shape vertex *(Place data points to define the vertices of the area to be measured. When the area is completely defined, click the Reset button to initiate area measurement.)*

As the data points are specified, a closed dynamic image of the area appears on the screen. When the Reset button is pressed, the area and the perimeter length appear in the Status bar/Tool Settings window.

ELEMENT INFORMATION

The Element Information dialog box is used to review or modify the properties of an element(s), such as its type, attributes, and geometry. The properties displayed in the dialog box automatically update as elements are selected and deselected using the Element Selection tool. The dialog box can be docked to the edge of the application window. Open the Element Information dialog box from:

Primary tool box	Select the Element Information tool (see Figure 9–14).
Key-in window	**element info dialog open** (or **el I d o**) (ENTER)

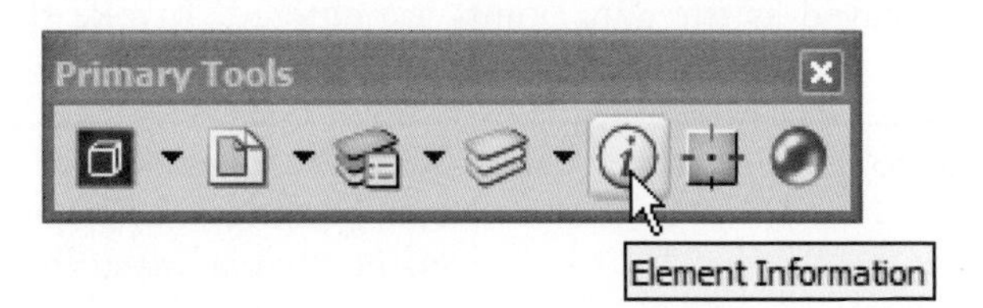

Figure 9–14 *Invoking the Element Information tool from the Primary tool box*

MicroStation opens the Element Information dialog box (see Figure 9-15).

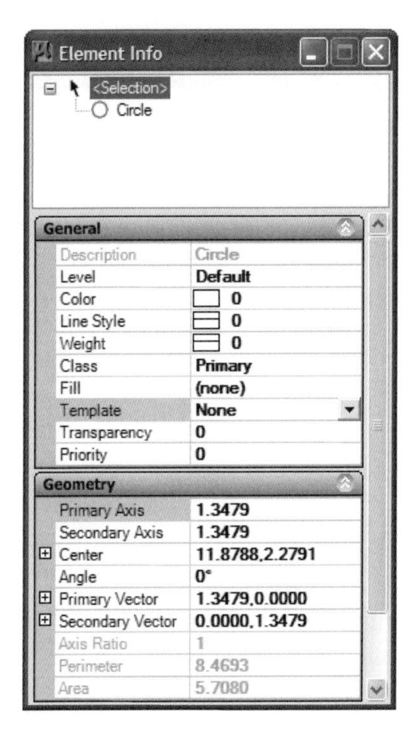

Figure 9–15 *Element Information dialog box*

The identified or selected element(s) is listed in the top frame. Depending on the type of element selected, the General, Geometry, Material, Extended, Raw Data, Text Formatting, Text Contents, Symbol Properties, Annotation Links, Image, Color, Display Print, Pattern Parameters and Groups tabs appear in the bottom frame. Information displayed on each of these tabs pertains to the element whose list entry is selected in the top frame (element list box).

To review one of multiple selected or identified elements, select the list entry for that element. Expand the tree to display the information of the selected element. If necessary make the necessary changes to the properties of the selected element. If the field and its setting display in gray text, the value is read-only and cannot be modified.

The Element Info dialog box can also be opened by right-clicking on an element with the *Element Selection* pointer and choosing Properties from the pop-up menu. As an alternative to opening the Element Information dialog box to review and change, open the Quick Info (compact version) of the dialog box by placing the cursor on an element, hold the ALT key and press the Reset button. The Quick Info dialog box (see Figure 9-16) can be used to change the general properties of an element, such as level and color.

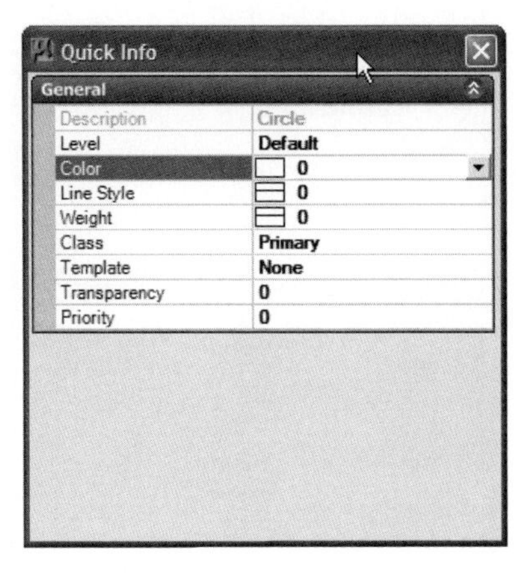

Figure 9–16 *Compact version of the Element Information dialog box*

DIMENSIONING

MicroStation's dimensioning features provide an excellent way to add dimensional information to a design, such as lengths, widths, angles, tolerances, and clearances.

Dimensioning of any drawing is generally one of the last steps in manual drawing; however, it does not need to be the last step in a MicroStation drawing. If you place the dimensions and find out later they must be changed because the size of the objects they are related to have changed, MicroStation allows you to stretch or extend the objects and have the dimensions change automatically to the new size. MicroStation provides three basic types of dimensions: linear, angular, and radial dimensioning. Figure 9–17 shows examples of these three basic types of dimensions.

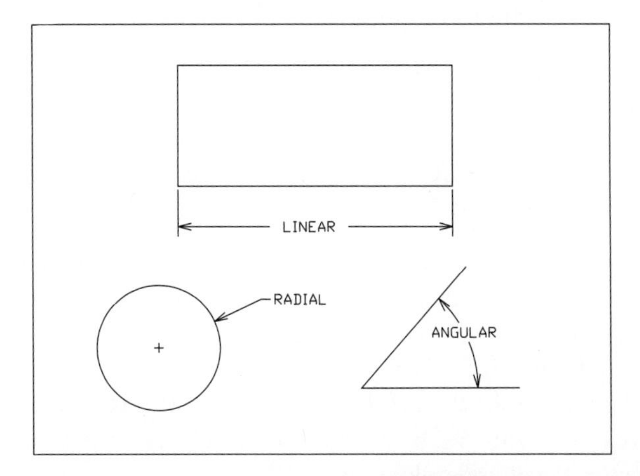

Figure 9–17 *Examples of the three basic types of dimensions*

DIMENSIONING TERMINOLOGY

The following terms occur commonly in the MicroStation dimensioning procedures.

Dimension Line

This is a line with markers at each end (arrows, dots, tick marks, etc.). The dimensioning text is located along this line; it may be placeed above the line or in a break in the dimension line. Usually, the dimension line is inside the measured area. If there is insufficient space, MicroStation places the dimensions and draws two short lines outside the measured area with arrows pointing inward.

Extension Lines

The extension lines (also called *witness lines*) are the lines that extend from the object to the dimension line. Extension lines normally are drawn perpendicular to the dimension line. (Several options that are associated with this element will be reviewed later in this chapter.) Also, one or both of the extension lines can be suppressed.

Terminators

Terminators are placed at one or both ends of the dimension line, depending on the type of dimension line placed. Arrows, tick marks, or arbitrary user-defined symbolsmay be used for the terminators. Terminators can also be sized.

Dimension Text

This is a text string that usually indicates the actual measurement. The default measurement computed automatically by MicroStation can be used, or user-defined text may be substituted.

Figure 9–18 shows the different components of a typical dimension.

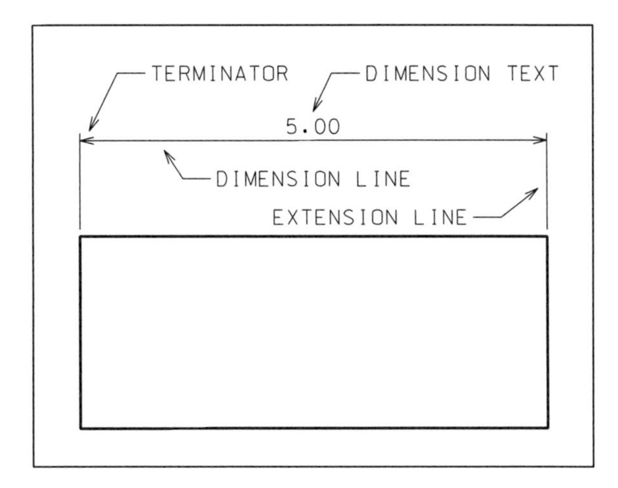

Figure 9–18 *Different components of a typical dimension*

Leader

The leader line is a line from text to an object on the design, as shown in Figure 9–19. For some dimensioning, the text may not fit next to the object it describes; hence, it is customary to place the text nearby and draw a leader from it to the object.

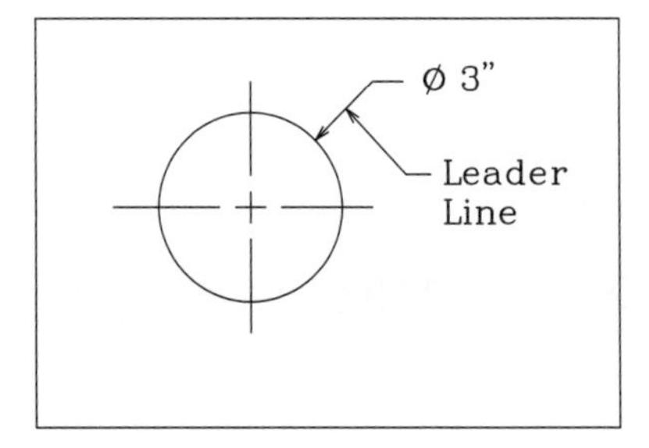

Figure 9–19 *Example of placing a leader line*

ASSOCIATIVE DIMENSIONING

Dimensions can be placed as either associative dimensions or normal (non-associative) dimensions. Associative dimensioning links dimension elements to the objects being dimensioned. An association point does not have its own coordinates, but is positioned by the coordinates of the point with which it is associated. When an element's size, shape, or position is changed, MicroStation modifies the associated dimension automatically to reflect the change. It also draws the dimension entity at its new location, size, and rotation.

To place associative dimensions, turn ON the **Association** check box in the Tool Settings window when invoking one of the dimensioning tools, as shown in Figure

9–20. If a dimension is placed when the **Association** check box is OFF, the dimension will not associate with the object dimensioned, and if the object is modified, the dimension is not changed.

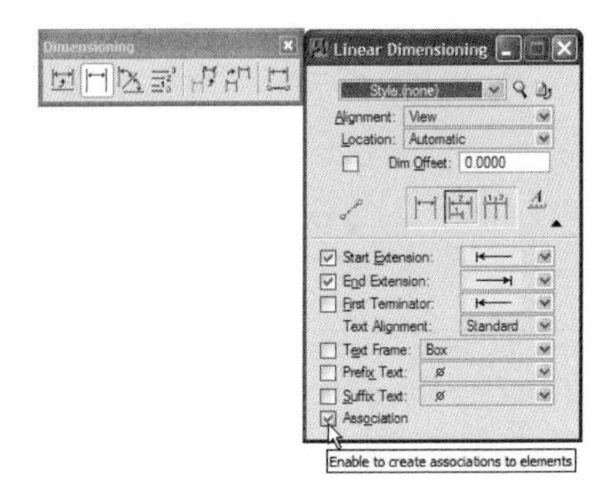

Figure 9–20 *Status of the check box for the Association in the Tool Settings window*

Placing associative dimensions can significantly reduce the size of a design file that has many dimensions, since a dimension element is usually smaller than its corresponding individual elements.

ALIGNMENT CONTROLS

The alignment controls the orientation of linear dimensions. **View**, **Drawing**, **True**, and **Arbitrary** are the available options. The options are selected from the **Alignment** option menu in the Tool Settings window when invoking one of the linear dimensioning tools.

- The **View** option aligns linear dimensions parallel to the view *X* or *Y* axis. This is useful when dimensioning *three dimensional* reference files with dimensions parallel to the viewing plane.
- The **Drawing** option aligns linear dimensions parallel to the design plane *X* or *Y* axis.
- The **True** option aligns linear dimensions parallel to the element being dimensioned. The extension lines are constrained to be at right angles to the dimension line.
- The **Arbitrary** option places linear dimensions parallel to the element being dimensioned. The extension lines are not constrained to be at right angles to the dimension line. This is useful when dimensioning elements in *2D* isometric drawings. The **Iso Lock** check box must be set to ON.

Figure 9–21 shows examples of placing linear dimensions with different alignment controls.

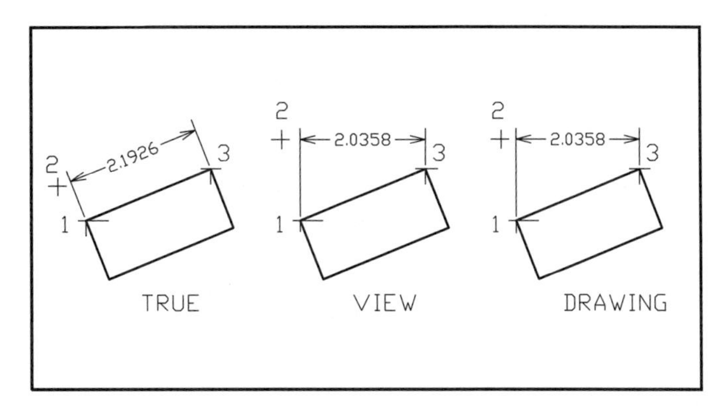

Figure 9–21 *Examples of placing linear dimensions with different alignment controls*

DIMENSION STYLES

A dimension style is a saved set of dimensioning settings. Dimension styles can be defined and applied to dimension elements during placement. The active dimension style can be set from the drop-down list box located in the Tool Settings window when invoking one of the Dimension commands. If the active dimension style is set to (none), the active dimensioning settings are applied. If the dimension style is changed, any dimension with that style also changes to those settings. Dimension styles can be created, customized and saved for easy recall.

Refer to the Dimension Settings section for detailed explanations for defining and modifying dimension styles.

LINEAR DIMENSIONING

Linear dimensioning tools dimension such linear elements as lines and line strings. Following are the tools available to dimension the linear elements:

- Dimension Linear—Dimensions the linear distance between two points (length).
- Dimension Location—Dimensions linear distances from an origin (datum), with the dimensions placed in line (chained).
- Dimension Location (Stacked)—Dimensions linear distances from an origin (datum), but with the dimensions stacked.
- Dimension Size Perpendicular to Points—Dimensions the linear distance between two points. The first two data points entered define the dimension's *Y* axis.
- Dimension Size Perpendicular to Line—Dimensions the linear distance perpendicular from an element to another element or point. The dimension's *Y* axis is defined by the element identified.

- Dimension Symmetric—Creates symmetric dimensions by defining a center point (non-associative).
- Dimension Half—Creates one-sided dimension by defining a center.
- Dimension Chamfer—Places a dimension on a chamfer.

Dimension Linear

To place linear dimension between two points (length), invoke the Dimension Size tool from:

Task Navigation tool box (active task set to Dimensioning)	Select the Dimension Linear tool and select Linear size from the Tool Settings window (see Figure 9–22). Select active dimension style, Alignment, and Location in the Tool Settings window.
Keyboard Navigation (Task Navigation tool box with active task set to Dimensioning)	**w** (Select Linear Size from the Tool Settings windows. Select active dimension style, Alignment, and Location in the Tool Settings window.)

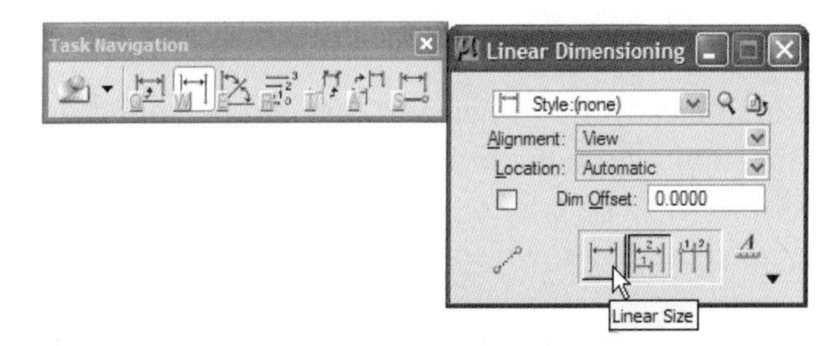

Figure 9–22 *Invoking the Dimension Linear size tool*

MicroStation prompts:

Dimension Linear Size > Select start of dimension *(Place a data point as shown in Figure 9–23 (point 1) to define the starting point of the dimension.)*

Dimension Linear Size > Select dimension endpoint *(Place a data point as shown in Figure 9–23 (point 3) to define the end point of the dimension.)*

Dimension Size with Arrow > Define length of extension line *(Place a data point as shown in Figure 9–23 (point 2) to define the length of the extension line.)*

Dimension Size with Arrow > Select dimension endpoint (*Continue placing data points as shown in Figure 9–23 (points 4 and 5) to continue linear dimensioning in the same direction, or click the Reset button to start all over again.)*

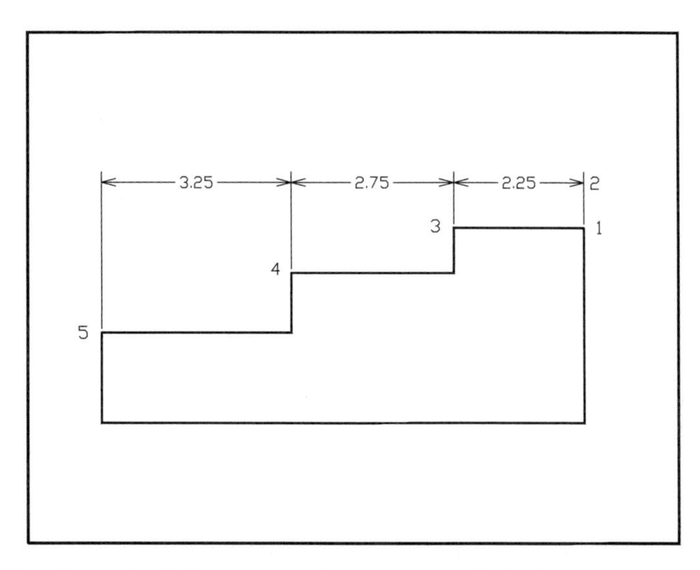

Figure 9–23 *Example of placing linear dimensioning with the Dimension Linear tool*

Note: *Use the appropriate Snap Lock when placing the data points for the starting point and end point of the dimension line.*

After dimensions are placed, the dimension text can be edited using the Edit Text tool in the Text tool box. To edit dimension text, select the Edit Text tool, then select the dimension text to be edited. MicroStation displays the Dimension Text dialog box similar to the one shown in Figure 9–24.

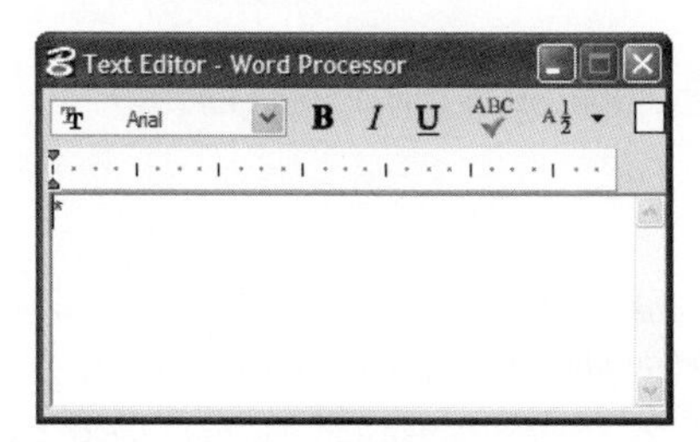

Figure 9–24 *Text dialog box*

The asterisk (*) in the Primary Text edit field indicates the current default dimension text string.

- To change the default dimension string, delete the asterisk (*) and key-in the new dimension text string.

- To add any prefix and/or suffix text string to the default dimension text string, keep the asterisk (*) and type in the appropriate text string in the Text edit field.

Click the data button to accept the changes or click the Reset button to disregard the changes.

The **Alignment** selection in the Tool Settings window determines the axis along which the dimension is aligned. The **Location** selection controls the location of dimension text: Automatic selection places the Dimension text according to the Justification setting; Semi-Automatic selection places the Dimension text according to the Justification setting if the text fits between the extension lines. If the text does not fit, the text is placed anywhere in response to the prompt; and Manual selection allows to place the Dimension text anywhere in response to the prompt.

Dimension Location

The Dimension Location tool dimensions linear distances from an origin (datum), as shown in Figure 9–25. The dimensions are placed in a line (chain). All dimensions are measured on an element originating from a common surface, centerline, or center plane. The Dimension Location tool is commonly used in mechanical drafting.

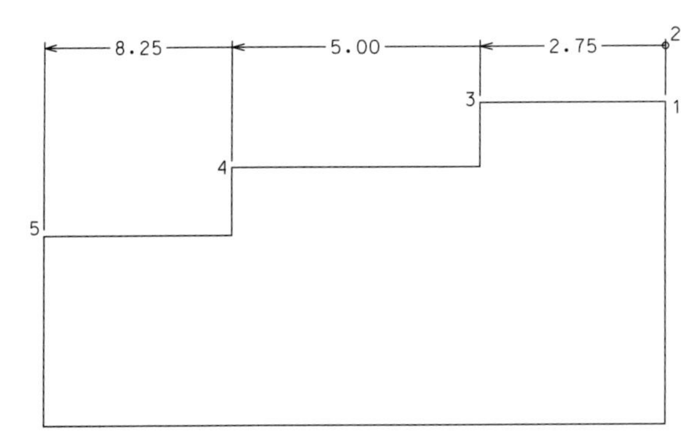

Figure 9–25 *Example of placing linear dimensioning from an origin (datum) with the Dimension Location tool*

To place linear dimensions in a chain, invoke the Dimension Location tool from:

Task Navigation tool box (active task set to Dimensioning)	Select the Dimension Linear tool and then select Linear Single from the tool settings window (see Figure 9–26). Select active dimension style, Alignment, and Location in the Tool Settings window.
Keyboard Navigation (Task Navigation tool box with active task set to Dimensioning)	**w** (Select Linear Single from the tool settings window. Select active dimension style, Alignment, and Location in the Tool Settings window.)

Figure 9–26 *Invoking the Dimension Location tool*

MicroStation prompts:

Dimension Location > Select start of dimension *(Place a data point to define the origin.)*

Dimension Location > Select dimension endpoint *(Place a data point to define the end point of the dimension.)*

Dimension Location > Define length of extension line *(Place a data point to define the length of the extension line. If necessary, press* (ENTER) *to edit the dimension text.)*

Dimension Location > Select dimension endpoint *(Continue placing data points to continue linear dimensioning in the same direction, or click the Reset button to start all over again.)*

Note: *Use the appropriate Snap Lock when placing the data points for the starting point and end point of the dimension line.*

Dimension Linear (Stacked)

The Dimension Location (Stacked) tool dimensions linear distances from an origin (datum), as shown in Figure 9–27. The dimensions are stacked. All dimensions are measured on an element originating from a common surface, centerline, or center plane. The Dimension Location (Stacked) tool is commonly used in mechanical drafting because all dimensions are independent, even though they are taken from a common datum. If necessary, the stack offset distance can be changed; see the later section on Dimension Settings.

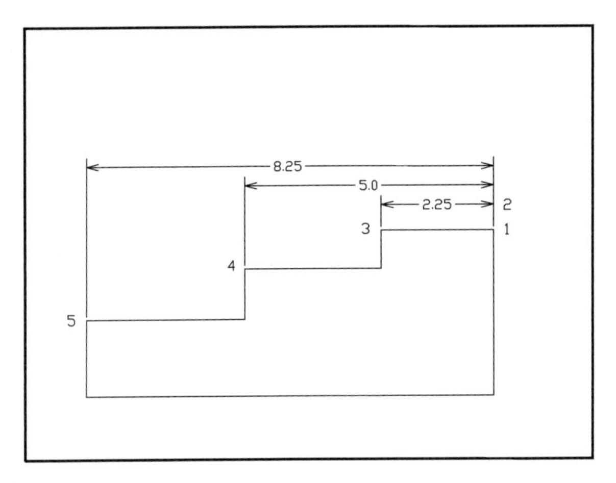

Figure 9–27 *Example of placing linear dimensioning (stacked) from an origin (datum) via the Dimension Location (Stacked) tool*

To place linear dimensions in a stack, invoke the Dimension Location (Stacked) tool from:

Task Navigation tool box (active task set to Dimensioning)	Select the Dimension Linear tool and then select Linear Stacked from the tool settings window (see Figure 9–28). Select active dimension style, Alignment, and Location in the Tool Settings window.
Keyboard Navigation (Task Navigation tool box with active task set to Dimensioning)	**w** (Select Linear Stacked from the Tool Settings window. Select active dimension style, Alignment, and Location in the Tool Settings window.)

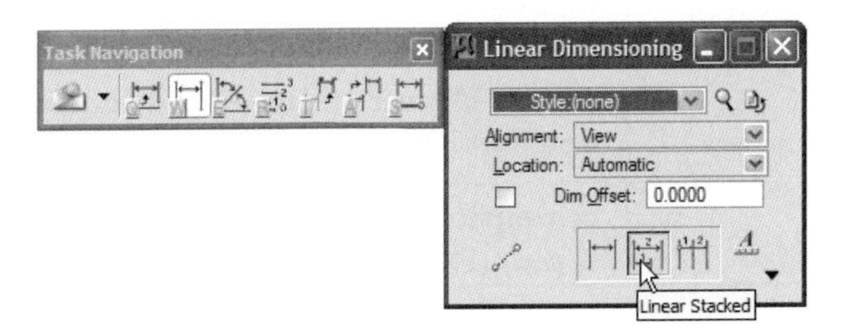

Figure 9–28 *Invoking the Dimension Location (Stacked) tool*

MicroStation prompts:

> Dimension Location (Stacked) > Select start of dimension *(Place a data point to define the origin.)*

Dimension Location (Stacked) > Select dimension endpoint *(Place a data point to define the end point of the dimension line.)*

Dimension Location (Stacked) > Define length of extension line *(Place a data point to define the length of the extension line. If necessary, press* (ENTER) *to edit the dimension text.)*

Dimension Location (Stacked) > Select dimension endpoint *(Continue placing data points to continue linear dimensioning in the same direction, or click the Reset button to start all over again.)*

Note: *Use the appropriate Snap Lock when placing the data points for the starting point and end point of the dimension line.*

Dimension Size Perpendicular to Points

The Dimension Size Perpendicular to Points tool can dimension the linear distance between two points. The first two data points entered define the dimension's *Y* axis, as shown in Figure 9–29.

Figure 9–29 *Example of placing linear dimensioning with the Dimension Size Perpendicular to Points tool*

To place linear dimensions between two points, invoke the Dimension Size Perpendicular to Points tool from:

Linear Dimensions tool box	Select the Dimension Size Perpendicular to Points tool. If necessary, set the Association check box to ON and select a Dimension Style in the Tool Settings window (see Figure 9–30).
Key-in window	**dimension size perpendicular to points** (or **dime si p p**) (ENTER)

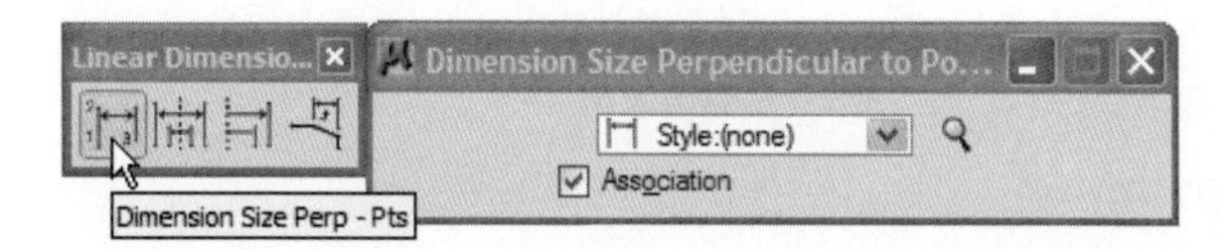

Figure 9–30 *Invoking the Dimension Size Perpendicular to Points tool from the Linear Dimensions tool box*

MicroStation prompts:

Dimension Size Perpendicular to Points > Select base of first dimension line *(Place a data point to define the base point of the first dimension line.)*

Dimension Size Perpendicular to Points > Select end of extension line *(Place a data point to define the length of the extension line.)*

Dimension Size Perpendicular to Points > Select dimension endpoint *(Place a data point to define the end point of the dimension line.)*

***Note:** Use the appropriate Snap Lock when placing the data points for the starting point and end point of the dimension line.*

Dimension Symmetric

The Dimension Symmetric tool is used to create symmetric dimensions by defining a center point (non-associative). To dimension symmetrically, invoke the Dimension Symmetric tool from:

Linear Dimensions tool box	Select the Dimension Symmetric tool. Select an Alignment and Dimension Style in the Tool Settings window (see Figure 9–31).
Key-in window	**dimension symmetric** (or **dime sy**) (ENTER)

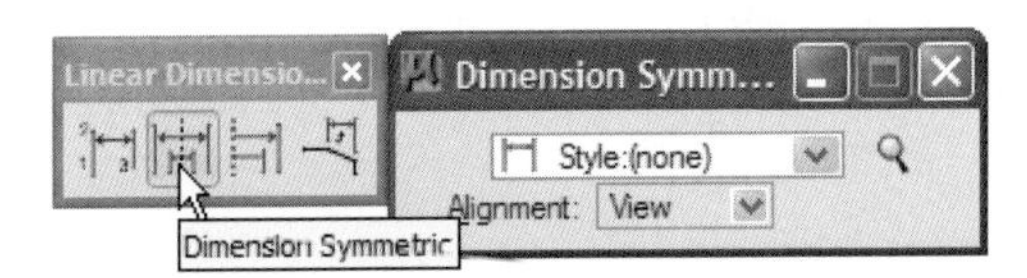

Figure 9–31 *Invoking the Dimension Symmetric tool from the Linear Dimensions tool box*

MicroStation prompts:

Dimension Symmetric > Select point on center line *(Place a data point to define the center of the dimension.)*

Dimension Symmetric > Select point to define dimension location *(Place a data point to define the length of the extension line.)*

Dimension Symmetric > Select dimension endpoint *(Place a data point to define the end point of the dimension.)*

***Note:** Use the appropriate Snap Lock when placing the data points for the starting point and end point of the dimension line.*

Dimension Half

The Dimension Half tool is used to create one-sided dimensions by defining a center. To dimension one-side of an element, invoke the Dimension Half tool from:

Linear Dimensions tool box	Select the Dimension Half tool. Select an Alignment and Dimension Style in the Tool Settings window (see Figure 9–32).
Key-in window	**dimension half** (or **dime h**) (ENTER)

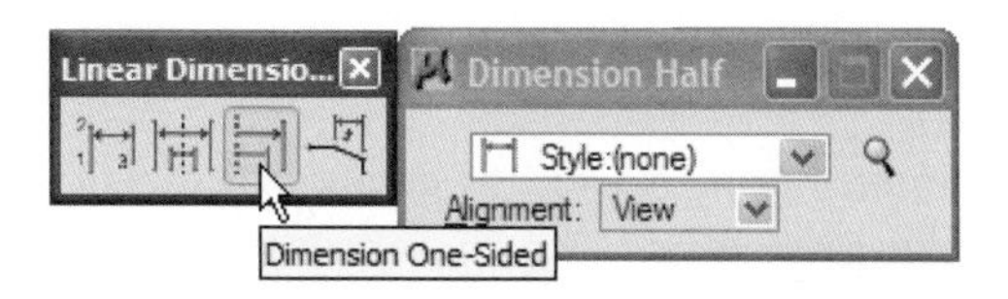

Figure 9–32 *Invoking the Dimension Half tool from the Linear Dimensions tool box*

MicroStation prompts:

Dimension Half > Select point on center line *(Place a data point to define the center of the dimension.)*

Dimension Half > Select point to define dimension location *(Place a data point to define the length of the extension line.)*

Dimension Half > Select dimension endpoint *(Place a data point to define the end point of the dimension.)*

Note: *Use the appropriate Snap Lock when placing the data points for the starting point and end point of the dimension line.*

Dimension Chamfer

The Dimension Chamfer tool is used to place a dimension on chamfer. To dimension a chamfer, invoke the Dimension Chamfer tool from:

Linear Dimensions tool box	Select the Dimension Chamfer tool. Select an Alignment and Dimension Style in the Tool Settings window (see Figure 9–33).
Key-in window	**dimension chamfer** (or **dime ch**) (ENTER)

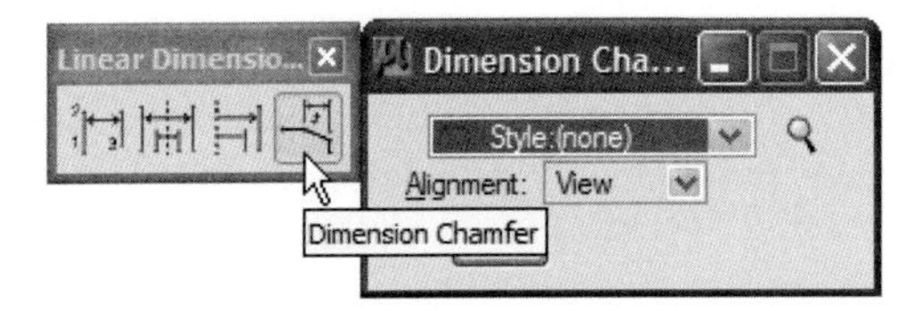

Figure 9–33 *Invoking the Dimension Chamfer tool from the Linear Dimensions tool box*

MicroStation prompts:

Dimension Chamfer Angle> Select chamfer element *(Identify chamfer element.)*

Dimension Chamfer Angle> Select second element to measure chamfer from *(Identify a second element from which to measure the chamfer.)*

Dimension Chamfer Angle > Define Length of extension line *(Place a data point to define the length of extension line or click* **Next** *button in the Tool Settings window to cycle through the available tools to dimension the chamfer.)*

ANGULAR DIMENSIONING

The angular dimensioning tools create dimensions for the angle between two non-parallel lines, using the conventions that conform to the current dimension variable settings. "Angle" is defined as "a measure of an angle or of the amount of turning necessary to bring one line or plane into coincidence with or parallel to another." Following are the eight different tools available to create angular dimensions:

- Dimension Angle Size—Each dimension (except the first) is computed from the end point of the previous dimension.
- Dimension Angle Location—Each dimension is computed from the dimension rigin (datum).
- Dimension Angle Between Lines—Dimensions the angle between two lines, two segments of a line string, or two sides of a shape.
- Dimension Angle from *X* axis—Dimensions the angle between a line, a side of a shape, or a segment of a line string and the view *X* axis.
- Dimension Angle from *Y* axis—Dimensions the angle between a line, a side of a shape, or a segment of a line string and the view *Y* axis.
- Dimension Arc Size—Dimensions an arc or circle. Each dimension is computed from the endpoint of the previous dimension, except the first one.
- Dimension Arc Stacked—Dimensions an arc or circle. Each dimension is computed from the dimension origin (datum). The dimensions are stacked.
- Dimension Angle Chamfer—Dimension the angle of a chamfer.

Dimension Angle Size

To place angular dimensioning with the Dimension Angle Size tool, invoke the tool from:

Task Navigation tool box (active task set to Dimensioning)	Select the Dimension Angular tool and then select Angle Size from the tool settings window (see Figure 9–34). Select active dimension style, Alignment, and Location in the Tool Settings window.
Keyboard Navigation (Task Navigation tool box with active task set to Dimensioning)	**e** (Select Angle Size from the tool settings window. Select active dimension style, Alignment, and Location in the Tool Settings window.)

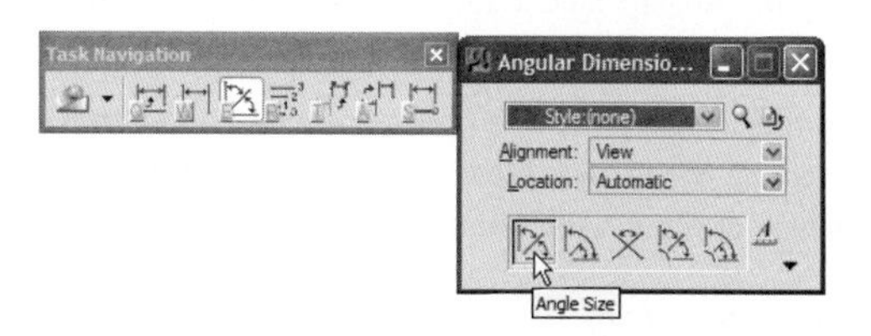

Figure 9–34 *Invoking the Angle size tool*

MicroStation prompts:

Dimension Angle Size > Select start of dimension *(Place a data point as shown in Figure 9–35 (point P1) to define the start of the dimension, which is measured counterclockwise from this point.)*

Dimension Angle Size > Enter point on axis *(Place a data point as shown in Figure 9–35 (point P3) to define the vertex of the angle.)*

Dimension Angle Size > Select dimension endpoint *(Place a data point as shown in Figure 9–35 (point P4) to define the end point of the dimension line. If necessary, press* (ENTER) *to edit the dimension.)*

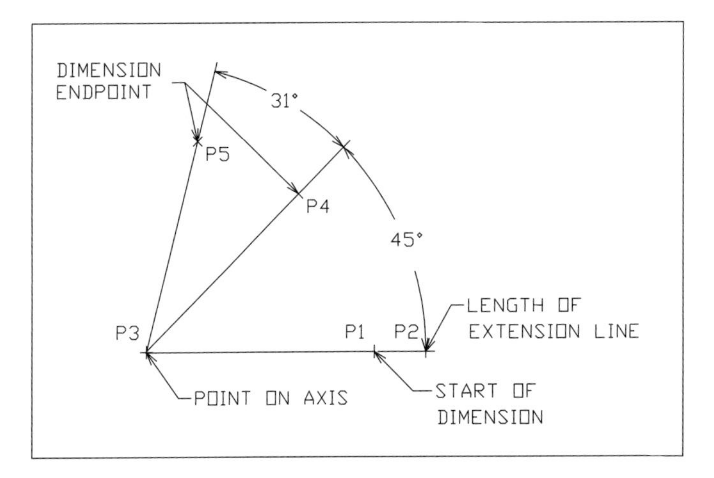

Figure 9–35 *Example of placing angular dimensioning via the Dimension Angle Size tool*

Dimension Angle Size > Define length of extension line *(Place a data point as shown in Figure 9–35 (point P2) to define length of extension line..)*

Dimension Angle Size > Select dimension endpoint *(Continue placing data points for angular dimensioning, and click the Reset button to complete the tool sequence.)*

Dimension Angle Location

To place angular dimensioning with the Dimension Angle Location tool, invoke the tool from:

Task Navigation tool box (active task set to Dimensioning)	Select the Dimension Angular tool and then select Angle Location from the tool settings window (see Figure 9–36). Select active dimension style, Alignment, and Location in the Tool Settings window.
Keyboard Navigation (Task Navigation tool box with active task set to Dimensioning)	**e** (Select Angle Location from the tool settings window. Select active dimension style, Alignment, and Location in the Tool Settings window.)

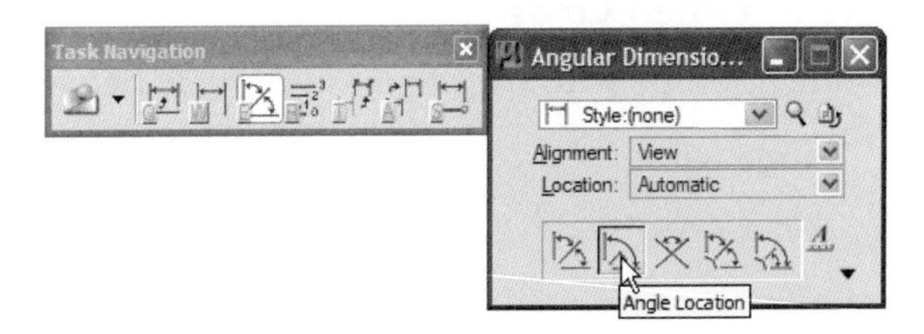

Figure 9–36 *Invoking the Dimension Angle Location tool*

MicroStation prompts:

> Dimension Angle Location > Select start of dimension *(Place a data point as shown in Figure 9–37 (point P1) to define the start of the dimension, which is measured counterclockwise from this point.)*
>
> Dimension Angle Location > Enter point on axis *(Place a data point as shown in Figure 9–37 (point P3) to define the vertex of the angle.)*
>
> Dimension Angle Location > Select dimension endpoint *(Place a data point as shown in Figure 9–37 (point P4) to define the end point of the dimension. Press* (ENTER) *to edit the dimension text.)*
>
> Dimension Angle Location > Define length of extension line *(Place a data point as shown in Figure 9–37 (point P2) to define the length of the extension line.)*
>
> Dimension Angle Location > Select dimension endpoint *(Continue placing data points for angular dimensioning, and click the Reset button to complete the tool sequence.)*

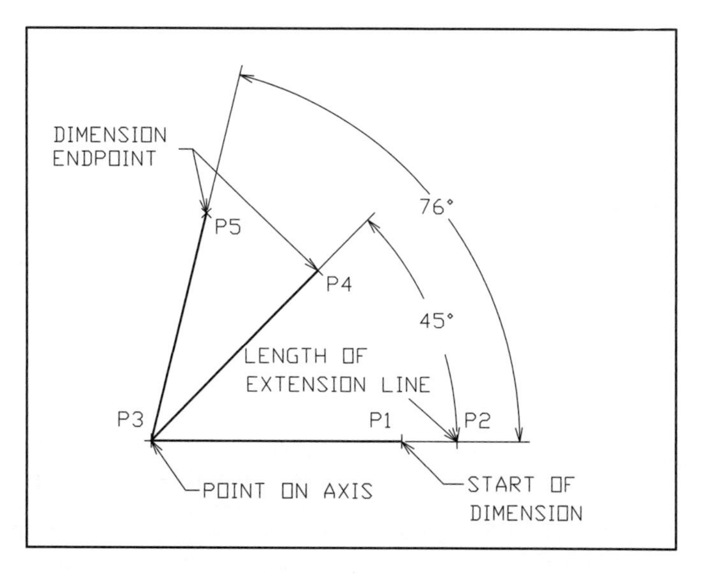

Figure 9–37 *Example of placing angular dimensioning via the Dimension Angle Location tool*

Dimension Angle Between Lines

To place angular dimensioning with the Dimension Angle Between Lines tool, invoke the tool from:

Task Navigation tool box (active task set to Dimensioning)	Select the Dimension Angular tool and then select Angle Between Lines from the tool settings window (see Figure 9–38). Select active dimension style, Alignment, and Location in the Tool Settings window.
Keyboard Navigation (Task Navigation tool box with active task set to Dimensioning)	**e** (Select Angle Between Lines from the tool settings window. Select active dimension style, Alignment, and Location in the Tool Settings window.)

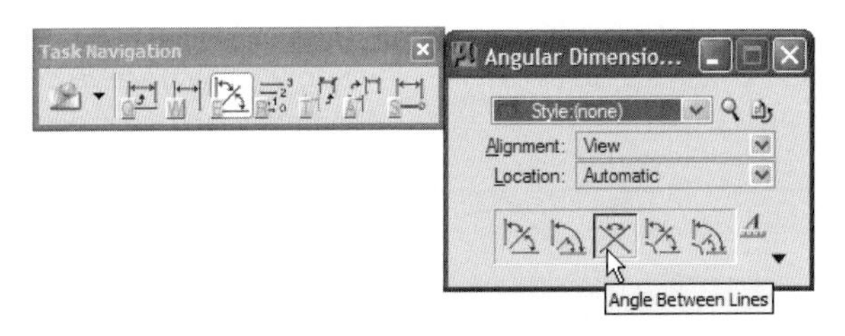

Figure 9–38 *Invoking the Dimension Angle Between Lines tool*

MicroStation prompts:

Dimension Angle Between Lines > Select first line *(Identify the first line or segment, as shown in Figure 9–39 (point P1).)*

Dimension Angle Between Lines > Select second line *(Identify the second line or segment, as shown in Figure 9–39 (point P2).)*

Dimension Angle Between Lines > Select location of dimension, Accept/Reset *(Place a data point to define the location of the dimension line, as shown in Figure 9–39 (point P3).)*

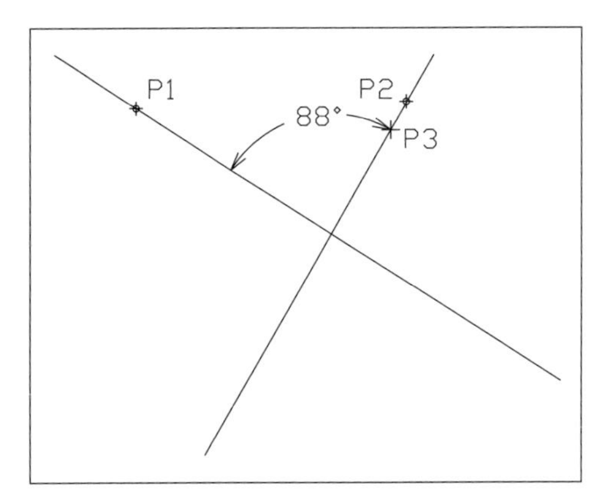

Figure 9–39 *Example of placing angular dimensioning with the Dimension Angle Between Lines tool*

Dimension Arc Size

The Dimension Arc Size tool dimensions a circle or circular arc. Each dimension is computed from the end point of the previous dimension, except the first one, similar to the Dimension Angle Size tool. To dimension arcs with the Dimension Arc Size tool, invoke the tool from:

Task Navigation tool box (active task set to Dimensioning)	Select the Dimension Angular tool and then select Arc Size from the tool settings window (see Figure 9–40). Select active dimension style, Alignment, and Location in the Tool Settings window.
Keyboard Navigation (Task Navigation tool box with active task set to Dimensioning)	**e** (Select Arc Size from the tool settings window. Select active dimension style, Alignment, and Location in the Tool Settings window.)

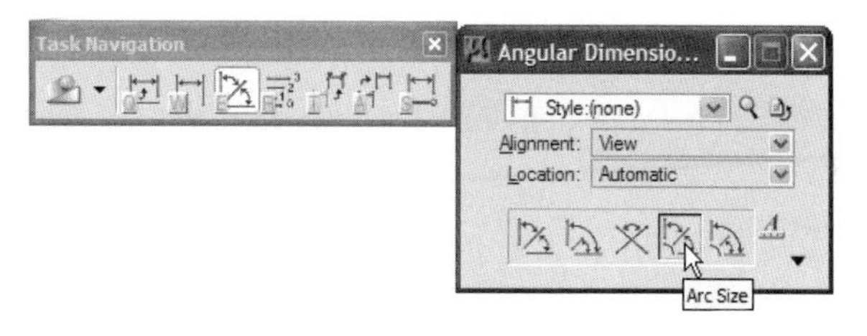

Figure 9–40 *Invoking the Dimension Arc Size tool*

MicroStation prompts:

> Dimension Arc Size > Select start of dimension *(Place a data point, as shown in Figure 9–41 (point P1), to define the origin point. The dimension is measured counterclockwise from this point. This point must select an arc, circle, or ellipse.)*
>
> Dimension Arc Size > Select dimension endpoint *(Place a data point, as shown in Figure 9–41 (point P3), to define the dimension end point. If necessary, press* (ENTER) *to edit the dimension text.)*
>
> Dimension Arc Size > Define length of extension line *(Place a data point, as shown in Figure 9–41 (point P2), to define the length of the extension line.)*
>
> Dimension Arc Size > Select dimension endpoint *(Continue placing data points to continue angular dimensioning, and click the Reset button to terminate the tool sequence.)*

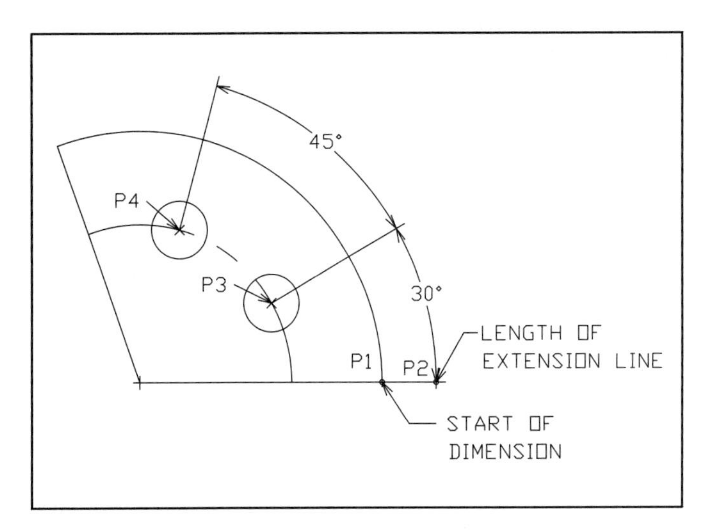

Figure 9–41 *Example of placing arc dimensioning with the Dimension Arc Size tool*

Dimension Arc Stacked

The Dimension Arc Stacked tool dimensions a circle or circular arc. Each dimension is computed from the dimension origin (datum), as shown in Figure 9–42.

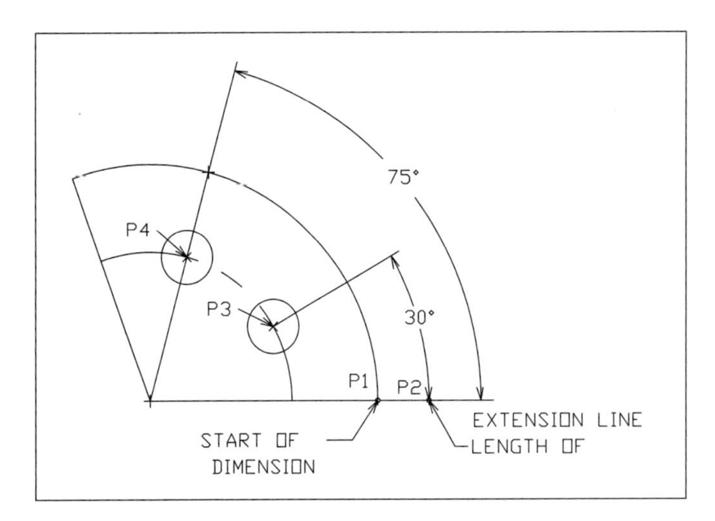

Figure 9–42 *Example of placing arc dimensioning via the Dimension Arc Location tool*

To dimension arcs with the Dimension Arc Location tool, invoke the tool from:

Task Navigation tool box (active task set to Dimensioning)	Select the Dimension Angular tool and then select Arc Stacked from the tool settings window (see Figure 9–43). Select active dimension style, Alignment, and Location in the Tool Settings window.
Keyboard Navigation (Task Navigation tool box with active task set to Dimensioning)	**e** (Select Arc Stacked from the tool settings window. Select active dimension style, Alignment, and Location in the Tool Settings window.)

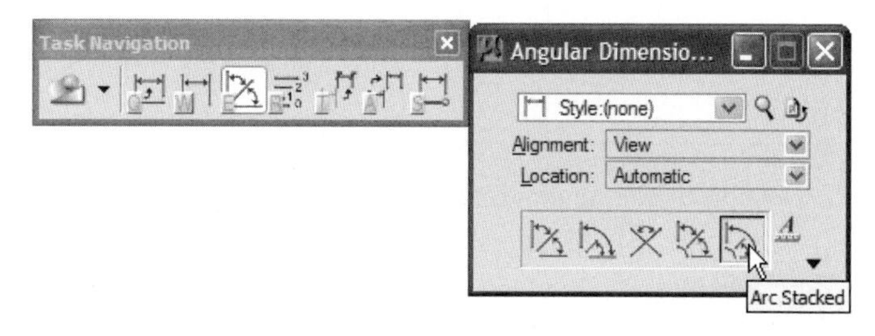

Figure 9–43 *Invoking the Dimension Arc Stacked tool from the Angular Dimensions tool box*

MicroStation prompts:

> Dimension Arc Location > Select start of dimension *(Place a data point to define the origin point. The dimension is measured counterclockwise from this point. This point must select an arc, circle, or ellipse.)*
>
> Dimension Arc Location > Select dimension endpoint *(Place a data point to define the dimension end point. If necessary, press* (ENTER) *to edit the dimension text.)*
>
> Dimension Arc Size > Define length of extension line *(Place a data point to define the length of the extension line.)*
>
> Dimension Arc Size > Select dimension endpoint *(Continue placing data points to continue angular dimensioning, and click the Reset button to terminate the tool sequence.)*

Dimension Angle from X-Axis

To place angular dimensioning with the Dimension Angle from X-axis tool, invoke the tool from:

Angular Dimensions tool box	Select the Dimension Angle from X-axis tool. If necessary, set the Association check box to ON and select a dimension style in the Tool Settings window (see Figure 9–44).
Key-in window	**dimension angle x** (or **dime a x**) (ENTER)

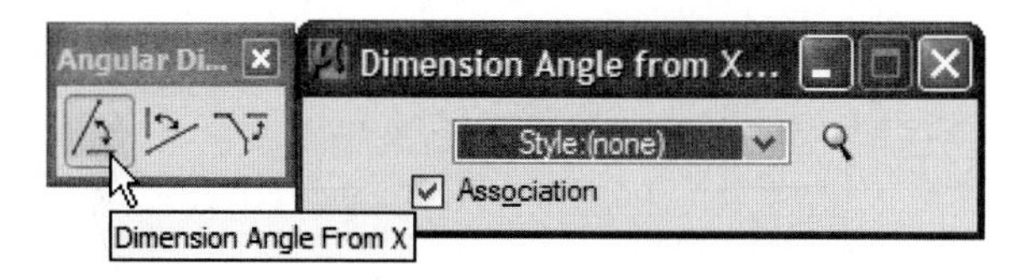

Figure 9–44 *Invoking the Dimension Angle from X-Axis tool from the Angular Dimensions tool box*

MicroStation prompts:

> Dimension Angle from X Axis > Identify element *(Identify the element, as shown in Figure 9–45 (point P1).)*
>
> Dimension Angle from X Axis > Accept, define dimension axis *(Place a data point, as shown in Figure 9–45 (point P2), to specify the location and direction of the dimension.)*

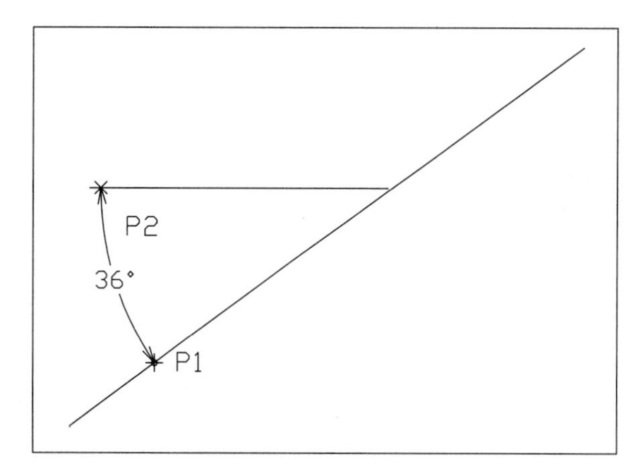

Figure 9–45 *Example of placing angular dimensioning with the Dimension Angle from X Axis tool*

Dimension Angle from Y-Axis

To place angular dimensioning with the Dimension Angle from Y-axis tool, invoke the tool from:

Angular Dimensions tool box	Select the Dimension Angle from Y-axis tool. If necessary, set the Association check box to ON and select a dimension style in the Tool Settings window (see Figure 9–46).
Key-in window	**dimension angle y** (or **dime a y**) (ENTER)

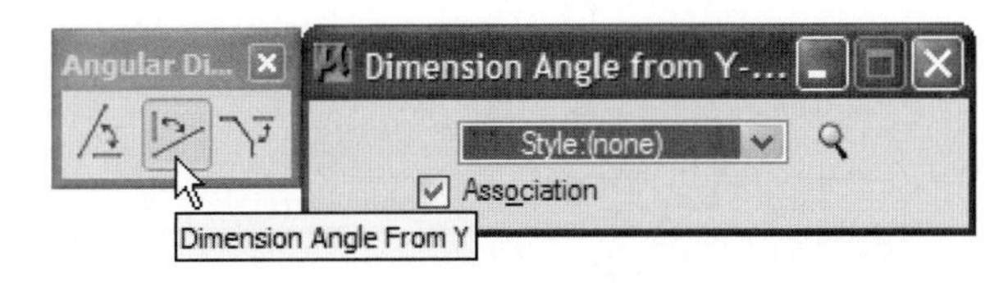

Figure 9–46 *Invoking the Dimension Angle from Y-Axis tool from the Angular Dimensions tool box*

MicroStation prompts:

Dimension Angle from Y Axis > Identify element *(Identify the element.)*

Dimension Angle from Y Axis > Accept, define dimension axis *(Place a data point to specify the location and direction of the dimension.)*

Dimension Angle Chamfer

The Dimension Angle Chamfer tool is used to dimension the angle of chamfer. To dimension chamfer with Dimension Angle Chamfer tool, invoke the tool from:

Angular Dimensions tool box	Select the Dimension Angle Chamfer tool. Select a dimension style from the Tool Settings window (see Figure 9–47).
Key-in window	**dimension angle chamfer** (or **dime ang ch**) (ENTER)

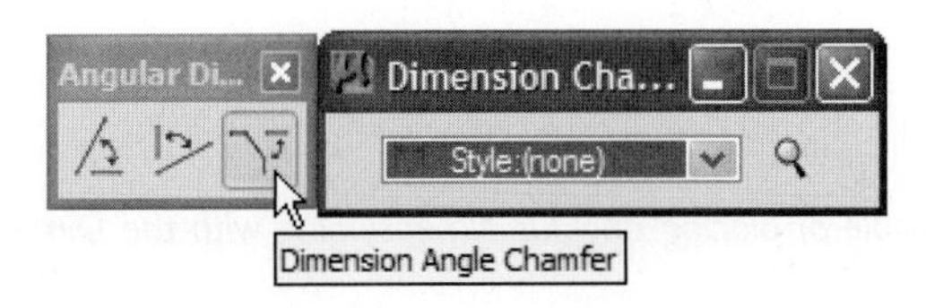

Figure 9–47 *Invoking Dimension Angle Chamfer tool from the Angular Dimensions tool box*

MicroStation prompts:

Dimension Chamfer > Select first line *(Identify chamfer line.)*

Dimension Chamfer > Select second line (*Identify the base line.*)

Dimension Chamfer (*Enter the data position the dimension.*)

DIMENSION RADIAL

The Dimension Radial feature provides tools to create dimensions for the radius or diameter of a circle or arc and to place a center mark. Following are the tools available for radial dimensioning:

- Dimension Radial—Dimensions the radius of a circle or a circular arc, dimensions the diameter of a circle or circular arc, and place a center mark of a circle or a circular arc.
- Dimension Radius (Extended Leader)—Identical to the Radius tool, except the leader line continues across the center of the circle, with terminators that point outward.
- Dimension Diameter—Dimensions the diameter of a circle or a circular arc.
- Dimension Diameter Perpendicular—Dimensions the diameter of a circle or circular arc, with the dimension placed parallel to the plane of the circle or arc and with extension lines extending to the circle or arc.
- Dimension Center—Places a center mark at the center of a circle or circular arc.
- Dimension Radius/Dimension Note—Places a dimension note on a circle or circular arc.
- Dimension Arc Distance—Dimensions the distance between two arcs that have the same center.

Dimension Radius

To place a radial dimension with the Dimension Radial tool, invoke the tool from:

Radial Dimensions tool box	Select the Dimension Radial tool (see Figure 9-48). Select one of the dimension types from the Mode menu: Radius, Radius Extended, Diameter, Diameter Extended, or Center Mark. Select alignment from the Alignment menu.
Key-in window	**dimension radius** (or **dime radiu**) (ENTER)

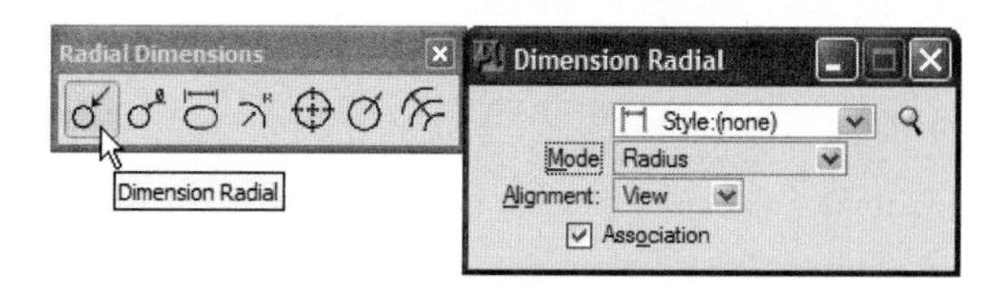

Figure 9–48 *Invoking the Dimension Radial tool from the Radial Dimensions tool box*

MicroStation prompts when Radius mode is selected:

> Dimension Radius > Identify element *(Identify a circle or arc, as shown in Figure 9–49.)*
>
> Dimension Radius > Select dimension endpoint *(Place a data point inside or outside the circle or arc to place the dimension line, as shown in Figure 9–49.)*

***Note:** The placement of the dimension text (horizontal or in-line) is set by selecting the Text **Orientation** in the Dimension Settings dialog box.*

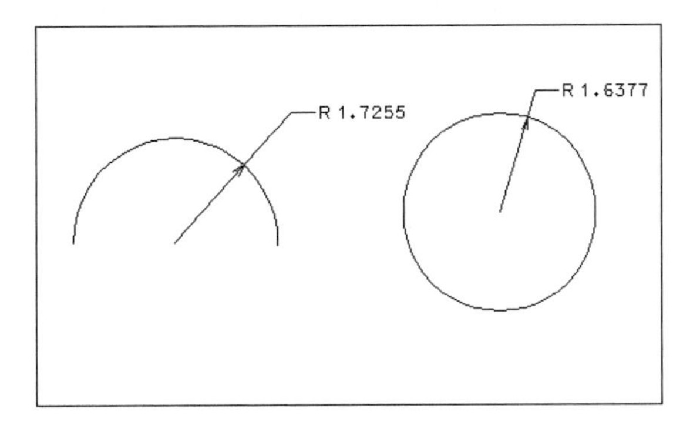

Figure 9–49 *Examples of placing radial dimensions with the Dimension Radius tool*

Dimension Diameter

To place diameter dimensioning with the Dimension Diameter tool, invoke the tool from:

Radial Dimensions tool box	Select the Dimension Diameter tool (see Figure 9-50). If necessary, set the Association check box to ON and select a dimension style from the Tool Settings window.
Key-in window	**dimension diameter** (or **dime d**) (ENTER)

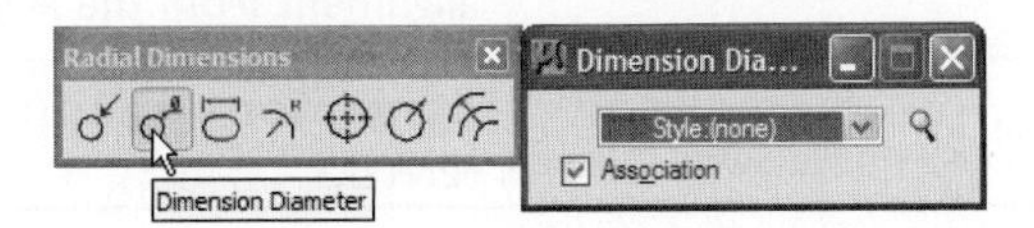

Figure 9–50 *Invoking the Dimension Diameter tool from the Radial Dimensions tool box*

MicroStation prompts:

Dimension Diameter > Identify element *(Identify a circle or arc, as shown in Figure 9–51.)*

Dimension Diameter > Select dimension endpoint *(Place a data point inside or outside the circle or arc to place the dimension line, as shown in Figure 9–51.)*

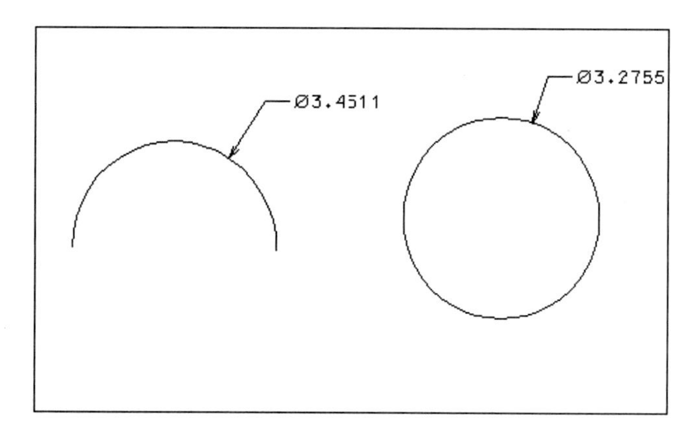

Figure 9–51 *Examples of placing diameter dimensions with the Dimension Diameter tool*

Dimension Diameter Perpendicular

The Dimension Diameter Perpendicular tool is used to dimension the diameter of a circle, with the dimension placed inside the circle. Invoke the Dimension Diameter Perpendicular tool from:

Radial Dimensions tool box	Select the Dimension Diameter Perpedicular tool (see Figure 9-52). If necessary, set the Association check box to ON, and select a dimension style from the Tool Settings window.
Key-in window	**dimension diameter perpedicular** (or **dime dia per**) (ENTER)

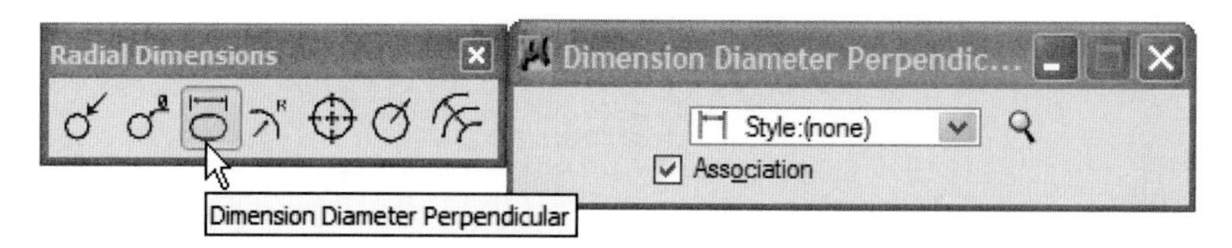

Figure 9–52 *Invoking the Dimension Diameter Perpendicular tool from the Radial Dimensions tool box*

MicroStation prompts:

Dimension Diameter Perpendicular > Identify element *(Identify a circle or arc.)*

Dimension Diameter Perpendicular > Select dimension endpoint *(Select dimension endpoint.)*

Dimension Radius (Extended Leader)

To place radial dimensioning with the Dimension Radius (Extended Leader) tool, invoke the tool from:

Radial Dimensions tool box	Select the Dimension Radius (Extended Leader) tool (see Figure 9-53). If necessary, set the Association check box to ON, and select a dimension style from the Tool Settings window.
Key-in window	**dimension radius extended** (or **dime radiu e**) (ENTER)

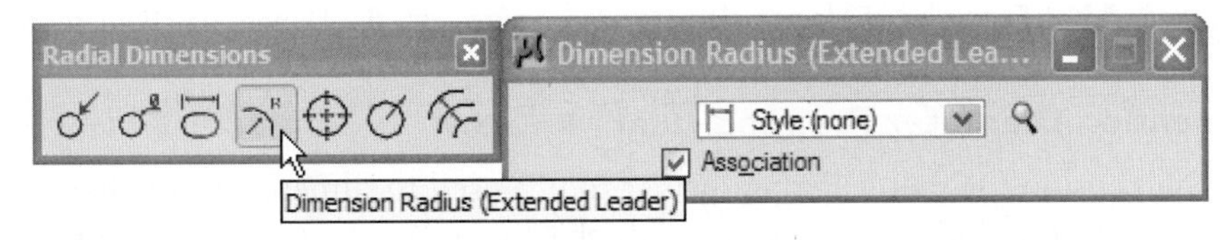

Figure 9–53 *Invoking the Dimension Radial (Extended Leader) tool from the Radial Dimensions tool box*

MicroStation prompts:

Dimension Radius (Extended Leader) > Identify element *(Identify a circle or arc, as shown in Figure 9–54.)*

Dimension Radius (Extended Leader) > Select dimension endpoint *(Place a data point inside or outside the circle or arc to place the dimension line as shown in Figure 9–54.)*

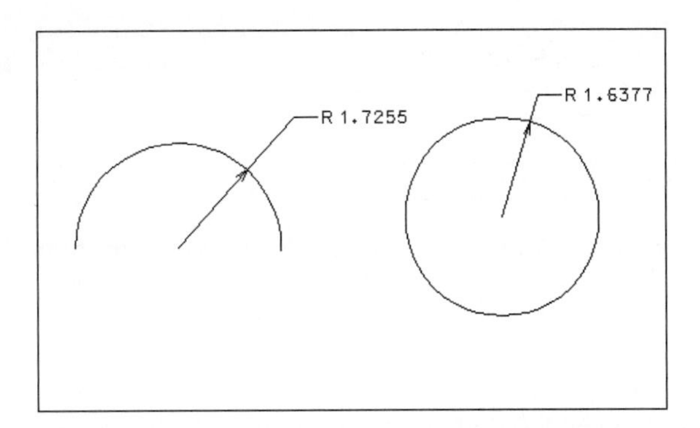

Figure 9–54 *Examples of placing radial dimensions with the Dimension Radius (Extended Leader) tool*

Place Center Mark

The Place Center Mark tool can place a center mark at the center of a circle or circular arc, as shown in Figure 9–55.

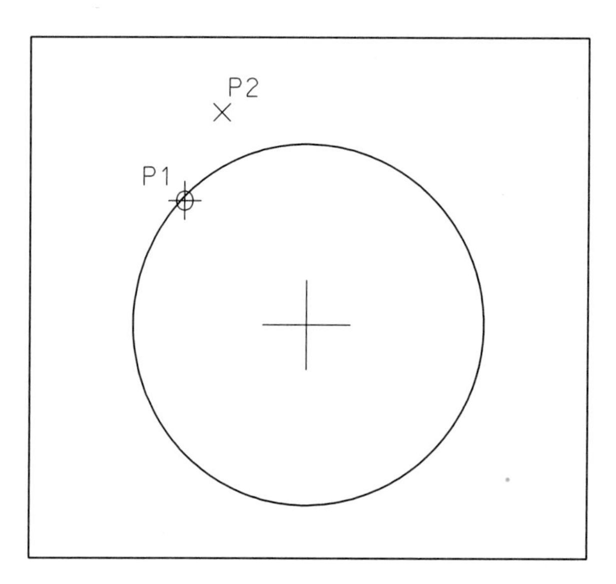

Figure 9–55 *Example of placing a center mark by the Place Center Mark tool*

To place a center mark with the Place Center Mark tool, invoke the tool from:

Radial Dimensions tool box	Select the Dimension Center tool (see Figure 9-56). If necessary, set Association Lock to ON, select a Dimension Style and key-in the appropriate size of the center mark in the Center Size text field in the Tool Settings window.
Key-in window	**dimension center mark** (or **dim c m**) (ENTER)

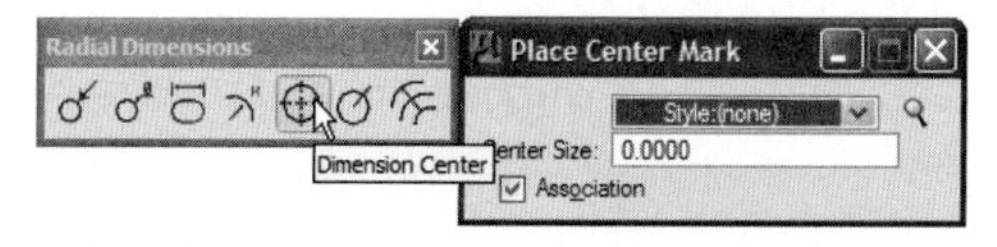

Figure 9–56 *Invoking the Dimension Center tool from the Radial Dimensions tool box*

MicroStation prompts:

Place Center Mark > Identify element *(Identify a circle or arc to place a center mark on.)*

Place Center Mark > Accept (next input) *(Click the Accept button, or identify another circle or arc to place the center mark.)*

Note: *If the Center Size text is set to 0.0000, then it applies the text size set in the Text settings box.*

Dimension Radius/Diameter Note

The Dimension Radius/Diameter Note tool is used to place a dimension note for a circle or arc indicating either the radius or diameter of the selected element. To place a note, invoke the tool from:

Radial Dimensions tool box	Select the Dimension Radius/Diameter Note tool (see Figure 9-57). In the Tool Settings window, select a dimension style and choose either Radius or Diameter information to be included as part of the note.
Key-in window	**dimension radius/diameter note** (or **dim rad/dia no**) (ENTER)

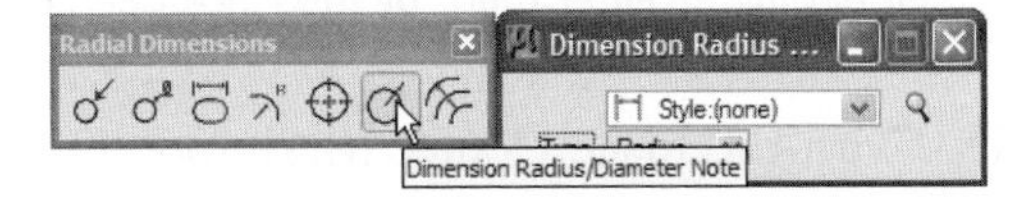

Figure 9–57 *Invoking the Dimension Radius/Diameter Note tool from the Radial Dimensions tool box*

MicroStation prompt:

Dimension Radius > Select circular element to dimension *(Identify circle or arc.)*

Dimension Radius > Define length of extension line *(Enter a data point to define the length of the extension line.)*

Dimension Arc Distance

The Dimension Arc Distance tool is used to dimension the distance between two arcs that have the same center. To dimension distance between two arcs, invoke the tool from:

Radial Dimensions tool box	Select the Dimension Arc Distance tool (see Figure 9-58). In the Tool Settings window, select a Dimension Style and View Alignment.
Key-in window	**dimension arc distance** (or **dim ar dist**) (ENTER)

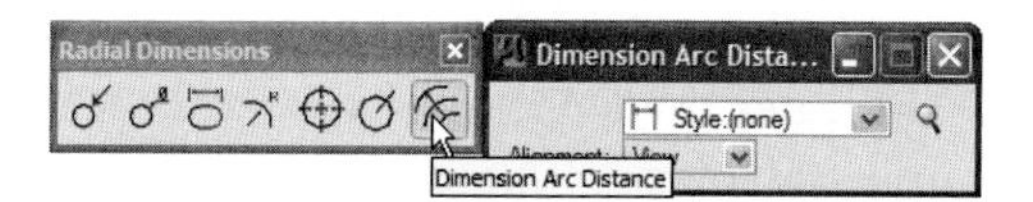

Figure 9–58 *Invoking the Dimension Arc Distance tool from the Radial Dimensions tool box*

MicroStation prompt:

> Dimension Arc Distance > Select first arc *(Identify the first circle or arc.)*
>
> Dimension Arc Distance > Select next arc *(Identify the second circle or arc.)*
>
> Dimension Arc Distance > Define length of extension line *(Enter a data point to define the length of the extension line.)*

DIMENSION ELEMENT

The Dimension Element tool provides a fast way to dimension an element. Simply identify the element and MicroStation selects the type of dimensioning tool it thinks is best. For linear elements it selects one of the linear dimensioning tools; for circles, arcs, and ellipses it selects one of the radial dimensioning tools.

When the Dimension Element tool is invoked, the tool name Dimension Element appears in the Status bar. If the selected element is a linear (non-circular) shape, MicroStation provides available dimension tools applicable to the linear (non-circular) shape in the Tool Settings window and choose the appropriate one. If the selected element is a circle, arc, ellipse, or b-spline, MicroStation provides available dimension tools applicable to the selected element in the Tool Settings window and choose the appropriate one. As soon as an element is selected, the name of the dimensioning tool MicroStation intends to use appears in the Status bar. Invoke the Dimension Element tool from:

Task Navigation tool box (active task set to Dimensioning)	Select the Dimension Element tool (see Figure 9-59). If necessary, set the Association check box to ON, and select an Alignment option and Dimension Style.
Keyboard Navigation (Task Navigation tool box with active task set to Dimensioning)	**q**

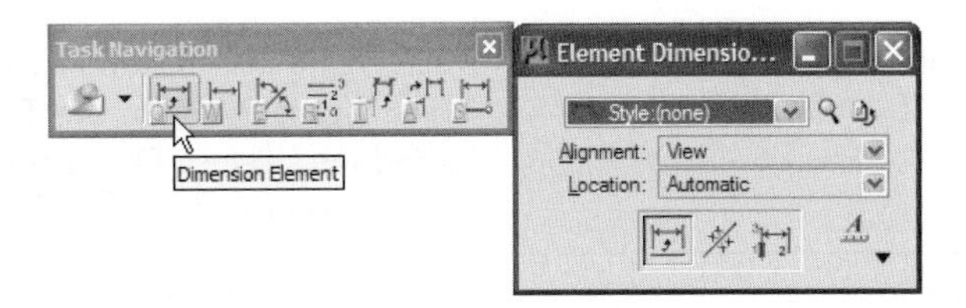

Figure 9–59 *Invoking the Dimension Element tool from the Linear Dimensions tool box*

MicroStation prompts depend on the selected element. For most dimension tools, the second data point indicates the length and direction of the extension line. See Figure 9–60 for examples of placing dimensions with the Dimension Element tool.

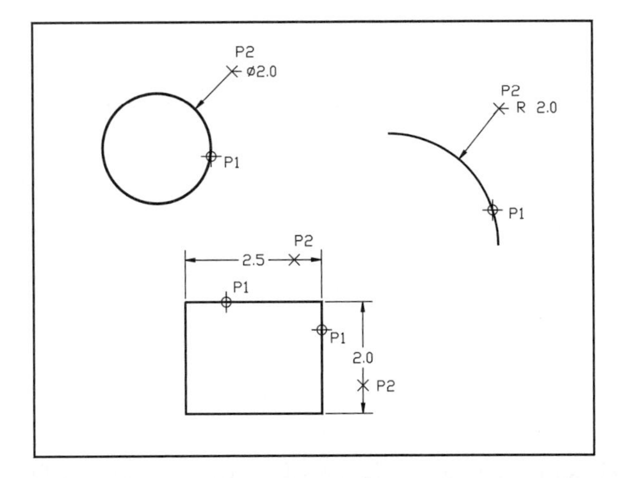

Figure 9–60 *Examples of placing dimensions with the Dimension Element tool*

MISCELLANEOUS DIMENSIONS

The Misc Dimensions tools are used to perform dimensioning that is not specific to linear, angular, or radial dimensioning. Following are the tools categorized as the Misc Dimensions tools:

- Dimension Ordinates—Labels distances along an axis from a common point of origin.
- Insert Dimension Vertex—Adds an extension line to a dimension element.
- Delete Dimension Vertex—Removes an extension line from a dimension element.
- Modify Dimension Location—Moves dimension text or modifies the extension line length of a dimension element.
- Reassociate dimension—Recreates a dimension's association to an element.

- Geometric Tolerance—Builds a feature control with geometric tolerance symbols.

Dimension Ordinates

Ordinate dimensioning is common in mechanical designs. It labels distances along an axis from a point of origin on the axis along which the distances are measured. See Figure 9–61 for an example of ordinate dimensioning.

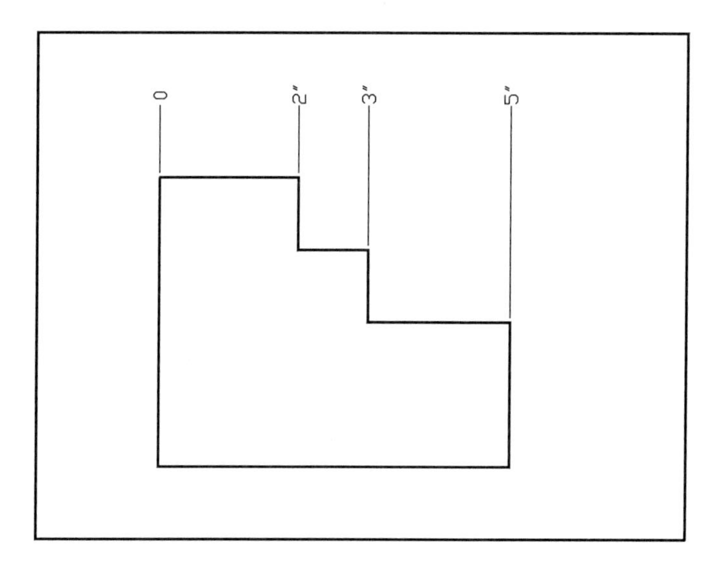

Figure 9–61 *Example of ordinate dimensioning*

To place ordinate dimensions, invoke the Dimension Ordinates tool from:

Task Navigation tool box (active task set to Dimensioning)	Select the Dimension Ordinates tool (see Figure 9-62). If necessary, set the Association check box to ON, and select an Alignment option and Dimension Style in the Tool Settings window.
Keyboard Navigation (Task Navigation tool box with active task set to Dimensioning)	**r**

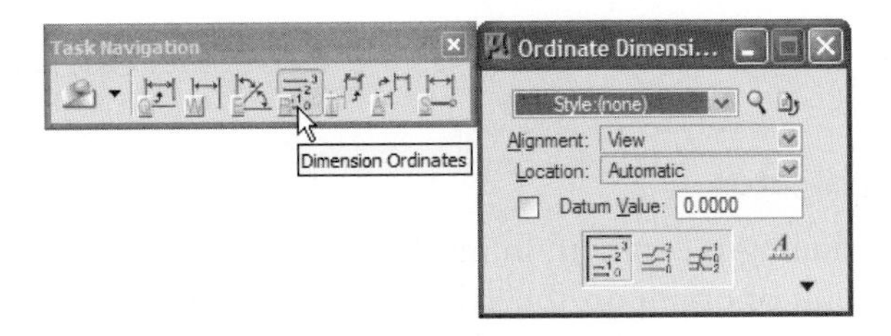

Figure 9–62 *Invoking the Dimension Ordinates tool*

If necessary, change the default base value of the origin when labeling distances along an axis in the **Datum Value** text field. Three methods are available for ordinate dimensioning which can be chosen in the Tool Settings window: The first icon sets the tool's mode to Ordinate Unstacked, which is used to label distances along an axis from an origin (datum) along the ordinate axis (the line along which distances are measured). The second icon sets the tool's mode to Ordinate Stacked, which is used to label distances along an axis from an origin (datum) along the ordinate axis (the line along which distances are measured). The third icon sets the tool's mode to Ordinate Free Location, which is used to label distances along an axis from an origin (datum) along the ordinate axis (the line along which distances are measured).

MicroStation prompts:

Dimension Ordinates > Select ordinate origin *(Place a data point from which all ordinate labels are to be measured.)*

Dimension Ordinates > Select ordinate direction *(Place a data point to indicate the rotation of the ordinate axis.)*

Dimension Ordinates > Select dimension endpoint *(Place a data point to define the end point of the dimension line. If necessary, press* (ENTER) *to edit the dimension text.)*

Dimension Ordinates > Select start of dimension *(Place data points at each place where an ordinate dimension is desired. These points define the base of the extension line. After placing all the points, click the Reset button to terminate the tool sequence.)*

Note: *To prevent the text for a higher ordinate from overlapping the text for a lower ordinate, turn ON the* ***Stack Dimensions*** *check box in the Dimension Settings dialog box—Tool Settings category.*

Insert Dimension Vertex

To add an extension line to a dimension element, invoke the Insert Dimension Vertex tool from:

Misc Dimensions tool box	Select the Insert Dimension Vertex tool (see Figure 9–63).
Key-in window	**insert dimvertex** (or **ins dimv**) (ENTER)

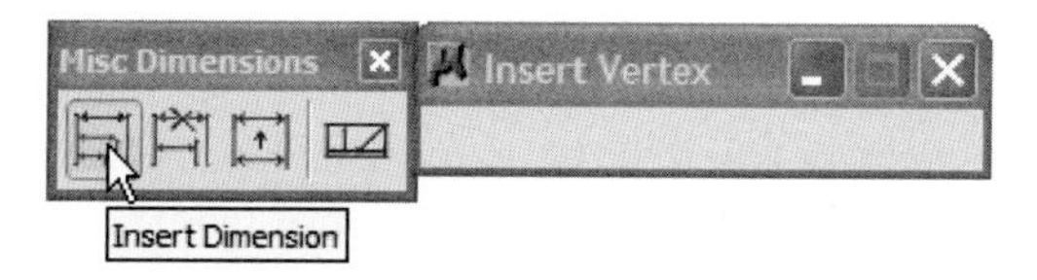

Figure 9–63 *Invoking the Insert Dimension Vertex tool from the Misc Dimensions tool box*

MicroStation prompts:

> Insert Vertex > Identify element *(Identify the dimension line near the desired extension line location.)*
>
> Insert Vertex > Accept/Reject (select next input) *(Enter a data point to position the end of the extension line as shown in Figure 9–64.)*

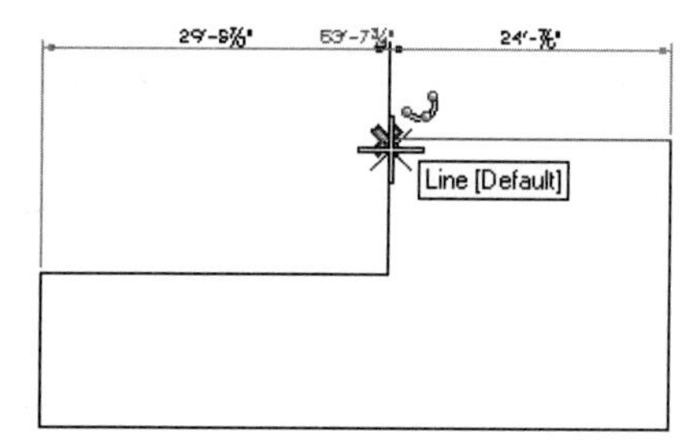

Figure 9–64 *An example of inserting a dimension vertex*

Delete Dimension Vertex

To delete an extension line from a dimension element, invoke the Delete Vertex tool from:

Misc Dimensions tool box	Select the Remove Dimension tool (see Figure 9–65).
Key-in window	**delete dimvertext** (or **del dimv**) (ENTER)

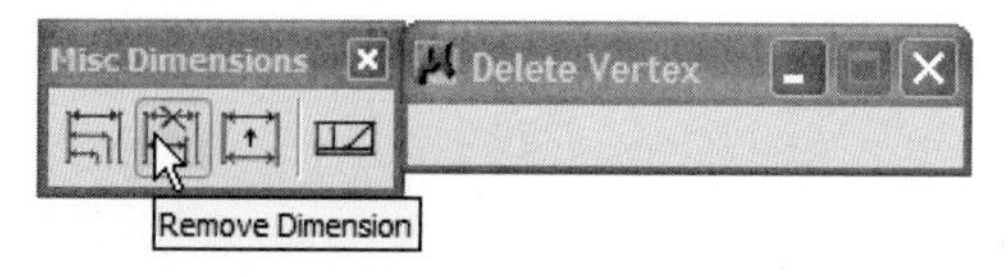

Figure 9–65 *Invoking the Delete Vertex tool from Misc Dimensions tool box*

> Delete Vertex > Identify element (*Identify the extension element.*)
>
> Delete Vertex > Accept/Reject (select next input) *(Click the accept button to accept the deletion.)*

Modify Dimension Location

To move dimension text or modify the extension line length of a dimension element, invoke the Modify Dimension Location tool from:

Misc Dimensions tool box	Select the Modify Dimension tool (see Figure 9–66).
Key-in window	**modify dimension loc** (or **modi d l**) (ENTER)

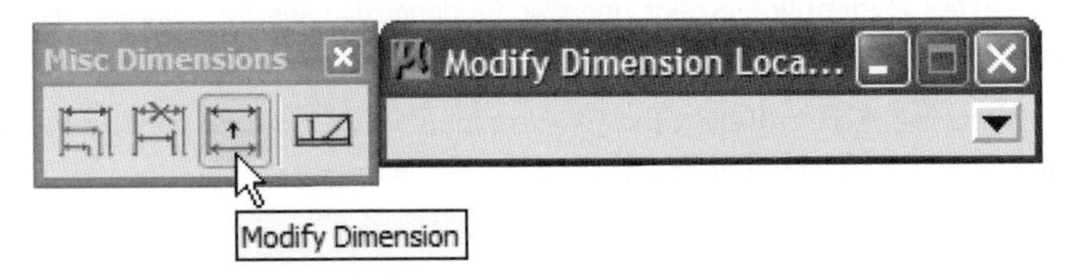

Figure 9–66 *Invoking the Modify Dimension Location tool from the Misc Dimensions tool box*

Modify element > Identify element (*identify the dimension to modify*)

Modify element > Accept/Reject (*select next input*) (*specify a data point to define the dimension's new position*).

Reassociate Dimension

The Reassociate Dimension tool is used to recreate a linear or radial dimension's association to an element. Dimensions can be reassociated to elements individually, or by using a fence and selection set. The intended elements must appear in the view window for the reassociation of their dimensions to occur. Invoke the Reassociate Dimension tool from:

Task Navigation tool box (active task set to Dimensioning)	Select the Reassociate Dimension tool (see Figure 9-67).
Keyboard Navigation (Task Navigation tool box with active task set to Dimensioning)	**s**

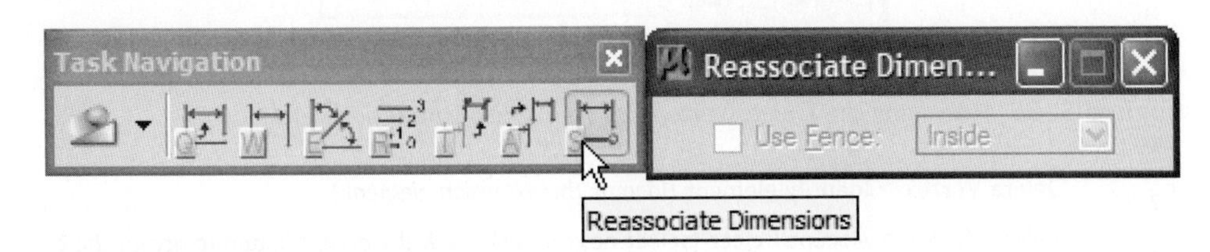

Figure 9–67 *Invoking the Reassociate Dimensions tool*

Reassociate Dimension > Select dimension (*Identify the dimension to reassociate.*)

Reassociate Dimension > Accept/Reject (*Enter the data point to accept the reassociation.*)

Geometric Tolerance

The Geometric Tolerance settings box (see Figure 9–68) helps build feature control frames with geometric tolerance symbols. The feature control frames are used with the Place Note tool and Place Text tool. The **Fonts** menu available in the settings box provides two fonts, 100—Ansi symbols, and 101—Feature Control Symbols. When a specific font is chosen, the buttons in the settings box reflect the availability of the symbols in the selected font. To add the symbols, click on the buttons as part of placing the text to the Place Note tool and Place Text tool.

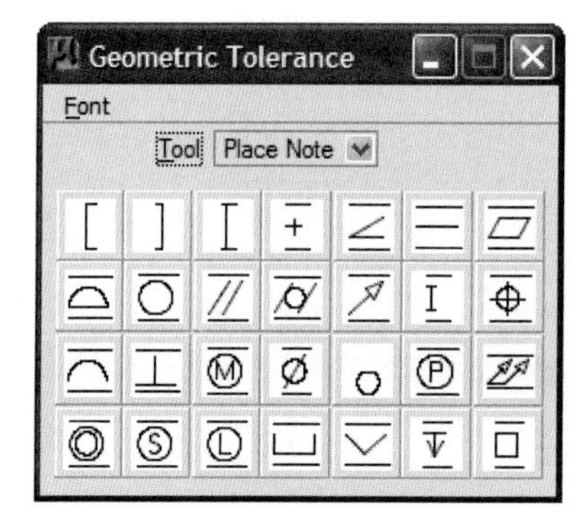

Figure 9–68 *Geometric Tolerance settings box*

The left bracket ([) and right bracket (]) form the ends of compartments; the vertical line (|) separates compartments. Invoke the Geometric Tolerance settings box from:

Misc Dimensions tool box	Select the Geometric Tolerance tool (see Figure 9–69).
Key-in window	**mdl load geomtol** (ENTER)

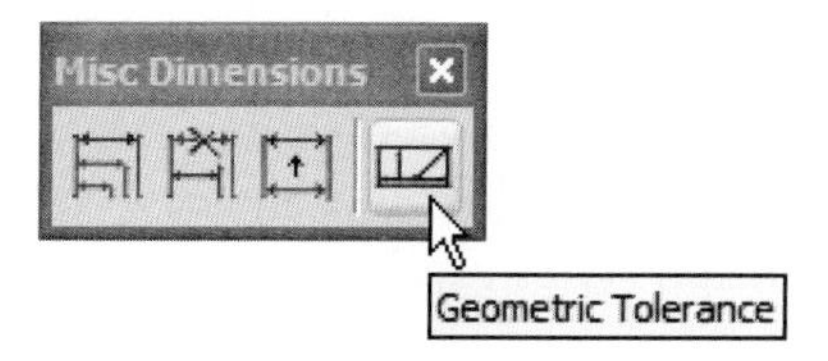

Figure 9–69 *Invoking the Geometric Tolerance tool from the Misc Dimensions tool box*

Select one of the two tools from the Geometric Tolerance settings box, **Place Note** or **Place Text**. MicroStation sets up the appropriate tool settings in the Tool Settings window. Follow the prompts to place text.

DIMENSION SETTINGS

MicroStation provides dimension settings that allow customization of dimension tools. The settings are flexible enough to accommodate users in all engineering disciplines. The entire set of dimensioning settings can be saved to a given dimension style name with in the design. Dimensions styles can be defined and applied to dimension elements during placement. Dimension styles can be created, customized and saved for easy recall, or imported from another design file to the currently working design. To change the active dimension settings, open the Dimension Styles box from:

Menu	Element > Dimensions Styles

MicroStation displays the Dimension Styles box, as shown in Figure 9–70, which serves to control the settings for dimensioning.

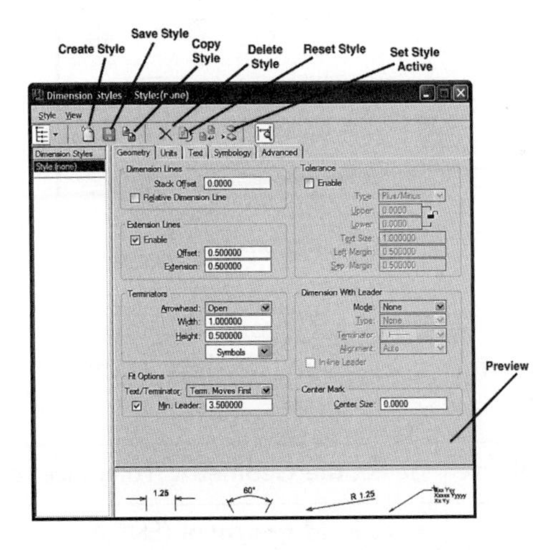

Figure 9–70 *Dimension Styles box*

The Dimension Styles dialog box is used to manage dimension styles. The options are organized under tabs, and a preview area shows style modifications. Under the Advanced tab, two dimension styles can be compared.

A dimension style is a saved set of dimensioning settings. Define dimension styles and apply them to dimension elements during placement. Dimension styles can be created, customized, and saved for easy recall. Dimension styles can be stored in either the open DGN file or a DGN library.

MicroStation lists all the available dimension lists in the Dimension Styles column of the Dimension Styles settings box. To create a new "Untitled" dimension style in the Dimension Style list, click the **Create Style** icon. If necessary, rename the

dimension style with an appropriate name. This style inherits all of the attributes of the active dimension style.

To create a copy of a selected dimension style, click the **Copy Style** icon. MicroStation adds newly created dimension style to the Dimension Style list.

To set a dimension style active, click the **Set Active Style** icon. The active dimension style name is displayed in the dialog box title bar and it is used by the Dimensioning tools.

To delete a selected dimension style, click the **Delete Style** icon.

To save the selected dimension style, and all its settings, click the **Save Style** icon. When an existing style is modified, the changes are displayed in blue until it is saved. Once saved, all dimensions created in that style are updated.

To reset the selected dimension style to previously saved version, click the **Reset Style** icon.

The Preview pane as shown in Figure 9-70 displays how dimensions will look using the active settings in the active dimension style. Right-click to display a menu that controls visibility of the Preview pane, as well as sthe type of dimensions displayed: Linear, Angular, Radial, or Note.

The Dimension Styles settings are available in four categories: Geometry, Units, Text, and Symbology.

GEOMETRY RELATED SETTINGS

The Geometry tab contains controls that affect the appearance of the dimension's geometry as shown in Figure 9-71.

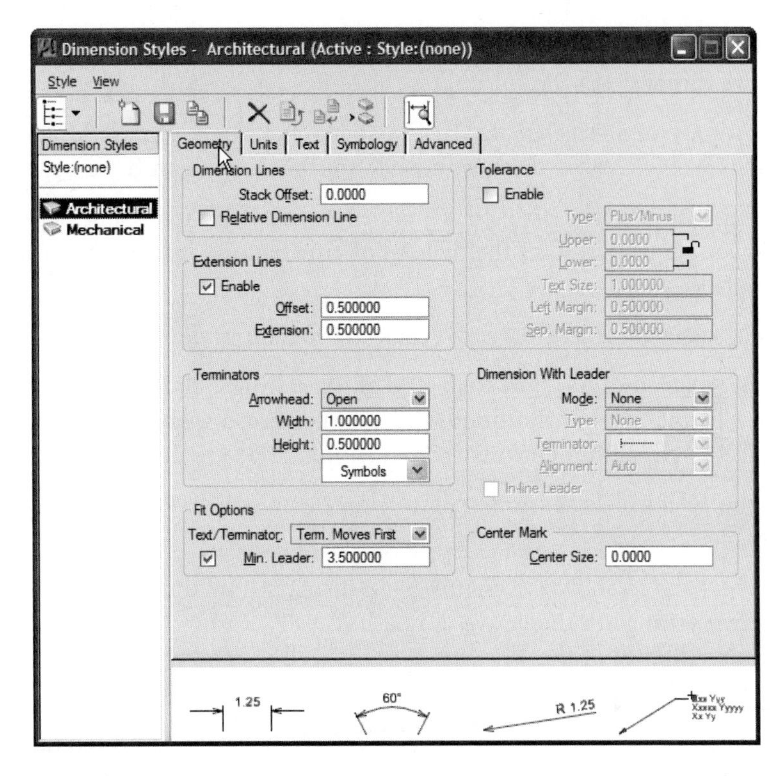

Figure 9–71 *Dimension Styles – Geometry tab selected*

Dimension Lines

The Dimension Lines section (see Figure 9-72) consists of controls that affect the general appearance of dimension lines.

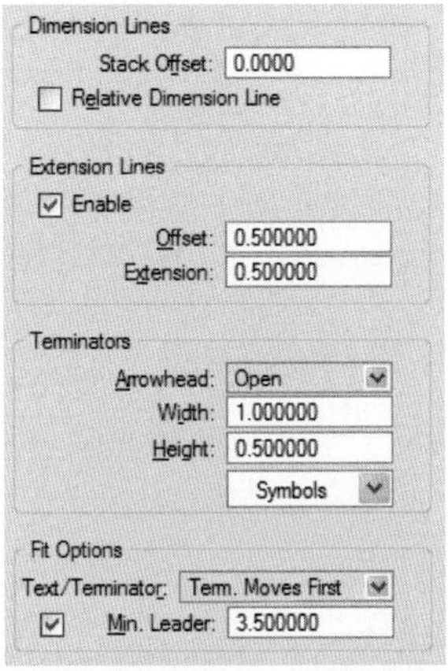

Figure 9–72 *Dimension Styles – Geometry tab – Dimension, Extension and Terminators sections*

The **Stack Offset** in the Dimension Lines section sets the space, in working units, between dimension lines if dimensions are stacked. If set to 0.0 (the default), It Stack Offset is set to a value other than 0 only if constant spacing between dimension lines is needed.

- If the Stack Offset is set to 0.0 (the default), a reasonable value based on text size and orientation is computed for space between the dimensions for stacked dimensioning.
- If the Stack Offset is set to a value other than 0 then the space between the dimensions for stacked dimensioning will be based on the specified value.

The **Relative Dimension Line** check box comes into play when the dimension associated with the element is modified. If the check box is set to:

- ON, the dimension will be moved as necessary to keep the extension line the same length.
- OFF, the dimension stays in the same position and the length of the extension line is varied as necessary to maintain the dimension's relationship to the element.

Extension Lines

The Extension Lines section (see Figure 9-72) consists of controls that affect the placement of extension lines (also called witness lines or projection lines).

- The **Enable** setting controls the display of the extension lines that are drawing the dimension.
- If the **Enable** check box is set to ON, the extension lines are drawn with the dimension. If the check box is set to OFF, the dimension is placed in the same way, but the extension lines are not drawn.

The **Offset** value sets the gap between the object being dimensioned and the start of the extension line. The gap is calculated by multiplying the text height by the value in this field. For example, if the value is 1.0, the offset is equal to the text height.

The **Extension** value sets the distance the extension line extends beyond the dimension line end. The extension is calculated by multiplying the text height by the value in this field. For example, if the value is 0.5, the offset is equal to half the text height.

Terminators

The Terminators section (see Figure 9-72) consists of controls that set the appearance and geometry of the default dimension terminators.

The **Arrowhead** options menu controls the style of arrowhead used for the dimension terminators:

- **Open**—The arrowheads look like angle brackets.
- **Closed**—The arrowheads look like triangles.
- **Filled**—The arrowheads look like triangles and are filled with the outline color.

Figure 9–73 shows examples of each terminator placement position and each style of arrowhead.

Note: *To make the filled arrowheads actually appear filled, turn on the Fill View Attribute for the current view window. The View Attributes settings box is opened from the **Settings** menu.*

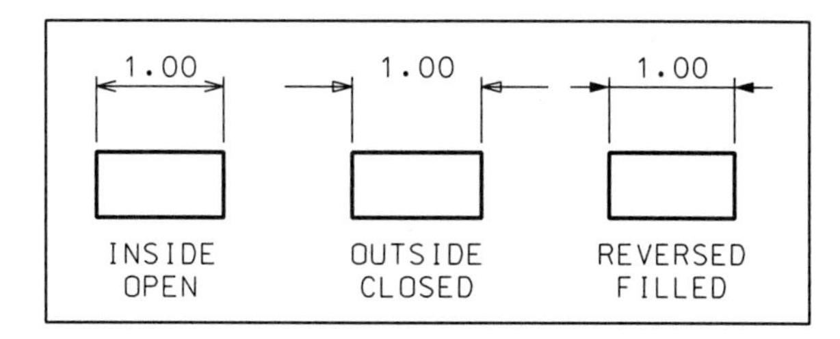

Figure 9–73 *Examples of terminator placement positions and styles*

Two Geometry text fields in the Terminators section control the size of arrowhead terminators. Each size is the product of the dimension text height multiplied by the number entered in the text field. For example, if the text size is one inch and 2.0 is entered in the width field, the arrowhead width will be two inches.

- **Width**—Sets the width of the arrowhead from tip to back (parallel to the dimension line).
- **Height**—Sets the height of the arrow head at its widest part (at right angles to the dimension line).

The **Symbols** option menu consists of controls that are used to specify alternate symbols (characters from symbol fonts or cells) for each of the default dimension terminators.

- The **Arrow** selection specifies an alternate to the default Arrowhead.
- The **Stroke** selection specifies an alternate to MicroStation's default stroke terminator symbol.
- The **Origin** selection specifies an alternate to MicroStation's default origin terminator symbol.
- The **Dot** selection specifies an alternate to MicroStation's default dot terminator symbol.

- The **Note** selection specifies an alternate to MicroStation's default Place Note terminator symbol.

Fit Options

The Fit Options section consists of controls that affect the general appearance or placement of text and terminators.

The Text/Terminator selection selects the minimum fit dependant on the text, terminator or combination.

The Min. Leader - Sets the space, in text width units, between extension lines and dimension text.

Tolerance

The Tolerance section (see Figure 9-74) consists of controls that affect the generation of tolerance dimensions.

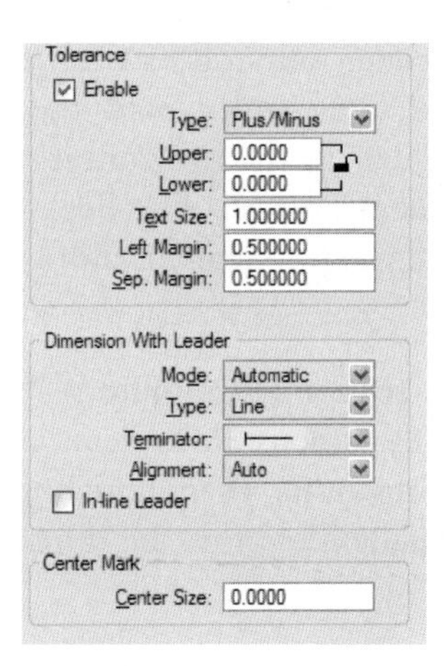

Figure 9–74 *Dimension Styles – Geometry tab – Tolerance, Dimension with Leader and Center Mark sections*

Tolerance values are added to the dimensions when the **Enable** check box is set to ON.

The Type option menu sets the format for the tolerance:

- The **Plus/Minus** selection sets the dimension with the upper and lower limits are expressed as positive and negative limits.
- The **Limit** selection sets the dimension expressed as the upper and lower limits.

Figure 9-75 shows examples of each tolerance type with the upper and lower tolerance values equal to 0.002.

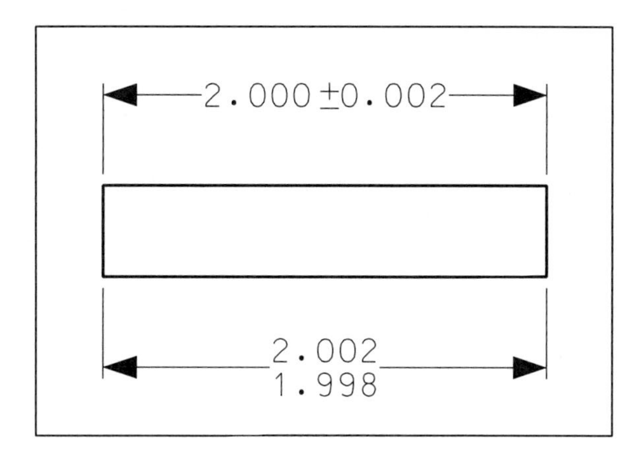

Figure 9–75 *Examples of Plus/Minus and Limit tolerance attributes*

The **Upper** and **Lower** values sets the upper tolerance limit, in working units and lower tolerance limit, in working units respectively. The **Text Size** value sets the tolerance text size, specified as a multiple of the dimension text Height and Width. The **Left Margin** value sets the horizontal space, in text height units, between tolerance text and dimension text. The **Sep. Margin** value sets the vertical space, in text height units, between tolerance values.

Dimension with Leader

The Dimension with Leader section (see Figure 9-74) consists of controls that affect the general appearance of dimension leaders. Location of a dimension value can be modified to something other than horizontal along the dimension line. How the dimension text visually identifies the dimension with which it is associated can also be modified.

The **Mode** option menu controls the display of the dimensions with leader. If it is set to None, then the options are disabled. If it is set to On, then MicroStation displays a leader with dimensions. And if it is set to Automatic, then MicroStation automatically places a leader with a line dependent on the Fit Option.

The **Type** option menu is used to select type of leader line to display: None, Line, Arc or Bspline.

The **Terminator** option menu is used to select the terminator type: None, Arrow, Slash, Empty ball or Filled ball.

The **Alignment** option menu is used to select the alignment type: Auto, Left or Right for the placement of the leader.

The **In-Leader** check box, when set to ON, adds a horizontal line to the leader.

Center Mark

The **Center Size** (see Figure 9-74) value sets the size of the center mark in radial dimensions. If set to 0, the center mark size is the active text height. If set to a number other than 0, the center mark size uses the working units.

UNITS RELATED SETTINGS

The Units tab contains controls that affect the display format for units in dimension text as shown in Figure 9-76.

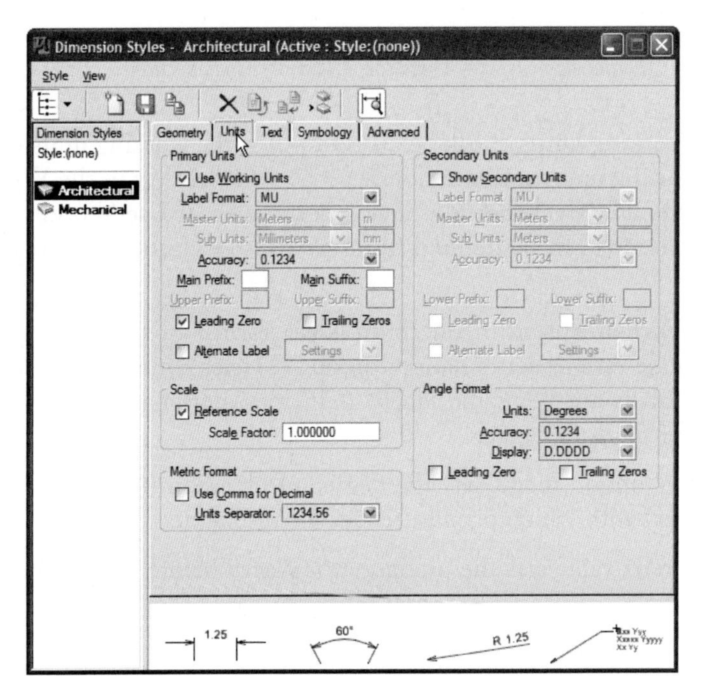

Figure 9–76 *Dimension Styles – Units tab selected*

Primary Units

The Primary Units section (see Figure 9-77) consists of controls that adjust the display of primary linear dimension text. In dual dimensions (for example, English/metric), the primary units are the upper units and the secondary units are the lower units.

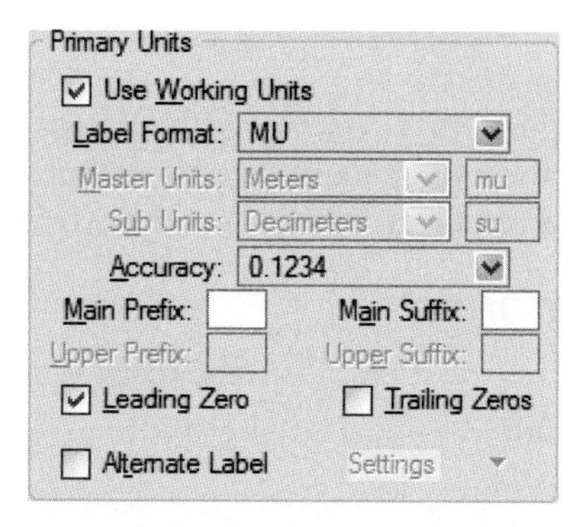

Figure 9–77 *Dimension Styles – Units tab – Primary Units section*

If the **Use Working Units** check box is set to ON, the working units of the design file is used for the units of dimensioning. The same label as in the Working Units category of the DGN File Settings dialog box is used. If **Use Working Units** is set to OFF, then units can be defined from the **Master Units** and **Sub Units** option menus.

The **Label Format** option menu controls the format of the primary dimension. Options include various combinations of MU (Master Units), SU (Sub Units), and labels. The labels are the same labels as in the Working Units category of the DGN File Settings dialog box.

The **Accuracy** option menu sets the degree of precision (for instance, number of decimal places) with which primary values are displayed.

The **Main Prefix** value sets the prefix that is placed before a single line of primary dimension text.

The **Main Suffix** value sets the suffix that is placed after a single line of primary dimension text.

The **Upper Prefix** value (available when **Show Secondary Units** is set to ON), sets the prefix that is placed before the primary (upper) dimension text.

The **Upper Suffix value** (available when **Show Secondary Units** is set to ON), sets the suffix that is placed after the primary (upper) dimension text.

The **Leading Zero** check when set to ON, adds a leading zero to the primary dimension text for a value of less than 1.0.

The **Trailing Zeros** check box when set to ON, displays values filled with zeros, if necessary, as specified by the Accuracy.

Figure 9–78 shows an example of a dimension that is set to three decimal places of accuracy, with the zero display check boxes ON and OFF.

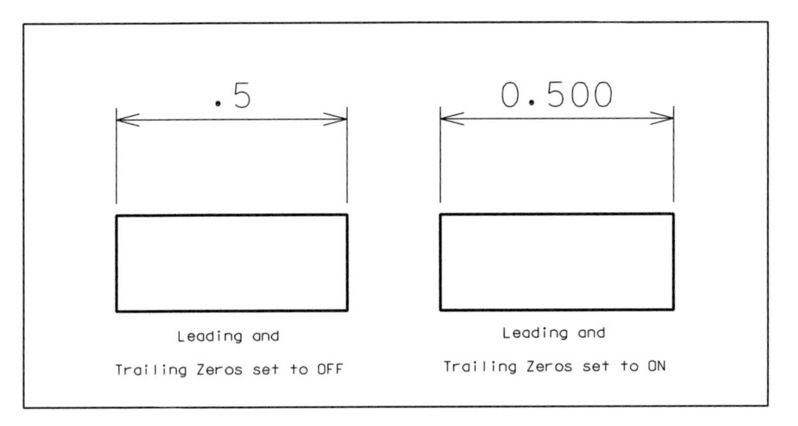

Figure 9–78 *Dimension example with leading and trailing zero display ON and OFF*

The **Alternate Label** check box is used to set up alternate dimensions based on the criteria that are defined from the **Settings** menu. For example, if master units are set to feet and dimensions of less than one foot are to display as inches. To do so, set **Alternate Label** check box to ON and from the Settings menu set the distance less than (<) to 1, and select MU.

Secondary Units

The Secondary Units section (see Figure 9-79) consists of controls that adjust the display of secondary linear dimension text. In dual dimensions (for example, English/metric), the primary units are the upper units and the secondary units are the lower units. The **Show Secondary Units** check box when set to ON, then secondary dimension text is displayed. The controls available are similar to the controls for Primary Units.

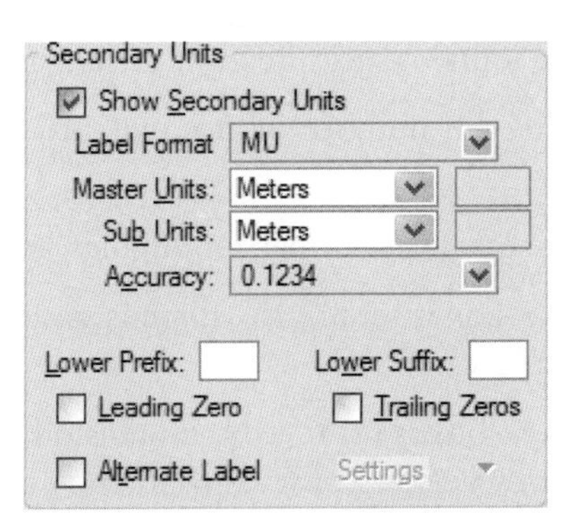

Figure 9–79 *Dimension Styles – Units tab – Secondary Units section*

Scale

The Scale section (see Figure 9-80) consists of controls for dimension scale.

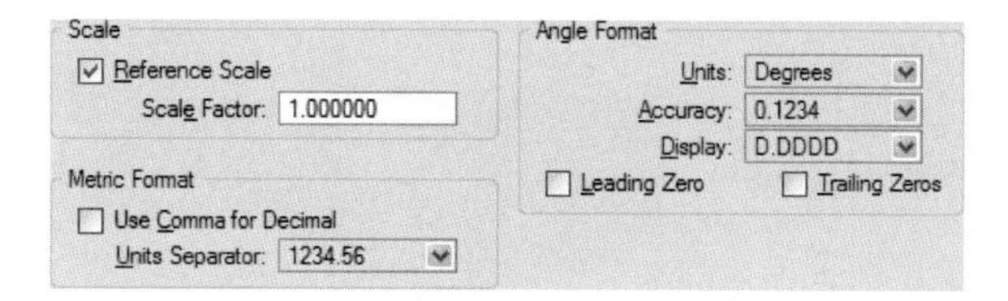

Figure 9–80 *Dimension Styles – Units tab – Scale, Metric and Angle sections*

The **Reference Scale** check box when set to ON, dimensions display the true size of referenced geometry. This option applies only when dimensioning elements in a scaled reference.

The **Scale Factor** value sets the scale factor that is applied to the dimension value. For instance, if the value is set to 2.0, then all the dimensions will be multiplied by a factor of 2.0.

Metric Format

The Metric Format section (see Figure 9-80) consists of controls that adjust the display of dimension values.

The **Use Comma for Decimal** check box when set to ON, the decimal point in dimension text displayed in the metric format are replaced with commas to conform to European standards.

The **Units Separator** menu sets the format for the separator used after the thousandths and millionths places.

Angle Format

The Angle Format section (see Figure 9-80) contains controls that adjust the display of angular dimension text.

The **Units** menu sets the units of measurement for angular dimensions.

The **Accuracy** menu sets the number of decimal places to display for angular dimensions.

The **Display** menu sets the display format for angular dimension text.

If the **Leading Zero** check box is set to ON, dimension text for a dimension of less than 1.0 is preceded by a leading zero.

If the **Trailing Zeros** check box is set to ON, dimension text is filled with zeros, if necessary, to the number of decimal places specified by the Accuracy.

TEXT RELATED SETTINGS

The Text tab contains controls that affect the placement and appearance of dimension text as shown in Figure 9-81.

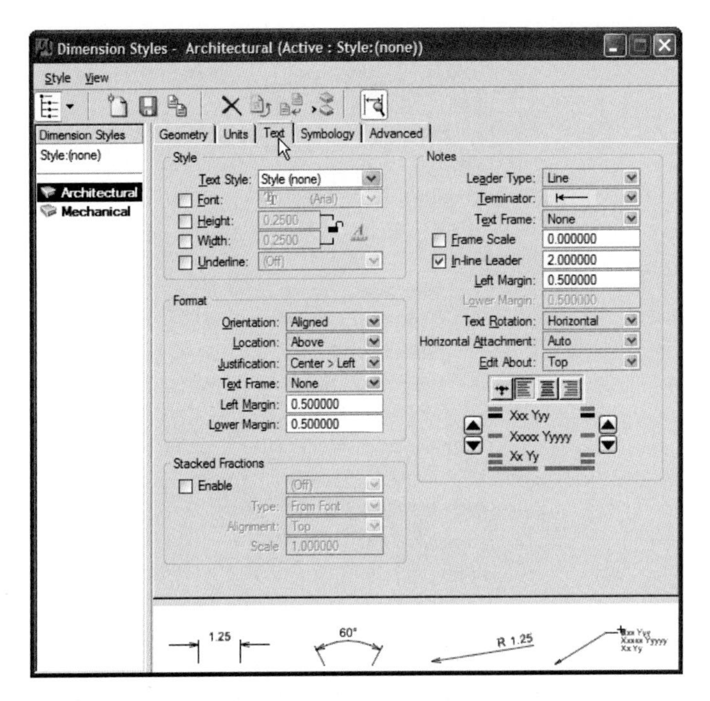

Figure 9–81 *Dimension Styles – Text tab selected*

Style

The Style section (see Figure 9-82) consists of controls that can be used to set active Text Style, Font, Height, Width, and Underline.

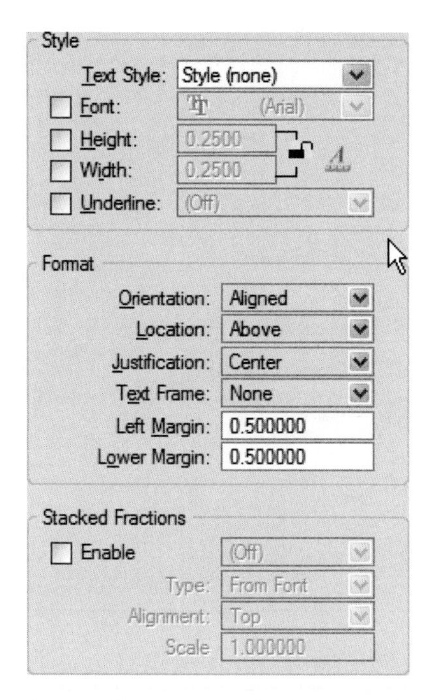

Figure 9–82 *Dimension Styles – Text tab – Style, Format and Stacked Fractions section*

The **Text Style** menu sets the text style that controls the presentation of dimension text. Dimensions placed using a text style are automatically updated when the text style is modified. Using a text style enables use of advanced text style properties that are not otherwise available for dimensions, such as italics and background color. Some text attributes can be directly controlled by the dimension style. For example, if the text height and width are set on the dimension style, these values are used instead of the height and width of the text style.

The **Font, Height, Width, and Underline** check boxes when set to ON, allows setting appropriate values. If the settings are set to OFF, then the settings will be set as in the active text style.

Format

The Format section (see Figure 9-82) controls the placement of dimension text.

The **Orientation** menu sets the orientation of dimension text relative to the dimension line. Aligned selection aligns the text with the dimension line and Horizontal selection displays the text horizontally, regardless of the orientation of the dimension line.

The **Location** menu sets the location of the text relative to the dimension line: Inline selection aligns the text with the dimension line; Above selection places the text above the dimension line; Outside selection places the text to the opposite side

of the extension origin; and Top Left selection places the text to the top left of the origin of the extension line.

The **Justification** menu sets the justification of dimension text and available options include: Left, Center, or Right justified.

The **Test Frame** menu sets the framing of dimension text: None selection places with no frame around the text; Box selection places dimension in a box; and Capsule selection places dimension in a capsule. Figure 9-93 shows examples of placing linear dimensions with different Text frame modes.

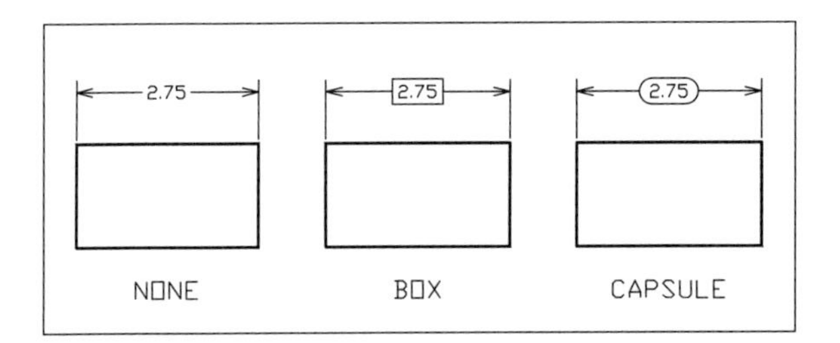

Figure 9–83 *Examples of placing linear dimensioning in different Text Frame modes*

The **Left Margin** text field sets the space, in text height units, between the leader line and the dimension text.

The **Lower Margin** text field sets the space, in text height units, between the dimension line and the bottom of the dimension text.

Stacked Fractions

The Stacked Fractions section (see Figure 9-82) controls the format for stacked fractions.

To change the format for the stacked fractions, set the **Enable** check box to ON and select On from the menu to modify the stacked fraction settings. By default, the stacked fractions settings are controlled by the text settings.

Notes

The Notes section (see Figure 9-84) contains controls that are used to set the appearance of the leader and text used by the Place Note tool. The controls available include Leader Type, Terminator type, Text Frame type, Frame Scale, placement of In-Line Leader, Left Margin setting, Lower Margin setting, Text Rotation, Horizontal Rotation of the text, and position of the vertical origin to edit text about.

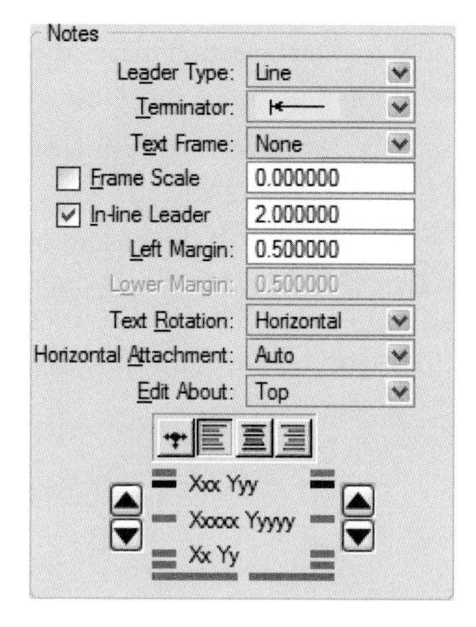

Figure 9–84 *Dimension Styles – Text tab – Notes section*

SYMBOLOGY RELATED SETTINGS

The Symbology tab contains controls for dimension lines, extension lines, text, and terminator symbology as shown in Figure 9-85.

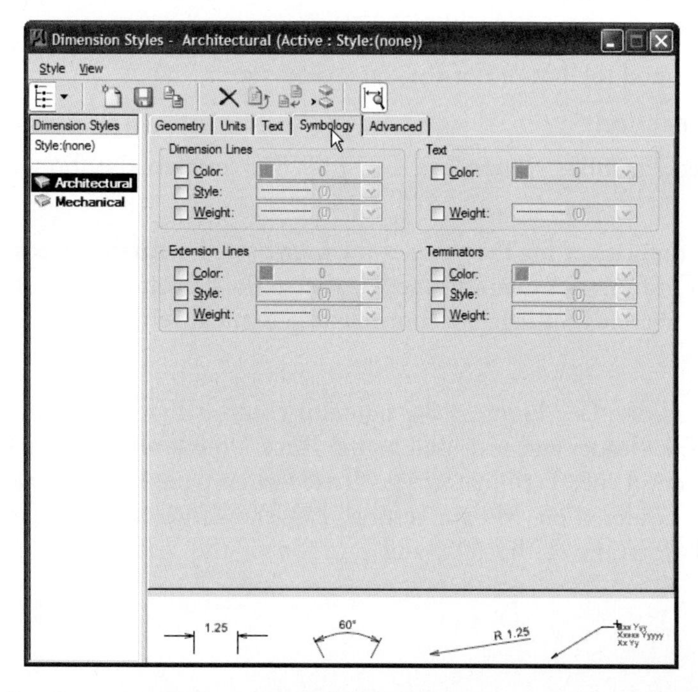

Figure 9–85 *Dimension Styles – Symbology tab selected*

Dimensions Lines

The Dimension Lines section contains controls for symbology (Color, Style, and Weight) settings for dimension lines. Set the appropriate control to ON and make necessary changes to the symbology for dimension lines. If the controls are set to OFF, then the symbology for the dimension lines is set to the active symbology.

Text

The Text section contains controls for symbology (Color and Weight) settings for dimension text. Set the appropriate control to ON and make necessary changes to Color and/or Weight for dimension text. If the controls are set to OFF, then the symbology for dimension text is set to the active symbology.

Terminators

The Terminators section contains controls for symbology (Color, Style, and Weight) settings for dimension terminators. Set the appropriate control to ON and make necessary changes to the symbology for dimension terminators. If the controls are set to OFF, then the symbology for the terminators is set to the active symbology.

ADVANCED SETTINGS

The Advanced tab lists all the available settings for the selected dimension style when the **Edit** mode is selected as shown in Figure 9-86.

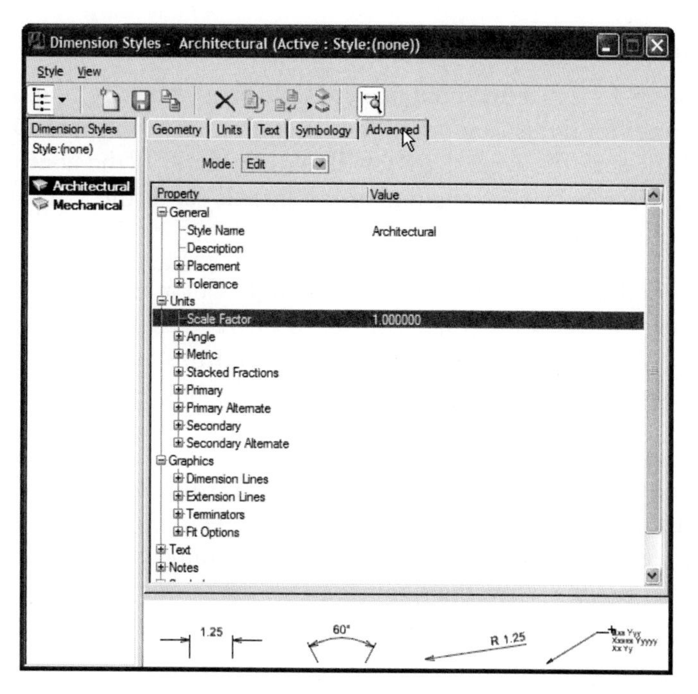

Figure 9–86 *Dimension Styles – Advanced tab selected*

To modify a specific setting for the selected dimenion style, click in the Value column and make the necessary changes.

The **Comparison** mode selection compares the values for dimension styles as shown in Figure 9-87. Turn on **Compare with Library** to compare a dimension style and a dimension style library. All dimension style properties are listed. The properties with different values are displayed in bold, and the values are listed.

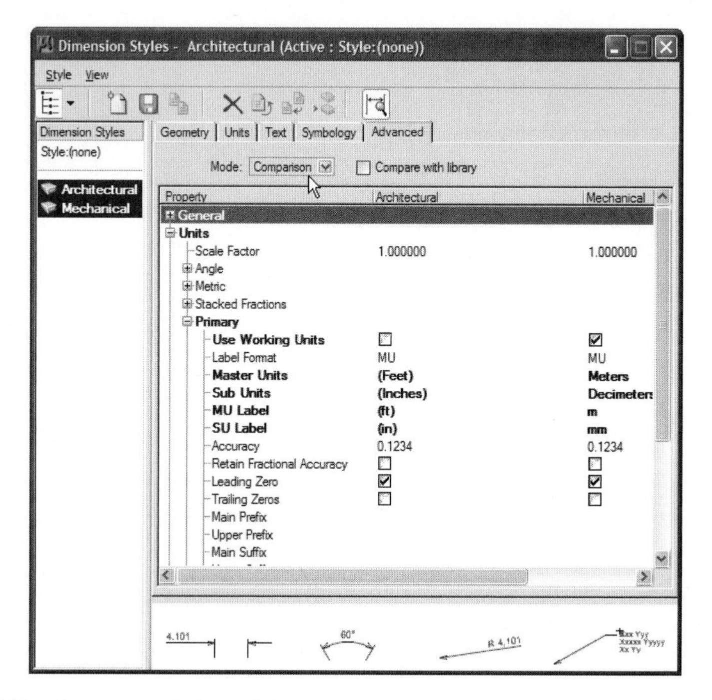

Figure 9–87 *Dimension Styles – Advanced tab with comparison mode selected*

The Differences mode selection compares the values for two dimension styles as shown in Figure 9-88. Turn on **Compare with Library** to compare a dimension style and a dimension style library. This option displays only the properties with different values.

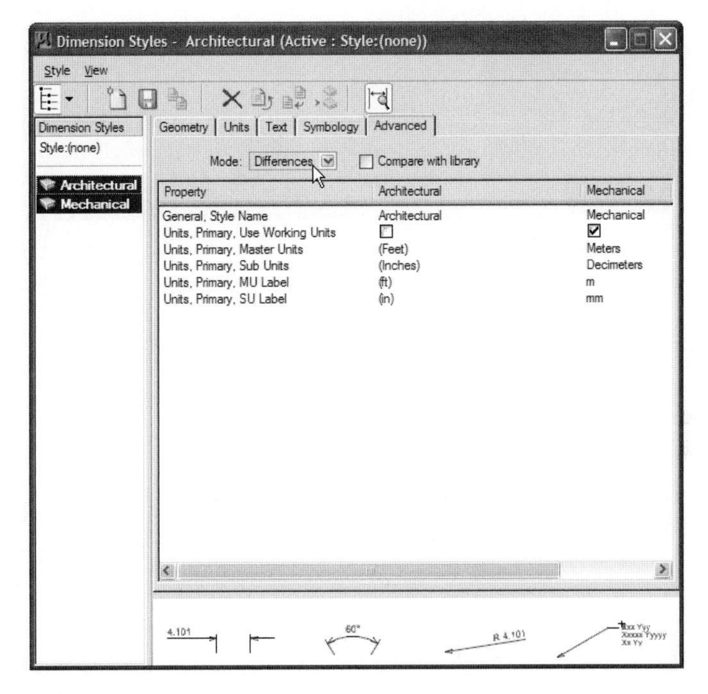

Figure 9–88 *Dimension Styles – Advanced tab with Differences mode selected*

MATCH DIMENSION SETTINGS

The steps required to set up dimensioning involve several settings from various categories and are time consuming. After setting everything up, it is all too easy to forget to save it as part of the current design file by using the Save Settings option. If dimension settings are lost, but there are dimension elements with appropriate settings in the design, the Match Dimension tool can set the current settings by matching them to the settings in effect when the dimensions were placed. Invoke the Match Dimension tool from:

Task Navigation tool box (active task set to Dimensioning)	Select the Match Dimension Attributes tool (see Figure 9–89).
Keyboard Navigation (Task Navigation tool box with active task set to Dimensioning)	**a**

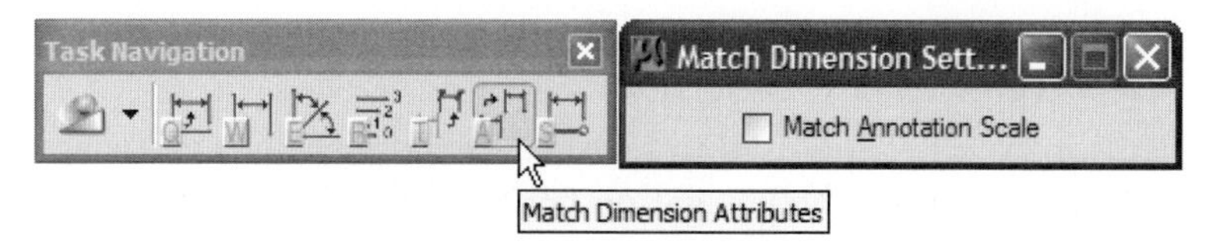

Figure 9–89 Invoking the Match Dimension Attributes tool from the Task Navigation tool box (active task set to Dimensioning)

MicroStation prompts:

Match Dimension Settings > Identify element *(Identify the dimension element whose dimension settings will be matched.)*

Match Dimension Settings > Accept/Reject (Select next input) *(Click the Data button to set the selected dimension element settings as the current dimension settings, or click the Reject button to reject the settings.)*

CHANGE DIMENSIONS

The Change Dimension tool is used to change a dimension element to the active dimension attributes, which can be set in the Dimension Style settings box or selected one of the available dimension styles. Invoke the Change Dimension tool from:

Task Navigation tool box (active task set to Dimensioning)	Select the Change Dimension tool (see Figure 9–90) and select one of the available dimension style to apply from the tool settings window.
Keyboard Navigation (Task Navigation tool box with active task set to Dimensioning)	**t** (Select one of the available dimension style to apply from the tool settings window.)

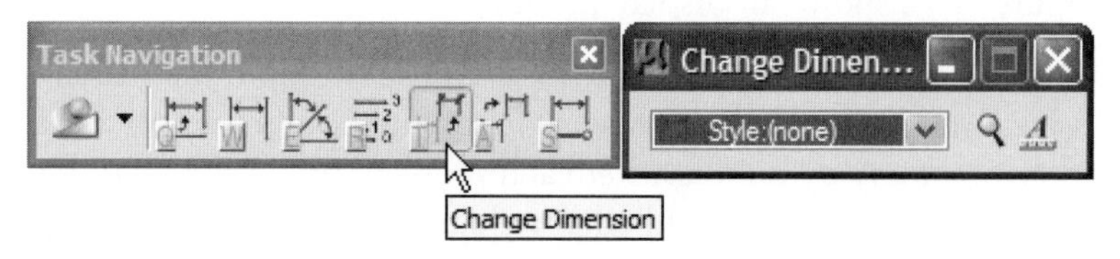

Figure 9–90 *Invoking the Change Dimension tool from the Dimension tool box*

MicroStation prompts:

Change Dimension > Identify element *(Identify the dimension element to set it to the active dimension attributes.)*

Change Dimension > Accept/Reject (Select next input) *(Click the Data button to update the dimension attributes to the selected element, or click the Reject button to disregard the selection of the element.)*

DIMENSION AUDIT

The Dimension Audit tool is used to search all the dimensions in the active model and report any problems. The navigation controls highlight each problem dimension with a red ellipse and zoom in on the area in the active view. If the problem is corrected, the red ellipse changes to green. Invoke the Dimension Audit tool from:

Utilities menu	Select the Dimension Audit tool

MicroStation displays the Dimension Audit tool box with various tool with navigation controls (see Figure 9-91).

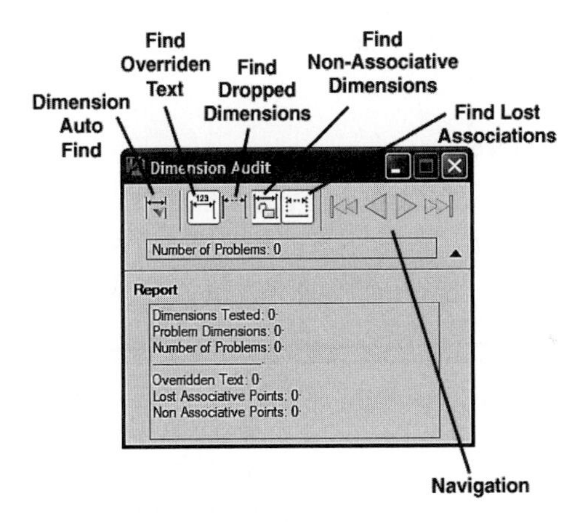

Figure 9–91 *Dimension Audit tool box*

The Hide/Show button is used to display or hide a detailed report of the current search (Figure 9-91 shown in display mode).

Choose **Dimension Audit Find** (first icon from left) to begin the search for the dimensions in an area of the active view. MicroStation lists number of dimensions tested and errors if any by highlighting each problem dimension with a red ellipse. If the problem is corrected, the red ellipse changes to green.

Choose **Find Overridden Text** (second icon from left) to display dimensions whose text has been edited.

Choose **Find Dropped Dimensions** (third icon from left) to display dimensions that have been dropped to elements.

Choose **Find Non-Associative Dimensions** (fourth icon from left) to display dimensions that were placed with the association lock off or were not snapped to an element correctly.

Choose **Find Lost Associations** (fifth icon from left) to display dimensions that have failed associations which are displayed with a heavy dashed line.

Open the Exercise Manual PDF file for Chapter 9 on the accompanying CD for project and discipline specific exercises.

REVIEW QUESTIONS

Write your answers in the spaces provided.

1. The dimension line is a
   ____.
2. The extension lines are
   ____.
3. The leader line is a ____________________.
4. Associative dimensioning links
   ____.
5. To place associative dimensions, the Association Lock must be
   ____________________ .
6. What are the three options available for orientation of dimension text relative to the dimension
   line?____
7. The justification field setting for dimension text does not apply when the __________ placement location mode is selected.
8. Name the two options that are available with dimension text length format.____
9. What are the three types of linear dimensioning available in MicroStation?____
10. The linear dimensioning tools are provided in the __________ tool box.
11. Name the two types of dimensioning included in circular dimensioning.____
12. The Dimension Angle Size tool is used to
    ____ .
13. The Place Center Mark tool serves to
    ____ .
14. List the four options available with the Measure Distance tool.____
15. The Measure Radius tool provides information on such elements as
    ____
16. List the seven area options available with the Measure Area tool.____

17. The Measure Area Flood option measures the area enclosed by ____________________________.

18. The Geometric Tolerance settings box is used to build _________________________________.

19. Ordinate dimensions are used to label ____.

20. The Label Line tool places the ____________________________ and ____________________________.

CHAPTER 10

Printing

One task has not changed much in the transition from board drafting to CAD, and that is obtaining a hard copy. The term *hard copy* describes a tangible reproduction of a screen image. The hard copy is usually a reproducible medium from which prints are made, and it can take many forms, including slides, videotape, prints, and plots. This chapter describes the most common process used to create a hard copy: printing.

OBJECTIVES

Topics explored in this chapter:

- How the printing process works
- What components are involved in the process
- How to create a plot file
- How to create a hard copy
- How to use the Batch Print utility

OVERVIEW OF THE PRINTING PROCESS

In manual drafting, if output is needed at two different scales, two physical drawings must be created. In CAD, on the other hand, with minor modifications one design can be printed at many different scales on different-sized paper.

To print a design with MicroStation :

1. Set up the view to be printed or place a fence around the part of the design to be printed.
2. Use the Print dialog box to set the necessary print settings.
3. If the selected printer is the Windows system printer, the printed output is sent to the printer. If the selected printer is not the Window system printer, as when using the Bentley driver, the Save Print As dialog box opens, so the print file can be saved to disk for later submission to the printer.

The plot file describes all the elements in the print area in a language the printing device can understand, and provides commands to control the printing device. It is separate from the design file, and contains the design as it existed when the plot

file was created. If changes are made to the design after creating the plot file and a new printout is needed, a new plot file must be created.

MicroStation stores plot files in the directory path contained in the MS_PLTFILES configuration variable. By default, that path is:

> *<disk>:\Documents and Settings\All Users\Application Data\Bentley\Workspace\projects\untitled\out\.*

Replace <disk> with the letter of the disk that contains the MicroStation program. (See Chapter 16 for a detailed explanation of configuration variables.)

Printing devices print the information contained in the plot file on the hard copy page. MicroStation supports many types and models of printing devices. There are electrostatic plotters that provide only shades of gray and more expensive models that print in color. Pen plotters use ink pens contained in a movable rack. A mechanical control mechanism selects pens and moves them across the page under program control.

MicroStation provides a plotter driver file for each supported plotting device. The information contained in this file (combined with user-defined information from the Print settings box) tells MicroStation how to create the plot file and send it to the plotting device.

The plotter driver file specifies the following:

- Printer model
- Number of pens the printer can use
- Resolution and units of distance on the printer
- Pen change criteria
- Name, size, offset, and number for all paper sizes
- Stroking tolerance for arcs and circles
- Border around the print and information about the border comment
- Pen speeds, accelerations, and force, where applicable
- Pen-to-element color or weight mapping
- Spacing between multiple strokes on a weighted line
- Number of strokes generated for each line weight
- Definitions for user-defined line styles (for printing only)
- Method by which prints are generated
- Actions to be taken at print's start and end and on pen changes

Plotter driver files can be edited with any text editor. For more information on the contents of these files, and how to change them, consult the *MicroStation User Guide*.

Note: *If a sample plotter driver file needs to be changed, it is a good idea to retain the original file and to save the modified file as a new file with a different name.*

PRINTING FROM MICROSTATION

All printing functions can be performed from the Print dialog box. Use the options in this dialog box to select a printer and adjust various settings that affect printing. Additionally, the printed output can be previewed. To Open the Print dialog box from:

Menu	File > Print
Key-in window	**print** (or **pri**) (ENTER)

MicroStation displays the Print dialog box, as shown in Figure 10–1.

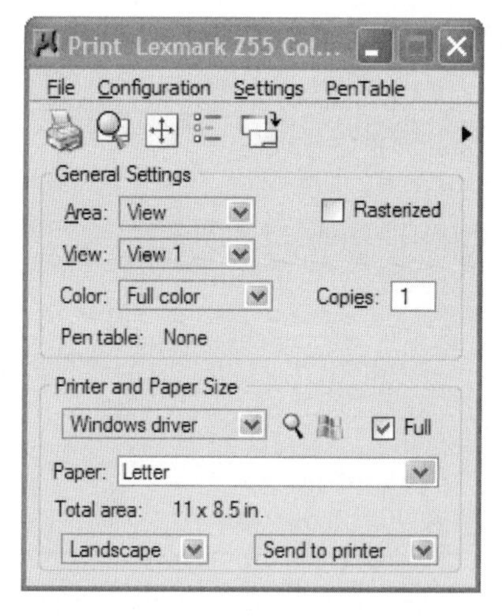

Figure 10–1 Print dialog box

All options for adjusting printing settings are contained in the menu bar at the top of this dialog box and via the icon bar directly below it. The currently selected printer driver displays in the title bar of the dialog box. By default, the printed output is maximized. It will be printed to the largest scale that will fit on the selected paper size. Click the **Show Preview** arrow to the right of the icon bar to display the print preview window.

The Print dialog box expands to display the print preview window as shown in Figure 10–2. The blue rectangle represents the size of the printed output on the selected sheet. To see the part of the drawing that is to be printed, set the **Show design in preview** check box to ON.

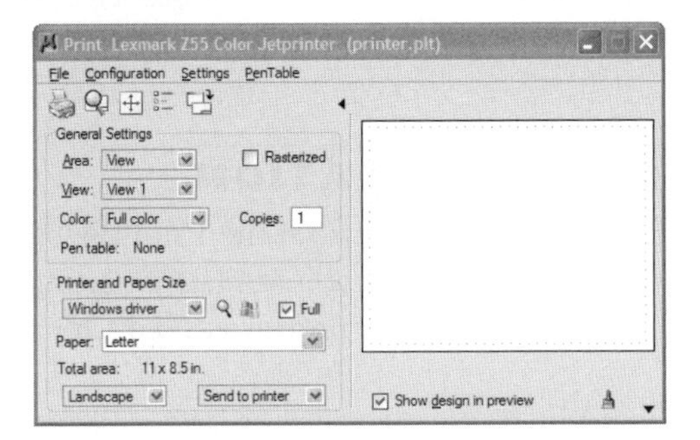

Figure 10–2 The Print dialog box expanded to show the preview window

In the expanded dialog box, click the **Show Details** arrow at bottom right. The Print dialog box expands to display further printing parameters as shown in Figure 10–3.

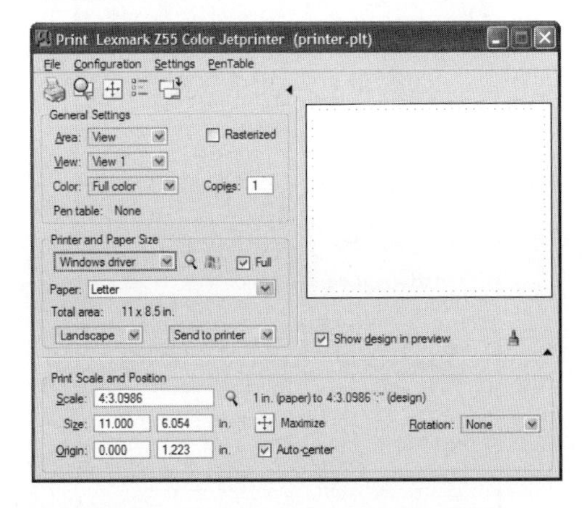

Figure 10–3 The Print dialog box fully expanded to show all settings

SELECTING THE AREA OF THE DESIGN TO PRINT

By default, the initial print area in the Print dialog box is determined as follows:

- If the active model contains a sheet definition, the print area is obtained from the sheet definition.
- If no sheet definition exists, and a fence exists, the fence defines the print area.

- If no sheet definition or fence exists, the print area is set to the first open view.

If there is no sheet or fence print area defined in the current design, the current view is set to the print area. Any open view can be selected from the **View** menu in the **General Settings** section to use as the print area. Alternately, Fit Master or Fit All can be selected from the **Area** option menu in the **General Settings** section of the Print dialog box to encompass all the elements in either the master file or the master file and all its references respectively.

SETTING THE VECTOR OUTPUT COLOR

MicroStation can print vectors in grayscale or monochrome, using the **Vector** output setting in the Print dialog box. Often, it is advantageous to display the vector information in grayscale or monochrome, rather than the element colors. The **Color** option menu in the **General Settings** section provides these options:

- **Monochrome**—Output is black and white.
- **Grayscale**—Design file colors are output as grayscale.
- **Color**—Design file colors are used.

Set the **Rasterized** check box to ON to plot the selected view output as a single raster image.

SELECTING A PRINTER AND PAPER SIZE

MicroStation can work with either of two types of printer drivers: **Windows Printer** or **Bentley Driver**. An option menu in the **Printer and Paper Size** section of the Print dialog box toggles between the two types of drivers.

Windows Printer selection automatically loads the Windows printer driver file (default is *printer.plt*). Predefined paper size from the **Paper** menu with an appropriate selection of Orientation Portrait or Landscape can be selected. Select one of the available options for output destination: Send to printer, Create plot file and Create metafile.

- **Send to printer** selection sends the print to the selected printer, using the selected printer driver.
- **Create plot file** selection creates a plot file. If the *.plt* file is configured to write directly to an LPT port; this is equivalent to Create plot file.
- **Create metafile** selection creates a Windows enhanced metafile (*.emf*).

Bentley Driver selection, by default, loads the Bentley printer driver file that was last used. If desired, another Bentley Driver can be selected from the Select Printer Driver File dialog box which can be opened by clicking the **Select Printer Driver** icon in the **Printer and Paper Size** section of the Print dialog box.. The configuration variable MS_PLOTDLG_DEF_PLTFILE can be used to define a default

printer driver file to be selected each time the Print dialog box is opened. That is, the defined printer driver file will be selected rather than the printer driver file last used.

When printing with a Bentley printer driver, **Create plot file** is the only available choice. MicroStation's printing system generates output in formats supported by most printing devices. All delivered MicroStation printer drivers reference drivers that create print information in industry-recognized formats (such as HPGL/2, HPGL/RTL and TIF). In addition, drivers are provided to create Acrobat Reader (PDF) file format, JPG and PNG graphic file formats.

PRINT SCALE AND POSITION

Settings in the Print Scale and Position section sets the scale for the print, and positions the print on the selected sheet.

The **Scale** edit field defines the number of design units (in working units) that equate to printer output unit (printer units). Key-in this value in the **Scale** edit field, or click the **Scale Assistant** icon, and MicroStation displays the Scale Assistant dialog box as shown in Figure 10–4. Define the scale criteria either as Design to Paper or Paper to Design.

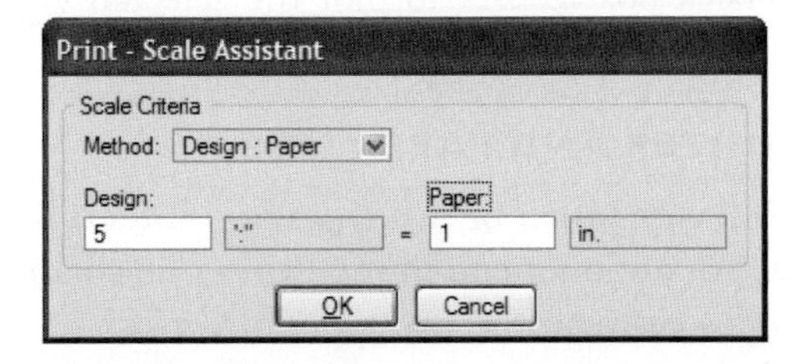

Figure 10–4 Scale Assistant dialog box

As an alternative to setting the scale for the print, the X (width) and Y (height) dimensions can be set for the print. Changing the **Scale**, or either dimension (X or Y), automatically results in changes to the remaining parameters to maintain the aspect ratio of the print.

If necessary, the printer's units can be changed using the **Settings** drop-down menu of the Print dialog box. The settings will remain until they are changed again.

The **Print Position** section of the Print dialog box controls the position of the printable area of the print. It specifies the position of the lower left corner of the print relative to the lower left corner of the page. The **Origin** value defines the distance horizontally and vertically in reference to the lower left corner.

To center the print on the page, set the **Auto-center** check box to ON. Choose the **Maximize** button to automatically fit the selected view or fenced area into as much

of the printable area as possible. The **Rotation** menu is used to set the rotation of the printed output.

The **Preview** section of the Print dialog box displays preview of the print for quickly checking the printing parameters. For more accurate previewing, open the resizable Preview window by selecting **Preview** from the **File** drop-down menu.

SETTING PRINT ATTRIBUTES

Use the Print Attributes dialog box to change aspects of the printed output's default appearance. Check boxes in the Print Attributes dialog box vary the relevant settings for printing purposes. Additionally, the display of the Fence Boundary and/or the default Print Border can be turned on/off for output. Open the Print Attributes dialog box from:

Print dialog box tool bar	Select the Print Attributes tool (see Figure 10–5).
Print dialog box menu	Settings > Print Attributes

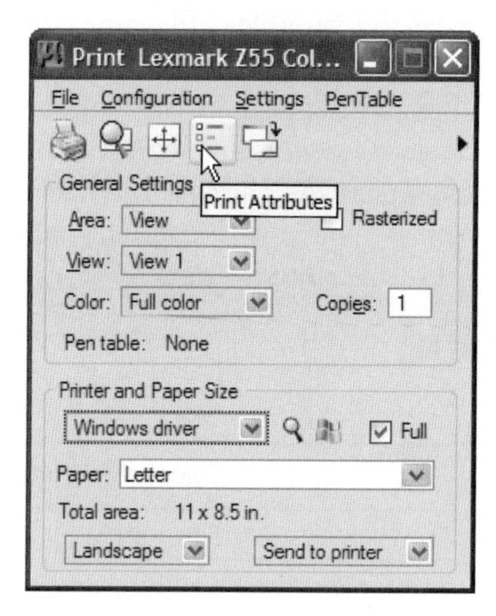

Figure 10–5 Invoking the Print Attributes tool from Print dialog box tool bar

MicroStation displays the Print Attributes dialog box as shown in Figure 10–6.

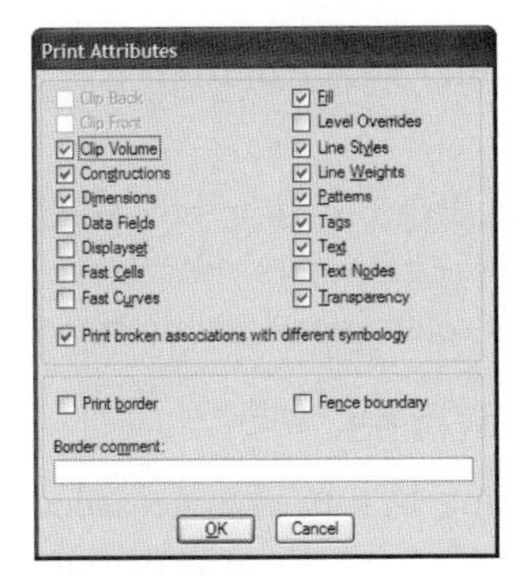

Figure 10–6 Print Attributes dialog box

The options provided in the Print Attributes dialog box are similar to the View Attributes settings box. Make the necessary changes to the available print output options. If the **Fence Boundary** check box is set to ON, then the printed output includes the fence shape. If the **Print Border** check box is set to ON, the printed output includes a default border, which can include a label giving information such as the name of the design file and the time and date of the print. By default, the supplied printer drivers have the variables "filename" and "time" included in the border record. Additionally, if **Print Border** is turned on, text can be added in the **Border Comment** field. This will appear in the label outside the border and can include configuration variable references, which are expanded in the printed output. For example, if the configuration variable NAME was defined as "Joe Doe" and the keyed-in description is "Name=$(NAME)," the printed output would expand to "Name=Joe Doe" in the printed output.

PRINTING THE DESIGN

To print a design file or to create a plot file, invoke the Print tool from:

Print dialog box tool bar	Select the Print tool (see Figure 10–7).
Print dialog box menu	File > Print

Figure 10–7 Invoking the Print tool from the Print dialog box tool bar

What happens at this stage depends on the system configuration and the selected printer driver. For a standard configuration, with no modifications to printer driver files or configuration variables, the print will either go directly to a printer, or will be saved to disk for later submission to a printer. If **Windows Printer** is selected, then output is sent directly to the Windows system printer. If **Bentley Driver** is selected, the Save Print As dialog box opens to specify a name and location for the print to be saved to disk.

The default plot file name is the same as the design file name, with the .000 extension added to the file name. If necessary, change the plot file name.

By default, the plot file is saved in the *<disk>:\Documents and Settings\ All Users\ Application Data\Bentley\Workspace\projects\untitled\out*. Replace <disk> with the letter of the disk that contains the MicroStation program.

If the selected file name already exists, an Alert window opens. Click the **OK** button to overwrite the file, or click the Cancel button to return to the Save Print As dialog box and enter a new file name.

SAVING A PRINT CONFIGURATION

A print configuration file saves the print information specific to a design file. Print configuration files are a way to streamline repetitive printing tasks. Information saved in a print configuration file includes the following:

- Printing area
- Print option settings
- Fence location
- Displayed levels

- Page size, margin, and scale
- Pen table if attached

Before a print configuration file is created, set the appropriate controls in the Print dialog box. To create a print configuration file, open the Save Print Configuration File As dialog box from:

Print dialog box menu	Configuration > Save

MicroStation displays the Save Print INI File As dialog box. Specify the name of the configuration file in the **Files** edit field and click the OK button. MicroStation saves the configuration to the given file name with the *.ini* extension, and by default it is saved in the *<disk>:Documents and Settings\All Users\Application Data\Bentley\Workspace\System\Data\.* directory.

To open an existing configuration file, invoke the **Open** tool from the **Configuration** menu in the Print dialog box. MicroStation displays the Select Print INI File dialog box. Select the appropriate configuration file and click the OK button. MicroStation makes the necessary changes to the print settings.

PEN TABLES

The pen table is a data structure used to modify the appearance of a print without modifying the design file, by performing one or more of the following at print-creation time:

- Changing the appearance of elements
- Determining the printing order of the active design file and its attached reference files
- Specifying text string substitutions

A pen table is stored in a pen table file. The pen table consists of sections that are tested against each element in the design. When a match is found, the output action is applied to the element. The modified element is then converted into print data, which in turn is printed or written to the plot file. At no time are the elements of the design file or its reference files modified.

CREATING A PEN TABLE

To create a pen table, invoke the New tool from:

Print dialog box menu	PenTable > New

MicroStation displays the Create New Pen Table file dialog box. Specify the pen table file name and click the OK button. MicroStation displays the Modify Pen Table settings box, as shown in Figure 10–8.

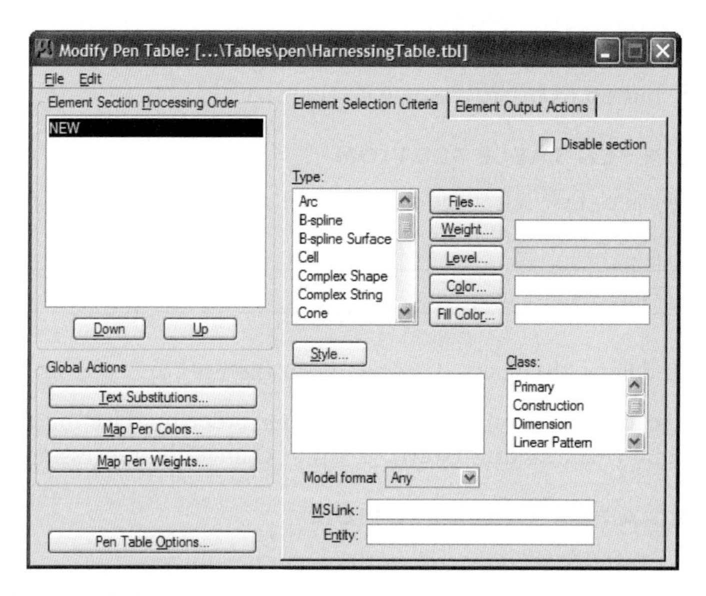

Figure 10–8 Modify Pen Table settings box

By default, MicroStation adds the section called NEW in the **Element Section Processing Order** list box. Either rename the section or insert a new one and delete the NEW section.

RENAMING A PEN TABLE SECTION

To rename a section, first select the name of the section in the list box, and double-click the selected section name. MicroStation displays the Rename Section dialog box. Specify the new name and click the OK button to rename the selected section.

INSERTING A NEW PEN TABLE SECTION

To insert a new section above an existing section, first select the name of the section in the list box, and then invoke the Insert New Section Above tool from:

Modify Pen Table settings box menu	Edit > Insert New Section Above

MicroStation displays the Insert Section dialog box. Specify the new name and click the OK button to insert the new section.

To insert a new section below an existing section, first select the name of the section in the list box, and then invoke the Insert New Section Below tool from:

Modify Pen Table settings box menu	Edit > Insert New Section Below

MicroStation displays the **Insert Section** dialog box. Specify the new name and click the OK button to insert the new section.

DELETING A PEN TABLE SECTION

To delete a section, first select the name of the section in the list box, and then invoke the Delete Section tool from:

Modify Pen Table settings box menu	Edit > Delete Section

MicroStation deletes the selected section from the list box.

Change the section's position in the processing order by selecting the section in the list box, and then clicking **Down** or **Up** to change the processing order.

MODIFYING A PEN TABLE SECTION

To modify a pen table section, follow these numbered steps.

STEP 1:

Highlight the name of the section to modify from the Sections list box.

STEP 2:

To set the element criteria, first select the **Element Criteria** tab (see Figure 10–9), located in the top right side of the Modify Pen Table settings box.

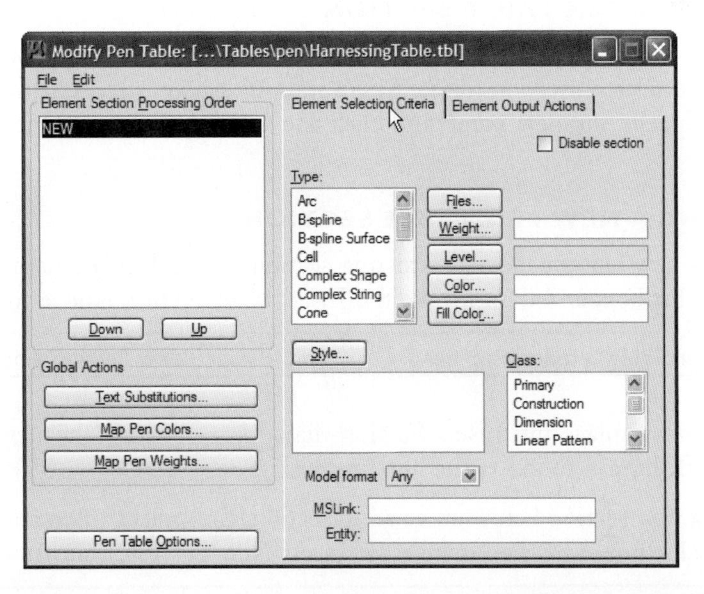

Figure 10–9 Selecting the Element Criteria tab in the Modify Pen settings box

Then select element type from the **Type** list box, and the element class from the **Class** list box to include in the selection criteria. To select multiple items from the

list box, hold down CTRL and then select the items. Click the **Files** button to select the file name and reference files, if any, attached to the active design file to select the order in which the elements will be identified.

To select all the available element types, choose **Select All Types** from **Edit** menu in the Modify Pen Table settings box. To deselect all the selected element types, choose **Clear Types** from **Edit** menu in the Modify Pen Table settings box. Similarly classes can be selected or deselected by choosing **Set All Classes** or **Clear Classes** respectively from **Edit** menu in the Modify Pen Table settings box.

In addition, selection criteria can be set based on weight, level, color, fill color, or style by specifying the appropriate values in the edit fields or by clicking the appropriate button in the Modify Pen Table settings box. MicroStation displays the appropriate dialog boxes. Use the controls in the dialog box to make the selections, and click the **OK** button to close the dialog box.

For example, if an Ellipse is specified as the element type, one or more types of modifications can be applied to all ellipses in the design file. However, if the specification is detailed, such as Ellipse, Level 40, Line Weight 2, then only ellipses on level 40 with a line weight of 2 will be affected by the output action. All other ellipses in the design will be ignored by this section. In addition, changes can be defined which will apply to other types of ellipses on other levels of the design file. Those changes will be applied without affecting the first section of changes.

The Model format menu is used to select the type of model (ANY, DGN and/or DWG/DWF) on which pen table sections will operate. The selection applies to all models.

STEP 3:

To set the output actions, first select the **Element Output Actions** tab (shown in Figure 10–10), located on the top right side of the Modify Pen Table settings box. Settings on the Element Output Actions tab are associated with each pen table section. Element output actions specify what is to be done once an element meets the section's Element Selection Criteria.

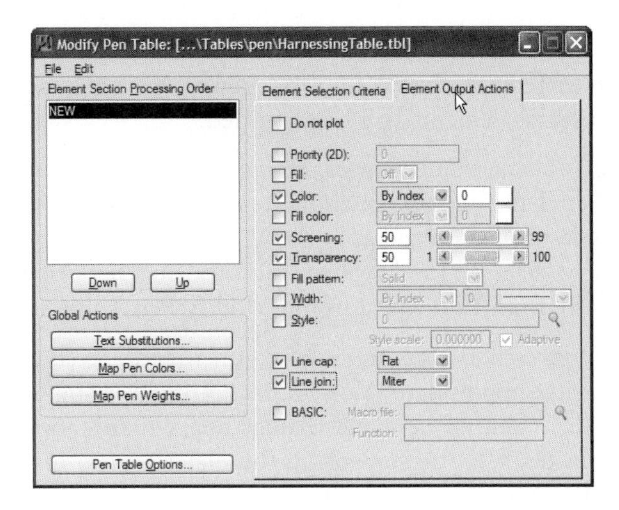

Figure 10–10 Selecting the Output Actions tab in the Modify Pen Table settings box

Optional: **Do not plot** check box controls the plotting of all the elements matching the current section. When selected, this option causes all other items on the Output Actions tab to be disabled and the elements that are matched to the current section will not be plotted.

Optional: **Priority (2D)** check box sets the priority. When set to ON, the desired priority value (range: –2147483648 to 2147483647) can be specified in the **Priority** edit field. Elements with a lower priority value are printed before elements with a higher priority value. Non-prioritized elements are always printed before all prioritized elements.

Optional: Set appropriate check boxes to ON to override the **Fill, Color, Fill Color, Fill Pattern, Screening, Transparency, Width, Line Cap, Line Join,** or **Style** appropriately for the elements that satisfy the selection criteria. Specify the values in the edit fields or choose the desired attributes from the pop-up palette. If the element is to be printed with a custom line style, specify the desired line style scale factor in the **Style Scale** edit field.

STEP 4:

Optional: Set Global actions that are applied across the entire design file and across all levels within the file. Global actions cannot be focused on any specific section. They provide three features for modifying a design's printed output:

- Text Substitutions
- Mapping Pen Colors
- Mapping Pen Weights

Text Substitutions is used to substitute text in the design with alternate text for printing. Click the Text Substitutions button to open the Text Substitutions settings box, as shown in Figure 10–11.

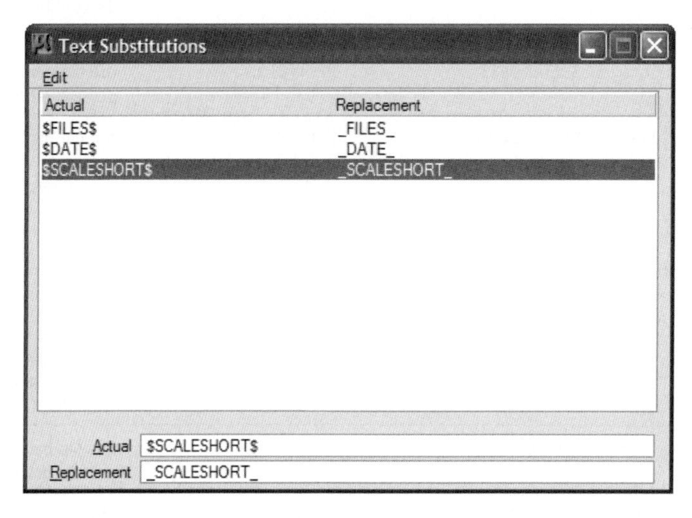

Figure 10–11 Text Substitutions settings box

To insert a text substitution entry, invoke the Insert New tool from:

Text Substitutions settings box menu	Edit > Insert New

An entry labeled "Original" appears in the list box and in the **Actual** edit field. Replace "Original" with the string in the design to be replaced for printing purposes. Type the replacement text string in the **Replacement** edit field and press (ENTER).

Note: *The defined text string substitutions apply universally to all text elements in the print and only to exact matches of the specified strings.*

In addition, a text string in a design can be replaced with a file name, current date, or time.

To replace a text string in a design with a file name, pen table file name, print driver file name, date, or time, select options from the **Edit** menu (Text Substitutions settings box) as listed in Table 10–1.

Table 10–1 Replacing a Text String with a File Name, Date, or Time

Edit Menu Item	Actual string in the Design*	Replacement string for printing	Replaces the actual text string with...
Insert Design File Short	$FILES$	_FILES_	the file name of the active design file.

Insert Design File Short Abbrev	$FILENAME$	_FILEA_	the file name of the active design file with the name of the folder.
Insert Design File Long	$FILEL$	_FILEL_	the file name of the active design file name with full path.
Insert Pen Table Short	$PENTBLS$	_PENTBLS_	the file name of the attached pen table file.
Insert Pen Table Abbrev	$PENTBLA$	_PENTBLA_	the file name of the attached pen table with the name of the folder.
Insert Pen Table Long	$PENTBLL$	_PENTBLL_	the name of the attached pen table with full path.
Insert Print Driver Short	$PLTDRVS$	_PLTDRVS_	the file name of the selected printer driver.
Insert Print Driver Abbrev	$PLTDRVA$	_PLTDRVA_	the file name of the selected printer driver and name of the folder.
Insert Print Driver Long	$PLTDRVL$	_PLTDRVL_	the file name of the selected printer driver with full path.
Insert Date	$DATE$	_DATE_	the current date.
Insert Time	$TIME$	_TIME_	the current time.
Insert Scale Long	$SCALE$	_SCALE_	the scale used for print output.
Insert Scale Short	$SCALESHORT$	_SCALESHORT_	the shortened version of the scale use for print output.

* The actual text string is shown with the dollar sign character ($) as the delimiter character just to differentiate it from normal text. It is not necessary to have the delimiter character as part of the text string in the design file.

To delete a text substitution entry, first highlight the text string substitution, and then invoke the Delete tool from:

Text Substitutions settings box menu	Edit > Delete

The selected text string substitution is deleted from the settings box.

Note: *Preview the substitutions by clicking the* **Preview Refresh** *icon in the Print Preview settings box.*

Mapping Pen Colors is used to apply multiple output color and width symbology to different parts of the same element, based on the component colors. For example, an element with a multicolored custom line style or an associative hatch linkage definition with a different color may be assigned unique widths for the specific colors using pen color maps. Pen color maps correspond to printer driver pen records, and may be overridden using element-based output actions.

Mapping Pen Widths is used to specify print output widths for each of the MicroStation weight values which corresponds to the *weight_strokes* record in the printer driver (*.plt*) file. Weight maps have lower priority than both pen color maps and element-based output actions.

STEP 5:

Before the design is printed, click the Preview Refresh icon in the Print Preview settings box. MicroStation displays all the elements that are set to print in the Preview box, with appropriate changes as per the settings of the Output Actions. When satisfied with the changes, print the design file.

STEP 6:

To save the pen table, invoke the Save tool from:

Modify Pen Table settings box menu	File > Save

MicroStation saves the modifications made to the existing pen table.

To save the modifications to a different pen table file, invoke the Save As tool from:

Modify Pen Table settings box menu	File > Save As

MicroStation displays the Create Pen Table File dialog box. Key-in the name of the file to save the settings, and click the **OK** button.

STEP 7:

To disable pen table processing, unload the pen table. To unload the pen table, invoke the Unload tool from:

Modify Pen Table settings box menu	File > Exit/Unload

MicroStation unloads the pen table.

The pen table can also be unloaded from the **Pen Table** menu located in the Print settings box.

BATCH PRINTING

MicroStation provides a utility program called Batch Print to print sets of design files. This utility program allows printing of multiple design files. Job sets that identify design files to be printed, and the specifications that describe how they should be printed, can be composed and re-used. Individual files or subsets of the files in large job sets can be printed for spot-checking.

To invoke the Batch Print program, select from:

Menu	File > Batch Print

MicroStation displays the Batch Print dialog box, as shown in Figure 10–12.

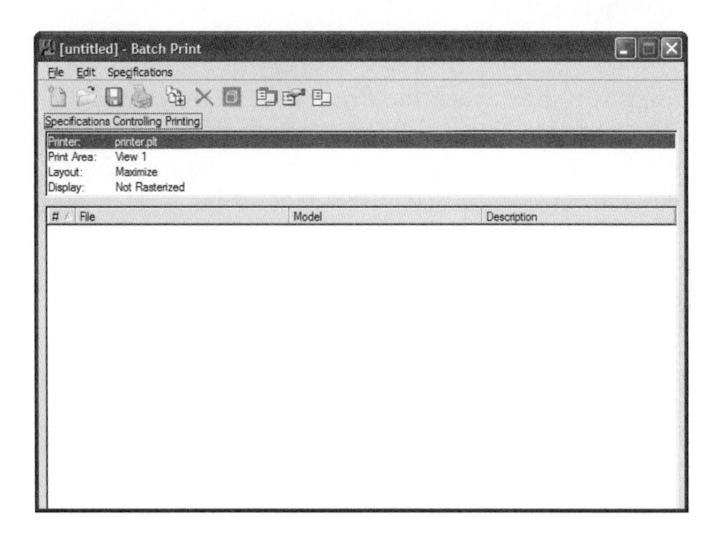

Figure 10–12 Batch Print dialog box

SETTING PRINT SPECIFICATIONS

A print specification is a named group of instructions describing certain steps in the printing process. The Batch Print utility program provides four specification types: Printer, Print Area, Layout, and Display. The Printer specification type describes the printer, paper size, and post-processing options. The Print Area specification type selects the portions of the design file to print. The Layout specification type places a representation of the given print area on the paper at the specified size and position. The Display specification type can be used to select a pen table and set the attributes for printing.

Printer Specification

The Printer specification type describes the printer, paper size, and post-processing options. The utility can be used to create a new printer specification, and modify or delete an existing specification. To create a new, modify or delete a printer specification, open the Batch Print Specification Manager from:

Batch Print dialog box menu	Specifications > Manage

MicroStation displays Batch Print Specification Manager as shown in Figure 10–13.

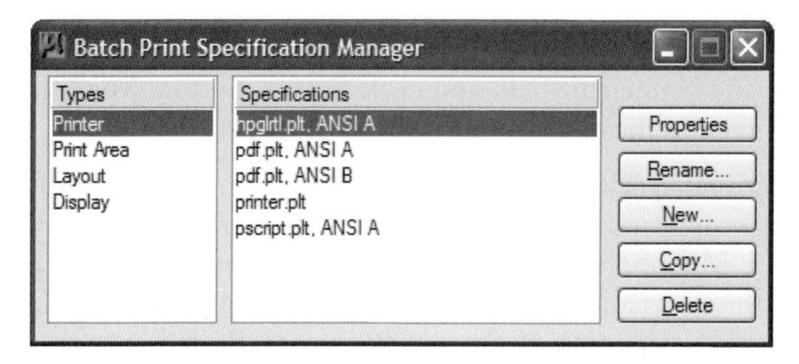

Figure 10–13 Batch Print Specification Manager dialog box

To change the properties of an existing printer specification, select the appropriate specification from the **Specifications** list box, and then click the **Properties** button. MicroStation displays a Printer Specification dialog box similar to Figure 10–14.

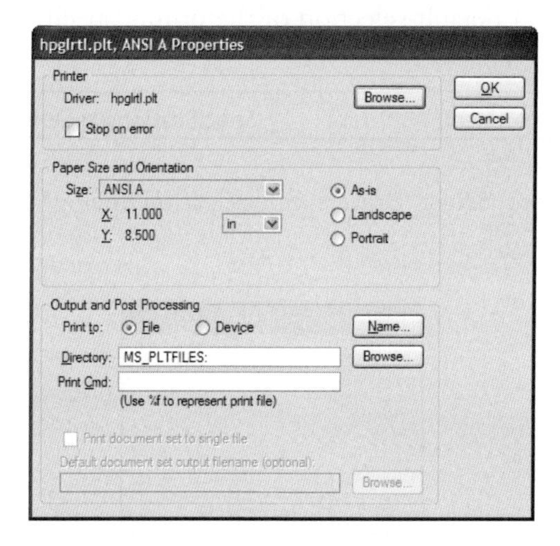

Figure 10–14 Printer Specification dialog box

The Printer Specification dialog box is used to select a specific printer driver, paper size, orientation, and output and post processing settings. Click the OK button to accept the changes and close the dialog box.

To create a new printer specification, click the New button in the Batch Print Specification Manager. MicroStation displays the New Printer Specification Name dialog box. Specify the name for the new Batch Print Specification in the **Name** edit field and click the OK button. If desired, change the properties of the selected Printer specifications. MicroStation lists the newly created printer specification in the **Specifications** list box.

To rename a new printer specification, first select the specification to be renamed, and then click the Rename button in the Batch Print Specification Manager.

MicroStation displays the Rename Printer Specification dialog box. Specify the new name in the **Name** edit field and click the OK button. MicroStation lists the renamed printer specification in the **Specifications** list box.

To create a printer specification from an existing one, first select the specification from which to copy, and then click the Copy button in the Batch Print Specification Manager. MicroStation displays the New Printer Specification Name dialog box. Specify the new name in the **Name** edit field and click the **OK** button. If desired change the properties of the newly created Printer specifications. MicroStation lists the newly created printer specification in the **Specifications** list box.

To delete a printer specification, first select the specification to delete from the **Specifications** list box, and then click the Delete button in the Batch Print Specification Manager. MicroStation deletes the selected printer specification from the **Specifications** list box.

To change the current default selection of the printer specification, invoke the Select tool from:

Batch Print dialog box menu	Specifications > Select

MicroStation displays Select Printer Specification dialog box shown in Figure 10–15.

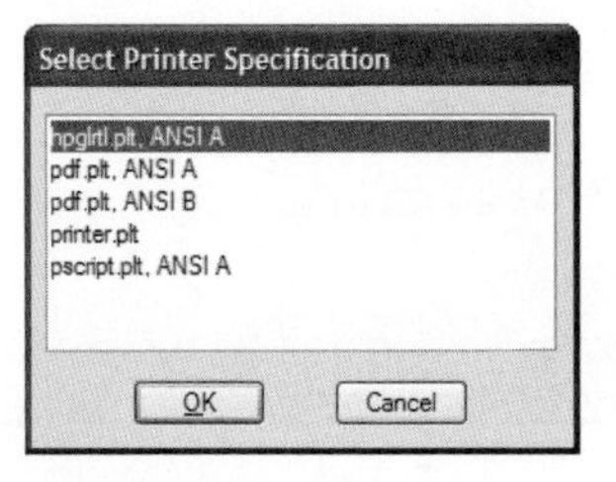

Figure 10–15 Select Printer Specification dialog box

Select the appropriate printer specification from the list box and click the OK button.

The default printer specification can also be set by double-clicking the name of the specification from the **Specifications** list box in the Batch Print Specifications Manager dialog box. MicroStation displays the Select Printer Specifications dialog box, similar to Figure 10–15, from which to select the appropriate printer specification.

Print Area

The Print Area specification type is used to select the portions of the design file to print. Use the utility to create a new Print Area specification, or modify or delete

an existing specification. To create, modify or delete a Print Area specification, open the Batch Print Specification Manager from:

Batch Print dialog box menu	Specifications > Manage

MicroStation displays the Batch Print Specification Manager, shown in Figure 10–13. To change the properties of an existing Print Area specification, first select the **Print Area** from the **Types** list box, and then select the appropriate specification from the **Specifications** list box and click the Properties button. MicroStation displays the Print Area Properties dialog box, similar to Figure 10–16.

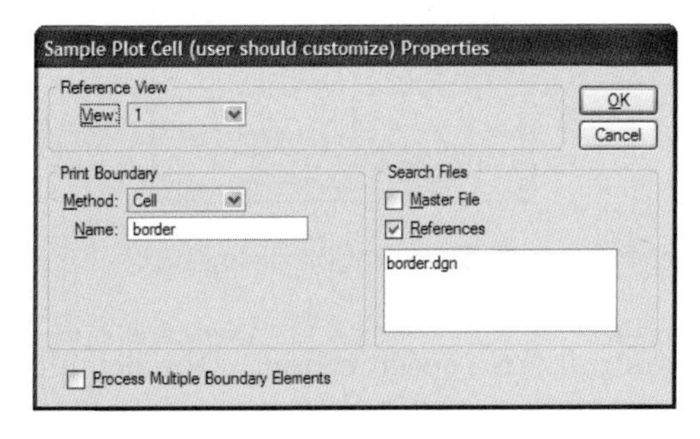

Figure 10–16 Print Area Properties dialog box

Use the **View** options menu to select the view number or saved view that will be printed.

The **Print Boundary** section of the dialog box is used to select the boundary-defining elements (similar to placing a fence by snapping to the vertices of a shape). The option menu in the **Print Boundary** section includes three options: **View, Shape,** and **Cell.** The **View** selection prints to the extent of the view window. The **Shape** selection prins an area bounded by a particular shape. Specify the attributes of the shape in the appropriate fields in the **Boundary** section of the dialog box. The **Cell** selection prints an area bounded by a cell. Specify the name of the cell in the **Name** edit field in the Boundary section of the dialog box.

The **Master File** and **References** check boxes set the limit for the search for the boundary-defining shape or cell. By default, the utility searches each master file and all of its reference files to find the boundary-defining shape or cell. If necessary, restrict the search to specific reference files by typing their logical names or file names in the **References** edit field.

The **Process Multiple Boundary Elements** check box controls whether to generate the prints for each boundary element found in the design file. Set the check box to ON to generate a print for each boundary element found and OFF to generate a print for the first boundary element found only.

Click the OK button to accept the changes and close the dialog box.

To set default Print Area specification, invoke the Select tool from:

Batch Print dialog box menu	Specifications > Select

MicroStation displays the Select Print Area Specification dialog box, shown in Figure 10–17.

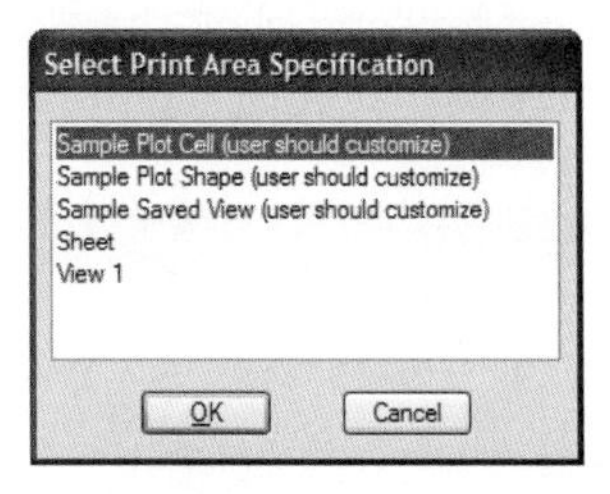

Figure 10–17 Select Print Area Specification dialog box

Select the appropriate Print Area specification as the default from the list box and click the **OK** button.

The default Area specification can also be set by double-clicking the name of the specification from the **Specifications** list box in the Batch Print Specifications Manager dialog box. MicroStation displays the Select Printer Specifications dialog box, similar to Figure 10–17 and select the appropriate Print Area specification.

Layout

The Layout specification type describes how the utility program determines the size and position of each print. The utility is used to create a new Layout specification, or modify or delete an existing specification. To create, modify or delete a Layout specification, open the Batch Print Specification Manager from:

Batch Print dialog box menu	Specifications > Manage

MicroStation displays the Batch Print Specification Manager, as shown in Figure 10–13. To change the properties of an existing Layout specification, first select the Layout from the **Types** list box, and then select the appropriate specification from the **Specifications** list box, and click the **Properties** button. MicroStation displays the Layout Properties dialog box, similar to Figure 10–18.

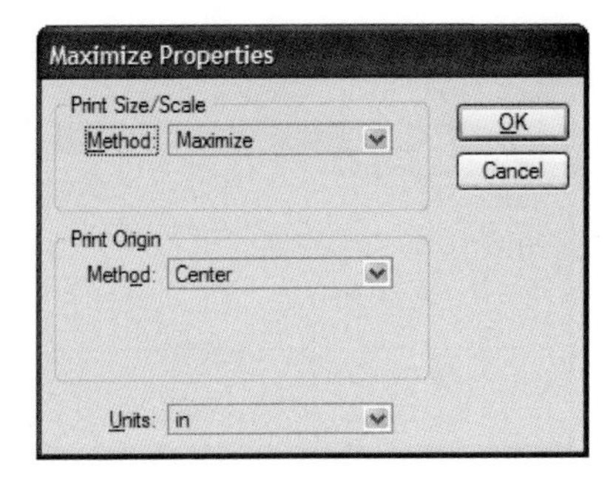

Figure 10–18 Layout Properties dialog box

The **Print Size/Scale Method** menu provides five print size options:

- The **Maximize** selection makes each plot as large as possible, given the paper size and orientation in the job set's printer specification.
- The **Scale** selection is used to specify a scale factor in terms of master units in the design file to physical units of the output media.
- The **% of Maximum Size** selection is used to specify an integer value between 10 and 100 percent of its maximum possible size.
- The **X Size** selection is used to specify an explicit X size (width) for the print.
- The **Y Size** selection is used to specify an explicit Y size (height) for the print.

The **Plot Origin Method** menu provides two options:

- The **Center** selection centers each print on the output media.
- The **Manual Offset** selection is used to specify explicit *X* and *Y* offsets for the print. The offsets provided are relative to the media's lower-left margin.

The **Units** menu sets the method by which the X size and Y sizes are measured.

Click the OK button to accept the changes and close the dialog box.

To change the current selection of the Layout specification, invoke the Select tool from:

Batch Print dialog box menu	Specifications > Select

MicroStation displays the Select Layout Specification dialog boxes shown in Figure 10–19.

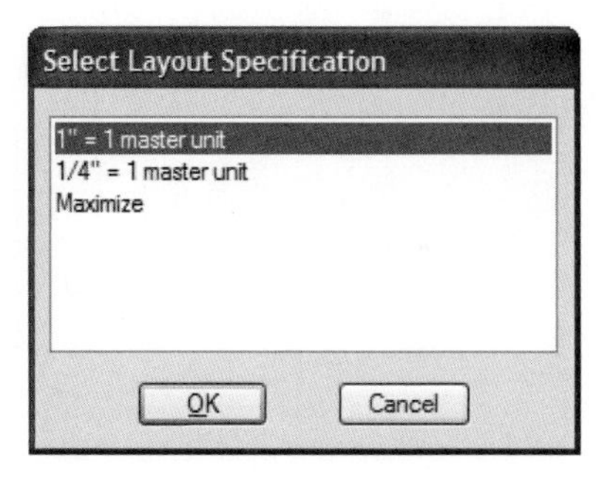

Figure 10–19 Select Layout Specification dialog box

Select the appropriate Layout specification from the list box and click the OK button.

The default Layout specification can also be set by double-clicking the name of the specification from the **Specifications** list box in the Batch Print Specifications Manager dialog box. MicroStation displays the Select Layout Specifications dialog box, similar to Figure 10–19; select the appropriate Print Layout specification.

Display

The Display specification controls the appearance of printed elements. It is used to control the printing equivalents of the view attributes, and to specify a pen table that will resymbolize the print. To create, modify or delete a Display specification, open the BatchPrint Specification Manager from:

Batch Print dialog box menu	Specifications > Manage

MicroStation displays the Batch Print Specification Manager shown in Figure 10–13. To change the properties of an existing Display specification, first select **Display** from the **Types** list box, and then select the appropriate specification from the **Specifications** list box, and click the **Properties** button. MicroStation displays the Display Properties dialog box, similar to Figure 10–20.

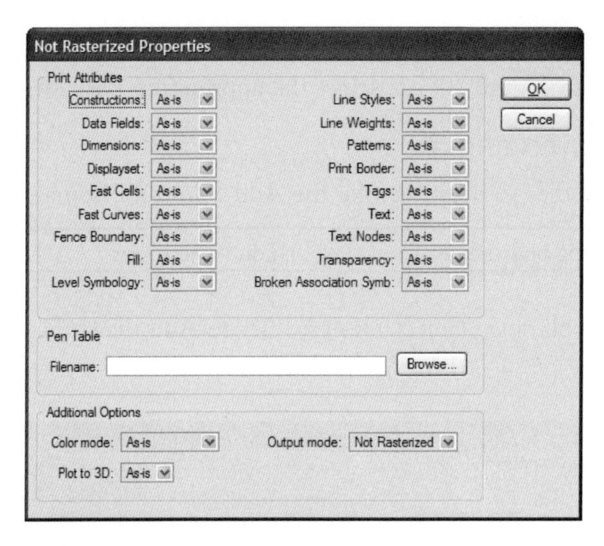

Figure 10–20 Display Properties dialog box

The options provided are similar to the View Attributes setting options. Make necessary changes to the display options. Specify the name of the pen table file to apply for printing the selected design file, if desired. Click the OK button to accept the changes and close the dialog box.

To change the current selection of the Display specification, invoke the Select tool from:

Batch Print dialog box menu	Specifications > Select

MicroStation displays the Select Display Specification dialog box shown in Figure 10–21.

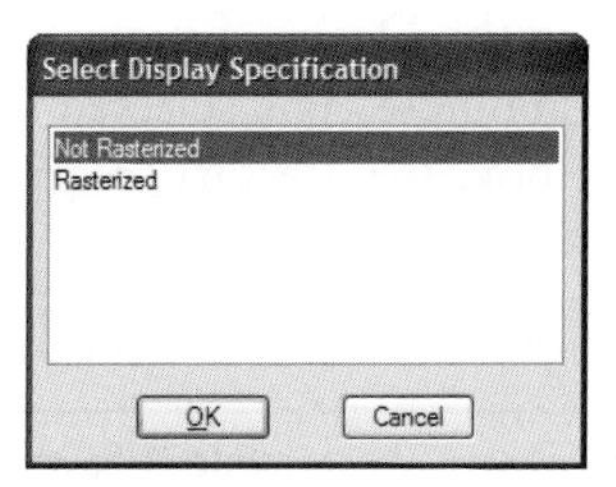

Figure 10–21 Select Display Specification dialog box

Select the appropriate Display specification from the list box and click the OK button.

The default Display specification can also be set by double-clicking the name of the specification from the **Specifications** list box in the Batch Print Specifications

Manager dialog box. MicroStation displays the Select Display Specifications dialog box, similar to Figure 10–21; select the appropriate Display specification.

DESIGN FILES TO PRINT

To select design files to print, invoke the Add Files tool from:

Batch Print dialog box menu	Edit > Add Files

MicroStation displays the Select Design Files to Add dialog box shown in Figure 10–22.

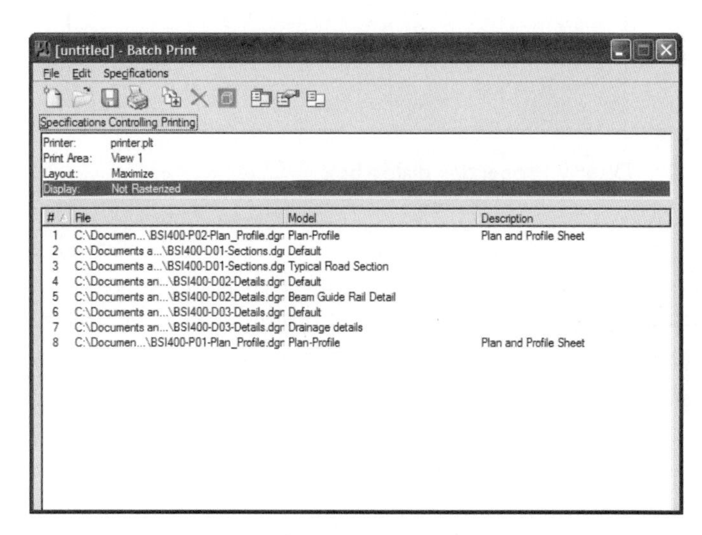

Figure 10–22 Select Design Files to Add dialog box

Select the design files from the appropriate directory and then click the Add button. Selected files are added to the **Design Files to Print** list box. Click the Done button when the selection is complete. The selected files are printed in the order in which they appear in the list box. If necessary, rearrange the files by highlighting their names and invoking one of the modify tools available in the **Edit** menu.

To save the current selection of the design files set to print, invoke Save tool from:

Batch Print dialog box menu	File > Save

MicroStation displays the Save Job Set File dialog box. Specify the name of the file in the **Files** edit field to which the current job set will be saved. Click the OK button to save the current job set and close the dialog box.

To print the current selection of the design files, invoke the Print tool from:

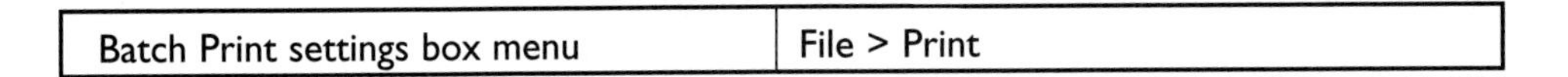

Batch Print settings box menu	File > Print

MicroStation displays Print Batch dialog box similar to Figure 10–23.

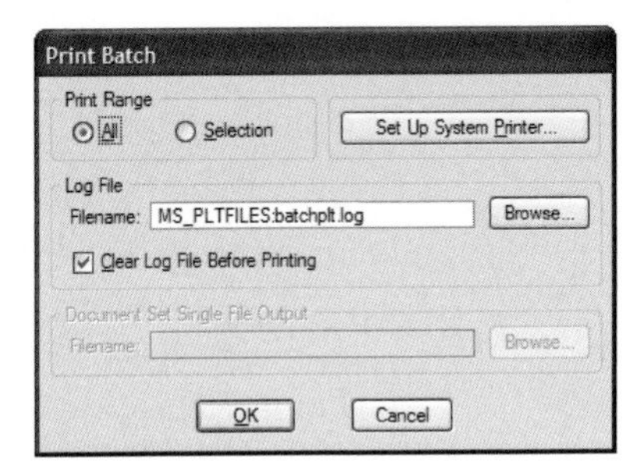

Figure 10–23 Print Batch dialog box

Select one of the two radio buttons available in the **Print Range** section of the dialog box. The **All** option allows MicroStation to print all the design files in the current job set. The **Selection** option prints only the design files that are selected explicitly in the current job set. Click the OK button to print the selection and close the dialog box.

Open the Exercise Manual PDF file for Chapter 10 on the accompanying CD for project and discipline specific exercises.

CHAPTER 11

Cells and Cell Libraries

OBJECTIVES

Topics explored in this chapter:

- Creating cell libraries
- Attaching cell libraries
- Creating cells
- Selecting active cells
- Placing cells
- Placing line terminators
- Placing point elements, characters, and cells
- Maintaining cells and cell libraries
- Placing and maintaining shared cells
- Using and modifying cells from the cell selector

CELLS

For paper drawings, metal or plastic templates with cutouts provide a means of drawing standard symbols. In MicroStation, cell libraries act as the templates for collections of standard symbols. Cells in the libraries hold the standard symbols. Figure 11–1 shows some common uses of cells in various engineering disciplines.

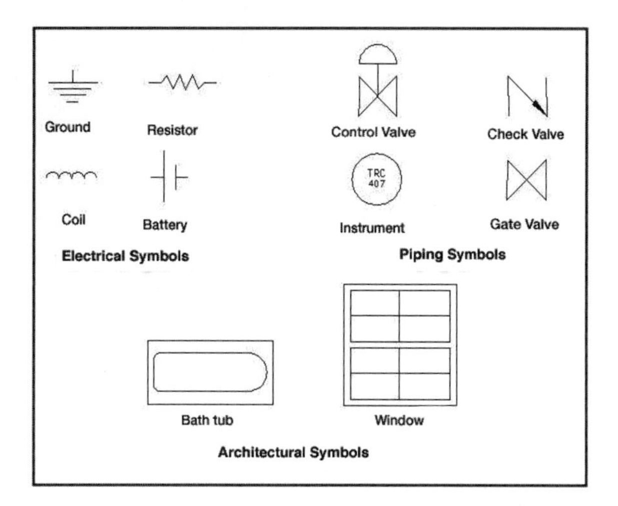

Figure 11–1 Common uses of cells in various engineering disciplines

Even though a cell contains separate elements, copies of a cell placed in a design file act as a single element when manipulated with tools such as Delete, Rotate, Array, and Mirror. Cells save time by eliminating the need to draw the same thing more than once and they promote standardization.

CELL LIBRARIES

If cells are like the holes in a plastic template, cell libraries are the template. A cell library is a file that holds cells. Most engineering companies that use MicroStation have several cell libraries to provide standard symbols for all of their design files.

Cell placement tools access cells from cell libraries and put copies of selected cells on the active design file. MicroStation can also place cells that are not in the attached cell library. Typing a cell name with a cell placement tool causes MicroStation to search for the cell in the cell libraries specified by the Cell Library List configuration variable (MS_CELLIST). (Refer to Chapter 16 for information on setting up the configuration variables).

Although there is no limit to the number of cells in a library, very large libraries are hard to manage. It is advisable to create separate cell libraries for specific disciplines, such as electrical and plumbing.

CREATING A NEW CELL LIBRARY

To start the process of creating a new cell library, invoke the Cell Library box from:

Menu	**Element > Cells** (see Figure 11–2).
Key-in window	**dialog cellmaintenance** (or **di ce**) (ENTER)

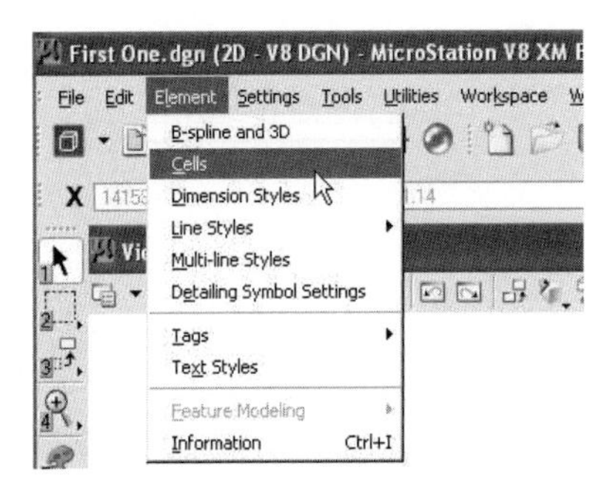

Figure 11–2 Invoking the Cell Library box from the Element menu

The Cell Library settings box appears, as shown in Figure 11–3.

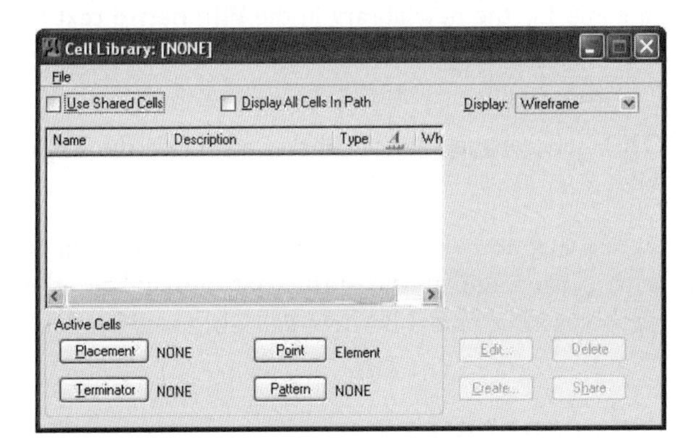

Figure 11–3 Cell Library settings box

To create a new cell library file, open the Create Cell Library dialog box from:

Cell Library box	**File > New**

MicroStation displays the Create Cell Library dialog box, as shown in Figure 11–4.

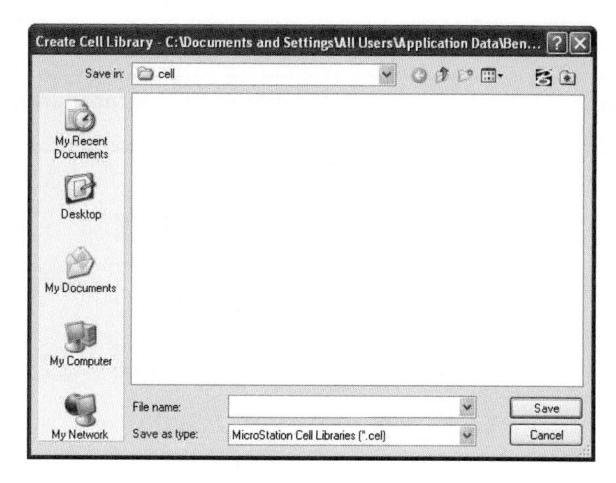

Figure 11–4 Create Cell Library dialog box

The following steps create a new cell library file:

1. In the **Save in** menu, select a folder to hold the new cell library.
2. Key-in a name for the new library in the **File name** text field. Type only the file name, not a period for file type code. MicroStation appends the ".cel" file type to the file name.
3. Click **OK** to create the new cell library and close the Create Cell Library dialog box.

MicroStation attaches the new cell library to the active design file, acknowledges the attachment with a message in the status bar, and places the filename in the Cell Library dialog box's title bar. The new cell library is available for use in storing new cells.

ATTACHING CELL LIBRARIES

Attaching a library to the active design file makes the cells in the library available for use. MicroStation provides tools to attach a single cell library or all the cell libraries in a selected folder.

Attach a Cell Library

Invoke the Cell Library box from:

Menu	**Element > Cells**

The Cell Library settings box opens, as shown in Figure 11–3. To attach a cell library from the Cell Library box, invoke the Attach Cell Library dialog box from:

Cell Library box	**File > Attach File**
Key-in window	**rc=**<name of the library> (ENTER)

MicroStation displays the Attach Cell Library dialog box, as shown in Figure 11–5.

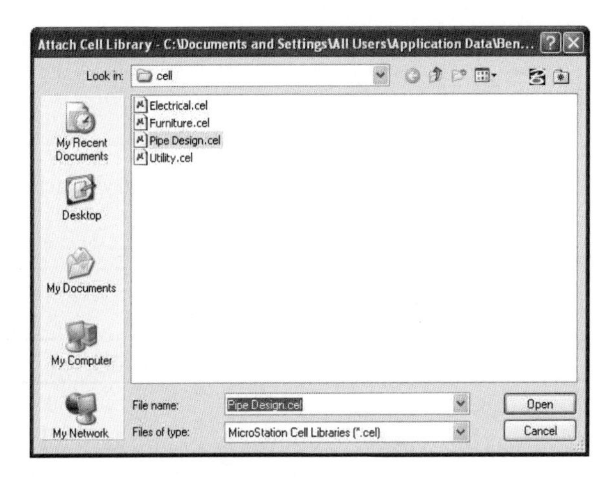

Figure 11–5 Attach Cell Library dialog box

The following steps attach a cell library:

1. In the **Save in** menu, select the folder that holds the cell library.
2. In the Files area, select the cell library name.
3. Click **Open** to attach the cell library to the active design file.

MicroStation attaches the selected cell library to the active design file and displays the cell library path and file name on the Cell Library settings box's title bar. The name and description of cells in the library appear on the cells area of the settings box. Figure 11–6 shows the Cell Library settings box with a typical cell library attached.

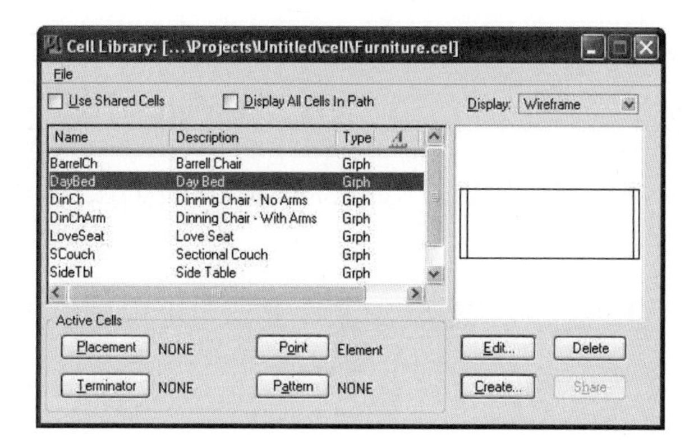

Figure 11–6 The Cell Library settings box with a cell library attached

Attach all Libraries in a Folder

Invoke the Cell Library settings box from:

Menu	**Element > Cells**

The Cell Library settings box opens, as shown in Figure 11–3. To attach all cell libraries in a folder, invoke the Attach Cell Library dialog box from:

Cell Library box	**File > Attach Folder**
Key-in window	**rc=**<name of the library> (ENTER)

MicroStation displays the Browse for Folder dialog box, as shown in Figure 11–7.

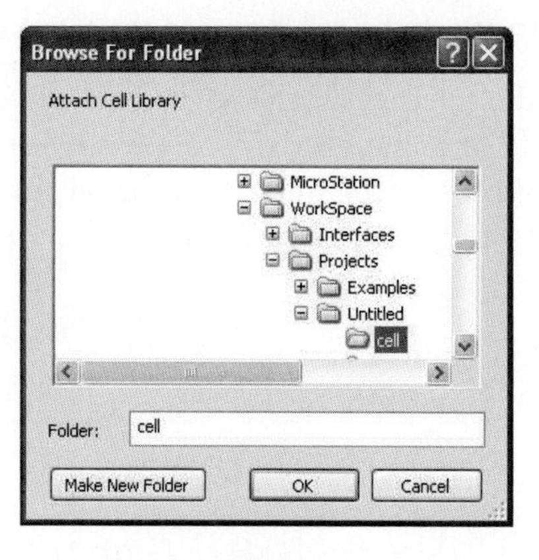

Figure 11–7 The Browse for Folder dialog box

The following steps attach all cell libraries in a folder:

1. In the folder selection box, find and select the folder to attach.

2. Click **OK** to attach the cell libraries in the folder to the active design file.

MicroStation attaches all cell libraries in the selected folder to the active design file and displays the folder path on the Cell Library settings box's title bar. The name and description of all cells in the attached libraries appear on the cells area of the settings box. Figure 11–8 shows the Cell Library box with a typical set of cell libraries attached.

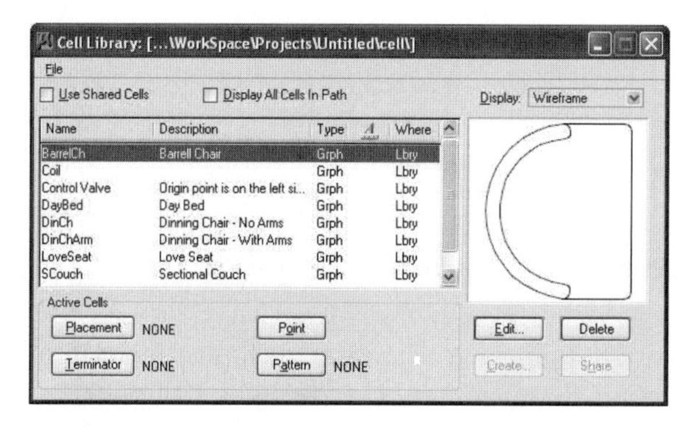

Figure 11–8 The Cell Library settings box with all cell libraries in a folder attached

Things to Remember about Cell Library Attachments

- The attachment is permanent as long as MicroStation can find the library file or folder.
- Only one cell library or folder of cell libraries can be attached. If the attach options are used to attach a different library or folder, the original attachment is lost.
- Cells cannot be edited or created when all cell libraries in a folder are attached, but cells can be deleted.

CREATING CELLS

The Cell Library settings box provides tools for creating new cells and loading the new cells into the attached cell library. The cell library can be attached to other design files, making its cells available for placement in the other design files.

Here are a few things to consider before drawing the elements for the new cell.

A cell is a group of elements that are loaded into the attached cell library. There is no limit on the size of cells; all element types can be part of a cell; cells can be on any level; and they can be any color, weight, or style.

The tool creates the new cell from copies of the elements. The originals remain in the design as separate elements.

Draw the object that is to be copied into a cell library with zero degrees of rotation. That means the object is upright and facing to the right, if it has a direction (see Figure 11–9). Drawing objects with zero rotation makes it easier to understand what happens to the cell when it is rotated.

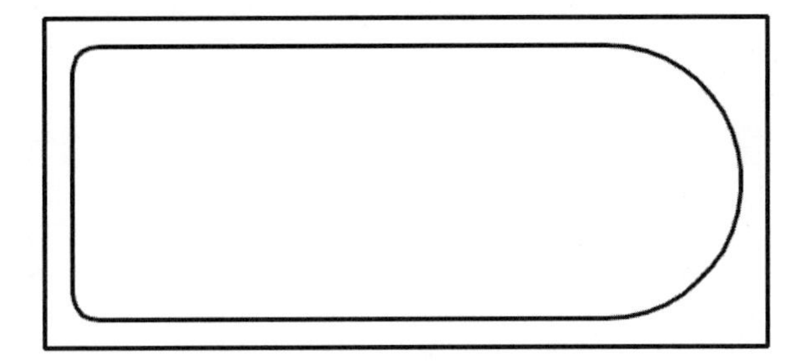

Figure 11–9 Example of a bathtub drawn left to right

Each cell has an origin point that determines cell placement in relation to the location of the data point that places it. The data point that places the cell is the cell origin. For example, if the origin point is in the center of the cell, the center of the cell is on data point that places it. Defining the location of the origin is part of cell creation.

Consider the cell's purpose when deciding where to place the origin while creating the cell. If the cell is to be placed connected to other elements, place the origin point at the connection point. For example, if the cell is a control valve, place the origin at pipeline attachment point, as shown in Figure 11–10.

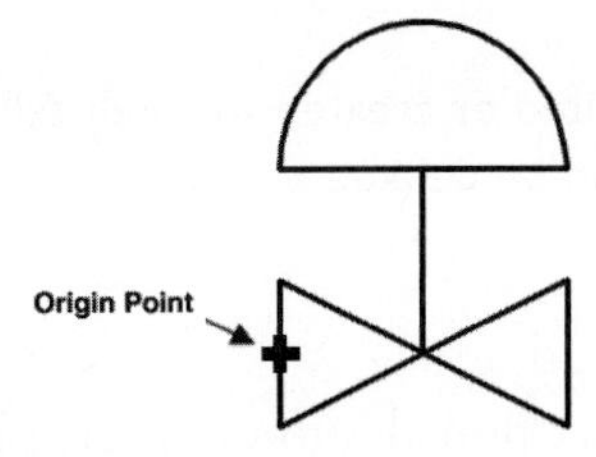

Figure 11–10 Example of defining a cell origin

STEPS FOR CREATING A CELL

The following steps create a new cell:

1. Draw the elements that create the new cell.
2. Group the elements by placing a fence around them or by using Element Selection.

Note: *Use any of the Fence modes to define the cell using the contents of the fence. Be aware that if the fence mode is Void or Void-Clip, MicroStation creates the cell from all elements in the design that are outside of the fence.*

3. Open the Cell Library settings box from:

Menu	**Element > Cells**

4. On the cell Library settings box, either attach an existing cell library to the design or create a new cell library.
5. Define the cell's origin point by invoking Define Cell Origin from:

Task Navigation tool box (active task set to Cells)	Select the Define Cell Origin tool (see Figure 11–11).
Keyboard Navigation (Task Navigation tool box with active task set to Cells)	**r**

MicroStation prompts:

Define Cell Origin > Define origin

Place a data point where the origin is to be located. Snap, if necessary, to place the point precisely.

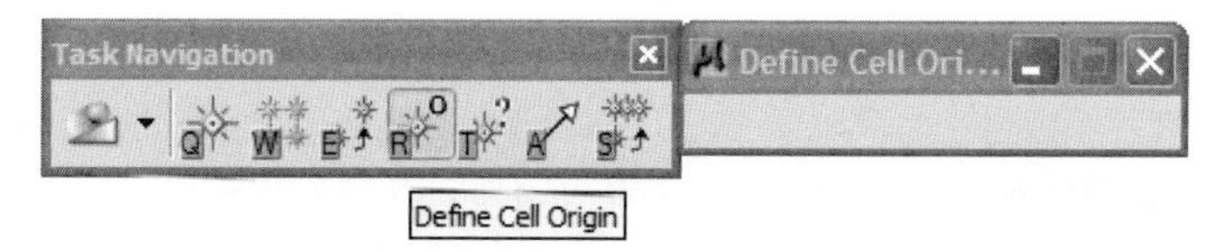

Figure 11–11 Invoking the Define Cell Origin tool from the Cells tool box

Note: *A cross indicates the location of the origin point on the design. The cross is not an element, just an indication of the origin location on the view. If the origin point is in the wrong place, define another one.*

6. Click **Create** (located in the bottom right corner of the Cell Library settings box) to open the New Cell dialog box, as shown Figure 11–12.

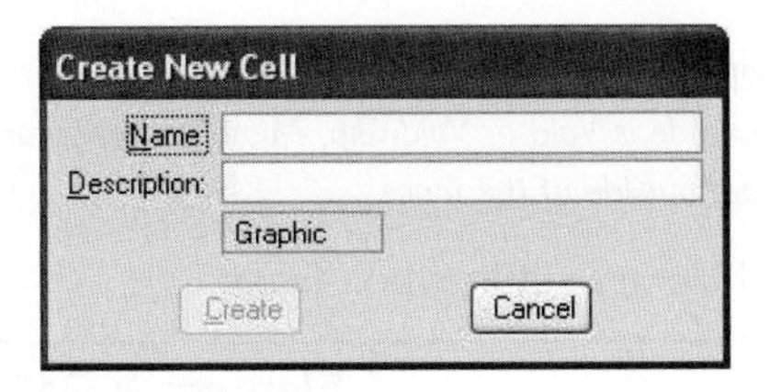

Figure 11–12 Create New Cell dialog box

7. Click the **Name** field in the Create New Cell dialog box and type a name for the new cell. There is no restriction on the length of the cell name and it can contain any combination of numbers, characters, and symbols (except: & * = < > ?/ : "\ |). The name can also contain space characters, but it cannot begin with a space. Give the cell a name that will indicate its purpose, making it easier for the user to pick appropriate cells from the libraries.
8. Optionally, click the **Description** field and type a description of the cell. There is no restriction on the length of the description and it can contain any combination of numbers, characters, and symbols. Use the description to add information that will help users understand when and how to use the cell.
9. Select the appropriate cell type from the menu located just below the **Description** field. Cells that can be placed in a design can be one of two types—**Graphic** or **Point**. The difference between the two types is described later.
10. Click **Create** to create a new cell from a copy of the selected elements.

Note: *The new cell is placed in the attached library and it appears in the cells list on the Cell Library box, as shown in Figure 11–13. The cell is available for placement in the current design file and in any other design files to which the cell library is attached.*

11. If the original elements are no longer needed, delete them.
12. To create another new cell, repeat steps 3 though 11.

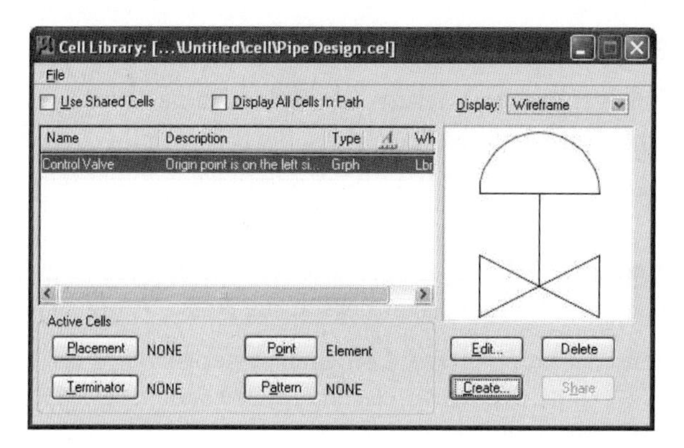

Figure 11–13 Displaying the name and description of a new cell in the Cell Library box

GRAPHIC AND POINT CELL TYPES

Two cell types, Graphic and Point, can be placed in design files. The default cell type is Graphic (also called Normal). Here are the differences between Graphic and Point cells.

A Graphic cell, when placed in a design file:

- Keeps the symbology (color, weight, style, and levels) of its elements.
- Remembers the levels on which its elements were drawn. (MicroStation provides two methods for placing the cells—Absolute and Relative modes, discussed in detail later in the chapter.)
- Retains the keypoints of each cell element.

A Point cell, when placed in a design file:

- Takes on the current active color, weight, and style.
- Places all cell elements on the current active level, regardless of what level they were drawn on.
- Has only one keypoint—the cell's origin point. The individual cell elements do not have keypoints.

While creating the new cell, select either **Graphic** or **Point** from cell type menu on the Create New Cell dialog box.

ACTIVE CELLS

To place a copy of a cell in the active design file, select the cell's name in the Cell Library box, click one of the **Active Cells** buttons to select a placement method, and select a cell placement tool from the Cells tool box.

PLACE CELLS

To place copies of a cell in the design file, select the cell to use from the Cell Library box, click the Active Cells Placement button to tell MicroStation which cell to use, and select a placement tool. Three tools place a cell at data points in the design file:

- **Place Active Cell** places copies of the Active Placement Cell at data points in the design. The tool places cells at the Active Angle and Active Scale.
- **Select and Place Active Cell** enables designating a cell already placed in the design to be the Active Placement Cell and places copies of the selected cell at the data points in the design.
- **Place Active Cell Matrix** places a rectangular matrix of copies of the active placement cell, with the lower left corner of the matrix at the data point.

Note: *MicroStation provides a shortcut for placing individual copies of a cell. Double-clicking the cell's name in the Cell Library box places the cell name next to the Active Cells Placement button and invokes the* ***Place Active Cell*** *tool.*

Place Active Cell

The **Place Active Cell** tool places one copy of the Active Placement Cell at the location defined by a data point. Invoke Place Active Cell from:

Task Navigation tool box (active task set to Cells)	Select the Place Active Cell tool (see Figure 11–14).
Keyboard Navigation (Task Navigation tool box with active task set to Cells)	**q**

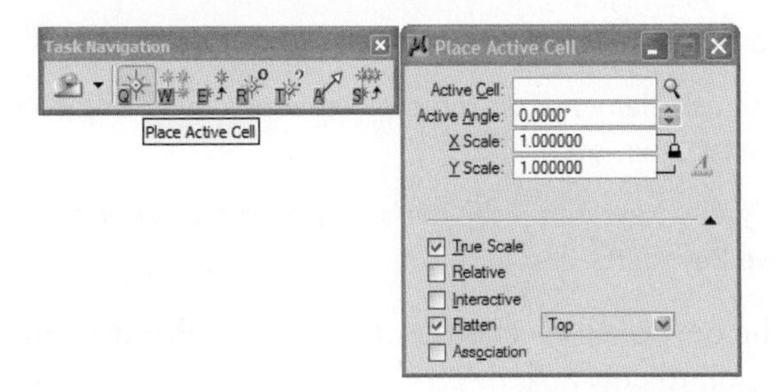

Figure 11–14 The Place Active Cell tool with its tool box expanded to show all options

MicroStation prompts:

Place Active Cell > Enter cell origin

Define the origin point at each location where a copy of the cell is to be placed.

The tool places a copy of the cell at each data point with the cell's origin is on the data point. While this tool is active, the screen cursor drags a dynamic image of the cell. The cell's origin point is at the screen cursor position.

Note: *Invoke the Place Active Cell tool with no active cell defined and MicroStation displays the message "No Active Cell" in the Status bar. Declare an active cell before using this tool (discussed earlier in this section).*

The tool's Settings window provides several options for modifying active cell placement.

- The **Active Cell** text field displays the name of the Active Placement Cell. To change to a different cell, type the cell's name in the field or click the **Browse Cell(s)** magnifying glass symbol to open a selection box that displays all cells in the attached library. Select an active cell by clicking the cell name in the selection box. If the selection box is already open, clicking the magnifying glass switches focus to the settings box.
- The **Active Angle** field allows rotating placements of the Active Placement Cell by the number of degrees keyed into the field. For example, keying in 90 flips the cell over on its left side. Positive angles rotate the cell in a counter-clockwise direction and negative angles rotate the cell clockwise. The scroll bar on the right side of the field allows selecting standard rotation angles such as 45, 90, and 180 degrees.
- The **X Scale** and **Y Scale** fields are multipliers that determine the size of the cell when placed in the design. If the scale factor is 2, the placed cell is twice its actual size. If the scale factor is 0.5, the placed cell is half its actual size. Turn OFF the lock next to the two fields to allow entering different numbers in the fields. Turn ON the lock to keep the aspect ratio: the same number is used for both X and Y scaling.

Additional settings are available on the tool settings box. If they are not visible, click the **Show Cell Placement Options** arrow on the lower right side of the box to expand it and reveal more settings fields.

- An active cell created in a design file that has different working units than the active design file is adjusted to the correct size when placed if the **True Scale** check box ON. For example, the Active Placement Cell is a twelve-inch square box that was created in a design whose Master Unit is Inches. The active design file's Master Unit is Feet. If the check box is ON, the cell is placed as a twelve-inch square box, but if the check box is OFF, the cell is placed as a twelve-foot square box.
- The actions of the **Relative** and **Interactive** check boxes are described later.
- The **Flatten** check box and menu are for 3D designs only.

- Turning the **Association** check box ON associates the placed cell with an element in the design when the cell is placed by first snapping to the element and then accepting it. Move an element in the design and all associated cells move with it.

Place Active Cell using Absolute or Relative Mode

When the **Relative** check box in the Place Active Cell tool's Tool Settings window is OFF, the tool places cells in Absolute placement mode. When the check box is ON, the tool places cells in Relative placement mode.

In **Absolute** placement mode, the tool places a Graphic cell's elements on the same levels on which they were drawn regardless of the Active Level setting. For example, if a graphic cell contains elements on levels 1 and 3, they are placed in the design file on levels 1 and 3 (see Figure 11–15a).

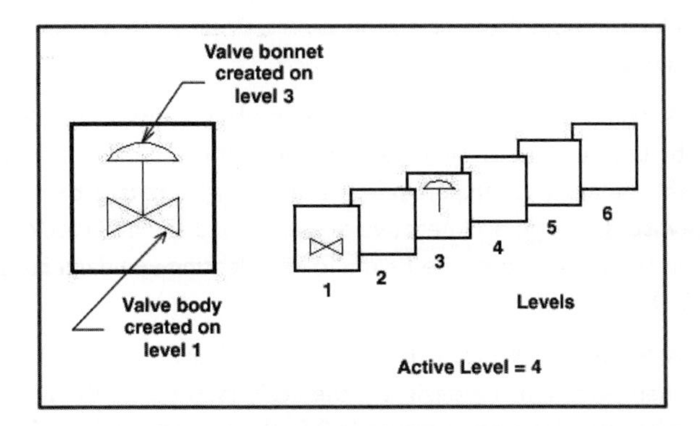

Figure 11–15a Example of placing a cell by Absolute placement mode

In **Relative** placement mode, MicroStation shifts all levels used in the cell such that the lowest level in the cell is on the Active Level in the design. For example, if a Graphic cell has elements on levels 1 and 2 and the active level is 4, the cell elements on level 1 are moved three levels up to level 4 in the design file and the cell elements on level 3 are moved three levels up to level 6 (see Figure 11–15b).

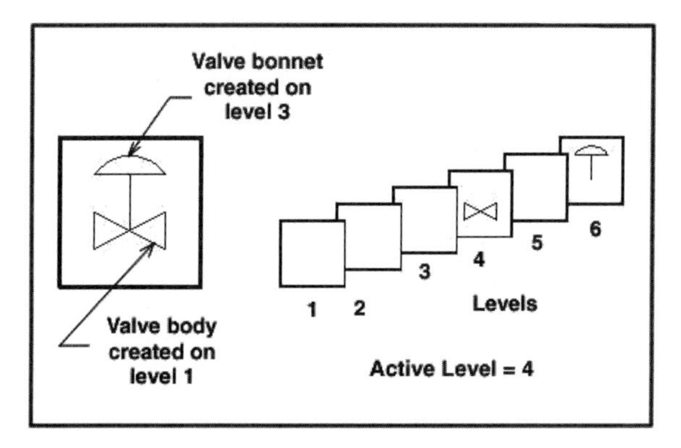

Figure 11–15b Example of placing a cell by Relative placement mode

Place Active Cell using Interactive Placement

Interactive placement is helpful for aligning a new cell with existing elements when the required angle and scale are not known. The tool allows defining the angle and scale graphically while placing the cell.

To place the active cell interactively, turn the **Interactive** check box ON in the settings box. When a cell is placed interactively, MicroStation prompts:

Place Active Cell (Interactive) > Enter cell origin

Define the origin point of the cell.

Place Active Cell (Interactive) > Enter Scale or Corner Point

Either place a data point to scale the cell or key-in the scale factor in the key-in window.

Place Active Cell (Interactive) > Enter Rotation by Angle or Point

Either place a data point to rotate the cell or key-in the rotation angle in the key-in window.

Select and Place Cell

Often, when working in an existing design, additional copies are needed of a cell that was placed earlier but the cell is no longer the Active Placement Cell. The Select and Place Cell tool allows selecting the cell for placement simply by clicking on a copy of it in the design file. The selected cell becomes the active placement cell and dynamic update shows it at the pointer position. Additional data points place copies of the cell.

The Select and Place Cell tool's Tool settings box provides the **Active Angle, X Scale,** and **Y Scale** fields, the **Browse Cells** magnifying glass, and the **Relative** check box. These fields were described earlier in the discussion of the Place Active Cell tool. Invoke the Select and Place Cell tool from:

Task Navigation tool box (active task set to Cells)	Select the Select and Place Cell tool (see Figure 11–16).
Keyboard Navigation (Task Navigation tool box with active task set to Cells)	**e**

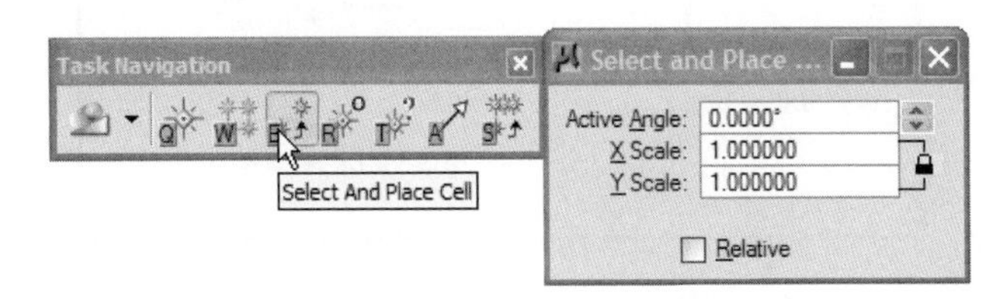

Figure 11–16 Invoking the Select and Place Cell tool from the Cells tool box

MicroStation prompts:

Select and Place Cell > Identify element

Select the cell to be copied.

Select and Place Cell > Accept/Reject (Select next input)

Accept the cell by placing a data point at the location where the first copy is to be placed.

Select and Place Cell > Enter cell origin

Place a data point at the location of each additional copy of the cell, or click the Reset button to cancel the operation.

This tool only works with the copy of the cell in the design, so the tool can select copies of cells that are not in the currently attached cell library. In fact, the tool works when there is no cell library attached.

Place Active Cell Matrix

The Place Active Cell Matrix tool places copies of the Active Placement Cell in a rectangular matrix using parameters defined in the settings box (see Figure 11–17).

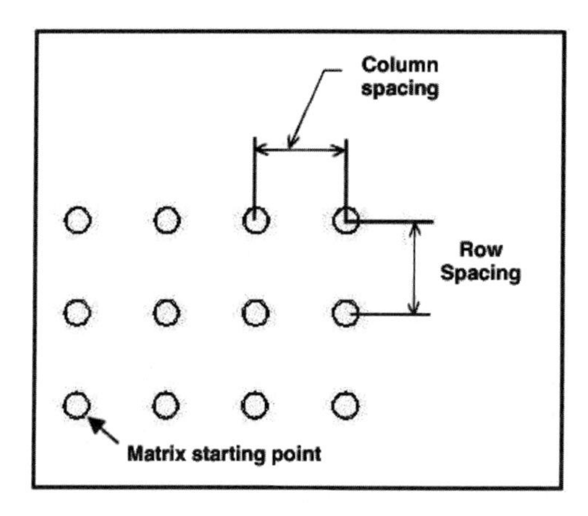

Figure 11–17 Example of placing a cell in a rectangular array

The Place Active Cell Matrix tool's Tool Settings window provides the following fields for defining the size and shape of the matrix:

- **Active Cell**—The active cell name
- **Browse Cells**—The magnifying glass that opens the Cell Library box
- **Rows**—The number of rows in the matrix
- **Columns**—The number of columns in the matrix
- **Row Spacing**—The space, in Working Units, between the rows
- **Column Spacing**—The space, in Working Units, between the columns

Note: *The row and column spacing is from origin point to origin point, not the space between the cells.*

Invoke the Place Active Cell Matrix tool from:

Task Navigation tool box (active task set to Cells)	Select the Place Active Cell Matrix tool (see Figure 11–18).
Keyboard Navigation (Task Navigation tool box with active task set to Cells)	**w**

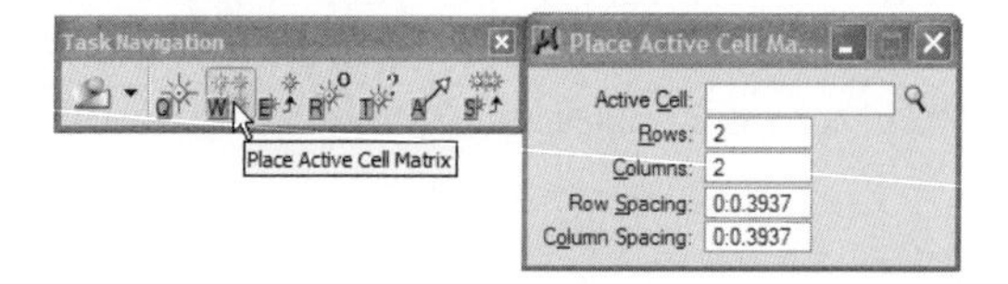

Figure 11–18 Invoking the Place Active Cell Matrix tool from the Cells tool box

MicroStation prompts:

Place Active Cell Matrix > Enter lower left corner of matrix

Place a data point to define the origin location of the cell in the lower left corner. The data point placed to start the matrix designates the position of the origin of the lower left cell in the matrix (see Figure 11–17).

The Place Active Cell Matrix tool places each cell in the matrix at the active angle and active X and Y scales, but these settings are not available in the Tool Settings window. Invoke the Place Active Cell tool or the Select and Place Cell tool to gain access to the **Active Angle**, **X Scale**, and **Y Scale** settings.

PLACE ACTIVE LINE TERMINATOR

The Place Active Line Terminator tool places the active terminator cell at the end of the selected element and automatically rotates the cell to match the rotation of the element at the point of connection. See Figure 11–19 for examples of placing line terminators.

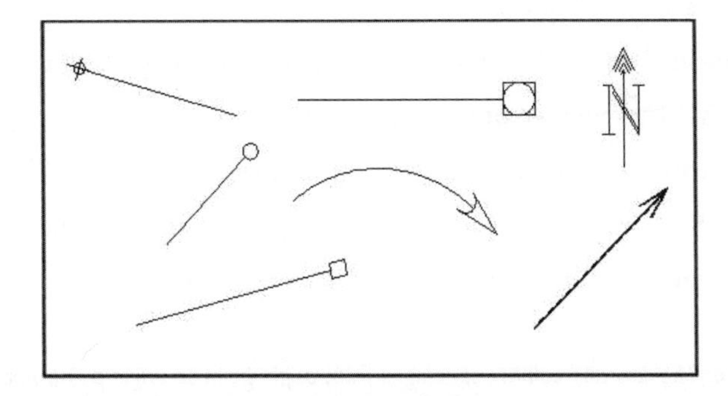

Figure 11–19 Examples of placing line terminators

Place the Active Terminator Cell selected on the Cell Settings box by selecting a cell and clicking the Terminator button.

The Tool Settings window provides fields for modifying the tool's action. The **Terminator** text field displays the name of the Active Terminator Cell. To switch to a different active cell, type the cell name in the field. The **Browse Cells** magnifying glass displays the Cell Library box. The **Scale** field allows keying in a scale factor that is applied to the terminator cell when it is placed. The field sets both the X and Y scale values. A number greater than zero increases the size of the cell and a number less than zero decreases its size. Invoke the Place Active Line Terminator tool from:

Task Navigation tool box (active task set to Cells)	Select the Place Active Line Terminator tool (see Figure 11–20).
Keyboard Navigation (Task Navigation tool box with active task set to Cells)	**a**

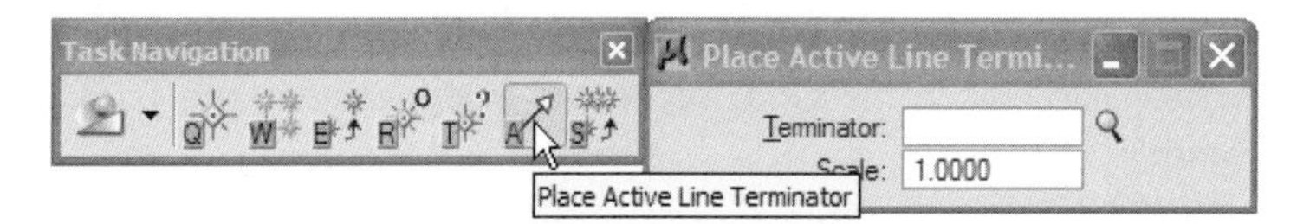

Figure 11–20 Invoking the Place Active Line Terminator tool from the Cells tool box

MicroStation prompts:

Place Active Line Terminator > Identify element

Select the element near the end where the terminator is to be placed.

Place Active Line Terminator > Accept/Reject (Select next input)

Click the Data button again to place the terminator, and, optionally, select another element to place a terminator.

Note: *Beginning users of MicroStation often select the element to place a terminator by first snapping to the element with the Tentative button. That is not necessary, because MicroStation finds the end of the element automatically. Just identify the element near the end where the terminator is to be placed.*

POINT PLACEMENT

MicroStation provides six point tools that place a dot, character, or cell in the active design file:

- The **Place Active Point** tool places a single point at the data point.
- The **Construct Points Between Data Points** tool places a set of equally spaced points between two data points.
- The **Project Point Onto Element** tool places a point on an element at the point on the element nearest to a data point.
- The **Construct Point at Intersection** tool places a point at the intersection of two elements.
- The **Construct Points Along Element** tool places a set of equally spaced points along an element between two data points on the element.
- The **Construct Point at @Dist Along Element** places one point at a keyed-in distance along an element from a data point.

Types of Points

The settings box for each point tool provides the **Point Type** menu for selecting the type of point to place. The available point types are as follows:

- **Element** places dots (0-length lines). To make the dots more noticeable, increase the Active Line Weight before placing them.
- **Character** places a text character using the Active Font. Key-in the appropriate character in the **Character** edit field located in the Tool Settings window.
- **Cell** places the Active Point Cell as the point. Select the Active Point Cell from the Cell Library box by selecting a cell name and clicking the Point button. The settings box for the point tools provides a **Cell** field for typing in a cell name to select the Active Point Cell and the **Browse Cell(s)** magnifying glass. Click the magnifying class to open the Cell Library box and select an Active Point Cell.

Note: *Character and cell points are rotated to the Active Angle. Cell points are rotated to the Active Angle and scaled to the Active Scale. Angle and Scale fields are not provided in the Tool Settings window for the Point tools.*

Do not confuse using a cell as the active point with point and graphic cells. A point or a graphic cell can be the active point cell. The Point tools place graphic cells in Absolute mode only.

Place Active Point

The Place Active Point tool places a single point (element, character, or cell) at the location of the data point. Invoke the Place Active Point tool from:

Task Navigation tool box (active task set to Points)	Select the Place Active Point tool (see Figure 11–21).
Keyboard Navigation (Task Navigation tool box with active task set to Points)	**q**

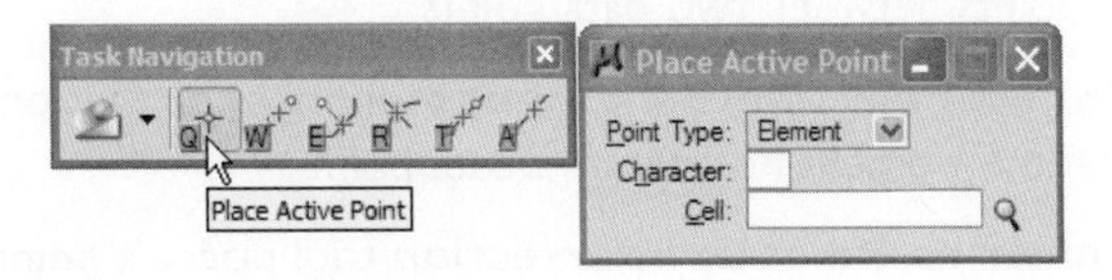

Figure 11–21 Invoking the Place Active Point tool from the Points tool box

MicroStation prompts:

Place Active Point > Enter point origin

Place a data point to place one active point in the design or click Reset to drop the point.

Construct Points Between Data Points

The Construct Points Between Data Points tool places a specified number of points (element, character, or cell) between two data points. In addition to the fields in the settings box for all point tools, the box provides the **Points** field for keying in the number of points to place between the two data points. The number includes the points placed to identify the position of the points. Invoke the Construct Points Between Data Points tool from:

Task Navigation tool box (active task set to Points)	Select the Construct Points Between Data Points tool (see Figure 11–22).
Keyboard Navigation (Task Navigation tool box with active task set to Points)	**w**

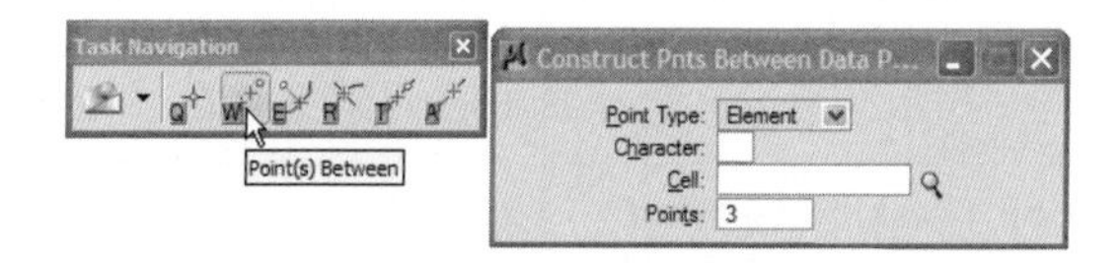

Figure 11–22 Invoking the Construct Points Between Data Points tool from the Points tool box

MicroStation prompts:

Construct Pnts Between Data Points > Enter first point

Key-in the number of points in the **Points** field and place a data point to define the location of the first point in the series.

Construct Pnts Between Data Points > Enter endpoint

Place a data point to define the location of the last point in the series or click the Reset button to cancel the tool without placing any points.

After placing the first set of points, additional sets can be placed. Each additional set uses the last data point of the previous set as its starting point. To start over with a new first data point, click the Reset button. For an example of placing 6 points between two data points, see Figure 11–23.

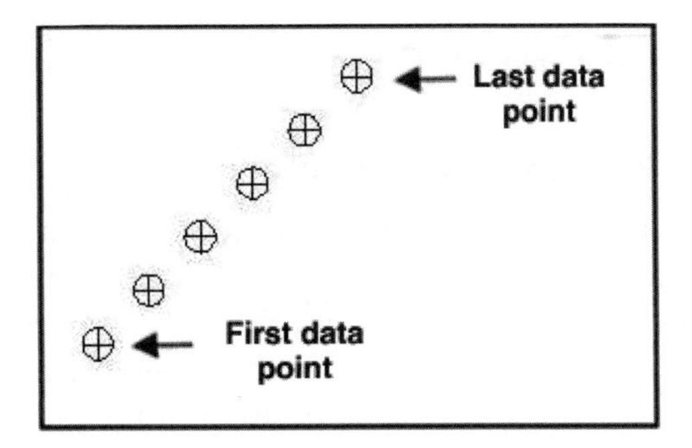

Figure 11–23 Example of placing 10 points between two data points

Project Point Onto Element

The Project Point Onto Element tool places the active point (element, character, or cell) on the selected element at a location projected from the acceptance data point. Invoke the Project Point Onto Element tool from:

Task Navigation tool box (active task set to Points)	Select the Project Point Onto Element tool (see Figure 11–24).
Keyboard Navigation (Task Navigation tool box with active task set to Points)	**e**

Figure 11–24 Invoking the Project Point Onto Element tool from the Points tool box

MicroStation prompts:

Construct Active Point Onto Element > Identify element

Select the element.

Construct Active Point Onto Element > Accept/Reject (Select next input)

Select the point in the design from which to project the point.

For an example of projecting the active point onto an element, see Figure 11–25.

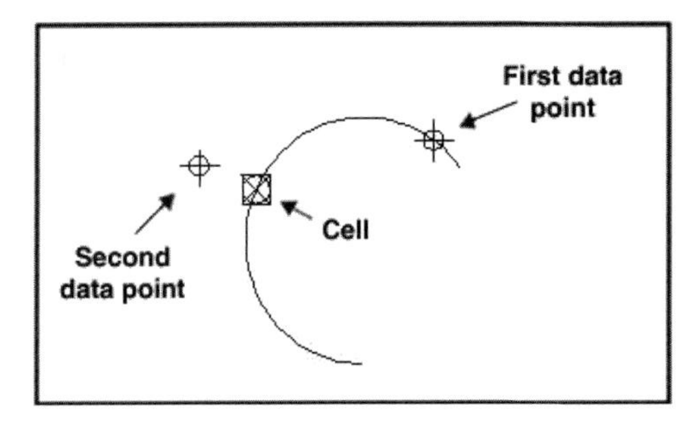

Figure 11–25 Example of placing an active point projected onto an element

Note: *The only purpose of the first data point is to identify the element. There is no need to use the Tentative button when identifying the element.*

Construct Point at Intersection

The Construct Point at Intersection tool places an active point (element, character, or cell) at the intersection of two elements. Invoke the Construct Point at Intersection tool from:

Task Navigation tool box (active task set to Points)	Select the Construct Point at Intersection tool (see Figure 11–26).
Keyboard Navigation (Task Navigation tool box with active task set to Points)	**r**

Figure 11–26 Invoking the Construct Point at Intersection tool from the Points tool box

MicroStation prompts:

Construct Active Point at Intersection > Select element for intersection

Select one of the elements.

Construct Active Point at Intersection > Select element for intersection

Select the other element.

Construct Active Point at Intersection > Accept - Initiate intersection

Place a data point to place the point.

The acceptance data point only initiates placement of the point; it does not identify another element. For examples of placing an active point at an intersection of two elements, see Figure 11–27.

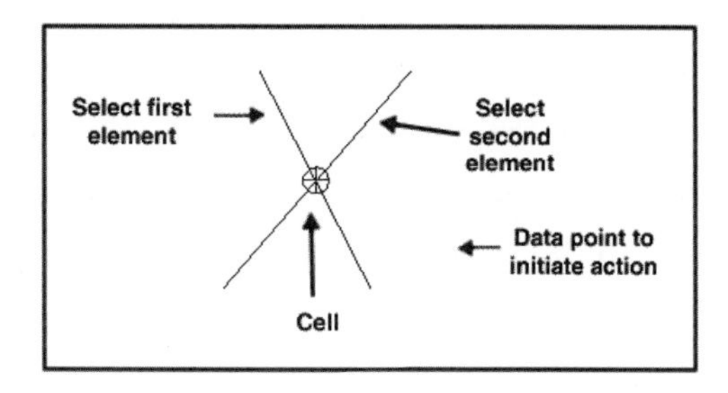

Figure 11–27 Example of placing an active point at an intersection of two elements

Note: *If the two elements intersect more than once (such as a line passing through a circle), identify the two elements close to the intersection where the point is to be placed. There is no need to use the Tentative button. Just place the Data button close to the intersection.*

Construct Points Along Element

The Construct Points Along Element tool places a set of active points (elements, characters, or cells) equally spaced along an element between two data points on the element. In addition to the fields in the settings box for all point tools, the settings box also provides the **Points** tool for keying in the number of points to place between the two data points. The number includes the two data points. Invoke the Construct Points Along Element tool from:

Task Navigation tool box (active task set to Points)	Select the Construct Points Along Element tool (see Figure 11–28).
Keyboard Navigation (Task Navigation tool box with active task set to Points)	**t**

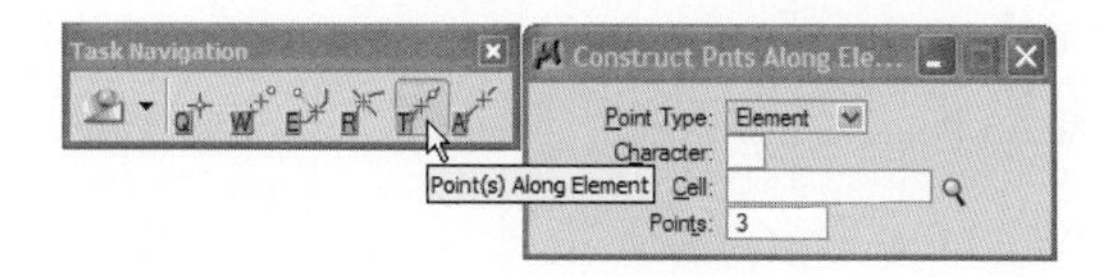

Figure 11–28 Invoking the Construct Points Along Element tool from the Points tool box

MicroStation prompts:

Construct Pnts Along Element > Enter first point

Key-in the number of points in the **Points** field and select the element at the location of the first point.

Construct Pnts Along Element > Enter endpoint

Define the location on the element at the location of the last point or click the Reset button to cancel the tool without placing any points.

The tool remains active so that additional sets of points can be placed along elements. While continuing to place sets, change the active point the number of points as necessary. For an example of placing 6 points along an element between two data points, see Figure 11–29.

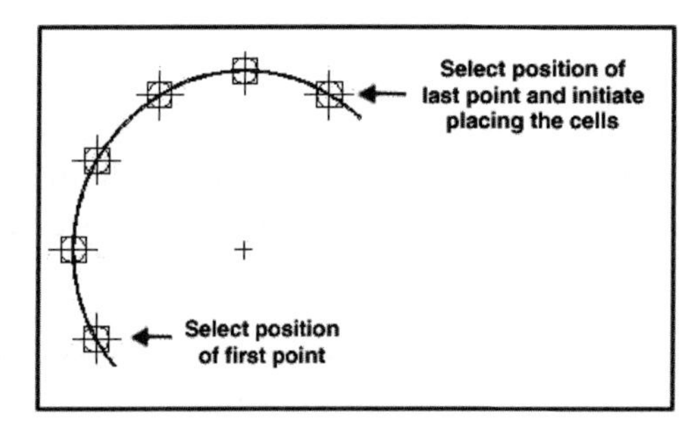

Figure 11–29 Example of placing 10 points along an element between two data points

Point at Distance Along

The Point at Distance Along tool places the active point (element, character, or cell) at a keyed-in distance along an element from the first data point that identified the element. In addition to the fields on the settings box for all point tools, the window also provides the **Distance** tool for keying in the distance along the element in working units. Invoke the Construct Point at Distance Along tool from:

Task Navigation tool box (active task set to Points)	Select the Point at Distance Along tool (see Figure 11–30).
Keyboard Navigation (Task Navigation tool box with active task set to Points)	**a**

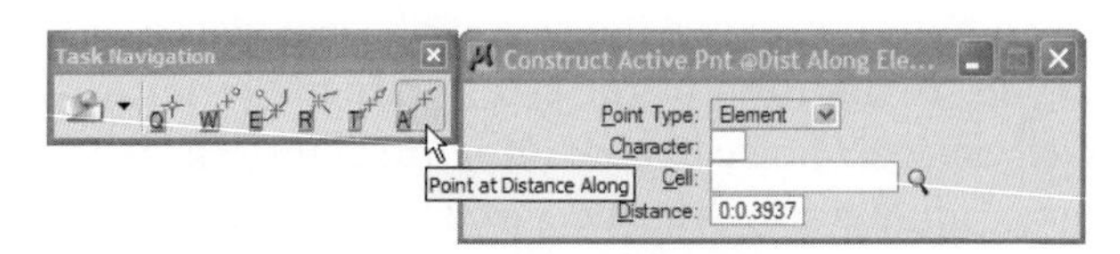

Figure 11–30 Invoking the Point at Distance Along tool from the Points tool box

MicroStation prompts:

Construct Active Pnt @Dist Along Element > Identify element

Key-in the distance in the **Distance** field and identify the element at the starting point of the distance along calculation.

Construct Active Pnt @Dist Along Element > Accept/Reject (Select next input)

Place a data point to accept the construction or click the Reset button to cancel the tool sequence without placing a point. The position of the second data point, relative to the first data point, determines the direction of the distance along calculation. If, for example, the second data point is to the right of the first data point, the distance calculation places the point on the element to the right of the first element.

The tool remains active to allow continuing to selecting elements for point placement. Change the distance and Active Point Cell as required while continuing to select elements. For an example of placing a point at a specified distance along an element, see Figure 11–31.

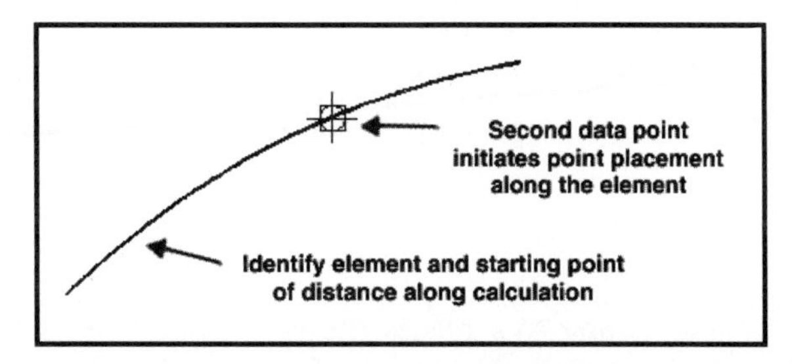

Figure 11–31 Example of placing a point at a specified distance along an element

CELL SELECTOR

In addition to the Cell Library box, MicroStation provides the Cell Selector box for previewing and selecting cells. The Cell Selector box has a button for each cell in the attached library. Clicking one of the buttons selects a placement action.

The Cell Selector box provides several advantages over the Cell Library box. The visual representation of all cells in the attached library allows quickly finding a specific cell. Clicking a button initiates the action of a cell placement tool and the tool used can be changed. In addition to the cell library file attached from the Cell Library box, the Cell Selector box can contain buttons for cells from other cell libraries. Cell Selector configurations can be saved to a file and previously saved Cell Selector configurations can be opened on the design. The ability to save configurations allows the Cell Selector to be used in multiple designs.

Tools on the Cell Library menu allow customizing the actions of each cell button.

OPENING THE CELL SELECTOR BOX

The type of library attachment controls the way the Cell Selector opens the first time after the cell library attachment is changed. Selecting the Cell Selector open option with a single cell library attached to the design opens the Cell Selector box and displays the cell library's cells on the Cell Selector box's buttons. Selecting the Cell Selector open option with all cell libraries in a folder attached to the design, displays a dialog box for selecting a cell library to use with the Cell Selector box.

Note: *The action of the Cell Selector box, when opened, can be changed by the button configuration specified in the cell selector file (.csf) pointed to by the MS_CELLSELECTOR configuration variable. For more information on configuration variables, see Chapter 16.*

Open the cell Selector When One Cell Library is Attached

Open the Cell Selector box from:

Menu	Utilities > Cell Selector

MicroStation displays the Cell Selector box, as shown in Figure 11–32.

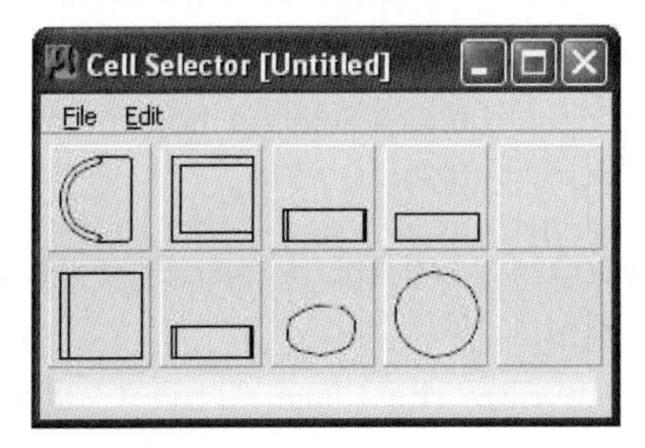

Figure 11–32 Cell Selector box

When the Cell Selector box opens the first time, it may only display two rows by two columns of buttons. Drag one edge of the box to show more cell buttons.

Open the Cell Selector When a Folder is Attached

Open the Cell Selector box from:

Menu	Utilities > Cell Selector

If this is the first opening after a folder was attached from the Cell Library settings box, the Select Cell Library to Load dialog box appears, as shown in Figure 3–33.

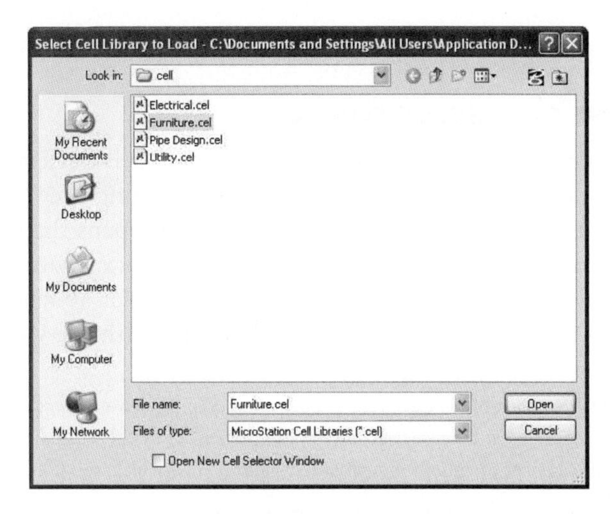

Figure 11–33 Select Cell Library to Load dialog box

The following procedure selects a library for the Cell Selector box:

1. Use the **Look in** menu to navigate to the folder that contains the library to be attached.
2. Select the library from the folder contents list area directly below the **Look in** menu.
3. Click **Open** to open the Cell Selector box and put the selected library's cells on the Cell Selector buttons.

INVOKING TOOLS FROM THE CELL SELECTOR BOX

By default, clicking a cell's button on the Cell Selector Box sets the cell as the Active Placement Cell and invokes the Place Active Cell tool. As will be discussed later, unique tool actions can be assigned to each button.

CUSTOMIZING THE CELL SELECTOR

The **Edit** menu on the Cell Selector box's menu bar provides options for customizing the action, appearance, and content of the cell buttons.

Editing A Button's Content

Use the **Button** option to edit the content and appearance of a selected cell button. Click the button to be edited and invoke the **Button** option from:

Cell Selector settings box	Edit > Button

MicroStation displays the Configure Cell Selector Button settings box, as shown in Figure 11–34.

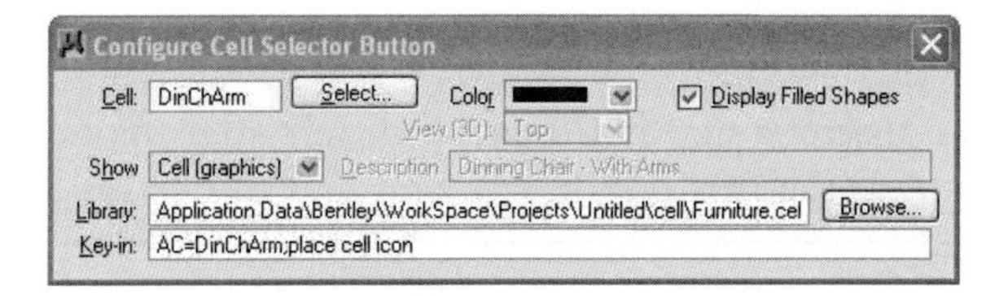

Figure 11–34 Configure Cell Selector Button dialog box

Make the required settings changes and close the dialog box. Following are descriptions of the customization features available on the Configure Cell Selector Button dialog box.

- Click the Color button to display a menu of button colors. Click one of the colors to change the button's color. This changes the color of the button, not the color of the cell itself. As will be discussed later, the same cell can be placed on additional buttons and different click actions can be assigned to each button. Using unique colors for each type of action can be an aid to finding the correct button. For example, black buttons invoke the Place Active Cell tool and green buttons invoke the Place Terminator tool.
- If the **Display Filled Shapes** check box is ON, and a cell contains filled elements, the cell's elements are displayed on the button with their fill color. If the check box is OFF, no fill colors are displayed on the button. For more information on the fill tools, Chapter 12.
- The **Show** menu provides options that control the information displayed on the selected button. Select **Cell (Graphics)** to display a picture of the cell, **Cell Name** to display only the cell's name, **Description** to display the cell's description, or **Cell and Name** to display a picture of the cell on the button and the cell name below the button.
- The **Description** field is available only when the **Description** option is selected on the Show menu. It allows editing the cell description that appears on the button.
- The **Library** field displays the name of the cell library associated with the selected button. To change to a different library, type the location and file name in the field or click the Browse button next to the field to open the **Select Cell Library** dialog box. If using the dialog box, use the **Look in** menu to navigate to the folder containing the library, select the library's name in the field under **Look in**, and click **OK** to select the library.
- The **Key-in** field shows the MicroStation key-in that is initiated by clicking the cell button. Enter multiple key-ins by separating them with a semi-colon. For example, AC=CHAIR1 makes CHAIR1 the Active Placement Cell and activates the Place Active Cell tool.

Inserting a New Button

The **Insert** option inserts a new button immediately after the selected cell button. Select the button that the new button to follow and invoke Insert from:

Cell Selector box	Edit > Insert

MicroStation displays the Define Button dialog box, as shown in Figure 11–35.

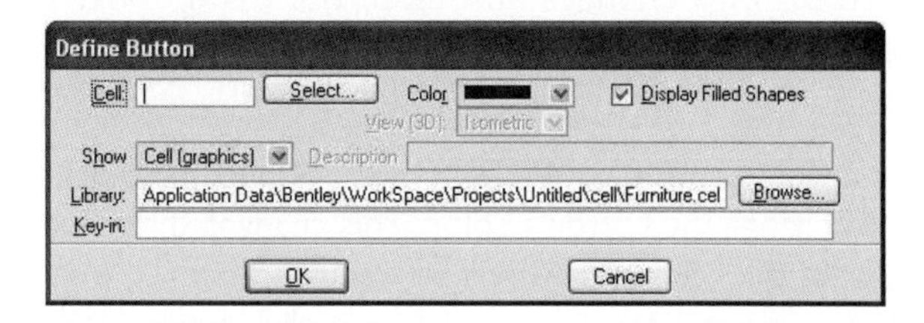

Figure 11–35 Define Button dialog box

The Define Button dialog box contains the same fields as the Configure Cell Selector Button settings box. Make the required settings for the new button and click **OK** to save the settings or click **Cancel** to cancel the operation without creating a new button.

Copying and Pasting a Button

The **Copy** option copies the content of a selected cell button and the **Paste** option pastes the copied content into another selected cell button. To copy the content of a cell button, select the button and invoke the **Copy** option from:

Cell Selector box	Edit > Copy

MicroStation places the contents of the selected button on a temporary "clipboard." To paste the contents of the clipboard onto another cell button, select the button and invoke the **Paste** option from:

Cell Selector box	Edit > Paste

MicroStation places the contents of the clipboard onto the selected cell button.

This procedure is useful when using the same cell for more than one tool action. For example, to use the Arrow cell for placing copies of the active cell and for placing line terminators, set the action in the first Arrow button to activate the Place Active Cell tool, copy and paste the Arrow button to a second button, and set that button's action to activate the Place Terminator tool.

The **Cut** option removes the contents from the selected cell button and places the content onto the temporary clipboard so it can be pasted onto another button. Select the button whose content is to be cut and invoke the **Cut** option from:

Cell Selector box	Edit > Cut

MicroStation immediately removes the selected button's content.

Used with the **Paste** option, **Cut** is useful for moving a cell to a new location on the Cell Selector box.

Deleting a Button

The **Delete** option deletes the selected cell button from the Cell Selector box. Click the button to be deleted and invoke the Delete tool from:

Cell Selector box	Edit > Delete

MicroStation immediately deletes the configuration of the selected button.

Deleting all Buttons

The **Clear** option deletes all buttons from the Cell Selector box. Invoke the Clear Configuration tool from:

Cell Selector box	Edit > Clear

MicroStation immediately deletes the content of all cell buttons on the Cell Selector box.

Changing Button and Gap Size

The **Button Size** option allows changing the size of the button and the gap between buttons. To see more buttons without scrolling the Cell Selector box, make the size and gap smaller. To improve readability of the buttons, make them larger. Invoke the Button Size tool from:

Cell Selector box	Edit > Button Size

MicroStation displays the Define Button Size dialog box, as shown in Figure 11–36.

Figure 11–36 The Define Button Size dialog box

The **Button Size (Pixels)** and **Gap Size (Pixels)** text fields set the size in pixels of the buttons and of the gap between the buttons. After keying-in the desired pixel values in the two fields, click **OK** to apply the new sizes or click **Cancel** to cancel the operation without changing either size.

Setting Button Defaults

The **Defaults** option selects default settings for creating new buttons in the Cell Selector box. Invoke the Define Defaults tool from:

Cell Selector box	Edit > Defaults

MicroStation displays the Define Defaults dialog box, as shown in Figure 11–37.

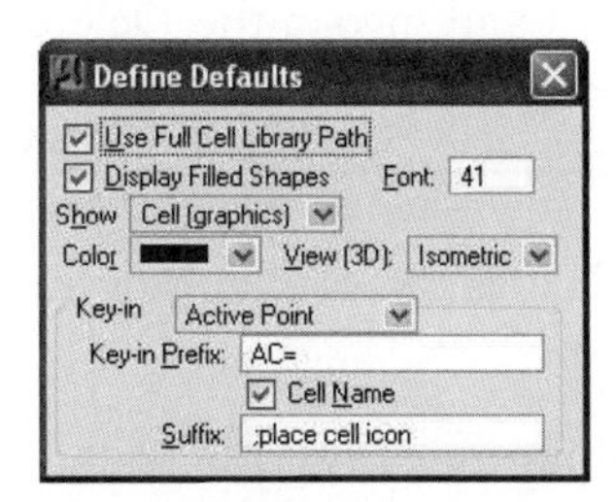

Figure 11–37 Define Defaults dialog box

Make the required settings changes and close the dialog box. Most of the settings are identical to settings in the Configure Cell Selector Button dialog box described earlier in this section. Following are descriptions of additional default settings:

- If the **Use Full Cell Library Path** check box is ON, the full path is displayed in the **Library** field of the Define Button dialog box when inserting a new button. If the check box is OFF, only the cell library file name is displayed.
- The **Font** text field displays the number of the MicroStation font used for the cell name or description on the new button.
- The **View (3D)** menu sets the default display method for 3D cells and is not used in 2D design files.
- The **Key-in** menu selects the Active Cell key-in that is placed in the **Key-in** field when the new button is created. The options are **Active Cell** (AC=), **Active Point** (PT=), **Active Pattern** (AP=), **Active Terminator** (LT=), and **User Defined** (when typing the key-in in the **Key-in Prefix** field).
- The **Key-in Prefix** contains the first key-in placed in the **Key-in** field for new buttons. If one of the active cell options is selected in the **Key-in** menu, the key-in characters for the type of active cell are placed in the **Key-in Pre-**

fix field. If the **User Defined** option is selected, the field is empty. Text can be added, changed, or replaced in this field.

- If the **Cell Name** check box is ON, the name of the cell assigned to the new button is added to the **Key-in** field for new buttons. This check box is automatically turned on when one of the active cell options is selected in the **Key-in** menu.
- The **Suffix** field contains additional key-in sequences for new buttons. This key-in is placed after the key-in listed in the **Key-in Prefix** field and is separated from it by a semicolon. To place more than one key-in in this field, separate them with a semi-colon.

CREATING AND USING CELL SELECTOR FILES

The **File** options in the Cell Selector box allows adding cells from other libraries to the Cell Selector box and saving the box's configuration to a file that can be opened in other design files.

Adding Buttons from Other Cell Libraries

In the cell selection descriptions thus far, the cells on the buttons have all been from the cell library currently attached to the design file. The Cell Selector's **Load Cell Library** option allows placing cells from additional cell libraries in the active design file's Cell Selector box. The cell buttons from loaded libraries have the same features as buttons from the attached library. For example, they can be used with the Place Active Cell tools. Invoke the Load Cell Library tool from:

Cell Selector box	File > Load Cell Library

MicroStation displays the Select Cell Library to Load dialog box, as shown in Figure 11–38.

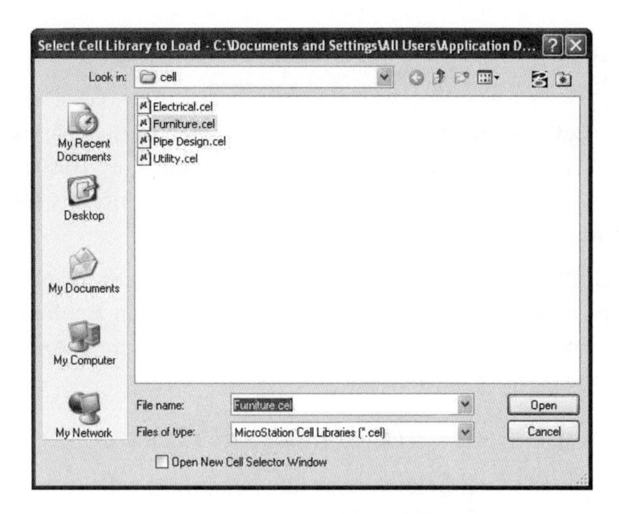

Figure 11–38 The Select Cell Library to Load dialog box

Select the folder that contains the file to be loaded from the **Look in** menu and select the file name from the file names field directly under **Look in**. Click **OK** to load the selected library or click **Cancel** to cancel the operation without loading the library.

Creating a New File

Use the **New** option to create a file that contains the Cell Selector configuration. Invoke the **New** option from:

Cell Selector box	File > New

MicroStation displays the Define Cell Selector File dialog box, as shown in Figure 11–39.

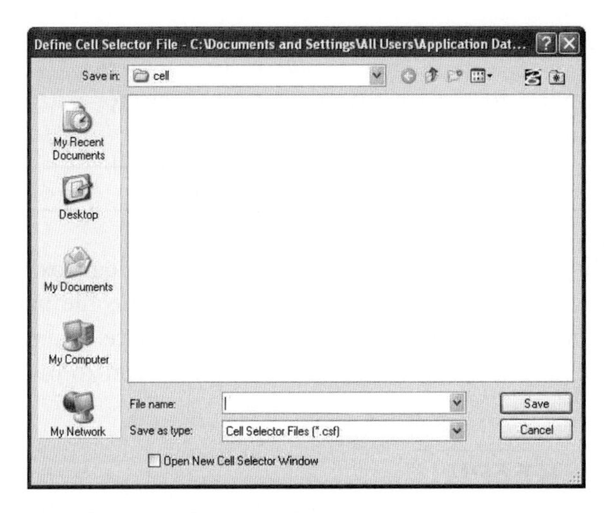

Figure 11–39 The Define Cell Selector dialog box

Select the folder where the file is to be placed in the **Save in** menu and key-in the file name in the **File name** field. Click **OK** to create the file or click **Cancel** to cancel the operation without creating the file. Clicking **OK** displays the file location and name on the Cell Selector box.

Saving the Current Configuration

To save changes made to the Cell Selector configuration after creating a new configuration file, use the **Save** option to save the changes to the file. Invoke the **Save** option from:

Cell Selector box	File > Save

If there is a file location and name in the Cell Selector title bar, MicroStation saves the Cell Selector changes in that file. Otherwise, it invokes the **New** option for creating a new file to store the configuration.

Saving the Current Configuration to a Different Cell Selector File

Use the **Save As** option to save the current Cell Selector content and configuration to a different file from the one displayed in the title bar. Invoke the **Save As** option from:

Cell Selector box	File > Save As

MicroStation displays the Define Cell Selector File dialog box. Select a folder for the file in the **Save in** menu and key-in the file name in the **File name** field. Click **OK** to create the file or click **Cancel** to cancel the operation without creating the file.

Opening a Cell Selector File

Use the **Open** option to replace the current Cell Selector with the Cell Selector stored in a file. Invoke the **Open** option from:

Cell Selector box	File > Open

MicroStation displays the Select Cell Selector File dialog box. Select the folder that contains the library on the **Look in** menu and the file name from the file names field directly below **Look in.** Click **OK** to replace active design file's Cell Selector with the one in the selected file or click **Cancel** to cancel the operation without opening the file.

CELL HOUSEKEEPING

Following are descriptions of additional tools that work on cells already placed in the design. The available tools include:

- **Identify Cell**—Displays the name and other related information about the selected cell.
- **Replace Cell**—Replaces a cell in the design file with another cell from the currently-attached cell library or libraries.
- **Drop Complex Status**—Breaks a cell into its individual elements. The elements lose their identities as being part of a cell.
- **Drop Fence Contents**—Breaks all cells selected by a fence to their individual elements. The elements lose their identities as being part of cells.
- **Fast Cells View**—Speeds up view updates by displaying only a box showing the location of cells in the view, rather than the cell elements.

IDENTIFY CELL

The Identify Cell tool displays the name of a selected cell. This tool is useful when a cell already placed in the design file needs to be used as a terminator cell or point cell but the cell's name is unknown. The tool displays the information in the Status bar. Invoke the Identify Cell tool from:

Task Navigation tool box (active task set to Cells)	Select the Identify Cell tool (see Figure 11–40).
Keyboard Navigation (Task Navigation tool box with active task set to Cells)	**t**

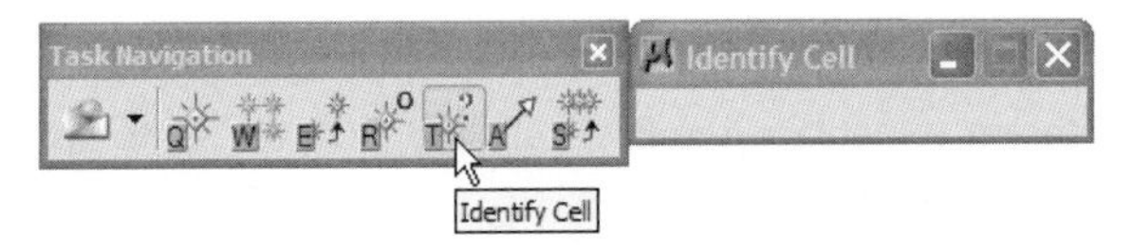

Figure 11–40 Invoking the Identify Cell tool from the Cells tool box

MicroStation prompts:

Identify Cell > Identify element

Select the cell.

Identify Cell > Accept/Reject (Select next input)

Place a data point to accept the cell or click the Reset button to cancel the selection.

The name of the cell appears in the Status bar after the first data point and again after the second data point.

REPLACE CELLS

The Replace Cells tool updates or replaces cells in the design file. It is useful when the design of a cell is changed and there are copies of the old cell in the design, and when design changes require different cells, such as a different type of valve.

The Replace Cells Tool Settings Window

The Tool Settings window provides several settings to control the way the Replace Cells tool operates, but some of them may not be visible on the settings box when the tool is selected. Click the **Show Cell Replacement Options** arrow on the lower right of the box to display all the options. Figure 11–41 shows the expanded Replace Cells settings box.

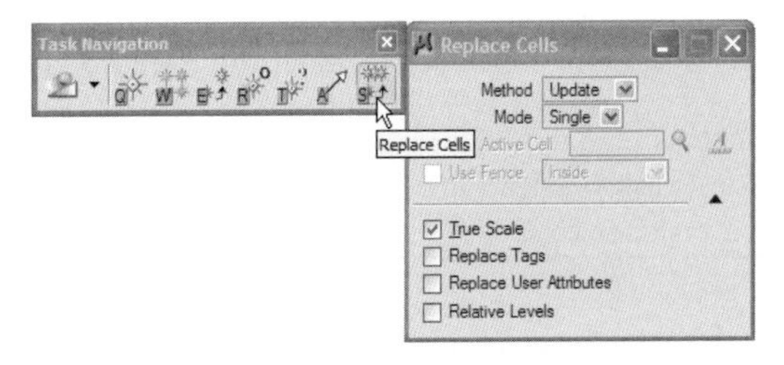

Figure 11–41 Invoking the Replace Cells tool from the Cells tool box

The **Method** menu allows choosing between updating and replacing selected cells. The **Update Method** replaces the selected cell in the active design with the cell of the same name in the attached cell library. The **Replace Method** action depends on the **Use Active Cell** check box (which is only available when the **Replace Method** is selected). If the **Use Active Cell** check box is:

- OFF, identify the cell in the design that is to be replaced, identify another cell in the design to replace the first cell, and click the Data button a third time to initiate the replacement.
- ON, identify and accept a cell to have it replaced by the Active Placement Cell. Set the Active Placement Cell by keying-in its name in the **Use Active Cell** text field or by clicking the **Browse Cell(s)** magnifying glass and selecting it from the Cell Library settings box.

The **Mode** menu provides two options that control cell updating or replacement. Select the **Single Method** to update individual cells by selecting and accepting them or select **Global Update** to update or replace all cells with the same name as the selected cell.

If the **Use** Fence check box is ON, and a fence is defined in the design, the tool manipulates all cells selected by the fence. When this check box is ON, the menu next to the check box allows selecting a **Fence Mode** to use for the tool operation.

If the **True Scale** check box is ON, and the replacement cell was created in a design file that used different working units from those in the active design file, the cell is adjusted to its true size using the units of the active file. If the check box is OFF, the size is not adjusted.

The **Replace Tags** and **Replace User Attributes** check boxes control the sets of tags and attributes associated with the new cell. Turn the check boxes ON to use the replacement cell's tags and attributes. Turn the check boxes OFF to retain the tags and attributes of the old cell.

If the **Relative Levels** check box is ON, the lowest level in the new cell is set equal to the lowest level in the old cell and the other levels in the new cell are adjusted up or down by the same number of levels as the lowest level. If the check box is OFF, no level adjustment is made. Invoke the Replace Cells tool from:

Task Navigation tool box (active task set to Cells)	Select the Replace Cells tool (see Figure 11–41).
Keyboard Navigation (Task Navigation tool box with active task set to Cells)	**s**

If no fence defined in the design, or the **Use Fence** check box is OFF, MicroStation's first prompt is:

Replace Cell > Identify Cell

Identify the cell to update or replace.

If the **Update Method** is selected or the **Replace method** is selected, and the **Use Active Cell** check box is ON, MicroStation' second prompt is:

Replace Cell > Accept/Reject

Click the Data button to initiate the replacement action or click the Reset button to cancel the operation.

If the **Replace Method** is selected and the **Use Active Cell** check box is OFF, MicroStation's second and third prompts are:

Replace Cell > Accept/Reject Replacement Cell

Select the cell that is to replace the first selected cell.

Replace Cell > Identify Replacement Cell

Click the Data button to initiate the replacement action or click the Reset button to cancel the operation.

Note: *To replace a shared cell (discussed later in this chapter), identify one of the shared cells and MicroStation replaces all instances of the shared cell with the same name.*

Note: *The replaced cell may shift position in the design file if the new cell's origin point does not have in the same relationship to the cell elements as the old cell's origin.*

If a fence is defined in the design and the **Use Fence** check box is ON, the fence identifies the cells to be replaced, so MicroStation skips the cell identification prompt.

If the **Update Method** is selected, or the **Replace Method** is selected and the **Use Active Cell** check box is off, MicroStation prompts:

Replace Cell > Accept/Reject Fence

Click the Data button anywhere in the design to initiate the replacement action or click the Reset button to cancel the operation.

If the **Replace Method** is selected and the **Use Active Cell** check box is OFF, MicroStation prompts:

Replace Cell > Accept/Reject Replacement Cell

Select the cell that is to replace the cell selected first.

Replace Cell > Identify Replacement Cell

Click the Data button to initiate the replacement action or click the Reset button to cancel the operation.

DROP COMPLEX STATUS

Cells placed in the design file are "complex shapes" that act like one element when manipulated. Before the shape of a cell in the design can be changed, it must be "dropped" to break the cell into separate elements. A dropped cell loses its identity

as a cell and becomes separate, unrelated elements. To drop a cell, invoke the Drop Element tool from:

Groups tool box	Select the Drop Element tool (see Figure 11–42).
Keyboard Navigation	**6 1**

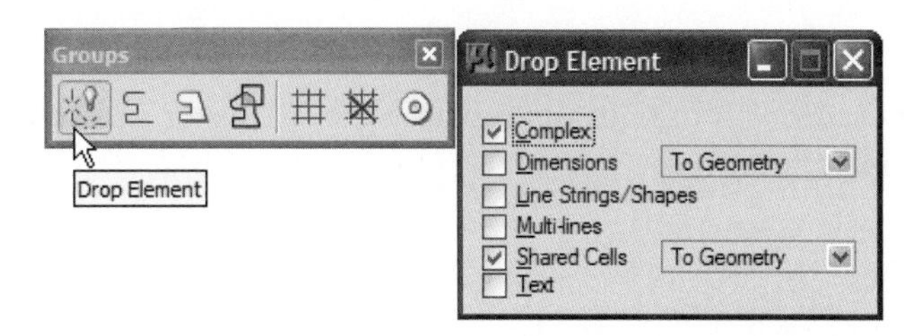

Figure 11–42 Invoking the Drop Element tool from the Groups tool box

MicroStation prompts:

Drop Element > Identify element

Turn the **Complex** check box ON, turn the other check boxes OFF, and select the cell to drop.

Drop Element > Accept/Reject (select next input)

Click the Data button to drop the cell or click the Reset button to cancel the operation.

DROP FENCE CONTENTS

The Drop Fence Contents tool breaks all the cells selected by a fence into separate elements. Before selecting the tool, group the cells to be dropped by placing a fence and selecting the appropriate fence mode. Invoke the Drop Fence Contents tool from:

Fence tool box	Select the Drop Fence Contents tool (see Figure 11–43).
Keyboard Navigation	**2 5**

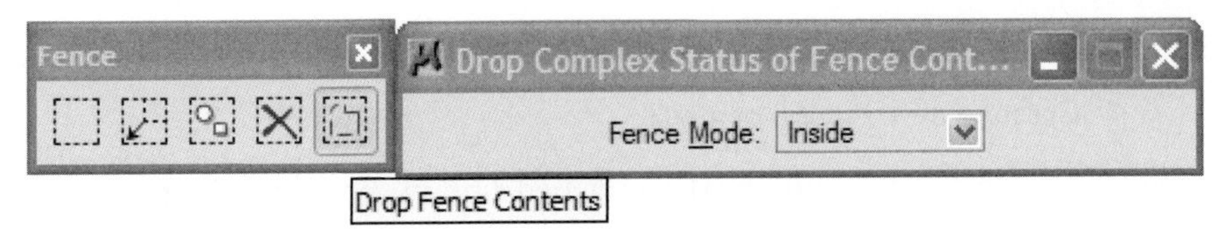

Figure 11–43 Invoking the Drop Fence Contents tool from the Fence tool box

MicroStation prompts:

Drop Complex Status of Fence Contents > Accept/Reject fence contents

Click the Data button to initiate the drop action or click the Reset button to cancel the operation.

FAST CELLS VIEW

A large number of complex cells in a view may cause the Update View tool to take an unacceptable amount of time to complete refreshing the view. If that is happening, turn the **Fast Cells** check box ON in the View attributes settings box to reduce the update time. When the check box is ON for a view, MicroStation displays a box at each cell location, rather than the cell elements.

To display boxes at the locations of cells in the view, do the following:

1. Select Settings > View Attributes to open the View Attributes settings box.
2. Check the **View Number** menu and, if necessary, change it to the number of the view where boxes are to be displayed in place of the actual cells.
3. Turn the **Fast Cells** check box ON.
4. If finished with the View Attributes settings box, close it.

The view attribute settings stay in effect until change or the design is closed.. To make the settings permanent, select **File** > **Save Settings**.

Note: *Only the boxes print when* ***Fast Cells*** *is ON in the printed view.*

LIBRARY HOUSEKEEPING

Thus far, the chapter described how to place cells in the design and maintain them. This topic looks at some housekeeping tools that take care of the cells in the attached cell library. The discussion includes explanations of how to do the following:

- Edit a cell's name and description
- Delete a cell from the Cell Library

- Compress the attached Cell Library
- Create a new version of a cell

All these tools affect the cells in the attached library, not the cells placed in the design.

EDIT A CELL'S NAME AND DESCRIPTION

The Edit Cell Information dialog box, which is available from the Cell Library box, allows changing a cell's name and description. To create a new version of a cell and keep the original cell, rename it before creating the new version. If the person who designed the cell failed to provide a description, provide one to help other users of the cell library understand the cell's purpose. This procedure renames a cell and changes the cell description.

1. If the Cell Library box is not already open, select **Element > Cells** to open it.
2. Select the cell to edit from the list of cells in the Cell Library box.
3. Click the Edit button to open the Edit Cell Information dialog box (see Figure 11–44).
4. Make the necessary changes to the cell name and description.
5. Click **Modify** to make the necessary changes and close the Edit Cell Information dialog box or click **Cancel** to cancel the operation.

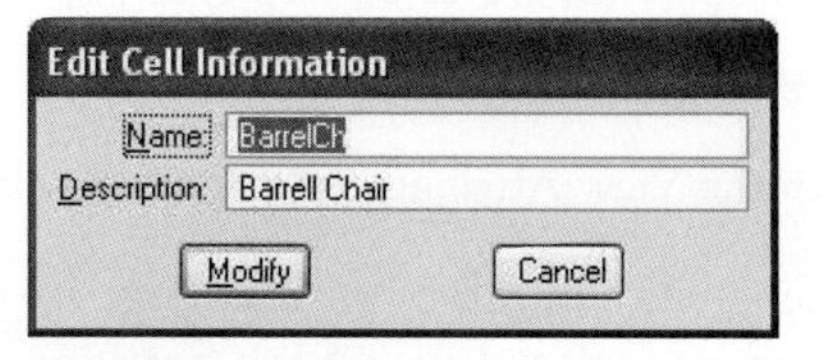

Figure 11–44 Edit Cell Information dialog box

Note: *The Edit button only works for a single cell library attachment. It does not work for a folder attachment.*

Note: *Changing the cell name and description in the library does not affect the cells already in the design file (or any other design file). The cells in the design keep their old names.*

Alternate Method

To rename a cell from the key-in window, type CR=<old>, <new>, and press ENTER. Replace <old> with the cell's current name and <new> with the new cell name. The key-in only changes the cell names. There is no key-in to change a description.

DELETE A CELL FROM THE LIBRARY

If a cell becomes obsolete, or if it was drawn incorrectly, it can be deleted from the cell library using the Delete button on the Cell Library box. Clicking **Delete** with a cell selected causes MicroStation to open an Alert dialog box that asks for confirmation of deleting the cell. This procedure deletes a cell from the Cell Library:

1. If the Cell Library box is not already open, select **Element > Cells** to open it.
2. Select the cell to delete from the cells list in the dialog box.
3. Click **Delete**.
4. When the Alert box appears, click **OK** to delete it or click **Cancel** to cancel the operation.

Note: *The* **Delete** *tool works for both single cell library attachments and folder attachments.*

Note: *This procedure deletes a cell from the cell library. It does not delete copies of the cell already placed in the design file (or any other design file).*

Alternate Method

To delete a cell from the key-in window, type CD=<name> and press ENTER. Replace <name> with the cell's name. When the key-in is used to delete a cell, MicroStation does not open the Alert window.

Note: *The Undo tool cannot undo the deletion of a cell from the attached library.*

COMPRESS THE ATTACHED CELL LIBRARY

Deleting a cell from a cell library does not really delete it. The cell is marked as deleted in the library and no longer appears in the cells list, but its elements still take up space in the cell library file. To get rid of the no-longer-usable cell elements, compress the cell library with the **Compress** option in the Cell Library box's **File** menu. This procedure compresses the attached Cell Library:

1. If the Cell Library box is not already open, select **Element > Cells** to open it.
2. Select **File > Compress** on the Cell Library box.

The attached cell library is compressed, and the Status bar indicates that the Cell Library is compressed.

***Note:** The **Compress** tool only works for a single cell library attachment. It does not work for a folder attachment.*

CREATE A NEW VERSION OF A CELL

Occasionally there is a need to replace a cell with an updated version of that cell. The geometric layout of the object represented by the cell may have changed or a mistake may have been made when the cell elements were originally drawn. Cell elements cannot be edited in the library. A new cell must be created and placed in the library.

This procedure replaces a cell in the library with a new version of the cell:

1. Place a copy of the cell in a design file that has the same Working Units as the design file in which the cell was created.
2. Drop the cell.
3. Make the required changes to the cell elements in the design.
4. If possible, place the cell origin at the same place as in the old cell.
5. Delete the old cell from the library or rename it.
6. Create the new cell from the modified elements.
7. If the old cell was deleted, compress the library.
8. If there are copies of the cell in design files, invoke the Replace Cell tool in each design file to update the copies with the new version of the cell.

SHARED CELLS

The cell placement tools described thus far place a separate copy of the cell's elements in the design file. That uses up a lot of disk space in a design file that contains many copies of the same cell.

Shared cells reduce the size of design files because Each time a shared cell is placed, the placement refers to the shared copy of the cell in the design file rather than placing another copy of the cell's elements in the design file. No matter how many copies of a shared cell are placed, only the one shared copy is actually in the design file.

When a cell is declared to be shared, MicroStation places a copy of it in the design file. Shared cell's elements stored in the design file even though no copies of the cell have ever been placed in the design. Later, when the cell is placed in the design, the placements refer to the locally stored cell rather than being placed as a copy in the design. Shared cells can be placed even when no cell library is attached.

This section discusses the following shared cell procedures:

- Turn on the shared cell feature
- Determine which cells are shared
- Declare a cell to be shared
- Place shared cells
- Turn a shared cell into an unshared cell
- Delete the shared cell copy from the design file

TURN ON THE SHARED CELL FEATURE

The **Use Shared Cells** check box on the Cell Library box (see Figure 11–45) turns the shared cell feature ON and OFF.

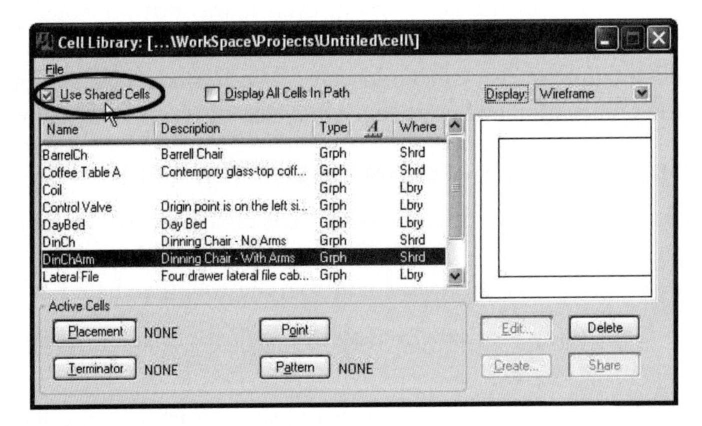

Figure 11–45 Use Shared Cells Check box is circled on the Cell Library box

DETERMINE WHICH CELLS ARE SHARED

When the shared cells feature is ON, all shared cells in the design file appear on the Cells List in the Cell Library box. Shared cells are indicated by "Shrd" in the **Where** column (see Figure 11–45). If the shared cell is also in the currently attached cell library, the list shows the shared copy in the design file—not the one in the library. Select a shared cell in the design file and the Status bar says it is a shared cell.

DECLARE A CELL TO BE SHARED

Use the Share button located in the bottom right in the Cell Library box to declare cells as being shared:

1. If the Cell Library box is not already open, select **Element > Cells** to open it.
2. Turn the **Use Shared Cells** check box ON.

3. Select the cell to be shared.
4. Click the Share button.

Clicking the Share button displays "Shrd" in the **Where** column of the selected cell. Sharing a cell stores a shared copy of the cell in the design file but does not display place the cell on the design. This provides a way to create a seed file for a group of design files that will use the same set of cells. Declare all required cells shared and create new design files by copying the seed file. Each copied file contains the shared cells.

PLACE SHARED CELLS

When the **Use Shared Cells** check box is ON, all cell placement tools place shared cells. Placing a cell makes the cell shared, even if was not previously shared. The tools that place the active placement, terminator, and point cells all work the same for shared cells as they do for unshared cells. The Replace Cell tool automatically replaces all the copies of the shared cell when one of them is identified.

TURN A SHARED CELL INTO AN UNSHARED CELL

The Drop Element tool provides a **Shared Cells** check box for turning shared cells into either unshared cells or individual elements in the design. The menu associated with the **Shared Cells** check box controls the way the shared cell is dropped:

- **To Geometry** drops the cells all the way back to individual elements in the design.
- **To Normal Cell** removes the shared status (each shared cell is replaced with a local copy of the cell).

Invoke the Drop Element tool from:

Groups tool box	Select the Drop Element tool (see Figure 11–46).
Keyboard Navigation	**6 1**

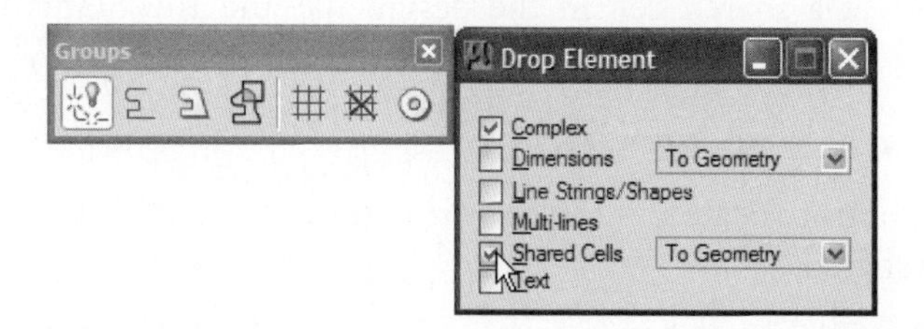

Figure 11–46 Invoking the Drop Element tool from the Fence tool box with Shared Cells selected

MicroStation prompts:

Drop Element > Identify element

Turn the **Shared Cells** check box ON and select the drop level (**To Geometry** or **To Normal** Cell and select the shared cell to drop.

Drop Element > Accept/Reject (select next input)

Click the Data button to convert the selected cell or click the Reset button to cancel the operation without dropping the cell.

DELETE THE SHARED CELL COPY FROM THE DESIGN FILE

The placement of shared cells can be deleted like any other element, but the actual shared cell copy takes a little more work to delete. It is done in the Cell Library box via the Delete button. Use this procedure to remove a shared cell from the design:

1. Either delete all placements of the shared cell or make them into unshared cells (see earlier).
2. If the Cell Library box is not already open, select **Element > Cells** to open it.
3. Turn the **Use Shared Cells** check box ON.
4. Select the shared cell to delete and make sure the cell has "Shrd" in the **Where** column.
5. Click the Delete button (located in the bottom right of the Cell Library box).
6. In the Alert box, click **OK** to delete the shared cell or click **Cancel** to cancel the operation.

This procedure deletes only the copy of the shared cell elements in the design file. It does not delete anything from the attached cell library.

Note: *If there are still placements of the shared cell in the design file, an Alert dialog box appears with a message stating that it cannot be deleted.*

Open the Exercise Manual PDF file for Chapter 11 on the accompanying CD for project and discipline specific exercises.

REVIEW QUESTIONS

Write your answers in the spaces provided.

1. Explain briefly the difference between a cell and cell library.

____

2. How many cell libraries can be attached at one time to a design file using the Attach Library option? ________________

3. List the steps involved in creating a cell.

____

____

4. What is the alternate key-in AC= used for?

____

5. What is the file that has an extension of .cel?

____

6. What does it mean to place a cell Absolute?

____

7. What does it mean to place a cell Relative?

____

8. Explain briefly the differences between a graphic cell and a point cell.

____

____

9. How many cells can be stored in a library?

____

10. What is the purpose of defining an active cell as a terminator?

____

11. What is the purpose of turning ON the Fast Cells View attribute

____

12. Explain briefly the benefits of declaring a cell to be shared.

____

____

CHAPTER 12

Patterning

Patterns are used in drawings for several reasons. Cutaways (cross sections) are hatched to help the viewer differentiate among components of an assembly and to indicate the material of construction. Patterns on surfaces depict material and add to the readability of the drawing. In general, patterns help communicate information about design. Because drawing patterns is a repetitive task, it is an ideal computer-aided drafting application.

OBJECTIVES

Topics explored in this chapter:

- Controlling the display of patterns in view windows
- Placing hatching, crosshatching, and area patterns via seven placement methods
- Manipulating patterns
- Filling elements

CONTROLLING THE VIEW OF PATTERNS

Patterns are complex elements that may take a long time to display. Placing numerous patterns in a design may cause view updates take longer, especially on slow workstations. To overcome that, MicroStation provides the **Patterns** view attribute check box to turn the display of pattern elements ON and OFF.

Turn the **Patterns** check box ON when placing new patterns so the placement results can be seen. When work with patterns is completed, the **Patterns** check box can be turned OFF to speed up view updates. Figure 12–1 shows a crosshatch pattern with the view attribute ON and with it OFF.

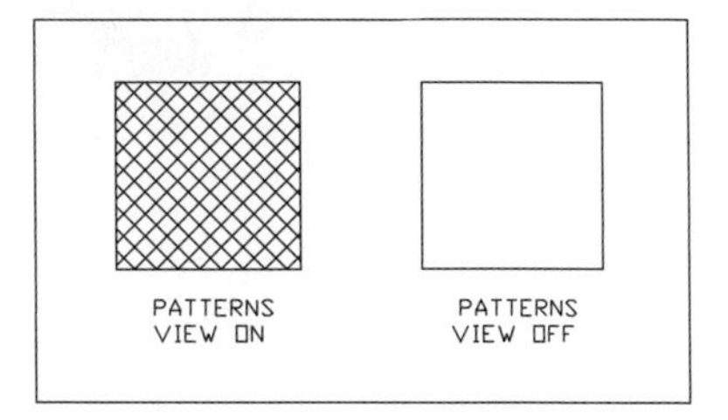

Figure 12–1 Example of the Patterns view attribute ON and OFF

To change the **Patterns** view status, invoke the View Attributes settings box from:

Menu	Select Settings > View Attributes
Key-in window	**dialog viewsettings** (or **di views**) (ENTER)

MicroStation displays the View Attributes settings box, as shown in Figure 12–2. Turn the **Patterns** check box ON or OFF and select a **View Number** from the top of the settings box. Click the **Apply** button to apply the settings to the selected view or set the check box Apply to **All** to ON to apply the settings to all view windows.

***Note:** If patterns are placed in a view that has the **Patterns**, check box turned OFF, the patterns are placed but they do not show up on the screen. To avoid confusion, turn the **Patterns**, check box ON before placing patterns.*

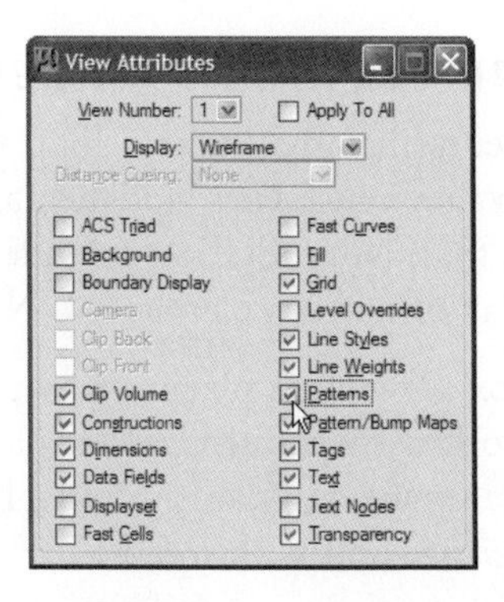

Figure 12–2 View Attributes settings box with the Patterns view attribute set to ON

PATTERNING TOOLS

MicroStation provides three tools for placing patterns in a design file:

- The Hatch Area tool places a set of parallel lines in the selected pattern area. Elements used to define the area to be hatched can be in the active model, or

in references. Where associative patterning is used, any modification to elements defining the hatched area results in an equivalent update to the hatching.

- The Crosshatch Area tool places two sets of parallel lines in the selected pattern area. Elements used to define the area to be crosshatched can be in the active model, or in references. Where associative patterning is used, any modification to elements defining the crosshatched area results in an equivalent update to the crosshatching.
- The Pattern Area tool fills the selected pattern area with tiled copies of the active pattern cell. Elements used to define the area to be patterned can be in the active file, or in references. Where associative patterning is used, any modification to elements defining the patterned area results in an equivalent update to the patterning.

Examples of the patterns placed by each tool are shown in Figure 12–3.

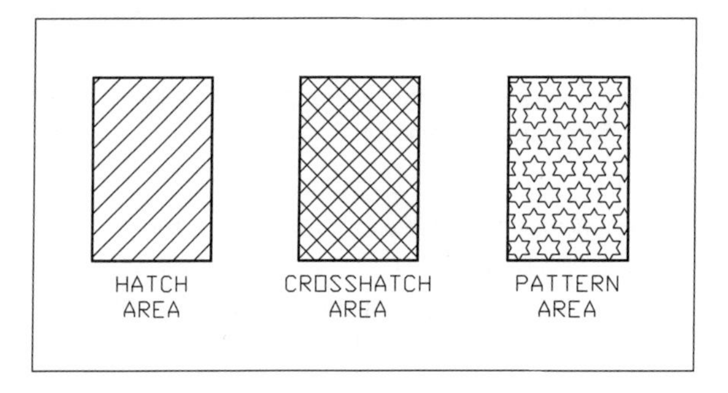

Figure 12–3 Examples of hatching, crosshatching, and patterning

HATCH AREA TOOL

Invoke the Hatch Area tool from:

Task Navigation tool box (active task set to Patterns)	Select the Hatch Area tool (see Figure 12–4).
Keyboard Navigation (Task Navigation tool box with active task set to Patterns)	**q**

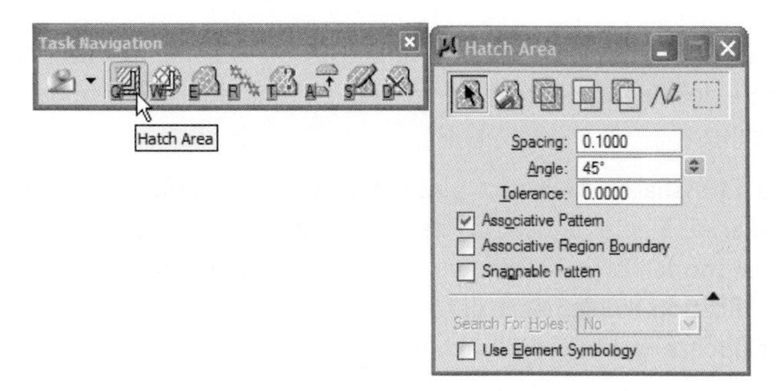

Figure 12–4 Invoking the Hatch Area tool from the Task Navigation tool box (active task set to Patterns)

Icons in the Tool Settings window are used to select the method for defining the area to be hatched. Element method is selected by default, and MicroStation prompts:

> Hatch Area > Identify element *(Select the closed element to be hatched as shown in Figure 12–5.)*
>
> Hatch Area > Accept @pattern intersection point *(Click the Data button again to accept the element and initiate hatch placement.)*

***Note:** For each method, the acceptance point defines a point through which pattern elements pass—one line for Hatch Area, the intersection of two lines for Crosshatch Area, and the origin point of one of the pattern cells for Pattern Area. If the acceptance point is not within the patterned area, the intersection will be on the edge of the area at the closest extension of the acceptance point.*

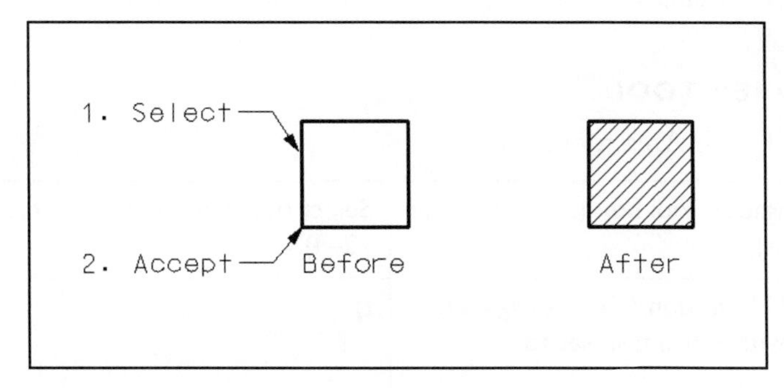

Figure 12–5 Example of patterning a closed element

The **Spacing** field in the Tool Settings window sets the interval between hatching lines.

The **Angle** field in the Tool Settings window sets the angle at which hatching lines are drawn. By default, the angle is relative to the view being used. Where AccuDraw is active, however, the angle is relative to the AccuDraw drawing plane.

The **Tolerance** field in the Tool Settings window sets the variance between the true element curve and the approximation of the curve when a curved element is patterned. The curve is approximated by a series of straight-line segments, and a low tolerance number increases the accuracy of the approximation by reducing the length of each segment. The low tolerance number also means a larger design file and slower pattern placement.

The **Associative Pattern** check box, when set to ON in the Tool Settings window, associates hatching with hatched geometry. Whenever changes are made to an associated element also affects the hatch. For example, if the element is stretched, the hatch pattern expands to fill the new size.

The **Associative Region Boundary** (available only when Associative Pattern is set to ON) check box, when set to ON in the Tool Settings window, places hatch on a level other than the level of the hatched geometry.

The **Snappable Pattern** check box, when set to ON in the Tool Settings window, enables snapping to, selecting, and manipulating individual elements in the hatch pattern. If the check box is set to OFF, individual hatch elements cannot snapped to or selected.

When creating a non-associative pattern, MicroStation provides additional options (available only for Element method) in the Tool Settings window for how hole elements are treated. Three options are available in the **Search for Hole** menu; **No** to ignore hole elements, **Element Level** to search for hole elements on the same level as the element to hatch and **View Levels** to search for hole elements on all levels displayed in the view.

The **Use Element Symbology** check box, when set to ON in the Tool Settings window, uses the active color, line weight, and line style for hatching of the element being patterned.

Hatch Methods

MicroStation provides six hatching methods in addition to Element method explained earlier for defining the area. Icons in the Tool Settings window select the method for defining the area.

Flood Method

The **Flood** method hatches an area enclosed by a set of elements, similar to flood fill tool common in painting programs. Figure 12–6 shows the available options in the Tool Settings window when the Flood method is selected to define the area.

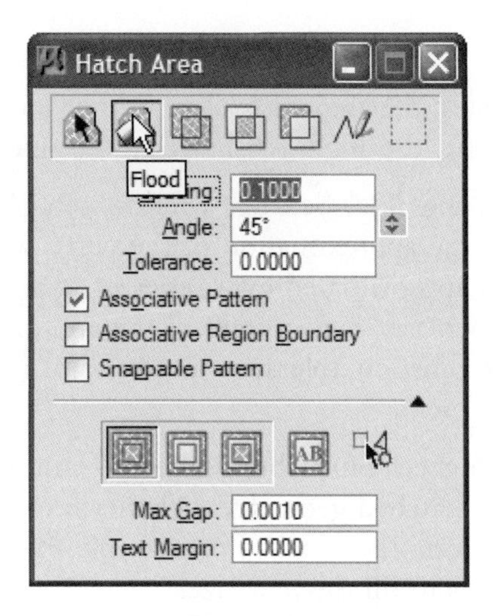

Figure 12–6 Tool Settings window (Flood Method selected)

MicroStation prompts:

> Hatch Area Enclosing Point > Enter data point inside area *(Place a data point inside the area to be hatched.)*
>
> Hatch Area Enclosing Point > Accept @pattern intersection point *(Click the Data button again to initiate hatched.)*

A dynamic image is drawn around the perimeter of the flood area.

Ignore Interior Shapes—ignores interior shapes when the region boundary is calculated.

Locate Interior Shapes—avoids hatching all closed elements (regardless of their Area attributes) inside the selected area when the area is hatched.

Identify Alternating Interior Shapes—hatches alternating areas where shapes are nested inside one another.

Locate Interior Text—avoids hatching any text or dimension text inside or overlapping the selected area when the area is hatched.

Dynamic Area Locate—the hatching is displayed dynamically in the area to be hatched as the screen pointer is moved over the shapes. To select several areas to hatch, hold down CTRL while selecting the areas. After all areas are selected, release CTRL and click the Data button to initiate hatching all selected areas.

Max Gap—sets the maximum distance, in working units, between the endpoints of enclosing elements.

Text Margin—sets the size of the margin between hatching lines and existing text elements in the area to hatch.

Associative Pattern, Associate Region Boundary, and **Snappable Pattern** settings are the same as in Element method.

Union Method

The Union method hatches the areas of the edges bound by the union of two or more closed planar elements. Where more than two elements are involved, use CTRL + Data points to select the extra elements. Figure 12–7 shows the available options in the Tool Settings window when the Union method is selected to define the area.

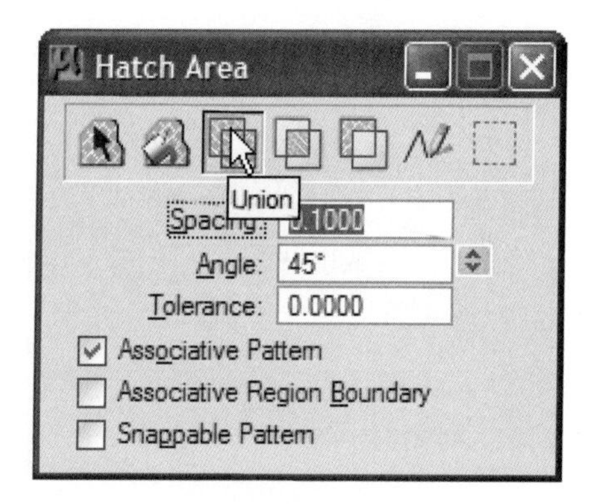

Figure 12–7 Tool Settings window (Union Method selected)

MicroStation prompts:

Hatch Element Union > Identify element *(Select the first element, as shown in Figure 12–8.)*

Hatch Element Union > Identify additional/Reset to complete *(Select the second element.)*

Hatch Element Intersection > Accept/Reject (CTRL+Data to select) *(Click the Data button to initiate the hatching or CTRL + Data button to select additional element.)*

As each element is accepted, the dynamic image changes the appearance of the selected elements to show the union area with interior segments removed. The elements are not affected and they reappear after the completion of hatching.

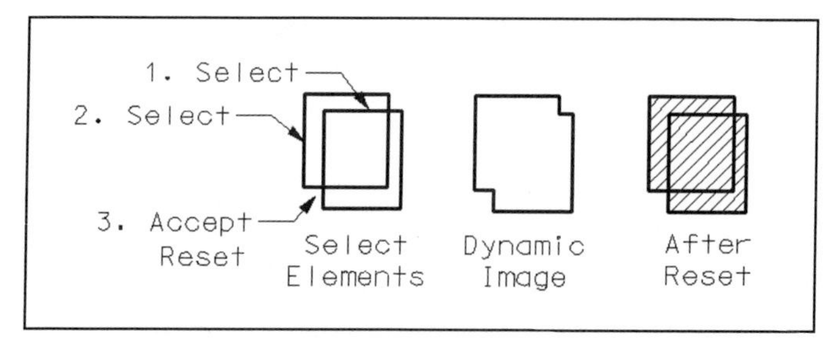

Figure 12–8 Example of union patterning

Associative Pattern, **Associate Region Boundary**, and **Snappable Pattern** settings are the same as in Element method.

Intersection Method

The Intersection method hatches the areas of the edges bound by the intersection (common area) of two or more closed planar elements. Where more than two elements are involved, use CTRL + Data points to select the extra elements. Figure 12–9 shows the available options in the Tool Settings window when the Intersection method is selected to define the area.

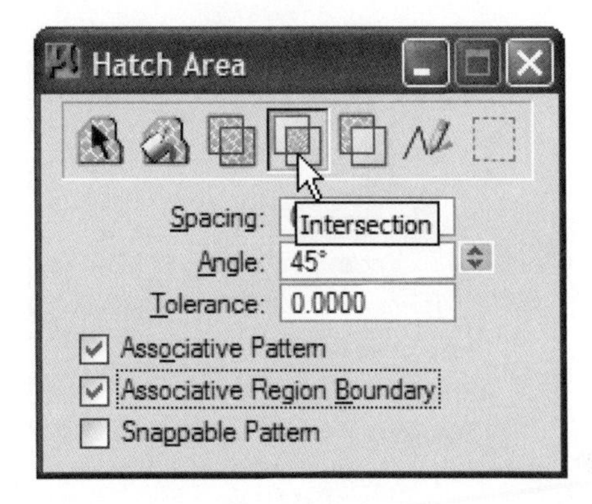

Figure 12–9 Tool Settings window (Intersection Method selected)

MicroStation prompts:

Hatch Element Intersection > Identify element *(Select the first element, as shown in Figure 12–10.)*

Hatch Element Intersection > Identify additional/Reset to complete *(Select the second element)*

Hatch Element Intersection > Accept/Reject (CTRL+Data to select) *(Click the Data button to initiate the hatching or CTRL + Data button to select additional element.)*

As each element is accepted, the dynamic image changes the appearance of the selected elements to show only the intersection area. The elements are not affected and they reappear after the completion of hatching.

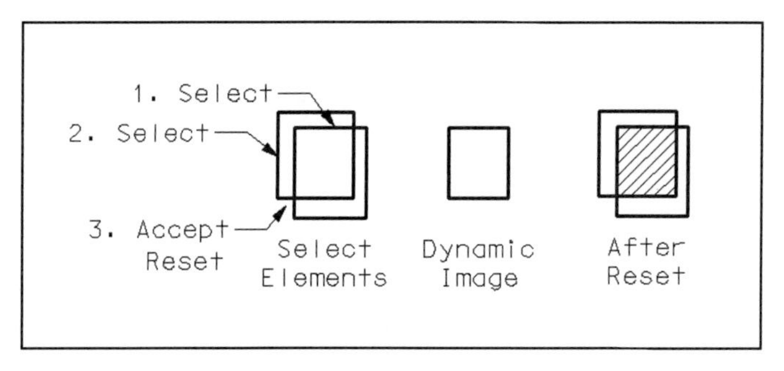

Figure 12–10 Example of intersection patterning

Associative Pattern, **Associate Region Boundary**, and **Snappable Pattern** settings are the same as in Element method.

Difference Method

The Difference method hatches the areas of the edges bound by the difference of two or more closed planar elements. The part of the first selected element that does not overlap the second element is hatched. Where more than two elements are involved, use CTRL + Data points to select the extra elements. Figure 12–11 shows the available options in the Tool Settings window when the Difference method is selected to define the area and MicroStation prompts:

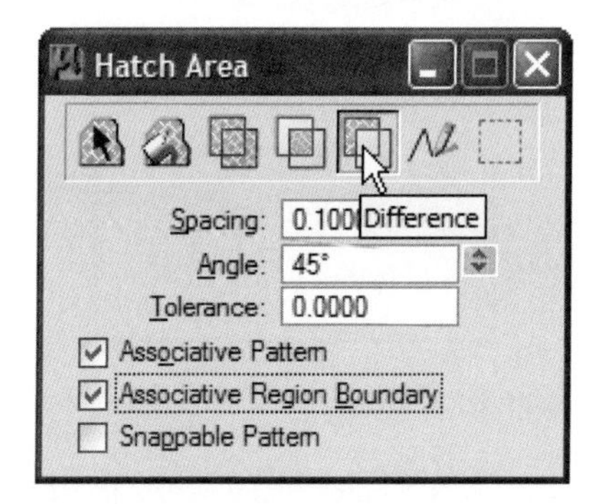

Figure 12–11 Tool Settings window (Difference Method selected)

> Hatch Element Difference > Identify element *(Select the first element, as shown in Figure 12–12.)*
>
> Hatch Element Difference > Identify additional/Reset to complete *(Select the second element)*
>
> Hatch Element Difference > Accept/Reject (CTRL+Data to select) *(Click the Data button to initiate the hatching or* CTRL *+ Data button to select additional element.)*

As each element is accepted, the dynamic image changes the appearance of the selected elements to show only the difference area. The elements are not affected and they reappear after the completion of hatching.

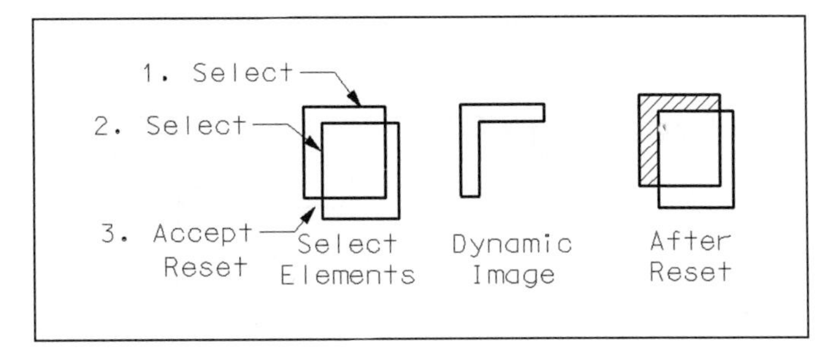

Figure 12–12 Example of difference patterning

Associative Pattern, Associate Region Boundary, and **Snappable Pattern** settings are the same as in Element method.

Points Method

The Points method hatches an area defined by a series of data points, each of which defines a vertex. Figure 12–13 shows the available options in the Tool Settings window when the Points method is selected to define the area and MicroStation prompts:

Figure 12–13 Tool Settings window (Points Method selected)

Hatch Area Defined By Points > Enter shape vertex *(Place the first three vertex points of the shape, as shown in Figure 12–14.)*

Hatch Area Defined By Points > Enter point or Reset to complete *(Continue entering shape vertex points, or click the Reset button to complete the shape and initiate hatching.)*

After the third point, the dynamic image shows the shape of the hatch area if the Reset button were to be clicked.

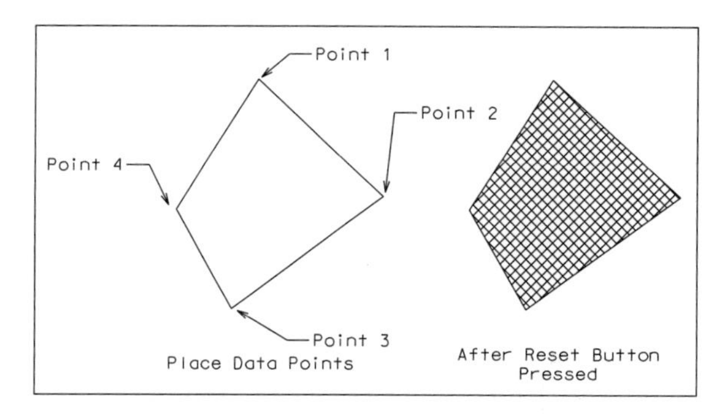

Figure 12–14 Example of patterning an area enclosed by points

Associative Pattern and **Snappable Pattern** settings are the same as in Element method.

Fence Method

The Fence method hatches the area inside the fence. Figure 12–15 shows the available options in the Tool Settings window when the Fence method is selected to define the area and MicroStation prompts:

Figure 12–15 Tool Settings window (Fence Method selected)

Hatch Fence > Accept/Reject fence contents *(Place a data point to define an intersection point for the hatch, as shown in Figure 12–16, and initiate hatching.)*

The example in Figure 12–16 is a circular fence. The fence must be placed before it can be hatched. The fence is not part of the hatch and removing it does not affect the hatch.

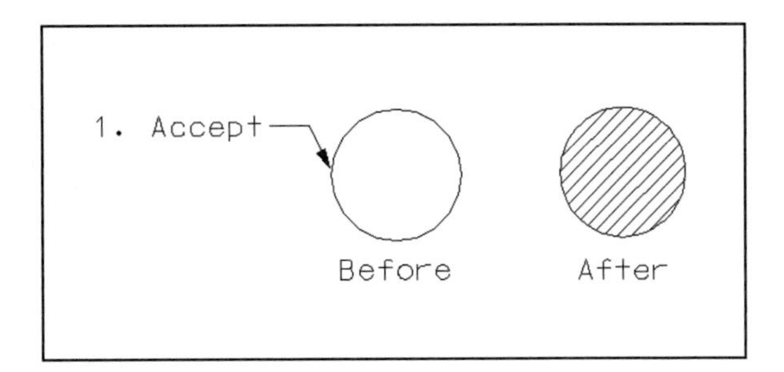

Figure 12–16 Example of hatching a fence

Associative Pattern, Search for Holes, and **Snappable Pattern** settings are the same as in Element method.

CROSSHATCH AREA TOOL

Invoke the Crosshatch Area tool from:

Task Navigation tool box (active task set to Patterns)	Select the Crosshatch Area tool (see Figure 12–17).
Keyboard Navigation (Task Navigation tool box with active task set to Patterns)	**w**

Figure 12–17 Invoking the Crosshatch Area tool from the Task Navigation tool box (active task set to Patterns)

Icons in the Tool Settings window select the method for defining the area to be cross hatched. By default, MicroStation selects Element method, and MicroStation prompts:

> Crosshatch Area > Identify element *(Select the closed element to be crosshatched as shown in Figure 12–18.)*
>
> Crosshatch Area > Accept @pattern intersection point *(Click the Data button again to accept the element and initiate crosshatch placement.)*

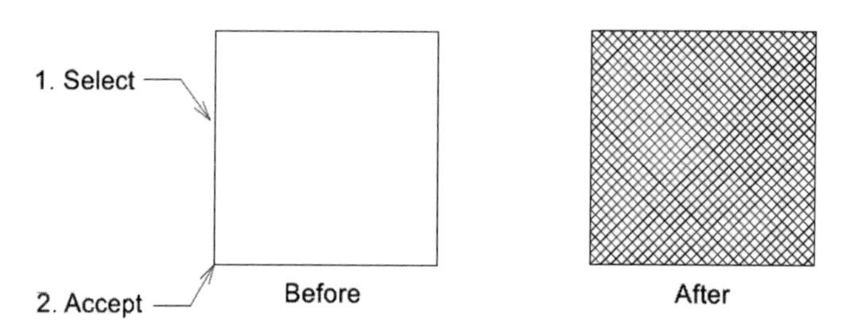

Figure 12–18 Example of Crosshatching a closed element

The tool settings and procedures are the same as those for the Hatch Area tool, except that there are additional fields to specify the Spacing and Angle of the crosshatch lines (for second set of hatch lines) in addition to those for the hatch lines.

The available methods for Crosshatching to definc the area are the same as in Hatch Area tool. Refer to the Hatch Area section for a detailed explanation for defining the area for crosshatching.

PATTERN AREA TOOL

Invoke the Pattern Area tool from:

Task Navigation tool box (active task set to Patterns)	Select the Pattern Area tool (see Figure 12–19).
Keyboard Navigation (Task Navigation tool box with active task set to Patterns)	**e**

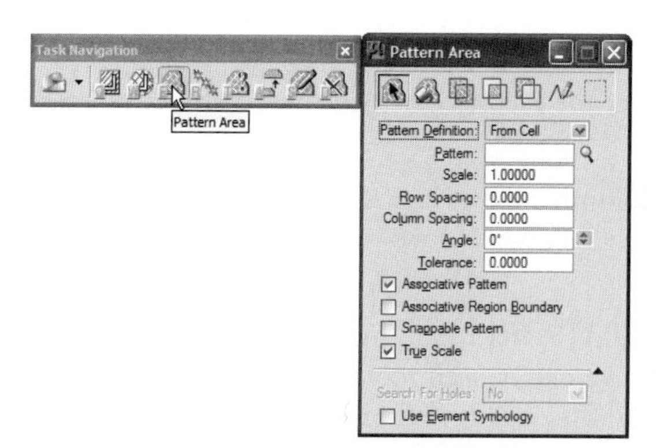

Figure 12–19 Invoking the Pattern Area tool from the Task Navigation tool box (active task set to Patterns)

Icons in the Tool Settings window select the method for defining the area to be patterned. By default, MicroStation selects Element method, and MicroStation prompts:

> Pattern Area > Identify element *(Select the closed element to be the pattern of the selected cell as shown in Figure 12–20.)*
>
> Pattern Area > Accept @pattern intersection point *(Click the Data button again to accept the element and initiate patterning.)*

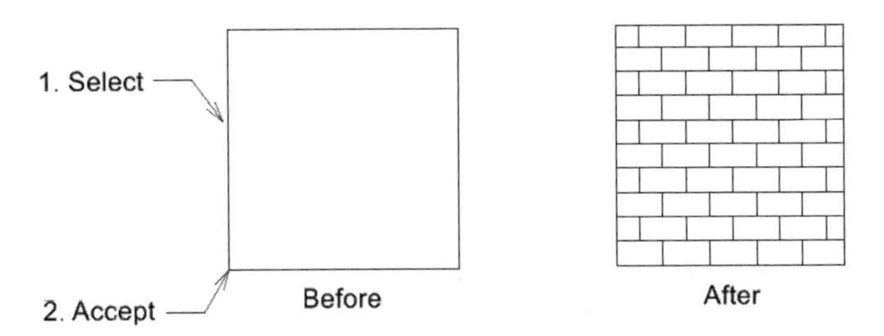

Figure 12–20 Example of Patterning a closed element

The **Pattern Definition** menu in the Tool Settings window selects the active cell pattern or the active pattern file (**.pat*) from an AutoCAD file. Select the **From Cell** option to key in the name of the cell that is tiled to create the pattern. Optionally, click the Browse Cells button to the right of the input field, which opens the Cell Library dialog box to browse the cells in the attached library and set one of the cell as an active pattern. Select the **From File** option to select a hatch pattern (**.pat*) from an AutoCAD pattern file. To select the pattern file, click the Browse button.

The **Pattern** field displays the name of the current pattern cell or file.

The **Scale** field sets the factor by which the active pattern cell or active pattern is scaled.

The **Row** and **Column** field (available only Pattern Definition set to **From Cell** only) sets the interval between rows and columns respectively for the tiling of the pattern.

The **True Scale** field (Pattern Definition set to From Cell only), when it is set to ON scales, the active pattern cell to adjust it to the units of the active model. The scaling occurs only if the cell is shared and the units of the model in which the cell was created differ from those of the active model.

The **Angle, Tolerance, Associative Pattern, Associative Region Boundary, Search for Holes, Snappable Pattern** and **Use Element Symbology** settings are the same as in Element method (Hatch).

The available methods for Patterning to define the area are the same as in Hatch Area tool. Refer to the Hatch Area section for a detailed explanation for defining the area for patterning.

USING THE AREA SETTINGS TO CREATE HOLES

When a shape is drawn that represents a hole in a solid element, the interior of that shape cannot be hatched or patterned with Element and Fence methods (Associative Pattern should be set to off), and the background will show "through" the hole. Closed shapes, such as blocks, circles, and ellipses can be created with either a **Solid Area** or **Hole Area**. Figure 12–21 shows an example of a brick pattern in which wall was drawn in solid mode and windows were drawn in hole mode.

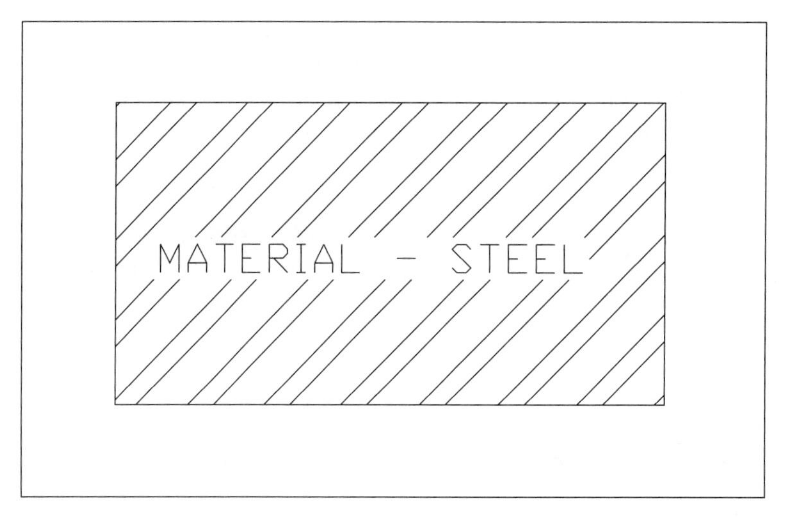

Figure 12–21 Example of the effect of Hole elements within a patterned element

Figure 12–22 shows a pattern where Text is placed in hole mode.

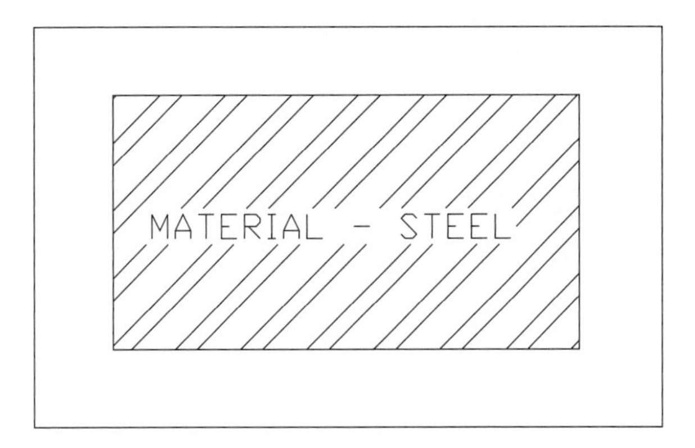

Figure 12–22 Example of patterning over a text string placed in Hole Area mode

The Tool Settings window for each tool that places closed shapes includes the **Area** menu. The **Area** option is used to select the active setting (see Figure 12–23 for an example) and all closed elements (and text) will be created with the Active area mode.

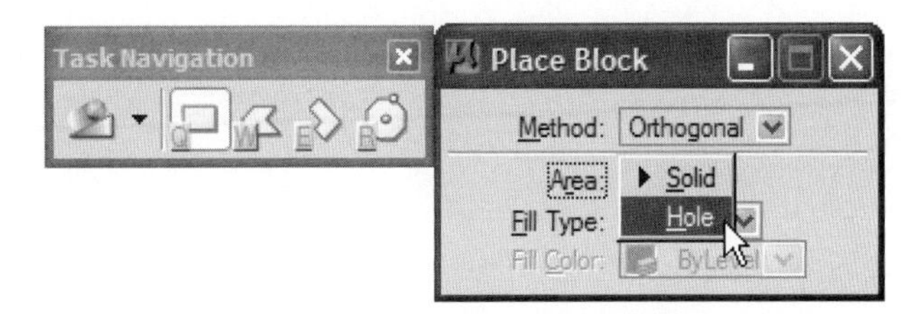

Figure 12–23 Example of the Area mode tool setting for a closed element placement tool

CHANGING AN ELEMENT'S AREA

MicroStation allows for changing the area attribute of a closed element(s) (shapes, ellipses, complex shapes, or B-spline curves) to the Active Area with Change to Active Area tool. Invoke the tool from:

Change Attributes tool box	Select the Change to Active Area tool and select the required Area (Solid or Hole) in the Tool Settings window (see Figure 12–24).
Key-in window	**change area** (or **chan a**) (ENTER)

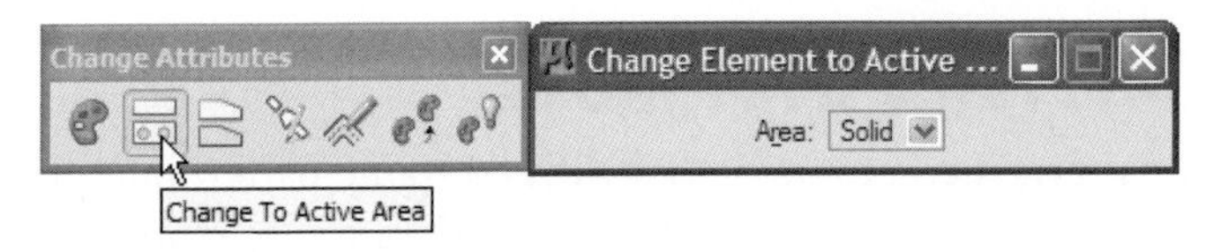

Figure 12–24 Invoking the Change Element to Active Area tool from the Change Attributes tool box

MicroStation prompts:

Change Element to Active Area > Identify element *(Identify the element.)*

Change Element to Active Area > Accept/Reject (Select next input) *(Click the Data button to initiate the area change or click Reset to reject the selected element. Optionally, also select the next element to be changed.)*

DELETE PATTERNS

A special delete tool for patterns deletes all pattern elements but not the element that contains the pattern. Invoke the Delete Pattern tool from:

Task Navigation tool box (active task set to Patterns)	Select the Delete Pattern tool (see Figure 12–25).
Keyboard Navigation (Task Navigation tool box with active task set to Patterns)	**d**

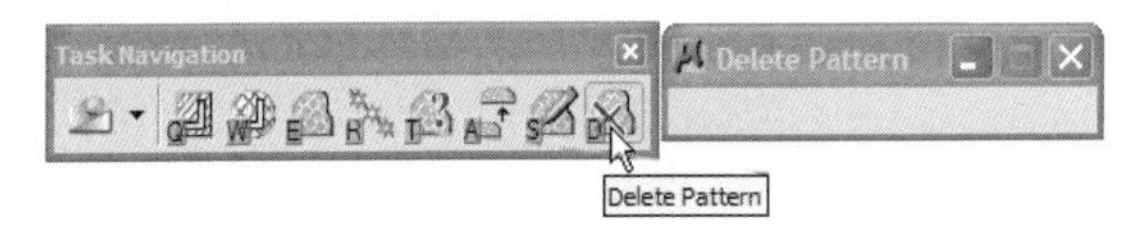

Figure 12–25 Invoking the Delete Pattern tool from the Task Navigation tool box (active task set to Patterns)

MicroStation prompts:

Delete Pattern > Identify element *(Identify the pattern to be deleted.)*

Delete Pattern > Accept/Reject (select next input) *(Click the Data button again to accept and delete the pattern or click the Reset button to drop the selected pattern.)*

Note: *Patterns can contain a large number of elements. After deleting patterns, select* ***File > Compress Design*** *to remove the deleted elements from the design file. Remember though, that when the design is compressed, the Undo buffer is cleared.*

MATCH PATTERN ATTRIBUTES

If additional patterns need to be placed with the same patterning attribute settings as an existing pattern, the Match Pattern Attributes tool provides a fast way to set the attributes. The tool sets the active patterning attributes to match those of a selected pattern.

Invoke the Match Pattern Attributes tool from:

Task Navigation tool box (active task set to Patterns)	Select the Match Pattern Attributes tool (see Figure 12–26).
Keyboard Navigation (Task Navigation tool box with active task set to Patterns)	**a**

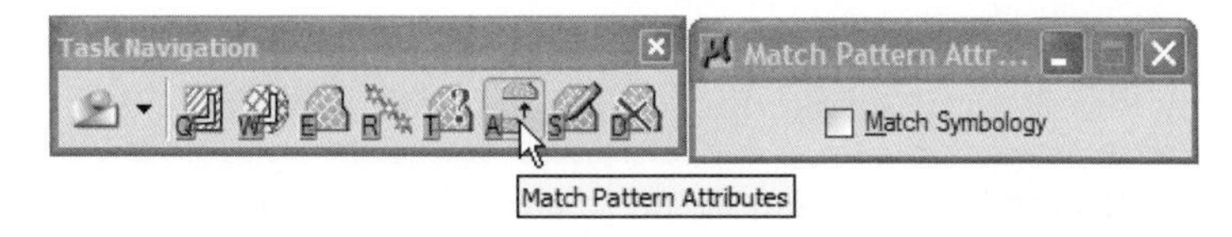

Figure 12–26 Invoking the Match Pattern Attributes tool from the Task Navigation tool box (active task set to Patterns)

MicroStation prompts:

> Match Pattern Attributes > Identify element *(Identify the element containing the pattern to be matched.)*
>
> Match Pattern Attributes > Accept/Reject (Select next input) *(Click the Data button to accept the element or click the Reject button to reject it.)*

When the selected pattern is accepted, the active patterning elements are set to match those of the selected pattern and the active settings appear in the Status bar.

CHANGE PATTERN

The Change Pattern tool is used to change an existing pattern to match the current attributes and/or pattern parameters, redefine the intersection point of the patterning and re-flood the area. Invoke the Change Pattern tool from:

Task Navigation tool box (active task set to Patterns)	Select the Change Pattern tool (see Figure 12–27).
Keyboard Navigation (Task Navigation tool box with active task set to Patterns)	**s**

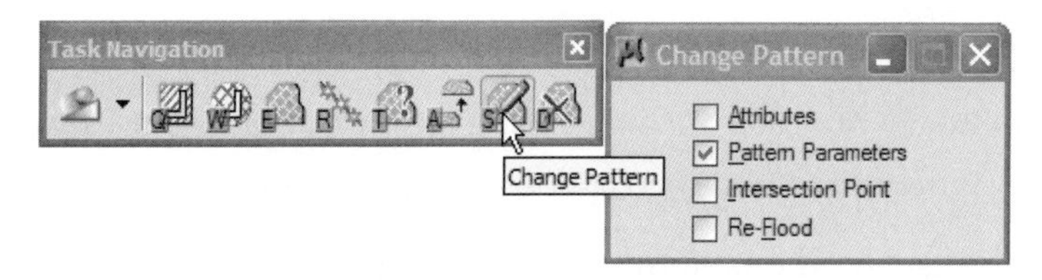

Figure 12–27 Invoking the Change Pattern tool from the Task Navigation tool box (active task set to Patterns)

MicroStation prompts:

> Change Pattern > Identify element *(Identify the element containing the pattern to be changed.)*

The **Attributes** check box when set to ON in the Tool Settings window applies the current element attributes (color, line weight, and line style) to the selected pattern.

The **Pattern Parameters** check box when set to ON in the Tool Settings window applies the current pattern parameters to the selected pattern.

The **Intersection Point** check box when set to ON in the Tool Settings window, the pointer position defines the new intersection point of the pattern.

The **Re-flood** check box when set to ON in the Tool Settings window, the selected region is re-flooded.

SHOW PATTERN ATTRIBUTES

The Show Pattern Attributes tool is used to display the Pattern Angle and Pattern Scale attributes of the selected pattern element. Invoke the Show Pattern Attributes tool from:

Task Navigation tool box (active task set to Patterns)	Select the Show Pattern Attributes tool (see Figure 12–28).
Keyboard Navigation (Task Navigation tool box with active task set to Patterns)	**t**

Figure 12–28 Invoking the Show Pattern Attributes tool from the Task Navigation tool box (active task set to Patterns)

MicroStation prompts:

Show Pattern Attributes > Identify element *(Identify the pattern to display the attributes.)*

Show Pattern Attributes > Accept/Reject (Select next input) *(Accept the selected element.)*

MicroStation displays the pattern angle and scale in the status bar.

FILLING AN ELEMENT

The Tool Settings window for tools that place closed elements (such as Circle and Block) include the **Fill Type** and **Fill Color** menus to control what happens to the area enclosed by the element.

The **Fill Type** options are:

- **None**—The element area is transparent and the outline color is set to the **Active Color**.
- **Opaque**—The element area and outline are set to the color selected in the **Fill Color** menu (which is also the **Active Color**).
- **Outlined**—The element area is set to the color selected in the **Fill Color** menu and the element outline is set to the **Active Color**.

The **Fill Color** menu opens a menu that contains all of the colors supported by MicroStation.

- When the **Opaque Fill Type** is selected, the default **Fill Color** is equal to the **Active Color**. Selecting a **Fill Color** also selects the **Active Color**.
- When the **Outlined Fill Type** is selected, the **Fill Color** sets the color of the element area, and the **Active Color** sets the color of the element outline. Changing the **Fill Color** does not change the **Active Color**.

Figure 12–29 shows a typical Tool Settings window with the Fill Type menu open; Figure 12–30 shows examples of opaque and outlined fill types.

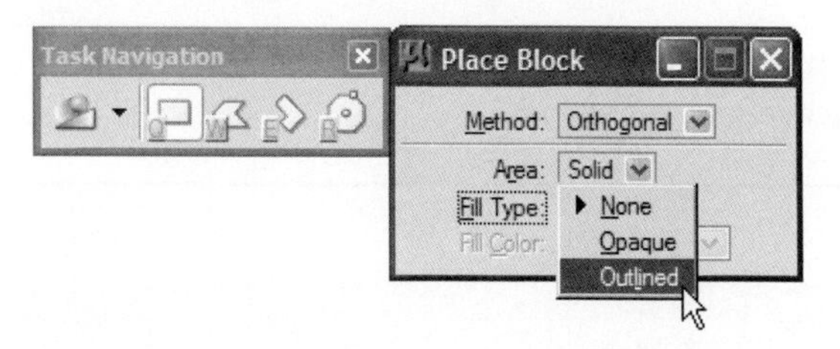

Figure 12–29 Fill Type menu in the Tool Settings window

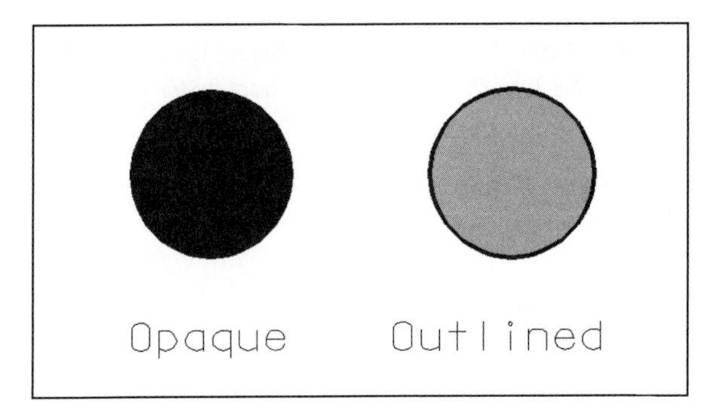

Figure 12–30 Examples of Opaque and Outlined Fill Types

To place filled closed elements, just select the desired placement tool, and then select the desired **Fill Type** and **Fill Color** before placing the element. The selected Fill settings become the active setting, and they are applied to all closed elements created subsequently.

Note: *Before placing filled closed elements, make sure the* ***Fill*** *check box is set to ON in the View Attributes settings box for the view window being used to place the elements. If the check box is set to OFF, filled elements appear to be transparent, and they print that way.*

CHANGING THE FILL TYPE OF AN EXISTING ELEMENT

The Change to Active Fill Type tool changes the fill type of existing elements. Invoke the tool from:

Change Attributes tool box	Select the Change to Active Fill Type tool and, if necessary, set the desired Fill Type and Fill Color in the Tool Settings window (see Figure 12–31).
Key-in window	**dialog change fill** (or **chan f**) (ENTER)

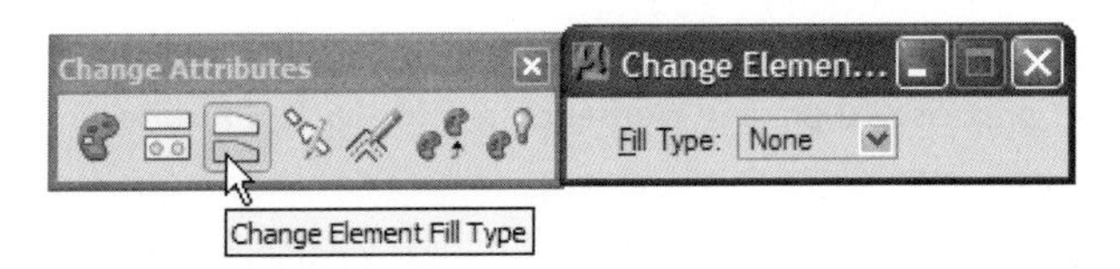

Figure 12–31 Invoking the Change Element to Active Fill Type tool from the Change Attributes tool box

MicroStation prompts:

Change Element to Active Fill Type > Identify element *(Select the element to be changed.)*

Change Element to Active Fill Type > Accept/Reject (Select next input) *(Click the Data button to initiate the Fill Type change or click the Reset button to drop the element.)*

Open the Exercise Manual PDF file for Chapter 12 on the accompanying CD for project and discipline specific exercises.

REVIEW QUESTIONS

Write your answers in the spaces provided.

1. List the three tools MicroStation provides for area patterning.
____
2. Explain briefly the difference between Hatch Area and Crosshatch Area patterning.
____
____
3. List the methods that are available to set the area patterning.
____
4. Explain with illustrations the differences between the **Intersection**, **Union**, and **Difference** options in area patterning.
____
____
5. Explain the difference between the **Element** and **Points** options in area patterning.
____
____
6. What is the purpose of providing a second data point in hatching a closed element?
____
7. The Pattern Area tool places copies of the **Active Pattern** ____________________ in the area you select.
8. What is the purpose of turning ON the **Associative Pattern** check box?
____
9. What is the purpose of placing elements in a **Hole Area** mode?
____
10. Explain briefly the purpose of invoking the Match Pattern Attributes tool.
____
11. Explain briefly the three options available for area **Fill** placement.
____
12. Explain the steps involved in switching an existing element between area **Fill** modes.
____
____

CHAPTER 13

Attaching References

OBJECTIVES

Topics explored in this chapter:

- Describing references
- Creating new reference attachments
- Using the References list box
- Using the reference manipulation tools
- Using standard manipulation tools to manipulate references
- Copying elements from a reference onto the active design file

OVERVIEW OF REFERENCES

A reference attachment is a model attached to and displayed with the active model for plotting or construction purposes. A reference cannot be modified. An attached reference can be a model that resides in either the open DGN file, some other DGN file, or a DWG file. Elements in a reference display as though they were in the active model. Although elements in a reference cannot be manipulated they can be used to snap to, and even copied into the active model.

MicroStation allows for attaching an unlimited number of models from one or more design files to the active design file as references. The only information about referenced model that becomes part of the active design file is the name of each referenced design file, its location, and attachment settings.

In addition to models, raster images can also be referenced. MicroStation supports referencing Monochrome, continuous-tone (gray-scale), and color images in a variety of image formats.

EXAMPLES OF USING REFERENCES

Borders and title blocks are an excellent example of design files that are useful as references. For example, a design team creates a standard border for their project and attaches it as a reference to all the design files created for the project. Any changes made to the border design file will automatically appear in all the design files that reference the border.

When combined with the networking capability of MicroStation, external references give the project manager powerful new tools for file management. By combining drawings through the referencing tools, the project manager can quickly see the work of the various departments working on different aspects of the project. The manager can overlay a drawing where appropriate, track the progress, and maintain document integrity. At the same time, departments need not lose control over individual designs and details.

REFERENCE MANIPULATIONS

Referenced models can be viewed but not modified from the active file (but there are options that allow for making a reference the active design file). The tentative button snaps to all elements in the references.

The view of reference models can be scaled, moved, and rotated by tools specifically programmed to work with references and by standard element manipulation tools. Each referenced model acts as one element when manipulated. The Copy tool can copy elements from a referenced model to the active model.

ACCESS THE REFERENCE TOOLS

A set of reference manipulation tools are available on the Task Navigation tool bar. Access the tools from:

Task Navigation tool box	Select Drawing Composition > References (see Figure 13–1).

MicroStation displays the References tools on the Task Navigation tool box (see Figure 13–2).

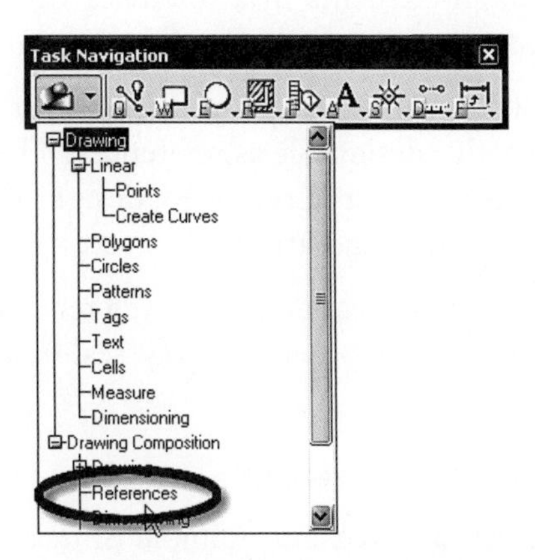

Figure 13–1 Accessing the references manipulation tools

Note: *The reference manipulation tools are also available on the References settings box.*

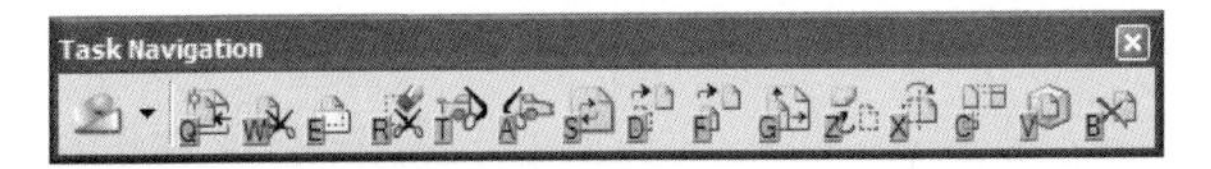

Figure 13–2 The references manipulation tools on the Task Navigation tool bar

LIST THE ATTACHED REFERENCES

Tools for maintaining reference attachments are also available on the References settings box. Open the Reference settings box from:

Menu	File > Reference.
Primary tool box	Click References (See Figure 13–3).

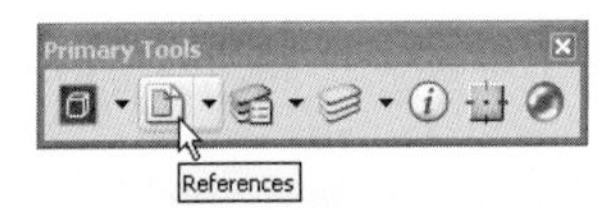

Figure 13–3 References icon on the Primary Tools tool bar

MicroStation displays the References list box, as shown in Figure 13–4.

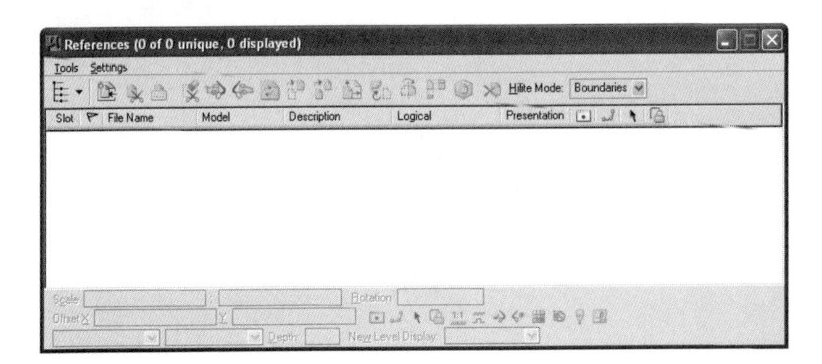

Figure 13–4 The References list box

The options on the list box are described later in this chapter.

ATTACHING A REFERENCE

To attach a model as a reference to the active model, invoke the Attach Reference tool from one of the following:

Task Navigation tool box (active task set to References)	Select the Attach Reference tool (see Figure 13–5).
Keyboard Navigation (Task Navigation tool box with active task set to References)	**Q**
References settings box	Tools > Attach
References settings box	Click the Attach Reference icon on the tool bar.

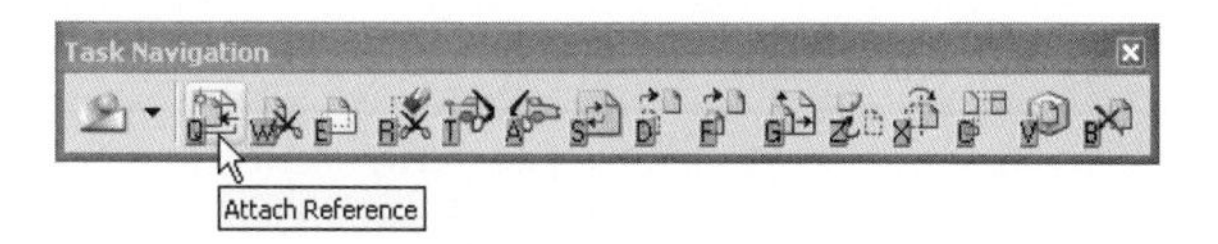

Figure 13–5 Invoking the Attach Reference tool from the Task Navigation tool bar

MicroStation displays the Attach Reference dialog box, as shown in Figure 13–6.

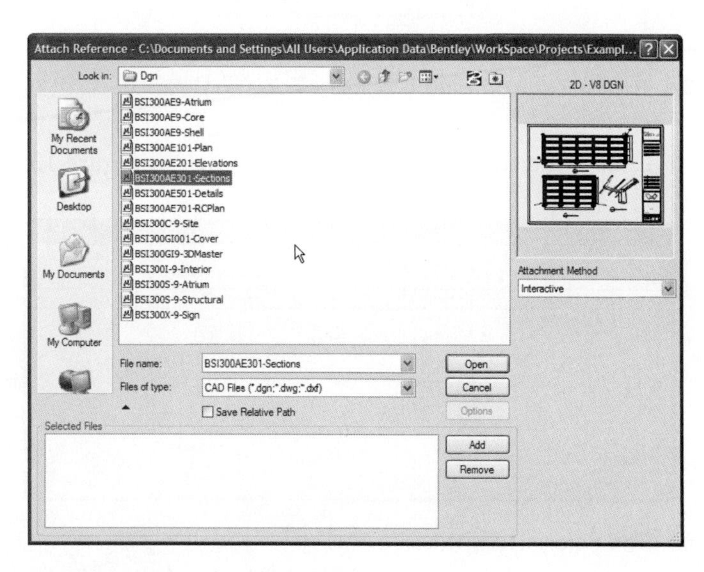

Figure 13–6 Attach Reference dialog box

***Note:** In Figure 13–6 the dialog box is expanded to show the **Selected Files** field. To expand the dialog box, click the up/down arrow on the lower left of the dialog box.*

The following procedure creates a reference to a single design file using default settings in the Attach Reference dialog box.

1. In the **Look in** menu, navigate to the folder that holds the design file to reference.
2. Select the design file's name from the filenames list field directly below the **Look in** menu.
3. Select **Coincident** from the **Attachment Method** menu (discussed later).
4. Click **Add** and MicroStation adds the full path and filename of each selected file in the **Selected Files** list box. Continue navigating to folders and selecting design files until all required design files are selected. If a design file is selected by mistake, select the design file in the **Selected Files** list field and click **Remove** to remove it from the field.
5. Click **Open** to attach the selected design file(s) as reference(s) or click **Cancel** to cancel the attachment.

Figure 13–7 shows a design before and after attaching a reference (title block and border).

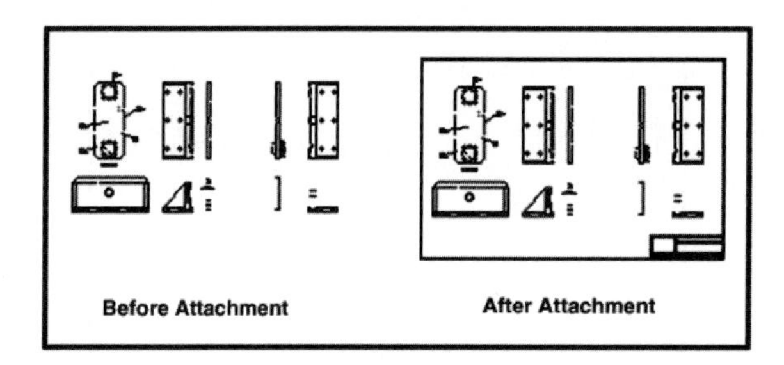

Figure 13–7 Design before and after attaching a reference model

ATTACH REFERENCE DIALOG BOX OPTIONS

The Attach Reference dialog box provides various options for modifying the view of the attached reference(s).

Attachment Method

The **Attachment Method** menu controls the method used to attach the selected design files as references. Three options (**Interactive**, **Coincident**, and **Coincident World**) are for 2D design files, **Top** is for both 2D and 3D design files, and several additional options are for 3D only.

- With **Interactive** selected, clicking **Open** opens the Reference Attachment Settings dialog box, which provides more attachment settings. For more information on the dialog box, see the next section.
- **Coincident** attaches the reference aligned with the active design file with regard to design plane coordinates only. This option is available only when referencing a DGN file. Clicking **Open** completes the attachment.

- **Coincident World** attaches the reference aligned with regard to both Global Origin and design plane coordinates. Clicking **Open** completes the attachment.
- **Top** allows for placing the referenced design file anywhere on the design plane of the active design file. Clicking **Open** displays a dynamic image of the outline of the area occupied by the referenced design file. The dynamic image follows the screen pointer. Place a data point to define the location of the center of the reference.

The Reference Attachment Settings Dialog Box

Selecting the **Interactive Attachment Method** and clicking **Open** on the Reference Attachment dialog box opens the Reference Attachment Settings dialog box, as shown in Figure 13–8.

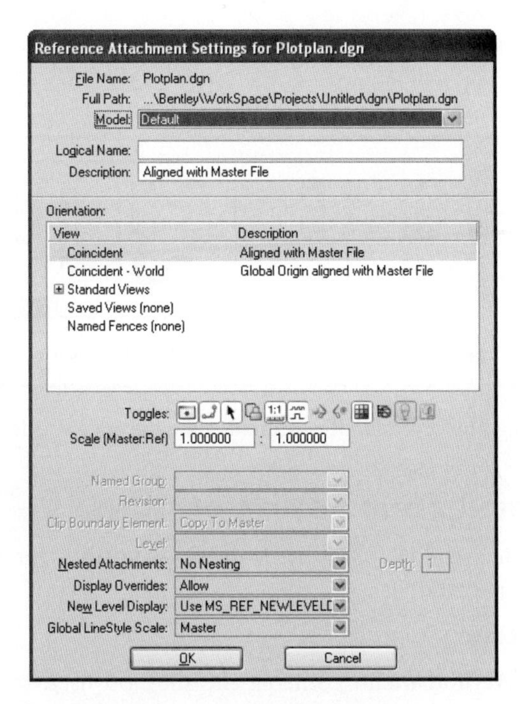

Figure 13–8 Reference Attachment Settings dialog box

The Reference Attachment Settings dialog box provides additional control over the attachment of each design file selected on the Reference Attachment dialog box. When the dialog box opens, it displays the name and path of the first selected file at the top of the dialog box. When all settings are made for that file and **OK** is clicked, the dialog box displays settings for the second selected design file. The process continues until settings are set for all selected design files.

Following are descriptions of the options on the Reference Attachment Settings dialog box.

The **Model** menu lists the models within the selected design file and is used to select the model to attach as a reference to the active design file. If there is only one model in the design, the menu displays only the **Default** option.

The **Logical Name** field provides a place to type a 40-character maximum identifier that can be used to identify the reference file for manipulation. The logical name is optional unless there is more than one attachment of the same model to the active design file.

The **Description** field provides an optional 40-character maximum description of the attachment. If several reference files are attached, the descriptions can be an aid to determining the purpose of each reference.

The **Orientation** options field lists possible views of the model when it is attached:

- **Coincident** (also available on the Attach Reference dialog box) attaches the reference aligned with the active design file with regard to design plane coordinates only.
- **Coincident World** (also available on the Attach Reference dialog box) attaches the reference aligned with regard to both Global Origin and design plane coordinates.
- **Standard Views** attaches one of the standard views of a design. For 2D designs, only the **Top** view is available (Attach Reference dialog box also provides the **Top** option). To expand **Shared Views**, click the plus sign on the left side of **Standard Views**.
- **Saved Views** allows for selecting one of the saved views in the referenced model. If there are no saved views in the referenced model, "none" appears next to **Saved Views**. If there are saved views, click the plus sign next to **Saved Views** to display the view names. For more information on saved views, see Chapter 3.
- **Named Fences** allows for selecting one of the named fences in the referenced model. If there are no named fences in the referenced model, "none" appears next to **Saved Views**. If there are named fences, click the plus sign next to **Saved Views** to display the view names. For more information on named fences, see Chapter 6.

The **Toggles** are a set of icons that, when clicked, turn reference settings ON and OFF. Figure 13–9 identifies each toggle icon.

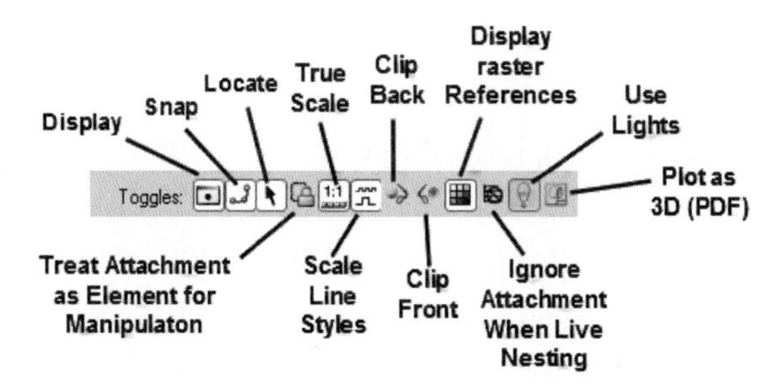

Figure 13–9 The Toggles on the Reference Attachment Settings dialog box

- If **Display** is ON, MicroStation displays the reference. If it is OFF, MicroStation hides the reference and the model is not displayed.
- If **Snap** is ON, the Tentative button can snap to elements in the referenced model.
- If **Locate** is ON, elements in the referenced model can be located for construction purposes (such as extending a line to its intersection with an element in the referenced model).
- If **Treat Attachment as Element for Manipulation** is ON, the standard manipulation tools can manipulate the reference as a single element.
- If **True Scale** is ON, units in the active model and those in the referenced model are used to calculate an appropriate scaling factor so that the reference scale reflects a true real-world scale.
- If **Scale Line Styles** is ON, custom line style components (for example, dashes) are scaled by the reference scale.
- If **Clip Back** is ON, a back-clipping plane can be applied to a referenced 3D model.
- If **Clip Front** is ON, a front-clipping plane can be applied to a referenced 3D model.
- If **Display Raster References** is ON, custom line style components (for example, dashes) are scaled by the **Scale (Master:Ref)** factor.
- If **Ignore Attachment When Live Nesting** is ON, when the active model is referenced with nesting, this attachment is not displayed.
- **Use Lights** is a 3D-only tool. During rendering, the source lighting cells present in the active file are always considered. Any source lighting cells located in the reference are ignored unless the **Use Lights** setting is ON for the reference.
- **Plot as 3D** controls the printing of 3D models.

The Scale (Master:Ref) edit fields allows for specifying the scale factor as a ratio of the Master Units in the active model to the Master Units in the attached model.

The **True Scale** check box, when ON, causes one-to-one alignment with units in the referenced model. For example, if the units in the active model are feet and the units in the referenced model are meters, meters are converted to feet so the elements are referenced at the same relative size.

The **Named Group** menu is used to select a named group to limit the view of the reference to only elements in the named group. For more information on named groups, see Chapter 14.

If **Design History** is on in the referenced model, the **Revision** menu allows for selecting one of the model's revisions to attach as a reference. For more information on Design History, see Chapter 14.

The **Clip Boundary Element** menu provides two options to control a clip boundary element that is part of a saved view in the referenced model. Clip boundaries limit the view of a reference model to the area within the clip boundary.

- The **Copy to Master** option copies the reference clip boundary to the master file. The clip boundary can be modified directly in the active design file, but it will not change if the boundary in the reference is updated.
- The **Associate to Saved View** associates the clip boundary directly to the boundary element in the reference's saved view. The reference clip boundary automatically reflects any changes to the clip boundary in the reference but it cannot be modified directly in the active design file.

The **Level** menu is used when referencing DWG files that were created using the AutoCAD application.

The **Nested Attachments** menu comes into play when the directly-attached reference also contains reference attachments. It determines how nested references (the references attached to the model selected as the reference) are handled.

- Select **No Nesting** to display only the directly-attached reference.
- Select **Live Nesting** to display the nested references in the directly-attached reference, except for the following situations:
    - MicroStation does not display nested references that are at a greater depth than the **Depth** setting.
    - MicroStation does not display nested references that have the **Ignore Attachment When Live Nesting** toggle switch turned ON.
- If **Copy Attachments** is selected, nested references in the attached reference are copied into the active design file.

The **Depth** edit field sets the number of levels of nested references that are recognized. If the **Depth** is set to zero, only the directly-attached reference is visible on the active design file and nested references in the directly-attached reference are ignored.

The **Display Overrides** menu controls how overrides are saved for nested references. For each nested reference, overrides allow for controlling the settings for reference display, locate, snap, raster reference display, and level display.

- Select **Allow** to create overrides automatically when the corresponding settings for nested references are modified.
- Select **Always** to save the current toggle and level display state for every nested attachment. This option locks in the settings for all nested attachments.
- Select **Never** to never create overrides for any of the nested references. The nested references are displayed the same as they are when the reference is opened as the master file.

The **New Level Display** menu controls the display of newly created levels in the attached reference and in nested references attached to the reference.

- Select **Use MS_REF_NEWLEVELDISPLAY Configuration Variable** to display new levels in the reference according to the setting for the MS_REF_NEWLEVELDISPLAY configuration variable. For more information on configuration variables, see Chapter 16.
- Select **Always** to always display new levels in the attached reference.
- Select **Never** to never display new levels in the attached reference.

The **Global LineStyle Scale** menu controls the scaling of cosmetic custom line styles. Every model can have a global line style scale factor that is applied to every line style within the model. The scale of line styles within a reference can be affected by the global line style scale of the active model, the referenced model, both, or neither.

- Select **None** to not use the active model's nor the reference model's global line style scale to scale the cosmetic custom line styles.
- Select **Master** to use the active model's global line style scale to scale the cosmetic custom line styles.
- Select **Reference** to use the referenced model's global line style scale to scale the cosmetic custom line styles.
- Use **Master * Reference** to multiply the active global line style scale by the referenced model's global line style scale to scale the cosmetic custom line styles.

Click **OK** to close the dialog box and attach the referenced model or click **Cancel** to close the dialog box without attaching the referenced model.

REFERENCES SETTINGS BOX

To list the references attached to the active design file, open the References settings box from:

Menu	File > Reference
Primary tool box	Click References

MicroStation displays the References settings box, as shown in Figure 13–10.

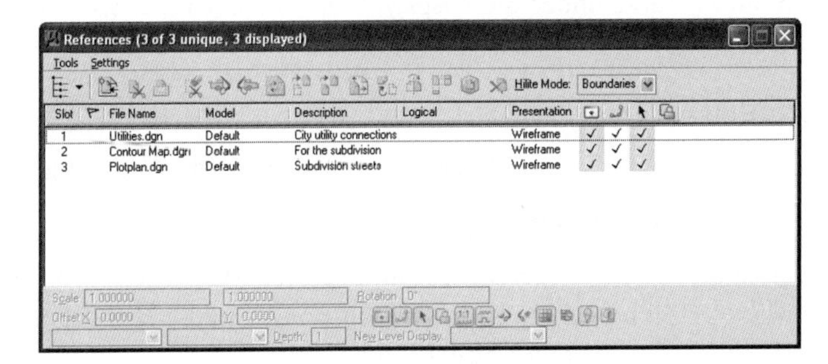

Figure 13–10 References settings box

The References settings box lists all the references attached to the active design file and provides options for maintaining the reference attachments. Following are descriptions of options on the list box.

TOOLS MENU

The **Tools** menu provides another way of selecting the same reference manipulation tools that are on the Task Navigator tool bar and the tool bar of this list box (the tool are presented later in this chapter). In addition to the shared tools, the **Tools** menu provides the following options that are only available on the **Tools** menu:

- **Detach All** detaches all references attached to the active design file. Before MicroStation detaches them, it displays a confirmation message asking if the references should be deleted. Click **Yes** to delete them or **Cancel** to cancel detaching the reference attachments.
- **Reload All** updates the attachment with the current content of all attached references to make sure they display the latest information.
- **Exchange** closes the active design file and opens the selected reference attachment in the same MicroStation session using the current view window settings so that necessary modifications can be made to the design file.

- **Open in New Session** opens the selected reference attachment in a separate MicroStation session.
- **Merge into Master** copies the content of the selected reference attachment into the active design file and detaches the reference. The contents of the referenced model are not changed but copies of the elements in the reference are placed in the active design file.
- **Make Direct Attachment** promotes a nested attachment to a direct attachment so its attachment settings can be manipulated. The nested attachment becomes redundant with the newly created direct attachment. Quotation marks under the Display, Snap, and Locate tools indicate a redundant attachment.

SETTINGS MENU

The Settings menu provides housekeeping options for attached references.

Changes to Attachment Settings

Select a reference on the list of references and select **Attachment** from the **Settings** menu to open the Attachment Settings dialog box to make the changes to the attachment settings, as shown in Figure 13–11.

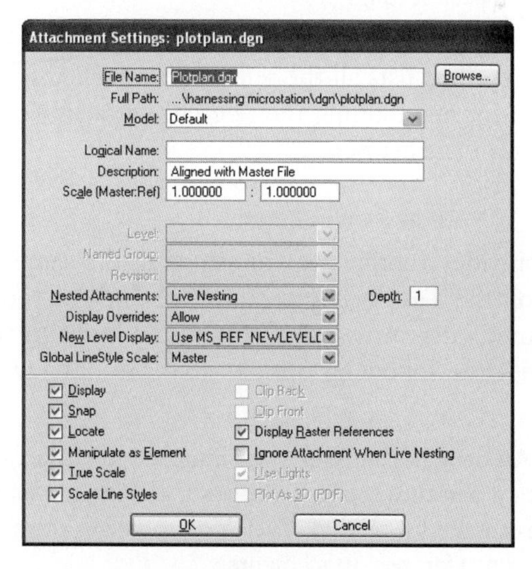

Figure 13–11 Attachment Settings dialog box

The dialog box contains many of the same settings as the Reference Attachment Settings dialog box. Some options that were menu options or toggle icons on the other dialog box are check boxes on this dialog box. The dialog box also contains a **Browse** button that, when clicked, opens the Reattach Reference dialog box, to al-

low changing the reference to a different design file. For more information on the options, see the discussion of attaching references earlier in this chapter.

***Note:** The **Attachment** option is only available when one reference is selected on the References list box. The option is unavailable if no references are selected or more than one reference is selected.*

***Note:** The Attachment Settings options are also available for the selected reference across the bottom of the References list box and double-clicking a reference on the References list box opens the Attachment Settings dialog box.*

Display Attributes settings

Select **Presentation** to open the Hidden Line Settings dialog box, which sets the display attributes of the selected referenced models, as shown in Figure 13–12.

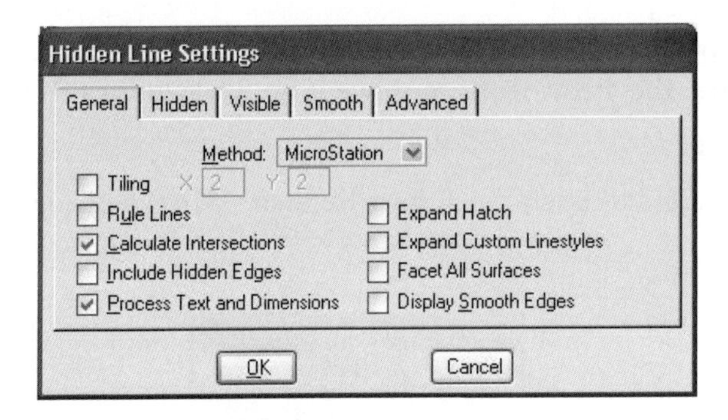

Figure 13–12 Hidden Line Settings dialog box

***Note:** The **Presentation** option is only available when the active model is a sheet model and the presentation of a selected reference is set to True Hidden Line.*

Changing the Display Order

Select **Update Sequence** to open the Update Sequence dialog box to change the display order, as shown in Figure 13–13. The dialog box allows for changing the display order for all operations involving the updating of the view windows. This is useful when references overlap because it allows for controlling which one paints on the screen last.

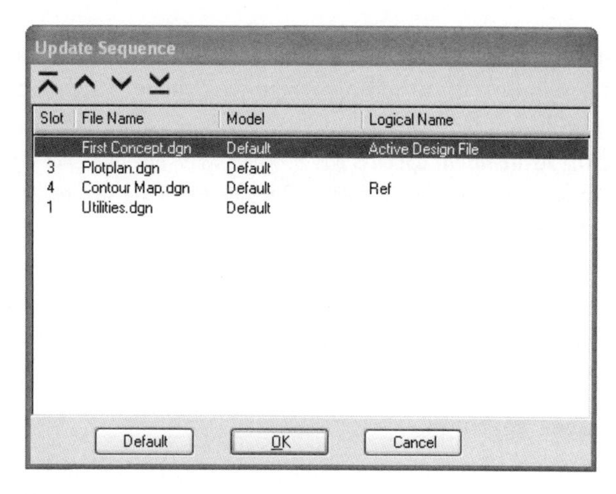

Figure 13–13 Update Sequence dialog box

To change the position of a reference in the update sequence, select the reference's file name and use the arrow icons to move it up or down in the sequence.

Adjusting Colors

Select **Adjust Colors** to open the Adjust Reference Colors dialog box, as shown in Figure 13–14.

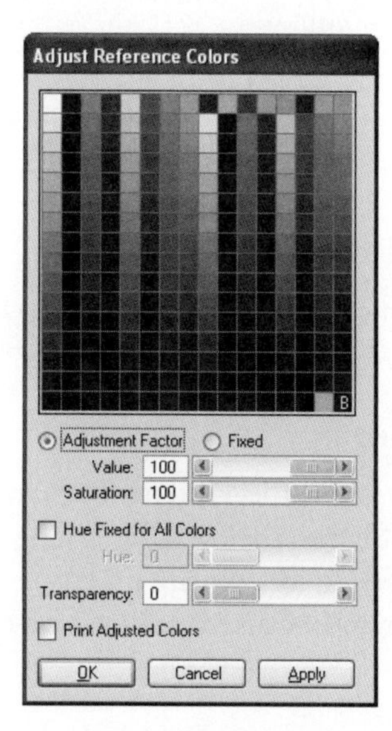

Figure 13–14 Adjust Reference Colors dialog box

The dialog box is used to modify or adjust the element colors in an attached reference. Each color in the color table is defined using the HSV (hue, saturation, value) color model.

Hilite Settings

Select **Hilite** to open a sub-menu containing two options for controlling the way references are identified when selected from the References list box.

- **Boundaries** identifies selected references by placing a dashed border around them.
- **Hilite** identifies selected references by highlighting all the elements in the reference with a color.

Note: *It is helpful to have one or both of the two reference highlighting methods turned on when manipulating elements with the standard manipulation tools or the reference manipulation tools. For information on the highlighting options, see the Reference list box topic earlier in this chapter.*

Auto Arrange Icons

If **Auto Arrange Icons** is ON, icons on the References list box's tool bar wrap when the dialog box is resized so that they are always visible. A check mark to the left of the option on the menu indicates that it is ON.

Level Manager for Reference

Select **Level Manager** to open the Level Manager dialog box, as shown in Figure 13 –15. The dialog box manages levels for the reference selected on the References list box.

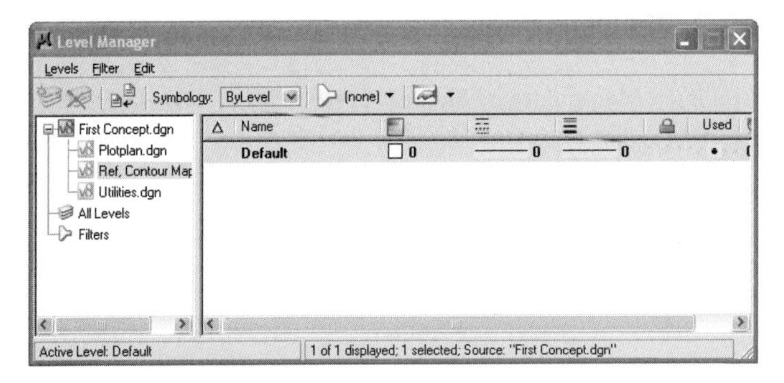

Figure 13–15 Level Manager dialog box

Select **Level Display** to open the Level Display dialog box, as shown in Figure 13–16). The dialog box turns ON and OFF the display of levels in the reference selected on the References list box.

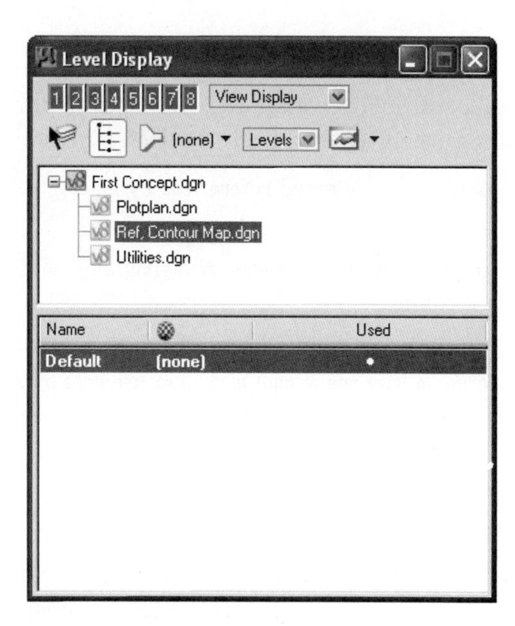

Figure 13–16 Level Display dialog box

REFERENCE MANIPULATION TOOLS

MicroStation provides a set of manipulation tools specifically designed to manipulate references. As mentioned earlier, the reference behaves as one element, and there is no way to drop it.

Note: *If the **Manipulate as Element** option is ON for the reference attachment, the standard manipulation tools also manipulate the reference as one element, as discussed later in this chapter.*

Note: *The Copy/Fold Reference tool and the Set Reference Presentation tool are for 3D designs only.*

SELECTING THE REFERENCE

Most of the reference manipulation tools on the Task Navigation tool box manipulate a selected reference attachment. There are three ways to select the reference attachments to manipulate:

- Select the reference to manipulate by clicking an element in the reference.
- Manipulate the references selected by the active fence in the design. To use a fence, place it before invoking the reference manipulation tool. Each tool that can use the active fence has a **Use Fence** check box and Fence Mode menu on the tool's settings box. The check box and menu are only available when there is an active fence in the design.

- Manipulate the references selected in the References list box. Open the list box and select the references before invoking the reference manipulation tool. Each tool that can use the References list box has a Use Reference dialog list check box on the tool's settings box. Turn the check box on to use the references selected in the References list box. The check box is only available when the References list box is open.

CLIPPING A REFERENCE

The Clip Reference tool clips an attached reference so that only a portion of a reference is visible. The clip boundary can be a closed element, the active fence, or a named fence.

Note: *MicroStation displays and plots nonrectangular clipping boundaries in a view only if the* ***Fast Ref Clipping*** *View Attribute is set to OFF for the selected view.*

Invoke Clip Reference from:

Task Navigation tool box (active task set to References)	Select the Clip Reference tool (see Figure 13–17).
Keyboard Navigation (Task Navigation tool box with active task set to References)	**W**

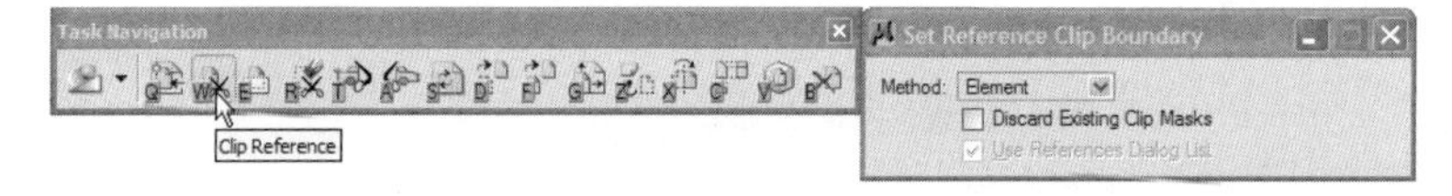

Figure 13–17 The Clip Reference tool on the Task Navigation tool box

After invoking the tool, adjust the settings on the tool's settings box as needed.

From the **Method** menu, select the type of clip boundary.

- Select **Element** to use a closed element as the clip boundary.
- Select **Active Fence** to use the active fence as the clip boundary. The **Fence** option is only available when there is an active fence on the design.
- Select **Named Fence** to use a named fence as the clip boundary. The **Named Fence** option is only available when there are named fences in the design. Selecting this option displays the list of named fences on the settings box.

If the reference to be clipped already has a clip boundary, turning the **Discard Existing Clip Masks** check box ON, removes the existing clip boundary.

If the **Method** is **Element,** MicroStation prompts:

Set Reference Clip Element > Identify Clipping Element

Select the closed element to use as the clip boundary.

Note: *If the Use Reference dialog bo x check box is ON, MicroStation initiates the clipping immediately after the element is selecting. There are no more prompts.*

Set Reference Clip Boundary > Select Reference

Identify the reference to clip by clicking an element in the reference.

Set Reference Clip Boundary > Accept/Reject Reference

Click the Data button to initiate the clipping the selected references at the boundary of the selected element.

If the **Method** is **Active Fence** or **Named Fence,** MicroStation prompts:

Set Reference Clip Boundary > Select Reference

If **Named** Fence is selected, select the named fence to use from the list on the settings box. Identify the reference to clip by clicking an element in the reference.

Note: *If the Use Reference dialog box check box is ON, MicroStation skips the first prompt and only asks to accept or reject the reference.*

Set Reference Clip Boundary > Accept/Reject Reference

Click the Data button to initiate the clipping the selected references at the boundary of the active fence.

MASKING A REFERENCE

The Mask Reference tool, like the Clip Reference tool, allows for displaying a portion of a reference. Clip Reference displays the part inside a clip boundary, whereas the Mask Reference tool displays the part outside the clip boundary. Place a Fence on the reference to define the desired clipping boundary before invoking this tool. Invoke Mask Reference from:

Task Navigation tool box (active task set to References)	Select the Mask Reference tool (see Figure 13–18).
Keyboard Navigation (Task Navigation tool box with active task set to References)	**E**

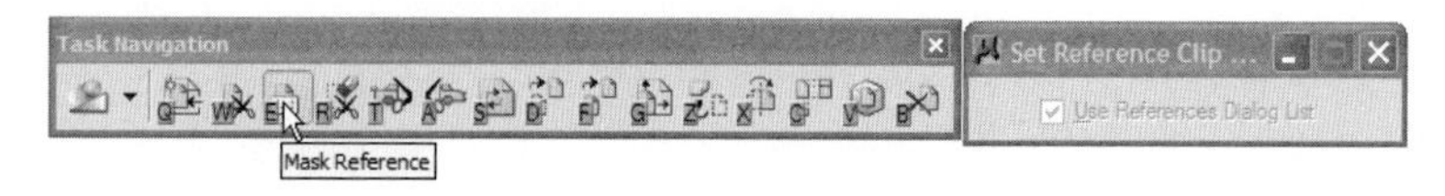

Figure 13–18 The Mask Reference tool on the Task Navigation tool box

After invoking the tool, make the required setting on the tool's settings box.

MicroStation prompts:

> Set Reference Clip Boundary > Select Reference

Identify the reference to clip mask by clicking an element in the reference.

> Set Reference Clip Boundary > Accept/Reject Reference

Click the Data button to initiate the clipping mask for the selected reference at the boundary of the active fence.

DELETING A REFERENCE CLIP

The Delete Clip tool can delete a reference clipping mask for a selected reference or the references selected by a fence. Invoke Delete Clip from:

Task Navigation tool box (active task set to References)	Select the Delete Clip tool (see Figure 13–19).
Keyboard Navigation (Task Navigation tool box with active task set to References)	**R**

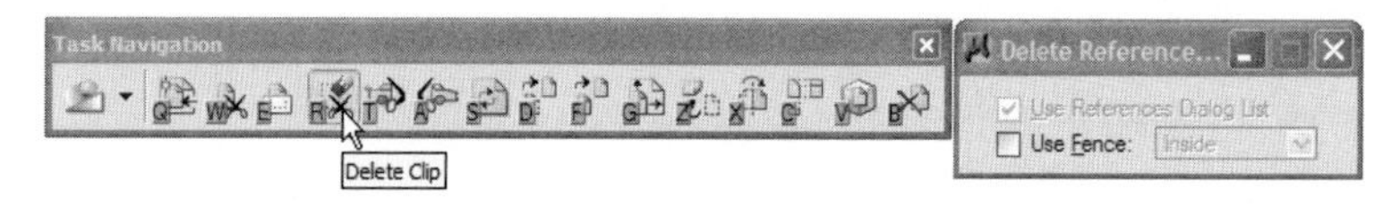

Figure 13–19 The Delete Clip tool on the Task Navigation tool box

After invoking the tool, adjust the settings on the tool's settings box as needed.

MicroStation prompts:

> Delete Reference Clip Component > Select Reference

Select the reference from which the clip or clip mask boundary is to be removed.

Note: *If the Use Reference dialog box check box is ON, MicroStation skips the first prompt and only asks to accept or reject the reference.*

> Delete Reference Clip Component > Accept/Reject Delete Clip Boundary

Click the Data button to accept the selected reference and initiate deleting the boundary.

RELOADING A REFERENCE

MicroStation automatically loads all references attached to a design file only when the design file is opened. To get the latest version of a reference after the design is opened, use the Reload tool. Invoke Reload Reference from:

Task Navigation tool box (active task set to References)	Select the Reload Reference tool (see Figure 13–20).
Keyboard Navigation (Task Navigation tool box with active task set to References)	**S**

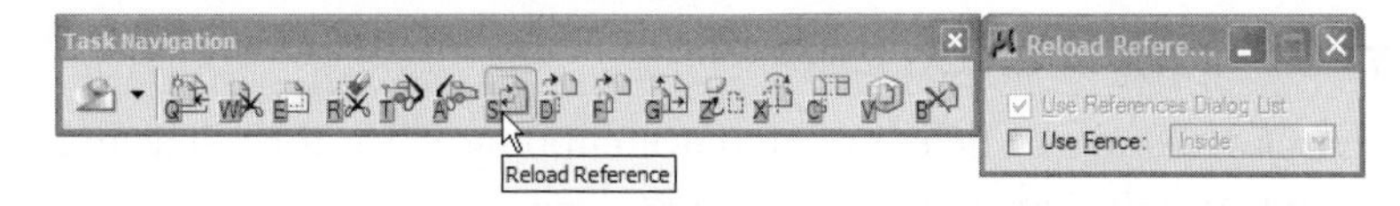

Figure 13–20 The Reload Reference tool on the Task Navigation tool box

After invoking the tool, adjust the settings on the tool's settings box as needed.

MicroStation prompts:

Reload Reference > Select Reference

Select the reference to reload.

Delete Reference Clip Component > Accept/Reject Delete Clip Boundary Reload Reference > Accept/Reject Reference

Click the Data button to accept the selected reference and initiate reloading it.

MOVING A REFERENCE

The Move References tool moves a reference from one location to another. To move a reference, invoke Move References from:

Task Navigation tool box (active task set to References)	Select the Move References tool (see Figure 13–21).
Keyboard Navigation (Task Navigation tool box with active task set to References)	**D**

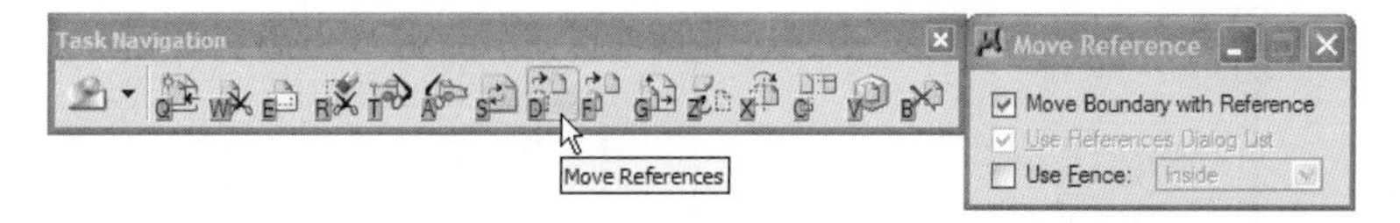

Figure 13–21 The Move References tool on the Task Navigation tool bar

After invoking the tool, adjust the settings on the tool's settings box as needed.

If the reference has a clip or clip mask boundary, turn the **Move Boundary with Reference** check box ON to move the boundary with the reference. If the check box is OFF, the reference slides through the boundary when it is moved.

If the Use Fence and Use Reference dialog box check boxes are both OFF, MicroStation prompts:

Move Reference > Select Reference

Select the reference at the point from which the move is to start.

Move Reference > Enter point to move to

Place a data point at the move to location. The reference is moved such that the location of the data point that identified the reference is placed at the location of the move to data point. The tool continues to drag an image of the reference and additional data points continue moving it. Click Reset or select another tool to drop the dynamic image.

If the Use Fence or Use Reference dialog box check box is ON, MicroStation prompts:

Move Reference > Enter point to move from

Place a data point at the location from which the move is to start.

Move Reference > Enter point to move to

Place a data point at the move to location. The reference is moved such that the location of the data point that identified the reference is placed at the location of the move to data point.

Figure 13–22 shows a view before and after moving a reference border.

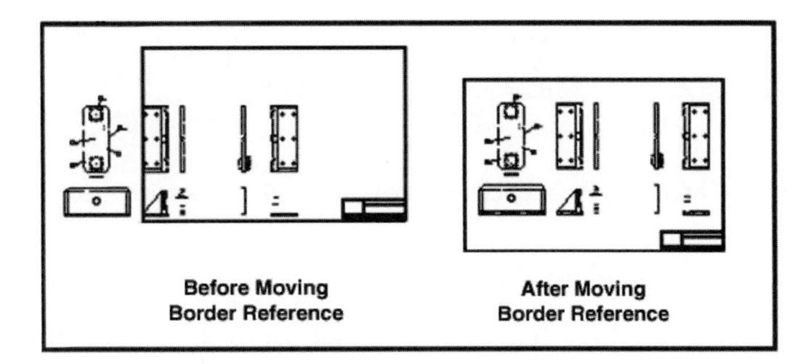

Figure 13–22 The design view before and after moving a reference

COPYING A REFERENCE

The Copy References tool makes additional attachments of the selected reference. Everything about the attachment is copied and a unique logical name is provided for each copy. Invoke Copy References from:

Task Navigation tool box (active task set to References)	Select the Copy References tool (see Figure 13–23).
Keyboard Navigation (Task Navigation tool box with active task set to References)	**F**

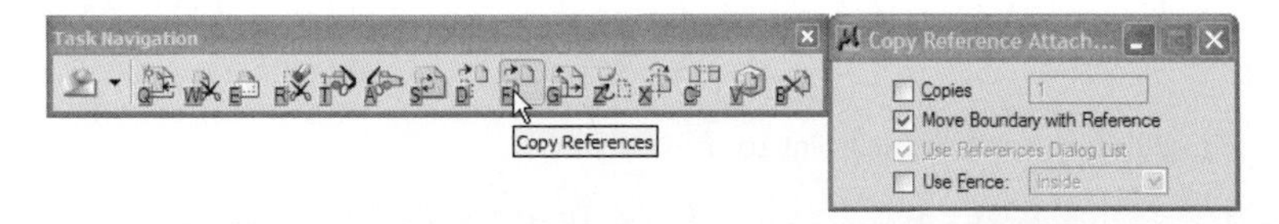

Figure 13–23 Invoking the Copy References tool from the References tool box

After invoking the tool, adjust the settings on the tool's settings box as needed.

Type the number of times to copy the attachment in the **Copies** field. The default value for the field is one. If more than one copy is to be placed, they are placed along an imaginary line from the selection point to the placement data point and the space between each copy is equal to the space between the original and first copy.

If the Use Fence and Use Reference dialog box check boxes are both OFF, MicroStation prompts:

Copy Reference > Select Reference

Select the reference at the point from which the copy is to start.

Copy Reference > Enter point to move to

Place a data point at the copy to location. The reference is copied such that the location of the data point that identified the reference is placed at the location of the copy to data point.

If the Use Fence or Use Reference dialog box check box is ON, MicroStation prompts:

Copy Reference > Enter point to move from

Place a data point at the location from which the copy is to start.

Copy Reference > Enter point to move to

Place a data point at the copy to location. The reference is copied such that the location of the data point that identified the reference is placed at the location of the copy to data point. The tool continues to drag an image of the reference and additional data points continue placing copies. Click Reset or select another tool to drop the dynamic image.

SCALING A REFERENCE

The Scale References tool enlarges or reduces the viewed size the reference attachment. Invoke the Scale References tool from:

Task Navigation tool box (active task set to References)	Select the Scale References tool (see Figure 13–24).
Keyboard Navigation (Task Navigation tool box with active task set to References)	**G**

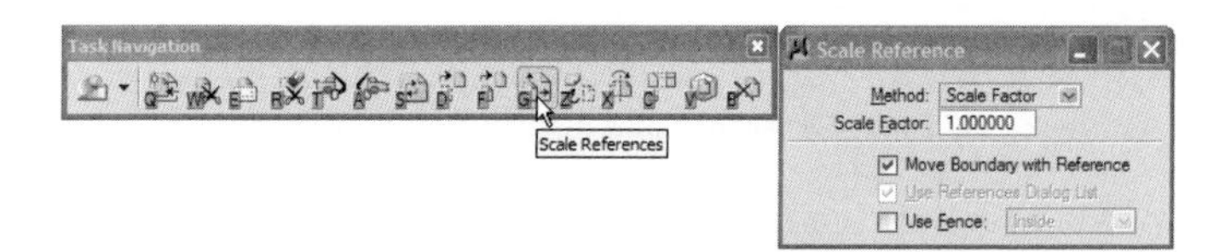

Figure 13–24 The Scale References tool on the Task Navigation tool box

After invoking the tool, adjust the settings on the tool's settings box as needed.

In the Scale Reference settings box, the **Method** option menu sets the method used to scale the reference attachment:

- **Scale Factor** scales the model by a specified factor. Selecting this method displays the **Scale Factor** field on the settings box. Type the scaling factor in the field. Typing a number greater than one increases the size or typing a number less than one decreases the size.

- **Absolute Ratio** scales the model by a ratio of Master Units in the active design file to Master Units in the reference. Selecting this method displays the **Scale (Master:Ref)** fields on the settings box. Type the ratio of the active design file's Master Units to the reference's Master Units in the two fields.

Note: *The scale factor ratio between the active design file and the reference is not cumulative. For instance, if a scale of 3:1 is specified, followed by 6:1, the result will be 6:1.*

- **By Points** scales the model by data points entered on the design. There is no additional field for this method.

If the reference has a clip or clip mask boundary, turn the **Move Boundary with Reference** check box ON to scale the boundary with the reference. If the check box is OFF, the reference slides through the boundary when it is scaled.

If the Use Fence and Use Reference dialog box check boxes are both OFF, and the **Method** is **Scale Factor** or **Absolute Ratio**, MicroStation prompts:

Scale Reference > Select Reference

Select the reference at the point about which the reference attachment is to be scaled.

Scale Reference > Accept/Reject Reference

Place a data point to scale the image and define its new location.

If the Use Fence and Use Reference dialog box check boxes are both OFF, and the **Method** is **By Points**, MicroStation prompts:

Scale Reference By Points > Select Reference

Select the reference.

Scale Reference By Points > Accept and identify origin point/Reject

Place a data point to begin scaling the reference by points.

Scale Reference By Points > Enter reference point

Place a data point to define the base for scaling the reference.

Scale Reference By Points > Enter point to define amount of scaling

Move the pointer until the reference is the correct new size and place a data point to complete the scaling.

If the Use Fence or Use Reference dialog box check box is ON, and the **Method** is **Scale Factor** or **Absolute Ratio**, MicroStation prompts:

Scale Reference > Enter point to scale about

Place a data point to scale the image and define it new location.

If the Use Fence or Use Reference dialog box check box is ON, and the **Method** is **By Points** or **Absolute Ratio**, MicroStation prompts:

Scale Reference By Points > Enter origin point for reference scale

Place a data point to begin scaling the reference by points.

Scale Reference By Points > Enter reference point

Place a data point to define the base for scaling the reference.

Scale Reference By Points > Enter point to define amount of scaling

Move the pointer until the reference is the correct new size and place a data point to complete the scaling.

Figure 13–25 shows a view before and after scaling the reference border.

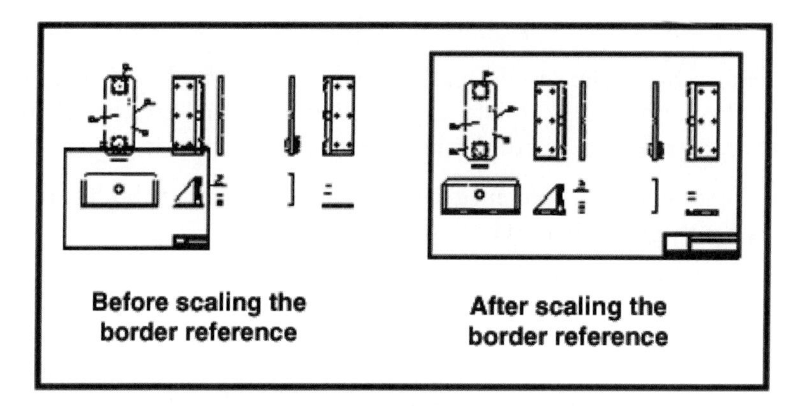

Figure 13–25 The design view before and after scaling a reference

ROTATING A REFERENCE

The Rotate References tool rotates a reference to any angle around a pivot point. Invoke the Rotate Reference tool from:

Task Navigation tool box (active task set to References)	Select the Rotate References tool (see Figure 13–26).
Keyboard Navigation (Task Navigation tool box with active task set to References)	**Z**

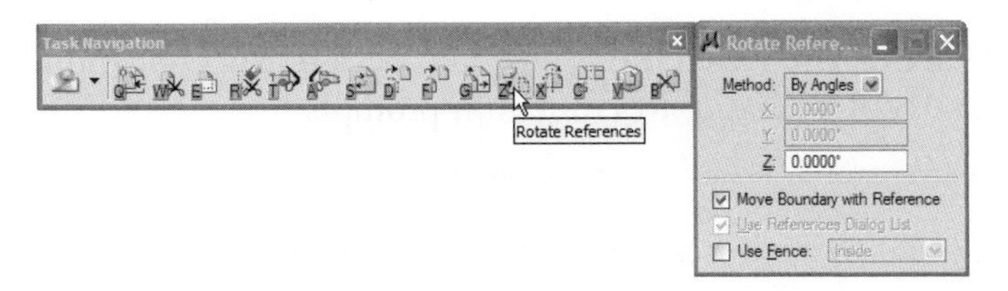

Figure 13–26 Invoking the Rotate References tool from the References tool box

After invoking the tool, adjust the settings on the tool's settings box as needed.

The **Method** menu sets the method by which the reference is rotated:

- **By Angles** uses the angles in the **X**, **Y**, and **Z**-axis fields to rotate the reference. Only the **Z**-axis field is used for 2D designs. The **X** and **Y**-axis fields are 3D only.
- **By Points** graphically rotates the reference.

If the reference has a clip or clip mask boundary, turn the **Move Boundary with Reference** check box ON to rotate the boundary with the reference. If the check box is OFF, the reference slides through the boundary when it is rotated.

If the Use Fence and Use Reference dialog box check boxes are both OFF, and the **Method** is **By Angles**, MicroStation prompts:

Rotate Reference > Select Reference

Select the reference at the point about which the reference attachment is to be rotated.

Rotate Reference > Accept/Reject Reference

Place a data point to rotate the image and define it new location.

If the Use Fence and Use Reference dialog box check boxes are both OFF, and the **Method** is **By Points,** MicroStation prompts:

Rotate Reference By Points > Select Reference

Select the reference.

Rotate Reference By Points > Accept and identify pivot point/Reject

Place a data point to identify the point about which the references will be rotated.

Rotate Reference By Points > Enter point to rotate ref file about

Place a data point to complete rotation of the reference.

Rotate Reference By Points > Enter point to define amount of rotation

Move the pointer until the reference is the correct rotation and place a data point to complete the rotation.

If the Use Fence or Use Reference dialog box check box is ON, and the **Method** is **By Angles**, MicroStation prompts:

Rotate Reference > Enter point to rotate ref file about

Place a data point to move and rotate the reference.

If the Use Fence or Use Reference dialog box check box is ON, and the **Method** is **By Points** or **Absolute Ratio**, MicroStation prompts:

Rotate Reference By Points > Enter pivot point for reference rotation

Place a data point to begin scaling the reference by points.

Rotate Reference By Points > Enter point to define start of rotation

Place a data point to define the point about which the reference will be rotated.

Rotate Reference By Points > Enter point to define amount of rotation

Move the pointer until the reference is the correct new rotation and place a data point to complete the rotation.

MIRRORING A REFERENCE

The Mirror Reference tool mirrors a reference about the X or Y-axis. Invoke Mirror Reference from:

Task Navigation tool box (active task set to References)	Select the Mirror References tool (see Figure 13–27).
Keyboard Navigation (Task Navigation tool box with active task set to References)	**X**

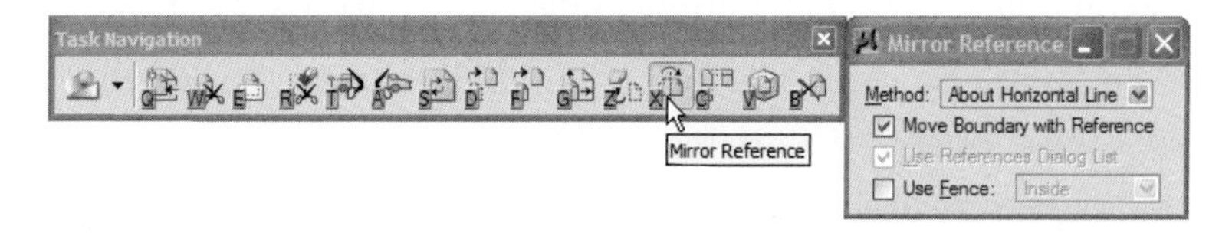

Figure 13–27 The Mirror Reference tool on the Task Navigation tool box

After invoking the tool, adjust the settings on the tool's settings box as needed.

The **Method** menu determines which axis is used to mirror the reference:

- **About Horizontal Line** mirrors the reference about the X–axis.
- **About Vertical Line** mirrors the reference about the Y–axis.

If the reference has a clip or clip mask boundary, turn the **Move Boundary with Reference** check box ON to mirror the boundary with the reference. If the check box is OFF, the reference slides through the boundary when it is mirrored.

MicroStation prompts:

If the Use Fence and Use Reference dialog box check boxes are both OFF, and the **Method** is **About Horizontal Line**, MicroStation prompts:

Mirror Reference About Horizontal > Select Reference

Select the reference at the point that is to be mirrored.

Mirror Reference About Horizontal > Accept/Reject Reference

Move the pointer up or down until the mirror image of the reference is at the correct location and click the Data button to place the mirrored reference.

If the Use Fence and Use Reference dialog box check boxes are both OFF, and the **Method** is **About Vertical Line**, MicroStation prompts:

Mirror Reference About Vertical > Select Reference

Select the reference at the point that is to be mirrored.

Mirror Reference About Vertical > Accept/Reject Reference

Move the pointer left or right until the mirror image of the reference is at the correct location and click the Data button to place the mirrored reference.

If the Use Fence or Use Reference dialog list check box is ON, and the **Method** is **About Horizontal Line**, MicroStation prompts:

Mirror Reference About Horizontal > Enter point to mirror about

Move the pointer up or down until the mirror image of the reference is at the correct location and click the Data button to place the mirrored reference.

If the Use Fence or Use Reference dialog box check box is ON, and the **Method** is **About Horizontal Line**, MicroStation prompts:

Mirror Reference About Vertical > Enter point to mirror about

Move the pointer left or right until the mirror image of the reference is at the correct location and click the Data button to place the mirrored reference.

DETACHING A REFERENCE

The Detach Reference tool detaches references from the active design file. Invoke Detach Reference from:

Task Navigation tool box (active task set to References)	Select the Detach Reference tool (see Figure 13–28).
Keyboard Navigation (Task Navigation tool box with active task set to References)	**B**

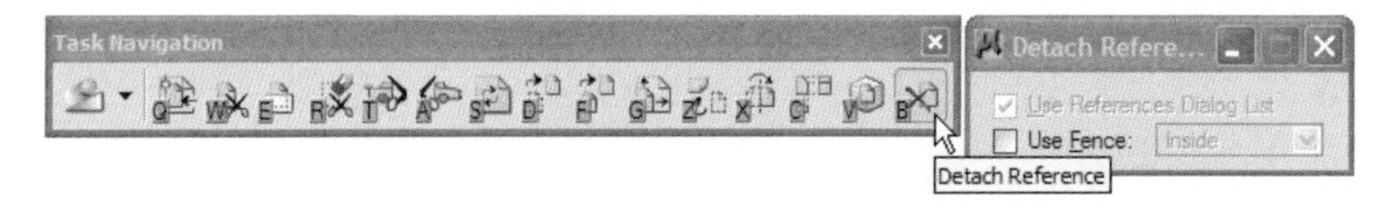

Figure 13–28 The Detach Reference tool on the Task navigator tool box

After invoking the tool, adjust the settings on the tool's settings box as needed.

If the Use Fence and Use Reference dialog box check boxes are both OFF, MicroStation prompts:

Detach Reference > Select Reference

Select the reference to detach.

Detach Reference > Accept/Reject Reference

Click the Data button to detach the reference. The reference is immediately detached.

If the **Use Fence or Use Reference** dialog box check box is ON, MicroStation displays an Alert box (see Figure 13–29). Click **OK** to detach the selected reference or click **Cancel** to cancel the operation.

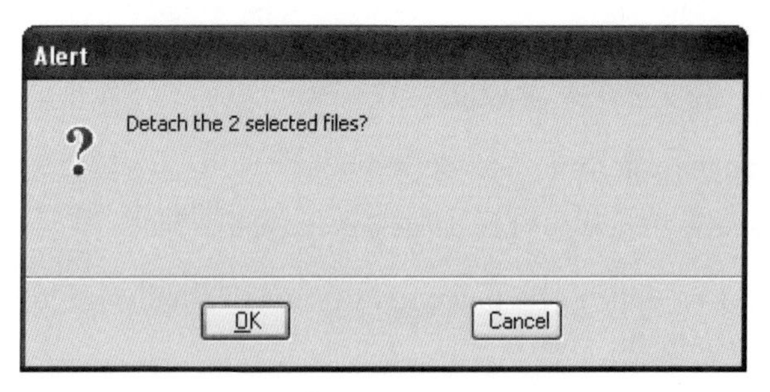

Figure 13–29 Alert box to confirm detaching a reference

USING STANDARD MANIPULATION TOOLS

If **Treat Attachment as Element for Manipulation** is ON for a reference attachment, the standard element manipulation tools can be used to manipulate the reference. All manipulations treat the reference as one element, and manipulations

can be preformed on individual references or references selected by the active fence.

If a reference is set to allow manipulation as an element, it highlights when selected. If the reference is not set to allow manipulation as an element, attempting to select it changes the appearance of the pointer changes as shown in Figure 13–30.

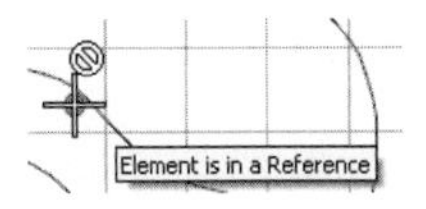

Figure 13–30 Pointer appearance when element manipulation is not allowed for a reference

To move a reference attachment to a new location using a standard element manipulation tool, invoke Move from:

Manipulate tool box	Select the Move tool (see Figure 13–31).
Keyboard Navigation (active task set to Main tool box)	**3 2**

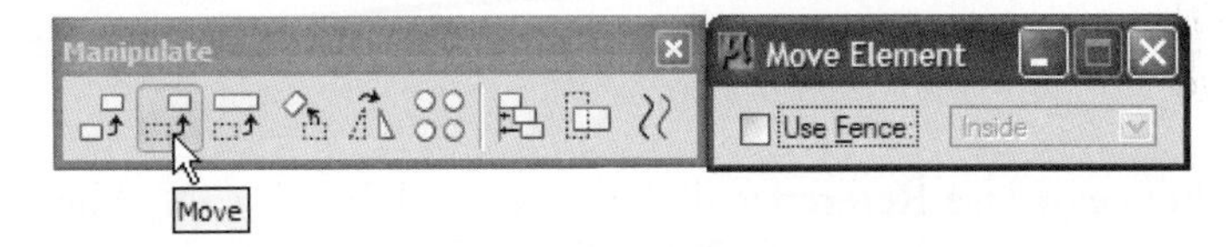

Figure 13–31 The Move tool selected on the Manipulate tool box

MicroStation prompts:

Move Element > Identify element

Select the reference attachment to be moved.

Move Element > Enter point to define distance and direction

Place a data point to define the placement point in the design for the reference attachment and initiate the move.

When a reference attachment's **Treat Attachment as Element for Manipulation** check box is ON, the element Delete tool can remove the reference attachment from the active design file.

The interaction of the Copy tool with references is different than the other standard manipulation tools. The other tools can only manipulate references as an element when **Treat Attachment as Element for Manipulation** is ON for the reference. The copy can work with references when the check box ON or OFF:

- When **Treat Attachment as Element for Manipulation** is ON, the Copy tool copies the reference as one element and the copy is a new reference attachment.
- When **Treat Attachment as Element for Manipulation** is OFF, the Copy tool copies individual elements from the reference into the active design file. The Copy tool can also copy elements in the reference selected by the active fence. The copied elements are part of the active design file, not a reference attachment.

Being able to copy elements from the reference is useful when, for example, an existing design file contains design information needed, with minor changes, in a new design file.

To copy an element from a reference to the active design, make sure **Treat Attachment as Element for Manipulation** is OFF for the reference and invoke Copy from:

Manipulate tool box	Select the Copy tool (see Figure 13–32).
Keyboard Navigation (active task set to Main tool box)	**3 1**

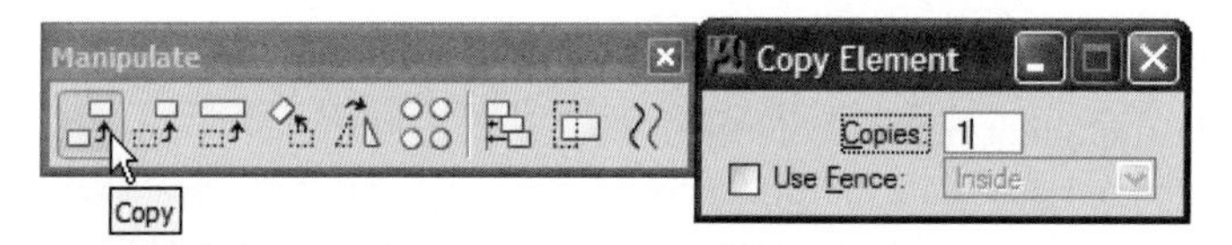

Figure 13–32 The Copy tool selected on the Manipulate tool box

MicroStation prompts:

Move Element > Identify element

Select the reference attachment to be moved.

Move Element > Enter point to define distance and direction

Place a data point to define the placement point in the design for the reference attachment and initiate the move.

CREATING SHEET MODELS

Creating Sheet models automates the creation of drawing sheets for printing of design drawings. This process is similar to the manual drafting process. Where it differs, however, is that instead of redrawing the model's geometry for each view, like the manual system requires, views of the design model can be attached as references.

Sheet models can be thought of as virtual sheets of paper, which are printed to produce hard-copy documents. Design geometry that appears in a Sheet model is attached as references, either from models in the same DGN file, or from models in other DGN files, or a combination of both. The reference tools can be used to manipulate the references as needed to compose the sheet. Typically, this may include folding, copying, scaling, clipping, rotating, masking, and/or mirroring the references. When working with 3D references, true hidden line representation can be used to display the reference correctly for the drawing.

To simplify this process, when a Sheet model is created, an Annotation Scale may be associated to it, as well as a sheet layout size. Additionally, if required, the origin of the sheet layout and its rotation can be specified.

Following are the steps for creating a sheet model (Drawing Composition):

1. Create a DGN file and draw the design model using references as required.
2. Create a Sheet model (this can be in the same DGN file as the design geometry, or in a separate DGN file, as required). Select appropriate paper size and set the scale where the drawing border is scaled up (or down) to cover the required area in the design. All text and dimensioning must be scaled the same amount by setting the Annotation scale lock to ON. Refer to Chapter 1 for a detailed explanation on creating sheet models.
3. In the Sheet model, reference the border file at the required scale.
4. In the Sheet model, reference all views of the design model at the appropriate scale factor to compose the sheet. If the scale was set to 1:1 in Step 2 then scale the references to fit the sheet. Set nesting to the level necessary to include all required sub references.
5. For any views that are to be at a different scale from the main scale, reference them at a scale factor of Main Scale/Required Scale.
6. If necessary, use the Set Reference Presentation tool, in the References dialog box, to change the display format for the references. This applies particularly to 3D models.
7. Place text and dimensions at their required size. Annotation Scale lock will scale the size of the text so that it prints at the required size.
8. Plot the sheet model at the scale set in Step 2.

Open the Exercise Manual PDF file for Chapter 13 on the accompanying CD for project and discipline specific exercises.

REVIEW QUESTIONS

Write your answers in the spaces provided.

1. List at least two benefits of using references.
   ____
2. By default, how many references can you attach to a design file?
   ____
3. What effect does attaching references have on a design file size?
   ____
4. List the tools that are specifically provided to manipulate references.
   ____
5. List the steps involved in attaching a reference.
   ____
6. Explain the difference between the Reference Clip Boundary tool and the Reference Clip Mask tool.
   ____
7. List the steps involved in detaching a reference using a reference manipulation tool.
   ____

Special Features

MicroStation provides special features that, although used less often than the tools described earlier in this book, provide added power and versatility. This chapter introduces several such features.

OBJECTIVES

Topics explored in this chapter:

- Creating and using graphic groups
- Creating and using named groups
- Using the Merge utility
- Selecting groups of elements using element attributes
- Cleaning up a design file
- Changing the highlight and vector cursor colors
- Importing and exporting drawings
- Working with AutoCAD drawings
- Manipulating graphic images
- Annotating designs
- Creating dimension-driven designs
- Using object linking and embedding

GRAPHIC GROUPS

Graphic groups provide a quick way to group elements. Turn the Graphic Group lock ON to manipulate groups as one element and turn the lock OFF to add, manipulate, or delete individual elements in the group without breaking up the group. Add more versatility to groups by naming them.

Note: *The elements of a pattern are part of one graphic group, but the elements that contain the pattern are not part of the group. See Chapter 12 for more information on patterning.*

CREATE OR ADD TO A GRAPHIC GROUP

The Add To Graphic Group tool creates new groups and adds elements to an existing group. Its action is the same whether the Graphic Group lock is OFF or ON.

- If the first element selected is not part of an existing graphic group, a new graphic group is created and all elements selected after first element become part of the new graphic group.
- If the first element selected is already in a group, and following element selections add the elements to the existing group.

Invoke Add To Graphic Group from:

Groups tool box	Select the Add To Group tool (see Figure 14–1).
Keyboard Navigation (Main tool box active)	**6 5**

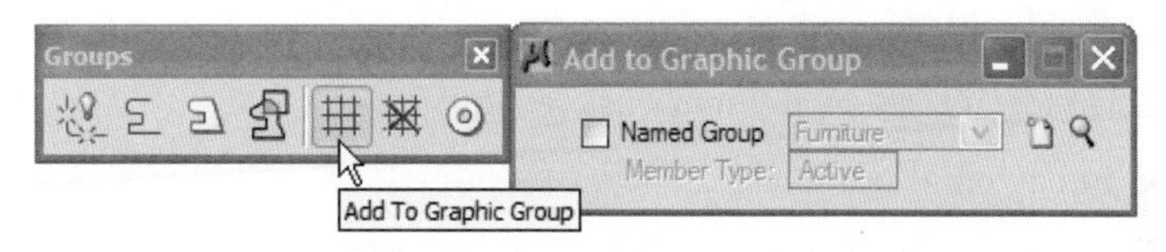

Figure 14–1 The Add To Graphic Group tool on the Main tool box

Note: *The options on the tool's settings box are for Named Groups that are discussed later in this chapter.*

MicroStation prompts:

Add to Graphic Group > Identify element

Identify the first element.

If the selected element is not already in a graphic group, a new group is created and MicroStation prompts:

Add to Graphic Group > Add to new group (Accept/Reject)

Identify the next element to add to the new group or click the Reset button to reject the first selected element.)

If the first selected element is already in a graphic group, MicroStation prompts:

Add to Graphic Group > Add to existing group (Accept/Reject)

Identify the first element to add to the existing graphic group or click the Reset button to reject the selected group. If the cursor is on another element when the Data button is clicked to accept the selected element, MicroStation selects the element under the pointer and prompts:

Add to Graphic Group > Accept/Reject (select next input)

Continue selecting elements until all required elements are in the graphic group. After selecting the last element to be added to the group, accept it without accepting another element.

CREATE A NEW GRAPHIC GROUP FROM SELECTED ELEMENTS

The element selection tools can select elements for a new graphic group before invoking the Add To Graphic Group tool. Use element selection tools to select the elements for the new graphic group and then invoke Add To Graphic Group from:

Groups tool box	Select the Add To Group tool (see Figure 14–1).
Keyboard Navigation	**6 5**

MicroStation prompts:

Add to Graphic Group > Accept to add selected elements to group

Click the Data button to initiate creating a group from the selected elements.

USING THE ELEMENT MANIPULATION TOOLS WITH GRAPHIC GROUPS

The element manipulation tools can manipulate a graphic group as one element when the Graphic Group lock is ON. Only the selected elements are manipulated when the Graphic Group lock is OFF. Turn the **Graphic Group** lock ON and OFF from any of the Lock settings boxes and menus (see Figure 14–2).

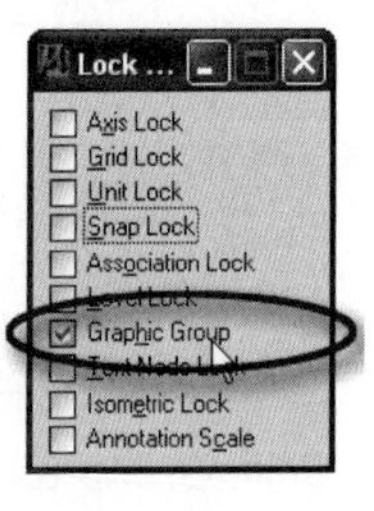

Figure 14–2 The Graphic Group lock on the Lock Toggles box

For example, invoke the Rotate element tool and select an element in a graphic group when prompted and the following happens:

- If Graphic Group lock is OFF, only the selected element rotates and it remains part of the graphic group.
- If Graphic Group lock is ON, the entire graphic group rotates.

If the Copy element tool is invoked and an element in a graphic group is selected after the tool is invoked, the tool's action depends on the Graphic Group lock status.

- If the lock is OFF, the tool copies only the selected element and puts the copied element in a new graphic group.
- If the lock is ON, the tool copies all elements in the graphic group and places the copied elements in a new graphic group.

If the element selection tools are used to select the elements before invoking the Copy element tool, the Copy tool copies only the selected elements and puts them in a new graphic group, regardless of the Graphic Group lock setting.

If the Copy Fence Contents tool is used to copy the elements selected by the fence, and the elements are in a graphic group, the Copy tool copies only the selected elements and puts the copies in a new graphic group, regardless of the Graphic Group lock setting.

DROPPING ELEMENTS FROM A GRAPHIC GROUP

The Drop From Graphic Group tool drops elements from a graphic group. When the Graphic Group lock is OFF, the tool drops selected elements from the graphic group. When the Graphic Group lock is ON, the tool drops the entire graphic group. Invoke Drop From Graphic Group from:

Groups tool box	Select the Drop From Group tool (see Figure 14–3).
Keyboard Navigation	**6 6**

Figure 14–3 The Drop From Graphic Group tool on the Main tool box

MicroStation prompts:

Drop From Graphic Group > Identify element

Identify an element in the graphic group.

Drop From Graphic Group > Accept/Reject (Select next input)

Click the Data button to initiate the drop.

- If Graphic Group Lock is OFF, the tool drops only the selected element. The graphic group still exists and the remaining elements are all still a part of it.
- If Graphic Group Lock is ON, the tool drops the entire graphic group.

Note: *The Drop From Graphic Group tool can only select elements that are part of a graphic group, regardless of the lock setting. Attempting to select an element not in a graphic group displays a pop-up message and changes the appearance of the pointer, as shown in Figure 14–4.*

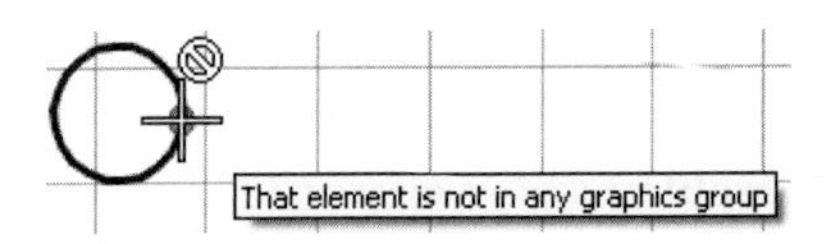

Figure 14–4 Pointer appearance for illegal element selection with the Drop From Graphic Group tool active

The element selection tools can select elements for dropping before invoking the Drop From Graphic Group tool. Selecting elements before invoking the tool always drops only the selected elements, regardless of the Graphic Group lock status. Use element selection tools to select the elements to drop and then invoke Add to Graphic Group from:

Groups tool box	Select the Drop From Group tool (see Figure 14–3).
Keyboard Navigation	**6 6**

MicroStation prompts:

Drop From Graphic Group > Accept/Reject (Select next input)

Click the Data button to initiate the drop.

Because the elements are selected before the tool is invoked, there is no error for selecting an element that is not in a graphic group. When elements not in a graphic group are selected, the tool proceeds normally and there is no indication that one or more of the selected elements was not in a group.

NAMED GROUPS

Named groups provide a new method for grouping elements in a design. Named groups are similar to graphic groups but are more versatile. Named groups can be

grouped in a hierarchy with "parent" named groups that contains "child" named groups. Parent named group manipulations include the child named groups. Elements in individual named groups and in the hierarchy can be added, dropped and manipulated.

The elements in a named group can be in the active design file or in a directly attached reference. Member elements in a directly attached reference are not copied into the active design file. Rather, the named group refers to their locations in the reference. Detaching the reference removes the reference elements from the named group.

Named groups consist of a name, an optional description, and a selection setting. The selection setting determines how elements in the named group (the members) are selected by other tools:

- When the **Selectable** toggle switch is OFF, the selection tools can select individual elements in the named group.
- When the **Selectable** toggle switch is ON, using the selection tools to select one element in the named group selects all elements in the group.

CHANGE PROPAGATION

Each element in a named group has three propagation settings that control when changes applied to the element can be applied to other elements in the group, when changes applied to other elements in the group can be applied to the element, and when changes applied to the element can be applied to other groups.

- In the **Group Lock** state, element changes are only applied when Graphic Group lock is ON.
- In the **Never** state, element changes are never applied.
- In the **Always** state, element changes are always applied.

The propagation settings are applied to an element when it is added to a named group by selecting an option form the **Member Type** menu. The type options are:

- **Active**, which has the following propagation settings:
  - To Other Members = When Graphic Group Lock is ON
  - From Other Members = When Graphic Group Lock is ON
  - To Other Groups = Never
- **Passive**, which has the following propagation settings:
  - To Other Members = Never
  - From Other Members = When Graphic Group Lock is ON
  - To Other Groups: Never

- **Custom**, which sets the three propagation settings as desired.

For example, with the Graphic Group Lock is ON, the color is changed for an element that has the Active Member Type. The color change is applied to all other elements in the group that are of the Active or Passive Member type. The only elements in the group to which the color change is not applied are those that have the Custom Member Type with From Other Members set to Never.

NAMED GROUPS LIST BOX

The Named Groups list box lists all named groups in the current design and provides options for creating and maintaining graphic groups. The box is opened from the MicroStation menu bar or from the Add to Graphic Group settings box.

Open from the Menu

Invoke the Named Groups list box from:

Menu	Utilities > Named Groups

MicroStation displays the Named Groups list box as shown in Figure 14–5.

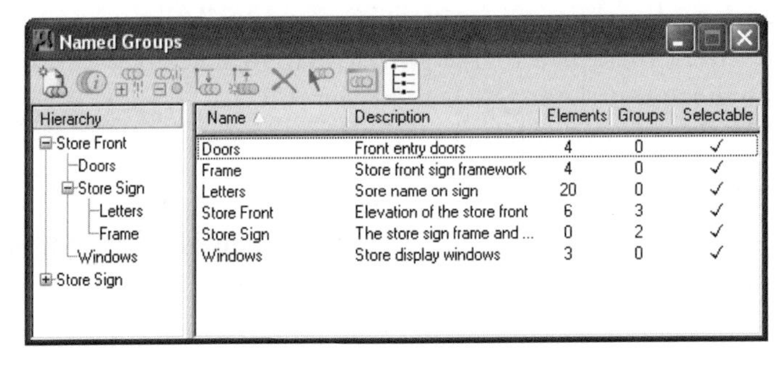

Figure 14–5 Named Groups list box

Note: *If the list of named groups on the settings box does not have all the columns shown in Figure 14–5, right-click the column headings row to display a list of the columns that can be added to the list. Click a column name to add it.*

Open from the Add to Graphic Group Tool

Invoke Add To Graphic Group from:

Groups tool box	Select the Add To Group tool
Keyboard Navigation	**6 5**

Click the magnifying glass icon, as shown in Figure 14–6.

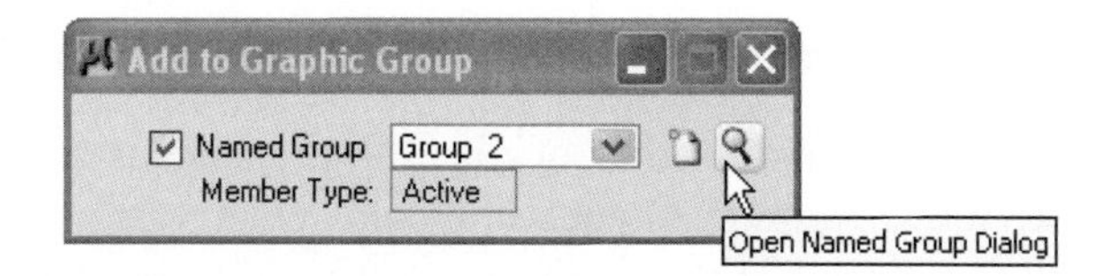

Figure 14–6 The Named Groups list box option

MicroStation displays the Named Groups list box as shown in Figure 14–5.

Named Groups List

The area of the window below the tool bar icons lists all existing named groups. In Figure 14–5, two named group list areas are displayed. The list on the right is a simple list of all named groups and the list on the left shows the named groups in a hierarchical tree with "Parent" and "Child" named groups. The hierarchical tree display appears when the Show Hierarch button is clicked on the setting box's tool bar. The initial display of the settings box only shows the simple list of named groups.

A check mark in the **Selectable** column on the simple list of named groups indicates that the named group is selectable, which means that when one element in the group is selected with an element selection tool, all elements in the group are selected. The check mark is a toggle switch that switches between ON and OFF each time it is clicked. If the toggle is ON when clicked, it turns OFF and individual elements in the group can be selected.

Named Groups Tools

A set of icons on a tool bar across the top of the settings box provide all the named group creation and maintenance tools. Figure 14–7 names the tools and following the figure are descriptions of each tool.

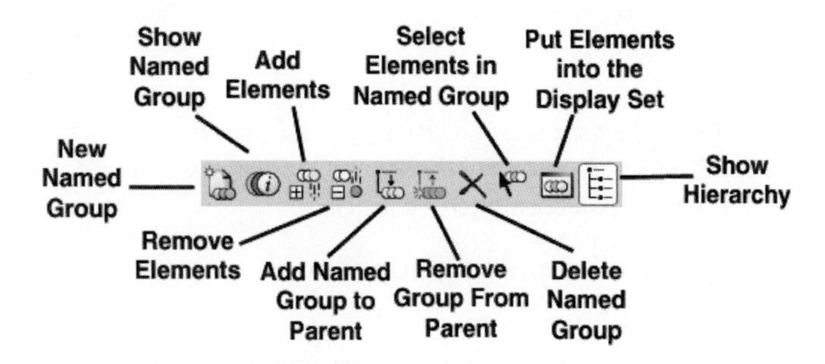

Figure 14–7 The Named Groups tools

- **New Named Group** creates a new named group using a default name of "Group #" where # is a number. Creating new named groups is discussed later in this chapter.

- **Show Named Group** opens the Element Info dialog box for a selected named group. For an example of the box, see Figure 14–8.

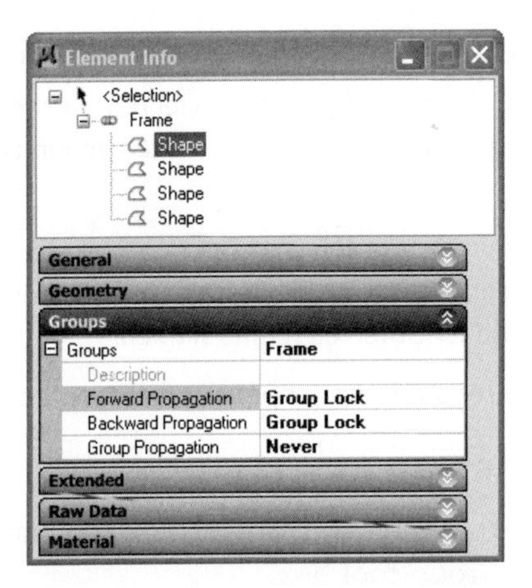

Figure 14–8 The Element Info box for a selected named group

- **Add Elements** invokes the Add to Graphic Group tool for adding elements to the selected named group. Adding elements is discussed later in this chapter.
- **Remove Elements** invokes the Drop From Graphic Group tool for removing elements from the named group selected on the simple list. Removing elements is discussed later in this chapter.
- **Add Named Group to Parent** adds a named group as a child of another named group:
  1. If the hierarchical tree is not displayed, click **Show Hierarchy** to display it.
  2. On the simple list, select the named group that is to be the child.
  3. On the hierarchical tree, select the parent named group.
  4. Click **Add Named Group to Parent** to add the selected named group as a child.
- **Remove Group from Parent** removes a child named group from the parent named group:
  1. If the hierarchical tree is not displayed, click **Show Hierarchy** to display it.
  2. On the hierarchical tree, select the child named group that is to be removed.
  3. Click **Remove Group from Parent** to remove the selected named group from the parent.

- **Delete Named Group** deletes the selected named group. The named group is deleted but the elements remain in the design and are no longer grouped.
- **Select Elements in Named Group** selects all the elements in the selected named group. If the named group is selected on the hierarchical tree, clicking this tool selects all elements in the selected group and all elements in its child named groups.
- **Put Elements into the Displayset** puts the elements in a selected named group into a displayset. A displayset is a group of elements that display in selected views with all other elements hidden.
- **Show Hierarchy** alternately opens and closes the hierarchical tree pane.

CREATE A NAMED GROUP

Tools to create new named groups are provided on the Named Groups list box and on the Add to Graphic Group tool box.

Use the Named Groups List Box

Invoke the New Named Group option from:

Named Groups list box	Click the **New Named Group** tool

MicroStation creates a new named group, displays it in the simple named groups list on the Named Groups list box, and opens the group name for editing, as shown in Figure 14–9.

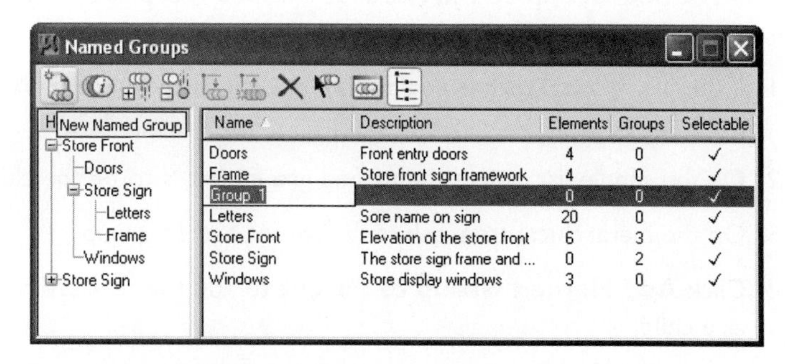

Figure 14–9 A new named group with its name open for editing

Type a descriptive name for the new group and press ENTER.

Optionally, click the Description field for the new named group, type a description, and press ENTER.

Optionally, turn the Selectable check box OFF to allow selecting individual elements in the graphic group.

***Note:** The new named group contains no elements. Adding elements to a named group is discussed later in this chapter.*

Use the Add to Graphic Group Tool

Invoke Add to Graphic Group from:

Groups tool box	Select the Add To Group tool.
Keyboard Navigation	**6 5**

On the Add to Graphic Group settings box, click the **Create New Named Group** icon, as shown in Figure 14–10.

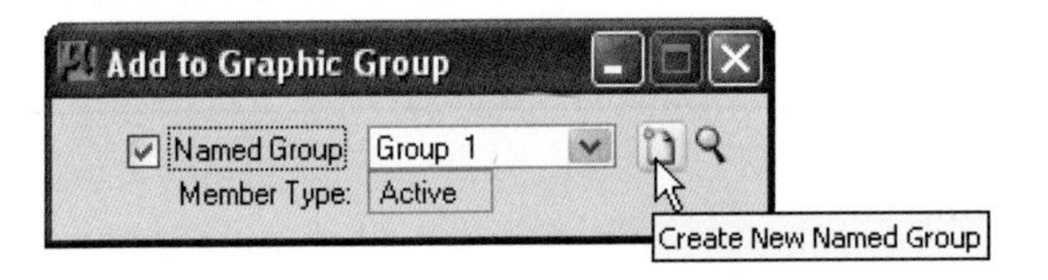

Figure 14–10 The Create New Named Group icon on the Add to Graphic Group settings box

MicroStation opens the Create Named Group dialog box, as shown in Figure 14–11.

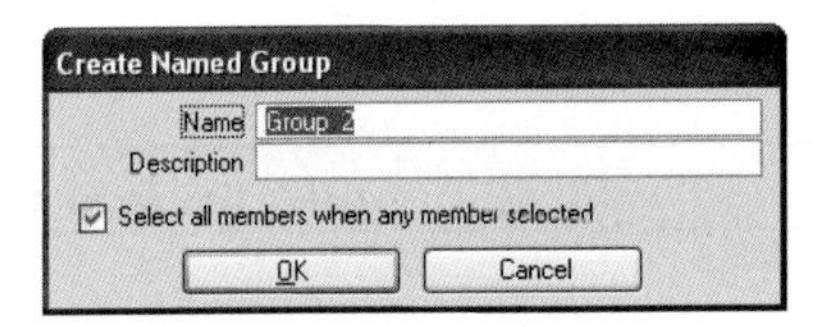

Figure 14–11 The Create Named Group dialog box opened from the Add to Graphic Group settings box

Type a descriptive name for the named group in the **Name** field.

Optionally, type a description for the named group in the **Description** field.

Optionally, turn the **Select all members when any member selected** check box OFF to allow selecting individual elements in the graphic group.

***Note:** The new named group contains no elements. Adding elements to a named group is discussed later in this chapter.*

ADDING ELEMENTS TO A NAMED GROUP

The Add to Graphic Group tool adds elements to a selected named group. The tool can be invoked from the Named Groups list box or the Main tool box. When the tool is invoked from the Named Groups list box, the elements to be added must have already been selected using element selection tools. When the tool is invoked from the Main tool box, it can add previously-selected elements or elements selected when prompted by the tool.

The following procedure starts by selecting elements using the element selection tools.

Invoke Element Selection from:

Main tool box	Select the Element Selection tool
Keyboard Navigation	1

MicroStation prompts:

> Element Selection > Identify element to add to set (The prompt varies depending on the last element selection tool used.)

Use the element selection tools as required to select elements to add to the named group.

After selecting all required elements, invoke the Add to Graphic Group tool using either of the following two methods:

Invoke Add Elements from:

Named Groups list box	Click the **Add Elements** icon

MicroStation invokes the Add to Graphic Group tool.

Invoke the Add to Graphic Group tool from:

Groups tool box	Select the Add To Graphic Group tool.
Keyboard Navigation	**6 5**

MicroStation prompts:

> Add to Named Group > Accept to add selected elements to group

Do the following on the Add to Graphic Group settings box (see Figure 14–12a):

1. Turn the Named Group check box ON.
2. Select a named group from the Named Group menu.

3. Select a type (**Active**, **Passive**, or **Custom**) from the **Member Type** menu.

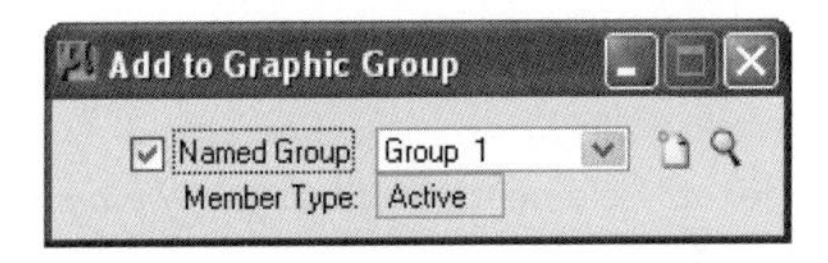

Figure 14–12a The Add to Graphic Group settings box

Note: *Selecting the* **Custom Member Type** *expands the settings box to display three* **Change Propagation** *options (see Figure 14 –12b).*

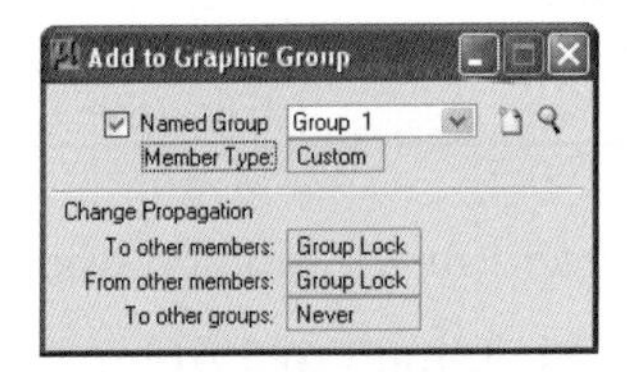

Figure 14–12b The expanded Add to Graphic Group settings box

If **Custom** is the selected on the **Member Type** menu, select the required propagation settings (**Group Lock, Never,** or **Always**) from the three **Change Propagation** options in the expanded settings box.

Click the Data button to add the selected elements to the named group.

Note: *A column in the list of named groups on the Named Groups list box provides the number of elements in each group.*

REMOVE ELEMENTS FROM A NAMED GROUP

The Drop From Graphic Group tool drops elements from a selected named group. The tool can be invoked from the Named Groups list box or the Main tool box. When the tool is invoked from the Named Groups list box, the elements to be dropped must have already been selected using element selection tools. When the tool is invoked from the Main tool box, it can drop previously-selected elements or elements selected when prompted by the tool.

The following procedure starts by selecting elements using the element selection tools.

Invoke Element Selection from:

Main tool box	Select the Element Selection tool
Keyboard Navigation	**1**

MicroStation prompts:

Element Selection > Identify element to add to set *(the prompt varies depending on the last element selection tool used.)*

Use the element selection tools as required to select elements to add to the named group.

After selecting all required elements, invoke the **Drop From Graphic Group** tool using either of the following two methods:

Invoke Remove Elements from:

Named Groups list box	Click the **Remove Elements** icon

MicroStation invokes the Drop From Graphic Group tool.

Invoke the Drop From Graphic Group tool from:

Groups tool box	Select the Drop From Graphic Group tool.
Keyboard Navigation	**6 6**

MicroStation prompts:

Drop from Named Group > Accept/Reject (select next input)

On the Drop From Graphic Group settings box, turn the **Drop From Named Groups** check box ON.

Click the Data button to drop the selected elements from the named group.

Note: *It was not necessary to specify the named group that contained the elements because the group identification is stored with the elements.*

SELECTION BY ATTRIBUTES

The **Select By Attributes** option on the **Edit** menu limits element selection for manipulation to elements with certain element attributes. For example, if all red ellipses on the Border level need to be changed to the color green, use the Select By Attributes dialog box to select only those elements before applying the color change. The selected elements are highlighted and can be manipulated by the element manipulation tools.

SELECT BY ATTRIBUTES DIALOG BOX

Open the Select By Attributes dialog box from:

Menu	Edit > Select By Attributes

MicroStation opens the Select By Attributes dialog box (see Figure 14–13).

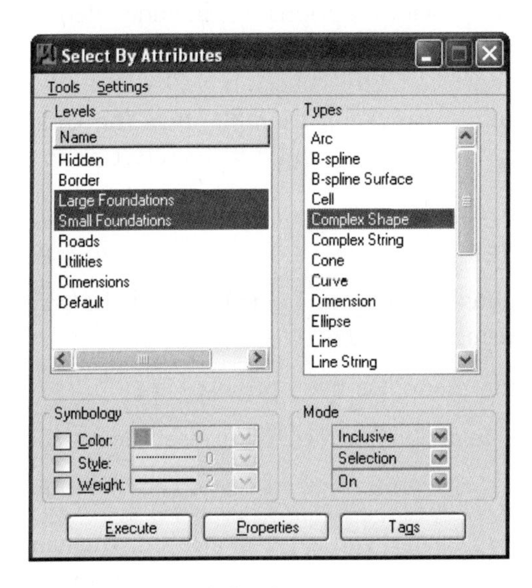

Figure 14–13 Select By Attributes dialog box

SELECTION BY LEVELS

The **Levels** box lists all level names defined in the design file (see Figure 14–13). If a name is shown with a dark background, it is part of the selection criteria. A light background means the level is not part of the selection criteria.

To select a specific level name and turn off all others, click the Data button on the desired name. To select additional names, hold down CTRL while clicking the Data button on level names. To select a contiguous group of names, drag the screen pointer across them while holding down the Data button. For example, to select only the Border and Roads levels, click the Data button on the Border and CTRL click the Data button on Roads.

SELECTION BY TYPES

The **Types** box contains a list of all element types (see Figure 14–13). Type names shown with a dark background are part of the selection criteria. A light background means that type is not part of the selection criteria.

To select a specific element type and turn off all others, click the Data button on the desired type. To select additional types, hold down CTRL while clicking the

Data button on type names. To select a contiguous group of types, drag the screen pointer across them while holding down the Data button. For example, to select only ellipses and line strings, click the Data button on the Ellipse type, and then CTRL click the Data button on the Line String type.

SELECTION BY SYMBOLOGY

The Symbology area provides options to include element **Color**, **Style**, and **Weight** in the selection criteria (see Figure 14–13). To add a symbology item to the selection criteria, turn the item's check box ON and select a value in the menu to the right of the item's name. For example, turn the **Color** check box ON and select the color **Red** (3) in the menu to limit the selection to only elements that are red.

SELECTION MODES

The three **Mode** menus (see Figure 14–13) control the way the selection criteria are applied when **Execute** is clicked.

The first **Mode** option menu has two options:

- **Inclusive**—Select only elements that meet the selection criteria.
- **Exclusive**—Select only elements that do not meet the selection criteria.

For example, if the selection **Type** is set to Ellipse, **Inclusive** mode causes all ellipses to be selected and **Exclusive** mode causes all elements except ellipses to be selected.

The second **Mode** option menu has three options:

- **Selection**—Clicking **Execute** immediately selects and highlights all elements that meet the selection criteria. Element manipulations act on all selected elements.
- **Location**—Clicking **Execute** turns on the selection criteria but does not select any elements. Elements that do not meet the criteria cannot be selected.
- **Display**—Clicking **Execute** turns on the selection criteria and makes all elements that do not meet the criteria disappear from the view. The remaining elements are not selected so the Element Selection tool must be used to select elements for manipulation.

The third **Mode** option menu has two options:

- **Off**—Clicking **Execute** turns off the previously set selection criteria so it has no effect on element selection.
- **On**—Clicking Execute button turns on the current selection criteria so it can be used.

Note: *Closing the Select By Attributes dialog box with a selection criteria in effect displays an Alert box that asks what to do with the current selection criteria. Click **OK** to keep the selection criterion in effect or **Cancel** to turn off the criterion.*

SELECT BY PROPERTIES DIALOG BOX

Use the Select By Properties dialog box to include additional element attributes in the selection criteria. Open the box by clicking **Properties** on the Select By Attributes dialog box.

On the left side of the box, select element **Properties** settings. Each property has an options menu for selecting the property to be included in the selection criteria. To select a property, turn the property's check box ON and select a setting from the menu. For example, the area property options menu has options to select either **Solid** or **Hole** elements as shown in Figure 14–14.

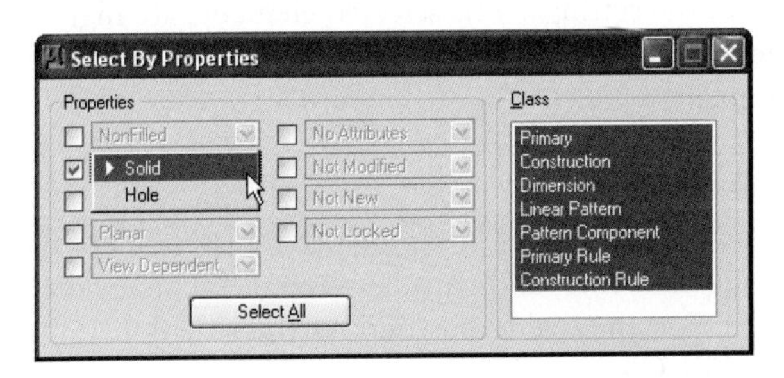

Figure 14–14 Selecting an area property in the Select By Properties dialog box

The Command button below the **Properties** area turns all the **Properties** check boxes ON and OFF. The name of the button says what happens next. If it says **Select All**, it turns ON all the check boxes and if it says **Clear All**, it turns OFF all the check boxes.

On the right side of the box is a field for selecting what class or classes of elements to include in the selection criteria (see Figure 14–14). If the class has a dark background, it is selected. Clicking the Data button on the class name adds it to, or removes it from the selection criteria. For example, to select only elements in the **Construction** class, set the word **Construction** in the menu to have a dark background and set all the others to have a light background.

SELECT BY TAGS DIALOG BOX

The Select By Tags dialog box (see Figure 14–15) specifies criteria based on tag values. If selection criteria based on tag values are specified, elements that do not have attached tags with the specified tag names are not selected, located, or displayed. To open the Select By Tags dialog box, click the Tags button in the Select By Attributes dialog box.

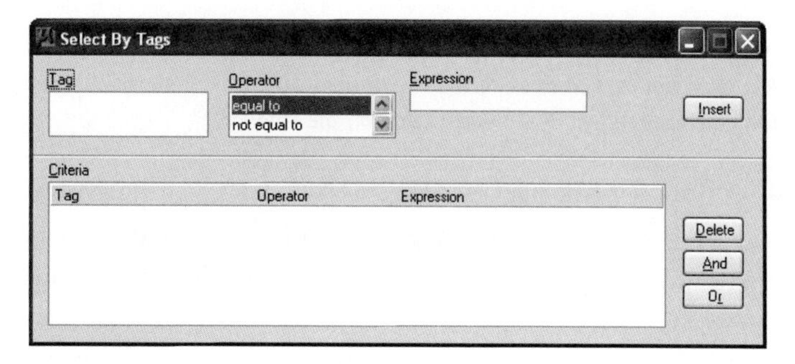

Figure 14–15 Select by Tags dialog box

EXECUTING SELECTION CRITERIA

Click the Execute button in the Select By Attributes dialog box to select the elements according to the settings. The selected elements are highlighted. After completing the selection criteria, invoke the appropriate element manipulation tool and follow the prompts.

DATA CLEANUP

The Design File Cleanup dialog box provides tools for identifying duplicate elements, overlapping elements, and gaps between elements in the active design file. The **Overview** area on the dialog box defines the situations to be cleaned up. Invoke the Design File Cleanup dialog box from:

Menu	Utilities > Data Cleanup
Keyboard Shortcut	**ALT+U,C**

MicroStation displays Design File Cleanup settings box (see Figure 14–16).

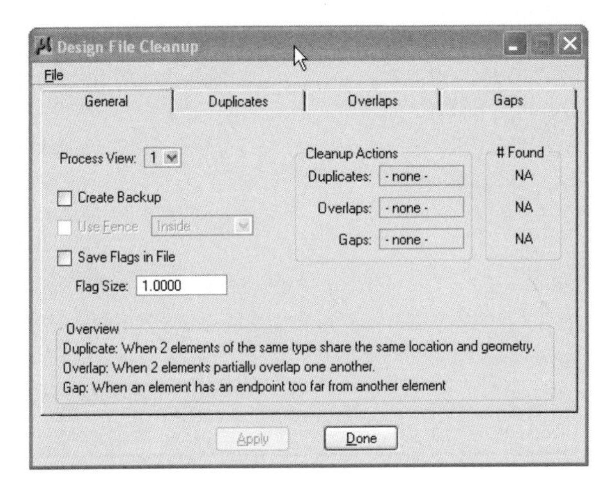

Figure 14–16 The General tab on the Design File Cleanup dialog box

The dialog box contains four tabs for specifying cleanup actions, for modifying the search criteria, and for displaying cleanup results.

The following three options appear on all four tabs:

The **File** menu contains options for retrieving and storing cleanup settings, and for returning to the default "out–of-the-box" cleanup settings.

- **Load Settings** retrieves saved cleanup settings and loads the settings in the active design file. Select the option to open the Load Cleanup Settings dialog box (see Figure 14–17). To retrieve saved settings, use the **Look In** menu to navigate to the folder that contains the saved cleanup settings file (**.rsc*), select the file, and click **Open**. The retrieved settings are loaded in the fields on the Design File Cleanup dialog box's tabs.
- **Save Settings** opens the Save Cleanup Settings dialog box, which contains options for saving the settings currently in the Design File Cleanup dialog box to a file. Use the **Look in** menu to navigate to the folder that is to hold the saved settings file, type a file name in the **File name** field (or select an existing file to be overwritten), and click **Save**. The current cleanup settings are saved to the file. If an existing file was selected, an Alert message box asks for confirmation that the file is to be overwritten. Click **Yes** to overwrite the file with the new settings or click **No** to cancel the operation.

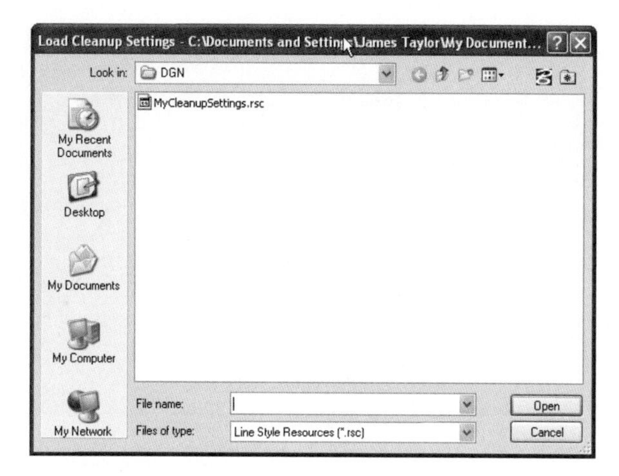

Figure 14–17 The Load Cleanup Settings dialog box

- **Default Settings** returns the Design File Cleanup dialog box's fields to MicroStation's default settings.

Click **Apply** to accept the current settings in the dialog box an initiate the search for duplicates, overlaps,and gaps.

Click **Done** to close the dialog box after cleanup is complete.

GENERAL SETTINGS

The **General** tab contains options for specifying the search scope and the main search items from the other tabs (see Figure 14–16).

- **Process View** menu selects a view window. The cleanup tool only searches for duplicates, overlaps, and gaps on the levels turned on (displayed) in the selected view.
- **Create Backup** creates a backup copy of the active design file and stored it on the folder specified in the MS_BACKUP configuration variable. For more information on configuration variables, see Chapter 16.
- Turn the **Use Fence** check box ON to limit the cleanup search to the contents of the active fence and select a fence mode from the **User Fence** menu.
- Turn the **Save Flags in File** check box on to add flags to the active design file as elements, or turn it OFF to clear the displayed flags. For more information, see the "Annotation Tools" topic in this chapter.
- The **Flag** Size field sets the size of the flag displayed on overlapping elements or gaps between elements. The tool places flags when the **Cleanup Actions** are set to **Flag** or **Interactive**.

- The **Cleanup Actions** menus allow selecting a specific action to take for **Duplicates**, **Overlaps**, and **Gaps**.
    - The **Duplicates** menu provides choices to do nothing (**None**), **Select** the duplicates, or **Remove** the duplicates.
    - The **Overlaps** and **Gaps** menus provide choices to do nothing (**None**), **Select** the overlapping elements and gaps, **Flag** overlapping elements and gaps, or show overlapping elements and gaps (**Interactive**)
- The **# Found** fields are read-only and show the number of elements that are duplicates, overlap, or have gaps.

DUPLICATES SETTINGS

The **Duplicates** tab provides options for setting the duplicates selection criteria (see Figure 14–18).

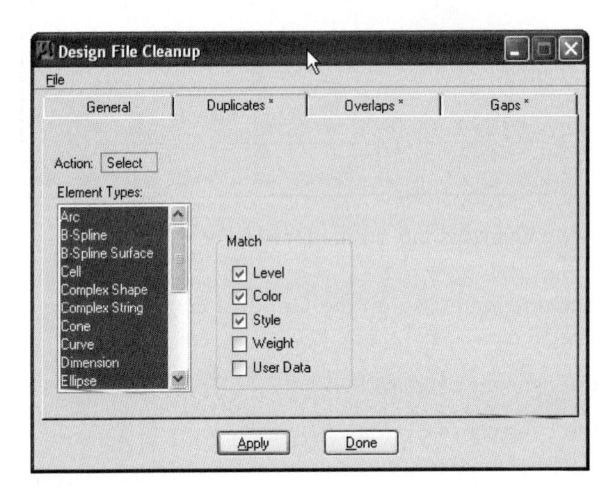

Figure 14–18 The General tab on the Design File Cleanup dialog box

- The **Action** menu (also on the **General** tab), selects the action to perform on all found duplicates:
    - **None** ignores duplicates.
    - **Delete** deletes the duplicates.
    - **Select** selects all duplicates found.
- The **Element Types** menu contains the names of all element types for which duplicates can be identified. Select the element types to include in the search.
- The **Match** check boxes limit the search for duplicates to elements with the attributes whose check boxes are ON. If no check boxes are ON, the clean-

up tool ignores attributes and finds all duplicate elements that meet the other criteria.

OVERLAPS SETTINGS

The **Overlaps** tab provides options for setting the overlap selection criteria (see Figure 14–19).

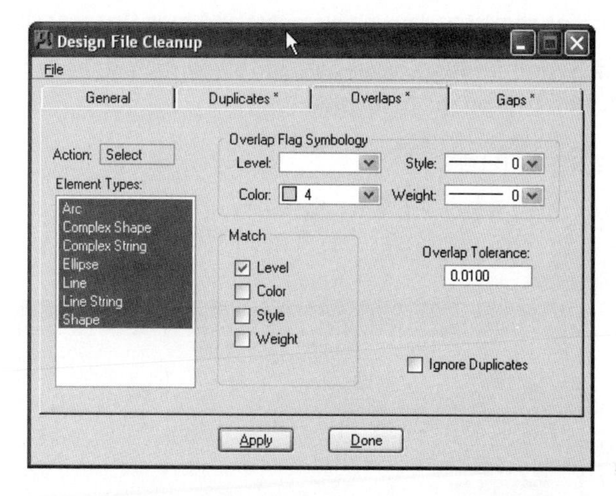

Figure 14–19 The Overlaps tab on the Design File Cleanup dialog box

- The **Action** menu (also on the **General** tab), selects the action to perform on all found overlaps:
  - **None** ignores overlaps.
  - **Select** selects all overlaps found.
  - **Flag** flags overlapping elements in the design file. The flag is a line, with a circle at each end, crossing over each overlapping element.
  - **Interactive** opens the Interactive Cleanup dialog box (see Figure 14–20), which steps through and displays each overlap one at a time.

Figure 14–20 The Interactive Cleanup dialog box

- The **Element Types** menu contains the names of all element types for which overlaps can be identified. Select the element types to include in the search.
- The **Overlap Flag Symbology** sets the attributes of flags placed on overlaps when the **Action** is **Flag**:

- The **Level** menu selects the level for placing flags.
- The **Style** menu sets the style of the flag line.
- The **Color** menu sets the color of the flag line and circles.
- The **Weight** menu sets the weight of the flag line and circles.

- The **Match** check boxes limit the search for overlaps to the elements with the attributes whose check boxes are ON. If no check boxes are ON, the cleanup tool ignores attributes and finds all overlapping elements that meet the other criteria.
- Use the **Tolerance** field to enter the allowable distance between elements before they are considered to be overlapping.
- Duplicate elements also overlap. Turn the **Ignore Duplicates** check box ON to ignore duplicates when searching for overlapping elements.

GAPS SETTINGS

The **Gaps** tab provides options for setting the gaps selection criteria (see figure 14–21).

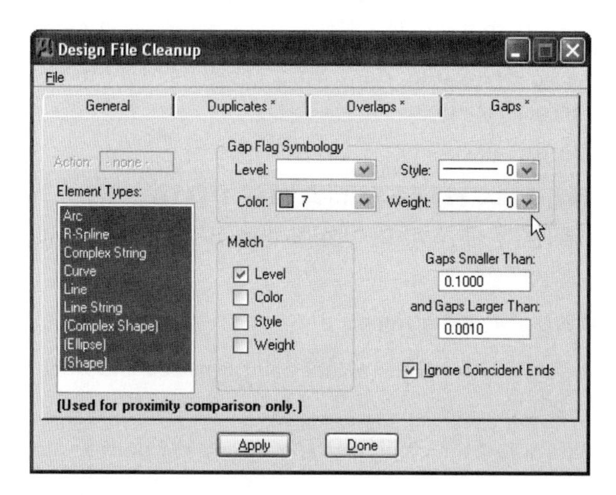

Figure 14–21 The Gaps tab on the Design File Cleanup dialog box

- The **Action** menu (also on the **General** tab), selects the action to perform on all found gaps:
  - **None** ignores gaps.
  - **Select** selects all gaps found.
  - **Flag** flags gaps in the design file. The flag is a circle inside a hexagon that is placed around each element endpoint that is part of a gap.

- **Interactive** opens the Interactive Cleanup dialog box (see Figure 14–20), which steps through and displays each gap one at a time.

- The **Element Types** menu contains the names of all element types for which gaps can be identified. Select the element types to include in the search.
- The **Gap Flag Symbology** sets the attributes of flags placed on gap endpoints when the **Action** is **Flag**:
  - The **Level** menu selects the level for placing flags.
  - The **Style** menu sets the style of the flag line.
  - The **Color** menu sets the color o the flag line and circles.
  - The **Weight** menu sets the weight of the flag line and circles.
- The **Match** check boxes limit the search for gaps to the elements with the attributes whose check boxes are ON. If no check boxes are ON, the cleanup tool ignores attributes and finds all gap endpoints that meet the other criteria.
- The **Gaps Smaller Than** and **Gaps Larger Than** fields set the range of gap endpoints to flag.
- Turn the **Ignore Coincident Ends** check box ON to ignore endpoints shared across elements.

CHANGE THE HIGHLIGHT AND POINTER COLOR

The highlight and drawing pointer colors can be changed from the Design File settings window. These colors are used for:

- Highlighting selected elements
- The drawing pointer when a data point is placed and when an element is manipulated
- The locate tolerance circle that appears on the pointer during manipulations

The highlight colors may need to be changed if many elements in the design are the same colors as the selection and pointer. To change the colors, invoke the Design File dialog box from:

Menu	Settings > Design File (see Figure 14–22).
Key-in window	**mdl load dgnset** (or **md l dgnset**) (ENTER)

MicroStation opens the Design File dialog box. Click the **Color Category** to display the highlight, pointer, and selection colors.

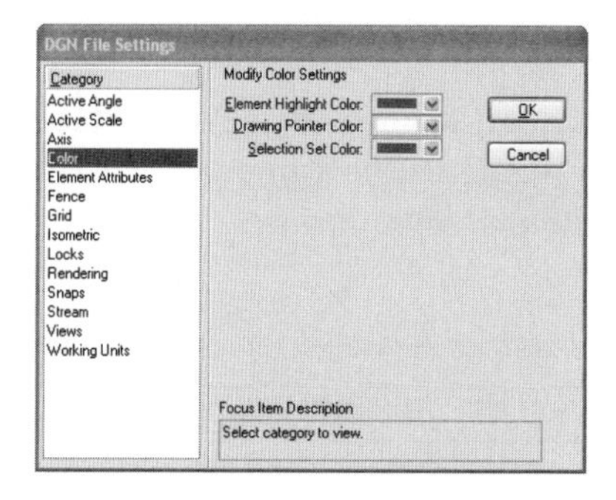

Figure 14–22 Design File dialog box with the Color category selected

On the Design File Settings window:

- Open the **Element Highlight Color** menu and select the desired color for highlighting identified elements.
- Open the **Drawing Pointer Color** menu and select the desired color for the drawing pointer.
- Open the **Selection Set Color** menu and select the desired color for elements selected by the element selection tools.
- Click **OK** to make the new color settings active.
- To make the changes permanent in this design file, select **File > Save Settings**.

Note: *The color changes apply only to the design file in which they were changed. Each design file has its own highlight and pointer color settings.*

Do not set the colors the same as the view window's background color. If the color is the same, highlighted elements and the pointer cannot be seen.

IMPORT AND EXPORT DRAWINGS IN OTHER FORMATS

MicroStation can import and export graphic files in several file formats.

- Several formats can be imported from the MicroStation Manager and File Open windows.
- Design files can be exported to several formats from the File Save As window.
- **Export** and **Import** options are also provided in the **File** menu.

This support for other formats allows MicroStation users to share design files with clients and vendors using other CAD and graphic applications.

SUPPORTED FORMATS

Table 14–1 lists file exchange formats that MicroStation can open directly and save, and that are also available in the **Import** and **Export** submenus on the **File** menu. Table 14–2 lists file exchange formats that are available only from the **Import** and **Export** submenus.

Table 14–1 File Exchange Formats Available for Opening and Saving Files

FORMAT	OPEN	SAVE AS	DESCRIPTION
DGN	Yes	Yes	Native format MicroStation version 7 and version 8 design files
CEL	Yes	No	Cell files
DGNLIB	Yes	Yes	DGN Library files
S	Yes	No	Sheet files
H	Yes	No	Hidden Line files
DWG	Yes	Yes	Native format AutoCAD drawings
DXF	Yes	Yes	Drawing Interchange Format—developed by Autodesk, Inc., to exchange graphic data among many CAD and graphics applications. DWG and DXF imports are handled identically.
CGM	Yes	No	Computer Graphics Metafile Format—an ANSI standard for the exchange of picture data between different graphics applications; device and environment independent
RDL	Yes	Yes	Redline files are created by the Bentley Redline application
D	Yes	No	MicroStation TriForma Document Files. TriForma is another Bentley product.
3DS	Yes	No	3D Studio modeling Files

Table 14–2 File Exchange Formats Available in the Import and Export Submenus

FORMAT	IMPORT	EXPORT	DESCRIPTION
DGN, DWG, DXF	No	Yes	MicroStation and AutoCAD design file formats
IGES	Yes	Yes	A public domain, neutral file format that serves as an international standard for the exchange of data between different CAD/CAM systems
XMT	Yes	Yes	Parasolid files are mathematical definitions of engineering parts and assemblies
ACIS SAT	Yes	Yes	ACIS digital imaging files
CGM	Yes	Yes	Computer Graphics Metafile Format—an ANSI standard for the exchange of picture data between different graphics applications; device and environment independent

FORMAT	IMPORT	EXPORT	DESCRIPTION
STEP AP203/ AP214	Yes	Yes	Application Protocols (APs) used to exchange data. AP203 applies to representations of mechanical parts and assemblies. AP214 applies to representations of data relating to automotive design
VRML	No	Yes	VRML (Virtual Reality Modeling Language) is an open, extensible, industry-standard scene description language for 3D scenes, or worlds, on the Internet
STL	Yes	Yes	Stereolithography machines produce 3D drawings of objects files
KML	No	Yes	Google Earth provides a 3D interface to view imagery from anywhere on Earth. MicroStation provides data to Google Earth as KML documents
Image	Yes	No	Several graphics formats used by text processing and publication graphics packages
Text	Yes	No	ASCII text files (discussed in Chapter 6)
3D	No	Yes	MicroStation's 3-dimensional design file format. If the open design file is 3D, there will be an option to save it as a 2D (2-dimensional) drawing

OPENING A FILE OF ANOTHER FILE FORMAT

The file formats shown in Table 14–1 open directly in MicroStation. MicroStation converts the files before opening them.

To open a drawing created in another format using the MicroStation Manager or the Open dialog box:

1. Select **Files > Open**.
2. Open the **Files of type** menu on the Open dialog box (see Figure 14–23).
3. Select the format of the drawing to be opened.
4. Follow the usual procedure for opening a file.

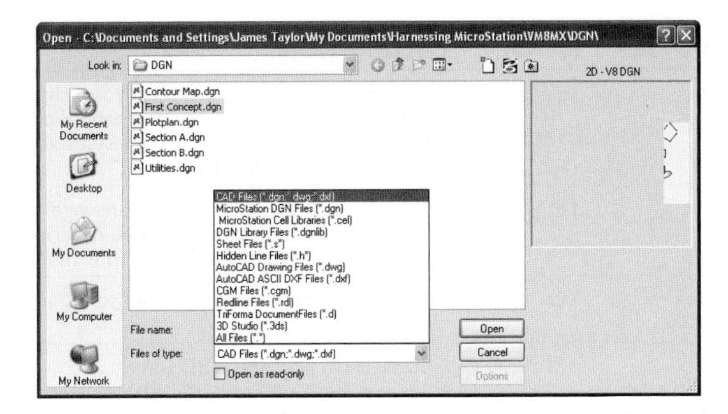

Figure 14–23 Selecting the format of a file to be opened

SAVING A DESIGN FILE USING ANOTHER FILE FORMAT

The file formats shown in Table 14–1 save directly from the File Save As dialog box. MicroStation converts the file and then saves it.

1. Select File > Save As.
2. Open the Files of type menu (see Figure 14–24).
3. Select the format to be used.
4. Supply a directory path and file name.
5. Click OK to initiate the conversion and close the dialog box.

MicroStation converts the active design file to the other format and saves it. The design file remains open.

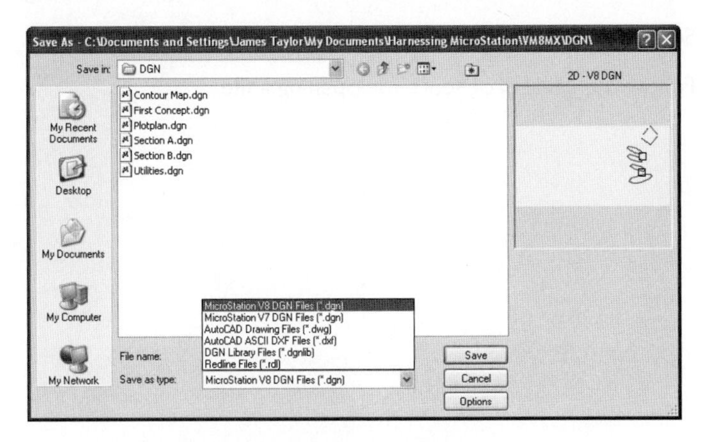

Figure 14–24 Selecting the format of a file to be saved

IMPORT AND EXPORT OTHER FORMATS INTO AN OPEN DESIGN FILE

The file formats described in Table 14–2 are available for import and export from an open design file. The **Import** and **Export** options are available from the **File** menu.

- The **Import** option inserts the imported file contents into the open design file and converts the elements in the imported file to design file elements. Image files remain as images in the design file and do not become elements.
- The action of the **Export** tool varies with the chosen format. In some cases it opens the Save As dialog box and follows that window with an Export dialog box containing options to control the way the design file opens. In other cases, the Export dialog box opens directly.

Note: *For detailed information on importing and exporting, refer to the technical documentation furnished with MicroStation.*

AUTOCAD INTEROPERABILITY

The MX version of MicroStation expands support for opening and saving documents in AutoCAD's DWG format. An important change to DWG support is AutoCAD's addition of a unit setting to its "Design Center." Prior to this addition, the true size of objects in the DWG file could only be determined if the file had its units set to Architectural or Engineering, which implies feet and inches. The new value, if correctly set, makes it easy to determine the drawing's correct measurement units. Now, when MicroStation opens or creates a DWG file it sets the Design Center units correctly.

MicroStation provides DWG settings boxes that are available on the Open and Save As dialog boxes by clicking the Options button. Figure 14–25 shows the Save As DWG/DXF Options dialog box.

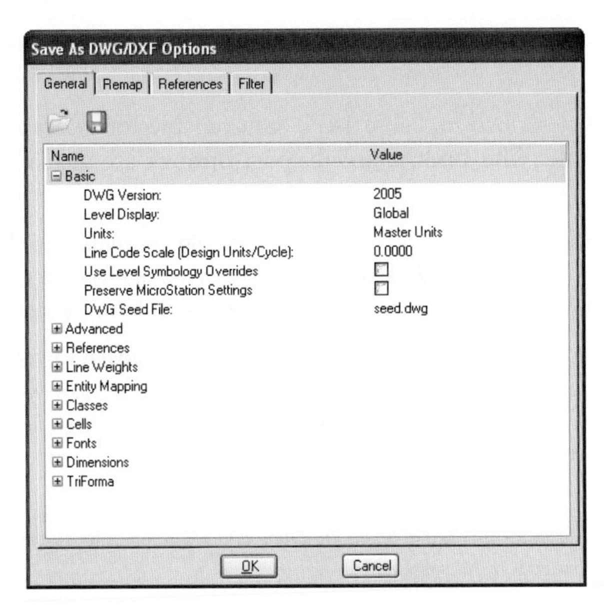

Figure 14–25 The Save As DWG/DXF Options dialog box

MANIPULATING IMAGES

In MicroStation, the term "image" refers to pixel-based graphics files such as those that can be inserted in word processing and graphics processing applications. The screen images of MicroStation used for the figures in this textbook are examples of such images. MicroStation provides a set of tools for creating, manipulating, and viewing images under the **Utilities** menu.

For example, a programmer creates an automated drawing procedure in MicroStation that involves several custom dialog boxes. A technical writer creates a training guide for the procedure in a Windows-based word processing package and needs pictures of the dialog boxes. The programmer uses the **Capture** option to capture the dialog boxes as bit-mapped image files that can be inserted in the training guide.

The **Utilities > Image > Save** option opens the Save Image dialog box (see Figure 14–26) from which the contents of one of the eight view windows can be saved as an image file using one of several available image formats.

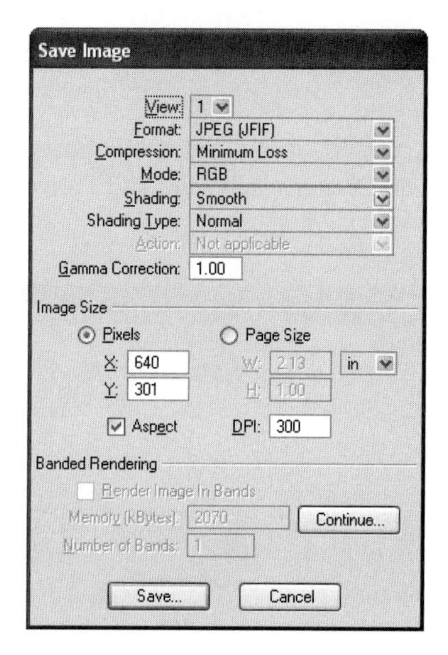

Figure 14–26 Save Image dialog box

The **Utilities > Image > Capture** option opens the Screen Capture dialog box (see Figure 14–27), which provides methods of capturing all or part of the image on the workstation screen. Each capture method opens a Capture Output dialog box for specifying a file name and folder for the captured image.

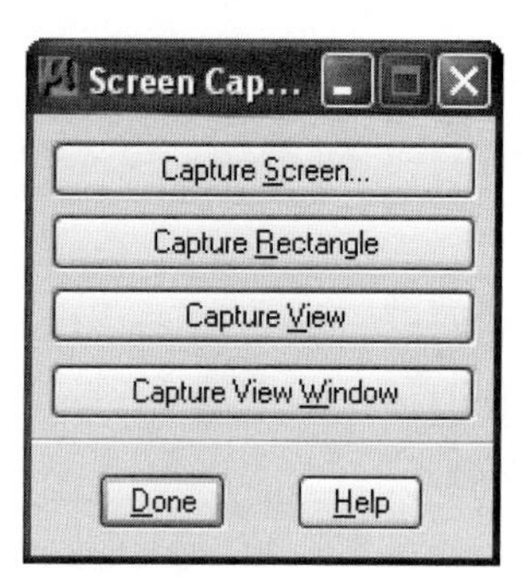

Figure 14–27 Screen Capture dialog box

The Screen Capture methods include:

- **Capture Screen**—Captures the entire screen.
- **Capture Rectangle**—Captures the contents of a rectangle of the desired area placed on the MicroStation workspace.
- **Capture View**—Captures the contents of a selected view. Windows on top of the view are also captured.

- **Capture View Window**—Captures the contents and window border of the selected view. Windows on top of the view are also captured.

The **Utilities** > **Image** > **Convert** option opens the Raster Convert dialog box (see Figure 14–28) for selecting a raster image file and converting it to an output format file.

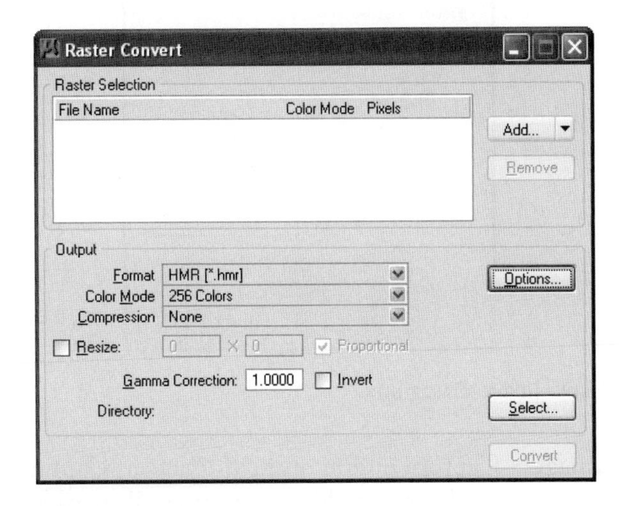

Figure 14–28 Raster Convert dialog box

The **Utilities** > **Image** > **Display** option opens the Display Image dialog box (see Figure 14–29) for selecting an image file to view in a separate dialog box.

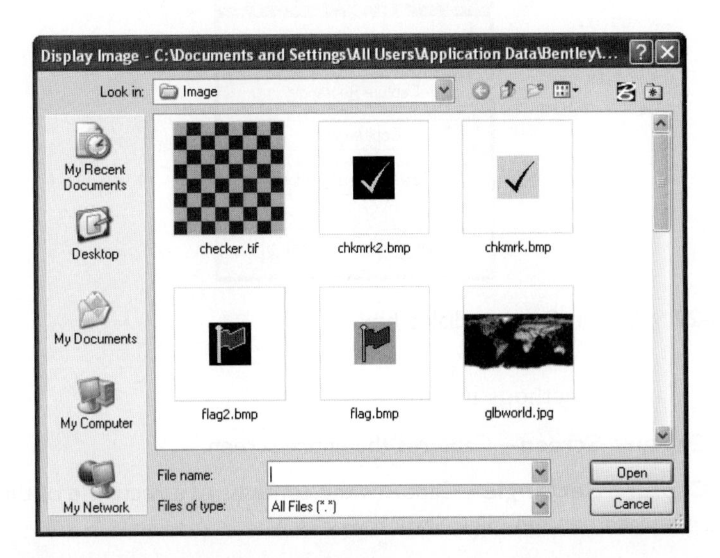

Figure 14–29 Display Image dialog box

The **Utilities** > **Image** > **Movies** option opens the Movies dialog box (see Figure 14–30) from which an animated sequence file can be selected for viewing in a separate window.

Figure 14–30 Movies dialog box

ANNOTATION TOOLS

MicroStation provides annotation tools that place annotation items in the design such as flags, callout markers, and leaders. Invoke the Annotation tools submenu from:

Menu	Tools > Annotation
Key-in window	**dialog toolbox annotation** (or **di to annotati**) (ENTER)

The **Annotation** submenu has options to open five tool boxes. Figure 14–31 shows the tool boxes open in submenu order from left to right and top to bottom.

Figure 14–31 The Annotation tool frame and tool boxes

The first tool bar, **Annotation,** is the main annotation tool bar and provides an alternate way to invoke the tools on the other four tool bars using the same method as the Main toolbar (see Figure 14–32).

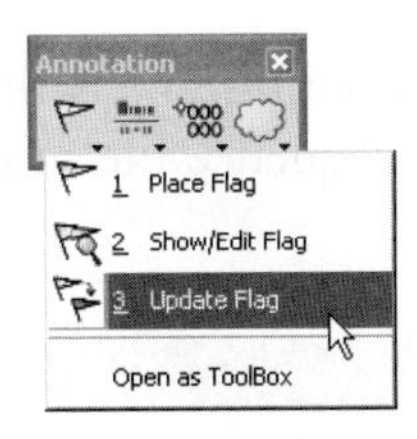

Figure 14–32 Selecting a tool from the main Annotation tool box

Note: *Like the Main tool box, the Annotation tool box always displays the last tool selected on each tool box.*

ANNOTATE TOOL BOX

The Annotate tool box provides tools for placing, editing, and updating design notes that are placed in the design file behind flag symbols. The flag holds the text of the note and provides a visual reminder that there is a note at the location of the flag. Invoke the Annotate tool box from:

Menu	Tools > Annotation > Annotate
Key-in window	**dialog toolbox annotate** (or **di to ann**) (ENTER)

Place Flag

The Place Flag tool inserts new annotation flags in the design file. Invoke the Place Flag tool from:

Annotate tool box	Select the Place Flag tool (see Figure 14–33).
Key-in window	**place flag** (or **pl fl**) (ENTER)

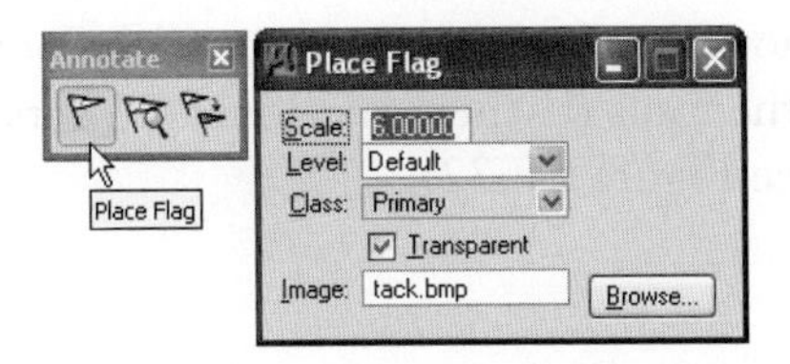

Figure 14–33 Invoking the Place Flag tool from the Annotate tool box

The Place Flag settings window provides options for setting up the new flag:

- Adjust the size of the placed flag by entering a number in the **Scale** field. Entering a number greater than one increases the size and entering a number less than one decreases the size.
- Select the level on which the flag is placed from the **Level** menu.
- Place the flag in the **Primary** or **Construction** class using the **Class** menu. If the **Construction View** attribute is turned OFF, elements in the **Construction** class do not appear in the view.
- Key-in the name of the image to use for the flag or click **Browse** to open the Select Image File for Flag dialog box. The dialog box opens displaying MicroStation's default images folder. To search for an image in a different folder, use the **Look in** menu to navigate to the folder and select an image filename from the filenames list directly below the **Look in** menu. Figure 14–34 shows examples of flag symbol images from MicroStation's default folder and the names of the files containing the images.

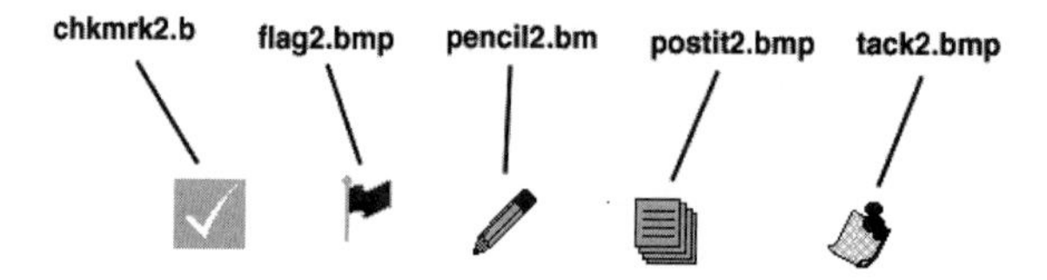

Figure 14–34 Typical flag images and their bitmap file names

MicroStation prompts:

Place Flag > Identify location

Place a data point to define the flag's location in the design.

Defining the flag location opens the Define Flag Information dialog box (see Figure 14–35). Type the annotation text in the field and click **OK** to place the flag and close the dialog box.

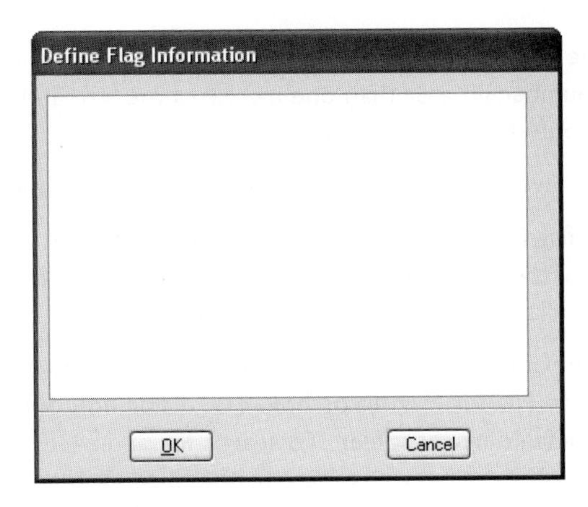

Figure 14–35 Define Flag Information dialog box

Note: *The flag's image appears in the design file but not the text. To view and edit the text, double-click the image to open the Define Flag Information dialog box or click the Show/Edit Flag icon in the Annotate tool box.*

Show/Edit Flag Tool

Use the Show/Edit Flag tool to view and edit the annotation text for a selected. Invoke the Show/Edit Flag tool from:

Annotate tool box	Select the Show/Edit Flag tool (see Figure 14–36).
Key-in window	**show flag** (or **sho f**) (ENTER)

Figure 14–36 Invoking the Show/Edit Flag tool from the Annotate tool box

MicroStation prompts:

Show/Edit Flag > Select flag

Click the Data button on the flag to be viewed or edited.

Show/Edit Flag > Accept/Reject

Click the Data button again to accept the selected flag. Accepting the flag opens the Define Flag Information dialog box, which displays the annotation text and makes it available for editing. After viewing or editing the flag text, click **OK** to close the dialog box and save the changes.

Update Flag Tool

The Update Flag tool changes the image of an existing flag to the image currently selected on the Place Flag tool settings box's **Image** field. Before invoking the Update Flag tool, invoke the Place Flag tool and use the settings box Browse button to find and select the correct image. Invoke the Update Flag command from:

Annotate tool box	Select the Update Flag tool (see Figure 14–37).
Key-in window	**flag update** (or **fl u**) (ENTER)

Figure 14–37 Invoking the Update Flag tool from the Annotate tool box

MicroStation prompts:

Update Flag > Select flag

Select the flag to be updated.

Update Flag > Accept/Reject

Click the Data button again to update the flag symbol with the image.

DETAILING SYMBOLS

The Detailing Symbols tool box provides tools that place callout markers and leaders, and a tool to customize the way the elements are placed by the other tools. Following are brief descriptions of each tool in the toolbox. Four tools place symbols and two tool work with symbol settings. Invoke the Detailing Symbols tool box from:

Menu	Tools > Annotation > Detailing Symbols
Key-in window	**dialog toolbox detailingsymbol** (or **di to de**) (ENTER)

The tool settings boxes for the four tools that place symbols provide similar settings. Following is a discussion of the settings:

- The **Text Style** menu is used to select one of the text styles for the symbol. For more information, see Chapter 7. The option is on all four symbol placement tool settings boxes.

- The magnifying glass icon opens the Text Styles list box from which a text style can be selected and text settings can be changed. The option is on all four symbol placement tool settings boxes.
- The **Link Target** menu selects a project link to associate with the title text. The option is on all four symbol placement tools.
- The Title and Subtitle buttons open the Text Editor box for entering title and subtitle text. These two buttons are on the Place Title Text and Place Section Marker settings boxes.
- The **Annotate Scale Lock** alternately locks and unlocks the annotation scale. The option is on all four symbol placement tools.
- The Ref # and Sheet # buttons open the Text Editor box for entering reference and sheet ID numbers. These two buttons are on the Place Section Marker, Place Arrow Marker, and Place Detail Marker settings boxes.
- The **Show/Hide Extended Settings** arrow expands the settings box to display more options on all four symbol placement tool settings boxes. The following options are on the extended part of the settings box.
- The **Color**, **Style**, and **Weight** menus set the element attributes for the graphic parts of the symbols (the lines and circles), but not the text.
- The **Level** menu sets the placement level of the symbol.
- The **Bubble Size** field is used to enter a size for the bubble (circle) placed as part of the symbol. This option is on the Place Section Marker, Place Arrow Marker, and Place Detail Marker settings boxes.

Place Title Text

The Place Title Text tool places an annotation symbol that consists of title and subtitle text on two lines with a horizontal line between them. Figure 14–38 includes an example of a title text symbol. Invoke Place Title Text from:

Detailing Symbols tool box	Select the Place Title Text tool (see Figure 14–38).
Key-in window	**place title text** (or **pl ti**) (ENTER)

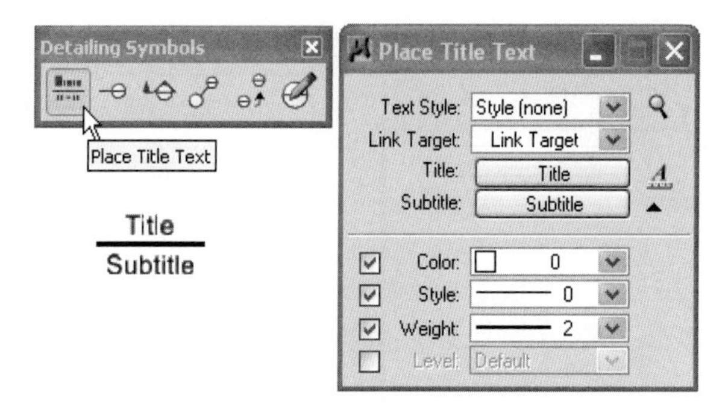

Figure 14–38 The Place Title Text tool and an example symbol

Make the required settings on the Place Title Text tool settings box.

MicroStation prompts:

> Place Title Text > Insert Text

Click the Data button at the symbol placement point on the design. The placement point is the left end of the horizontal line.

Place Section Marker

The Place Section Marker tool places an annotation symbol that consists of a title and subtitle on two lines separated by a horizontal line, and a reference ID number and sheet ID number on two lines separated by the same line with a circle enclosing the ID numbers. Figure 14–39 includes and example of a section marker symbol. Invoke Place Section marker from:

Detailing Symbols tool box	Select the Place Section Marker tool (see Figure 14–39).
Key-in window	**place section marker** (or **pl s**) (ENTER)

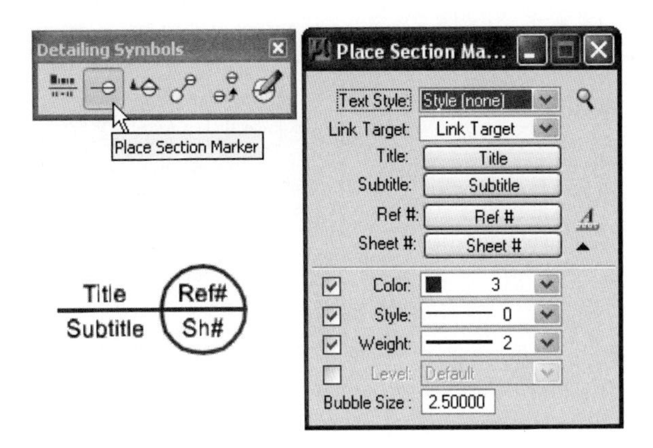

Figure 14–39 The Place Section Marker tool and an example symbol

Make the required settings on the Place Section Marker tool settings box.

MicroStation Prompts:

Place Section Marker > Insert Text

Click the Data button at the symbol placement point on the design. The placement point is the left end of the horizontal line.

Place Arrow Marker

The Place Arrow Marker places an annotation symbol that consists of a reference ID number and sheet ID number on two lines enclosed in a circle and arrow that has an extension line. Figure 14–40 includes an example of an arrow marker symbol.

Detailing Symbols tool box	Select the Place Arrow Marker tool (see Figure 14–40).
Key-in window	**place arrow marker** (or **pl arr**) (ENTER)

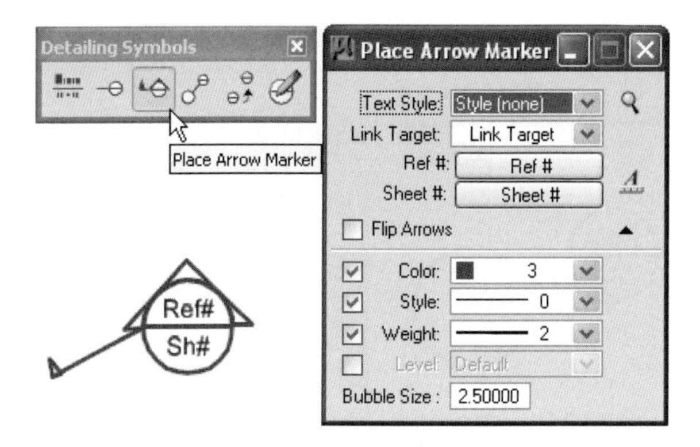

Figure 14–40 The Place Arrow Marker tool and an example symbol

Make the required settings on the Place Arrow Marker tool settings box.

MicroStation prompts:

> Place Arrow Marker > Insert text or define start point for marker

Click the Data button at the placement point of the center of the circle.

> Place Arrow Marker > Insert text or define marker end point

Click the Data button at to set the direction in which the arrow points.

> Place Arrow Marker > Insert text or define marker end point or reset to finish

Click the Data button at the end point of the extension line.

> Place Arrow Marker > Insert text or define marker end point or reset to finish

If the extension line is to be segmented, click the Data button at each segment line end point. When the extension line is complete, click the Reset button to complete symbol placement.

Place Detail Marker

The Place Detail Marker tool places a symbol that consists of a detail circle and a marker circle connected by a line. A reference ID number and sheet ID number display on two lines inside of the marker circle and are separated by a horizontal line. Figure 14–41 includes an example of the detail marker symbol. Invoke Place Detail Marker from:

Detailing Symbols tool box	Select the Place Detail Marker tool (see Figure 14–41).
Key-in window	**place Detail marker** (or **pl d**) (ENTER)

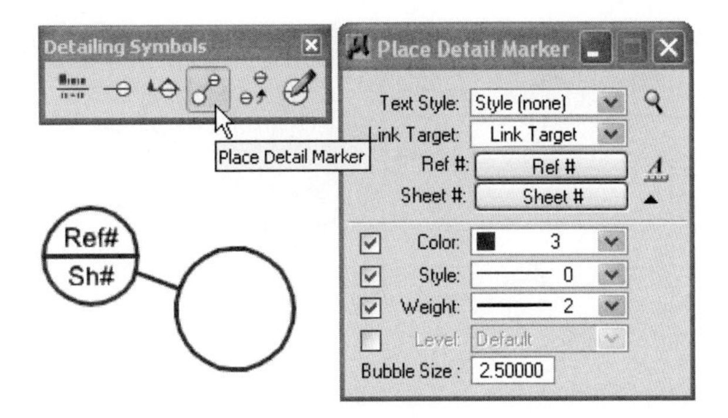

Figure 14–41 The Place Detail Marker tool and an example symbol

Make the required settings on the Place Detail Marker tool settings box.

MicroStation prompts:

> Place Detail Marker > Insert text or define center of detail circle

Click the Data button at the placement location of the center of the circle.

> Place Detail Marker > Insert text or define edge of detail circle

Click the Data button at the placement location of the circumference of the circle.

> Place Detail Marker > Insert text or define marker end point

Click the Data button at the placement location of the center of the circle.

> Place Detail Marker > Insert text or define marker end point or reset to finish

If the line connecting the detail and marker circles has to be segmented, continue clicking the Data button at the placement points of the ends of the line segments. When the line is complete, click the Reset button to complete placing the marker circle in the design.

Match Symbol Settings

The Match Symbol Settings tool applies the settings of a placed symbol in the design to the settings box of the tool that is used to placed the symbol. Invoke Match Symbol Settings from:

Detailing Symbols tool box	Select the Match Symbol Setting tool (see Figure 14–42).
Key-in window	**match detailingsymbol** (or **mat d**) (ENTER)

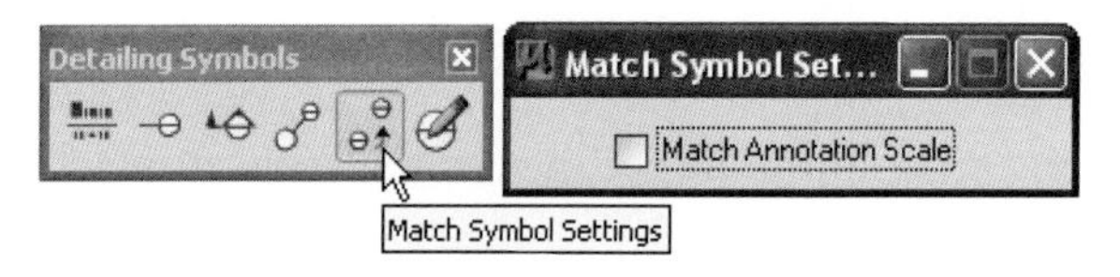

Figure 14–42 The Match Symbol Settings tool

Make the required settings on the Match Symbol Settings tool settings box.

MicroStation prompts:

Match Symbol Settings > Identify element

Select the symbol in the design whose settings are to be matched.

Match Symbol Settings > Accept/Reject (select next input)

Click the Data button again to accept the selected symbol element and apply the selected symbol's settings to the settings box of the tool that is used to that created the symbol. If, for example, the symbol is a section marker, the settings are applied to the Place Section Marker settings box. Thus, subsequent placement of symbols will have the same settings as the selected symbol.

Change Symbol Settings

The Change Symbol Settings tool changes the settings of a symbol in the design. Invoke Change Symbol Settings from:

Detailing Symbols tool box	Select the Change Symbol Settings tool (see Figure 14–43).
Key-in window	**change detailingsymbol** (or **chan d**) (ENTER)

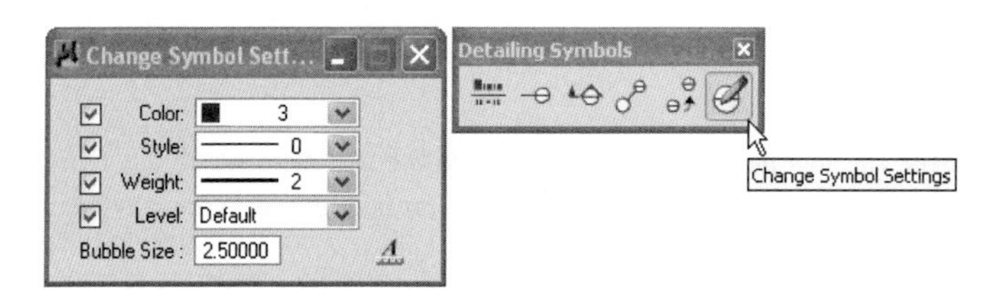

Figure 14–43 The Change Symbol Settings tool

Make the required settings on the Change Symbol Settings tool settings box.

MicroStation prompts:

Change Symbol Settings > Identify element

Select the symbol in the design whose settings are to be changed.

Change Symbol Settings > Accept/Reject (select next input)

Click the Data button to apply the settings in the tool settings box to the selected symbol element.

XYZ TEXT TOOL BOX

The XYZ Text tool box provides tools for placing coordinate labels in the design, exporting coordinate points to an ASCII file, and importing coordinate points from an ASCII file. Following are brief descriptions of each tool in the tool box. Invoke the Detailing Symbols tool box from:

Menu	Tools > Annotation > XYZ Text
Key-in window	**dialog toolbox xyztxt** (or **di to x**) (ENTER)

The settings boxes for the first three XYZ Text tools (Label Coordinates, Label Element, and Export Coordinates) all contain the following fields:

- The **Order** menu displays the X coordinate in **XYZ** OR **YXZ** order.
- The **Units** menu displays the coordinate values using **Master** Units, **Sub** Units, full **Working** Units, or **UORs** (Units of Resolution).
- The **Accuracy** menu sets the accuracy of the fractional part of the coordinate values.
- The **Separator** menu separates each coordinate value by placing each one on a separate line (**Newline**), separating the coordinates with a **Comma** character, or a **Space** character.
- The **View** menu is for 3D files only.
- **X**, **Y**, and **Z Prefix** text fields allow placing a prefix before each coordinate value.

***Note:** The tool places coordinates as multi-line text elements. If the **Text Node View Attribute** is ON, the Text Node number and cross display with the coordinate element.*

Label Coordinates

The Label Coordinates tool places a symbol showing the coordinates of the data point that places the symbol. In a 2D design, the coordinate places the X- and Y-coordinates, one above the other. In a 3D design, the tool also places the Y-coordinate. The tool places the coordinate text using the active text parameters. Figure 14–44 includes an example of a label coordinate placement. Invoke Label Coordinates from:

XYZ Text tool box	Select the Label Coordinates tool (see Figure 14–44).
Key-in window	**label point** (or **l p**) (ENTER)

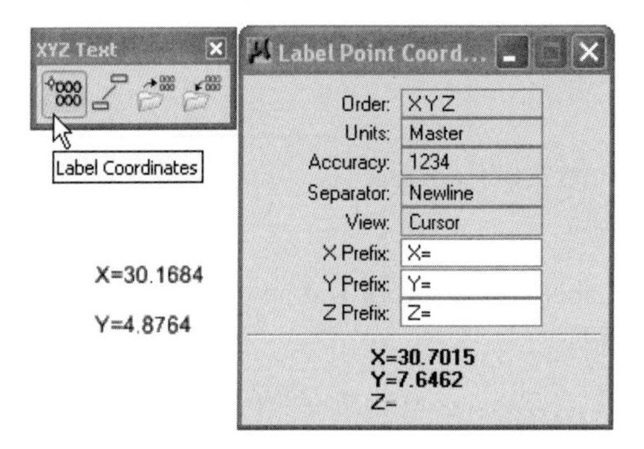

Figure 14–44 The Label Coordinates tool and an example of a label

Make the required settings on the Label Coordinates settings box.

***Note:** The tracking area at the bottom of the settings box displays the current coordinates of the cursor in read-only fields.*

MicroStation prompts:

Label Point Coordinate > Input Point to Label

Place a data point to define thc top, left corner of the coordinate label.

Label Element

The Label Element tool places coordinate labels on all of a selected elements key-points (the vertices and origin). Figure 14–45 includes an example of placing coordinates on a block element. Invoke Label Coordinates from:

XYZ Text tool box	Select the Label Element tool (see Figure 14–45).
Key-in window	**label point** (or **l e**) (ENTER)

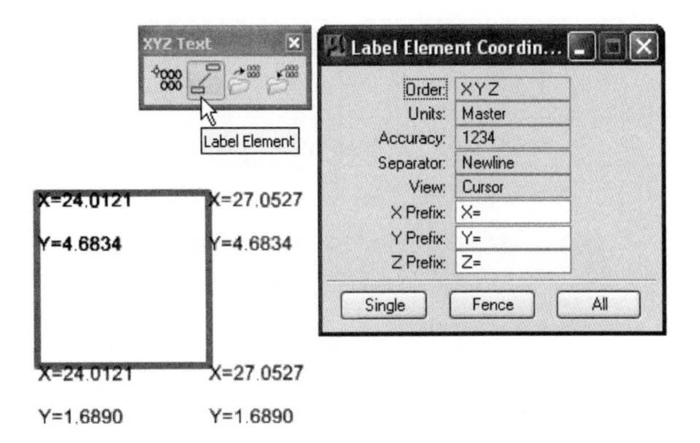

Figure 14–45 The Label Element tool and an example of placing coordinates on each vertex of a block element

Note: *The tool places coordinate labels at the center of circles and ellipses, at the center and ends of arcs, and at each vertex of linear elements.*

Make the required settings on the Label Element settings box and pick an element selection method by clicking one of the three buttons at the bottom on the settings box.

- **Single** places coordinate labels on a single selected element.
- **Fence** places coordinate labels on all elements selected by a fence. The fence must already be placed in the design.
- **All** places coordinate labels on all elements in a design.

Select **Single** and MicroStation Prompts:

Label Element Coordinates > Identify Element to Label

Select an element.

Label Point Coordinate > Input Point to Label

Click the Data button to place the coordinates.

Select **Fence** and MicroStation prompts:

Label Point Coordinate > Accept/Reject Fence Contents

Click the Data button to place coordinate labels on the keypoints of all elements selected by the active fence.

Select **All** and MicroStation displays an alert box that asks for confirmation that coordinate labels are to be placed on all elements in the design file. Click **OK** to place labels on all elements or click **Cancel** to cancel the tool without placing any coordinate labels.

Export Coordinates

The Export Coordinates tool exports the coordinates of elements in the design to a text file (*.*txt*). It can export the coordinates of one element, the coordinates of all elements selected by a fence, or the coordinates of all elements in the active design file. Figure 14–46 shows the Export Coordinates tool.

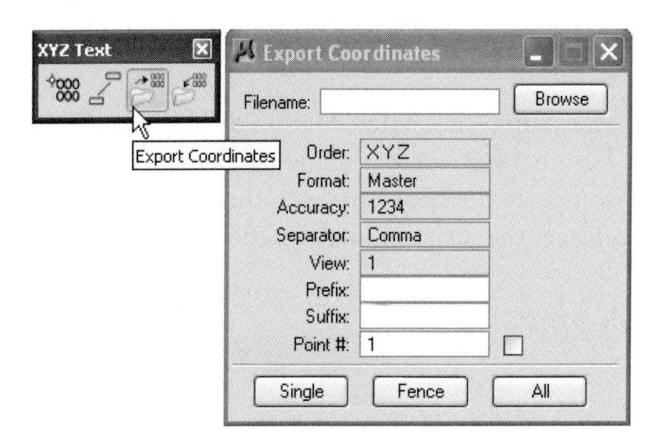

Figure 14–46 The Export Coordinates tool

The Export Coordinates settings box provides three options in addition to the options previously described:

- Type the name of the file to which the coordinates are to be exported in the **Filename** field or click **Browse** to open the Create Export File dialog box to create the file in a selected folder. If the file already exists, an alert message appears.
- Pick an element selection method by clicking one of the three buttons at the bottom on the settings box.
    - **Single** exports the coordinates of a single selected element.
    - **Fence** exports the coordinates of all elements selected by a fence. The fence must already be placed in the design.
    - **All** exports the coordinates of all elements in a design.

Select **Single** and MicroStation Prompts:

Export Coordinates > Identify Element

Select an element.

Export Coordinates > Identify Element

Click the Data button to export the coordinates of the selected element.

Select **Fence** and MicroStation prompts:

Export Coordinate > Accept/Reject Fence Contents

Click the Data button to export the coordinates of the elements selected by the active fence.

Select **All** and MicroStation displays an alert box that asks for confirmation that the coordinates of all elements in the design file are to be exported. Click **OK** to export the coordinates or click **Cancel** to cancel the tool without placing any coordinates.

If the export is to an existing file, MicroStation displays the Export File Exists dialog box shown in Figure 14–47 before allowing elements to be selected. Click **Append existing file** to export the coordinates to the end of the file or click **Overwrite existing file** to replace the existing information in the file with the exported coordinates.

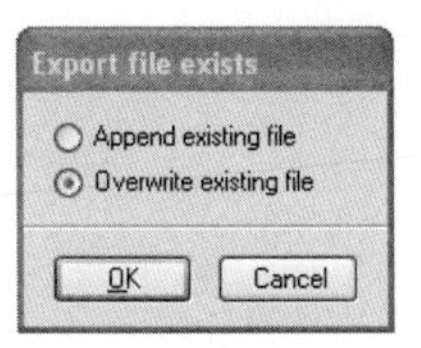

Figure 14–47 The Export File Exists dialog box

Figure 14–48 shows an example of coordinates exported to a text file.

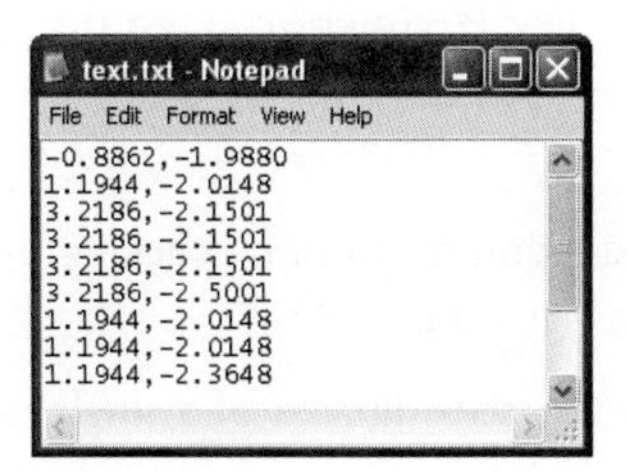

Figure 14–48 Example of exported coordinates

Import Coordinates

The Import Coordinates tool imports coordinate information from a text file and places a selected type of element at each coordinate location. The placed element can be a point, text string, or cell. Figure 14–49 shows the Import Coordinates tool.

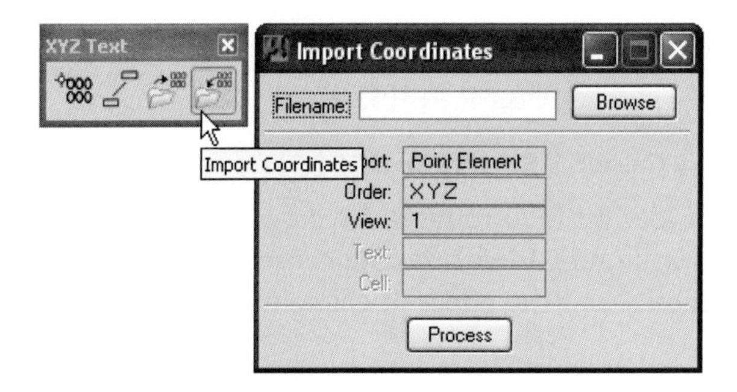

Figure 14–49 The Import Coordinates tool

Make the required settings on the Import Coordinates settings box.

- Type the name of the file to that contains the coordinates to be imported in the **Filename** field or click **Browse** to display the Open Import File dialog box from which the file containing the coordinates can be found and selected. The selected filename and path appears in the **Filename** field.
- Select an import method from the **Import** menu:
  - **Point Element** places a point element at each imported coordinate location.
  - **Text** places a text string at each imported coordinate location.
  - **Cell** places a cell from the attached cell library at each imported coordinate location.
- The **Order** menu sets the order in which the coordinates are imported (**XYZ** or **YXZ**).
- The **View** menu is for 3D designs.
- The **Text** field is available only when **Text** is selected on the **Import** menu. Type the text string that is to be placed at the coordinate locations.
- The **Cell** field is available only when **Cell** is selected on the **Import** menu. Type the name of the cell to place at each coordinate location. The cell library that contains the cell must be attached to the design.
- Click **Process** to import the coordinates from the file whose name is in the **Filename** field.

DIMENSION-DRIVEN DESIGN

Dimension-driven design provides a set of tools to define constraints on the elements that make up a design. These constraints control the size and shape of the design, so the design can be adjusted for changing requirements by simply entering new dimension values.

Dimension-driven cells are cells defined from constrained designs. The dimensions of such cells can be changed as they are placed.

EXAMPLE OF A CONSTRAINED DESIGN

Dimension-driven design is introduced by describing the steps required to create the model shown in Figure 14–50. Before constructing the model, the constraints on the example model and the effect of changing a constraint are examined.

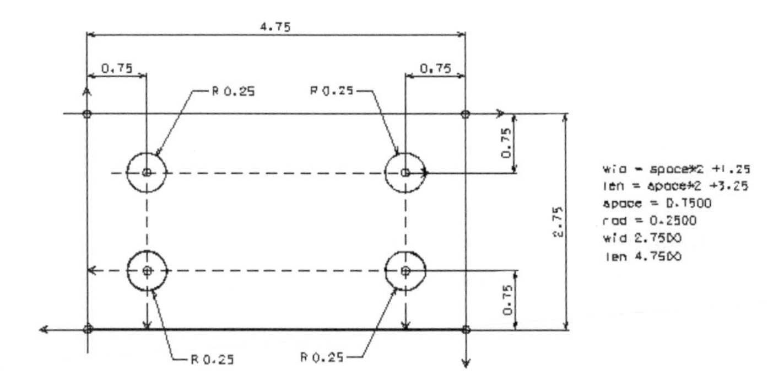

Figure 14–50 Example of a constrained design with the construction elements displayed

The Constraints

The design in the figure contains several "Construction" class elements that graphically represent the constraints:

- The dashed lines with arrows on one end are the construction lines to which the design is attached (some of the construction lines are covered by the lines and circles of the design).
- The equations and variables to the right of the model define the constraints.
- The Construction View attribute controls the display of construction elements. Figure 14–50 shows the view with Construction lines ON.
- The following constraints were placed on the design of Figure 14–50:
  - The angles of the construction lines are fixed, as indicated by the arrows.
  - The construction line intersections are constrained to always be connected, as indicated by the small circles.
  - The radius of the circles are set by the "rad = 0.2500" variable.
  - The circles are constrained always to be centered on the intersections of the inter-construction lines.
  - The space between the center of each circle and the adjacent design edges are set by the "space = 0.7500" variable.

- The overall length of the design is set by the "len = space*2 + 3.25" equation. The equation multiples the circle-center-to-edge space by 2 and adds 3.25 to that total (4.75 = 0.75*2 + 3.25).
- The overall width of the design is set by the "wid = space*2 + 1.25" equation. The equation multiplies the circle-center-to-edge space by 2 and adds 1.25 to that total (2.75 = 0.75*2 + 1.25).

Effect of Changing a Constraint

To illustrate the effect of changing a constraint, the Text Edit tool is used to change the add-on value in the wid (width) equation from 1.25 to 0.75 Master Units ("wid = space*2 + 0.75") The Re-solve Constraints tool is then used to solve the design for the new value. Figure 14–51 shows the design change caused by changing the wid add-on:

- The two rows of circles moved closer together.
- The overall width of the design decreased from 2.75 to 2.25 master units

The overall size was reduced because the changed value is part of the constraint on the overall width.

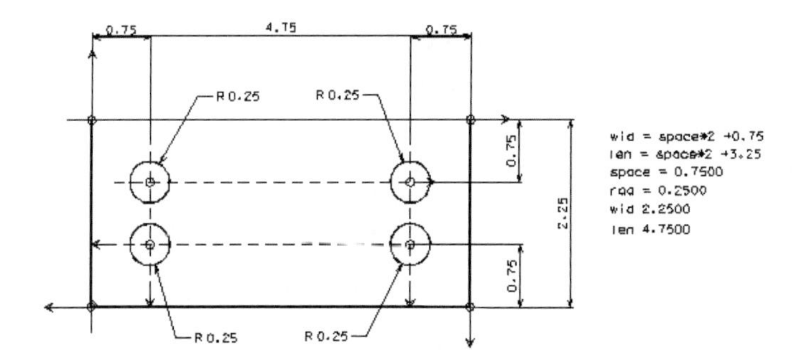

Figure 14–51 Effect of changing the width equation add-on from 1.25 to 0.75 Master Units

DIMENSION-DRIVEN DESIGN TERMS

Following are common terms used in dimension-driven design.

- **Constraint**—Information that controls how a construction is handled within a model. A constraint can be one of the following types:
  - **Location**—Fixes the location of a point in the design plane.
  - **Geometric**—Controls the position or orientation of two or more elements relative to each other.
  - **Dimensional**—Controlled by a dimension.

- **Algebraic**—Controlled by an equation that expresses a relationship among variables.
- **Construction**—An element, such as a line or circle, on which constraints can be placed to control its relation to other constructions in the model.
- **Well-Constrained**—A set of constructions that is completely defined by constraints and has no redundant constraints. It has what is needed to define it and no more.
- **Underconstrained**—A set of constructions that does not have enough constraints to define completely its geometric shape.
- **Redundant**—A constraint applied to a construction that is already well-constrained. It provides no useful information for the construction.
- **Degrees of Freedom**—A number that sums up a dimension-driven cell's ambiguity.
- **Solve**—Construction of the model from the given set of constraints. Each time constraints are modified or added the model is solved for the new set of constraints. If the model can be solved for the constraints, the model is updated; if not, an error message is displayed in the Status bar.

DIMENSION-DRIVEN DESIGN TOOLS

Dimension-driven tool boxes are available in the DD Design submenu. Invoke the submenu from:

Menu	Select Tools > DD Design > DD Design (see Figure 14–52)
Key-in window	**dialog toolbox dddtools** (or **di to dddt**) (ENTER)

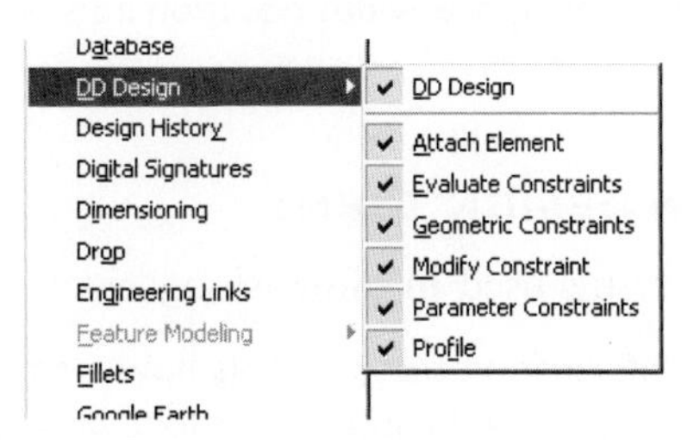

Figure 14–52 The DD Design submenu open from the Tools menu

The **DD Design** submenu has options to open seven tool boxes. Figure 14–53 shows the tool boxes opened in order left to right and top to bottom.

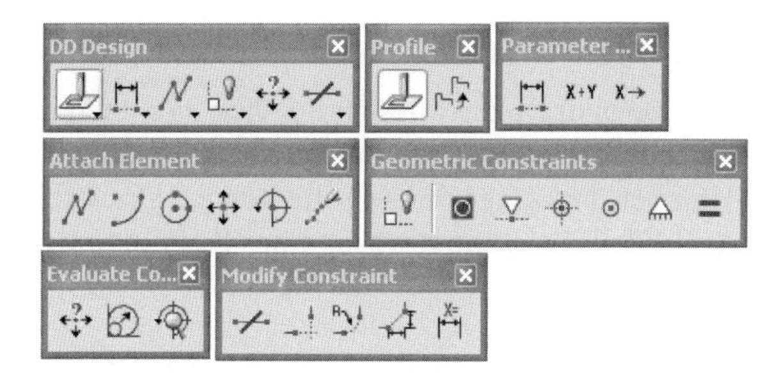

Figure 14–53 The dimension-driven design tool boxes

The first tool bar, **DD Design,** is the main dimension-driven design tool bar and provides an alternate way to invoke the tools on the other four tool bars using the same method as the Main toolbar (See Figure 14–54).

Figure 14–54 Invoking Dimension Driven Design tools from the DD Design tool bar

The tool boxes are:

- **Profile**—Tools for drawing a construction profile and for converting elements to a construction profile
- **Attach Elements**—Tools for attaching elements, such as lines and arcs, to the constrained construction elements
- **Evaluate Constraints**—Tools for solving and obtaining information about the construction
- **Parameter Constraints**—Tools for converting dimensions to constraints and for assigning equations and variables
- **Geometric Constraints**—Tools for placing constraints on the construction elements that are to define the model
- **Modify Constraint**—Tools for modifying constraints

CREATING A DIMENSION-DRIVEN DESIGN

A dimension-driven design requires careful planning to define the constraints. Numerous tools are available to constrain the design. A good introduction to the method is to walk through the creation of a design. The following discussion shows one way to create the model from the rough design shown in Figure 14–55.

Draw Construction Elements

Set the line weight to zero; draw the lines and circles, as shown in Figure 14–55. It is not necessary to draw the elements to specific dimensions because the constraints to added later set the dimensions of the design. Draw the elements in Primary mode. Use AccuDraw to insure drawing lines that are horizontal and vertical.

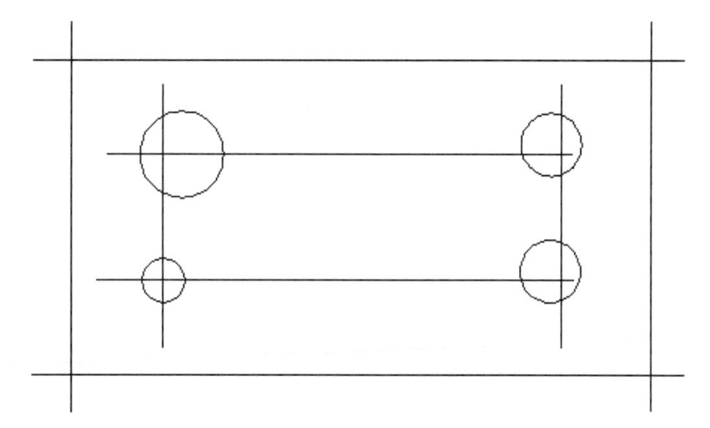

Figure 14–55 Rough construction lines and circles

Constrain the angle of the lines

The first constraint to apply to the design forces the lines to stay at the angles at which they were drawn. To constrain the angle of the lines shown in Figure 14–55, invoke the Constrain Elements tool from:

Geometric Constant tool box	Select the Constrain Elements tool. (see Figure 14–56).
Key-in window	**constrain element** (or **const e**) (ENTER)

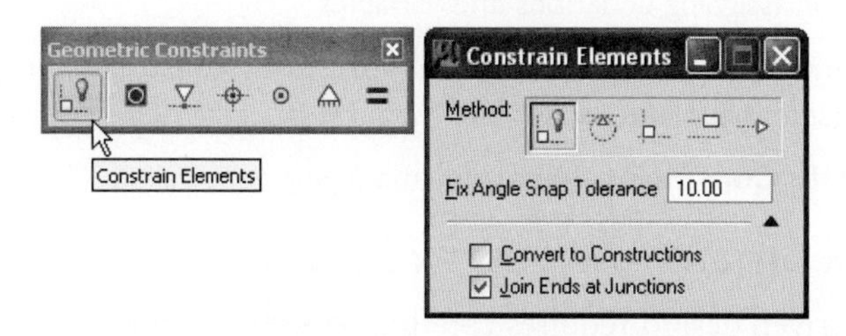

Figure 14–56 Invoking the Constrain Elements tool from the Geometric Constraints tool box

On the Constrain Elements settings box:

- Click the **Show Extended Information** arrow to expand the settings box and display more options.
- Select the **Smart Constrain Elements** method (the left-most **Method** tool).
- Turn the **Convert to Constructions** check box ON.

MicroStation prompts:

Smart Constrain Elements > Identify Construction

Identify one of the lines, and click the Data button again to accept it. The second click does not select another element.

Smart Constrain Elements > Identify Second Construction or Same Construction to Fix Angle

Identify the same line again to fix its angle.

Smart Constrain Elements > Accept

Click the Data button a third time to constrain the angle.

An arrow appears at one end of the line, the line's color changes to yellow and the line style to dashed, and the line changes to a construction element. This constraint ensures that the lines always remain at their original rotation angle.

Repeat the steps to constrain the angle of the rest of the elements in the design. Figure 14–57 shows the result of constraining the angle of all the lines.

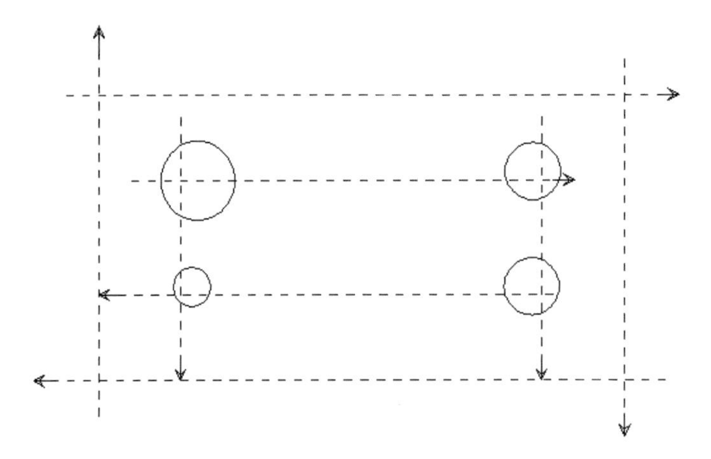

Figure 14–57 Result of constraining the angles of the lines

***Note:** If the **Constructions** View Attribute is OFF in the active view, the constrained elements disappear from the view.*

Constrain Point at Intersection

This tool either constrains a point to lie at the intersection of two constructions or forces two constructions to pass through a point. It works with any kind of construction except points. To constrain the line intersections as shown in Figure 14–59, invoke the Constrain Point At tool from:

Geometric Constraints tool box	Select the Constrain Point At tool (see Figure 14–58).
Key-in window	**constrain intersection** (or **const i**) (ENTER)

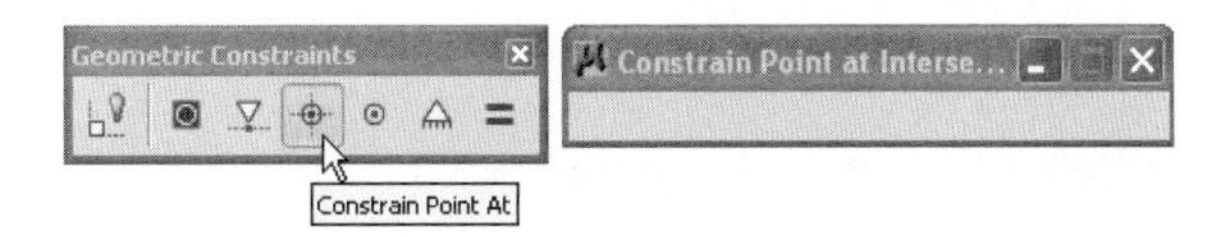

Figure 14–58 Invoking the Constrain Point At tool from the Geometric Constraints tool box

MicroStation prompts:

Constrain Point at Intersection > Identify Construction

Identify one of the intersecting lines near the point of intersection, then select the other intersecting line.

Constrain Point at Intersection > Identify point or Accept+Reset to Create

Click the Data button in the space, then click the Reset button to constrain the point.

A small circle appears at the constrained intersection. This constraint ensures that when one line is modified, the constrained lines always pass through the constraining point.

To constrain the remaining intersections, repeat the procedure for each section of the outer and inner set of lines. Figure 14–59 shows the result of constraining the intersections.

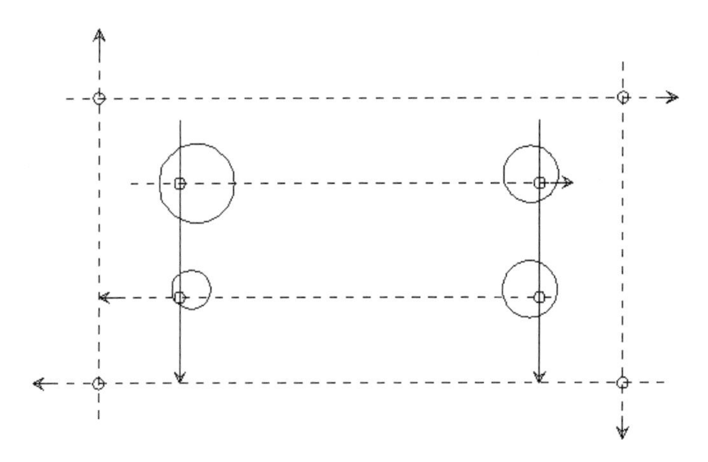

Figure 14–59 Result of constraining the construction intersections

Constrain Two Points to Be Coincident

This tool constrains two points to the same location (coincident), two circles to be concentric (have the same center), or a point to lie at the center of a circle. To constrain the circles to be centered at the intersection points as shown in Figure 14–61, invoke the Constrain Points Coincident tool from:

Geometric Constraints tool box	Select Constrain Points Coincident tool (see Figure 14–60).
Key-in window	**constrain concentric** (or **const c**) (ENTER)

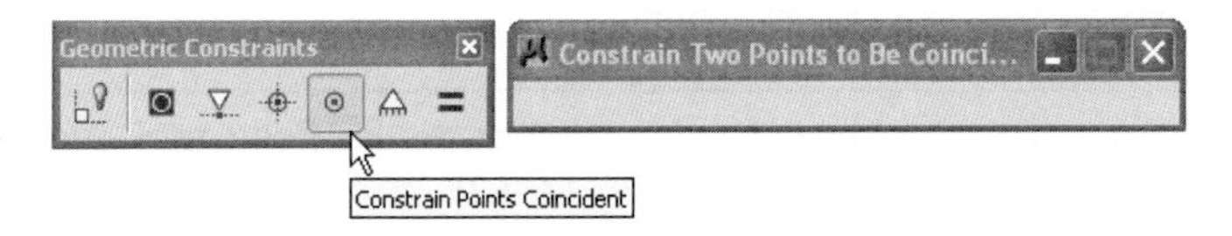

Figure 14–60 Invoking the Constrain Two Points to Be Coincident tool from the Geometric Constraints tool box

MicroStation prompts:

Constrain Two Points to Be Coincident > Identify Point (or Ellipse)

Identify the intersection point the circle is to be centered about.

Constrain Two Points to Be Coincident > Identify Next Point (or Ellipse)

Identify the circle.

Constrain Two Points to Be Coincident > Accept

Click the Data button in space to complete the constraint.

The constrained circle changes to construction elements that are yellow and dashed. The circles are constrained to stay centered over the intersections when the positions of the intersecting lines are modified.

Repeat the procedure to constrain the other three circles to the intersection points. Figure 14–61 shows the result of constraining the circles to the intersection points.

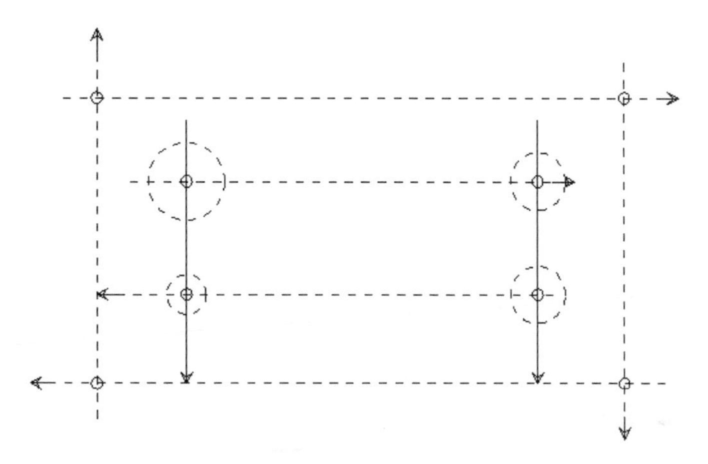

Figure 14–61 Result of constraining the circles

Fix Point at Location

This tool fixes the location of a point (or the center of a circle or ellipse) in the design plane. To attach the lower left intersection of the design to its current location in the design plane, invoke the Fix Point tool from:

Geometric Constraints tool box	Select the Fix Point tool (see Figure 14–62).
Key-in window	**constrain location** (or **const l**) (ENTER)

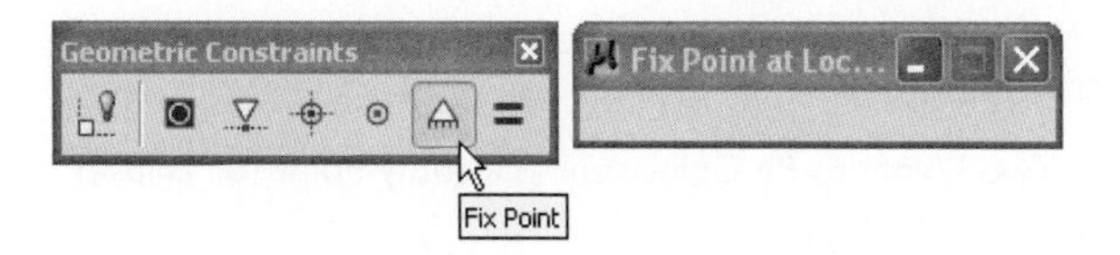

Figure 14–62 Invoking the Fix Point tool from the Geometric Constraints tool box

MicroStation prompts:

Fix Point at Location > Identify Point (or Ellipse)

Select the lower left intersection point.

Fix Point at Location > Accept/Reject

Click the Data button a second time to accept the point. This tool fixes the design to a location in the design plane that remains fixed when changes are made to the size of the constrained design.

Construct Attached Line-String or Shape

This tool creates a line-string or shape with the vertices attached to construction points, circles, or constraints. It is recommended to set a higher value for active line weight so that is easy to distinguish from the construction elements. To construct the shape as shown in Figure 14–64, invoke the Attach Line-String or Shape tool from:

Attach Element tool box	Select the Attach Line-String or Shape tool (see Figure 14–63).
Key-in window	**attach lstring** (or **at ls**) (ENTER)

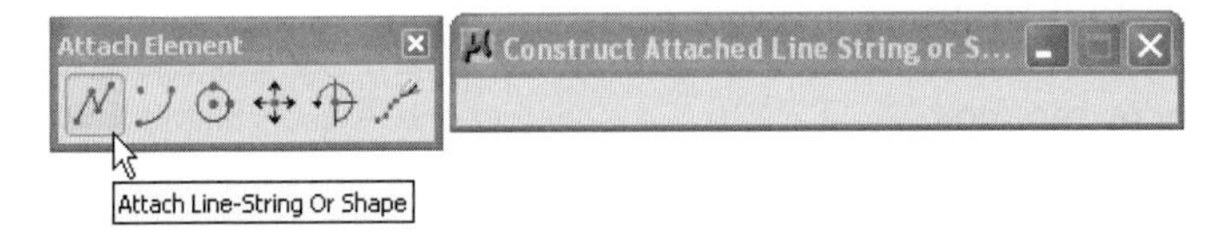

Figure 14–63 Invoking the Attach Line-String or Shape tool from the Attach Element tool box

MicroStation prompts:

Construct Attached Line String or Shape > Identify Point or Constraint

Select one of the constraint points on an intersection of a pair of the outer lines and select the constraint point at the other end of one of the lines.

Construct Attached Line String or Shape > Identify point, or RESET to finish

Select the other two constraint points on the outer lines and select the first constraint point again to complete the shape.

When the first constraint point is selected again, a closed shape is placed attached to the four points, as shown in Figure 14–64. Changes to the position of the constraint points change the shape of the attached shape element.

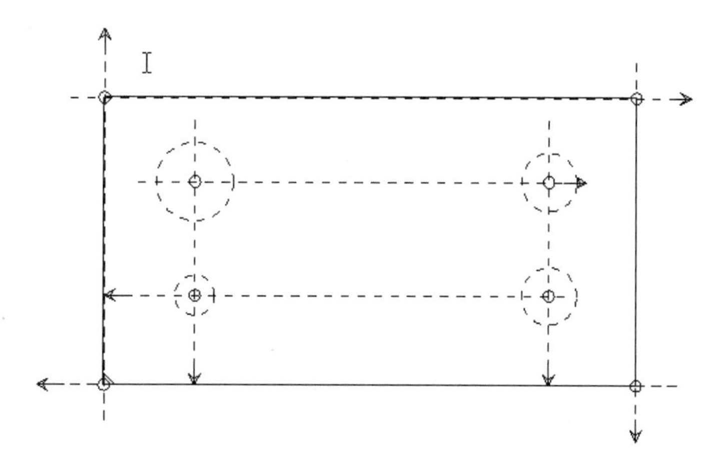

Figure 14–64 Result of drawing the design's outline shape

Construct Attached Ellipse or Circle

This tool creates and attaches a circle to a construction circle or point. To attach the design circles to the intersection points as shown in Figure 14–66, invoke the Attach Ellipse tool from:

Attach Element tool box	Select the Attach Ellipse tool (see Figure 14–65).
Key-in window	**attach circle** (or **at c**) (ENTER)

Figure 14–65 Invoking the Attach Ellipse tool from the Attach Element tool box

MicroStation prompts:

Construct Attached Ellipse or Circle > Identify Ellipse

Select one of the construction circles and click the Data button in space to accept the circle.

This attachment causes the design circle to always stay centered on the inner construction line intersections. Repeat the procedure for the other three circles. Figure 14–66 shows the result of attaching design circles to each of the construction circles.

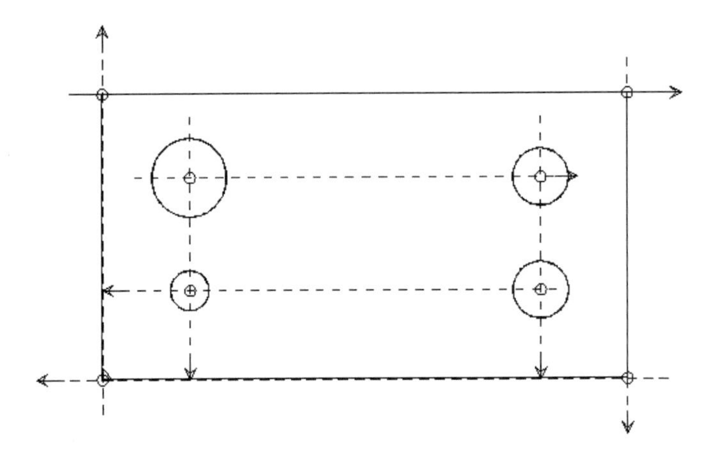

Figure 14–66 Result of drawing the design's circles

Add dimensions on the design as shown in Figure 14–67. Turn the Association check box ON before placing the dimensions. Associating the dimensions with the design elements allows using the dimensions as constraints. When constructing linear dimensions for the circles, start the dimension on the outside rectangle. Starting the dimension at the circles causes an error to appear.

Also, make sure the dimension extension lines connect to the design elements, not the constructions. To make sure of that, snap to the starting and ending points until the correct element is highlighted and accept it.

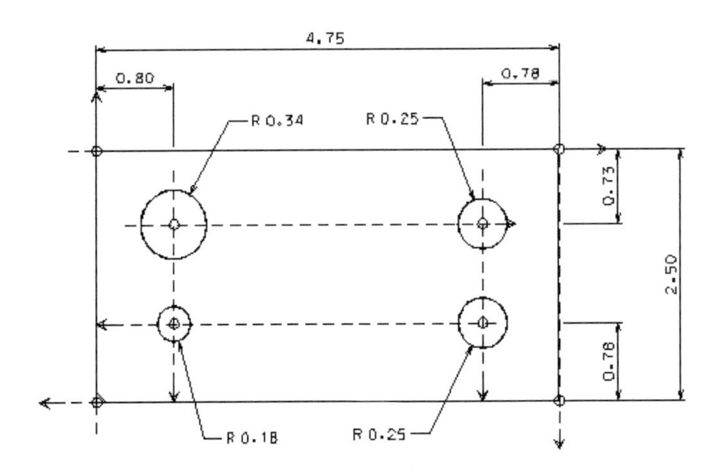

Figure 14–67 Result of dimensioning the design

Note: *Ignore the dimension values now. The dimensions are constrained and their values are set later.*

Place the following text strings for the equations and variables by using the Place Text at Origin tool, and place each line of text in separate, one-line text elements. Place the text to the right of the design.

- wid = space*2 + 1.25
- len = space*2 + 3.25
- space = 0.75
- rad = 0.25
- wid
- len

Figure 14–68 shows the design after adding the text.

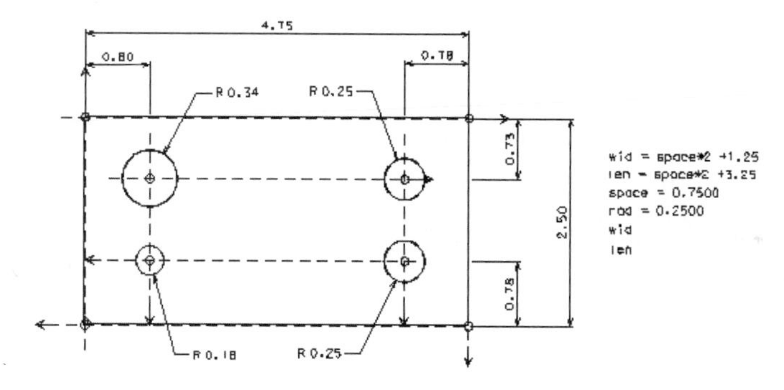

Figure 14–68 Result of adding the text strings to the design

Assign Variable to Dimensional Constraint

This tool assigns a constant or variable to a dimensional constraint. The constant or variable then represents the dimension's value in equations. Assign the "rad" variable as the radius of each circle. Invoke the Assign Variable tool from:

Parameter Constraint tool box	Select the Assign Variable tool (see Figure 14–69).
Key-in window	**assign variable** (or **as v**) (ENTER)

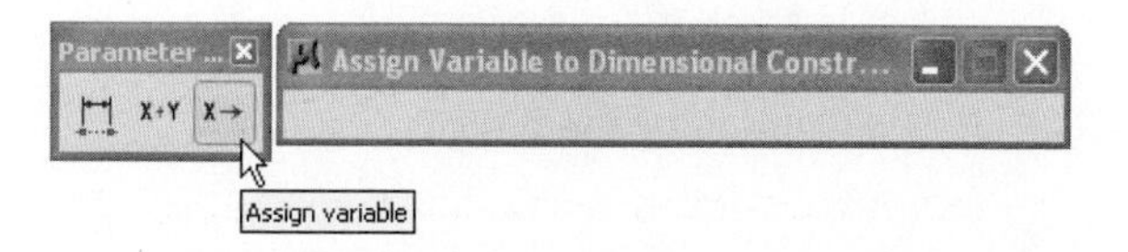

Figure 14–69 Invoking the Assign Variable tool from the Parameter Constraint tool box

MicroStation prompts:

Assign Variable to Dimensional Constraint > Identify variable

Select the "rad = 0.25" text string.

Assign Variable to Dimensional Constraint > Identify Constraint

Select the radial dimension of one of the circles.

Assign Variable to Dimensional Constraint > Accept

Click the Data button in space to assign the "rad" variable to the circle's dimension and set the circle's radius to the "rad" value.

Repeat the procedure for the radial dimensions of the other three circles.

Similarly, constrain the space between each circle center and the adjacent design edges with the Assign Variable. Assign the "Space" variable to each of the four circle-center-to-edge dimensions.

Assign Equation

This tool assigns an algebraic constraint—an equation that expresses a constraint relationship between variables, numerical constants, and built-in functions and constants—to a model. Create the "wid" and "len" equations by invoking the Assign Equation tool from:

Parameter Constraints tool box	Select the Assign Equation tool (see Figure 14–70).
Key-in window	**assign equation** (or **as e**) (ENTER)

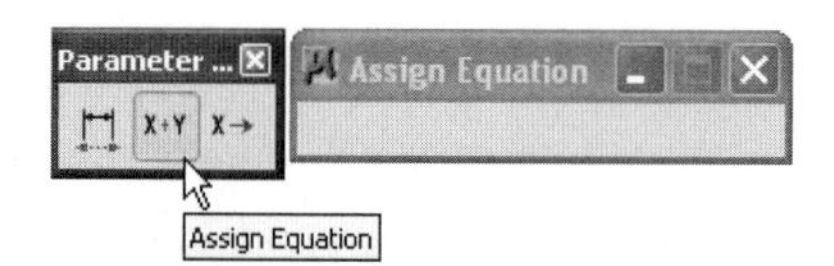

Figure 14–70 Invoking the Assign Equation tool from the Parameter Constraint tool box

MicroStation prompts:

Assign equation > Identify equation

Select the "wid = space*2 + 1.25" text string.

Assign equation > Identify variable, or RESET to finish

Select the "space = 0.75" text string, select the "wid" text string, then click the Data button in space to complete defining the "wid" equation.

Similarly, invoke the Assign equation tool again to create the "len" equation by selecting "len = space*2 + 3.25", "space = 0.75", and "len."

Constrain the design's outline dimensions by invoking the Assign Variable tool to assign "len" to the overall horizontal dimension and "wid" to the overall vertical dimension.

Figure 14–71 shows the completed design with the Construction view attribute turned OFF so that only the actual design is displayed.

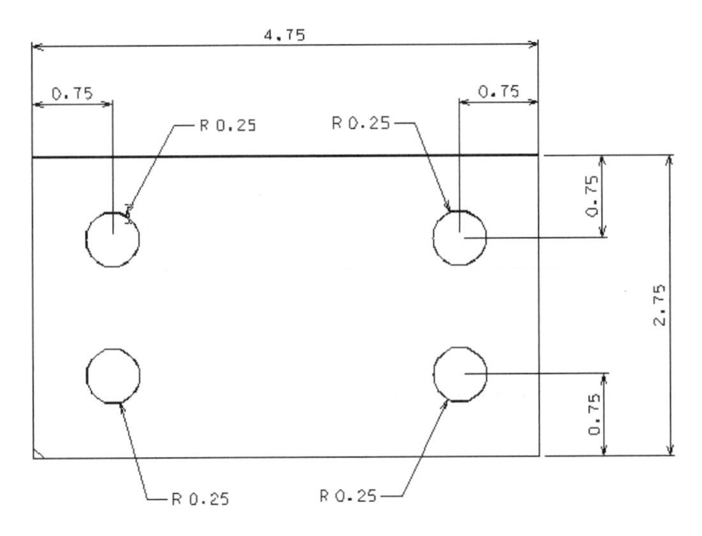

Figure 14–71 Completed design with construction elements turned off

Note: *After the design is completed, it may be necessary to invoke the Modify Element tool to adjust the position of some of the dimension elements and to move the text strings.*

MODIFYING A DIMENSION-DRIVEN DESIGN

Changing one of the variable values and re-solving the design modifies dimension-driven designs. Constraint equations and variables can be changed by editing the text strings or by invoking the Modify Value tool. For example, the design just constructed contains four variables:

- The "wid" and "len" variables are set equal to equations that contain constants (1.25 and 3.25). Editing the equations with the Edit Text tool can change the constants. These two constants control the horizontal and vertical space between the circles.
- The "space" variable is a constant (0.75) that is changed either by editing the text string with the Edit Text tool or with the Modify Value tool. Changing

this variable changes the position of the circles and, because it appears in the two equations, the overall width and length of the design.

- The "len" variable is a constant (0.25) that can be changed to control the size of the circles.

Modify Value of Dimension or Variable

This tool can edit the value of a dimensional constraint. To change the variable, invoke the Modify Value tool from:

Modify Constraint tool box	Select the Modify Value tool (see Figure 14–72).
Key-in window	**model edit_dimension** (or **mo e**) (ENTER)

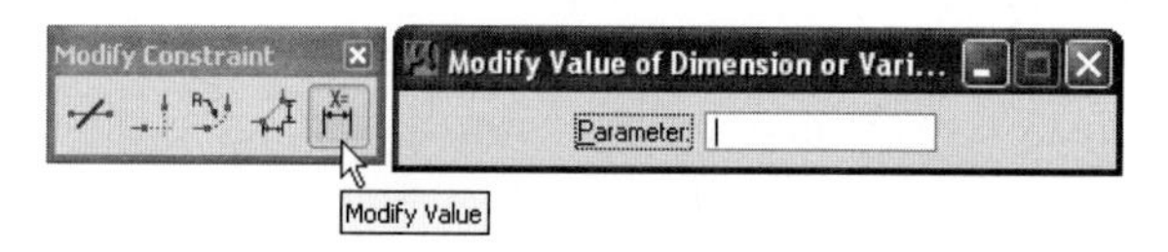

Figure 14–72 Invoking the Modify Value tool in the Modify Constraint tool box

MicroStation prompts:

Modify Value of Dimension or Variable > Identify element

Select the dimension or variable to be changed.

Modify Value of Dimension or Variable > Accept

Click the Data button in space to accept the selected element.

Modify Value of Dimension or Variable > Enter a value

Enter the new value in the Settings window edit field, then press (ENTER) to re-solve the design for the new value.

Re-solve Constraints

If the Edit Text tool changes a variable's value, the design must be "re-solved" to apply the new constraint value. To re-solve the design, invoke the Re-solve Constraints tool from:

Evaluate Constraints tool box	Select the Re-solve Constraints tool (see Figure 14–73).
Key-in window	**update model** (or **up m**) (ENTER)

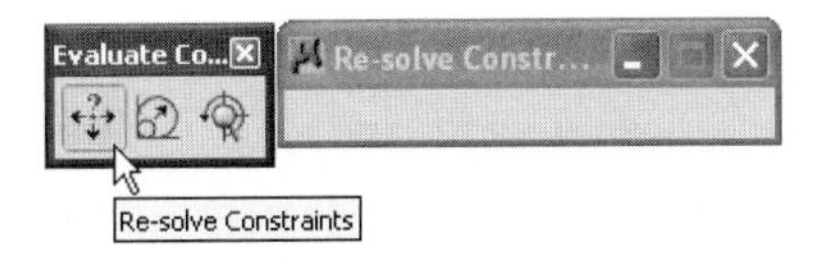

Figure 14–73 Invoking the Re-solve Constraints tool from the Evaluate tool box

MicroStation prompts:

Re-solve Constraints > Identify Element

Select the variable that has been changed or the text of a dimension that is constrained to the variable, then click the Data button in space to initiate re-solving of the design.

CREATING A DIMENSION-DRIVEN CELL

Dimension-driven cells are created from dimension-driven designs when the Construction view attribute is ON and all of the construction elements of the design are included in the design. When a dimension-driven cell is placed in a design file, its constraints can be changed.

Note: *If the Construction view attribute is OFF when a cell is created from a dimension-driven design, the cell is not created as a dimension-driven cell.*

OBJECT LINKING AND EMBEDDING (OLE)

Object linking and embedding (OLE) is a Microsoft Windows feature that allows a document to contain an object from other application. MicroStation supports OLE and can display information from other documents that also support OLE. For example, a MicroStation design file can contain a Microsoft Office Excel spreadsheet.

An object linked to a MicroStation design file is actually an image of the other document. It displays the current information from the other document and, when double-clicked it opens, the object's application to allow editing the object. For example, Double-click a linked Microsoft Office Excel spreadsheet object and the Excel application opens.

An object is a static snapshot of what was in the other document at the time it was embedded and there is no link to the original document. Changes made to the original document do not appear in the embedded object.

Linked and embedded objects appear in MicroStation view windows and print with the design file.

EMBEDDING MICROSTATION ELEMENTS IN ANOTHER DOCUMENT

Elements in a MicroStation design file can be copied into other documents that support graphic formats. As an illustration of the process, the following procedure copies elements from a design file into a Microsoft Office Word document:

1. Use the Selection Tool to select the elements.
2. Select **Edit > Copy** to copy the selected elements to the Windows clipboard.
3. In the Word document, go to the point in the Word document where the image of the MicroStation elements are to be placed.
4. Select **Edit > Paste Special** to open Word's Paste Special dialog box as shown in Figure 14–74.
5. In the **AS** field, select the paste method to use (such as **Picture**).
6. Click **OK** to paste the text.

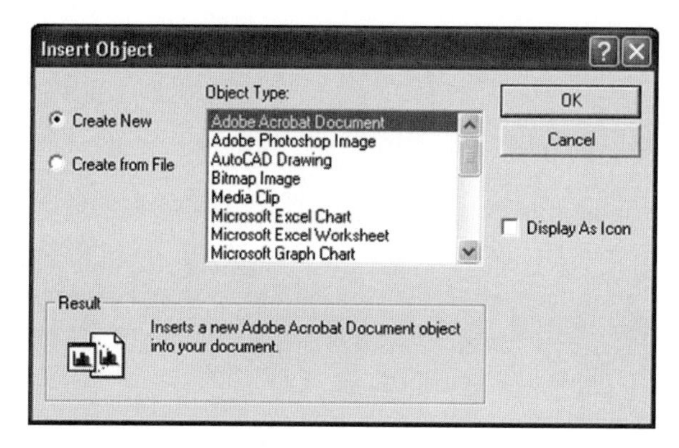

Figure 14–74 Word Paste Special dialog box

INSERTING OBJECTS IN A MICROSTATION DESIGN FILE

MicroStation provides options for inserting objects from other applications. The objects can be static copies of the documents (snapshots) or live links back to the documents that can be updated to display the current content of the documents. Insert objects into a MicroStation design file using the Insert Object dialog box. Invoke the dialog box from:

Menu	Edit > Insert Object

MicroStation displays the Insert Object dialog box, as shown in Figure 14–75a.

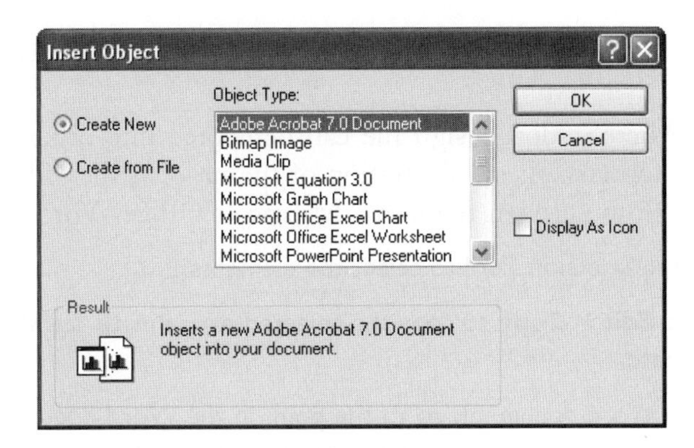

Figure 14–75a The Insert Object dialog box

The dialog box can create and insert a new object, or can insert an existing file as the object. Each insertion method has its own set of options. Figure 14–75a shows the options for creating a new file to insert and Figure 14–75b shows the options for selecting an existing file to insert.

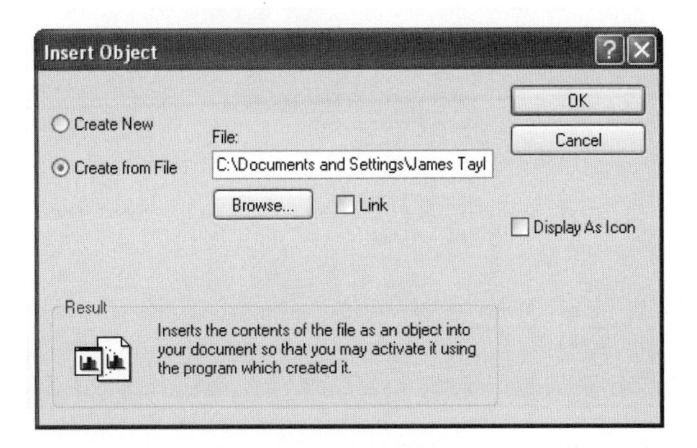

Figure 14–75b The Insert Object dialog box

To insert a new object and place it in the design, click **Create New** and select the type of object to create form the **Object Type** list box.

To insert an existing file as the object, click **Create from File**. Type the path and name of the file to insert in the **File** field or click **Browse** to open the Browse dialog box. On the Browse dialog box, use the **Look in** menu to find the file, click its filename in the list below the **Look in** menu, and click **Open** to put the selected file's path and name in the **File** field.

If the inserted object is an existing file, the **Link** check box controls the way the object is inserted:

- Turn the check box ON to create a live link to the file. When the design file is opened, MicroStation gets current contents of all live links to insure that the latest content is in the design file.
- Turn the check box OFF to insert a static copy (a "snapshot") of the current file content. MicroStation does not update static objects so they do not display content changes made after the object is inserted.

Both object insertion methods (new and existing) provide the **Display as Icon** check box to control the way the object content displays on the design file.

- Turn the check box OFF to display the content of the object on the design file.
- Turn the check box ON to display the object on the design file as an icon that is clicked to display the content of the object. Turning the check box ON expands the Insert Object dialog box to display the Change Icon button. Click **Change Icon** to open a dialog box from which a graphic symbol for the icon can be selected.

After specifying all required object settings, click **OK** to insert the object or click **Cancel** to cancel the operation. Clicking **OK** closes dialog box and displays the Insert OLE Object tool settings box. The box has options for placing and sizing the object on the design file (see Figure 14–76).

Figure 14–76 The Insert Object tool settings box

Expand the settings box to display all options by clicking the **Show/Hide Extended Options** arrow.

Select the placement **Method** as **By Corners** or **By Size**. Selecting **By Size** displays a **Scale** field that provides a place to enter a scaling factor. Type a number greater than one to increase the size of the object or type a number between zero and one to reduce the size of the object.

Turn the **Transparent Background** check box ON to allow elements behind the object to show through. Turn the **Rotate With View** check box ON to rotate the object when the view is rotated.

Place a data point to define the location of the object in the design file.

- If placing the object **By Center**, the data point defines the center of the object.
- If placing the object **By Corner**, the data point defines the upper, left corner of the object and the next data point defines the lower, right corner of the object.

If the **Link** check box is ON, MicroStation inserts the linked object in the design file and opens the objects application to allow editing the object.

Figure 14–77a shows a Microsoft Excel worksheet in the Excel application, and Figure 14–77b shows the worksheet as an object in a MicroStation design file.

AREA CALCULATIONS			
	WIDTH	LENGTH	AREA
ROOM 1	10	13	130
ROOM 2	12	11	132
ROOM 3	11	9	99
ROOM 4	12	12	144
TOTAL			505

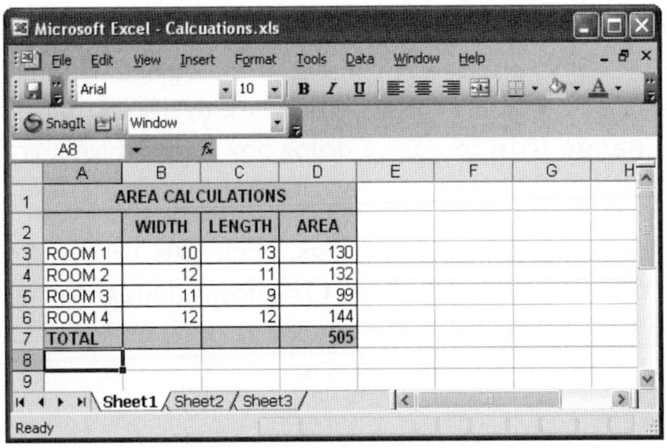

Figure 14–77a An Excel spreadsheet

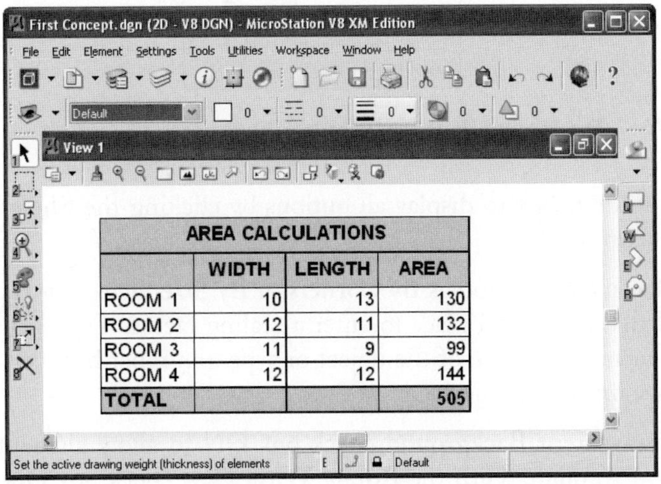

AREA CALCULATIONS			
	WIDTH	LENGTH	AREA
ROOM 1	10	13	130
ROOM 2	12	11	132
ROOM 3	11	9	99
ROOM 4	12	12	144
TOTAL			505

Figure 14–77b The Excel spreadsheet as an object in a MicroStation design file

UPDATING OBJECTS IN A MICROSTATION DESIGN FILE

Objects inserted with the **Link** check box ON are live links to the original document. Updating the links shows the current content of the linked document (if MicroStation can find it). To update all links in the design file, select Update Links from:

MicroStation menu bar	Select Edit > Update Links

MicroStation updates the links. While it is updating the links, it displays an Update Links message box that builds a percent complete bar as it completes its work. MicroStation closes the message box when it completes updating the links.

MAINTAINING OBJECTS IN A MICROSTATION DESIGN FILE

The Links dialog box provides options for maintaining links to objects. Invoke the dialog box from:

MicroStation menu bar	Select Edit > Links

MicroStation displays the Links dialog box, as shown in Figure 14–78.

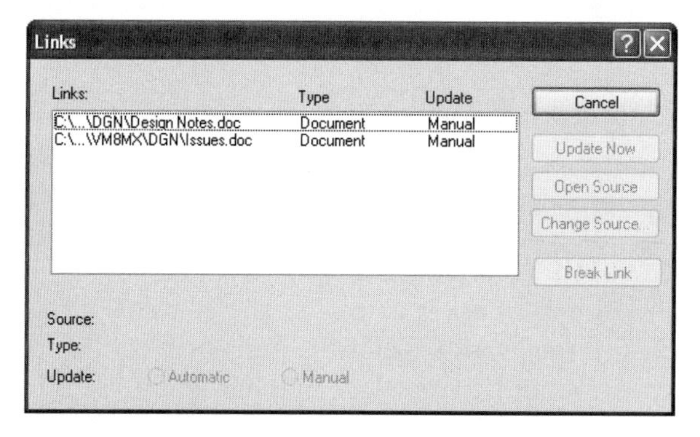

Figure 14–78 The Links dialog box

In the Links dialog box, the **Links** field displays the path and filename of all linked objects. To perform maintenance on a linked object, select the object from the **Links** field and click one of the maintenance buttons:

- Click **Update Now** to update the selected object with the latest information from the object.
- Click **Open Source** to open the object in its native application for editing.
- Click **Change Source** to open the Change Source dialog box and select a document to replace the current object.

- Click **Break Link** to break the link from the object to its native application and turn it into a static copy of the document. When the link is broken, the object cannot be updated from the original document. Before MicroStation breaks the link, it displays an alert message box that for confirmation of breaking the link. To break the link, click **OK**.

After completing all maintenance on the links, click **Cancel** to close the Links dialog box.

DESIGN MANAGEMENT FEATURES

Following are brief descriptions of MicroStation options for approving designs, checking to make sure a design meets design standards, tracking design changes, and using archive designs.

APPROVING DESIGNS

MicroStation provides features to help maintain design files in a secure environment and version V8 MX improves the security features.

Digital signatures indicate approval of a design and MicroStation allows attachment of multiple signatures. The signatures can also be applied in a hierarchy that forces signatures to be applied in a specific order. The signatures can apply only to a model in the design or to all of the models in the design including references to DGN and DWG design files. Users can detect the attached signatures, determine who owns them, who attached them, and if the design was changed after signature attachment.

Cells can indicate the attachment of signatures and signatures attachments can be limited to a specific area of the design. There are also features to allow determining that the digital signatures are authentic.

CHECKING DESIGN STANDARDS

The MicroStation Design Checker compares the content of the active design file to standards selected in the Standards Checker dialog box. To open the dialog box, select the **Utilities** menu and then select **Standards Checker** and **Configure.** To run the check, select the **Utilities** menu and then select **Standards Checker** and **Check.** A check box in the Standards Checker dialog box allows running the check interactively or in a batch that can contain multiple design files.

TRACK DESIGN CHANGES

MicroStation's Design History tracks changes to individual models in a design down to the element level and MicroStation V8 MX enhances the tracking features. Each change is identified by a unique ID number, change date, change author, and change description. For each change, MicroStation stores only enough information to undo or redo the change, not the entire design file. Groups of selected changes can be undone and redone in change order from the first selected change to the last.

All history tracking option are available in a sub-menu by selecting the **Utilities** menu and then selecting **Design History**.

ARCHIVE DESIGNS

The Archive dialog box (available from the Utilities menu) provides options for selecting and opening an archive file (compressed files), for selecting the path where an extracted file is to be stored, and for verifying the attached digital signatures.

REVIEW QUESTIONS

Write your answers in the spaces provided.

1. Explain briefly the purpose of creating a graphic group.

   ____

   ____

2. List the steps in creating a graphic group.

   ____

   ____

3. Explain the purpose of the Graphic Group lock setting.

   ____

   ____

4. Explain briefly the purpose of element selection by element type.

   ____

   ____

5. What is a parent named group? A child named group?

   ____

   ____

6. Describe how to set the symbology for a level.

   ____

   ____

7. List the steps involved in converting a MicroStation design file to an AutoCAD drawing file.

   ____

   ____

8. Explain briefly the benefits of dimension-driven design.

   ____

   ____

Customizing MicroStation

MicroStation is a powerful program off the shelf, but it also provides options to customize the core program to suit individual needs and applications.

Topics explored in this chapter:

- Settings groups
- Level filters
- Element Templates
- Custom line styles
- Workspaces
- Function keys
- Fonts
- MicroStation Environment Packages
- Project Explorer and Link Sets
- Associate files
- Scripts and Macros

SETTINGS GROUPS

Settings groups can be defined in MicroStation in three categories: Drawing, Scale, and Working Units. Any number of individual group components can be defined under each settings group of the category. As part of a group component, element attributes such as color, weight, line style, level, and class can be set, and the component can be associated with a tool (such as Place Line, Place Text). The current multi-line definition and active dimension settings can also be saved.

Settings groups and group components can be saved to an external file that, by default, has the extension *.stg*. This file has the same conceptual functionality as the cell library. Once created, the settings file can be attached to any design file, and one of the settings groups and corresponding group components can be activated. Selecting a component does the following:

- All element attributes associated with the component are set in the design.

- If a tool key-in is defined for the component, the tool is selected, enabling placement of an element or elements without invoking the tool from a tool box.

For example, MicroStation provides the **V40 – Dimension Styles** group that sets dimension attributes for several dimensioning standards. Selecting a dimension style from the components list is faster than manually setting up the many dimension attributes.

SELECTING A SETTINGS GROUP AND COMPONENT

Use the Select Settings dialog box to select a settings group and corresponding group component from the currently attached settings group file. The settings box also provides an option to attach another settings group file. To open the Select Settings dialog box, invoke the Manage tool from:

Menu	Settings > Manage

MicroStation displays the Select Settings dialog box, as shown in Figure 15–1.

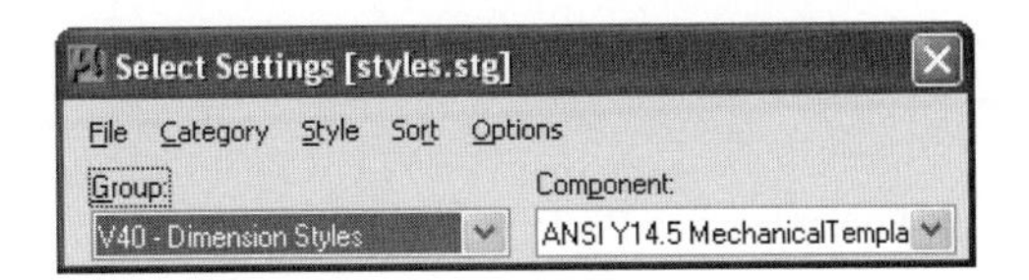

Figure 15–1 *Select Settings dialog box*

The name of the currently opened settings group file is displayed in the title bar. The dialog box is divided into two parts. The **Group** Menu lists the names of the available drawing settings groups, the **Component** Menu lists the name and type of each component in the selected settings **Group**.

To invoke one of the components, select the **Group** that contains it and then select the **Component**. If the component has a defined key-in, the corresponding tool is invoked.

The Select Settings dialog box can be docked on the top or bottom of the Microsoft desktop. It can also be displayed as a resizable large dialog box. Invoke the Large Dialog tool from:

Select Settings box	Options > Large Dialog

MicroStation displays the large Select Settings dialog box as shown in Figure 15–2.

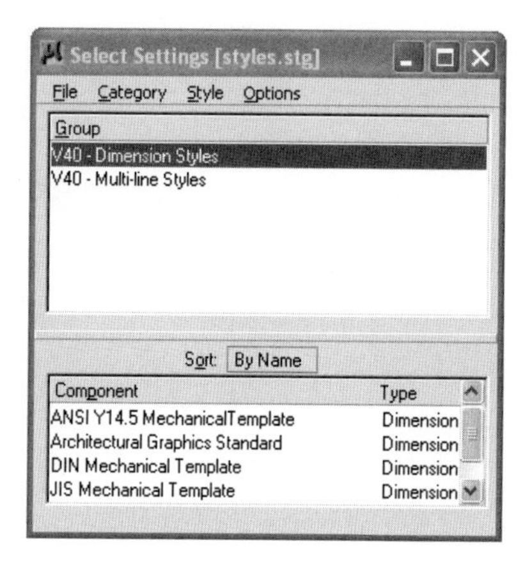

Figure 15–2 *Select Settings dialog box (Large Dialog)*

The top part lists the names of the available drawing settings groups and the bottom part lists the name and type of each component from the selected settings group. The **Category** Menu sets the category for the listing of groups in the **Group** list box. The **Sort** Menu allows sorting the components **By Name** or **By Type.**

ATTACHING A SETTINGS GROUP FILE

To attach a settings group file to a design file, invoke the Open tool from:

Select Settings box	File > Open

MicroStation opens the standard File Open dialog box. Select the location and name of the file and click **Open** to list the available groups and corresponding components in the Select Settings dialog box.

MANAGING A SETTINGS GROUP

The Edit Settings dialog box is used to define, modify, and delete settings groups and group components. To open the Edit Settings dialog box, invoke the Edit tool from:

Select Settings dialog box	File > Edit

MicroStation displays the Edit Settings dialog box, as shown in Figure 15–3.

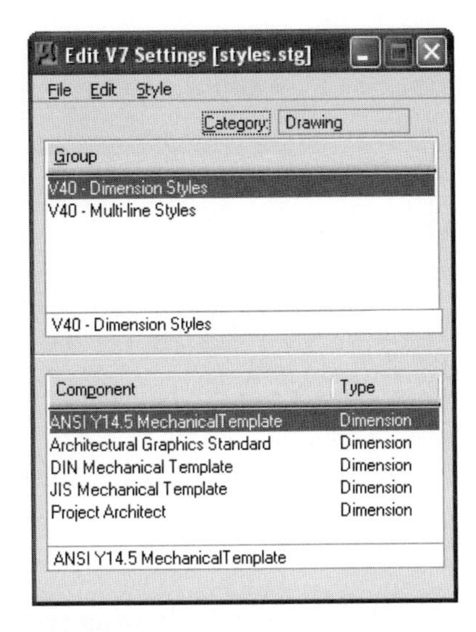

Figure 15–3 *Edit Settings dialog box*

The name of the currently opened settings group file is displayed in the title bar of the dialog box. The top part of the dialog box lists the names of the available drawing settings groups. The bottom part lists the name and type of each component from the selected settings group.

Creating a Settings Group File

To create a new settings group file, invoke the New tool from:

Edit Settings dialog box	File > New

MicroStation displays the standard file creation dialog box. To create the new file, select a folder to hold it, type the file name, and click **Save.**

Creating a Settings Group

To create a new settings group, invoke the Create Group tool from:

Edit Settings dialog box	Edit > Create > Group

MicroStation adds a new group to the **Group** list and names it "Unnamed." The name also appears in a text field at the bottom of the **Group** list. Replace the name in the text field with a descriptive name (32-character maximum).

Deleting a Settings Group

To delete a settings group, select the group's name from the **Group** list box and invoke the Delete tool from:

Edit Settings dialog box	Edit > Delete

MicroStation opens an Alert message box to confirm the deletion. Click **OK** to delete the group and all of its components.

Creating a New Component

To create a new component, select a settings group and select one of the seven available component types from:

Edit Settings dialog box	Edit > Create

MicroStation displays a menu of component types (see Table 15–1). Select a component type from the menu. MicroStation adds the new component to the **Component** list and names it "Unnamed." The name also appears in the text field at the bottom of the **Component** list. Replace the name in the text field with a descriptive name (32-character maximum).

Table 15–1 lists the component types and corresponding tools that can be used with each component type.

Table 15–1 Component Types and Corresponding Tools

COMPONENT TYPE	TOOL
Active Point	Points tool box
Area Pattern	Pattern tool box
Cell	Cells tool box
Dimension	Dimension tool box
Linear	Linear Elements tool box, Polygons tool box, Arcs tool box, Ellipses tool box, Curves tool box
Multi-line	The key-in PLACE MLINE CONSTRAINED corresponds to the Place Multi-line tool
Text	Text tool box

Modifying a Component

Select a **Group** and **Component** from the Edit Settings dialog box and then invoke the Modify tool from:

Edit Settings dialog box	Edit > Modify

Select a group from the **Group** area and a component from the **Component** area to open a settings box. Each component type has a unique settings box. In the settings boxes, the modifiable settings have check boxes. Turn ON a settings check box to modify the setting.

The following guidelines are common to all of the component settings boxes:

- To set the tool that MicroStation activates when the component is selected, turn the **Key-in** check box ON and type the appropriate key-in for the tool.

- Set the appropriate element attributes (**Level**, **Color**, **Weight**, and **Line Style**).
- Click **Save** to save the settings entered in the settings box fields.
- Click **Match** and select an existing element in the design of the same type as the component to match. The tool applies the settings of the element to the selected component. Click **Save** to save the settings.

For example, if the component type is a Point, selecting a point element in the design file fills in the settings fields with the settings that were in effect when the element was created.

- Click **Close** to close the settings box.

The following paragraphs show examples of the settings boxes and describe settings unique to each box.

The Modify Point Component settings box (see Figure 15–4) provides options for modifying settings related to Construct Point tools. Select the type of point (**Zero-Length Line, Cell,** or **Character**) from the **Type** menu. If **Cell** is selected, set the appropriate **Cell** settings. If **Character** is selected, set the appropriate **Character** settings.

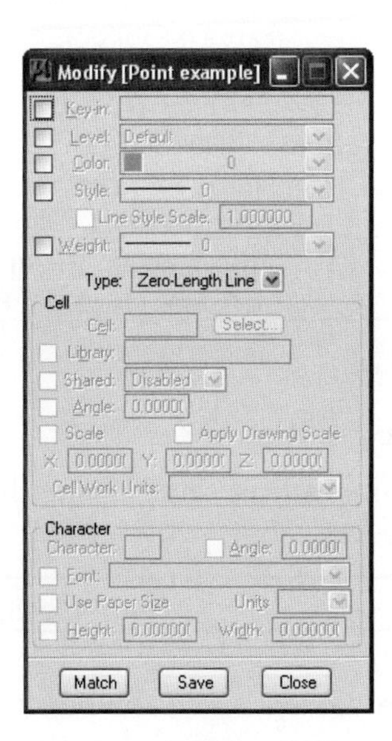

Figure 15–4 *Modify Point Component settings box*

The Modify Area Pattern Component settings box (see Figure 15–5) provides options for modifying settings related to hatching and patterning tools. Select either **Pattern** or **Hatch** from the **type** menu. If **Pattern** type is selected, set the appropriate **Pattern Settings.** If **Hatch** area is selected, set the appropriate **Hatch Settings.**

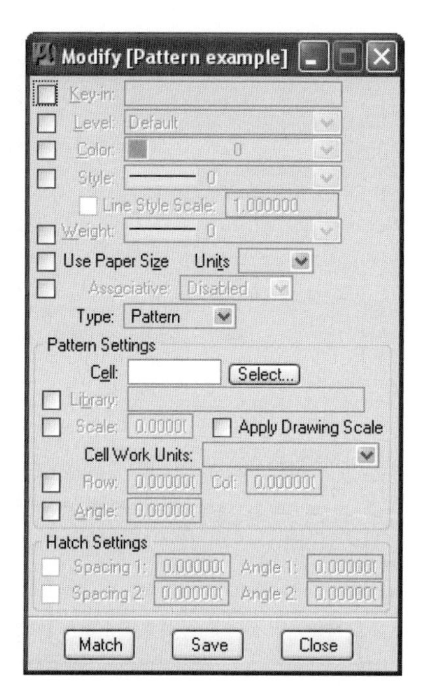

Figure 15–5 *Modify Area Pattern Component settings box*

The Modify Cell Component settings box (see Figure 15–6) provides options for modifying settings related to cell placement tools. Key-in the name of the placement cell in the **Cell** field or click **Select** to open the Select Cell dialog box and select a cell from the attached library. Select the placement method for the cell (**Placement** or **Terminator**) from the **Type** menu. If **Placement** is selected, the **Cell** selection becomes the Active Cell when the component is selected in the Select Settings dialog box. If **Terminator** is selected, the **Cell** selection becomes the Active Line Terminator.

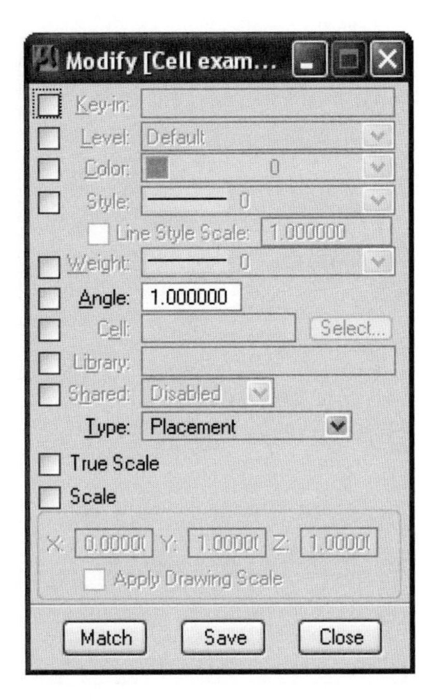

Figure 15–6 *Modify Cell Component settings box*

Note: *The **Level**, **Color**, **Style**, and **Weight** controls affect placement of a cell using the component only if the specified cell was created as a point cell.*

The Modify Dimension Component settings box (see Figure 15–7) provides options for modifying settings related to the Dimension component definition. To associate the dimensions with the elements they dimension, select **Enabled** from the **Associative** menu. Otherwise, select **Disabled.** Click **Select** to open the Select Dimension Definition dialog box and select the type of dimensions to place with this dimensioning component (such as "ISO Mechanical Template").

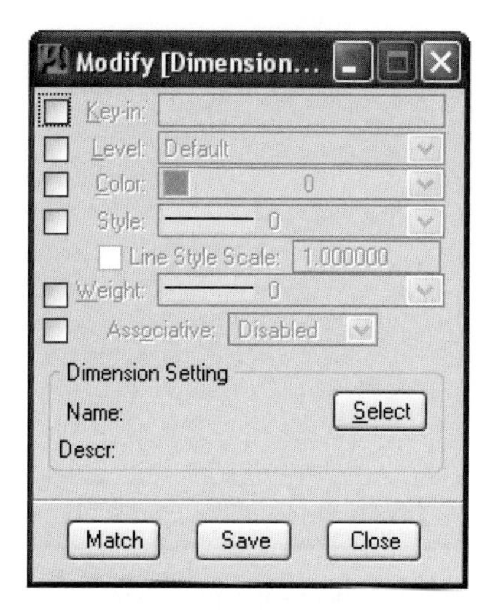

Figure 15–7 *Modify Dimension Component settings box*

The Modify Linear Component settings box (see Figure 15–8) provides options for modifying settings related to placement of lines, polygons, arcs, circles, ellipses, and curves.

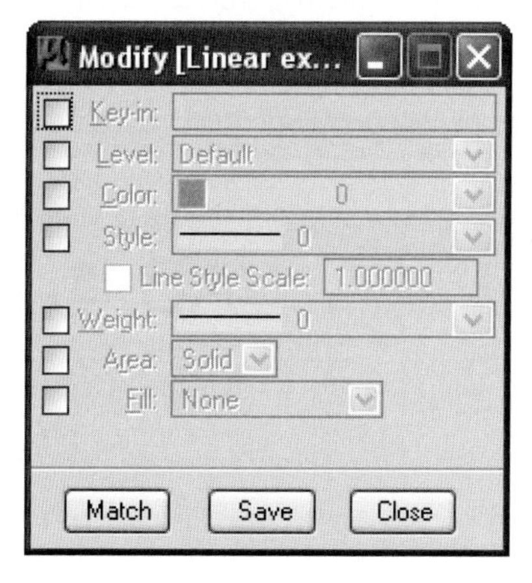

Figure 15–8 *Modify Linear Component settings box*

The Modify Multi-line Component settings box (see Figure 15–9) provides options for modifying settings related to multi-line component definitions. Click **Select** to open the Select Multi-line Definition dialog box for selecting multi-line definitions.

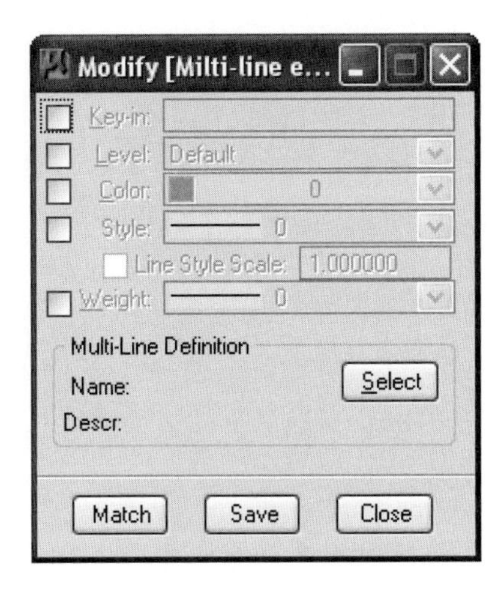

Figure 15–9 *Modify Multi-line Component settings box*

The Modify Text Component settings box (see Figure 15–10) provides options for modifying settings related to the placement of text. Set the appropriate text attributes (**Font, Slant Angle, Line Length, Fraction, Vertical,** and **Underline**), Justification (**Single-line** and **Multi-line**), and Size and Spacing (**Use Paper Size, Units, Height, Width, Intercharacter Spacing,** and **Line Spacing**). MicroStation makes the selected text settings the active settings when the component is selected.

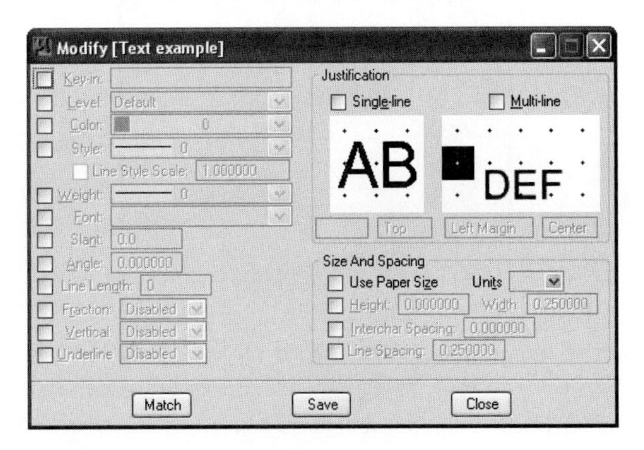

Figure 15–10 *Modify Text Component settings box*

Selecting Categories

Specify the relationship between printing units and design Master Units by selecting one of the available scale settings. Select the relationship from:

Select Settings dialog box	Category > Scale

MicroStation displays the Select Scale dialog box. Select a scale from the dialog box and click **OK** to apply the selection to the design.

Set the active Working Units from the Select Settings dialog box by selecting one of the available Working Units settings. Select the Working Units from:

Select Settings dialog box	Category > Working Units

MicroStation displays the Select Working Units dialog box, which lists a set of standard Working Units. Select a Working Units setup from the dialog box and click **OK** to apply it to the design file. An alert box opens requesting confirmation for the adjustment of the Working Units settings. Click **OK** to implement the selection or **Cancel** to cancel the selection.

LEVEL FILTERS

Chapter 3 introduced placing elements on different levels and naming levels. A practice of assigning meaningful names for levels can be further enhanced by creating filters to display sets of related levels by specifying what levels are displayed when the filter is active. For example, a seed file that that provides separate levels for each engineering discipline could have a filter for each discipline to help users find the levels for their discipline.

Filters are defined in the Level Manager settings box. The **Filters** option in the tree on the left side of the settings box is used to view the filters defined for the active design file. Click the **Filters** icon to display all defined filters. The filter criteria appear to the right of the tree area, as shown in Figure 15–11. Click one of the filter names in the tree area to list the names of the levels that meet the filter criteria on the right of the tree area.

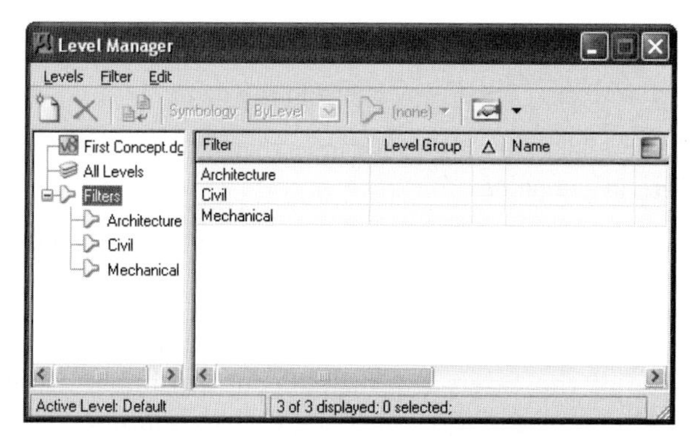

Figure 15–11 *Level Filters in the Level Manager settings box*

CREATING A NEW FILTER

To create a new filter, open the Level Manager settings box, click the **Filters** icon in the tree area of the settings box, and create the new filter from:

Level Manager settings box	Filter > New

MicroStation inserts a criteria line for the new filter in the area of the Level Manager settings box to the right of the tree area. The **Object Levels** column in the criteria area lists the filter names and the default name of the new filter ("New Filter") is selected for editing. Type a descriptive name for the new filter and enter the filter criteria in the adjacent columns.

CREATING FILTER CRITERIA

To define criteria for the filter, click the filter's name in the tree area. The area to the right of the tree displays a row for entering the filter criteria and all levels that currently match the filter criteria (initially all levels in the design). Filter criteria can be entered in the first row of each column. Examples of typical criteria follow.

Text strings entered in the **Name** column are compared to each level's name. Wild card symbols (* &) are accepted and evaluated. Following are examples of typical text strings:

- **obje**–Include only level names that contain the characters "obje" anywhere in the string.
- **obje***–Include only level names that start with "obje."
- ***obje**–Include only level names that end with "obje."
- **obje* *ing**–Include only level names that begin with "obje" or end with "ing."
- **obje bord**–Include only level names that contain either the characters "obje" or the characters "bord" anywhere in the name.
- **obje&bord**–Include only level names that contain both "obje" and "bord" anywhere in the name.
- **obje*&*ing**—Include only level names that begin with "obje" and end with "ing."

The **Color**, **Weight**, and **Style** columns are used to limit the display to levels that match one or more of the external filter element attributes. Enter the criteria by entering the attribute numbers. For example, type the number "3" in the **Color** criteria field to display only levels that are set to the color Red (3).

Note: *For additional information on entering filter criteria, refer to the MicroStation online help.*

ELEMENT TEMPLATES

Element templates are named sets of element properties (such as level, color, line style, and line weight) that can be applied to existing elements and can be used to set the active settings for placing new elements. Elements can retain an association to the template that was active when they were placed. Elements that contain an association to their template can be resymbolized when properties in the template are modified. Creating element templates in DGN libraries allows them to be shared with many users to promote drawing consistency within projects.

The Customize dialog box provides options for modifying, renaming, deleting, and organizing element templates for use in a DGN file. The Standards Checker is used to validate templates. The open DGN file's local templates can be checked against those defined in the DGN libraries. All elements in the open DGN file can be checked against the local templates with which they are associated.

CUSTOM LINE STYLES

In addition to the eight standard line styles introduced in Chapter 3, MicroStation provides tools for creating Custom line styles. Custom line styles can be selected from the **Active Line Style** menu in the Attributes tool box, where they are then listed below the eight standard line styles (Figure 15–12). Selecting a Custom line style makes it the Active Line Style that will be used by invoked placement tools.

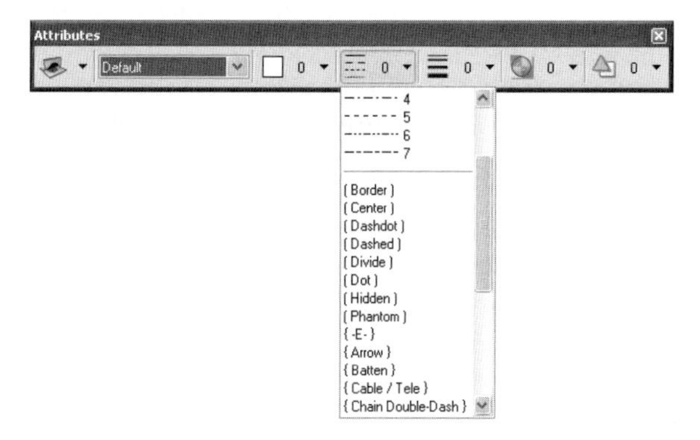

Figure 15–12 *Custom line styles in the Active Line Style menu*

Custom line styles are stored in style library files that have an "RSC" extension and must be attached to the design to make the Custom line styles available. Multiple style libraries can be attached to the active design at the same time.

THE LINE STYLES SETTINGS BOX

The Line Styles settings box can be used to preview the Custom line styles and modify stroke pattern. The settings box can also be used to designate a Custom line style as the Active Line Style. Invoke the Line Styles settings box from:

Menu bar	Element > Line Style > Custom
Key-in window	**linestyle settings** (or **lines s**) (ENTER)

MicroStation displays the Line Styles settings box, as shown in Figure 15–13.

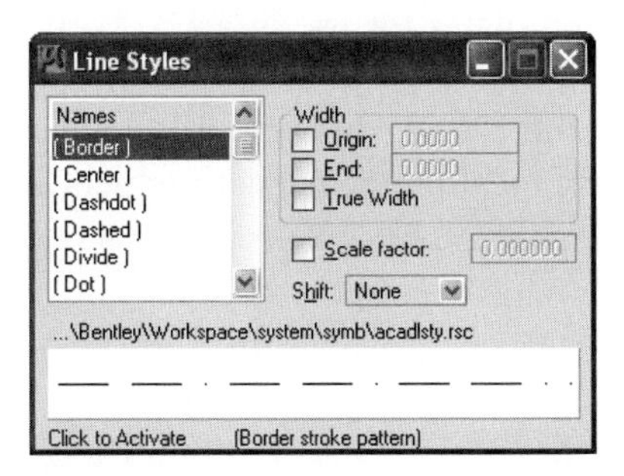

Figure 15–13 *Line Styles settings box*

The **Names** area on the left side of the settings box lists all available Custom line styles. When one of the styles is selected, a picture of the line style appears in the field near the bottom of the settings box. Make a Custom line style the Active Line Style by either clicking its picture or by double-clicking its name in the **Names** area.

When an Active Line Style is a Custom line style that contains dash strokes that have width, the **Origin** and **End** fields can be used to change the stroke width. To change one of the widths, turn its check box ON and key-in the width, in Working Units, in the associated text field.

The **Scale factor** sets the scale of all displayable characteristics (dash length and width, point symbol size) of the Active Line Style. To change the scale factor, turn the **Scale factor** check box ON and key-in the scale factor in the associated text field.

The **Shift** menu sets the distance or fraction by which each stroke pattern in the active line style is shifted or adjusted.

THE LINE STYLE EDITOR SETTINGS BOX

To create and modify Custom line styles, open the Line Style Editor settings box from:

Menu	Elements > Line Style > Edit
Key-in window	**linestyle edit** (or **lines e**) (ENTER)

MicroStation opens the Line Style Editor settings box. If no library file was previously opened in the settings box, the box's appearance is as shown in Figure 15–14.

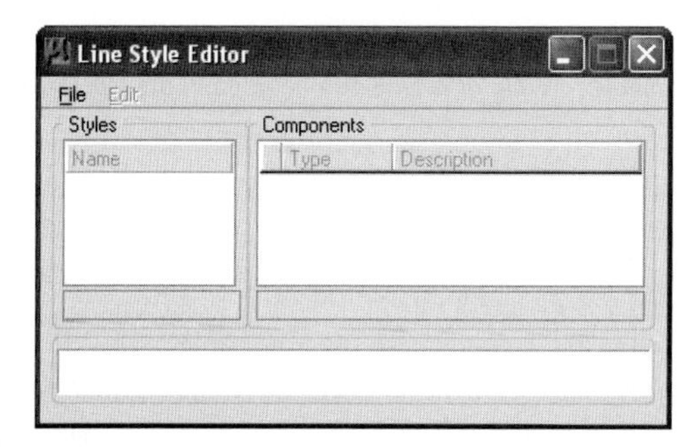

Figure 15–14 *The Line Style Editor settings box*

If a file was previously opened in the settings box, the size of the settings box varies depending on what type of line style was selected (Stroke Pattern, Point, or Compound).

The settings box provides two menus (**File** and **Edit**) in the menu bar. The **Edit** menu provides tools to create and maintain the line styles in a selected library file. If no file is opened in the box, only the **File** menu is available for use.

The File Menu

The Line Style Editor settings box's **File** menu provides options for creating and maintaining style library files, as shown in Figure 15–15.

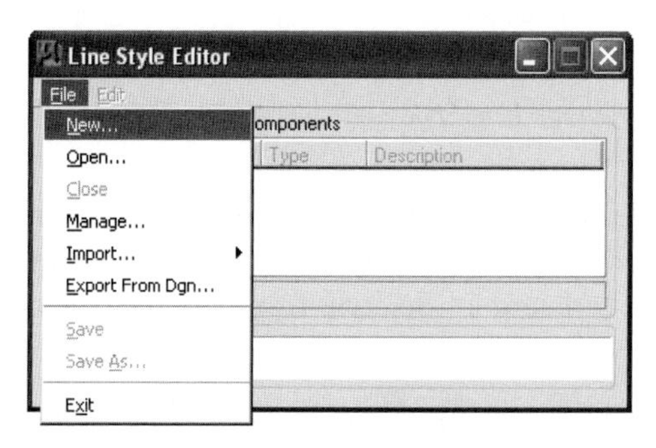

Figure 15–15 *The File Menu in the Line Style Editor settings box*

The **New** option opens the standard File Creation dialog box. Navigate to the folder that is to hold the new library and type a name for the library in the **File**

Name field. Click **Save** to create the library. The file path and name appear on the title bar of the Line Style Editor settings box, and the file's contents (MicroStation's default line styles) are listed in the **Components** area of the box.

The **Open** option opens the standard File Open dialog box. To open an existing library file for editing, navigate to the files location, select the file name, and click **Open**.

Note: *Only one library file at a time can be open in the Line Style Editor settings box.*

The **Close** option closes the file currently open in the Line Style Editor settings box, but leaves the settings box open. The file is immediately closed. There are no additional prompts.

The **Manage** tool Opens the Manage Line Style Definitions dialog box (see Figure 15–16), which contains buttons for renaming and editing styles from the open library If no library file is currently open in the Line Style Editor settings box when **Manage** is selected, MicroStation opens the standard File Open dialog box so that a library file can be selected and opened.

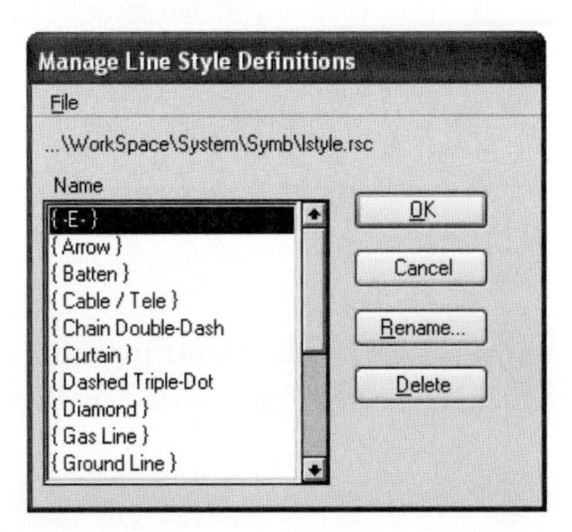

Figure 15–16 *The Manage Line Style Definitions dialog box*

To rename a line style, select the style's name in the **Name** field and click **Rename**. MicroStation opens the Rename Line Style dialog box, in which the new style name can be typed. To delete a line style from the file, select the style's name and click **Delete**.

Click **OK** to commit to the changes made in the Manager Line Style Definitions dialog box or click **Cancel** to discard all changes.

The **Import** tool provides options for importing line styles from an AutoCAD Line Style File (LIN) or a MicroStation Resource File (RSC). Clicking one of the options opens the standard File Open dialog box. Navigate to the location of the

file to be imported, select the file name, and click **Open** to import the selected file. When the selected file opens, MicroStation displays the Select Line Styles to Import dialog box that lists all custom line styles in the selected file. Select the line styles to import, and click **Import** to import the selected styles or **Cancel** to close the dialog box without importing any styles.

The **Export From Dgn** exports all custom line styles in the active design file to an RSC type library file. It opens a standard File Open dialog box from which the file that is to receive the exported styles can be opened or a new file can be created.

The **Save** option saves the changes made to the open library file. The changes do not become part of the design file unless they are saved. Closing the Line Style Editor dialog box without selecting **Save**, discards all changes.

The **Save As** tool opens the standard Save As dialog box. Navigate to the folder where the files is to be saved, type the new file name in the **File Name** field and click **Save**. Alternately, click **Cancel** to close the Save As dialog box without saving the file under a new file name.

The **Exit** option closes the Line Style Editor settings box. If the file contains no unsaved changes, the dialog box immediately closes. If the file contains unsaved changes, an Alert message appears that asks if the file should be saved. Click **Yes** to save the file, **No** to close the dialog box without saving the file, or **Cancel** to return to the dialog box.

The Edit Menu

The **Edit** menu on the Line Style Editor settings box provides options for creating and maintaining line style names and components in the open library file (see Figure 15–17).

Figure 15–17 *The Edit Menu in the Line Style Editor settings box*

The **Edit** menu contains the following options:

Create opens a submenu that provides options for creating new line style names and components. The options are discussed later.

Delete deletes the selected style name from the **Name** area and the associated style component from the **Components** area. To delete a style, select its name and select the **Delete** option. The style is immediately deleted. There are no additional prompts.

Note: *Style names that are associated with internal style components cannot be deleted.*

Duplicate creates a copy of the selected style component in the **Components** area. The copy has the same name and content as the original.

Link links the component selected in the **Components** field to the line style selected in the **Name** field.

Snappable is a check box that controls the ability to snap to the individual components of the Custom line style.

The Create Submenu

Selecting **Create** from the **Edit** menu opens a submenu, as shown in Figure 15–18.

Figure 15–18 *The Create submenu in the Line Style Editor settings box*

The options in the **Create** submenu create line style components and assign line style names to the components.

Name creates a new "Unnamed" line style. The new name is automatically selected and is linked to the component selected in the **Components** list box. Replace the default name by keying-in a new one in the text field below the **Names** list.

Stroke Pattern creates style components that consist of a series of strokes and gaps that are repeated along the length of elements.

Point creates style components that consist of a series of point symbols (such as a shared cell) that are repeated along the length of elements. Point components also contain an association with a stroke pattern component that determines the placement of the point symbol on the element.

Compound creates style components that are combinations of style components. For example, a stroke pattern might be combined with one of the internal line components.

Creating a Stroke Pattern Line Style Component

Selecting **Stroke Pattern** from the Create submenu expands the Line Style Editor settings box to show the settings fields for defining a Stroke Pattern, as shown in Figure 15–19.

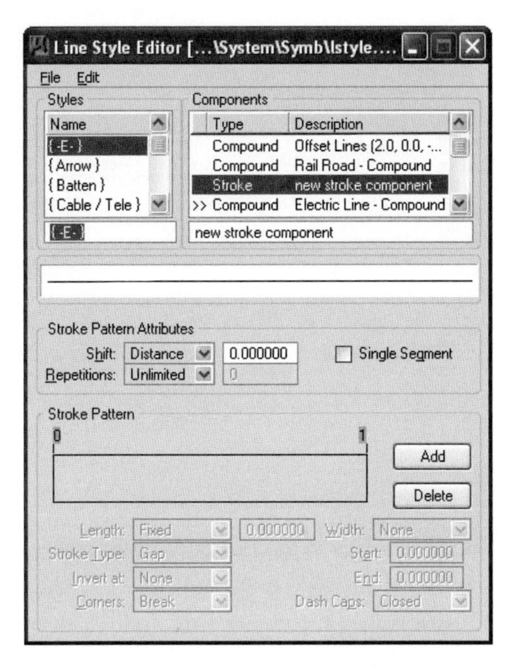

Figure 15–19 *Line Style Editor with Stroke Pattern settings*

Note: *The* ***Create*** *menu options create new components, but it is often easier to find and modify a similar component in one of the library files MicroStation furnishes.*

Below the **Name** and **Components** fields is a field that shows an example of the pattern component as it is defined. Below the picture are fields for defining the component.

The **Stroke Pattern Attributes** area of the settings box contains options that controls placement of the stroke pattern along elements.

- The **Shift** menu and text field provide options for settings the distance by which the stroke pattern is shifted from the starting point of the element. The shift can be a **Distance** from the beginning, a **Fraction** of the pattern length from the beginning, or the pattern can be **Centered** within the length of the element or element segment.
- The **Repetitions** menu and text field set the number of times the stroke pattern is repeated throughout the length of an element or element segment.
- The **Single Segment** check box controls the truncation of the stroke pattern at the end of each element segment. If it is turned ON, the pattern is shifted to end each segment with a complete stroke.

The **Stroke Pattern** area of the settings box contains options for defining the dashes and gaps that make up the stroke pattern. The pattern is a series of dash

and gap strokes, and the options in this area are used to add, delete, and define the strokes.

- The horizontal bar at the top of the area shows the relative length of each stroke in the pattern. A dark background indicates a dash and a clear background indicates a gap. The small blocks above and below the bar are handles that can be dragged to change the length of each stroke. To select a stroke for editing, click its part of the bar. Clicking a part changes the appearance of the stroke in the bar to indicate that it is selected.
- **Add** adds a new gap stroke to the end of the stroke pattern. After it is added, it can be changed from a gap to a dash.
- The **Delete** button removes a selected stroke from the pattern.

The following fields apply to both dash and gap strokes:

- The **Length** menu sets a **Fixed** or **Variable** length for the selected stroke. Enter the stroke length in the text field to the right of the menu. The length uses the working units of the design file. This controls how the pattern is shifted or repeated a fixed number of times.
- The **Stroke Type** menu defines the selected stroke as a **Dash** or **Gap**.
- The **Invert at** menu is used to invert the type of the selected stroke when it is used to place an element. The inversion options are **None**, **Origin**, **End** and **Both**.
- The **Corners** menu controls the behavior of the selected stroke at each element vertex. The stroke can **Break** at the vertex or **Bypass** the vertex. If it bypasses the vertex, the corner may not be sharp.

The following options are only available when the selected **Stroke Type** is **Dash**:

- The **Width** menu controls the width of the selected dash stroke. If the **Width** is **None**, the width of the stroke is determined by the Active Line Width. If the **Width** is **Full**, **Left**, or **Right**, the stroke width is controlled by the values entered in the **Start** and **End** fields. The **Left** option applies the width to the left (or above) the center point of the stroke, the **Right** option applies the width to the right (or bottom) of the center point, and the **Both** option applies the width to both sides of the center point. The **Start** option applies the **Width** to the starting point of the dash stroke. The **End** option applies the **Width** to the ending point of the dash stroke.
- The **Dash Caps** menu sets the shape of each end of the selected dash stroke when the ends have a width greater than zero. If the **None** option is selected, the dash stroke is only a line. If any other cap option is selected, the dash stroke is filled.

Figures 15–20a and 15–20b show the stroke settings for a simple stroke pattern that alternates between a dash with width and a gap. The field just above the

Stroke Pattern Attributes area of the settings box in the figures shows the way the line style appears when used in the design.

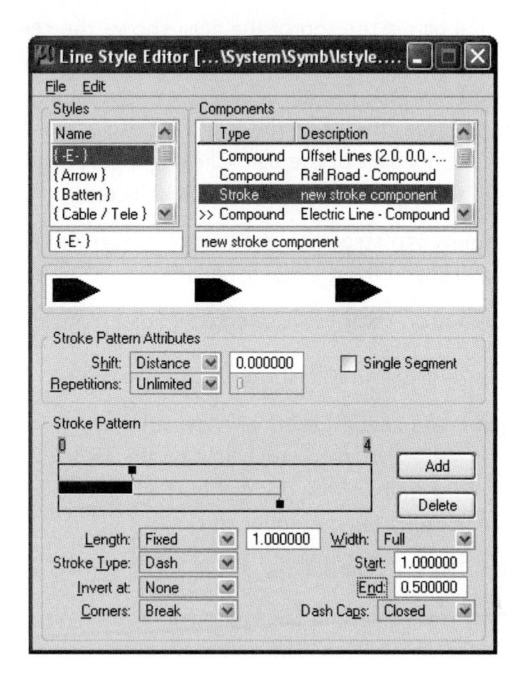

Figure 15–20a *Example of a dash stroke definition*

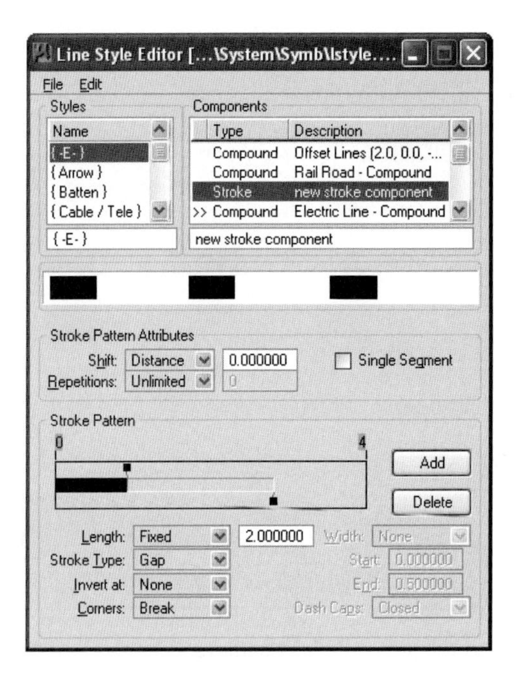

Figure 15–20b *Example of a gap stroke definition*

Creating a Point Line Style Component

Selecting **Point** from the **Create** submenu expands the Line Style Editor settings box to show the setting fields for defining a Point component, as shown in Figure 15–21.

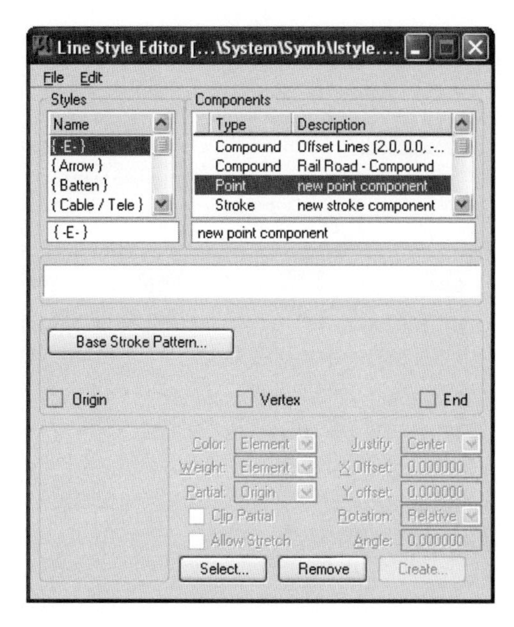

Figure 15–21 *Line Style Editor with Point settings*

To define a Point component, select one or more of the Point Symbols defined in the open style library file. If no Point Symbols are defined for the open file, they must be created. The procedure is presented later in this topic.

A Point component can be created that only places from one to three Point Symbols on each segment of an element or that places Point Symbols on one or more of the strokes in an associated stroke pattern.

- Below the **Name** and **Components** fields is a field that shows an example of the point component as it is defined.
- Below the picture are three check boxes that place selected Point Symbols at the **Origin**, at each **Vertex**, and **End** of elements.

Note: *The appearance of a slight depression of a check box indicates that it is selected. That is easy to miss.*

- Click the **Base Stroke Pattern** button to associate the new Point component with a Stroke Pattern. Selecting this option opens the Base Stroke Pattern dialog box, from which a stroke pattern can be selected. Selecting a pattern displays a bar graph that shows the strokes in the pattern.

The fields below the **Origin**, **Vertex**, and **End** check boxes are used to assign Point Symbols to positions within the Point component. Before using these commands, select the position by clicking the **Origin**, **Vertex**, or **End** check boxes, or by click-

ing one of the strokes in the stroke pattern bar graph. After selecting the position, the following Point Symbol fields become available:

- **Select** opens the Select Point Symbol dialog box. Select a symbol from the box's **Name** field and click **OK** to assign the selected symbol to the selected element position or pattern stroke.
- **Remove** removes a Point Symbol from the selected element location or pattern stroke.
- **Create** creates a new Point Symbol from the contents of a fence in the design file. This button is only available for use when an element position or pattern stroke is selected and a fence is defined in the design file. Here are the steps required to create a new Point Symbol:

1. Draw the symbol in the design file.
2. Place a fence around the elements. The **Fence Mode** setting is ignored by this command and all elements must be inside the fence.
3. Use the Define Cell Origin tool to define an origin for the Point Symbol.
4. Click **Create** to open the Create Point Symbol dialog box.
5. In the dialog box type the new symbol's name in the **Name** field and click **OK**. The symbol is created but not assigned to the Point component.

- The **Color** menu sets the color for the symbol. Choose **Element** to change the symbol color to the color of the element, or **Symbol** to use the color of the symbol.
- The **Weigh** menu sets the line weight to use for the symbol. Choose **Element** to use the weight of the element, or **Symbol** to use the weight of the symbol.
- The **Justify** menu sets the origin point on the **Left**, **Center**, or **Right** end of the selected pattern stroke.
- The **X Offset** and **Y Offset** text fields set the horizontal and vertical offset from the origin point.
- The **Rotation** menu sets the rotation calculation to be **Relative** to the element angle, **Absolute**, or **Adjusted**.
- The **Angle** text field sets the rotation angle value in degrees.

Creating a Compound Line Style Component

Selecting **Compound** from the **Create** submenu expands the Line Style Editor settings box to show the setting fields for defining a Compound component, as shown in Figure 15–22.

Figure 15–22 *Line Style Editor with Compound settings*

A Compound component consists of Stroke Pattern and Point sub-components. The fields in the settings box provide controls to insert and remove sub-components:

- **Sub-Components** lists each sub-component. The field shows the **Offset** from the origin, the **Type** of component, and the component **Description**.
- The **Offset** text field sets the distance, in working units measured perpendicular from the work line, by which the selected component is displayed parallel to the work line. If the **Offset** is zero, the selected component is displayed on the work line.
- **Insert** opens the Select Component dialog box, from which a sub-component can be selected to insert in the compound component selected in the **Components** field.
- **Remove** removes the sub-component selected in the **Sub-Components** field from the compound component selected in the **Components** list box.

WORKSPACES

A *workspace* is a customized drafting environment in which MicroStation can be set up for specific purposes. There is no limit on the number of workspaces. A workspace consists of "components" and "configuration files" for both the user and the project.

MicroStation comes with a set of workspaces for various disciplines, for example, a "civil" workspace. When the civil workspace is active, the files and tools needed for civil engineering design are available by default. Tools and tool boxes unrelated to civil engineering are removed from the interface so that they are out of the way.

SETTING UP THE ACTIVE WORKSPACE

Select the default workspace from the MicroStation Manager. At the bottom, right of the MicroStation Manager dialog box (see Figure 15–23) are menus (**User**, **Project**, and **Interface**) used to change the major components of the workspace before opening a design file.

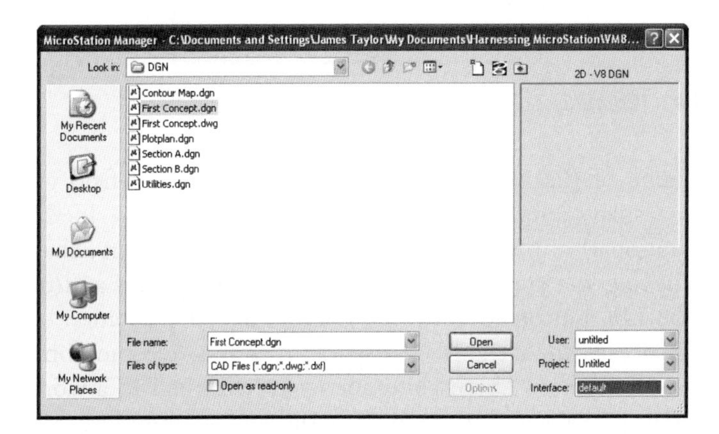

Figure 15–23 *MicroStation Manager*

User Options

The **User** menu sets the default workspace from the available workspaces. Selecting a workspace from the list reconfigures MicroStation to use that workspace's components. Selecting a workspace also resets the **Look In** path to a corresponding folder for loading design files. In addition, MicroStation sets the associated project and user interface.

When MicroStation starts with any workspace as the default workspace, a preference file is created for that workspace, unless one already exists. The settings in the preferences are set either to default settings or to AutoCAD Transition settings, depending on the active user interface component.

To create a new workspace, invoke **New** from the **User** menu. MicroStation displays the Create User Configuration File dialog box, similar to Figure 15–24.

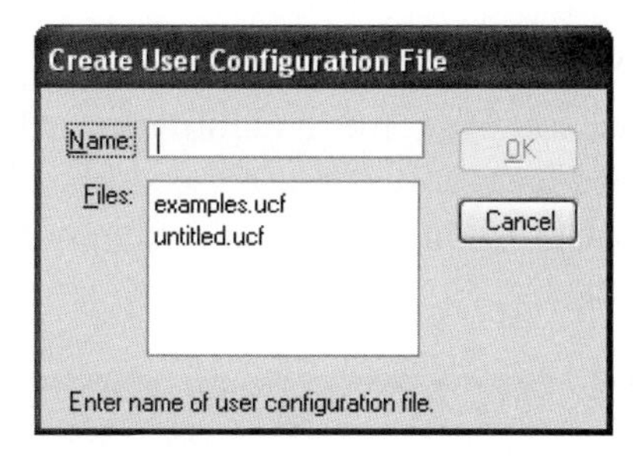

Figure 15–24 *Create User Configuration File dialog box*

Key-in the name of the new workspace in the **Name** text field and click **OK**. MicroStation displays another dialog box, as shown in Figure 15–25. Key-in the description (optional) in the **Description** text field. If necessary, change the components by clicking the appropriate **Select** button for **Project** and/or **User Interface**. Click the **OK** button to close the dialog box. MicroStation sets the newly created workspace as the default workspace. A workspace can contain only one project and one interface. These components are attached to a workspace, so, to use two different projects with the same interface or two interfaces for one project, you need to make additional workspaces.

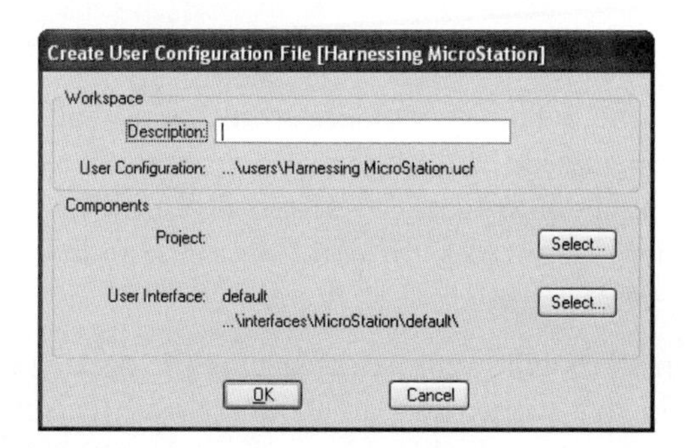

Figure 15–25 *Create User Configuration File dialog box*

Project Option Menu

The selection of the project sets the location and names of data files associated with a specific design project. Change the selected project from the **Project** menu. Create a new Project from the **Project** menu, similar to creating a workspace.

Interface Option Menu

The selection of the interface sets a specific look and feel for MicroStation's tools and on-screen operation. Change the selection of the interface from the **Interface** menu. MicroStation comes with discipline-specific interfaces—civil engineering, architecture, mechanical engineering, drafting, and mapping—and it also has in-

terfaces for previous versions of MicroStation (V. 4 and V. 5) and AutoCAD users.

To create a new interface, invoke **New** from the **Interface** menu. MicroStation displays the Create User Interface dialog box, as shown in Figure 15–26.

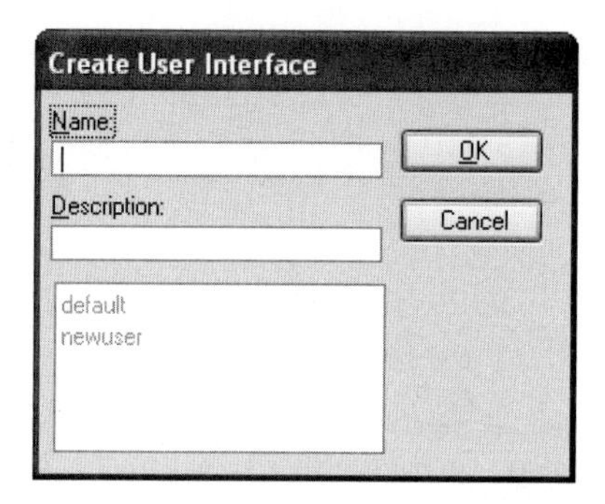

Figure 15–26 *Create User Interface dialog box*

Key-in the name of the new user interface in the **Name** text field and a description in the **Description** text field (optional). Click **OK** to close the dialog box. MicroStation creates an interface directory under the *Bentley\Workspace\interfaces* directory and sets the newly created user interface as the default user interface. The new interface takes the default interface as its starting point. Any changes made while using this new interface are written only to the new interface.

SETTING USER PREFERENCES

Preferences are settings that control the way MicroStation operates and the way its tool boxes appear on the screen. For example, they affect how MicroStation uses memory on a user's system, how windows are displayed, and how reference files are attached by default. The settings can be changed to suit specific needs. The user preferences are saved under the same name as the workspace, with the file extension *.ucf.* To set the user preferences, invoke Preferences from:

Menu	Workspace > Preferences

MicroStation displays the Preferences dialog box, as shown in Figure 15–27. MicroStation displays the name of the file under which preference settings are saved as part of the title bar.

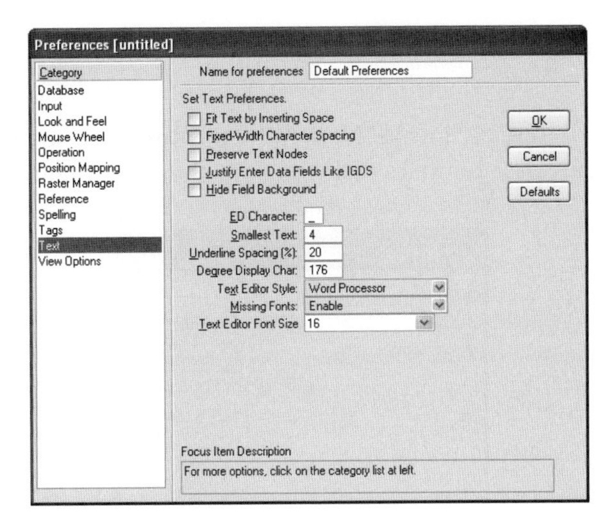

Figure 15–27 *Preferences dialog box*

Type a name for the current set of preferences in the **Name for preferences** text field at the top of the settings box.

User preferences are divided into areas. The **Category** list box lists all the available categories. Selecting a category causes the appropriate controls to appear to the right of the **Category** list. Each category controls a specific aspect of MicroStation's appearance or operation.

After making the required changes to the preferences settings, click **OK** to save the settings and close the Preferences dialog box or click **Cancel** to close the dialog box without saving any of the changes. Click **Default** to return all settings to the default values.

Note: *Some preferences changes do not go into effect until MicroStation is closed and re-opened.*

Note: *For descriptions of all user preferences, select* ***Contents*** *from the* ***Help*** *menu. On the On-line Help window, click the* ***Contents*** *tab and then select* ***Getting Started***, ***Fundamentals***, *and* ***User Preferences***. *Expand* ***Categories*** *to view a table listing all user preferences*

WORKING WITH CONFIGURATION VARIABLES

MicroStation provides configuration variables that control the way MicroStation works. For example, the library files that appear in the Open Line Style Library dialog box are in folders specified by the MS_SYMBRSCR configuration variable.

Create, edit, and delete configuration variables from the Configuration dialog box. Invoke the dialog box from:

Menu	Workspace > Configuration

MicroStation displays the Configuration dialog box as shown in Figure 15–28.

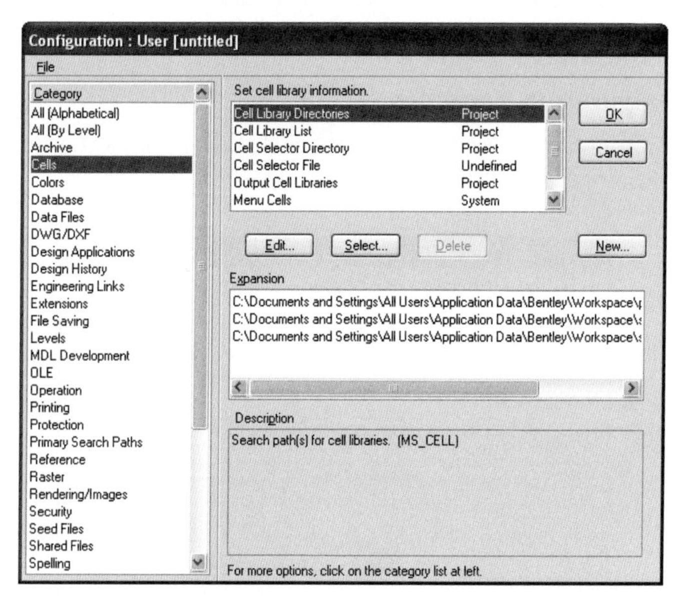

Figure 15–28 *Configuration dialog box*

The Configuration Dialog Box

The **Category** field lists the names of related groups of configuration variables. Click a category name to display the group's configuration variables. For example, click **Cells** to displays all variables related to cells. MicroStation lists all of the selected category's variables in the **View/modify all configuration variables** list box.

Select a configuration variable in the list box to display additional information about the variable in the **Expansion** list box and a description of the variable in the **Description** list box. For example, with the **Cells** category selected, select the **Cell Library Directories** variable and the **Expansion** field displays the paths that MicroStation searches for cell libraries when the open cell library dialog box is opened.

Control buttons on the dialog box provide variable maintenance options.

- **Edit** is used to change a variable's action.
- **Select** is used to add items to a variable's expansion.
- **Delete** deletes variables.
- **New** creates new variables.

If a control button is not available, that maintenance action is not allowed on the variable.

Note: *The content of the Configuration dialog box is different from that described above for the **Design Applications Category**.*

Editing Configuration Variables

To modify a configuration variable's action, click **Edit** to open the Edit Configuration Variable dialog box, as shown in Figure 15–29.

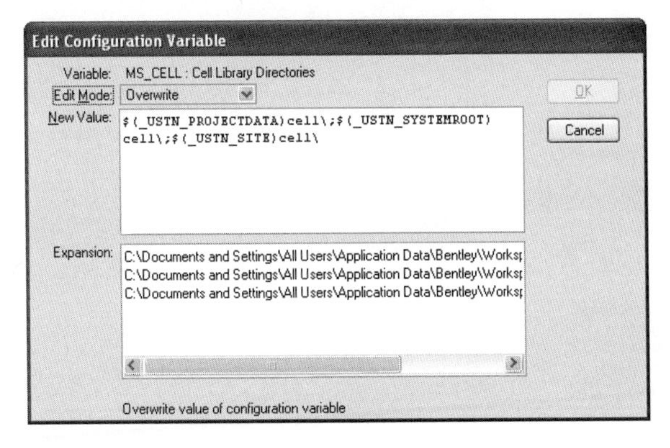

Figure 15–29 *Edit Configuration Variable dialog box*

The variable's name and descriptive name (if there is one) are displayed at the top of the dialog box. Below the names is the **Edit Mode** menu that contains options to **Overwrite** the current variable's action, **Append** new actions after the existing action, or **Prepend** new actions before the existing action. The **New Value** field provides a place to edit the variable's action, and the **Expansion** field shows the results of the variable's action.

The action for most, but not all, variables is a directory path definition that includes other variables and specific folder and file names. For example, the Cell Library Directories action shown in Figure 15–29, finds all cell library files in three paths.

The paths are formed from the contents of two other configuration variables (_USTN_SYSTEMROOT and _USTN_SYSTEMROOT), a specific folder (symb), and a specific type of file (*.*rsc*). The asterisk (*) is a wildcard that tells MicroStation to include all file names with the *.rsc* extension.

When **Overwrite** is selected, the variable's current action appears in the **New Value** field and the action can be edited or replaced. If **Append** or **Prepend** is selected, the **New Value** field is empty and only new information can be added. To view the action again after appending or prepending additional actions, select **Edit Mode** >

Overwrite. When editing is complete, click **OK** to save the changes and close the dialog box or click the **Cancel** button to close the dialog box without saving the changes.

Note: *Describing all possible variable action sequences is beyond the scope of this book. For more information on editing and creating configuration variables, consult the MicroStation on-line help.*

Selecting Configuration Variable Items

Clicking **Select** opens a dialog box for selecting an item to add to a configuration variable's expansion. The dialog box layout varies depending on what type of variable is selected. For example, if the action of the selected variable is to find files, the standard File Open dialog box appears, as shown in Figure 15–30. The dialog box is then used to add a file to the variable's expansion.

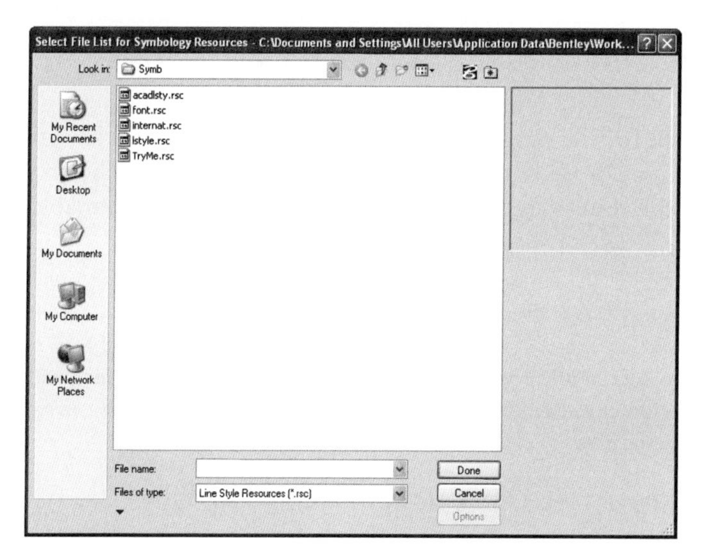

Figure 15–30 *Select File List dialog box*

Deleting Configuration Variables

To delete a configuration variable, select it and click **Delete**. The selected variable is immediately deleted but remains in the variables list with a different color background until **OK** is clicked to close the Configuration dialog box and save the changes.

Note: *Only user-defined configuration variables can be deleted.*

Creating New Configuration Variables

To create a new configuration variable click **New**, which opens the New Configuration Variable dialog box, as shown in Figure 15–31.

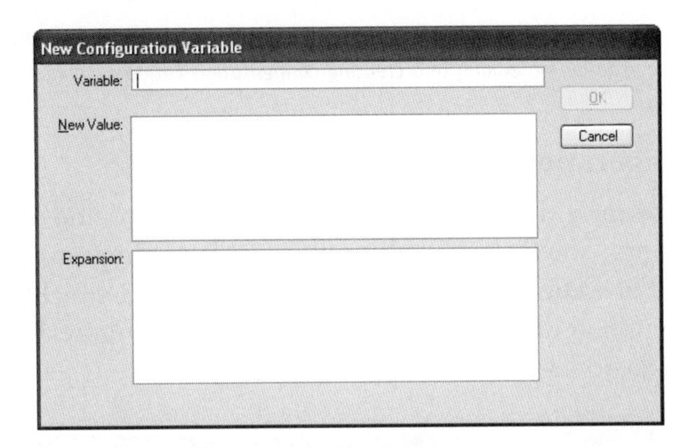

Figure 15–31 *New Configuration Variable dialog box*

Key-in a name for the new variable in the **Variable** field and key-in the variable's action in the **New Value** field. The results of your action are shown in the Expansion field. If the results are correct, click **OK** to save the new variable and close the dialog box.

Note: *For descriptions of all user configuration variables, select* ***Contents*** *from the* ***Help*** *menu. On the on-line help window, click the* ***Contents*** *tab and then select* ***Setting Up Projects, Workspaces, Workspace Configuration,*** *and* ***Working with Configuration Variables.*** *Expand* ***User Configuration Variables*** *to view a table listing all user configuration variables.*

CUSTOMIZING THE USER INTERFACE

MicroStation provides options for creating an interface with tools and tool boxes specifically designed to meet the work flow needs of a project. Custom tools can be modified, copied or created specifically for the design project. Related custom tool boxes can be grouped into tasks to create a task-based interface.

Custom tools, tool boxes, and tasks are managed from the Customize dialog box and are stored in DGN libraries. The use of DGN libraries allows many users to access the customized features. Open the Customize settings box from:

Menu	Workspace > Customize

MicroStation displays the Customize settings box, as shown in Figure 15–32.

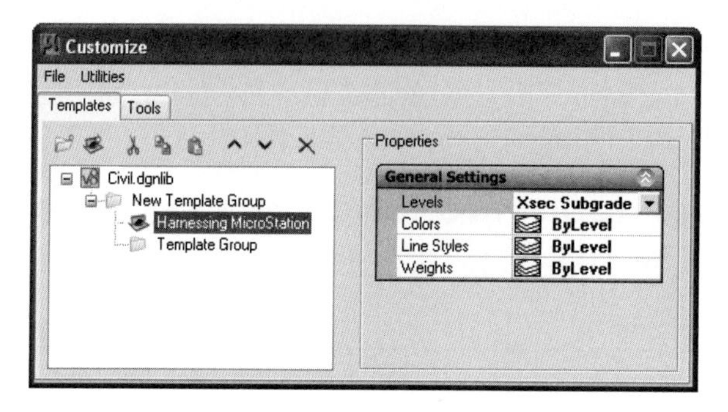

Figure 15–32 *Customize settings box*

***Note** : If the open DGN file is not a configured DGN library, the features on the Tools tab in the Customize dialog box are not available while that DGN file is open. The MS_GUIDGNLIBLIST configuration variable (set by default to point to MS_DGNLIBLIST and to the DGN libraries in the active interface component's folder) and the _USTN_SYSTEM_GUIDGNLIBLIST configuration variable specify which files can be used for task, tool box, tool, icon, and menu customizations.*

***Note**: For more information on customizing the user interface, see the MicroStation online help.*

FUNCTION KEYS

Personal computer keyboards have a set of function keys at the top of the keyboard. MicroStation has a utility that assigns actions to function keys F1 through F12. Tap a function key and the assigned action executes.

There are only 12 function keys on the keyboard, but MicroStation allows up to 96 function key assignments by combining the function keys with SHIFT, ALT, and/or ctrl. For example, F1 can have the following combinations:

- F1
- SHIFT + F1
- ALT + F1
- CTRL + F1
- SHIFT + ALT + F1
- SHIFT + CTRL + F1
- ALT + CTRL + F1
- SHIFT + ALT + CTRL + F1

Function key assignments are stored in an ASCII file, and by default the extension for the function key file is *.mnu.* This file follows the same concept as the cell li-

brary file. Once created, the function key file can be attached to any design file to activate the function keys. The default function key menu file is *funckey.mnu* and is stored in the */Bentley/Worskpace/interfaces/fkeys/* path.

CREATING AND MODIFYING FUNCTION KEY DEFINITIONS

To create and modify function key definitions, open the Function Keys dialog box from:

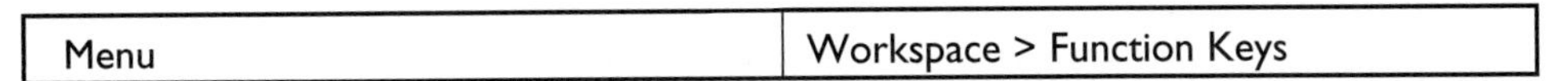

Menu	Workspace > Function Keys

MicroStation displays the Function Keys dialog box, as shown in Figure 15–33.

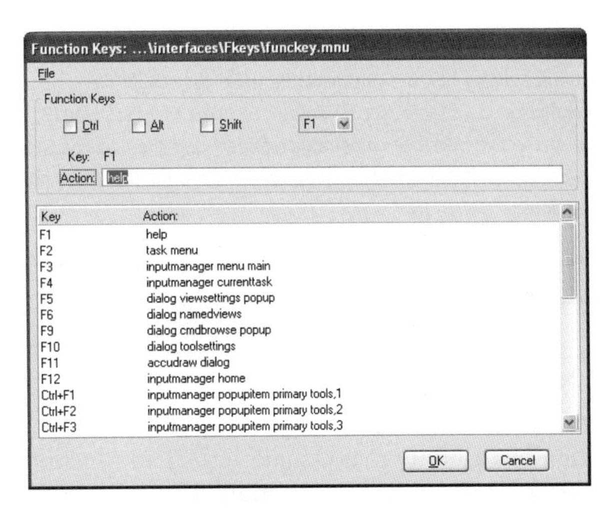

Figure 15–33 *Function Keys dialog box*

The title bar identifies the file name of the **Function Key** menu. The **Key/Action** list box lists all function key combinations that have an action assigned and the action.

The **File** menu on the Function Keys dialog box provides the following options:

- **Open** opens the standard File Open dialog box for opening a function key definitions file.
- **Save As** opens the standard Save As dialog box for saving the open function key definitions file using a different folder path and file name.
- **Save** saves the currently open function key definitions file.

To edit, delete, or add a function key action, click the function key in the **Key/Action** list box to place the action in the **Action** field and do the following:

- Change the action by editing the action text in the **Action** field

- Delete the action by removing the action text from the **Action** field.
- Add a new action by doing the following:
  - Use the key check boxes (**Ctrl**, **Alt**, and **Shift**) and the **Function Key** menu to select a key combination to use for the new action. For example, to assign an action to ALT+F8, turn the ALT check box ON and select F8 from the **Function Key** menu.
  - Type the action text in the **Action** Field.
  - Complete the edit, delete, or add by clicking in another field, tapping TAB, or tapping ENTER.
  - Click **OK** to close the dialog box and save all changes to the function key definitions file or click **Cancel** to close the dialog box without saving the changes.

***Note:** Selecting **Save** from the **File** menu saves the changes to the function definitions file and leaves the dialog box open. **Cancel** only throws out changes made after the last **Save**.*

FONTS

When MicroStation opens a design file, it also opens a font resource file that contains font definitions. The default resource file is named *FONT.RSC*. The Font Installer dialog box has options for importing fonts from different sources into the MicroStation font library, and to rename and renumber fonts. To open the Font Installer dialog box, invoke Install Fonts from:

Menu	Utilities > Install Fonts

MicroStation displays the Font Installer dialog box, as shown in Figure 15–34.

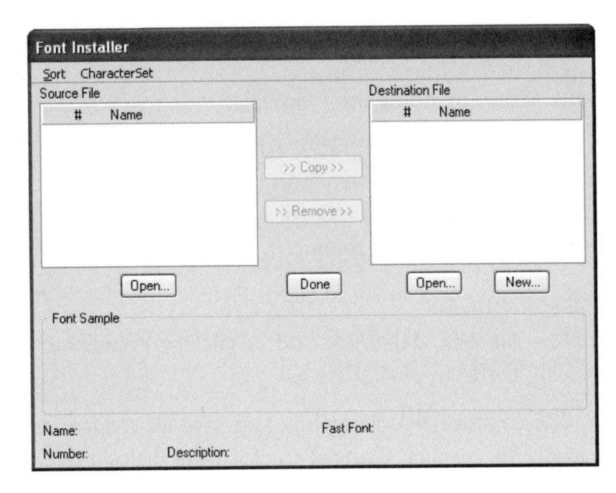

Figure 15–34 *Font Installer dialog box*

SELECTING THE SOURCE FONT FILES

To select the source fonts, click **Open,** located below the **Source File** field. MicroStation displays the Open Source Font Files dialog box, as shown in Figure 15–35.

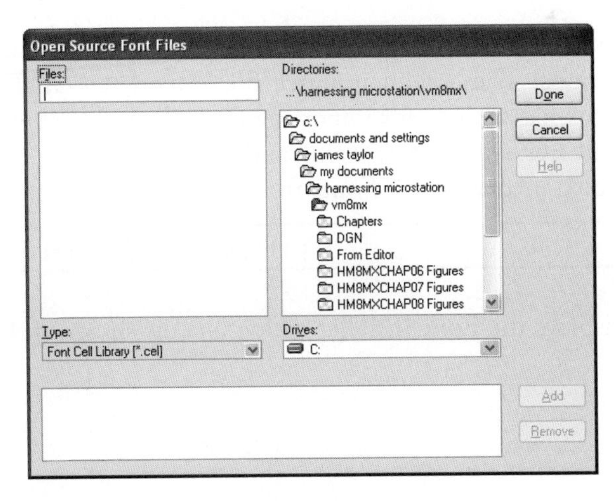

Figure 15–35 *Open Source Font Files dialog box*

The **Type** Menu provides options to set the file type to one of the following:

- **Font Cell Library**—MicroStation font cell library, a standard cell library that contains cells that define the characters and symbols in a traditional MicroStation font and the font's attributes (*.*cel*)
- **Font Library**—MicroStation Version 5 font library (*.*rsc*)
- **uSTN V4/IGDS Fontlib**—Version 4.1 or earlier (or IGDS) font library

Select the location of the font file to open from the **Directories** field, select the font file from the **Files** field, and click **Add** to open the selected file. Multiple files can be selected. To unselect selected files, click **Remove**. After selecting the files to open, click **Done** to open the files and close the dialog box. The **Source File** field in the Font Installer dialog box lists all the fonts contained in the opened source files.

SELECTING THE DESTINATION FONTS FILE

To select the destination fonts file in which to put fonts from the source files, click **Open** located beneath the **# Name** list box on the right side of the dialog box (see Figure 15–34). MicroStation opens the standard File Open dialog box. Use the **Look in** field to find and select the folder containing the font library file, and click **Open**. The **# Name** destination field in the Font Installer dialog box lists the fonts in the selected library.

IMPORTING FONTS

To import a font from the source to the destination file, select the font in the **Source File** field and click **Copy**. MicroStation copies the selected font onto the destination **# Name** field on the right side of the dialog box. To remove a font in the **# Name** destination field, select the font to be removed from the list and click **Remove**. MicroStation removes the selected font from the **Destination File** field.

To change a font's name, number, or description, select the font in the **# Name** destination field to display the font's information at the bottom of the dialog box. Make the necessary changes in the appropriate text fields. Make the selected font the default font for text tools by turning the **Default** check box on. Click **Done** to close the Font Installer dialog box.

MICROSTATION ENVIRONMENT PACKAGES

The Packaging utility provides options to package all the design files and other associated files in a workspace for sharing with users on other computers and to create a backup copy. The packaged files and operating environment can be applied to the software on another computer so that it has the same work environment and design files as the source computer. Invoke the Packager from:

Menu	Utilities > Packager

MicroStation opens the Create Package wizard, as shown in Figure 15–36.

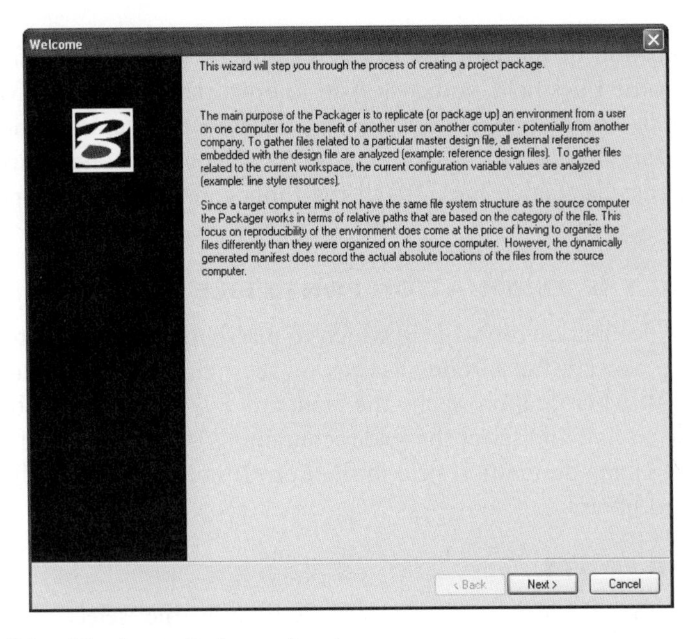

Figure 15–36 *The Create Package wizard*

The wizard provides a guide for the process of creating the package that contains the folders and files that make up the MicroStation work environment. The tool compresses the packaged files using a standard compressing format and places them in a "zip" file that has the PZIP extension. MicroStation also supplies an extractor tool for uncompressing and extracting the files on the target computer. Other commercial extraction applications can also uncompress and extract the files. Click **Next** to move through the packaging steps or click **Back** to back up to a previous step. To create a package, follow the instructions on each step of the wizard.

The person receiving the PZIP file double-clicks the file in Windows Explorer to open the Bentley Package Extractor dialog box, as shown in Figure 15–37.

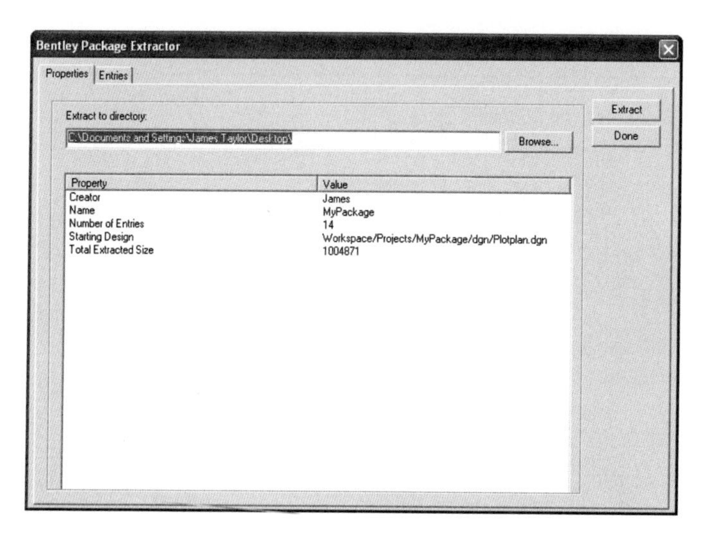

Figure 15–37 *The Bentley Package Extractor dialog box*

To extract the files and create the workspace, click **Extract**. MicroStation displays the PZIP Extractor alert box as shown in Figure 15–38.

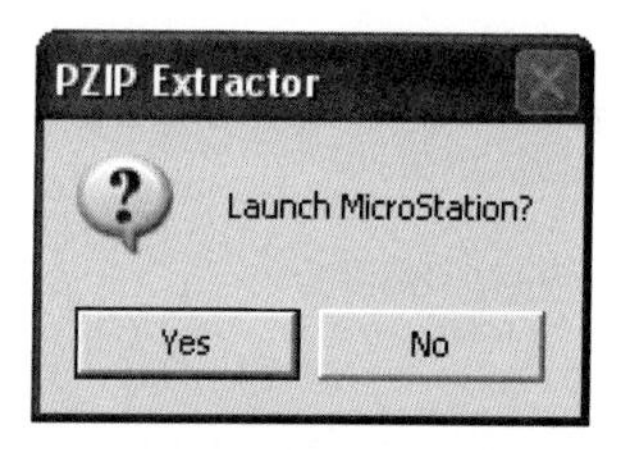

Figure 15–38 *The PZIP Extractor alert box*

Click **Yes** to extract and install the package, or click **No** to return to the Bentley Package Extractor dialog box without extracting the package. If **Yes** is clicked, MicroStation closes the Bentley Package Extractor dialog box after extracting the package.

PROJECT EXPLORER AND LINK SETS

The Project Explorer dialog box has options for managing project data by grouping data such as: design and sheet models and the files in which the models are stored; saved views and; references and documentation stored in other file formats such as Microsoft Word. Data is grouped by creating link sets that contains links to the various project data. The Link Sets dialog box contains options to create Link Sets that can be stored in design files and design library files.

Maintaining Links

Invoke the Project Explorer dialog box from:

Menu	Utilities > Packager

MicroStation opens the Project Explorer dialog box, as shown in Figure 15–39.

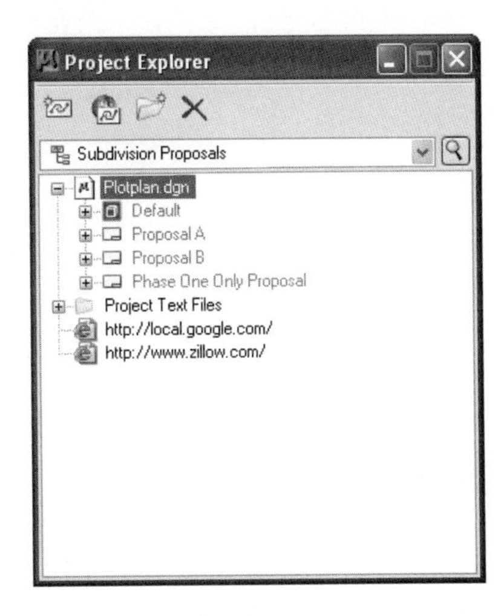

Figure 15–39 *The Project Explorer dialog box*

Select a link set from the menu located directly below the tool bar on the Project Explorer dialog box to display the links currently defined in the link set.

The dialog box's toolbar contains tools for adding and removing links:

- New **Link** adds new links to the selected link set. Click **New Link** to open the standard File Open dialog box for selecting a file. After a file is selected, MicroStation displays the Create Links dialog box, as shown in Figure 15–40. Select the file and all of its models by clicking the file name or select individual models. Click **OK** to add the selected links to the link set or click **Cancel** to close the Create Links dialog box without adding any new links.

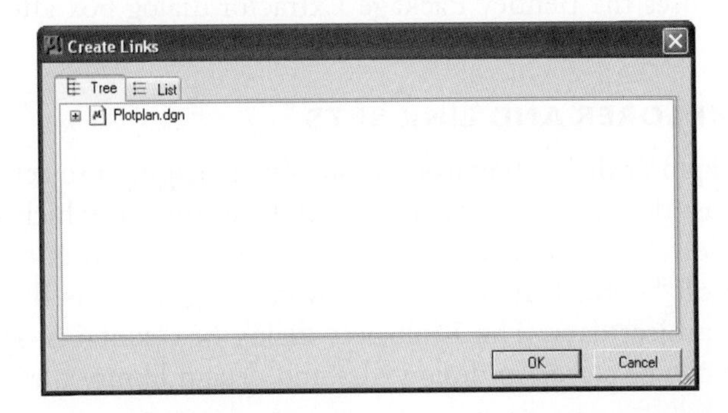

Figure 15–40 *The Create Links dialog box*

- **New URL** adds Uniform Resource Locator (URL) Internet addresses to the selected link set. Click the tool to open the Create URL Link dialog box, as shown in Figure 15–41. Type the URL address in the menu field and click **OK** to add the URL to the link set, or click **Cancel** to close the dialog box without adding a new URL.

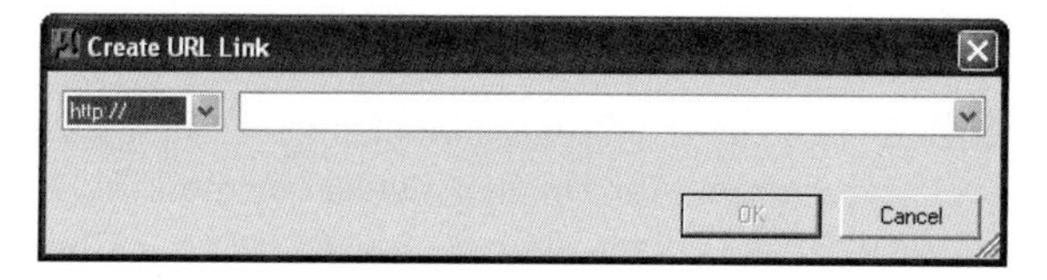

Figure 15–41 *The Create URL Link dialog box*

- **New Folder** adds folders to the selected link set. Click **New Folder** to insert a new folder in the link set and open the folder's name for editing. Type a name for the folder and click **Enter** to complete adding the folder.
- **Delete Folder or Link** deletes the selected link, URL, or folder.

MAINTAINING LINK SETS

To create a new link set, click the magnifying glass tool icon on the right side of the Project Explorer dialog box. Clicking the tool opens the Link Sets dialog box, as shown in Figure 15–42.

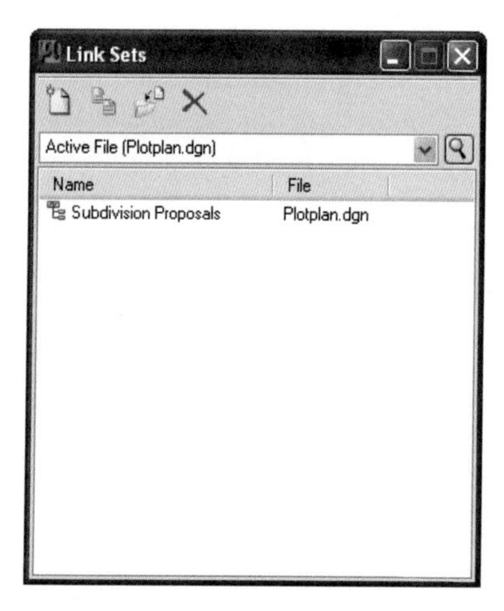

Figure 15–42 *The Link Sets dialog box*

Link Sets are created in design files and design library files. To select the design file or library, open the **File Selection** menu located directly below the dialog box's tool bar. The menu provides options to select the **Configured Libraries (MS_DGNLIBLIST)**, the **Active Design File,** or a **Selected File**. Clicking **Selected**

File opens the standard File Open dialog box from which a design file or design library file can be selected.

Note: *For more information on configured libraries, see the "Working with Configuration Variables" topic earlier in this chapter.*

The dialog box's toolbar provides tools for maintaining link sets in the selected file:

- **New Link Set** opens the Create New Link Set dialog box, as shown in Figure 15–43. Type a name for the new link set in the **Link Set Name** text field, and click **OK** to create the new link set or click **Cancel** to close the dialog box without creating a new link set. The name of the new link set appears on the Link Sets dialog box.

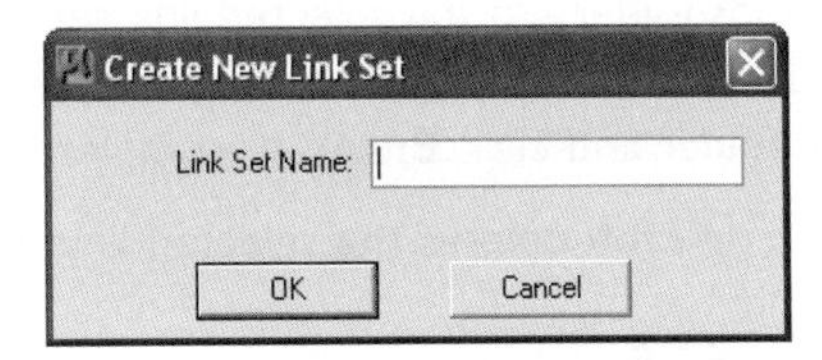

Figure 15–43 *The Create New Link Set dialog box*

- **Copy Link Set** creates a copy of the selected link set and opens the copy's name for editing. Type a name for the link set and tap ENTER to complete renaming it.
- **Import Link Set** opens the standard File Open dialog box. When a file is selected, the File Open dialog box closes and the Import Link Sets dialog box opens, as shown in Figure 15–44. The dialog box displays the names of all link sets in the selected file. Select the link sets to import and click **OK** to import them into the file selected in the Link Sets dialog box, or click **Cancel** to close the Import Link Sets dialog box with out importing any link sets.

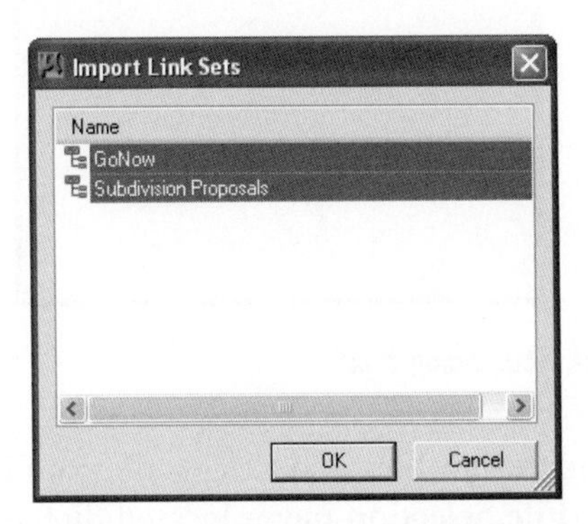

Figure 15–44 *The Import Link Sets dialog box*

- **Delete Link Set** deletes the selected link sets from the selected design file.

ASSOCIATE FILES

The Associate Files dialog box associates file extensions with file types. For example, if the extension "myp" is associated with the design file type, design files with the "myp" open in MicroStation. Open the Associate Files dialog box from:

Menu	File > Associate

MicroStation opens the Associate Files dialog box, as shown in Figure 15–45.

THE MICROSTATION DRAG/DROP TAB

The **MicroStation Drag/Drop** tab customizes the action taken by MicroStation when a file is dropped onto MicroStation from the desktop. Figure 15–45 shows the dialog box with the tab displayed.

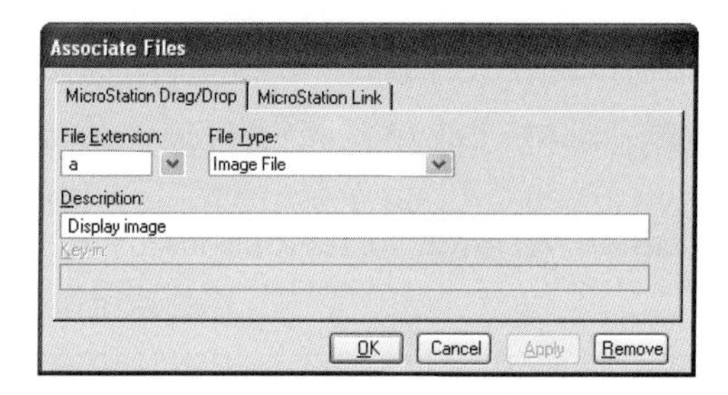

Figure 15–45 *The Associate Files dialog box with the MicroStation Drag/Drop tab selected*

To create an association:

1. Select an extension from the **File Extension** menu or type an extension in the associated field.
2. Select a type from the **File Type** menu.
3. Optionally, type a description of the association in the **Description** text field.
4. Click one of the command buttons to select the action to take for the new association:
   - **OK** creates the association and closes the dialog box.
   - **Cancel** closes the dialog box without creating the current association unless it was previously applied by clicking **Apply**.
   - **Apply** creates the association and leaves the dialog box open.

- **Remove** changes the file type for the selected extension to "Remove" but the association is not actually removed until **OK** or **Apply** is clicked.

The **Key-in** field is only available for use when the extension is associated with the **Custom** File Type. If it is available a command can be typed in the field.

THE MICROSTATION LINK TAB

The **MicroStation Link** tab customizes certain actions that occur when hypertext links are referenced from within the context of a MicroStation links-initiated Web browser. Figure 15–46 shows the dialog box with the tab displayed.

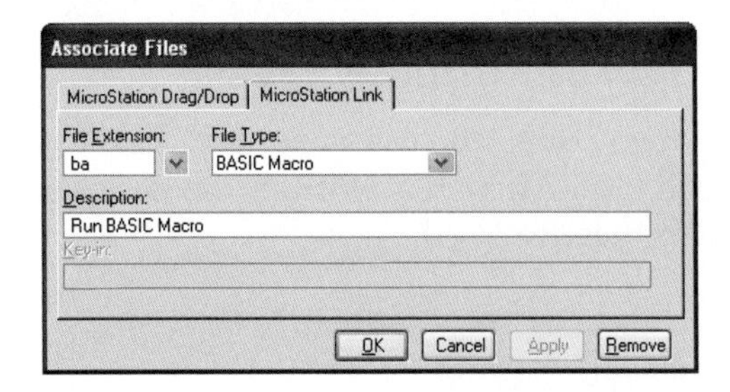

Figure 15–46 *The Associate Files dialog box with the MicroStation Link tab selected*

The options on the tab are identical to the options on the **MicroStation Drag/Drop** tab.

LAUNCHING THE WEB BROWSER AND OPENING FILES REMOTELY

Web browsers are applications designed to provide the user with direct, easy access to the diverse content of the World Wide Web. Highly intuitive in design and function, Web browsers provide a clean, easily navigable window through which to scan pages on the Web. Customizable to suit the user's aesthetics and interests, browsers typically support a number of secondary functions including sending and receiving email, monitoring newsgroups, or downloading files directly from Web sites.

MicroStation browsing tools provide everything users need for browsing all of the information on the Web. To open the Web browser, invoke the MicroStation Link from:

Menu	Utilities > Connect Web Browser
E-Links tool box	Select the Connect to Browser tool (see Figure 15–47).

Figure 15–47 *Invoking the Connect to Browser tool from the E-Links tool box*

MicroStation opens the default Web browser, if not open already, and connects to the Internet. The choice of browsers is defined with the variable MS_USEEXTERNALBROWSER. You can set the variable to open an external browser (such as Netscape Navigator or Microsoft Explorer) instead of the internal browser, if you prefer. Figure 15–48 shows an example of connecting to the Internet through the Internet Explorer Web browser.

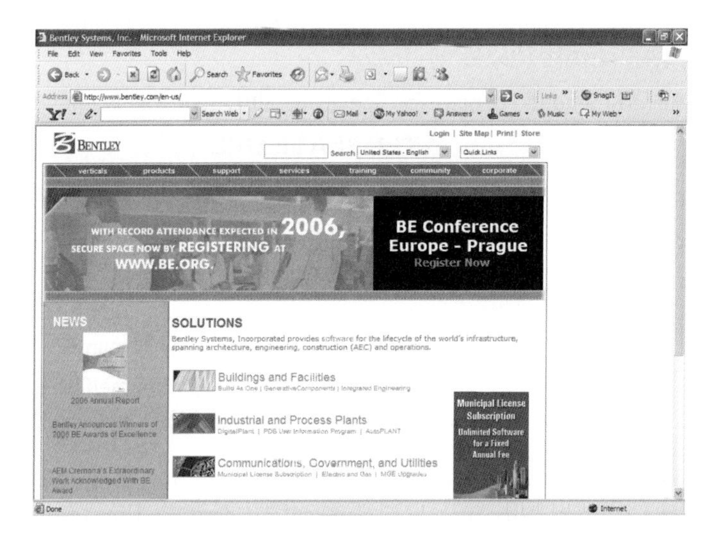

Figure 15–48 *Accessing the Internet through the Internet Explorer Web browser*

Specify the URL (short for "uniform resource locator") in the **Location** edit field in the browser. The URL is the Web site address, which usually follows the format http://www.bentley.com.

Note: *The "http://" prefix is not required. Most of today's Web browsers automatically add in this routing prefix, which saves you a few keystrokes. URLs can access several different kinds of resources—such as Web sites, e-mail, news groups—but always take on the following format:*

Scheme://netloc

The scheme accesses the specific resource on the Internet, including those listed in Table 15–2.

Table 15–2 URL Prefix Meanings

SCHEME PREFIX	MEANING
File://	Files on your computer's hard drive or local network
ftp://	File transfer protocol (downloading files)
http://	Hypertext transfer protocol (Web sites)
mailto://	Electronic mail (e-mail)
news://	Usenet news (news groups)
telnet://	Telnet protocol
gopher://	Gopher protocol

The "://" characters indicate a network address. Table 15–3 lists the formats recommended for specifying URL-style file names in the Location: edit field of the Browser.

Table 15–3 Recommended Formats for Specifying URL-Style File Names

DRAWING LOCATION	TEMPLATE URL
Web site	http://servername/pathname/filename
FTP site	ftp://servername/pathname/filename
Local file	file:///drive:/pathname/filename File:///drive /pathname/filename File://\\localPC\pathname\filename File:////localPC/pathname/filename
Network file	file://localhost/drive:/pathname/filename File://localhost/drive /pathname/filename

Servername is the name of the server, such as www.bentley.com. The *pathname* is the name of the subdirectory or folder name. The *drive:* is the drive letter, such as C: or D:. A *local file* is a file located on your computer. The *localhost* is the name of the network host computer.

MicroStation allows you to select a URL as a design file location instead of a specific local design file. You can also open files using URLs to open remote settings files, archives, reference files, or cell libraries. Downloaded files from a URL are stored in a directory specified by the configuration variable MS_WEBFILES.

MicroStation allows you to close the Web browser by selecting the same tool that opened the Web browser, whose icon has slightly changed and whose tool tip now says **Disconnect from Browser**. If you had originally opened the browser through the E-Links tool box, selecting **Disconnect from Browser** closes the browser in addition to disconnecting it from MicroStation. If you had opened the browser from outside MicroStation and then connected to it, selecting **Disconnect from Browser** only disconnects MicroStation from the browser without closing the browser.

REMOTE OPENING OF A FILE

To open a design file located at a remote server by selecting a URL, open the WEBLIB.MSL MDL application located in the *\Program Files\Bentley\MicroStation\mdlsys\asneeded* folder. Invoke the Open URL tool from:

Key-in window	WEBLIB GETURL

MicroStation displays a Select Remote File dialog box, similar to Figure 15–49.

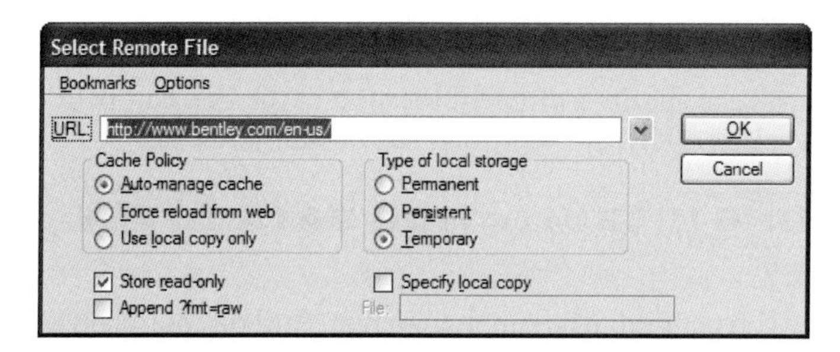

Figure 15–49 *Select Remote File dialog box*

Specify the URL in the edit field. Following are the available options in the dialog box.

Cache Policy

The Cache Policy section of the dialog box allows you to select whether the design files from selected URLs are downloaded or if a previously downloaded local copy is used. Select one of the three following options:

- **Auto-manage cache**—Downloads the file from the selected URL only if it is newer than the local version.
- **Force reload from web**—Reloads from the Web selection and automatically downloads the file from the selected URL.
- **Use local copy only**—Uses the local copy and no Internet request is sent.

Type of Local Storage

The Type of Local Storage section of the dialog box allows you to select how the downloaded file is stored. Select one of the three following options:

- **Permanent**—Stores the file that is downloaded and it is not deleted.
- **Persistent**—Deletes the file only when it exceeds the size requirements. By default, the total permitted size of local copies of remote files is 25MB.
- **Temporary**—Deletes the file at the end of the current session.

Store Read Only

The Store Read Only selection stores downloaded files as read-only. This is a useful reminder that the remote file is not being changed and local modifications could lead to confusion.

Append ?fmt=raw

The Append selection appends the string "?fmt=raw" to the end of the selected URL that is submitted for download. This is required only if the selected site is running ModelServer Publisher.

Specify Local Copy

If necessary, you can specify a different location in the **File** edit field for downloaded files from selected URLs rather than the default location.

PUBLISHING MICROSTATION DATA TO THE INTERNET

The HTML (HyperText Markup Language) Author tool allows you to create HTML files that can be viewed via MicroStation Link or any external browser (such as Netscape Navigator or Internet Explorer). HTML files can be created from a Design File Saved View, Design File Snapshot, Cell Library, or BASIC Macros.

CREATING HTML FILES FROM DESIGN FILE SAVED VIEWS

The Design File Saved Views option creates an HTML Web page with the selected view of the design file. To create an HTML file from a design file saved view, first open the HTML Author dialog box from:

Drop-down menu	Utilities > HTML Author

MicroStation displays the HTML Author dialog box, similar to Figure 15–50.

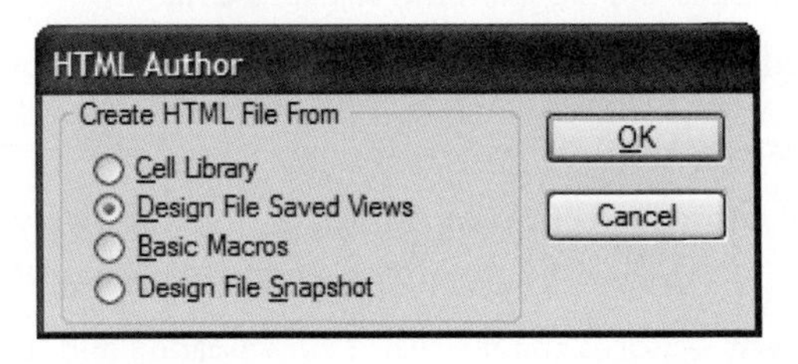

Figure 15–50 *HTML Author dialog box*

Select the **Design File Saved Views** radio button from the Create HTML File From section of the dialog box and click the **OK** button to close the HTML dialog box. The Select Design File dialog box opens. Select the design file from the

appropriate directory and click the **OK** button. MicroStation opens the Design File Walkthrough dialog box, similar to Figure 15–51.

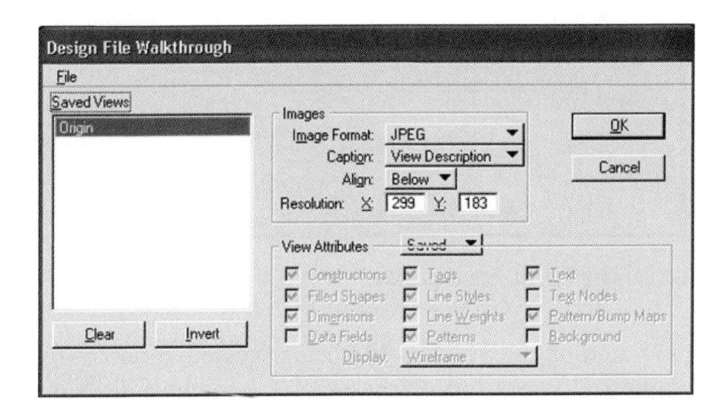

Figure 15–51 *Design File Walkthrough dialog box*

MicroStation lists all the saved views in the design file in the **Saved Views** list box. You can select any or all of the saved views to be included in the HTML page.

The Images section of the dialog box sets controls for image display. The **Caption** option menu controls whether to display **View Description**, **View Name**, or **None** as part of the image display on the HTML page. The **Align** option menu controls where the text (selected from the **Caption** option menu) appears with relation to the image on the HTML page. The available options include **Below, Left, Right,** and **Above.** The **Resolution** edit fields set the resolution of the generated image in pixels.

The View Attributes section of the dialog box lists all the available view attributes and the corresponding settings as saved in the design file. If you need to override the settings, select **Override** from the **View Attributes** option menu and make the necessary changes in the View Attributes settings box.

After making the necessary changes, click the **OK** button to close the Design File Walkthrough dialog box. MicroStation opens the Create HTML File dialog box, similar to Figure 15–52.

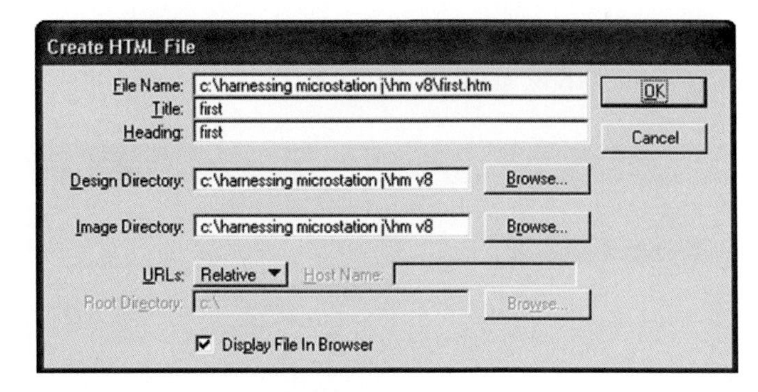

Figure 15–52 *Create HTML File dialog box*

Specify the file name of the HTML file being created, a descriptive title, and a heading description (appears top of the graphics) in the **File Name**, **Title**, and **Heading** edit fields. Specify the design and image directories in the **Design Directory** and **Image Directory** fields, respectively. Specify if the URLs being used are absolute or relative in the **URLs** option menu. If it is set to absolute, then specify the host name and root directory in the **Host Name** edit field and **Root Directory** edit fields. Click the **OK** button to create the HTML file from the design file saved view.

CREATING HTML FILE FROM DESIGN FILE SNAPSHOT

The Design File Snapshot option creates an HTML Web page with a view-only picture of the selected design file. To create an HTML file from a design file snapshot, first open the HTML Author dialog box from:

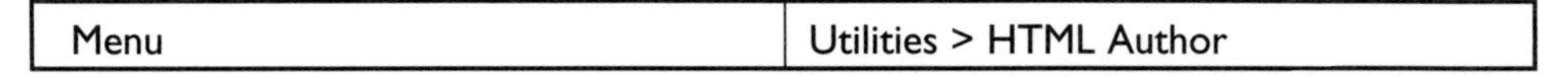

Menu	Utilities > HTML Author

MicroStation displays the HTML Author dialog box, similar to Figure 15–50.

Select the **Design File Snapshot** radio button from the Create HTML File From section of the dialog box and click the **OK** button to close the HTML dialog box. The Select Design File dialog box opens. Select the design file from the appropriate directory and click the **OK** button. MicroStation opens the Create HTML File dialog box, similar to Figure 15–53.

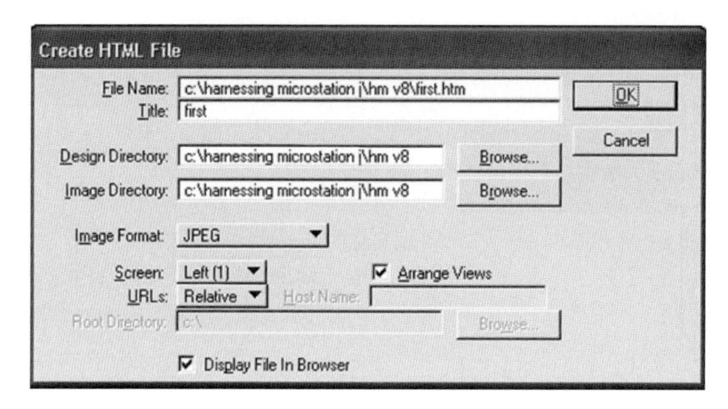

Figure 15–53 *Create HTML File dialog box*

Specify the file name of the HTML file being created, and the title in the **File Name** and **Title** edit fields, respectively. Specify the design and image directories in the **Design Directory** and **Image Directory** fields, respectively. Specify if the URLs being used are absolute or relative in the **URLs** option menu. If it is set to absolute, then specify the host name and root directory in the **Host Name** edit field and **Root Directory** edit field. Click the **OK** button to create the HTML file from the design file snapshot.

CREATING HTML FILE FROM CELL LIBRARY

The Cell Library option creates an HTML Web page from a cell library. To create an HTML file from a cell library, first open the HTML Author dialog box from:

Drop-down menu	Utilities > HTML Author

MicroStation displays the HTML Author dialog box, similar to Figure 15–50.

Select the **Cell Library** radio button from the Create HTML File From section of the dialog box and click the **OK** button to close the HTML dialog box. The Select Cell Library to Open dialog box opens. Select the cell library file from the appropriate directory and click the **OK** button. MicroStation opens the HTML Cell Page dialog box, similar to Figure 15–54.

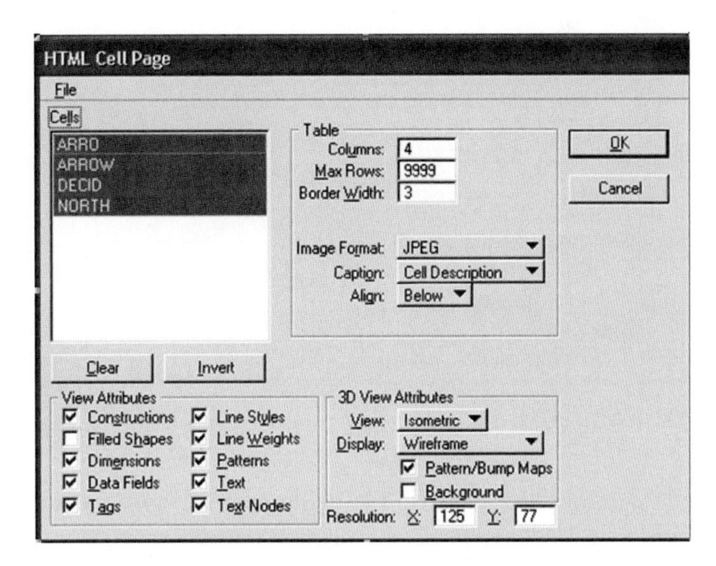

Figure 15–54 *HTML Cell Page dialog box*

MicroStation lists all the cells in the cell library file in the **Cells** list box. You can select any or all of the cells to be included in the HTML page.

The Tables section of the dialog box allows you to set the settings (such as the number of columns, maximum number of rows, and border width) for the table to be displayed on an HTML page.

The **Caption** option menu controls whether to display the **Cell Description**, **Cell Name**, or **None** as part of the image display on the HTML page. The **Align** option menu controls where the text (selected from Caption option menu) appears in relation to the image on the HTML page. The available options include **Below**, **Left**, **Right**, and **Above**.

The View Attributes section of the dialog box lists all the available view attributes and the corresponding settings as saved in the design file. If necessary, you can make changes in the View Attributes settings.

The 3D View Attributes section of the dialog box sets attributes for rendering generated images including view, display, patterns/bitmaps, and background.

The **Resolution** edit field sets the resolution of the generated image in pixels.

After making the necessary changes, click the **OK** button to close the HTML Cell Page dialog box. MicroStation opens the Create HTML File dialog box, similar to Figure 15–55.

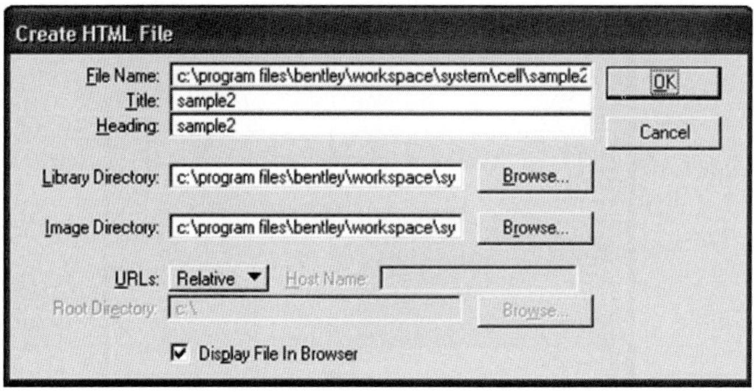

Figure 15–55 *Create HTML File dialog box*

Specify the file name of the HTML file being created, a descriptive title, and the heading description (appears top of the graphics) in the **File Name**, **Title**, and **Heading** edit fields, respectively. Specify the cell library directory location and image directories in the **Library Directory** and **Image Directory** edit fields, respectively. Specify if the URLs being used are absolute or relative in the **URLs** option menu. If it is set to absolute, then specify the host name and root directory in the **Host Name** edit field and **Root Directory** edit field. Click the **OK** button to create the HTML file from the cell library.

CREATING HTML FILE FROM BASIC MACROS

The Basic Macro HTML Page dialog box is used to create an HTML file that references a directory of MicroStation Basic macros. To create an HTML file from a directory of MicroStation Basic macros, first open the HTML Author dialog box from:

Drop-down menu	Utilities > HTML Author

MicroStation displays the HTML Author dialog box, similar to Figure 15–50.

Select the **Basic Macros** radio button from the Create HTML File From section of the dialog box and click the **OK** button to close the HTML dialog box. The Select Basic Macro Directory dialog box opens. Select the directory containing basic macro files and click the **OK** button. MicroStation opens the Basic Macro HTML Page dialog box, similar to Figure 15–56.

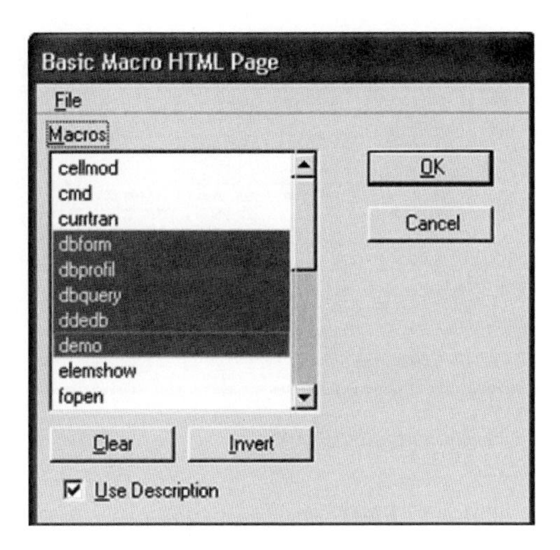

Figure 15–56 *Basic Macro HTML Page dialog box*

MicroStation lists all the basic macros in the selected directory in the **Macros** list box. You can select any or all of the basic macros to be included in the HTML page. Click the **OK** button to complete the selection.

MicroStation opens the Create HTML File dialog box, similar to Figure 15–57.

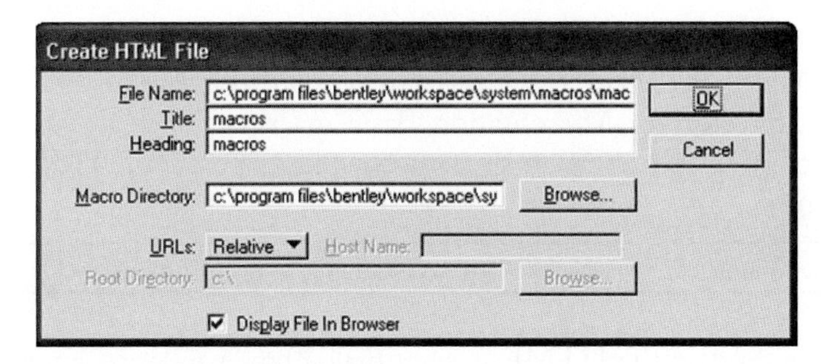

Figure 15–57 *Create HTML File dialog box*

Specify the file name of the HTML file being created, a descriptive title, and the heading description (appears top of the graphics) in the **File Name**, **Title**, and **Heading** edit fields, respectively. Specify the macro directory location in the **Macro Directory** edit field. Specify if the URLs being used are absolute or relative in the **URLs** option menu. If it is set to absolute, then specify the host name and root directory in the **Host Name** edit field and **Root Directory** edit field. Click the **OK** button to create the HTML file from Basic macros.

LINKING GEOMETRY TO INTERNET URLS

MicroStation provides tools to link Internet URLs to geometry. Most browsers highlight a linked entity when the cursor is placed on it and display the link title.

Clicking on the linked entity redirects the browser to the linked URL. This feature allows you to connect to component catalogs which can be accessed over the Internet, and, for example, engineers can execute a search over the Web to find a supplier whose components meet their specifications. Having URL links to geometry will also help when someone using the drawing needs more information on that component, since they can just click on the component and it can automatically connect them to the manufacturer's Web page.

ATTACHING LINK TO A GEOMETRY

To attach a URL to an element, invoke the Attach Engineering Link tool from:

E-Links tool box	Select the Attach Engineering Link tool (see Figure 15–58).

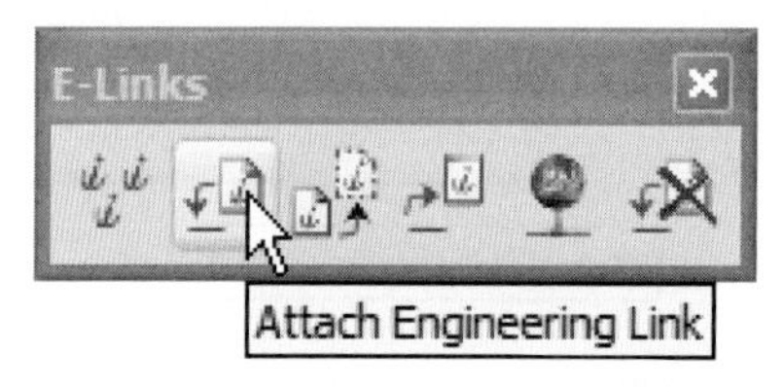

Figure 15–58 *Invoking the Attach Engineering Link tool from the E-Links tool box*

MicroStation displays the Attach Engineering Link settings box, similar to Figure 15–59.

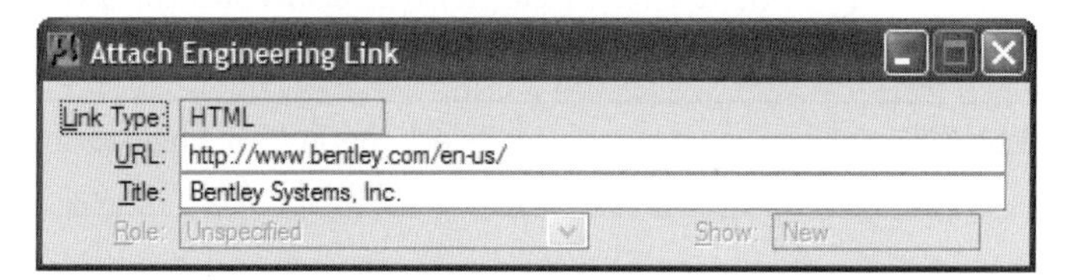

Figure 15–59 *Attach Engineering Link settings box*

The **Link Type** option menu sets the type of link you are creating: **HTML** or **XML:simple**. If you are creating an **XML:simple** link, you can specify parameters that further define how the link should operate. Specify the URL to attach to an element and the title in the **URL** and **Title** edit boxes. The **Role** option menu specifies what role the object of the link will specify, for example, that of a reference or cell library (enabled only when **Link Type** is set to **XML:simple**). The **Show** option menu determines whether any existing page should be replaced or if a new browser should be opened (enabled only when **Link Type** is set to **XML:simple**).

MicroStation prompts:

Attach Engineering Link > Identify element *(Identify an element to attach.)*

Attach Engineering Link > Accept/Reject *(Place a data point to accept the attachment of the link to the selected element.)*

You can attach multiple elements with the same URL.

DISPLAYING LINKS

To highlight the elements that have engineering links, invoke the Show Engineering Links tool from:

E-Links tool box	Select the Show Engineering Links tool (see Figure 15–60).

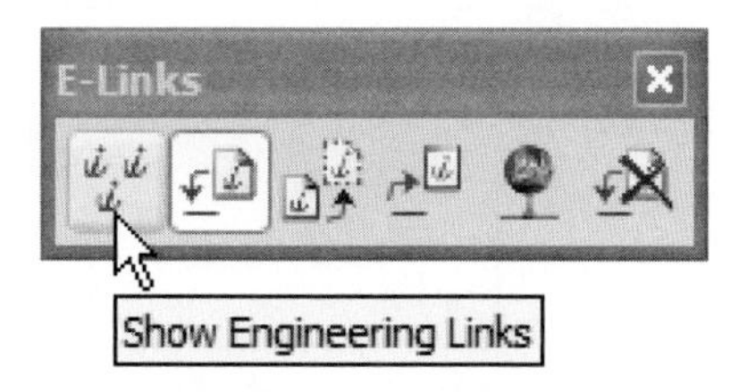

Figure 15–60 *Invoking the Show Engineering Links tool from the E-Links tool box*

MicroStation highlights the elements that have engineering links.

CONNECTING TO URLS

To connect to the URL attached to an element, first invoke the Follow Engineering Link tool from:

E-Links tool box	Select the Follow Engineering Link tool (see Figure 15–61).

Figure 15–61 *Invoking the Follow Engineering Link tool from the E-Links tool box*

MicroStation prompts:

Follow Engineering Link > Identify element *(Identify the element to which the URL is attached.)*

Attach Engineering Link > Accept/Reject *(Place a data point to accept the access to the URL.)*

MicroStation opens the default Web browser and connects to the URL.

If the Web browser is not able to access the specified URL, then it will display an error message. Make sure proper connection is established to access the Internet.

EDIT ENGINEERING LINK

To edit the URL attached to an element, first invoke the Edit Engineering Link tool from:

E-Links tool box	Select the Edit Engineering Link tool (see Figure 15–62).

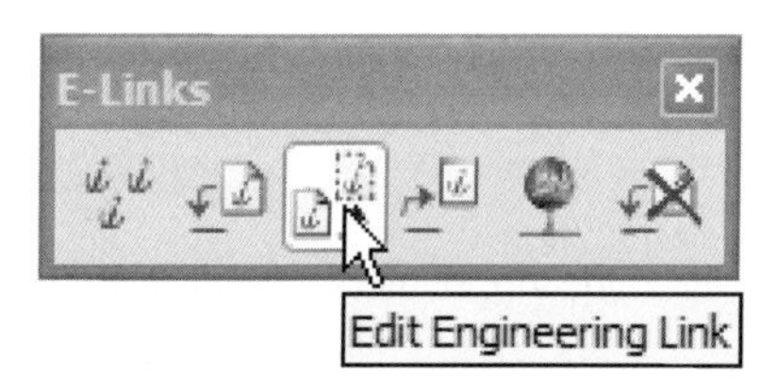

Figure 15–62 *Invoking the Edit Engineering Link tool from the E-Links tool box*

MicroStation prompts:

> Edit Engineering Tags > Identify element *(Identify the element to which the URL is attached.)*

MicroStation displays the Edit Tags [Internet] dialog box similar to Figure 15–63.

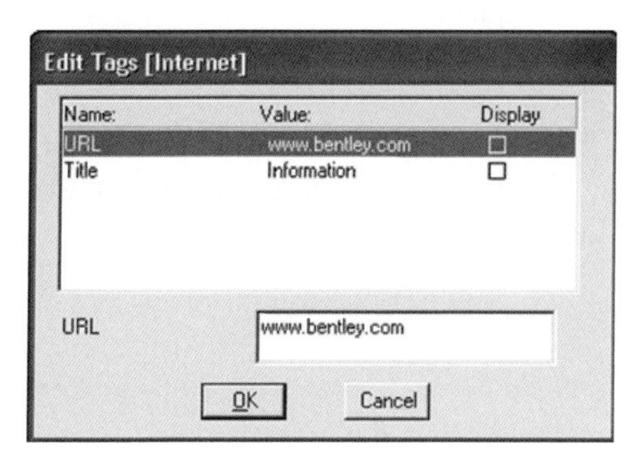

Figure 15–63 *Edit Tags [Internet] dialog box*

In the list box, select a line item to edit and in the edit text field at the bottom of the dialog box make the necessary changes. If necessary, the appropriate line item **Display** check box can be set to ON, to display the information in the design. Click the **OK** button to accept the changes and close the dialog box.

DELETE ENGINEERING LINK

To remove an attached engineering link from an element, invoke the Delete Engineering Link tool from:

E-links tool box	Select the Delete Engineering Link tool (see Figure 15–64).

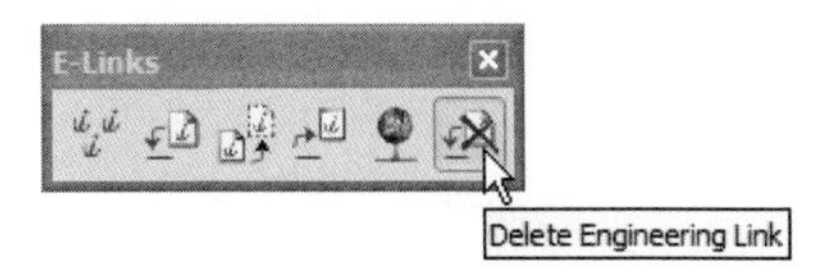

Figure 15–64 *Invoking the Delete Engineering Link tool from the E-Links tool box*

MicroStation prompts:

> Delete Engineering Link > Identify element *(Identify the element to which the URL is to be deleted.)*
>
> Delete Engineering Link > Accept/Reject *(Place a data point to accept the deletion of the Engineering Link.)*

MicroStation deletes the selected Engineering Link from the element

SCRIPTS AND MACROS

MicroStation provides the Scripts and Macros tools for automating often-used command sequences. Scripts can be used for simple command sequences and macros are used for complex sequences.

SCRIPTS

A script is an ASCII file containing the key-in sequence of the MicroStation commands. For example, the following script sets the Active Color, Active Line Weight, and Active Level:

active color red

active weight 6

active level Borders

To create a script, create a text file and type the commands onto it. To load and run a script, key-in the following in the Key-in window: **@<script_file>** and press (ENTER) (for example, **@attachRef**). MicroStation executes the command sequence.

MACROS

Macros are Microsoft Visual Basic for Applications (VBA) programs that automate often-used sequences of operations. Many MicroStation-specific extensions were added to the VBA language to customize it for the MicroStation environment. Macros select tools and view controls, send key-ins, manipulate dialog boxes, modify elements, and so forth.

For detailed information about creating macros, refer to the MicroStation VBA Guide that comes with the MicroStation software.

To load and run a macro, invoke the Macro tool from:

Menu	Utilities > Macro > Macros
Key-in window	**Macro** <macro_name> (ENTER)

MicroStation displays the Macros dialog box. Select a macro from the **Macro name** field and click the **Run** button to execute the macro. Macros can provide a variety of operations. Some may prompt for input, some may only display messages in the status bar as during execution, and others may complete their operations without any prompts or messages.

CHAPTER 16

3D Design and Rendering

WHAT IS 3D?

In two dimensional drawings, there are two axes, *X* and *Y*. In three dimensional drawings, in addition to the *X* and *Y* axes, there is a *Z* axis, as shown in Figure 16–1. Plan views, sections, and elevations represent only two dimensions. Isometric, perspective, and axonometric drawings, on the other hand, represent all three dimensions. For example, to create three dimensional views of a cube, the cube is simply drawn as a square with thickness. This is referred to as *extruded 2D*. Only objects that are extrudable can be drawn by this method. Any other views are achieved by simply rotating the viewpoint of the object. Isometric or perspective views can be created by simply changing the viewpoint.

Drawing objects in 3D provides three major advantages:

- An object can be drawn once and then can be viewed and plotted from any angle.
- A 3D object holds mathematical information that can be used in engineering analysis, such as finite-element analysis and computer numerical control (CNC) machinery.
- Shading can be added for visualization.

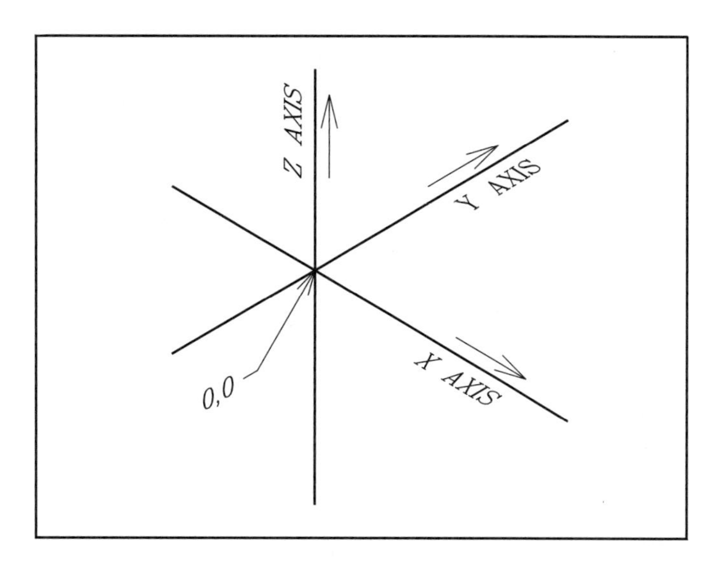

Figure 16–1 X, Y, *and* Z *axes for 3D Design*

MicroStation provides two types of 3D modeling: Surface and Solid. Surface modeling defines the edges of a 3D object in addition to surfaces, whereas solid models are the unambiguous and informationally complete representation of the shape of a physical object. Fundamentally, solid modeling differs from surface modeling in two ways:

- The information is more complete in the solid model.
- The method of construction of the model itself is inherently straightforward.

MicroStation's SmartSolids and SmartSurfaces tools can quickly construct complex 3D models. Finishing touches such as fillets and chamfers can be added to primitive solids or surfaces. The ShellSolid tool can create a hollow solid with defined wall thickness.

This chapter provides an overview of the tools and specific commands available for 3D design.

CREATING A 3D DESIGN FILE

The procedure for creating a new 3D design file is similar to that for creating a new 2D design file, except seed files must be designed specifically for 3D. To create a new 3D design file, invoke the New tool from:

Menu	File > New
Key-in window	**create drawing** (or **cr d**) (ENTER)

MicroStation opens the New dialog box. Click the Browse button, and MicroStation displays a list of seed files available, as shown in Figure 16–2. Select one of the 3D seed files from the **Files** list box and click the OK button. Key-in the name

for the new 3D design file in the **File name** edit field and click the OK button. MicroStation opens the new design file, the screen will look similar to the one shown in Figure 16–3. By default, MicroStation displays four views with Isometric orientation shown in View 2 of Figure 16–3. As part of the title of the view window, MicroStation displays the name of the view being displayed.

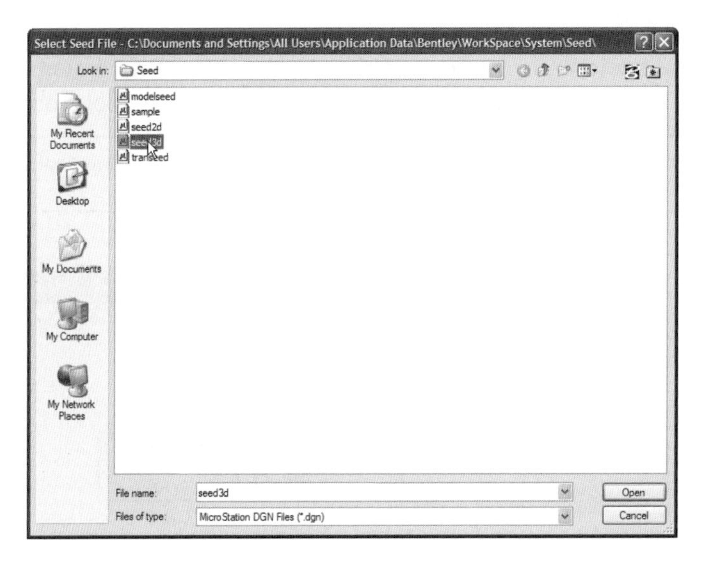

Figure 16–2 Select Seed File dialog box

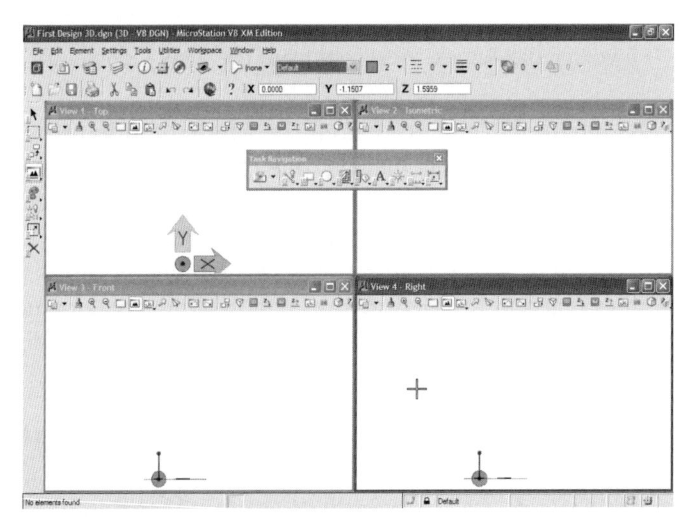

Figure 16–3 MicroStation screen display

VIEW ROTATION

There are seven standard view orientations defined in MicroStation: top, bottom, front, back, right, left, and isometric. Use one of the four tools available in MicroStation to display the standard view orientation in any of the view windows.

ROTATE VIEW

The Rotate View tool displays one of the standard view orientations. In addition, it can rotate the view dynamically. Invoke the Rotate View tool from:

View Control bar	Select the Rotate View tool (see Figure 16–4).

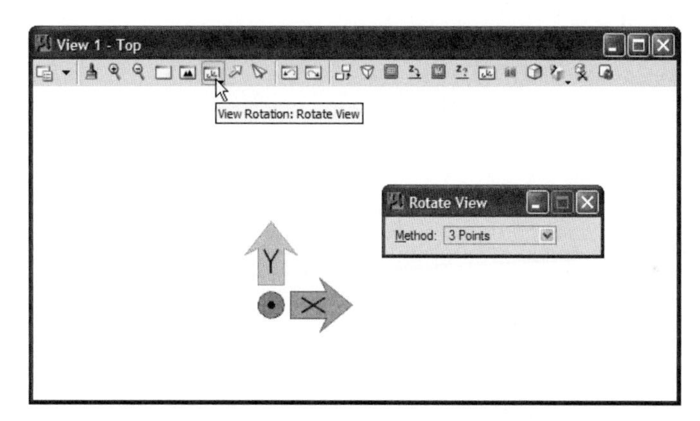

Figure 16–4 Invoking the Rotate View tool from the View Control bar

Select one of the ten available options from the **Method** option menu in the Tool Settings window. To position the view to one of the standard view orientations, first select the standard view orientation (Top, Bottom, Front, Back, Right, Left, Isometric [Top, Front, or Left] or Isometric [Top, Front, or Right]). Then place a data point in the view window where the view orientation will display. Use the Dynamic option to position the view at any angle around a data point. These three points rotate the view by defining three data points (origin, direction of the *X* axis, and a point defining the *Y* axis).

CHANGING VIEW ROTATION FROM THE VIEW ROTATION SETTINGS BOX

The view can also be rotated by specifying the angle of rotation in the View Rotation settings box. Open the View Rotation settings box from:

View Control bar	Select the Change View Rotation tool (see Figure 16–5).
Key-in window	**dialog view rotation** (or **di viewro**) (ENTER)

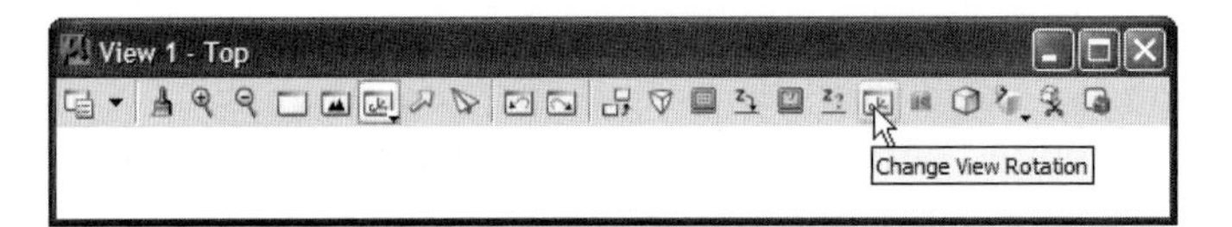

Figure 16–5 Invoking the Change View Rotation tool from the View Control bar

MicroStation displays the View Rotation settings box, similar to Figure 16–6.

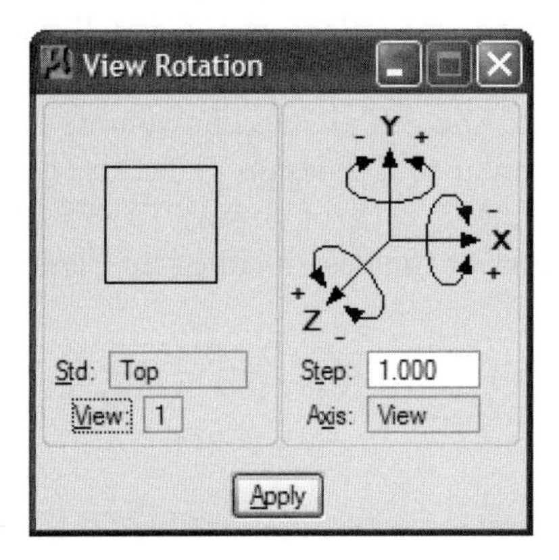

Figure 16–6 View Rotation settings box

The view to be manipulated is selected in the **View** option menu. Key-in the rotation increment in degrees in the **Step** text field. Click the "+" control to rotate the view in the positive direction by the Step amount around the specified Axis. Click the "–" to rotate the view in the negative direction by the Step amount around the specified Axis. To reposition the view to one of the standard view orientations, select the view orientation from the **Std** option menu. Click the Apply button to rotate the selected view to the specified rotation.

ROTATING VIEW BY KEY-IN

Rotate the view via one of the three key-in tools from:

Key-in window	**vi =** <name of the view> (ENTER) **rotate view absolute=**<xx,yy,zz> (ENTER) or **rotate view relative=**<xx,yy,zz> (ENTER)

Using the **VI** key-in, specify either one of the standard view rotations (top, bottom, front, back, right, left, or iso) or the name of the saved view.

In the Rotate View Absolute key-in, xx, yy, and zz are the rotations, in degrees, about the view *X*, *Y*, and *Z* axes (by default, 0 for each).

In the Rotate View Relative key-in, xx, yy, and zz are the relative, counter-clockwise rotations, in degrees, about the view *X*, *Y*, and *Z* axes. This key-in follows what is commonly called the "right-hand-rule." For example, if a positive *X* rotation is keyed in, point your right thumb in the view's positive *X* direction, then the way your fingers curl is the direction of the rotation.

DESIGN CUBE

Any new 2D design is created on a design plane—the electronic equivalent of a sheet of paper on a drafting table. The 2D design plane is a large, flat plane covered with an invisible matrix grid along the *X* and *Y* axes. A 3D design has that same *XY* plane plus a third dimension—the *Z* axis. The *Z* axis is the depth in the direction perpendicular to the *XY* plane. The volume defined by *X*, *Y*, and *Z* is called the *design cube.* Similar to the design plane, the design cube is covered with an invisible matrix grid along each of the *X*, *Y*, and *Z* axes. The global origin (0,0,0) is at the very center of the design cube (see Figure 16–7).

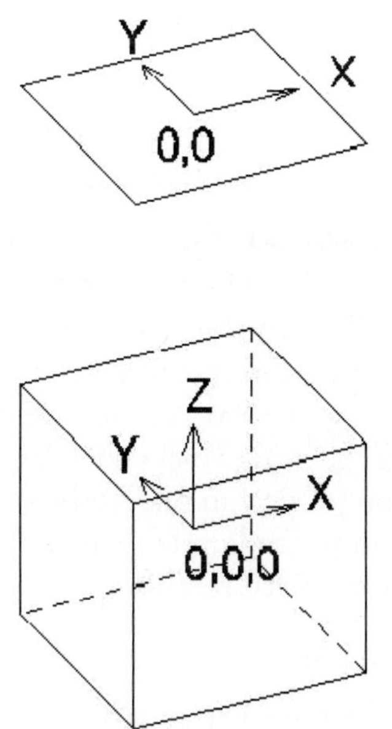

Figure 16–7 Design cube

DISPLAY DEPTH

Display depth is used to display a portion of the design rather than the entire design. The ability to look at only a portion of the depth comes in handy—especially if the design is complicated. Display depth settings define the front and back clipping planes for elements displayed in a view and is set for each view. Elements not contained in the display depth do not show up on the screen. To set the display depth, invoke the Set Display Depth tool from:

View Control bar	Select the Set Display Depth tool (see Figure 16–8).
Key-in window	**depth display** (or **dept d**) (ENTER)

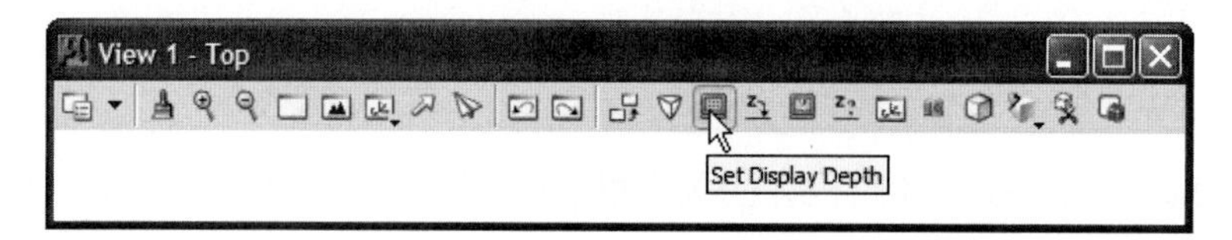

Figure 16–8 Invoking the Set Display Depth tool from the View Control bar

MicroStation prompts:

> Set Display Depth > Select view for display depth *(Place a data point in a view where you want to set the display depth.)*
>
> Set Display Depth > Define front clipping plane *(Place a data point in any view where you can identify the front clipping plane.)*
>
> Set Display Depth > Define back clipping plane *(Place a data point in any view where you can identify the back clipping plane.)*

The display depth can also be set by keying-in the distances in MU:SU:PU along the view *Z* axis by absolute or relative coordinates. To set the display depth by absolute coordinates, key-in:

Key-in window	**dp =** <front,back> (ENTER)

The <front,back> are the distances in working units along the view *Z* axis from the global origin to the desired front and back clipping planes. MicroStation prompts:

> Set Display Depth > Select view *(Identify the view with a data point to set the display depth.)*

To set the display depth by relative coordinates, key-in:

Key-in window	**dd =** <front,back> (ENTER)

The <front,back> are the distances in working units, and they add the keyed-in values to the current display depth settings. MicroStation prompts:

Set Display Depth > Select view *(Identify the view with a data point to set the display depth.)*

To determine the current setting for display depth, invoke the Show Display Depth tool from:

3D View Control tool box	Select the Show Display Depth tool (see Figure 16–9).
Key-in window	**dp = $** (or **dd = $**) (ENTER)

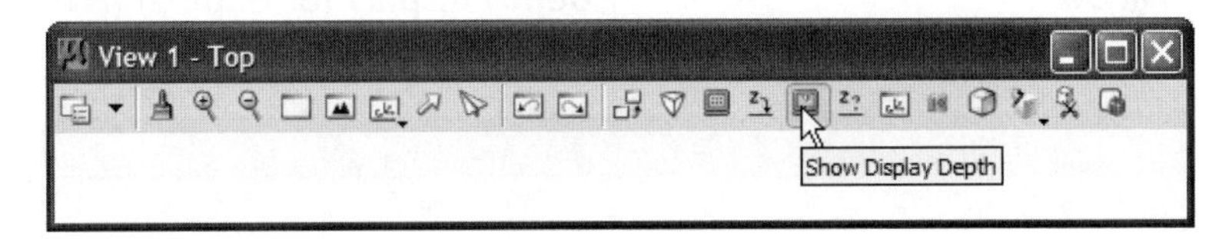

Figure 16–9 Invoking the Show Display Depth Tool from the View Control bar

MicroStation prompts:

Show Display Depth > Select view *(Place a data point anywhere in the view window.)*

MicroStation displays the current setting of the display depth in the Status bar.

Note: *When setting up the display depth in all views, dashed lines indicate the viewing parameters of the selected view. Both the display volume of the view and the active depth plane are dynamically displayed, with different-style dashed lines.*

In Figure 16–10, the display depth is set in such a way that only the square box is displayed but not the circles.

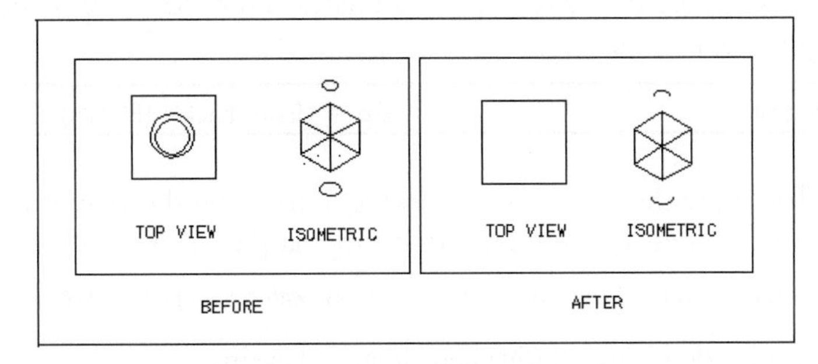

Figure 16–10 Example of setting up the display depth

FITTING DISPLAY DEPTH TO DESIGN FILE ELEMENTS

A fast way to display the entire design is to invoke the Fit View tool and select the appropriate view. In 2D design the Fit tool adjusts the view window to include all elements in the design file. Similarly, in 3D design the Fit View tool adjusts both the view window and the display depth to include all the elements in the design file. The Fit View tool automatically resets the display depth to the required amount to display the entire design file.

ACTIVE DEPTH

MicroStation has a feature that enables placement of an element in front of or behind the *XY* plane (front and back views), to the left or right of the *YZ* plane (right and left views), and above or below the *XY* plane (top and bottom views). This can be done by setting up the *active depth.* The active depth is a plane, parallel to the screen in each view, where elements will be placed by default. Each view has its own active depth plane, which can be changed at any time.

Elements are placed at the active depth by default. In the top and bottom views (*XY* plane), the depth value is along its *Z* axis. In the front and back views (*XZ* plane), the depth value is along its *Y* axis. And in the case of right and left views (*YZ* plane), the depth value is along its *X* axis.

SETTING ACTIVE DEPTH

To set the active depth, invoke the Set Active Depth tool from:

View Control bar	Select the Set Active Depth tool (see Figure 16–11).
Key-in window	**depth active** (or **dept a**) (ENTER)

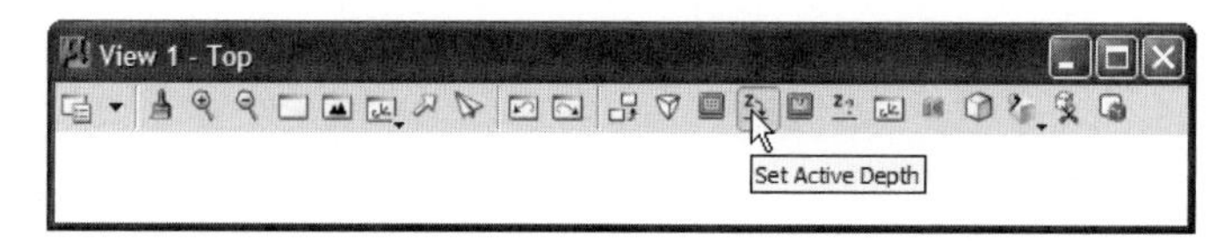

Figure 16–11 Invoking the Set Active Depth tool from the View Control bar

MicroStation prompts:

Set Active Depth > Select view *(Place a data point in a view to set the active depth.)*

Set Active Depth > Enter active depth point *(Place a data point in a different view to identify the location for setting up the active depth.)*

Note: *When setting up the active depth in all views, notice dashed lines indicating the viewing parameters of the selected view. Both the display volume of the view and the active depth plane are dynamically displayed, with different-style dashed lines.*

The active depth can also be set by keying-in the distances in MU:SU:PU along the view *Z* axis by absolute or relative coordinates. To set the active depth by absolute coordinates, key-in:

Key-in window	**az** = <depth> (ENTER)

The <depth> is the distance in working units along the view *Z* axis from the global origin to the desired active depth. MicroStation prompts:

Set Active Depth > Select view *(Identify the view with a data point to set the active depth.)*

To set the active depth by relative coordinates, key-in:

Key-in window	**dz** = <depth> (ENTER)

The <depth> is the distance in working units, and it adds the keyed-in values to the current active depth setting. MicroStation prompts:

Set Active Depth > Select view *(Identify the view with a data point to set the active depth.)*

To determine the current setting for active depth, invoke the Show Active Depth tool from:

View Control bar	Select the Show Active Depth tool (see Figure 16–12).
Key-in window	**az** = **$** (or **dz** = **$**) (ENTER)

Figure 16–12 Invoking the Show Active Depth tool in the View Control bar

MicroStation prompts:

Show Active Depth > Select view *(Place a data point anywhere in the view window.)*

MicroStation displays the current setting of the active depth in the Status bar.

Note: *Before placing elements, make sure the active depth and display depth are appropriate. If the active depth is set outside the range of the display depth, then MicroStation displays the following message in the error field:*

Active depth set to display depth

The active depth is set to the value closest to the display depth. Assume, for example, the current display depth is set to 100,450 and the active depth is set to 525. Since MicroStation sets the active depth to the closest value, in this case the active depth is set to 450. Make sure that MicroStation sets the value for the active depth to the value intended. If necessary, change the display depth and then set the active depth.

BORESITE LOCK

The Boresite lock controls the manipulation of the elements at different depths. If the Boresite lock is set to ON, elements can be identified or snapped at any depth in the view; elements being moved or copied will remain at their original depths. If it is set to OFF, elements can be identified at, or very near, the active depth of a view. Boresite lock can be set to ON or OFF from the Lock Toggles settings box, as shown in Figure 16–13.

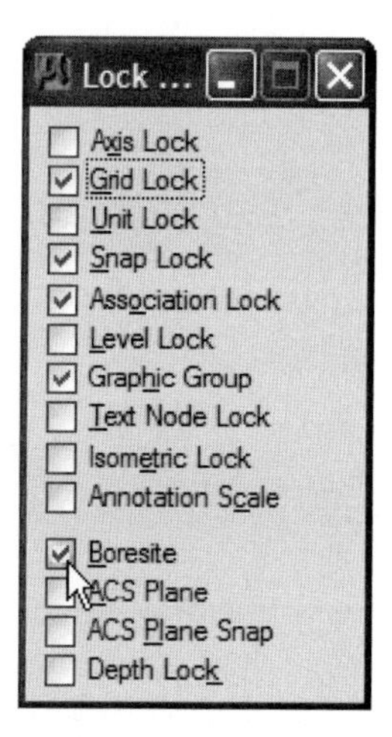

Figure 16–13 Lock Toggles settings box

Note: *Tentative points override the Boresite lock. Elements can be tentatively snapped at any depth regardless of the Boresite lock setting.*

PRECISION INPUTS

When MicroStation prompts for the location of a point, in addition to providing the data point with the pointing device, precision input tools can be used to place data points precisely. Similar to 2D placement tools, 3D tools also accept keyed-in coordinates. MicroStation provides two types of coordinate systems for 3D design: the drawing coordinate system and the view coordinate system.

DRAWING COORDINATE SYSTEM

The drawing coordinate system is the model coordinate system fixed relative to the design cube, as shown in Figure 16–14. For example, in the TOP view, *X* is to the right, *Y* is up, and *Z* is out of the screen (right-hand rule). In the RIGHT view, *Y* is to the right, *Z* is up, and *X* is out of the screen, and so on. Following are the two key-ins available for the drawing coordinate system:

XY=<X,Y,Z>

DL=<delta_x,delta_y,delta_z>

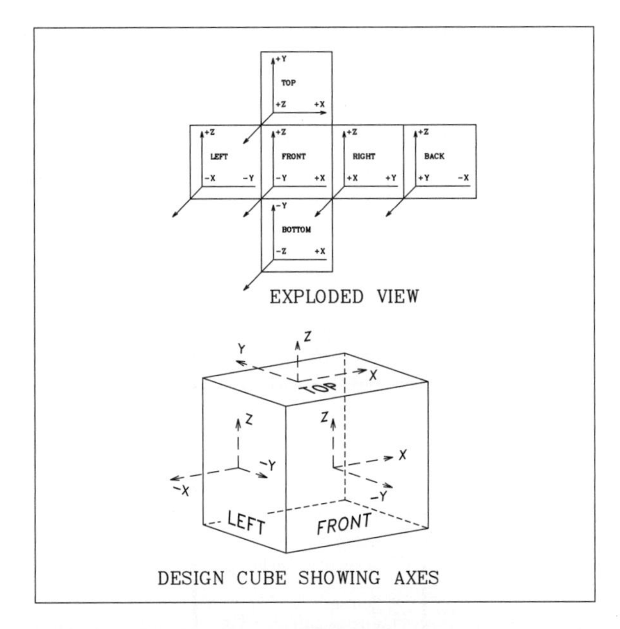

Figure 16–14 Design cube showing the drawing coordinate system

The XY= key-in places a data point measured from the global origin of the drawing coordinate system. The <X,Y,Z> are the *X*, *Y*, and *Z* values of the coordinates. The view being used at the time has no effect on them. The DL= places a data point to a distance along the drawing axes from a previous data (relative) or tentative point. The <delta_x,delta_y,delta_z> are the relative coordinates in the *X*, *Y*, and *Z* axes relative to the previous data point or tentative point.

VIEW COORDINATE SYSTEM

The view coordinate system inputs data relative to the screen, where *X* is to the right, *Y* is up, and *Z* comes directly out from the screen in all views, as shown in Figure 16–15. The view coordinate system is view-dependent, that is, depends on the orientation of the view for their direction. Following are the two key-ins available for the view coordinate system:

DX=<delta_x,delta_y,delta_z>

DI=<distance,direction>

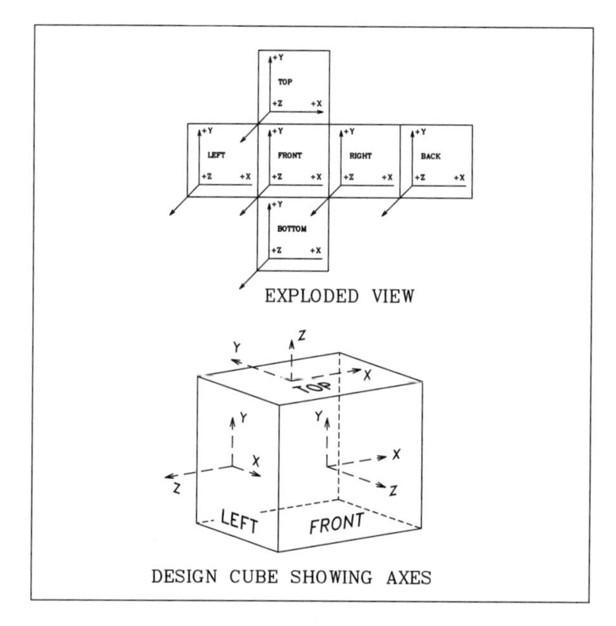

Figure 16–15 Design cube showing the view coordinate system

The DX= key-in places a data point to a distance from the previous data or tentative point (relative) in the same view where the previous point was defined. The <delta_x,delta_y,delta_z> are the relative coordinates in the *X*, *Y*, and *Z* axes relative to the previous data point or tentative point. The DI= key-in places a data point a certain distance and direction from a previous data or tentative point (relative polar) in the same view where the previous point was defined. The <distance,direction> are specified in relation to the last specified position or point. The distance is specified in current working units, and the direction is specified as an angle, in degrees, relative to the *X* axis.

Note: *With key-in precision inputs, MicroStation assumes that the desired view is the one last worked in—that is, the view in which the last tentative or data point was placed. The easiest way to make a view current is to place a tentative point and then press the Reset button. Updating a view is also another way to tell MicroStation that the selected view is the view last worked in.*

AUXILIARY COORDINATE SYSTEMS (ACS)

MicroStation provides a set of tools to define an infinite number of user-defined coordinate systems called *auxiliary coordinate systems.* An auxiliary coordinate system allows the user to change the location and orientation of the *X*, *Y*, and *Z* axes

to reduce the calculations needed to create 3D objects. The origin can be redefined in a drawing, and thus establish new positive *X* and *Y* axes. New users think of a coordinate system simply as the direction of positive *X* and positive *Y*. But once the directions *X* and *Y* are defined, the direction of *Z* will be defined as well. Thus, the user only has to be concerned with *X* and *Y*. For example, if a sloped roof of a house is drawn in detail using the drawing coordinate system, each end point of each element on the inclined roof plane must be calculated. On the other hand, if the auxiliary coordinate system is set to the same plane as the roof, each object can be drawn as if it were in the plan view. Any number of auxiliary coordinate systems can be defined, each with a user-determined name. But, at any given time, only one auxiliary coordinate system is current with the default system.

MicroStation provides a visual reminder of how the ACS axes are oriented and where the current ACS origin is located. The *X*, *Y*, and *Z* axis directions are displayed using arrows labeled appropriately. The display of the ACS axes is controlled by turning ON/OFF the ACS Triad in the View Attributes settings box, as shown in Figure 16–16.

MicroStation provides three types of coordinate systems for defining an ACS: rectangular, cylindrical, and spherical coordinate systems.

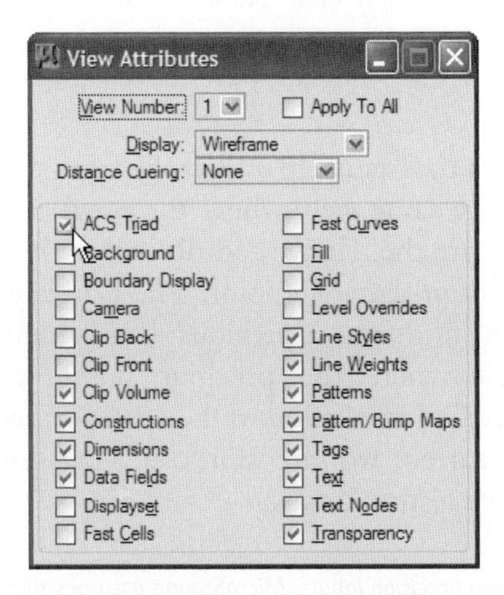

Figure 16–16 View Attributes settings box showing the ACS Triad set to ON

The rectangular coordinate system is the same one that is available for the design cube and is also the default type to define an ACS.

The cylindrical coordinate system is another 3D variant of the polar format. It describes a point by its distance from the origin, its angle in the *XY* plane from the *X* axis, and its *Z* value as shown in Figure 16–17. For example, to specify a point at a distance of 4.5 units from the origin, at an angle of 35 degrees relative to the *X* axis (in the *XY* plane), and with a *Z* coordinate of 7.5 units, enter **4.5,35,7.5**.

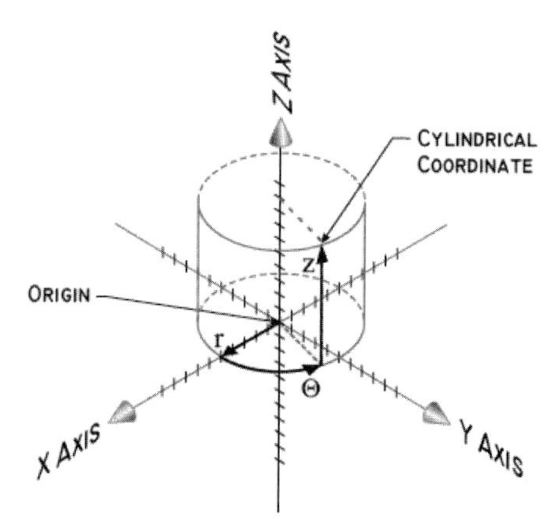

Figure 16–17 Cylindrical coordinate system

The spherical coordinate system is another 3D variant of the polar format. It describes a point by its distance from the current origin, its angle in the *XY* plane, and its angle up from the *XY* plane as shown in Figure 16–18. For example, to specify a point at a distance of 7 units from the origin, at an angle of 60 degrees from the *X* axis (in the *XY* plane), and at an angle 45 degrees up from the *XY* plane, enter 7,60,45.

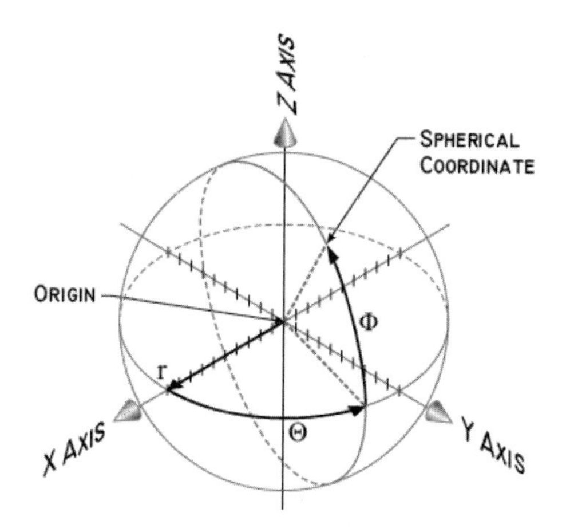

Figure 16–18 Spherical coordinate system

PRECISION INPUT KEY-IN

Similar to the key-ins available for drawing and view coordinates, MicroStation provides key-ins to input the coordinates in reference to the auxiliary coordinate system. Following are the two key-ins available for the auxiliary coordinate system:

AX=<X,Y,Z>

AD=<delta_x,delta_y,delta_z>

The AX= key-in places a data point measured from the ACS origin and is equivalent to the key-in XY=. The <X,Y,Z> are the *X*, *Y*, and *Z* values of the coordinates. The AD= places a data point to a distance along the drawing axes from a previous data (relative) or tentative point and is equivalent to the key-in DL=. The <delta_x,delta_y,delta_z> are the relative coordinates in the *X*, *Y*, and *Z* axes relative to the previous data point or tentative point.

DEFINING AN ACS

MicroStation provides three different tools to define an ACS. The tools are available in the ACS tool box. Before selecting one of the three tools, select the desired coordinate system for the new ACS from the **Type** option menu in the Tool Settings window. Control the ON/OFF check box for two locks in the Tool Settings window. When the **ACS Plane Lock** is set to ON, each data point is forced to lie on the active ACS's *XY* plane (Z=0). When the **ACS Plane Snap** Lock is set to ON, each tentative point is forced to lie on the active ACS's *XY* plane (Z=0).

Defining an ACS by Aligning with an Element

This option defines an ACS by identifying an element where the *XY* plane of the ACS is parallel to the plane of the selected planar element. The origin of the ACS is at the point of identification of the element. Upon definition, the ACS becomes the active ACS. To define an ACS aligned with an element, invoke the Define ACS (aligned with element) tool from:

ACS tool box	Select the Define ACS (Aligned with Element) tool (see Figure 16–19).
Key-in window	**define acs element** (or **de a e**) (ENTER)

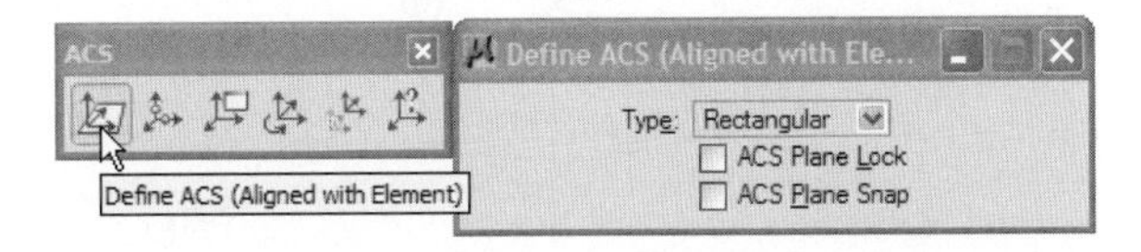

Figure 16–19 Invoking the Define ACS (Aligned with Element) tool from the ACS tool box

MicroStation prompts:

Define ACS (Aligned with Element) > Identify element *(Identify the element with which to align the ACS and define the ACS origin.)*

Define ACS (Aligned with Element) > Accept/Reject (Select next input) *(Place a data point to accept the element for defining an ACS or click the Reject button to cancel the operation.)*

Defining the ACS by Points

This option is the easiest and most used option for controlling the orientation of the ACS. It allows the user to place three data points to define the origin and the directions of the positive *X* and *Y* axes. The origin point acts as a base for the ACS rotation, and when a point is selected to define the direction of the positive *X* axis, the direction of the *Y* axis is limited because it is always perpendicular to the *X* axis. When the *X* and *Y* axes are defined, the *Z* axis is automatically placed perpendicular to the *XY* plane. Upon definition, the ACS becomes the active ACS. To define an ACS by Points, invoke the Define ACS (By Points) tool from:

ACS tool box	Select the Define ACS (By Points) tool (see Figure 16–20).
Key-in window	**define acs points** (or **de a p**) (ENTER)

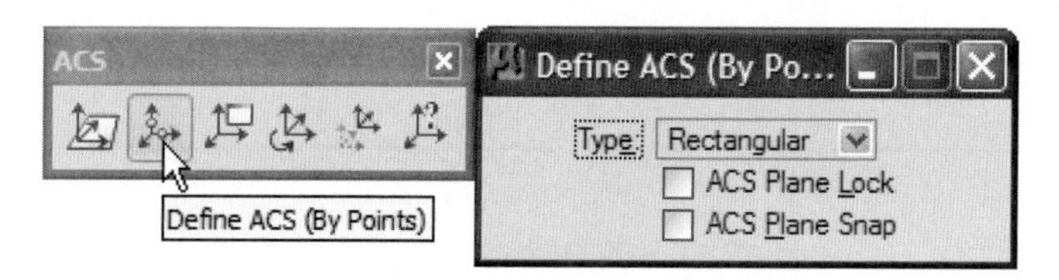

Figure 16–20 Invoking the Define ACS (By Points) tool from the ACS tool box

MicroStation prompts:

Define ACS (By Points) > Enter first point @x axis origin *(Place a data point to define the origin.)*

Define ACS (By Points) > Enter second point on X axis *(Place a data point to define the direction of the positive X axis, which extends from the origin through this point.)*

Define ACS (By Points) > Enter point to define Y axis *(Place a data point to define the direction of the positive Y axis.)*

Defining the ACS by Aligning with a View

In this option, the ACS takes the orientation of the selected view. That is, the ACS axes align exactly with those of the view selected. Upon definition, the ACS becomes the active ACS. To define an ACS by aligning with a view, invoke the Define ACS (Aligned with View) tool from:

ACS tool box	Select the Define ACS (Aligned with View) tool (see Figure 16–21).
Key-in window	**define acs view** (or **de a v**) (ENTER)

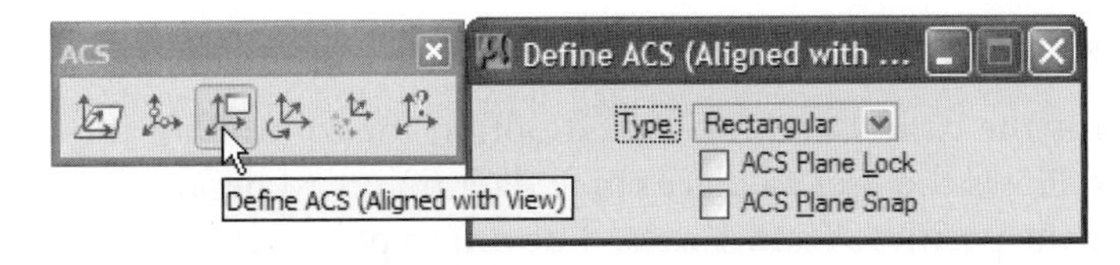

Figure 16–21 Invoking the Define ACS (Aligned with View) tool from the ACS tool box

MicroStation prompts:

> Define ACS (Aligned with View) > Select source view *(Place a data point to select the view with which the ACS is to be aligned and define the ACS origin.)*

ROTATING THE ACTIVE ACS

The Rotate Active ACS tool rotates the Active ACS. The origin of the ACS is not moved. To rotate the active ACS, invoke the Rotate Active ACS tool from:

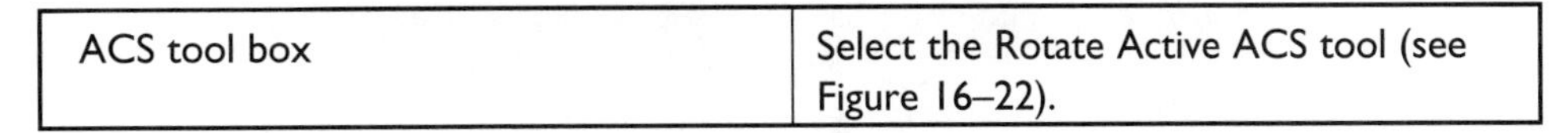

ACS tool box	Select the Rotate Active ACS tool (see Figure 16–22).

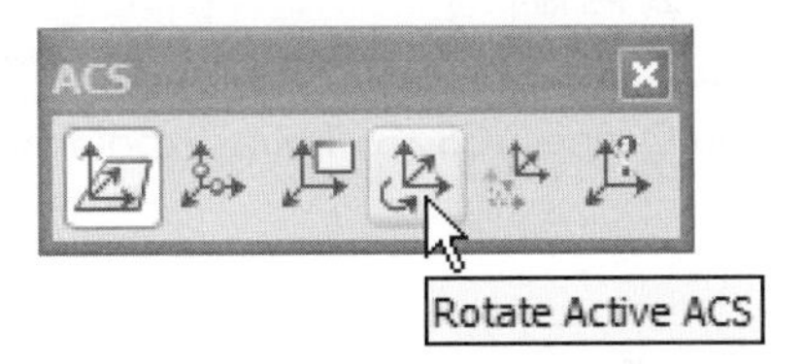

Figure 16–22 Invoking the Rotate Active ACS Tool from the ACS tool box

MicroStation displays the Rotate Active ACS dialog box, as shown in Figure 16–23.

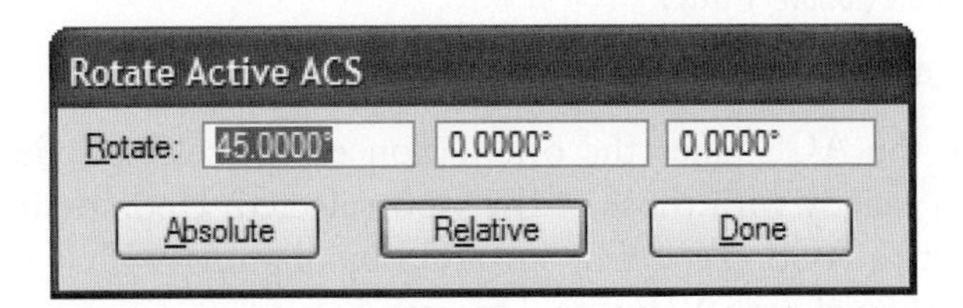

Figure 16–23 Rotate Active ACS dialog box

Key-in the rotation angles, in degrees, from left to right, for the *X*, *Y*, and *Z* axes. Click the Absolute button to rotate the ACS in relation to the unrotated (top) orientation. Click the Relative button to rotate the ACS in relation to the current orientation. Click the Done button to close the Rotate Active ACS dialog box.

MOVING THE ACTIVE ACS

Use the Move ACS tool to move the origin of the Active ACS, leaving the directions of the *X*, *Y*, and *Z* axes unchanged. To move the ACS, invoke the Move ACS tool from:

ACS tool box	Select the Move ACS tool (see Figure 16–24).
Key-in window	**move acs** (or **mov a**) (ENTER)

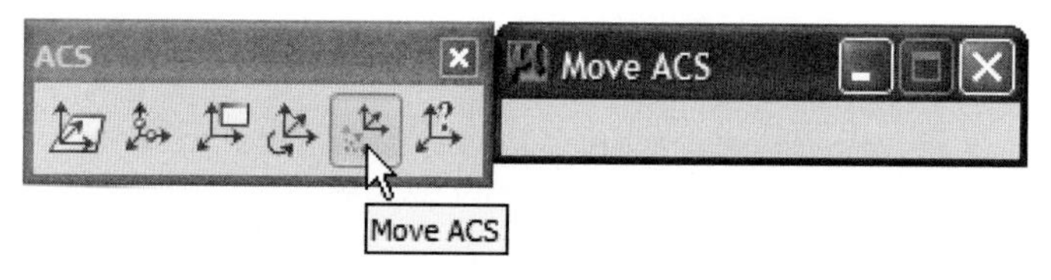

Figure 16–24 Invoking the Move ACS tool from the ACS tool box

MicroStation prompts:

Move ACS > Define origin *(Place a data point to define the new origin.)*

SELECTING THE ACTIVE ACS

Use the Select ACS tool to identify an ACS for attachment as the active ACS from the saved ACS in each view. To select the ACS, invoke the Select ACS tool from:

ACS tool box	Select the Select ACS tool (see Figure 16–25).

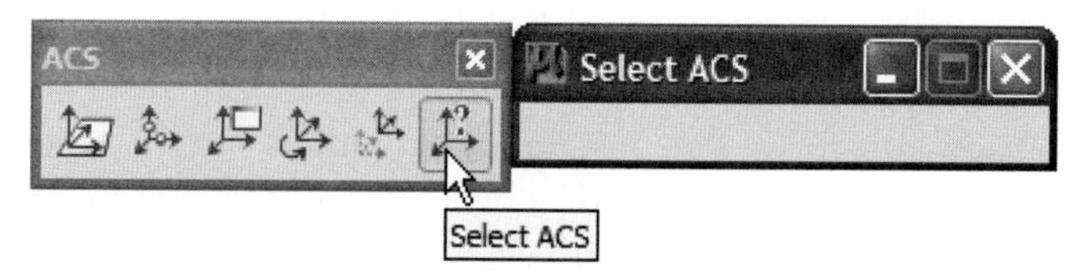

Figure 16–25 Invoking the Select ACS tool from the ACS tool box

MicroStation prompts:

Select ACS > Select auxiliary system @ origin *(Identify the ACS origin from the coordinate triad displayed.)*

SAVING AN ACS

Any number of ACSs can be defined in a design file. Of these, only one can be active at any time. Any ACS can be saved for future use. The Auxiliary Coordinate Systems settings box is used to name, save, attach, or delete an ACS. Open the Auxiliary Coordinate Systems settings box from:

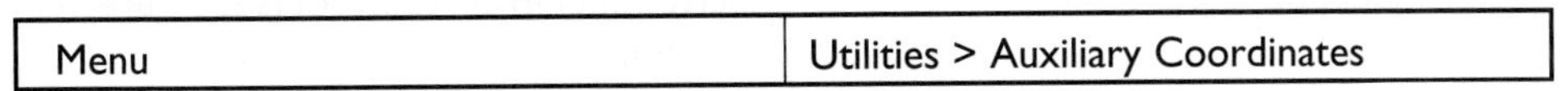

Menu	Utilities > Auxiliary Coordinates

MicroStation displays an Auxiliary Coordinate Systems settings box, similar to Figure 16–26.

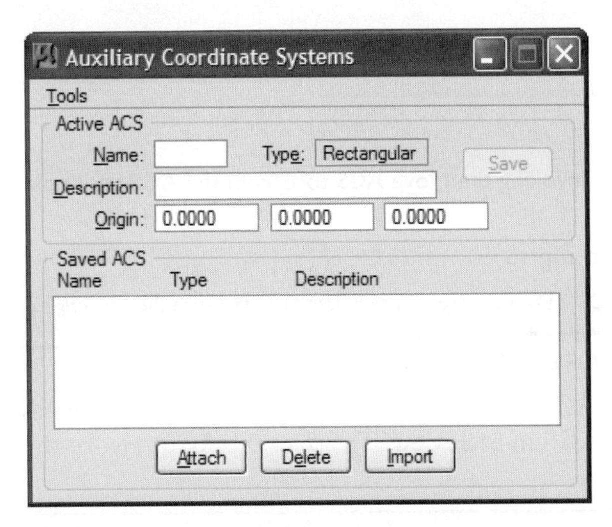

Figure 16–26 Auxiliary Coordinate Systems settings box

Key-in the name for the active ACS in the **Name** edit field. The name is limited to six characters. Select the coordinate system to be saved with the active ACS from the **Type** option menu. Key-in the description (optional) of the active ACS in the **Description** edit field. The description is limited to 27 characters. If necessary, change the origin of the active ACS by keying-in the coordinates in the **Origin** edit field. Click the Save button to save the active ACS for future attachment. MicroStation will display the ACS name, type, and description in the Saved ACS list box.

To attach an ACS as an active ACS, select the name of the ACS from the **Saved ACS** list box and click the Attach button. To delete an ACS, first highlight the ACS in the Saved ACS list box, and then click the Delete button.

Note: *All the tools available in the ACS tool box are also available in the Tools menu of the Auxiliary Coordinate Systems settings box.*

3D PRIMITIVES

MicroStation provides a set of tools to place simple 3D elements that can become the basic building blocks that make up the model. The primitive tools include slab, sphere, cylinder, cone, torus, and wedge.

Two important settings need to be taken into consideration before the primitives are created that control the way in which solids and surfaces are created and displayed on the screen. The settings include: display method and number of rule lines that represent a surface with a full 360-degree curvature and selection of solids and surfaces.

DISPLAY METHOD AND SURFACE RULE LINES

MicroStation provides two options: wireframe and surfaces, for on-screen display of the solids and surfaces. By default, it is set to wireframe, which is the more efficient mode for working with solids and surfaces in a design session. The surfaces display mode should be used only when the design is to be rendered with an earlier version of MicroStation.

Surface rule lines provide a visual indication of a surface's curvature. By default, it is set to 4—a full cylindrical solid is displayed with 4 surface rules lines. If necessary, increase the setting to display with additional surface rule lines. To change the display method and set the surface rule lines, open the Change SmartSolid Settings box from:

Task Navigation tool box (active task set to 3D Utility)	Select the Change SmartSolid Display tool (see Figure 16–27).
Keyboard Navigation (Task Navigation tool box with active task set to 3D Utility)	**w**

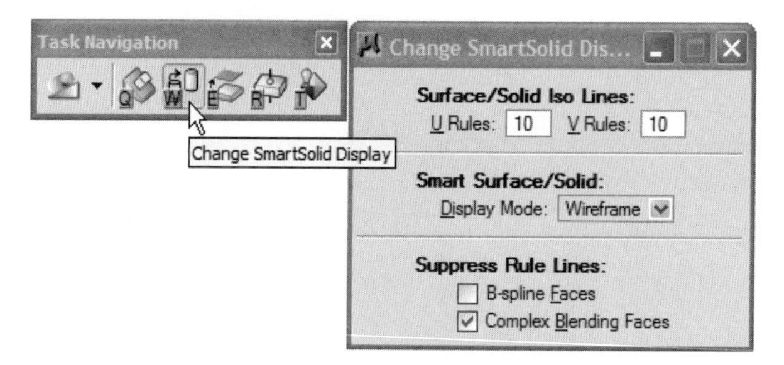

Figure 16–27 Invoking the Change SmartSolid Display tool from the Task Navigation tool box (active task set to 3D Utility)

The **Surface/Solid Iso Lines** section of the Tool Settings window sets the number of rule lines that represent a full 360° of curvature of curved surfaces for SmartSolids and SmartSurfaces. The **Smart Surface/Solid** section lets you set the Display Mode for SmartSurfaces and SmartSolids. The **Suppress Rule Lines** section lets you suppress, or turn off, the display of rule lines for particular faces on SmartSolids and SmartSurfaces.

SELECTION OF SOLIDS AND SURFACES

MicroStation provides three options in the selection of solids and surfaces. By default, surfaces and solids may be identified with a data point anywhere on their surface, not necessarily on an edge line or surface rule line. To change the selection mode, open the Preferences dialog box from:

Menu	Workspace > Preferences

MicroStation displays the Preferences dialog box Select Input category from the **Category** list box. The **Locate By Faces:** option menu provides the following three selection modes in the selection of solids and surfaces:

- **Never**—Solids and surfaces can only be identified with a data point on an edge or surface rule line.
- **Rendered Views Only**—Solids and surfaces rendered with any of the rendering options may be identified with a data point anywhere on their surface.
- **Always**—Solids and surfaces whether rendered or not may be identified with a data point anywhere on their surface.

PLACE SLAB

The Place Slab tool places a volume of projection with a rectangular cross section. To place a slab, invoke the Place Slab tool from:

Task Navigation tool box (active task set to 3D Primitives located in the Solids Modeling)	Select the Place Slab tool (see Figure 16–28).
Keyboard Navigation (Task Navigation tool box with active task set to 3D Primitives)	**q**

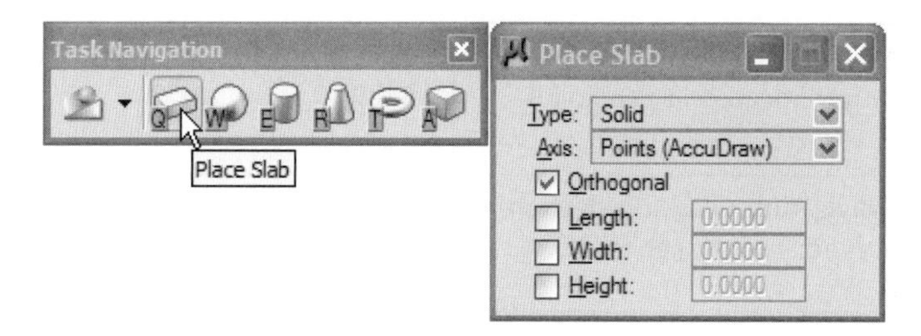

Figure 16–28 Invoking the Place Slab tool from the Task Navigation tool box (active task set to 3D Primitives)

Select the type of surface from the **Type** option menu in the Tool Settings window. The surface (not capped) option is considered to be open at the base and top, whereas the solid (capped) option is considered to enclose a volume completely.

From the **Axis** option menu in the Tool Settings window, select the direction in which the height is projected relative to the view or design file axes. If set to Screen X, Screen Y, or Screen Z, the height is projected with the selected screen (view) axis. If set to Drawing X, Drawing Y, or Drawing Z, the height is projected with the selected design file axis.

If necessary, turn ON the check boxes for **Orthogonal, Length, Width,** and **Height** in the Tool Settings window. If **Orthogonal** is set to ON, the edges are placed orthogonally. If constraints for **Length, Width,** and **Height** are ON, be sure to key-in appropriate values in the edit fields.

MicroStation prompts:

Place Slab > Enter start point *(Place a data point or key-in coordinates to define the origin.)*

Place Slab > Define Length *(Place a data point or key-in coordinates to define the length and rotation angle. If the Length constraint is set to ON, this data point defines the rotation angle.)*

Place Slab > Define Width *(Place a data point or key-in coordinates to define the width. If the Width constraint is set to ON, this data point accepts the width.)*

Place Slab > Define Height *(Place a data point or key-in coordinates to define the height. If the Height constraint is set to ON, this data point provides the direction.)*

Note: *To place a volume of projection with a nonrectangular cross section, use the Extrude tool in the 3D Construct tool box.*

PLACE SPHERE

The Place Sphere tool can place a sphere, in which all surface points are equidistant from the center. To place a sphere, invoke the Place Sphere tool from:

Task Navigation tool box (active task set to 3D Primitives located in the Solids Modeling)	Select the Place Sphere tool (see Figure 16–29).
Keyboard Navigation (Task Navigation tool box with active task set to 3D Primitives)	**w**

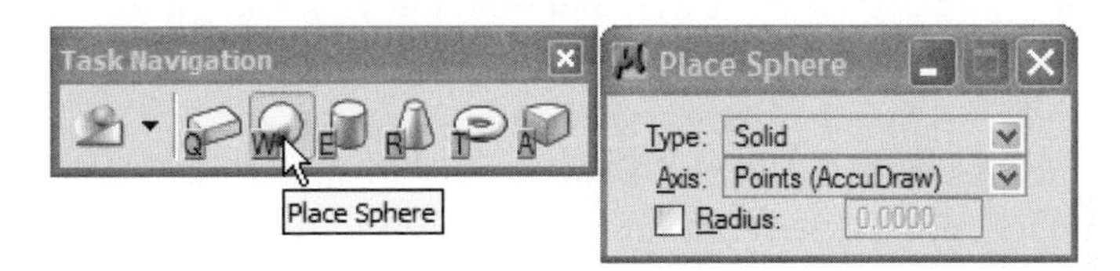

Figure 16–29 Invoking the Place Sphere tool from the Task Navigation tool box (active task set to 3D Primitives)

Select the type of surface from the **Type** option menu in the Tool Settings window.

From the **Axis** option menu in the Tool Settings window, select the direction of the sphere's axis relative to the view or design file axes. If set to Screen X, Screen Y, or Screen Z, the sphere's axis is set with the selected screen (view) axis. If set to Drawing X, Drawing Y, or Drawing Z, the sphere's axis is set with the selected design file axis.

If necessary, turn ON the check box for **Radius** constraint, and key-in the radius in the **Radius** edit field.

MicroStation prompts:

> Place Sphere > Enter center point *(Place a data point or key-in coordinates to define the sphere's center.)*
>
> Place Sphere > Define radius and axis *(Place a data point or key-in coordinates to define the radius. If Radius is set to ON, then the data point accepts the sphere.)*

Note: *To place a volume of revolution with a noncircular cross section, use the Extrude tool in the 3D Construct tool box.*

PLACE CYLINDER

The Place Cylinder tool places a cylinder of equal radius on each end and similar to an extruded circle. To place a cylinder, invoke the Place Cylinder tool from:

Task Navigation tool box (active task set to 3D Primitives located in the Solids Modeling)	Select the Place Cylinder tool (see Figure 16–30).
Keyboard Navigation (Task Navigation tool box with active task set to 3D Primitives)	e

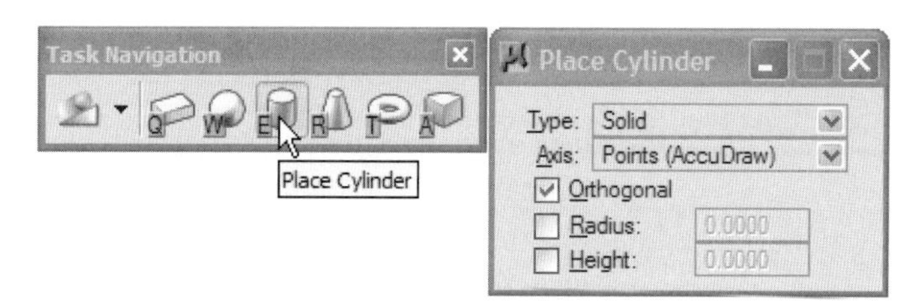

Figure 16–30 Invoking the Place Cylinder tool from the Task Navigation tool box (active task set to 3D Primitives)

Select the type of surface from the **Type** option menu in the Tool Settings window.

From the **Axis** option menu in the Tool Settings window, select the direction of the cylinder's axis or its height relative to the view or design file axes. If set to Screen X, Screen Y, or Screen Z, the direction of the cylinder's axis or height is set with the selected screen (view) axis. If set to Drawing X, Drawing Y, or Drawing Z, the direction of the cylinder's axis or height is set with the selected design file axis.

If necessary, turn ON the check boxes for **Orthogonal**, **Radius**, and **Height** in the Tool Settings window. If **Orthogonal** is set to ON, the cylinder is a right cylinder. If the constraints for **Radius** and **Height** are ON, be sure to key-in appropriate values in the edit fields.

MicroStation prompts:

> Place Cylinder > Enter center point *(Place a data point or key-in coordinates to define the center of the base.)*
>
> Place Cylinder > Define radius *(Place a data point or key-in coordinates to define the radius. If Radius is set to ON, then the data point accepts the base.)*
>
> Place Cylinder > Define height *(Place a data point or key-in coordinates to define the height. If Height is set to ON, then the data point accepts the cylinder.)*

PLACE CONE

The Place Cone tool places a cone of unequal radius on each end. To place a cone, invoke the Place Cone tool from:

Task Navigation tool box (active task set to 3D Primitives located in the Solids Modeling)	Select the Place Cone tool (see Figure 16–31).
Keyboard Navigation (Task Navigation tool box with active task set to 3D Primitives)	**r**

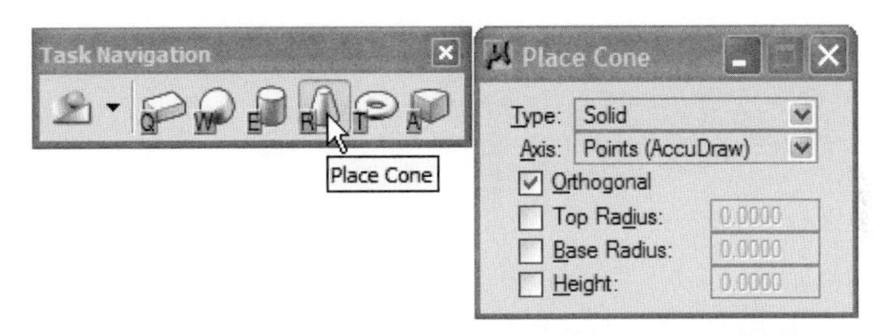

Figure 16–31 Invoking the Place Cone tool from the Task Navigation tool box (active task set to 3D Primitives)

Select the type of surface from the **Type** option menu in the Tool Settings window.

From the **Axis** option menu in the Tool Settings window, select the direction of the cone's axis or its height relative to the view or design file axes. If set to Screen X, Screen Y, or Screen Z, the direction of the cone's axis or height is set with the selected screen (view) axis. If set to Drawing X, Drawing Y, or Drawing Z, the direction of the cone's axis or height is set with the selected design file axis.

If necessary, turn ON the check boxes for **Orthogonal, Top Radius, Base Radius,** and **Height** in the Tool Settings window. If **Orthogonal** is set to ON, the cone is a right cone. If the constraints for **Top Radius, Base Radius**, and **Height** are ON, be sure to key-in appropriate values in the edit fields.

MicroStation prompts:

> Place Cone > Enter center point *(Place a data point or key-in coordinates to define the center of the base.)*
>
> Place Cone > Define base radius *(Place a data point or key-in coordinates to define the base radius. If Base Radius is set to ON, then the data point accepts the base.)*
>
> Place Cone > Define height *(Place a data point or key-in coordinates to define the height and top's center. If Height is set to ON, then the data point defines the top's center; if Orthogonal is set to ON, then the data point defines the direction of the height only.)*
>
> Place Cone > Define top radius *(Place a data point or key-in coordinates to define the top radius. If Top Radius is set to ON, then the data point accepts the cone.)*

PLACE TORUS

The Place Torus tool creates a solid or surface with a donut-like shape. To place a torus, invoke the Place Torus tool from:

Task Navigation tool box (active task set to 3D Primitives located in the Solids Modeling)	Select the Place Torus tool (see Figure 16–32).
Keyboard Navigation (Task Navigation tool box with active task set to 3D Primitives)	**t**

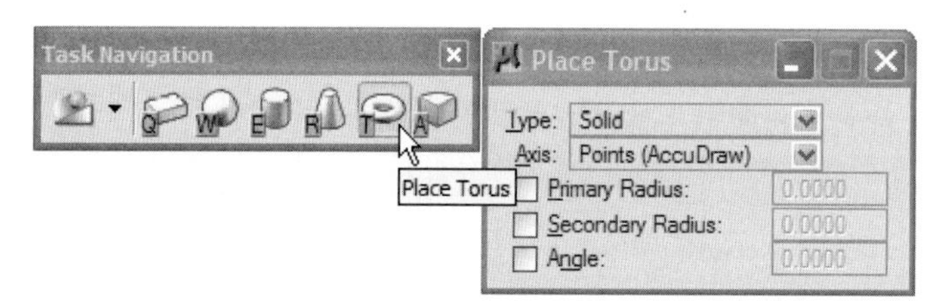

Figure 16–32 Invoking the Place Torus tool from the Task Navigation tool box (active task set to 3D Primitives)

Select the type of surface from the **Type** option menu in the Tool Settings window.

Select the direction of the axis of revolution relative to the view or design file axes from the **Axis** option menu in the Tool Settings window. If set to Screen X, Screen Y, or Screen Z, the axis or revolution is set with the selected screen (view) axis. If set to Drawing X, Drawing Y, or Drawing Z, the axis of revolution is set with the selected design file axis.

If necessary, turn ON the check boxes for **Primary Radius**, **Secondary Radius**, and **Angle** in the Tool Settings window. If the constraints for **Primary Radius**, **Secondary Radius**, and **Angle** are ON, be sure to key-in appropriate values in the edit fields.

MicroStation prompts:

> Place Torus > Enter start point *(Place a data point or key-in coordinates to define the start point.)*
>
> Place Torus > Define center point *(Place a data point or key-in coordinates to define the center point, primary radius, and start angle. If Primary Radius is set to ON, then the data point defines the center and the start angle.)*
>
> Place Torus > Define angle and secondary radius *(Place a data point or key-in coordinates to define the secondary radius and the sweep angle. If Secondary Radius is set to ON, then the data point defines the sweep angle; if Angle is set to ON, then the data point defines the secon-*

dary radius; and if both Secondary Radius and Angle are set to ON, then the data point defines the direction of the sweep angle rotation.)

PLACE WEDGE

The Place Wedge tool creates a wedge—a volume of revolution with a rectangular cross section. To place a wedge, invoke the Place Wedge tool from:

Task Navigation tool box (active task set to 3D Primitives located in the Solids Modeling)	Select the Place Wedge tool (see Figure 16–33).
Keyboard Navigation (Task Navigation tool box with active task set to 3D Primitives)	**a**

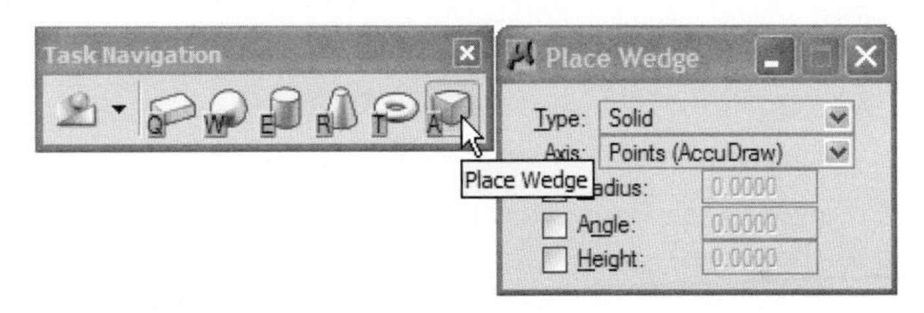

Figure 16–33 Invoking the Place Wedge tool from the Task Navigation tool box (active task set to 3D Primitives)

Select the type of surface from the **Type** option menu in the Tool Settings window.

Select the direction of the axis of revolution relative to the view or design file axes from the **Axis** option menu in the Tool Settings window. If set to Screen X, Screen Y, or Screen Z, the axis of revolution is set with the selected screen (view) axis. If set to Drawing X, Drawing Y, or Drawing Z, the axis of revolution is set with the selected design file axis.

If necessary, turn ON the check boxes for **Radius**, **Angle**, and **Height** in the Tool Settings window. If the constraints for **Radius**, **Angle**, and **Height** are ON, be sure to key-in appropriate values in the edit fields.

MicroStation prompts:

Place Wedge > Enter start point *(Place a data point or key-in coordinates to define the start point.)*

Place Wedge > Define center point *(Place a data point or key-in coordinates to define the center point and the start angle. If Radius is set to ON, then the data point defines the start angle.)*

Place Wedge > Define angle *(Place a data point or key-in coordinates to define the sweep angle. If Angle is set to ON, then the data point defines the direction of the rotation.)*

Place Wedge > Define Height *(Place a data point or key-in coordinates to define the height. If Height is set to ON, then the data point defines whether the wedge is projected up or down from the start plane.)*

CHANGING THE STATUS—SOLID OR SURFACE

The Convert 3D tool can change the status of an element from surface to solid, or vice versa. To change the status, invoke the Convert 3D tool from:

Task Navigation tool box (active task set to Modify Surfaces located in the Solids Modeling)	Select the Convert 3D tool (see Figure 16–34).
Keyboard Navigation (Task Navigation tool box with active task set to 3D Primitives)	**e**

Figure 16–34 Invoking the Convert 3D tool from the Task Navigation tool box (active task set to Modify Surfaces)

Select the type of surface to change from the **Convert To** option menu in the Tool Settings window.

MicroStation prompts:

Convert 3D > Identify solid or surface *(Identify the element whose status is to change.)*

Convert 3D > Accept/Reject *(Place a data point to accept the change in status, or click the Reject button to reject the operation.)*

USING ACCUDRAW IN 3D

AccuDraw 3D provides the ability to work in a pictorial view rather than the standard orthogonal views. AccuDraw automatically constrains data points to its drawing plane regardless of its orientation to the view.

Open the AccuDraw window from:

Primary Tools tool box	Select the Toggle AccuDraw tool (see Figure 16–35).
Key-in window	**accudraw activate** (or **acc a**) (ENTER)

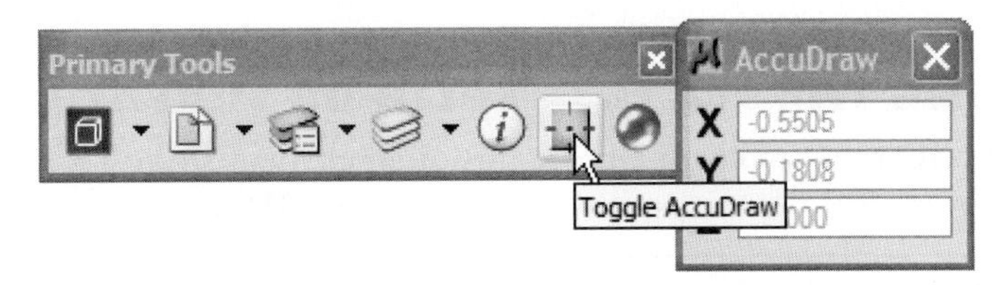

Figure 16–35 Invoking the Toggle AccuDraw tool from the Primary Tools tool box

The AccuDraw window opens, either as a floating window or docked at the top of the MicroStation workspace.

In 3D, when using rectangular coordinates, the AccuDraw window has an additional field for the *Z* axis. For polar coordinates in 3D, the AccuDraw window has the same two fields as in 2D.

The AccuDraw keyboard shortcuts can be used to rotate the drawing plane axes, making it convenient to draw in an isometric view. For example, it is easy with AccuDraw to place a nonplanar complex chain or complex shape in an isometric view in any direction without reverting to an orthogonal view.

AccuDraw's ability to adhere to the standard view axes while manipulating a drawing in a pictorial view is so important that it maintains the current orientation from tool to tool.

By default, AccuDraw orients the drawing plane to the view axes, similar to working with 2D design. AccuDraw can be returned to this orientation any time the focus is in the AccuDraw window by pressing the V key.

To rotate the drawing plane axes to align with the standard top view, focus in the AccuDraw window and press the T key. AccuDraw dynamically rotates the compass to indicate the orientation of the drawing plane.

To rotate the drawing plane axes to align with the standard front view, focus in the AccuDraw window and press the F key. AccuDraw dynamically rotates the compass to indicate the orientation of the drawing plane.

To rotate the drawing plane axes to align with the standard side (left or right) view, focus in the AccuDraw window and press the S key. AccuDraw dynamically rotates the compass to indicate the orientation of the drawing plane.

To rotate the drawing plane axes 90 degrees about an individual axis, focus in the AccuDraw window and press letters R and X to rotate 90 degrees about the *X* axis,

R and Y to rotate 90 degrees about the *Y* axis, and R and Z to rotate 90 degrees about the *Z* axis.

To rotate the drawing plane axes interactively, focus in the AccuDraw window and press letters R and A. Place data points to locate the *X* axis origin, the direction of the *X* axis, and the direction of the *Y* axis.

PROJECTED SURFACES

The Extrude tool creates a unique 3D object from 2D elements. Line, line string, arc, ellipse, text, multi-line, complex chain, complex shape, and B-spline curve are the elements that can be projected to a defined distance. Surfaces formed between the original boundary element and its projection are indicated by straight lines connecting the keypoints. To project a boundary element, invoke the Extrude tool from:

Task Navigation tool box (active task set to 3D Construct located in the Solids Modeling)	Select the Extrude tool (see Figure 16–36).
Keyboard Navigation (Task Navigation tool box with active task set to 3D Primitives)	**q**

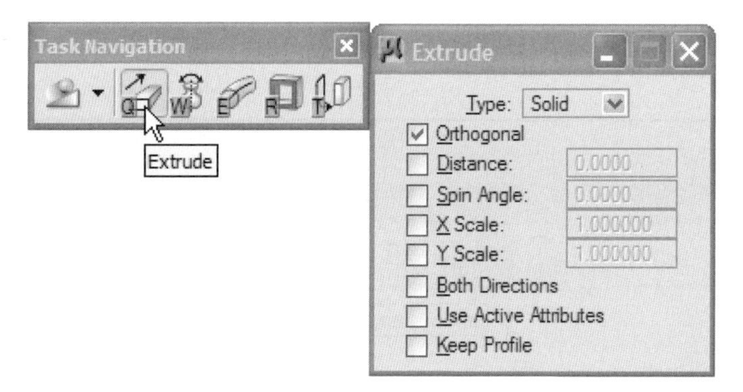

Figure 16–36 Invoking the Extrude tool from the Task Navigation tool box (active task set to 3D Construct)

Select appropriate options in the Tool Settings window. Select the type of surface from the **Type** option menu. When the **Orthogonal** check box is set to ON, the profile element is extruded orthogonally. The **Distance** check box, when set to ON, sets the distance in working units the element is extruded. The **Spin Angle** check box, when set to ON, sets the spin angle. The **X Scale** and **Y Scale** check boxes, when set to ON, sets the scale factor in the x-direction and y-direction respectively. When the **Both Directions** check box is set to ON, the profile element

is extruded in both directions. When **Use Active Attributes** check box is set to ON, the active attributes are applied. When the **Keep Profile** check box is set to ON, the original profile element is kept in the design.

MicroStation prompts:

Extrude > Identify profile *(Identify the boundary element.)*

Extrude > Define distance *(Place a data point to define the height. If Distance is set to ON, then the data point provides the direction.)*

EXTRUDE ALONG A PATH

The Extrude Along Path tool creates a tubular surface or solid extrusion along a path. Line, line string, arc, ellipse, text, multi-line, complex chain, complex shape, and B-spline curve are the elements that can be projected along a path. Straight lines connecting the keypoints indicate surfaces formed between the original boundary element and its projection.

To project a boundary element along a path, invoke the Extrude Along Path tool from:

Task Navigation tool box (active task set to 3D Construct located in the Solids Modeling)	Select the Extrude Along Path tool (see Figure 16–37).
Keyboard Navigation (Task Navigation tool box with active task set to 3D Primitives)	**e**

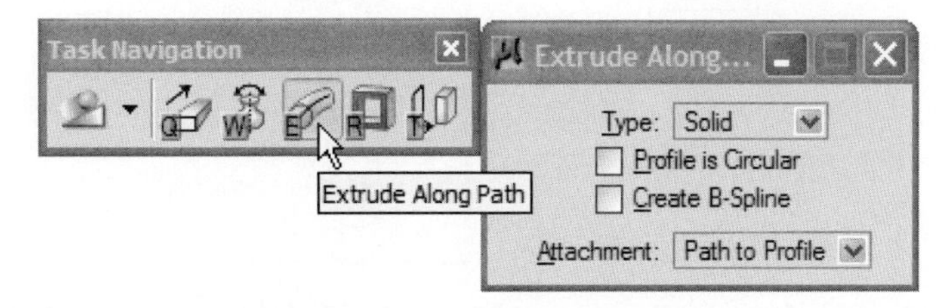

Figure 16–37 Invoking the Extrude Along Path tool from the Task Navigation tool box (active task set to 3D Construct)

Select the type of surface from the **Type** option menu in the Tool Settings window.

When the **Profile is Circular** check box is set to OFF, the surface/solid is constructed by extruding one element (the profile) along another element (the path). Orientation of the profile changes continually to follow the orientation of the path.

If it is set to ON, the Tool Settings window expands to display **Inside** and **Outside** radius settings for a tube with circular cross-section to be generated.

When the **Create B-Spline** check box is set ON, a B-Spline surface or solid is created.

The **Attachment** option (available only when **Profile is Circular** check box is set to OFF) sets the attachment mode for the profile to the path.

MicroStation prompts:

Extrude Along Path > Identify path *(Identify the boundary element.)*

Extrude Along Path > Identify profile or snap to profile at attachment point *(Identify profile or snap to profile at attachment point.)*

Extrude Along Path > Accept to create *(Click the Accept button to create the profile.)*

SURFACE OF REVOLUTION

The Construct Revolution tool is used to create a unique 3D surface or solid that is generated by rotating a boundary element about an axis of revolution. Line, line string, arc, ellipse, shape, complex chain, complex shape, and B-spline curve are the elements that can be used in creating a 3D surface or solid. Surfaces created by the boundary element as it is rotated are indicated by arcs connecting the key-points. To create a 3D surface or solid of revolution, invoke the Construct Revolution tool:

Task Navigation tool box (active task set to 3D Construct located in the Solids Modeling)	Select the Construct Revolution tool (see Figure 16–38).
Keyboard Navigation (Task Navigation tool box with active task set to 3D Primitives)	**w**

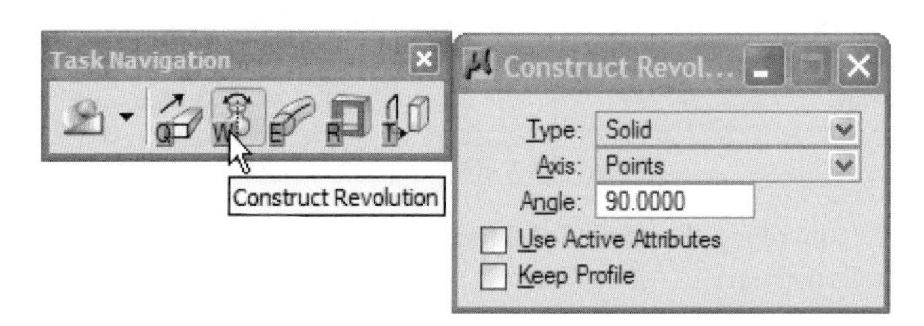

Figure 16–38 Invoking the Construct Revolution tool from the Task Navigation tool box (active task set to 3D Construct)

Select the type of surface from the **Type** option menu in the Tool Settings window.

Select the direction of the axis of revolution relative to the view or design file axes from the **Axis** option menu in the Tool Settings window. If set to Screen X, Screen Y, or Screen Z, the axis of revolution is set with the selected screen (view) axis. If set to Drawing X, Drawing Y, or Drawing Z, the axis of revolution is set with the selected design file axis.

If necessary, turn ON the check boxes for **Angle** and **Keep Profile** in the Tool Settings window. If the constraint for **Angle** is ON, be sure to key-in the appropriate value in the edit field.

MicroStation prompts:

Construct Revolution > Identify profile *(Identify the boundary element.)*

Construct Revolution > Define axis of revolution *(Place a data point or key-in coordinates. If Axis is set to Points, this data point defines one point on the axis of revolution and subsequently MicroStation prompts you for a second data point. If not, this data point defines the axis of revolution.)*

Construct Revolution > Accept, continue surface/reset to finish *(Place additional data points to continue, and/or press the Reset button to terminate the sequence.)*

SHELL SOLID

The Shell Solid tool creates a hollowed out solid for one or more selected faces of a defined thickness. To create a hollowed out solid, invoke the Shell Solid tool from:

Task Navigation tool box (active task set to 3D Construct located in the Solids Modeling)	Select the Shell Solid tool (see Figure 16–39).
Keyboard Navigation (Task Navigation tool box with active task set to 3D Primitives)	**r**

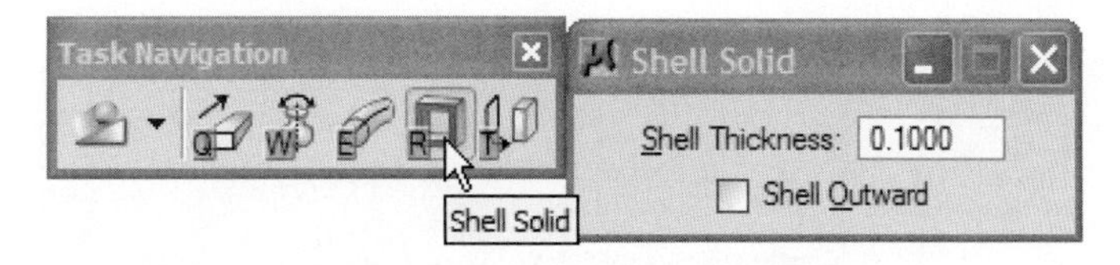

Figure 16–39 Invoking the Shell Solid tool from the Task Navigation tool box (active task set to 3D Construct)

Specify the shell thickness in the **Shell Thickness** edit field in the Tool Settings window. Set the **Shell Outward** check box to OFF to create a hollowed out solid for one or more selected faces. If set to ON, the material is added to the outside and the original solid defines the inside of the walls.

MicroStation prompts:

> Shell Solid > Identify target solid *(Identify the target solid.)*
>
> Shell Solid > Identify face to open *(Move the screen pointer over the solid; the face nearest the pointer highlights and data point selects the highlighted face.)*
>
> Shell Solid > Accept/Reject (select next face) *(Select additional faces, click the Reset button to deselect an incorrect face; to accept the selection of faces, click the Accept button to complete the selection.)*

THICKEN TO SOLID

The Thicken To Solid tool is used to add thickness to an existing surface to create a solid. Specify the thickness graphically or key-in a distance. To add thickness to an existing surface to create a solid, invoke the Thicken To Solid tool from:

Task Navigation tool box (active task set to 3D Construct located in the Solids Modeling)	Select the Thicken to Solid tool (see Figure 16–40).
Keyboard Navigation (Task Navigation tool box with active task set to 3D Primitives)	**t**

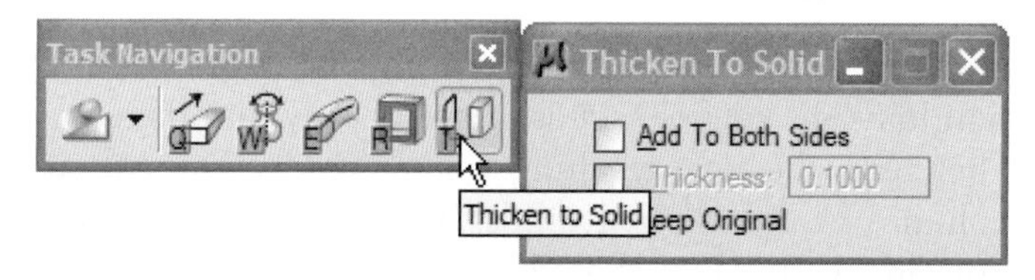

Figure 16–40 Invoking the Thicken to Solid tool from the Task Navigation tool box (active task set to 3D Construct)

To add the thickness to both sides of the selected surface, set the **Add To Both Sides** check box to ON. To increase the thickness to a specific value, key-in the values in the **Thickness** edit field and turn ON the check box. To keep the original profile element, turn ON the check box for **Keep Original**.

MicroStation prompts:

> Thicken to Solid > Identify surface *(Identify the surface.)*
>
> Thicken to Solid > Define thickness *(An arrow is displayed; move the pointer to the side you want to increase the thickness and click the Accept button.)*

PLACING 2D ELEMENTS

Any 2D elements (such as blocks and circles) that are placed with data points without snapping to existing elements will be placed at the active depth of the view. Also, they will be parallel to the screen. Elements that require fewer than three data points to define (such as blocks, circles with radius, circles with diameter/center, and polygons) take their orientation from the view being used. The points determine only their dimensions, not their orientation. Elements that require three or more data points to describe (shapes, circles by edge, ellipses, and rotated blocks) also provide their planar orientation. Once the first three points have been specified, any further points will fall on the same plane.

CREATING COMPOSITE SOLIDS

MicroStation provides three tools that can create a new composite solid by combining two solids using Boolean operations. There are three basic Boolean operations that can be performed in MicroStation:

- Union
- Intersection
- Difference

UNION OPERATION

The union is the process of creating a new composite solid from two solids. The union operation joins the original solids in such a way that there is no duplication of volume. Therefore, the total resulting volume can be equal to or less than the sum of the volumes in the original solids. To create a composite solid with the union operation, invoke the Construct Union tool from:

Task Navigation tool box (active task set to 3D Modify located in the Solids Modeling)	Select the Construct Union tool (see Figure 16–41).
Keyboard Navigation (Task Navigation tool box with active task set to 3D Primitives)	**r**

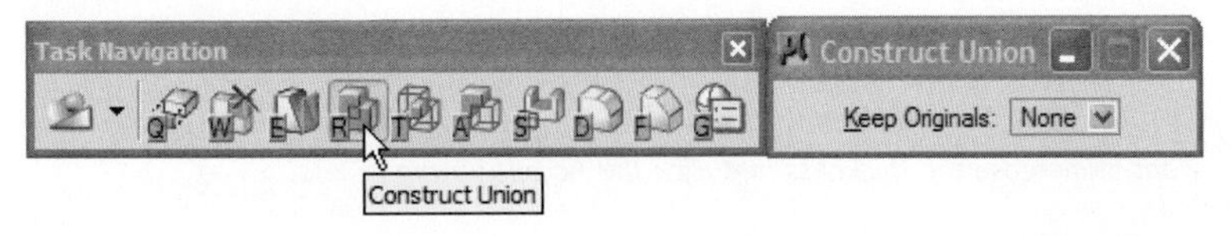

Figure 16–41 Invoking the Construct Union tool from the Task Navigation tool box (active task set to 3D Modify)

The **Keep Originals** options menu determines whether or not the original solids are retained after constructing the solid. The **None** selection does not retain any of the originals, **All** selection retains all the original solids, **First** selection retains first solid identified, and **Last** selection retains last solid identified.

MicroStation prompts:

> Construct Union > Identify first solid *(Identify the first solid element for union.)*
>
> Construct Union > Identify next solid *(Identify the second solid element for union.)*
>
> Construct Union > Identify next solid, or data point to finish *(Identify the third element for union or click the data point to accept the union of two selected solids.)*

See Figure 16–42 for an example of creating a composite solid by joining two cylinders with the Construct Union tool.

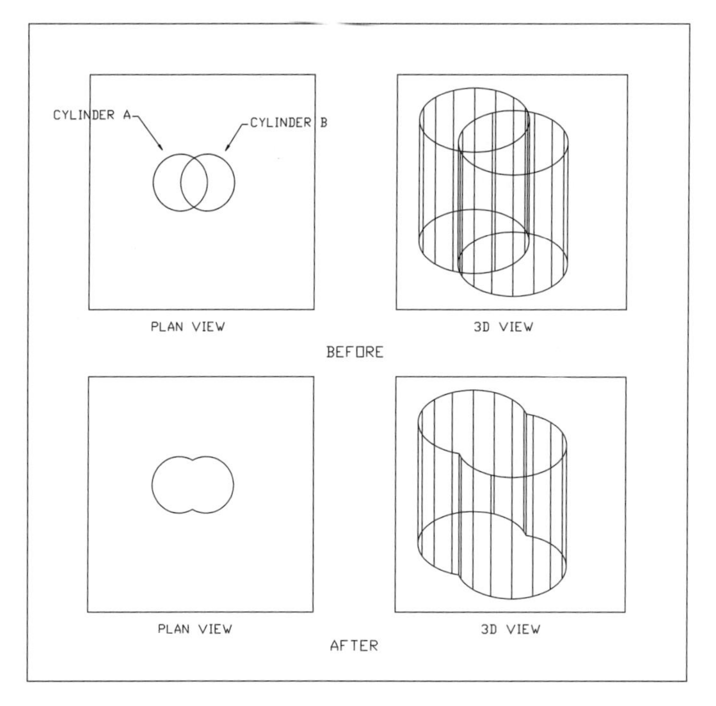

Figure 16–42 Creating a composite solid by joining two cylinders via the Construct Union tool

MicroStation also provides a tool to create a new composite solid from two surfaces. To combine two surfaces by union, invoke the Boolean Surface Union tool (available only by key-in) and select the surfaces to make it into a composite solid. The parts of the solids left are determined by their surface normal orientations. The surface normals can be changed by the Change Surface Normal tool.

INTERSECTION OPERATION

The intersection is the process of forming a composite solid from only the volume that is common to two solids. To create a composite solid with the intersection operation, invoke the Construct Intersection tool from:

Task Navigation tool box (active task set to 3D Modify located in the Solids Modeling)	Select the Construct Intersection tool (see Figure 16–43).
Keyboard Navigation (Task Navigation tool box with active task set to 3D Modify)	**t**

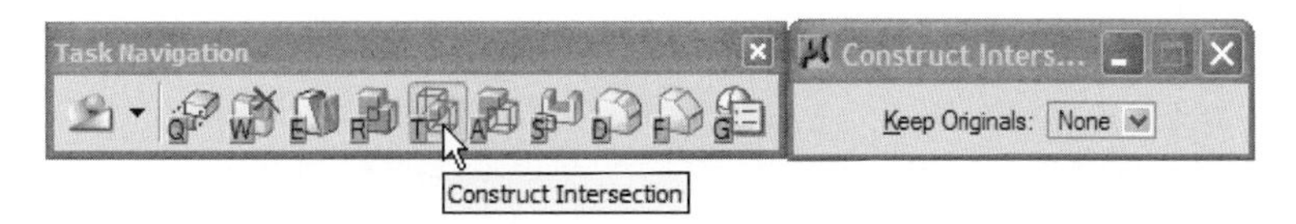

Figure 16–43 Invoking the Construct Intersection tool from the Task Navigation tool box (active task set to 3D Modify)

The **Keep Originals** options menu determines whether or not the original solids are retained after constructing the solid. The **None** selection does not retain any of the originals, **All** selection retains all the original solids, **First** selection retains first solid identified and **Last** selection retains last solid identified.

MicroStation prompts:

> Construct Intersection > Identify first solid *(Identify the first element for intersection.)*
>
> Construct Intersection > Identify next solid *(Identify the second element for intersection.)*
>
> Construct Intersection > Identify next solid, or data point to finish *(Identify the next element or click the Data button to complete the selection.)*

See Figure 16–44 for an example of creating a composite solid by joining two cylinders via the Construct Intersection tool.

MicroStation also provides a tool to create a new composite solid from two surfaces. To combine two surfaces by intersection, invoke the Boolean Surface Intersection tool (available only by key-in) and select the surfaces to make it to a composite solid. The parts of the solids left are determined by their surface normal orientations. The surface normals can be changed by the Change Surface Normal tool.

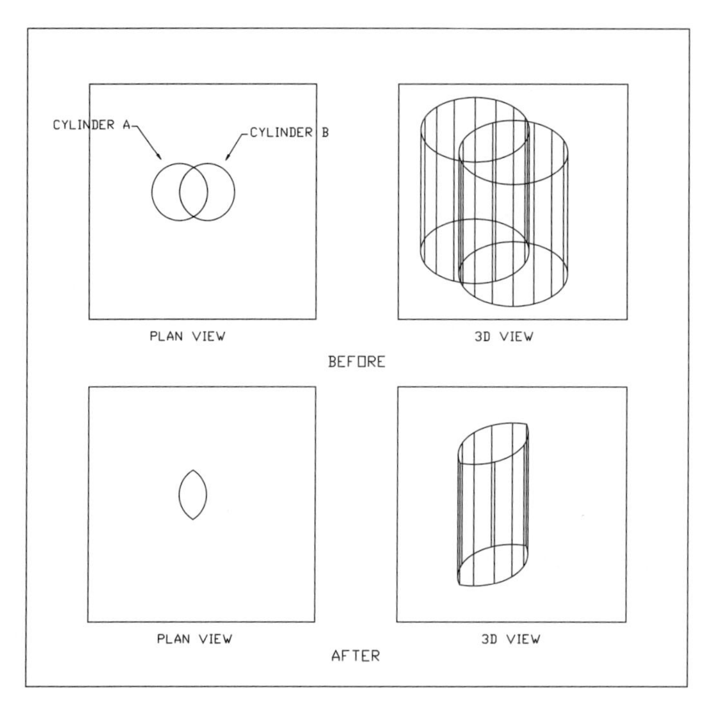

Figure 16–44 Creating a composite solid by intersecting two cylinders with the Construct Intersection tool

DIFFERENCE OPERATION

The difference operation is the process of forming a composite solid by starting with a solid and removing from it any volume it has in common with a second object. If the entire volume of the second solid is contained in the first solid, then what is left is the first solid minus the volume of the second solid. However, if only part of the volume of the second solid is contained within the first solid, then only the part that is duplicated in the two solids is subtracted. To create a composite solid with the difference operation, invoke the Construct Difference tool from:

Task Navigation tool box (active task set to 3D Modify located in the Solids Modeling)	Select the Construct Difference tool (see Figure 16–45).
Keyboard Navigation (Task Navigation tool box with active task set to 3D Modify)	**a**

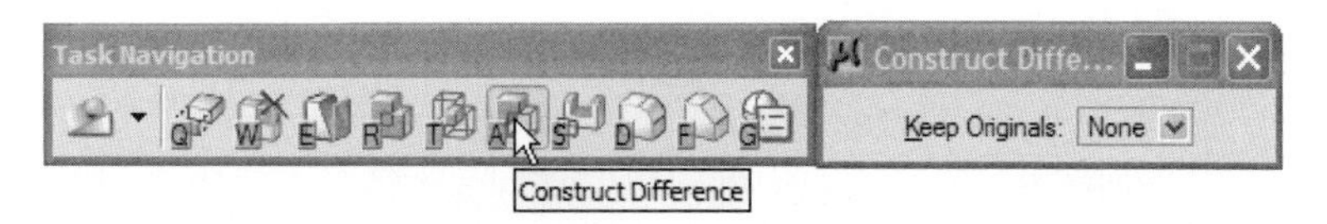

Figure 16–45 Invoking the Construct Difference tool from the Task Navigation tool box (active task set to 3D Modify)

The **Keep Originals** options menu determines whether or not the original solids are retained after constructing the solid. The **None** selection does not retain any of the originals, **All** selection retains all the original solids, **First** selection retains first solid identified and **Last** selection retains last solid identified.

MicroStation prompts:

> Construct Difference > Identify solid to subtract from *(Identify the first element for the difference operation.)*
>
> Construct Difference > Identify next solid or surface to subtract *(Identify the second solid or surface to subtract.)*
>
> Construct Difference > Identify next solid or surface to subtract, or data point to finish *(Identify the next element to subtract, or click the Data button to complete the selection.)*

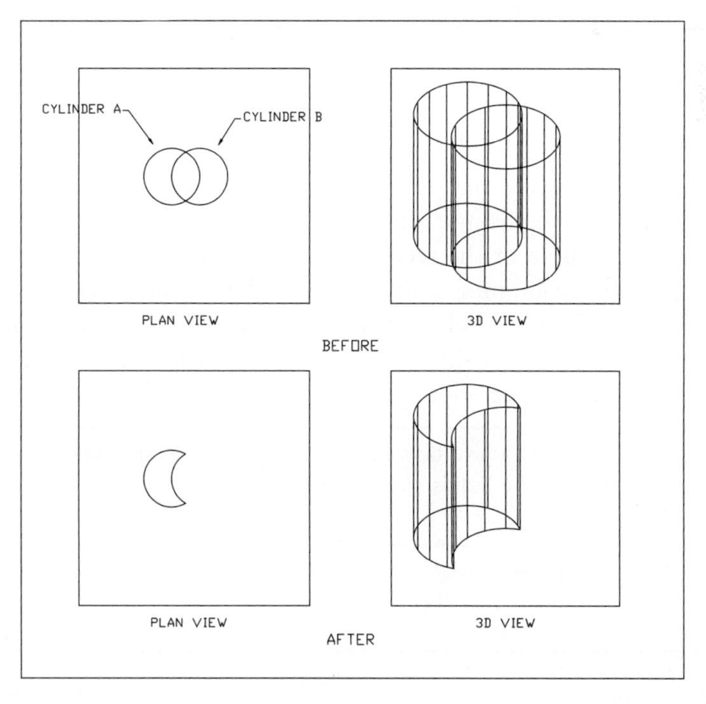

Figure 16–46 Creating a composite solid by subtracting Cylinder B from Cylinder A using the Construct Difference Between Surfaces tool

MicroStation also provides a tool to create a new composite solid from two surfaces. To combine two surfaces by the difference operation, invoke the Boolean Surface Difference tool (available only by key-in) and select the surfaces to make it into a composite solid. The parts of the solids left are determined by their surface normal orientations. The surface normals can be changed by Change Surface Normal tool.

CHANGE NORMAL

The Change Normal Direction tool can change the surface normal direction for a surface. This is useful to control the way the elements are treated while performing the Boolean operations. To change the surface normal of an element, invoke the Change Normal Direction tool from:

Task Navigation tool box (active task set to Modify Surfaces located in the Solids Modeling)	Select the Change Normal Direction tool (see Figure 16–47).
Keyboard Navigation (Task Navigation tool box with active task set to Modify Surfaces)	**t**

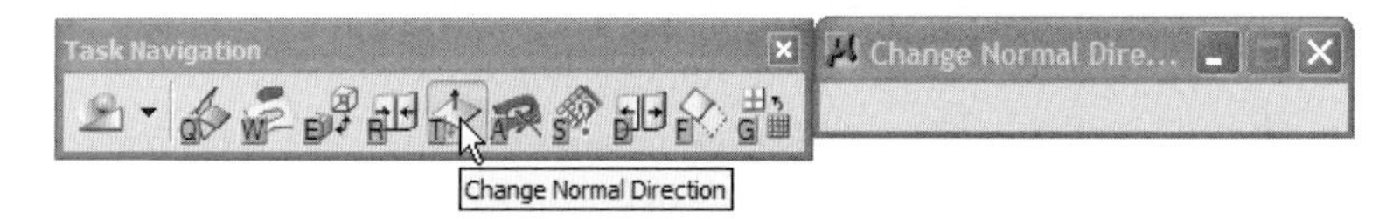

Figure 16–47 Invoking the Change Normal Direction tool from the Task Navigation tool box (active task set to Modify Surfaces)

MicroStation prompts:

Change Normal Direction > Identify element *(Identify the element; surface normals are displayed.)*

Change Normal Direction > Accept to change or Reject *(Place a data point to accept the change in the normal direction or reject the change.)*

MODIFY SOLID

The Modify Solid tool lets you relocate a face of a solid outward (positive) or inward (negative), relative to the center of the solid. To modify a solid, invoke the Modify Solid tool from:

Task Navigation tool box (active task set to 3D Modify located in the Solids Modeling)	Select the Modify solid tool (see Figure 16–48).
Keyboard Navigation (Task Navigation tool box with active task set to 3D Modify)	**q**

Figure 16–48 Invoking the Modify Solid tool from the Task Navigation tool box (active task set to 3D Modify)

To modify the selected solid face by key-in distance, set the **Distance** check box to ON and key-in the distance in the **Distance** edit field in the Tool Settings window.

MicroStation prompts:

Modify Solid > Identify target solid *(Identify the solid to modify.)*

Modify Solid > Select face to modify *(Select the face to modify.)*

Modify Solid > Define distance *(Using the arrow as the guide, move the pointer to define the distance dynamically when the Distance check box is set to OFF and direction.)*

REMOVE FACES

The Remove Faces tool is used to remove an existing face or a feature from a solid and then close the opening. It can also remove faces that are associated with a cut, a solid that has been added to or subtracted from the original, a shell solid, a fillet, or a chamfer. To remove a face or a feature, invoke the Remove Faces tool from:

Task Navigation tool box (active task set to 3D Modify located in the Solids Modeling)	Select the Remove Faces tool (see Figure 16–49).
Keyboard Navigation (Task Navigation tool box with active task set to 3D Modify)	**w**

Figure 16–49 Invoking the Remove Faces tool from the Task Navigation tool box (active task set to 3D Modify)

Select one of the two available methods to remove faces from the **Method** option menu in the Tool Settings window. The **Faces** selection removes one or more faces from a selected solid feature. The **Logical Groups** selection removes faces that are associated with a cut, a solid that has been added to or subtracted from the original, a shell solid, a fillet, or a chamfer. When the **Add Smooth Faces** check box is set to ON, any tangentially continuous faces are included with the selected face. If set to OFF, only the selected face is considered.

MicroStation prompts:

Remove Faces and Heal > Identify target solid *(Identify the solid to modify.)*

Remove Faces and Heal > Identify first face to remove *(Select the face to remove.)*

Remove Faces and Heal > Accept/Reject (select next face) *(Click the Accept button to remove the selected face, then select additional faces to continue the selection or click the Reject button to cancel the operation.)*

CUT SOLID

The Cut Solid tool splits a solid into two or more segments using a cutting profile. Cutting profiles may be open or closed elements. The open element profile must extend to the edge of the solid. To cut a solid, invoke the Cut Solid tool from:

Task Navigation tool box (active task set to 3D Modify located in the Solids Modeling)	Select the Cut Solid tool (see Figure 16–50).
Keyboard Navigation (Task Navigation tool box with active task set to 3D Modify)	**s**

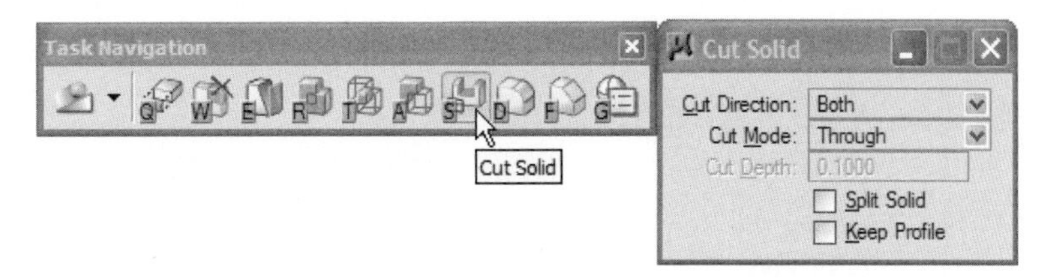

Figure 16–50 Invoking the Cut Solid tool from the Task Navigation tool box (active task set to 3D Modify)

The **Cut Direction** option menu in the Tool Settings window sets the direction of the cut relative to the cutting profile's Surface Normal. Available selections include:

- **Both**—Selects in both directions from the profile's plane.
- **Forward**—Selects from the forward direction from the profile's plane.
- **Back**—Selects from the backward direction from the profile's plane.

The **Cut Mode** option menu in the Tool Settings window sets the limits of the cut. Available selections include:

- **Through**—Cuts through all faces of the solid.
- **Define Depth**—Cuts into the solid a defined distance. Key-in the distance in the Cut Depth edit field.

The **Split Solid** check box controls whether the material is removed or not when it is split into segments. When it is set to ON, no material is removed from the solid, and it is split into two or more segments.

The **Keep Profile** check box controls whether the cutting profile remains in the design. When it is set to ON, the original cutting profile remains in the design.

MicroStation prompts:

Cut Solid > Identify target solid *(Identify the solid.)*

Cut Solid > Identify cutting profile *(Identify the cutting profile.)*

Cut Solid > Accept/Reject *(Click the Accept button to complete the cut or click the Reject button to cancel the operation.)*

CONSTRUCT A FILLET

The Fillet Edges tool is used to fillet or round for one or more edges of a solid, projected surface, or a surface of revolution. To construct a fillet, invoke the Fillet Edges tool from:

Task Navigation tool box (active task set to 3D Modify located in the Solids Modeling)	Select the Fillet Edges tool (see Figure 16–51).
Keyboard Navigation (Task Navigation tool box with active task set to 3D Modify)	**d**

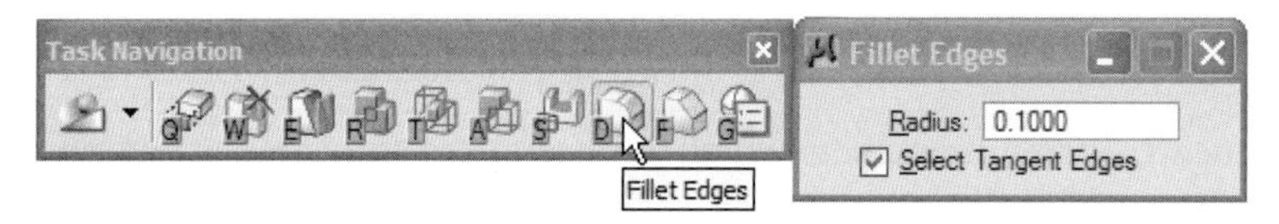

Figure 16–51 Invoking the Fillet Edges tool from the Task Navigation tool box (active task set to 3D Modify)

Key-in the radius for the fillet in the **Radius** edit field in the Tool Settings window. When the **Select Tangent Edges** check box is set to ON, the edges that are tangentially continuous are selected and rounded in one operation.

MicroStation prompts:

> Fillet Edges > Identify Edge to Fillet *(Identify the solid.)*
>
> Fillet Edges > Accept/Reject (select next edge) *(Click the Accept button to accept the fillet for the selected edge, and if necessary, select additional edges to fillet or click the Reject button to cancel the operation.)*

The Fillet Surfaces tool helps you construct a 3D fillet between two surfaces. The fillet is placed by sweeping an arc with a specified radius along the common intersecting curve. The fillet is created in the area pointed to by the surface normals of both surfaces.

To fillet between surfaces, invoke the Construct Fillet Between Surfaces tool from:

Task Navigation tool box (active task set to Fillet Surfaces located in the Solids Modeling)	Select the Fillet Surfaces tool (see Figure 16–52).
Keyboard Navigation (Task Navigation tool box with active task set to Fillet Surfaces)	**q**

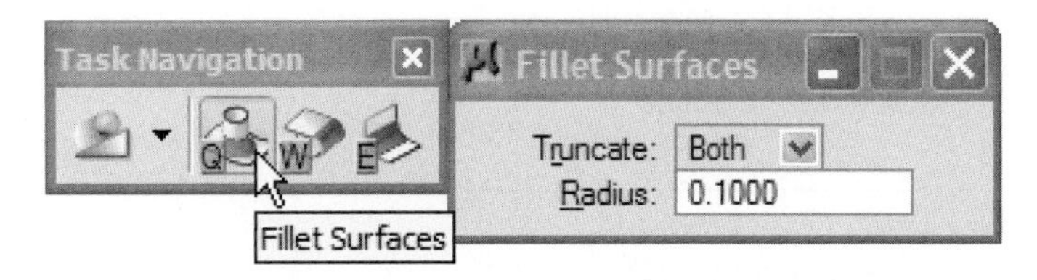

Figure 16–52 Invoking the Fillet Surfaces tool from the Task Navigation tool box (active task set to Fillet Surfaces)

Key-in the radius of the fillet to be drawn in the **Radius** edit field.

The **Truncate** option menu sets which surface(s) are truncated at the point of tangency with the fillet.

MicroStation prompts:

> Fillet Surfaces > Identify first surface *(Identify the first surface.)*
>
> Fillet Surfaces > Identify second surface *(Identify the second surface.)*
>
> Fillet Surfaces > Accept/Reject *(Click the Data button to accept the fillet, or click the Reject button to cancel the operation.)*

CONSTRUCT A CHAMFER

The Chamfer Edges tool is used to chamfer one or more edges of a solid, projected surface, or a surface of revolution. To construct a chamfer, invoke the Chamfer Edges tool from:

Task Navigation tool box (active task set to 3D Modify located in the Solids Modeling)	Select the Chamfer Edges tool (see Figure 16–53).
Keyboard Navigation (Task Navigation tool box with active task set to 3D Modify)	**f**

Figure 16–53 Invoking the Chamfer Edges tool from the Task Navigation tool box (active task set to 3D Modify)

Key-in chamfer distances in the **Distance 1** and **Distance 2** edit fields. The **Select Tangent Edges** check box, when set to ON, makes edges that are tangentially

continuous and are selected and rounded in one operation. The **Flip Direction** check box, when set to ON, reverses the direction of the chamfer and sets the values that the faces are trimmed.

MicroStation prompts:

> Chamfer Edges > Identify Edge to Chamfer *(Identify the solid.)*
>
> Fillet Edges > Accept/Reject (select next edge) *(Click the Accept button to accept the chamfer for the selected edge; if necessary, select additional edges to chamfer or click the Reject button to cancel the operation.)*

The Construct Chamfer Between Surfaces tool enables you to construct a 3D chamfer between two surfaces by a specified length along the common intersection curve. The chamfer is created in the area pointed to by the surface normals of both surfaces.

To chamfer between surfaces, invoke the Construct Chamfer Between Surfaces tool:

Key-in window	**chamfer surface** (or **cham s**) (ENTER)

The **Truncate** option menu sets which surface(s) are truncated at the point of tangency with the chamfer.

Key-in the Chamfer length in the **Chamfer Length** edit field.

The **Tolerance** check box sets the override for the system tolerance.

MicroStation prompts:

> Construct Chamfer Between Surfaces > Identify first surface *(Identify the first surface.)*
>
> Construct Chamfer Between Surfaces > Accept/Reject *(Click the Data button to accept the first surface selection.)*
>
> Construct Chamfer Between Surfaces > Identify second surface *(Identify the second surface.)*
>
> Construct Chamfer Between Surfaces > Accept/Reject *(Click the Data button to accept the second surface selection.)*
>
> Construct Fillet Between Surfaces > Accept/Reject *(Click the Data button to accept the chamfer, or click the Reject button to cancel the operation.)*

PLACING TEXT

MicroStation provides two options to place text in a 3D design: (1) placing text (view-dependent) in such a way that it appears planar to the screen in the view in which the data point is placed but rotated in the other views, or (2) placing text (view-independent) in such a way that it appears planar to the screen in all views.

To place text (view-dependent), click the Place Text icon in the Text tool box, select **By Origin** from the **Method** option menu, and follow the prompts. To place text (view-independent), click the Place Text icon in the Text tool box, select **View Ind** from the **Method** option menu, and follow the prompts. Text parameters are set up in the same way as in the 2D design.

FENCE MANIPULATIONS

Fences are used in a 3D design in much the same way as in a 2D design (see Chapter 6). The difference is that a 3D fence defines a volume. The volume is defined by the fenced area and the display depth of the view in which the fence is placed. The Fence lock options work the same way as in a 2D design.

CELL CREATION AND PLACEMENT

The procedure for creating and placing cells in 3D design is the same as in 2D design (see Chapter 11). Before creating a 3D cell, make sure the display depth is set to include all the elements to be used in the cell and the origin is defined at an appropriate active depth. If a normal (graphic) cell was created in the top view and then placed in the front view, it will appear as it did in the top view and rotated in other views. In other words, the normal cell is placed as view-dependent, whereas a point cell when placed will appear planar to the screen in all views. A point cell is placed view-independent.

A 2D cell library can be attached to a 3D design file, but the cells will have no depth and will be placed at the active depth of the working view.

DIMENSIONING

The procedure for dimensioning setup and placement in 3D design is similar to that for 2D design (see Chapter 9); however, consideration must be given to which plane will be used for the dimensioning. Before placing dimensions in a 3D design, make sure the appropriate option is selected from the **Alignment** option menu in the Linear Dimension tool box. The view measurement axis measures the projection of the element along the view's horizontal or vertical axis. The true measurement axis measures the actual distance between two points, not the projected distance. And the drawing measurement axis measures the projection of an element along the design cube coordinate system's axis.

RENDERING

Shading, or rendering, can turn a 3D model into a realistic, eye-catching image. MicroStation's rendering options provide complete control over the appearance of the final image. Add and control lighting in the design and define the reflective qualities of individual surfaces, making objects appear dull or shiny. The 3D model can be rendered entirely within MicroStation. This section provides an overview of

the various options available for rendering. Refer to the *MicroStation Reference Guide* for a more detailed description of various options.

SETTING UP CAMERAS

To establish a viewing position in MicroStation, imagine looking through a camera to see the image. By default, MicroStation places the camera at a right angle to a view's *XY* plane. If necessary, move or reposition the camera to view the model from a different viewing angle. To enable or disable the default camera setting and make changes to the camera setup, invoke the Camera Settings tool from:

Task Navigation tool box (active task set to Camera)	Select the Camera Settings tool (see Figure 16–54).
Keyboard Navigation (Task Navigation tool box with active task set to Camera)	**r**

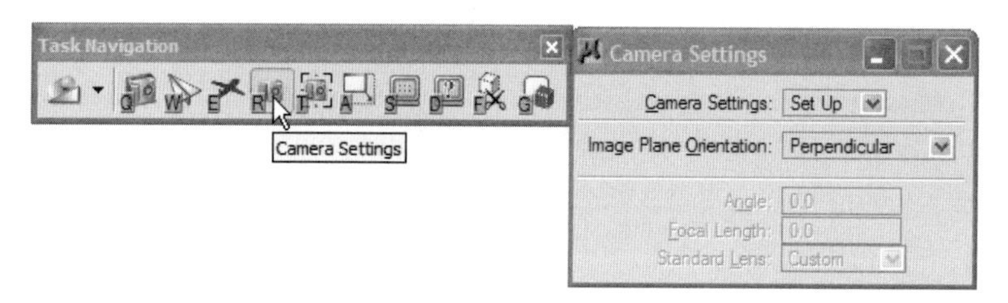

Figure 16–54 Invoking the Camera Settings tool from the Task Navigation tool box (active task set to Camera)

Select one of the available options from the **Camera Settings** option menu in the Tool Settings window:

- **Turn On**—Turns on the camera in a view or views.
- **Turn Off**—Turns off the camera in a view or views.
- **Set Up**—Turns on the camera in a view and sets the camera target and position. The target is the focal point (center) of a camera view. The position is the design cube location from which the model is viewed with the camera. Objects beyond the camera target appear smaller; objects in front of the camera target appear larger and may be outside of the viewing pyramid.
- **Move**—Moves the camera position.
- **Target**—Moves the target.

Select one of the available options from the **Image Plane Orientation** option menu in the Tool Settings window:

- **Perpendicular**—Perpendicular to the camera direction

- **Parallel to X axis**—Parallel to the view *X* axis; analogous to a bellows camera
- **Parallel to Y axis**—Parallel to the view *Y* axis; analogous to a bellows camera
- **Parallel to Z axis**—Parallel to the view *Z* axis; all vertical lines (along the axis) appear parallel

Set the lens angle in degrees and the lens focal length in millimeters in the **Angle** and **Focal Length** edit fields, respectively.

Select one of the available options from the **Standard Lens** option menu if you wish to use the standard lens type commonly used by photographers. MicroStation sets the appropriate lens angle and focal length.

MicroStation prompts depend on the options selected in the Tool Settings window.

PLACEMENT OF LIGHT SOURCES

The lighting setup is equally as important as setting the camera angle for producing a high-quality rendered image. MicroStation provides four types of lighting:

- Ambient lighting
- Flashbulb lighting
- Solar lighting
- Source lighting, including point, spot, and distant

Ambient lighting is a uniform light that surrounds the model. *Flashbulb lighting* is a localized, intense light that appears to emanate from the camera position. *Solar lighting* is sunlight. By defining a location on the earth in latitude, longitude, day, month, and time, MicroStation can simulate lighting for most exterior architectural projects. Ambient, flashbulb, and solar lighting are set in the Global settings box invoked from the Rendering submenu of the drop-down Settings menu.

Source lighting is achieved by placing light sources in the form of cells. The MicroStation program comes with three light source cells (point lights—PNTLT, spot lights—SPOTLT, and distant lights—DISTLT) provided in the LIGHTING.CEL cell library.

Point light can be thought of as a ball of light. It radiates beams of light in all directions. Such a light also has more natural characteristics. Its brilliance may be diminished as an object moves away from the source of light. An object that is near a point light will appear brighter; an object that is farther away will appear darker.

Spot lights are very much like the kind of spotlight you might be accustomed to seeing at a theater or auditorium. Spot lights produce a cone of light toward a target that you specify.

Distant light gives off a fairly straight beam of light that radiates in one direction, and its brilliance remains constant so that an object close to the light will receive as much light as a distant object.

Before placing the light cells, adjust the settings in the Source Lighting settings box invoked from the Rendering submenu of the Settings menu. MicroStation provides various options under the Tool drop-down menu, in the Settings box.

RENDERING METHODS

MicroStation provides seven different tools to render a view. To render a view, invoke the Render tool from:

View Control bar	Select the Render tool (see Figure 16–55).

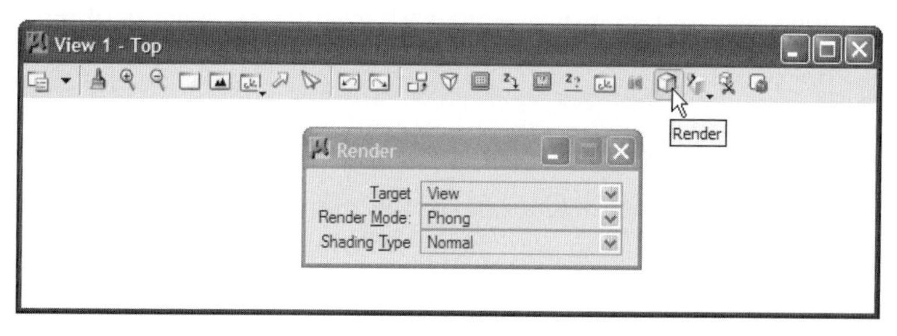

Figure 16–55 Invoking the Render tool from the View Control bar

Select the type of area or element to be rendered from the **Target** option menu in the Tool Settings window. The available options include View, Fence (contents), and Element.

Select one of the available options (Wiremesh, Hidden Line, Filled Hidden Line, Constant, Smooth, or Phong) from the **Render Mode** option menu in the Tool Settings window.

- **Wiremesh**—Similar to the default wireframe display, all elements are transparent and do not obscure other elements.
- **Hidden Line**—Displays only the element parts that would actually be visible.
- **Filled Hidden Line**—Identical to a Hidden Line option display except that the polygons are filled with the element color.
- **Constant**—Displays each element as one or more polygons filled with a single (constant) color. The color is computed once for each polygon, from the element color, material characteristics, and lighting configurations.

- **Smooth**—Displays the appearance of curved surfaces more realistically than in constant shaded models because polygon color is computed at polygon boundaries and color is blended across polygon interiors.
- **Phong**—Displays the image after re-computing the color of each pixel. Phong shading is useful for producing high-quality images when speed is not critical.

The **Shading Type** option menu sets the rendering method.

- **Anti-alias**—Displays the image with reduced jagged edges that are particularly noticeable on low-resolution displays. The additional time required for anti-aliasing is especially worthwhile when saving images for presentation, publication, or animated sequences.
- **Stereo**—Renders a view with a stereo effect that is visible when seen through 3D (red/blue) glasses. Stereo Phong shading takes twice as long as Phong because two images—one each from the perspective of the right and left eyes—are rendered and combined into one color-coded image.

DRAWING COMPOSITION

One of MicroStation's useful features is the ability to compose multiple views (standard and saved) on a drawing sheet. Multiple views can be plotted on one sheet of paper—what-you-see-is-what-you-get (WYSIWYG). Drawing Composition is used to attach multiple views. The views are attached as references. An attached view in a sheet file can be any standard (top, bottom, right, left, front, back, or isometric), fitted view, or any saved view of a model file. Standard views can be clipped or set to display only certain levels. A view of the model file can be attached in any position at any scale. MicroStation provides a tool that can group a set of views. A group of attached views can be moved, scaled, or detached as one. The tools in the Reference attachments (for a detailed explanation about References, refer to Chapter 13) simplify the process of creating sheet views.

Open the Exercise Manual PDF file for Chapter 16 on the accompanying CD for project and discipline specific exercises.

APPENDIX a

MicroStation Tool Boxes

Main

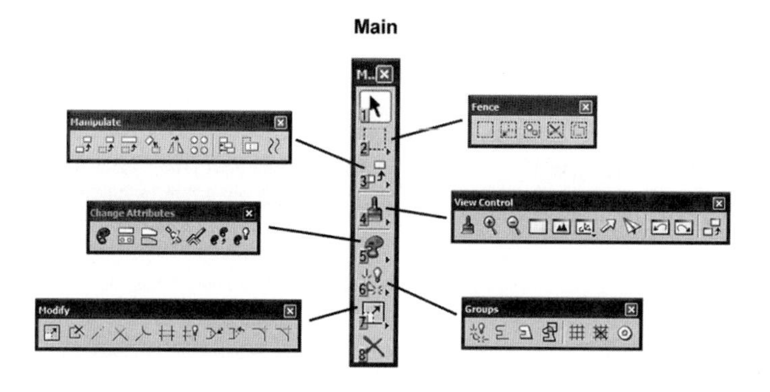

Standard

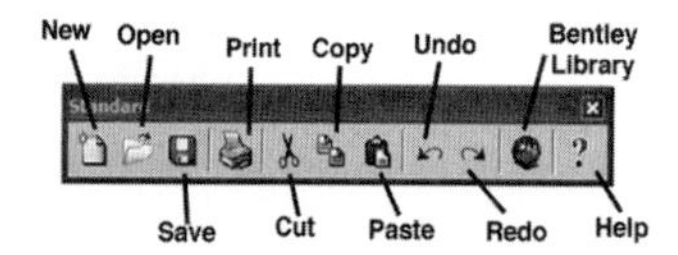

Primary

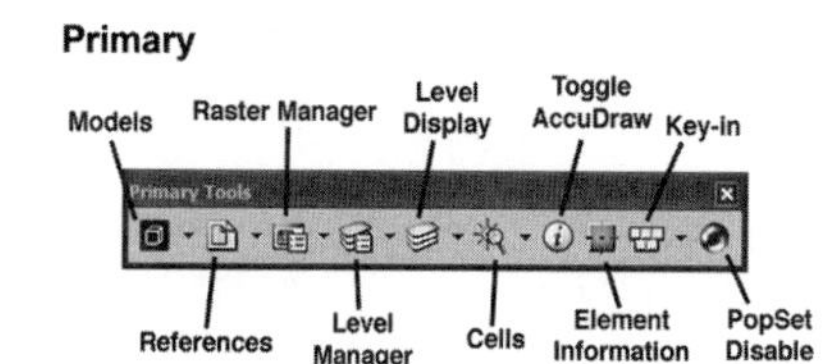

Attributes

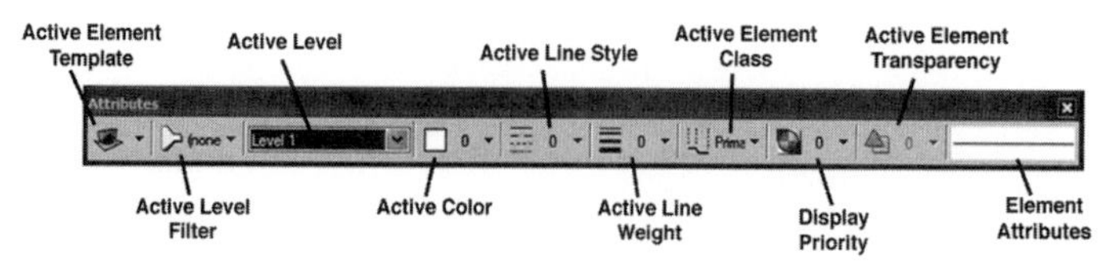

Task Navigation set to Drawing

Task Navigation set to Linear

Task Navigation set to Circles

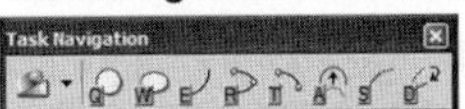

Task Navigation set to Polygons

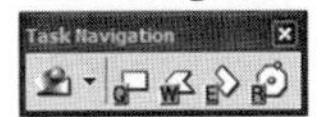

Task Navigation set to Text

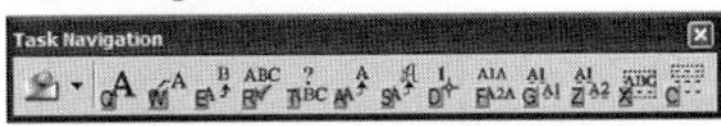

Task Navigation set to Tags

Task Navigation set to Cells

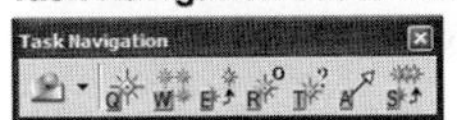

Task Navigation set to Patterns

Task Navigation set to Measure

Task Navigation set to Dimensioning

Task Navigation set to Drawing Composition

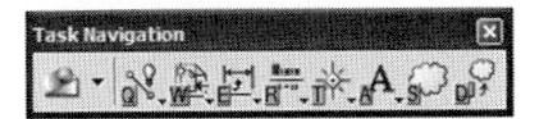

Task Navigation set to Basic 3D

Task Navigation set to Surface Modeling

Task Navigation set to Solid Modeling

Task Navigation set to Visualization

Task Navigation set to Animation

Key-in Commands

Note: Key-ins can be uppercase, lowercase, or combination.

TOOL NAME	KEY-IN
Add to Graphic Group	GROUP ADD
Attach Active Entity	ATTACH AE
Attach Active Entity to Fence Contents	FENCE ATTACH
Attach Displayable Attributes	ATTACH DA
Attach Reference File	REFERENCE ATTACH (RF=)
Automatic Create Complex Chain	CREATE CHAIN AUTOMATIC
Automatic Create Complex Shape	CREATE SHAPE AUTOMATIC
Automatic Fill in Enter Data Fields	EDIT AUTO
B-spline Polygon Display On/Off	MDL LOAD SPLINES; CHANGE BSPLINE POLYGON
Chamfer	CHAMFER
Change B-spline Surface to Active U-Order	MDL LOAD SPLINES; CHANGE BSPLINE UORDER
Change B-spline Surface to Active U-Rules	MDL LOAD SPLINES; CHANGE BSPLINE URULES
Change B-spline Surface to Active V-Order	MDL LOAD SPLINES; CHANGE BSPLINE VORDER
Change B-spline Surface to Active V-Rules	MDL LOAD SPLINES; CHANGE BSPLINE VRULES
Change B-spline to Active Order	MDL LOAD SPLINES; CHANGE BSPLINE ORDER

TOOL NAME	KEY-IN
Change Element to Active Class	CHANGE CLASS
Change Element to Active Color	CHANGE COLOR
Change Element to Active Level	CHANGE LEVEL
Change Element to Active Line Style	CHANGE STYLE
Change Element to Active Line Weight	CHANGE WEIGHT
Change Element to Active Symbol	CHANGE SYMBOLOGY
Change Fence Contents to Active Color	FENCE CHANGE COLOR
Change Fence Contents to Active Level	FENCE CHANGE LEVEL
Change Fence Contents to Active Style	FENCE CHANGE STYLE
Change Fence Contents to Active Symbology	FENCE CHANGE SYMBOLOGY
Change Fence Contents to Active Weight	FENCE CHANGE WEIGHT
Change Fill	CHANGE FILL
Change Text to Active Attributes	MODIFY TEXT
Circular Fillet (No Truncation)	FILLET NOMODIFY
Circular Fillet and Truncate Both	FILLET MODIFY
Circular Fillet and Truncate Single	FILLET SINGLE
Closed Cross Joint	MDL LOAD CUTTER; JOIN CROSS CLOSED
Closed Tee Joint	MDL LOAD CUTTER; JOIN TEE CLOSED
Complete Cycle Linear Pattern	PATTERN LINE SCALE
Construct Active Point at Distance Along an Element	CONSTRUCT POINT DISTANCE
Construct Active Point at Intersection	CONSTRUCT POINT INTERSECTION
Construct Active Points Between Data Points	CONSTRUCT POINT BETWEEN
Construct Angle Bisector	CONSTRUCT BISECTOR ANGLE
Construct Arc Tangent to Three Elements	CONSTRUCT TANGENT ARC 3

TOOL NAME	KEY-IN
Construct B-spline Curve by Least Squares	MDL LOAD SPLINES; CONSTRUCT BSPLINE CURVE LEAST SQUARE
Construct B-spline Curve by Points	MDL LOAD SPLINES; CONSTRUCT BSPLINE CURVE POINTS
Construct B-spline Curve by Poles	MDL LOAD SPLINES; CONSTRUCT BSPLINE CURVE POLES
Construct B-spline Surface by Cross-Section	MDL LOAD SPINES; CONSTRUCT BSPLINE SURFACE CROSS
Construct B-spline Surface by Edges	MDL LOAD SPLINES; CONSTRUCT BSPLINE SURFACE EDGE
Construct B-spline Surface by Least Squares	MDL LOAD SPLINES; CONSTRUCT BSPLINE SURFACE LEAST SQUARE
Construct B-spline Surface by Points	MDL LOAD SPLINES; CONSTRUCT BSPLINE SURFACE POINTS
Construct B-spline Surface by Poles	MDL LOAD SPLINES; CONSTRUCT BSPLINE SURFACE POLES
Construct B-spline Surface by Skin	MDL LOAD SPLINES; CONSTRUCT BSPLINE SURFACE SKIN
Construct B-spline Surface by Tube	MDL LOAD SPLINES; CONSTRUCT BSPLINE SURFACE TUBE
Construct B-spline Surface of Projection	MDL LOAD SPLINES; CONSTRUCT BSPLINE SURFACE PROJECTION
Construct B-spline Surface of Revolution	MDL LOAD SPLINES; CONSTRUCT BSPLINE SURFACE REVOLUTION
Construct Circle Tangent to Element	CONSTRUCT TANGENT CIRCLE 1
Construct Circle Tangent to Three Elements	CONSTRUCT TANGENT CIRCLE 3
Construct Line at Active Angle from Point (key-in)	CONSTRUCT LINE AA 4
Construct Line at Active Angle from Point	CONSTRUCT LINE AA 3
Construct Line at Active Angle to Point (key-in)	CONSTRUCT LINE AA 2
Construct Line at Active Angle to Point	CONSTRUCT LINE AA 1
Construct Line Bisector	CONSTRUCT BISECTOR LINE

TOOL NAME	KEY-IN
Construct Line Tangent to Two Elements	CONSTRUCT TANGENT BETWEEN
Construct Minimum Distance Line	CONSTRUCT LINE MINIMUM
Construct Perpendicular from Element	CONSTRUCT PERPENDICULAR FROM
Construct Perpendicular to Element	CONSTRUCT PERPENDICULAR TO
Construct Points Along Element	CONSTRUCT POINT ALONG
Construct Surface/Solid of Projection	SURFACE PROJECTION
Construct Surface/Solid of Revolution	SURFACE REVOLUTION
Construct Tangent Arc by Keyed-in Radius	CONSTRUCT TANGENT ARC 1
Construct Tangent from Element	CONSTRUCT TANGENT FROM
Construct Tangent to Circular Element and Perpendicular to Linear Element	CONSTRUCT TANGENT PERPENDICULAR
Construct Tangent to Element	CONSTRUCT TANGENT TO
Convert Element to B-spline (Copy)	MDL LOAD SPLINES; CONSTRUCT BSPLINE CONVERT COPY
Convert Element to B-spline Original	MDL LOAD SPLINES; CONSTRUCT BSPLINE CONVERT ORIGINAL
Copy Fence Content	FENCE COPY
Copy Parallel by Distance	COPY PARALLEL DISTANCE
Copy Parallel by Key-in	COPY PARALLEL KEYIN
Corner Joint	MDL LOAD CUTTER; JOIN CORNER
Create Complex Chain	CREATE CHAIN MANUAL
Create Complex Shape	CREATE SHAPE MANUAL
Crosshatch Element Area	CROSSHATCH
Cut All Component Lines	MDL LOAD CUTTER; CUT ALL
Cut Single Component Line	MDL LOAD CUTTER; CUT SINGLE
Define ACS (Aligned with Element)	DEFINE ACS ELEMENT
Define ACS (Aligned with View)	DEFINE ACS VIEW

TOOL NAME	KEY-IN
Define ACS (By Points)	DEFINE ACS POINTS
Define Active Entity Graphically	DEFINE AE
Define Cell Origin	DEFINE CELL ORIGIN
Define Reference File Back Clipping Plane	REFERENCE CLIP BACK
Define Reference File Clipping Boundary	REFERENCE CLIP BOUNDARY
Define Reference Clipping Mask	REFERENCE CLIP MASK
Define Reference File Front Clipping Plane	REFERENCE CLIP FRONT
Define True North	DEFINE NORTH
Delete Element	DELETE ELEMENT
Delete Fence Contents	FENCE DELETE
Delete Part of Element	DELETE PARTIAL
Delete Vertex	DELETE VERTEX
Detach Database Linkage	DETACH
Detach Database Linkage from Fence Contents	FENCE DETACH
Detach Reference File	REFERENCE DETACH
Dimension Angle Between Lines	DIMENSION ANGLE LINES
Dimension Angle from X-Axis	DIMENSION ANGLE X
Dimension Angle from Y-Axis	DIMENSION ANGLE Y
Dimension Angle Location	DIMENSION ANGLE LOCATION
Dimension Angle Size	DIMENSION ANGLE SIZE
Dimension Arc Location	DIMENSION ARC LOCATION
Dimension Arc Size	DIMENSION ARC SIZE
Dimension Diameter	DIMENSION DIAMETER
Dimension Diameter (Extended Leader)	DIMENSION DIAMETER EXTENDED
Dimension Diameter Parallel	DIMENSION DIAMETER PARALLEL

TOOL NAME	KEY-IN
Dimension Diameter Perpendicular	DIMENSION DIAMETER PERPENDICULAR
Dimension Diameter	DIMENSION DIAMETER
Dimension Element	DIMENSION ELEMENT
Dimension Location	DIMENSION LOCATION SINGLE
Dimension Location (Stacked)	DIMENSION LOCATION STACKED
Dimension Ordinates	DIMENSION ORDINATE
Dimension Radius	DIMENSION RADIUS
Dimension Radius (Extended Leader)	DIMENSION RADIUS EXTENDED
Dimension Size (Custom)	DIMENSION LINEAR
Dimension Size with Arrow	DIMENSION SIZE ARROW
Dimension Size with Strokes	DIMENSION SIZE STROKE
Display Attributes of Text Element	IDENTIFY TEXT
Drop Association	DROP ASSOCIATION
Drop Complex Status	DROP COMPLEX
Drop Complex Status of Fence Contents	FENCE DROP
Drop Dimension	DROP DIMENSION
Drop from Graphic Group	GROUP DROP
Drop Line String/Shape Status	DROP STRING
Drop Text	DROP TEXT
Edit Text	EDIT TEXT
Element Selection	CHOOSE ELEMENT
Extend 2 Elements to Intersection	EXTEND ELEMENT 2
Extend Element to Intersection	EXTEND ELEMENT INTERSECTION
Extend Line	EXTEND LINE DISTANCE
Extend Line By Key-in	EXTEND LINE KEYIN
Extract Bspline Surface Boundary	MDL LOAD SPLINES; EXTRACT BSPLINE SURFACE BOUNDARY
Fence Stretch	FENCE STRETCH

TOOL NAME	KEY-IN
Fill in Single Enter Data Field	EDIT SINGLE
Freeze Element	FREEZE
Freeze Elements in Fence	FENCE FREEZE
Generate Report Table	FENCE REPORT
Global Origin	ACTIVE ORIGIN (GO=)
Group Holes	GROUP HOLES
Hatch Element Area	HATCH
Horizontal Parabola (No Truncation)	PLACE PARABOLA HORIZONTAL NOMODIFY
Horizontal Parabola and Truncate Both	PLACE PARABOLA HORIZONTAL MODIFY
Identify Cell	IDENTIFY CELL
Impose Bspline Surface Boundary	MDL LOAD SPLINES; IMPOSE BSPLINE SURFACE BOUNDARY
Insert Vertex	INSERT VERTEX
Label Line	LABEL LINE
Load Displayable Attributes	LOAD DA
Load Displayable Attributes to Fence Contents	FENCE LOAD
Match Pattern Attributes	ACTIVE PATTERN MATCH
Match Text Attributes	ACTIVE TEXT
Measure Angle Between Lines	MEASURE ANGLE
Measure Area	MEASURE AREA
Measure Area of Element	MEASURE AREA ELEMENT
Measure Distance Along Element	MEASURE DISTANCE ALONG
Measure Distance Between Points	MEASURE DISTANCE POINTS
Measure Minimum Distance Between Elements	MEASURE DISTANCE MINIMUM
Measure Perpendicular Distance From Element	MEASURE DISTANCE PERPENDICULAR

TOOL NAME	KEY-IN
Measure Radius	MEASURE RADIUS
Merged Cross Joint	MDL LOAD CUTTER; JOIN CROSS MERGE
Merged Tee Joint	MDL LOAD CUTTER; JOIN TEE MERGE
Mirror Element About Horizontal (Copy)	MIRROR COPY HORIZONTAL
Mirror Element About Horizontal (Original)	MIRROR ORIGINAL HORIZONTAL
Mirror Element About Line Copy	MIRROR COPY LINE
Mirror Element About Line (Ordinal)	MIRROR ORIGINAL LINE
Mirror Element About Vertical (Copy)	MIRROR COPY VERTICAL
Mirror Element About Vertical (Original)	MIRROR ORIGINAL VERTICAL
Mirror Fence Contents About Horizontal (Copy)	FENCE MIRROR COPY HORIZONTAL
Mirror Fence Contents About Horizontal (Original)	FENCE MIRROR ORIGINAL HORIZONTAL
Mirror Fence Contents About Line (Copy)	FENCE MIRROR COPY LINE
Mirror Fence Contents About Line (Original)	FENCE MIRROR ORIGINAL LINE
Mirror Fence Contents About Vertical (Copy)	FENCE MIRROR COPY VERTICAL
Mirror Fence Contents About Vertical (Original)	FENCE MIRROR ORIGINAL VERTICAL
Mirror Reference File About Horizontal	REFERENCE MIRROR HORIZONTAL
Mirror Fence About Vertical	REFERENCE MIRROR VERTICAL
Modify Arc Angle	MODIFY ARC ANGLE
Modify Arc Axis	MODIFY ARC AXIS
Modify Arc Radius	MODIFY ARC RADIUS
Modify Element	MODIFY ELEMENT

TOOL NAME	KEY-IN
Modify Fence	MODIFY FENCE
Move ACS	MOVE ACS
Move Element	MOVE ELEMENT
Move Fence Block/Shape	MOVE FENCE
Move Fence Contents	FENCE MOVE
Move Reference File	REFERENCE MOVE
Multi-Cycle Segment Linear Pattern	PATTERN LINE MULTIPLE
Open Cross Joint	MDL LOAD CUTTER; JOIN CROSS OPEN
Open Tee Joint	MDL LOAD CUTTER; JOIN TEE OPEN
Pattern Element Area	PATTERN AREA ELEMENT
Pattern Fence Area	PATTERN AREA FENCE
Place Active Cell	PLACE CELL ABSOLUTE
Place Active Cell (Interactive)	PLACE CELL INTERACTIVE ABSOLUTE
Place Active Cell Matrix	MATRIX CELL (CM=)
Place Active Cell Relative	PLACE CELL RELATIVE
Place Active Cell Relative (Interactive)	PLACE CELL INTERACTIVE RELATIVE
Place Active Line Terminator	PLACE TERMINATOR
Place Active Point	PLACE POINT
Place Arc by Center	PLACE ARC CENTER
Place Arc by Edge	PLACE ARC EDGE
Place Arc by Keyed-in Radius	PLACE ARC RADIUS
Place B-spline Curve by Least Squares	MDL LOAD SPLINES; PLACE BSPLINE CURVE LEASTSQUARE
Place B-spline Curve by Points	MDL LOAD SPLINES; PLACE BSPLINE CURVE POINTS
Place B-spline Curve by Poles	MDL LOAD SPLINES; PLACE BSPLINE CURVE POLES

TOOL NAME	KEY-IN
Place B-spline Surface by Least Squares	MDL LOAD SPLINES; PLACE BSPLINE SURFACE LEASTSQUARES
Place B-spline Surface by Points	MDL LOAD SPLINES; PLACE BSPLINE SURFACE POINTS
Place B-spline Surface by Poles	MDL LOAD SPLINES; PLACE BSPLINE SURFACE POLES
Place Block	PLACE BLOCK ORTHOGONAL
Place Center Mark	DIMENSION CENTER MARK
Place Circle by Center	PLACE CIRCLE CENTER
Place Circle by Diameter	PLACE CIRCLE DIAMETER
Place Circle by Edge	PLACE CIRCLE EDGE
Place Circle by Keyed-in Radius	PLACE CIRCLE RADIUS
Place Circumscribed Polygon	PLACE POLYGON CIRCUMSCRIBED
Place Ellipse by Center and Edge	PLACE ELLIPSE CENTER
Place Ellipse by Edge Points	PLACE ELLIPSE EDGE
Place Fence Block	PLACE FENCE BLOCK
Place Fence Shape	PLACE FENCE SHAPE
Place Fitted Text	PLACE TEXT FITTED
Place Fitted View Independent Text	PLACE TEXT VI
Place Half Ellipse	PLACE ELLIPSE HALF
Place Helix	MDL LOAD SPLINES; PLACE HELIX
Place Inscribed Polygon	PLACE POLYGON INSCRIBED
Place Isometric Block	PLACE BLOCK ISOMETRIC
Place Isometric Circle	PLACE CIRCLE ISOMETRIC
Place Line	PLACE LINE
Place Line at Active Angle	PLACE LINE ANGLE
Place Line String	PLACE LSTRING POINT
Place Multi-line	PLACE MLINE

TOOL NAME	KEY-IN
Place Note	PLACE NOTE
Place Orthogonal Shape	PLACE SHAPE ORTHOGONAL
Place Parabola by End Points	MDL LOAD SPLINES; PLACE PARABOLA ENDPOINTS
Place Point Curve	PLACE CURVE POINT
Place Polygon by Edge	PLACE POLYGON EDGE
Place Quarter Ellipse	PLACE ELLIPSE QUARTER
Place Right Cone	PLACE CONE RIGHT
Place Right Cone by Keyed-in Radius	PLACE CONE RADIUS
Place Right Cylinder	PLACE CYLINDER RIGHT
Place Right Cylinder by Keyed-in Radius	PLACE CYLINDER RADIUS
Place Rotated Block	PLACE BLOCK ROTATED
Place Shape	PLACE SHAPE
Place Skewed Cone	PLACE CONE SKEWED
Place Skewed Cylinder	PLACE CYLINDER SKEWED
Place Slab	PLACE SLAB
Place Space Curve	PLACE CURVE SPACE
Place Space Line String	PLACE LSTRING SPACE
Place Sphere	PLACE SPHERE
Place Spiral By End Points	MDL LOAD SPLINES; PLACE SPIRAL ENDPOINTS
Place Spiral By Length	MDL LOAD SPLINES; PLACE SPIRAL LENGTH
Place Spiral by Sweep Angle	MDL LOAD SPLINES; PLACE SPIRAL ANGLE
Place Stream Curve	PLACE CURVE STREAM
Place Stream Line String	PLACE LSTRING STREAM
Place Text	PLACE TEXT
Place Text Above Element	PLACE TEXT ABOVE
Place Text Along Element	PLACE TEXT ALONG

TOOL NAME	KEY-IN
Place Text Below Element	PLACE TEXT BELOW
Place Text Node	PLACE NODE
Place Text On Element	PLACE TEXT ON
Place View Independent Text	PLACE TEXT VI
Place View Independent Text Node	PLACE NODE VIEW
Polar Array	ARRAY POLAR
Polar Array Fence Contents	FENCE ARRAY POLAR
Project Active Point Onto Element	CONSTRUCT POINT PROJECT
Rectangular Array	ARRAY RECTANGULAR
Rectangular Array Fence Contents	FENCE ARRAY RECTANGULAR
Reload Reference File	REFERENCE RELOAD
Replace Cell	REPLACE CELL
Review Database Attributes of Element	REVIEW
Rotate ACS Absolute	ROTATE ACS ABSOLUTE
Rotate ACS Relative	ROTATE ACS RELATIVE
Rotate Element Active Angle Copy	ROTATE COPY
Rotate Element Active Angle Original	ROTATE ORIGINAL
Rotate Fence Contents by Active Angle (Copy)	FENCE ROTATE COPY
Rotate Fence Contents by Active Angle (Original)	FENCE ROTATE ORIGINAL
Rotate Reference File	REFERENCE ROTATE
Scale Element (Copy)	SCALE COPY
Scale Element (Original)	SCALE ORIGINAL
Scale Fence Contents (Copy)	FENCE SCALE COPY
Scale Fence Contents (Original)	FENCE SCALE ORIGINAL
Scale Reference File	REFERENCE SCALE
Select ACS	ATTACH ACS
Select and Place Cell	SELECT CELL ABSOLUTE

TOOL NAME	KEY-IN
Select and Place Cell (Relative)	SELECT CELL RELATIVE
Set Active Depth	DEPTH ACTIVE
Show Active Depth	SHOW DEPTH ACTIVE
Show Active Entity	SHOW AE
Show Linkage Mode	ACTIVE LINKAGE
Show Pattern Attributes	SHOW PATTERN
Single Cycle Segment Linear Pattern	PATTERN LINE SINGLE
Spin Element (Copy)	SPIN COPY
Spin Element (Original)	SPIN ORIGINAL
Spin Fence Contents (Copy)	FENCE SPIN COPY
Spin Fence Contents (Original)	FENCE SPIN ORIGINAL
Symmetric Parabola (No Truncation)	PLACE PARABOLA NOMODIFY
Symmetric Parabola and Truncate Both	PLACE PARABOLA MODIFY
Thaw Element	THAW
Thaw Elements in Fence	FENCE THAW
Truncated Cycle Linear Pattern	PATTERN LINE ELEMENT
Uncut Component Lines	MDL LOAD CUTTER; UNCUT

Alternate Key-ins

Note: Key-ins can be uppercase, lowercase, or combination.

AA = ACTIVE ANGLE	set active angle
AC = ACTIVE CELL	set active cell; place absolute
AD = POINT ACSDELTA	data point—delta ACS
AE = ACTIVE ENTITY	define active entity
AM = ATTACH MENU	activate menu
AP = ACTIVE PATTERN CELL	set active pattern cell
AR = ACTIVE RCELL	set active cell; place relative
AS = ACTIVE SCALE	set active scale factors
AT = TUTORIAL	activate tutorial
AX = POINT ACSABSOLUTE	data point absolute ACS
AZ = ACTIVE ZDEPTH ABSOLUTE	set active depth
CC = CREATE CELL	create cell
CD = DELETE CELL	delete cell from cell library
CM = MATRIX CELL	place active cell matrix
CO = ACTIVE COLOR	set active color
CR = RENAME CELL	rename cell
CT = ATTACH COLORTABLE	attach color table
DA = ACTIVE DATYPE	set active displayable attribute type
DB = ACTIVE DATABASE	attach control file to design file
DD = SET DDEPTH RELATIVE	set display depth (relative)

DF = SHOW FONT	open Fonts settings box
DI = POINT DISTANCE	data point—distance, direction
DL = POINT DELTA	data point—delta coordinates
DP = DEPTH DISPLAY	set display depth
DR = TYPE	display text file
DS = SEARCH	specify fence filter
DV = VIEW	delete saved view
DX = POINT VDELTA	data point—delta view coordinates
DZ = ZDEPTH RELATIVE	set active depth (relative)
EL = ELEMENT LIST	create element list file
FF = FENCE FILE	copy fence contents to design file
FI = FIND	set database row as active entity
FT = ACTIVE FONT	set active font
GO = ACTIVE ORIGIN	Global Origin
GR = ACTIVE GRIDREF	set grid reference spacing
GU = ACTIVE GRIDUNIT	set horizontal grid spacing
KY = ACTIVE KEYPNT	set Snap divisor
LC = ACTIVE STYLE	set active line style
LD = DIMENSION LEVEL	set dimension level
LL = ACTIVE LINE LENGTH	set active text line length
LS = ACTIVE LINE SPACE	set active text node line spacing
LT = ACTIVE TERMINATOR	set active terminator
LV = ACTIVE LEVEL	set active level
NN = ACTIVE NODE	set active text node number
OF = SET LEVELS <level list> OFF	set level display off
ON = SET LEVELS <level list> ON	set level display on
OX = ACTIVE INDEX	retrieve user command index
PA = ACTIVE PATTERN ANGLE	set active pattern angle
PD = ACTIVE PATTERN DELTA	set active pattern delta (distance)
PS = ACTIVE PATTERN SCALE	set active pattern scale

PT = ACTIVE POINT	set active point
PX = DELETE ACS	delete ACS
RA = ACTIVE REVIEW	set attribute review selection criteria
RC = ATTACH LIBRARY	open cell library
RD = NEWFILE	open design file
RF = REFERENCE ATTACH	attach reference file
RS = ACTIVE REPORT	name report table
RV = ROTATE VIEW	rotate view (relative)
RX = ATTACH ACS	select ACS
SD = ACTIVE STREAM DELTA	set active stream delta
SF = FENCE SEPARATE	move fence contents to design file
ST = ACTIVE STREAM TOLERANCE	set active stream tolerance
SV = SAVE VIEW	save view
SX = SAVE ACS	save auxiliary coordinate system
TB = ACTIVE TAB	set tab spacing for importing text
TH = ACTIVE TXHEIGHT	set active text height
TI = ACTIVE TAG	set copy and increment value
TS = ACTIVE TSCALE	set active terminator scale
TV = DIMENSION TOLERANCE	set dimension tolerance limits
TW = ACTIVE TXWIDTH	set active text width
TX = ACTIVE TXSIZE	set active text size (height/width)
UC = USERCOMMAND	activate user command
UCC = UCC	compile user command
UCI = UCI	user command index
UR = ACTIVE UNITROUND	set unit distance
VI = VIEW	attach named view
WO = WINDOW ORIGIN	Window Orgin
WT = ACTIVE WEIGHT	set active line weight
XD = EXCHANGEFILE	open design file; keep view config.

XS = ACTIVE XSCALE	set active X scale
XY = POINT ABSOLUTE	data point absolute coordinates
YS = ACTIVE YSCALE	set active Y scale
ZS = ACTIVE ZSCALE	set active Z scale

INDEX

B

C

D

E

F

G

H

I

J

K

L

M

N

O

P

Q

R

S

T

U

V

W

X

Z